T0211770

Lecture Notes in Computer Science 12011

Advanced Research in Computing and Software Science

Subline of Lecture Notes in Computer Science

More information about this series at http://www.springer.com/series/7407

Alexander Chatzigeorgiou · Riccardo Dondi ·
Herodotos Herodotou · Christos Kapoutsis ·
Yannis Manolopoulos · George A. Papadopoulos ·
Florian Sikora (Eds.)

SOFSEM 2020: Theory and Practice of Computer Science

46th International Conference on Current Trends
in Theory and Practice of Informatics, SOFSEM 2020
Limassol, Cyprus, January 20–24, 2020
Proceedings

Editors
Alexander Chatzigeorgiou
University of Macedonia
Thessaloniki, Greece

Riccardo Dondi
University of Bergamo
Bergamo, Italy

Herodotos Herodotou
Cyprus University of Technology
Limassol, Cyprus

Christos Kapoutsis
Carnegie Mellon University Qatar
Doha, Qatar

Yannis Manolopoulos
Open University of Cyprus
Nicosia, Cyprus

George A. Papadopoulos
University of Cyprus
Nicosia, Cyprus

Florian Sikora
Paris Dauphine University
Paris, France

ISSN 0302-9743 ISSN 1611-3349 (electronic)
Lecture Notes in Computer Science
ISBN 978-3-030-38918-5 ISBN 978-3-030-38919-2 (eBook)
https://doi.org/10.1007/978-3-030-38919-2

LNCS Sublibrary: SL1 – Theoretical Computer Science and General Issues

This Springer imprint is published by the registered company Springer Nature Switzerland AG
The registered company address is: Gewerbestrasse 11, 6330 Cham, Switzerland

Preface

This volume contains the invited and contributed papers selected for presentation at SOFSEM 2020, the 46th International Conference on Current Trends in Theory and Practice of Computer Science, which was held during January 20–24, 2020, in Limassol, Cyprus.

SOFSEM (originally SOFtware SEMinar) is an annual international winter conference devoted to the theory and practice of computer science. Its aim is to present the latest developments in research for professionals from academia and industry working in leading areas of computer science. While being a well-established and fully international conference, SOFSEM also maintains the best of its original Winter School aspects, such as a high number of invited talks, in-depth coverage of selected research areas, and ample opportunity to discuss and exchange new ideas. SOFSEM 2020 is organized around the following four tracks:

- Foundations of Computer Science (chair: Christos Kapoutsis)
- Foundations of Data Science and Engineering (chair: Herodotos Herodotou)
- Foundations of Software Engineering (chair: Alexander Chatzigeorgiou)
- Foundations of Algorithmic Computational Biology (chairs: Ricardo Dondi and Florian Sikora)

Notably, the fourth track of the above takes place for the first time in the framework of SOFSEM and it is expected to be continued over the next years. With these tracks, SOFSEM 2020 covered the latest advances in both theoretical and applied research in leading areas of computer science.

An integral part of SOFSEM 2020 is the traditional Student Research Forum (chair: Theodoros Tzouramanis) organized with the aim of giving students feedback on both the originality of their scientific results and on their work in progress. The papers presented at the Student Research Forum will be available at the CEUR website (http://ceur-ws.org).

The SOFSEM 2020 Program Committee (PC) consisted of 95 international experts from 35 countries, representing the track areas with outstanding expertise. Another 72 external reviewers contributed as well. The committee undertook the task of assembling a scientific program for the SOFSEM audience by selecting from 125 submissions from 40 countries entered in the EasyChair system in response to the call for papers. The submissions were carefully reviewed with 2.9 reviews/paper, and thoroughly discussed. Following strict criteria of quality and originality, 40 papers were accepted for presentation as regular research papers, as well as 17 papers for presentation as short papers. These 57 papers were contributed by authors from 26 countries. Thus, the acceptance ratio for regular papers was 32%, plus another 14% for short papers. Additionally, based on the recommendation of the chair of the Student Research Forum, 9 papers were accepted for presentation in the Student Research Forum.

As editors of these proceedings, we are grateful to everyone who contributed to the scientific program of the conference. We would like to thank the invited speakers:

- Mikołaj Bojańczyk (University of Warsaw, Poland)
- Ernesto Damiani (Khalifa University, UAE)
- Erol Gelenbe (Polish Academy of Sciences, Poland)
- Gunnar Klau (Heinrich Heine University Düsseldorf, Germany)
- Elias Koutsoupias (University of Oxford, UK)

for presenting their work to the audience of SOFSEM 2020. We thank all authors who submitted their papers for consideration. Many thanks are due to the PC members, and to all external referees, for their precise and detailed reviewing of the submissions. The work of the PC was carried out using the EasyChair system, and we gratefully acknowledge this contribution.

Special thanks are due to the SOFSEM Steering Committee, headed by Július Štuller, for its support throughout the preparation of the conference. Finally, we want to thank Easyconferences.eu led by Petros Stratis, and the Deputy Ministry of Tourism of Cyprus, for their services and support.

January 2020

Alexander Chatzigeorgiou
Riccardo Dondi
Herodotos Herodotou
Christos Kapoutsis
Yannis Manolopoulos
George A. Papadopoulos
Florian Sikora

Organization

General Chairs

Yannis Manolopoulos	Open University of Cyprus, Cyprus
George-Angelos Papadopoulos	University of Cyprus, Cyprus

Program Committee Chairs

Alexander Chatzigeorgiou	University of Macedonia, Greece
Riccardo Dondi	University of Bergamo, Italy
Herodotos Herodotou	Cyprus University of Technology, Cyprus
Christos Kapoutsis	Carnegie Mellon University in Qatar, Qatar
Florian Sikora	University of Paris-Dauphine, France

Student Research Forum Chair

Theodoros Tzouramanis	University of Thessaly, Greece

Steering Committee

Barbara Catania	University of Genoa, Italy
Miroslaw Kutylowski	Wroclaw University of Technology, Poland
Tiziana Margaria-Steffen	University of Limerick, Ireland
Branislav Rovan	Comenius University, Slovakia
Petr Šaloun	Technical University of Ostrava, Czech Republic
Július Štuller (Chair)	Academy of Sciences in Prague, Czech Republic
Jan van Leeuwen	Utrecht University, The Netherlands

Program Committee

Andris Ambainis	University of Latvia, Latvia
Fabio Anselmi	Italian Institute of Technology, Italy
Mukul S. Bansal	University of Connecticut, USA
Ladjel Bellatreche	Poitiers University, France
Sadok Ben Yahia	University of Tunis El Manar, Tunisia
Maria Bielikova	Slovak University of Technology in Bratislava, Slovakia
Guillaume Blin	University of Bordeaux, France
Hans-Joachim Böckenhauer	ETH Zurich, Switzerland
Edouard Bonnet	ENS Lyon, CNRS, France
Stephane Bressan	National University of Singapore, Singapore

Francesco Buccafurri	University of Reggio Calabria, Italy
Laurent Bulteau	University of Paris-Est Marne-la-Vallée, CNRS, France
Cedric Chauve	Simon Fraser University, Canada
Zhi-Zhong Chen	Tokyo Denki University, Japan
Alfredo Cuzzocrea	ICAR-CNR, University of Calabria, Italy
Peter Damaschke	Chalmers University, Sweden
Bhaskar DasGupta	University of Illinois at Chicago, USA
Volker Diekert	University of Stuttgart, Germany
Martin Dietzfelbinger	Technical University Ilmenau, Germany
Johann Eder	Alpen Adria University of Klagenfurt, Austria
Mohammed El-Kebir	University of Illinois at Urbana-Champaign, USA
Oliver Eulenstein	Iowa State University, USA
Guillaume Fertin	University of Nantes, France
Pierre Fraigniaud	CNRS, Paris Diderot University, France
Johann Gamper	Free University of Bozen-Bolzano, Italy
Loukas Georgiadis	University of Ioannina, Greece
Pawel Gorecki	University of Warsaw, Poland
Giovanna Guerrini	University of Genova, Italy
Yo-Sub Han	Yonsei University, South Korea
Theo Härder	University of Kaiserslautern, Germany
Danny Hermelin	Ben-Gurion University, Israel
Irena Holubova	Charles University in Prague, Czech Republic
Markus Holzer	University of Giessen, Germany
Kazuo Iwama	Kyoto University, Japan
Jesper Jansson	The Hong Kong Polytechnic University, Hong Kong, China
Minghui Jiang	Utah State University, USA
Mark Jones	TU Delft, The Netherlands
Tomasz Jurdzinski	University of Wroclaw, Poland
Iyad A. Kanj	DePaul University, USA
Jarkko Kari	University of Turku, Finland
Selma Khouri	ESI, Algeria
Dennis Komm	ETH Zurich, Switzerland
Christian Komusiewicz	Philipps-University Marburg, Germany
Georgia Koutrika	Athena Research Center, Greece
Rastislav Kralovic	Comenius University, Slovakia
Evangelos Kranakis	Carleton University, Canada
Manuel Lafond	University of Sherbrooke, Canada
Michael Lampis	University of Paris-Dauphine, France
Sebastian Link	The University of Auckland, New Zealand
Zsuzsanna Lipták	University of Verona, Italy
Beatrice Markhoff	University of Tours, France
Giancarlo Mauri	University of Milano-Bicocca, Italy
Neeldhara Misra	Indian Institute of Technology, India
Elvira Mayordomo	University of Zaragoza, Spain
Carlo Mereghetti	University of Milan, Italy

Paolo Missier	Newcastle University, UK
Luc Moreau	King's College London, UK
Nelma Moreira	University of Porto, Portugal
Rim Moussa	ENICarthage, Tunisia
Alexander Okhotin	Saint Petersburg State University, Russia
Giovanni Pighizzini	University of Milan, Italy
Nadia Pisanti	University of Pisa, Italy
Alberto Policriti	University of Udine, Italy
Alex Popa	University of Bucharest, Romania
Andrei Popescu	Middlesex University London, UK
Igor Potapov	The University of Liverpool, UK
Eric Rivals	University of Montpellier, CNRS, France
Raffaella Rizzi	University of Milano-Bicocca, Italy
Gunter Saake	Otto-von-Guericke-University Magdeburg, Germany
Sherif Sakr	University of Tartu, Estonia
Kai Salomaa	Queen's University, Canada
Cem Say	Bogazici University, Turkey
Eike Schallehn	Otto von Guericke University Magdeburg, Germany
Marinella Sciortino	University of Palermo, Italy
Celine Scornavacca	University of Montpellier, CNRS, France
Shinnosuke Seki	The University of Electro-Communications, Japan
Ayumi Shinohara	Tohoku University, Japan
Ernest Teniente	Technical University of Catalonia, Spain
Olivier Teste	University of Toulouse, France
Martin Theobald	University of Luxemburg, Luxembourg
Eleftherios Tiakas	Aristotle University of Thessaloniki, Greece
Farouk Toumani	Blaise Pascal University, France
Gabriel Valiente	Technical University of Catalonia, Spain
Athena Vakali	Aristotle University of Thessaloniki, Greece
Stephane Vialette	University Paris-Est Marne-la-Vallée, CNRS, France
Peter Van Emde Boas	University of Amsterdam, The Netherlands
Leo Van Iersel	University of Twente, The Netherlands
Panos Vassiliadis	University of Ioannina, Greece
Mathias Weller	University Paris-Est Marne-la-Vallée, CNRS, France
Abuzer Yakaryilmaz	University of Latvia, Latvia
Tomoyuki Yamakami	University of Fukui, Japan
Christos Zaroliagis	CTI, University of Patras, Greece
Meirav Zehavi	Ben-Gurion University, Israel
Thomas Zeugmann	Hokkaido University, Japan
Louxin Zhang	National University of Singapore, Singapore

Additional Reviewers

Patrizio Angelini
Marcella Anselmo
Elvira-Maria Arvanitou
Christel Baier
Luca Bernardinello
Nicolas Bonichon
Sabine Broda
Elisabet Burjons
Cezar Campeanu
Costanza Catalano
Michele Chiari
Sarah Christensen
Ferdinando Cicalese
Maxime Crochemore
Federico Dassereto
Gianluca De Marco
Holger Dell
Emilio Di Giacomo
Mike Domaratzki
Gabriele Fici
Johannes Fischer
Marco Franceschetti
Fabian Frei
Janosch Fuchs
Esther Galbrun
Paweł Garncarek
Paweł Gawrychowski
Konstantinos Giannis
Szymon Grabowski
Massimiliano Goldwurm
Spyros Halkidis
Artur Jeż
Chris Keeler
Vasilios Kelefouras
Hwee Kim
Jetty Kleijn
Sang-Ki Ko
Athanasios Konstantinidis
Julius Köpke
Lukasz Kowalik
Tomas Kulik
Martin Kutrib
Markus Lohrey
Andreas Maletti
Florin Manea
Wim Martens
Radu-Stefan Mincu
Tobias Mömke
František Mráz
Reino Niskanen
Charis Papadopoulos
Matthew Patitz
Ilaria Pigazzini
Luca Prigioniero
Daniel Prusa
Simon Puglisi
Karol Rástočný
Rogério Reis
Traian Florin Serbanuta
Michiel Smid
Taylor Smith
Ana Paula Tomás
Spyridon Tzimas
Walter Unger
Diego Valota
Sergey Verlan
Christina Volioti
Kunihiro Wasa
David Wehner
Sebastian Wild
Petra Wolf
Viktor Zamaraev

Contents

Foundations of Data Science and Engineering – Regular Papers

Foundations of Software Engineering – Regular Papers

Foundations of Data Science and Engineering – Short Papers

Foundations of Software Engineering – Short Papers

Foundations of Algorithmic Computational Biology – Short Paper

Invited Papers

Certified Machine-Learning Models

Ernesto Damiani[1,2](✉) and Claudio A. Ardagna[2]

[1] Center on Cyber-Physical Systems, Khalifa University, Abu Dhabi, UAE
[2] Computer Science Department, Università degli Studi di Milano, Milan, Italy
{ernesto.damiani,claudio.ardagna}@unimi.it

Abstract. The massive adoption of Machine Learning (ML) has deeply changed the internal structure, the design and the operation of software systems. ML has shifted the focus from code to data, especially in application areas where it is easier to collect samples that embody correct solutions to individual instances of a problem, than to design and code a deterministic algorithm solving it for all instances. There is an increasing awareness of the need to verify key non-functional properties of ML-based software applications like fairness and privacy. However, the traditional approach trying to verify these properties by code inspection is pointless, since ML models' behavior mostly depends on the data and parameters used to train them. Classic software certification techniques cannot solve the issue as well. The Artificial Intelligence (AI) community has been working on the idea of preventing undesired behavior by controlling a priori the ML models' training sets and parameters. In this paper, we take a different, online approach to ML verification, where novel behavioral monitoring techniques based on statistical testing are used to support a dynamic certification framework enforcing the desired properties on black-box ML models in operation. Our aim is to deliver a novel framework suitable for practical certification of distributed ML-powered applications in heavily regulated domains like transport, energy, healthcare, even when the certifying authority is not privy to the model training. To achieve this goal, we rely on three key ideas: *(i)* use test suites to define desired non-functional properties of ML models, *(ii)* Use statistical monitoring of ML models' behavior at inference time to check that the desired behavioral properties are achieved, and *(iii)* compose monitors' outcome within dynamic, virtual certificates for composite software applications.

Keywords: Intelligent systems · Machine Learning · Certification

1 Introduction

Some time ago, TESLA's Artificial Intelligence (AI) Director Andrej Karpathy came up with an effective metaphor: "*Machine Learning models are eating software from within*". Indeed, Machine Learning (ML) has deeply changed the internal structure and operation of software systems. It has shifted the balance from code to data, especially for applications where it is easier to collect samples that embody correct solutions to instances of a problem than to design and code a deterministic algorithm solving it. Natural Language Processing (NLP) provides a good example of how support for the

A. Chatzigeorgiou et al. (Eds.): SOFSEM 2020, LNCS 12011, pp. 3–15, 2020.
https://doi.org/10.1007/978-3-030-38919-2_1

properties of an algorithm can move from code to data. Solving backward references (anaphora), that is, mapping pronouns to the noun they refer to in sentences like "*Sally was in the house when Alice arrived, but nobody saw her*" is a key functionality for automatic translation or text input prediction. The problem has been deeply researched for the English language [1]. Traditional algorithms used for English rely on distance tempered by context parsing. For instance, an algorithm could map a pronoun to the closest noun encountered going backward, but only if it refers to a person name whose gender matches the pronoun (to avoid classic mistakes that occur using distance only with sentences like "*Sally was in the house when the cat arrived, but nobody saw her*" or "*Sally was in the house when Bill arrived, but nobody saw her*"). The behavior of a traditional anaphora resolution algorithm based on context parsing and distance is hard-wired in the program code and can be analyzed via testing or formal modelling. Unfortunately, this is no longer the case. In their victorious Jeopardy! challenge in 2011, IBM managed to handle subtle anaphora that are frequent in Jeopardy! questions by using an ML model trained on the entire set of questions ever used in the game. The high accuracy of IBM ML-based anaphora solver was mostly due to its training set selection [2]. In terms of our example, an ML model for anaphora is fed via a training set like the one in Table 1.

Table 1. Feeding for ML model for anaphora.

Sentence	Anaphora
Sally was in the house when Alice arrived, but nobody saw ***her***	Alice
Sally was in the house when the cat arrived, but nobody saw ***her***	Sally
Sally was in the house when Bill arrived, but nobody saw ***her***	Bill

The point that we are trying to make is that substituting the traditional algorithm with an ML model may improve accuracy but makes it harder to verify the properties of the anaphora solver. As an example, let us assume we want to verify a functional property: the solver's capability to use presupposition to generate exceptions to the closest distance rule (e.g. mapping the adjective his to Robert rather than John in the sentence "*Robert's three children are with him; if Robert would argue loudly with John, his children could get frightened*"). Checking whether the presupposition property holds is straightforward if presupposition support (if any) is hard-wired in the algorithm's implementation code and verifiable via testing or formal modelling. However, this is clearly not the case when presupposition has been (or has not been) learnt from examples. Today, the capability to support presupposition of ML-based anaphora solvers depends on the solver's training dataset rather than on its code.

The paper is structured as follows. Section 2 presents the relevant related work. Section 3 presents our methodology for certified machine learning models. Section 4 describes a sample scenario on differential privacy. Section 5 gives our conclusions.

2 State of the Art

We now briefly discuss the state of the art in three main areas of interest for our research: AI Governance, AI Ethics by Design and Software Certification.

2.1 Governance of AI/ML Systems

Nowadays, industry and government agencies worldwide consider ML as the "next big thing" in applications and are keen on replacing the development and implementation of algorithms with ML models implemented using standard libraries. Large companies have quickly seen the interest of industrial-strength ML libraries, and most ML-powered applications use some stable, trustworthy open-source ML library (developed and managed by a large community) rather than relying on proprietary code. Consequently, many software developers no longer write code in the traditional way; rather, they build AI pipelines that include trained ML models. As discussed above, training ML models instead of implementing algorithms does not remove applications' behavioral variability. Rather, such variability is due to training data and to the circumstances of training. Randomization in the training process (e.g., neuron drop-off) and training data selection affect the behavior of ML models much more than their code.

The software engineering research community has started to realize that as ML models increasingly make key decisions for humans, the need arises for a new governance of ML models, allowing users to understand where and when (if not how) an ML inference came to be. The research agenda for AI governance is still under discussion [3], but it is centered around establishing policies regulating the use of AI and specifying the desired properties of AI systems. System properties come in many flavors [4], the three main traditional categories being architectural properties (how the system is structured), functional properties (what the system can do), and non-functional properties (how the system operates, e.g. in a fair or privacy-preserving way). Frameworks have been proposed to represent AI pipelines' architectural properties via symbolic representation [5, 6]. Recently, IEEE has launched an initiative aimed to foster standardization in Big Data management. The IEEE Big Data Governance and Metadata Management (BDGMM) group aims to control datasets consistency and provenance. Our own recent work on the TOREADOR platform for Big Data analytics [7] has led to a symbolic representation of AI pipelines using OWL/S [8] that can be used to reason about them. Recently, we have put forward the idea of using ML models together with symbolic reasoning to optimize the very structure of AI pipelines ("AI designing AI") [9].

Here, we want to discuss assessing the behavior of an ML-powered system by checking the functional and non-functional properties emerging by the ML model's training set and by other parameters of its training. A priori reasoning on architectural properties cannot guarantee that an ML-powered software system will hold some non-functional properties at run-time, including the key one of being harmless to its users or to third parties. In principle, we could think of generating a symbolic representation of the ML model and use it to prove the properties of interest, or to test the properties on the trained ML model.

Unfortunately, automatic reasoning on formal representations of ML models is still in its infancy [10], and we argue that no general approach to testing ML models' functional and non-functional properties is available, apart from testing robustness with respect to adversarial perturbations [11].

2.2 AI Ethics by Design

In the AI community, the debate on how AI can be made to behave ethically has initially focused on principles rather than on practices [12]. According to a recent paper by Morley, Floridi et al. [13], "*the AI community's ability to take action to mitigate [...] risks is still at its infancy*". We understand that excellent AI research has been done and is being done on methodologies for hardwiring non-functional properties into ML models at training time, especially by training set filtering [14, 15] and by bias shaping [16, 17]. From our standpoint, these approaches are the equivalent of trying to educate ML models toward the desired non-functional properties. Education requires enforcement authorities to be privy to the training process and to have access to the training sets. In this paper, we take an entirely different approach, investigating a direct monitoring mechanism to police black-box ML models to enforce the non-functional properties we want, even when the enforcing authority is not privy to the ML model training. We claim that enforcement can deliver ethical behavior as defined by the ethics initiative created in April 2016 by IEEE Standards Association, with the aim of embedding ethics into the design of processes for all AI and autonomous systems [18]. In our vision, education and enforcement will play a complementary role in defining and controlling AI behavior as they do for humans.

2.3 Software Certification

The software certification process "*demonstrates the non-functional properties of software systems so that they can be checked by an independent authority with minimal trust in the techniques and tools used in the certification process itself*" [19]. The result of certification process are software certificates, composed of assertions signed by the authority and stating the non-functional properties of the software component or service they refer to (Target-of-Certification or ToC). Certificates also contain (or point to) the evidence on which each assertion is based, usually in the digital form of a verification proof or of (static or dynamic) test results. Dynamic verification procedures can generate (and revoke) digital certificates at run time [20]. The book in [21] has introduced the notion of automatically selecting existing test cases based on the non-functional property one wants to validate. The notion of compositional certifications for systems composed of multiple ToCs (including cloud-based distributed application [22]) has been introduced and successfully tackled the problem of automatically composing certificates independently awarded to individual ToCs to asserting overall properties of composite systems. This approach identifies virtual active test cases to support system-wide "virtual certificates" based on certified properties of individual components [23]. It has been shown to scale nicely to the security certification of large software systems, including entire cloud stacks. Virtual certification can also be based on formal-proof evidence [24, 25]. There are several practical proof-based certification techniques [26–29] that provide a higher level of trust (corresponding to the highest level in the Common Criteria standard [30]) than the test-based ones. Using proof-based evidence for dynamic, composite application is currently a challenging and open research problem. As we have seen, we cannot directly apply existing software certification techniques to the new generation of ML-powered applications. Adopting ML models as building blocks of certified software applications challenges a key assumption of classic verification and testing "*applications' business logic, or at least its design, will be stable at verification time*" [21].

3 Methodology

In our quest for a way to enforce the desired non-functional properties of ML models, we do not start from scratch. Neuroscience has been using for decades statistical analysis of behavior to characterize the human decision-making process [5]. In the software engineering domain, techniques for statistics-based active testing of services can tell the tester (in a code-independent way) if variations in an application affect the users' experience or some other metrics [31].

Multi-Armed Bandit (MAB) problems [32] consist of K unknown probability distributions, $D_1, \ldots, D_k$, with expected values $\mu_1, \ldots, \mu_K$. These distributions are classically binary, modelling wins or losses (1 or 0) from various arms of slot machines (hence the name); in the active testing scenario the outcomes represent "success measures" of runs of different implementations of an algorithm, where the difference is some feature or variation in the application code. In our context, scores can be any (bounded) user-defined function.

A statistics test showing that $\mu_i \neq \mu_j$ (or $D_i \neq D_j$) can tell us if and how a difference in implementation makes (or does not make) a difference in success scores.

In a MAB test, the tester "pulls arms", that is, considers multiple variations to the module and tries to maximize a score function (Fig. 1). The foundation of MAB is Bayesian updating. Usually, each module is modelled as a Bernoulli process.

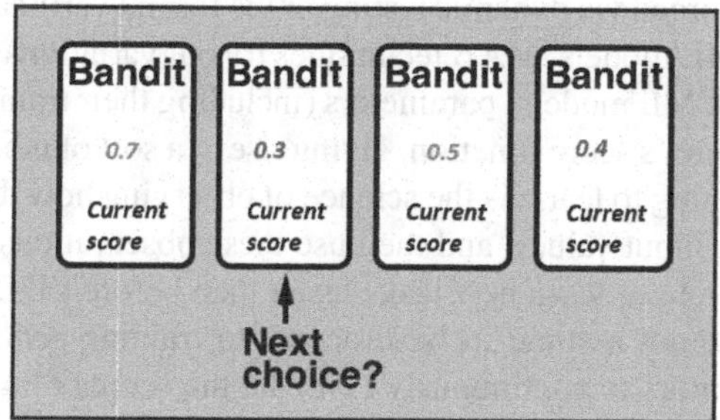

Fig. 1. The MAB experiment setting

The probability of success is unknown and is modelled by a Beta probability distribution. Each module receives input flows and the Beta distribution is updated accordingly. Modules are selected via Thompson sampling, a greedy method that always chooses the arm that maximizes expected reward.

In each iteration of the MAB experiment, Thompson sampling draws a sample score from each arm's Beta distribution and assign the next input to the arm with the highest score. In MAB, Thompson sampling and Bayesian update work together. If one of the modules is performing well, its Beta distribution parameters are updated to remember this, and Thompson sampling is more likely to draw a high score from it. In other words, the MAB setting permits testers to preferentially feed test cases to the best performing instances of a system: throughout the test, high-performing arms are rewarded with more inputs, whereas under-performing arms are punished with fewer inputs.

3.1 Using MAB to Test Behavior of ML Models: A Meta-Learning Process

Our key idea is using statistical active testing to explore ML model space at run-time, finding the variation that will keep a score function of interest in an acceptable area. Statistics has always played an important role in ML models' comparative assessment [33]. When comparing two binary ML models differing in their training sets, a simple Student t-test can be used to verify whether the difference in the training set translates in a statistically significant difference in the outputs [34]. More complex tests are often used to compare multi-category classifiers. ANOVA tests can point out significant differences among data grouped by classifiers, but do not report exactly where those differences lie. For instance, the Tukeys Honest Significant Difference (HSD) post-hoc test and its non-parametric equivalent, the Friedman-Nemenyi test, are often used to complement ANOVA for achieving a more detailed comparison between classifiers.

We rely on MAB to compare the behavior of an ML system to one or more desired properties. We argue that MAB is more suitable than statistical tests for supporting dynamic verification of non-functional properties of ML models. This is because, in principle, MAB can handle scenarios where the meta-learning process has competing objectives to the ones of the learning, as it often happens when an application's regulatory goals conflict with its commercial ones. We argue that our MAB-based line of attack will preserve classic duality between proof- and test-based certification, as the behavior of ML models can be enforced via dynamic testing or via formal verification. When applied to the verification of ML models, MAB techniques try out variations of the models. This way, the tester can tune ML models' parameters (including their training sets) to follow a gradient in the ML model's score function, giving rise to a sort of meta-learning process. Meta-learning, or learning to learn, is the science of observing how different ML models perform on a range of input values, and then use these observations to tune the models so they get more accurate or learn new tasks faster than before [35].

In practice, we cannot assume to have access to training sets or parameters. So, we rely on MAB monitors to continuously compute the scores of a pool of pre-trained ML models and output the inference of the pool member having the best score. The connection with meta-learning lies in the idea that the choice strategy can itself be learnt. The model pool plays the role of a grid in the model space. If no score from the pool is satisfactory, it can be (lazily) updated by requesting new variations from the model supplier. We envision that *model pools*, that is, sets of pre-trained ML model variations, can be supplied by ML model developers, who generate them via novel options of industrial ML model compilers or other software development environments. For designing and implementing execution time support, we can take advantage from approaches to the efficient simultaneous operation of multiple representations of models from FPGA to software platforms [36, 37]. Monitors work by applying a strategy for selecting within the available model pool the model likely to provide the best score. Monitored ML models whose score is unsatisfactory do not contribute to the inference; if all models in the pool are unsatisfactory for a while, a pool extension by adding new variations is triggered. Ideally, the selection strategy "keeps up" the score, minimizing the inevitable score loss from playing non-optimally in the ML model training, which is after all aimed to achieve accuracy. It is important to remark that we will be able to evaluate the quality of our solutions against theoretical bounds, as lower bound to this loss have been established already in the Eighties [38].

3.2 Open Issues

Our methodology faces four main issue briefly summarized below.

Develop Representation of Non-functional Properties using Score Functions. A key issue is how to express non-functional properties of interest in terms of a score function on the ML model's output. Let us consider an ML model estimating, say, the amount of bail to be posted by a crime suspect based on the suspect's ethnicity, gender, and age group [37]. As a high bail should be an incentive not to violate the corresponding restriction, the goal of the model is the resulting violations to bail restrictions. This suggests a game-theoretical definition for the effective behavior of the ML model as the one that minimizes such violations. However, this definition would probably make a bad score function, because it has high latency with respect to model output and requires costly large-scale monitoring [39]. Heuristics can be used to express the model's "fairness" quantitatively, for instance, as the ratio between output values corresponding to different inputs. Heuristics also guarantees low latency in the score computation. For example, we could argue that the ML model is fair if total amount of bails BS requested to any ethnicity, gender, or age group S (say, "people under 30") does not exceed $x_{S,T}$ times the amount requested to a benchmark group T (say, "Asian people over 50"). What is the right value of $x_{\mathrm{S,T}}$? Intuition suggests linking $x_{\mathrm{S,T}}$ to the group cardinalities, $x_{S,T}$ = |S|/|T|, which corresponds to imposing no calibration bias. At every estimate, one can compute the fairness score of each model in the grid. A lower bounded sub-optimality in meeting bail conditions can be tolerated to achieve an ethically acceptable proportion in the bail amounts requested to the different ethnic groups. This way, the model space can be explored by MAB testing to control the distance from the desired "fair" behavior.

In our approach, we define properties *by example*. Publicly available, shared examples of desired inferences could provide an operational consensus-based definition of properties easy to understand and acceptable for the community ("*I cannot say what fairness is, but I know it when I see it*").

A set of sample inferences can be used to define a *score function*, for instance as the distance (square error) between these inferences and the one performed by a ML model on the same inputs. Figure 2 hints to a more sophisticated definition of a property as the acceptable confusion matrix on the set of sample inferences.

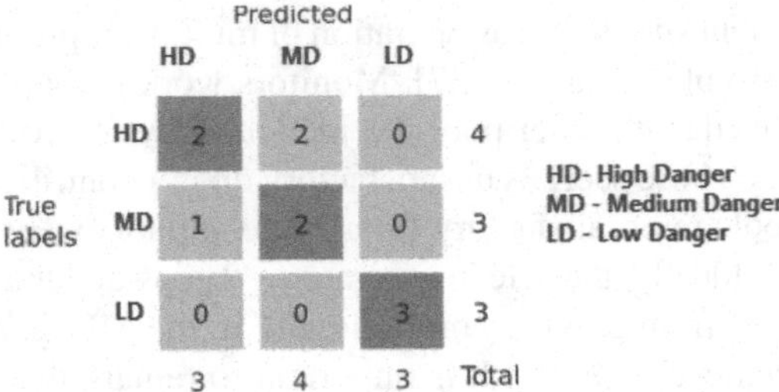

Fig. 2. Definition of a property via confusion matrix on a set of sample cases

Develop, Test and Validate Property-Dependent Choice Strategies for the MAB. The non-triviality of the MAB problem in our case lies in the fact that the agent operating the choice in the ML model pool cannot access the underlying probability distributions. Its strategy needs to be driven by trial-and-error only. So, the problem becomes, *how can we design a MAB choice strategy that can adapt to the score functions expressing the properties of interest?* Our first answer to this question is to include in the score function definition an exploratory budget ε that can be used to correct the "greediness" of the MAB basic greedy strategy, thus obtaining a "ε-greedy" strategy. We remark that how to spend this budget during the monitor operation (i.e., when to make a non-greedy choice, selecting a model in the pool that does not provide the best score) can itself be learnt, by supervised learning or by classical reinforcement learning should reward/penalties for holding/not holding the property be available.

Efficient Design and Implementation of the MAB Strategy. Executing the choice strategy at each inference can be time and resource consuming. Today's practice of using Monte Carlo simulation can be adopted to check termination of MAB experiments. The Monte Carlo procedure randomly draws samples from each of the K arms multiple times and computes how often each of the arms wins (highest score). If the winning arm is beating the second arm by a large enough margin, the procedure declares a winner and stops. More advanced criteria can be investigated, developing on the notion of residual value. Google Analytics made a first step in this direction by introducing the concept of "value remaining in experiment" [40]. In each Monte Carlo simulation, the remaining value is computed. The experiment terminates when 95% of the samples in a Monte Carlo simulation have a remaining value of less than 1% of the winning arm's value.

Build Trusted MAB Monitor. AMAB monitor (Fig. 3) and its ML model pool need to be digitally signed and certifiably tamper-proof (e.g., using tamper-proof probes for system assurance [41]). Any trusted implementation must rely on a chain of trust, where the root-of-trust is in a tamper-resistant hardware co-processor. The trusted hardware checks a part of the for integrity a part of the monitor that in turn checks other parts. Our system detects illegal modifications to monitors, loadable ML models and user applications.

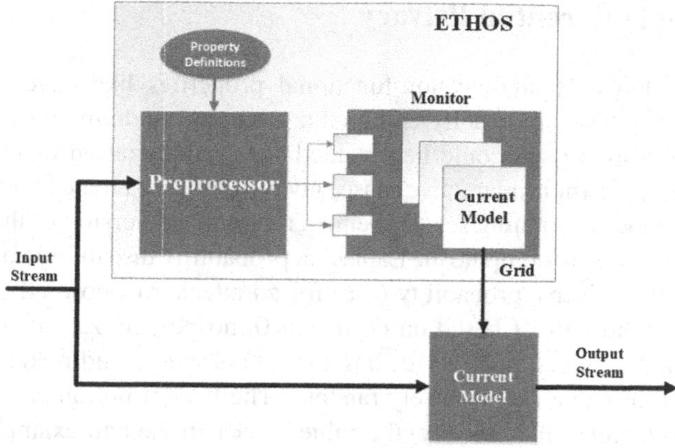

Fig. 3. A MAB monitor handling a ML model pool.

3.3 Ecosystem

Figure 4 provides a high-level view of the actors and of the complete eco-system. The definition of non-functional properties of interest (directly via suitable score functions or, as discussed above, via sets of desired inferences) can be carried out as an interactive process involving the entire community and the regulatory agencies interested in controlling ML models behavior in any given domain. Public property definitions, signed by the authority who released them, are made available to any platform (implemented as software, firmware or even hardware) capable of hosting MAB monitors. Suppliers of ML models use development environments capable of generating model pools and deploy them in environments suitable for monitoring. User organizations that run their ML models (e.g. for making healthcare, transport, telecommunication or FinTech decisions) exhibit to their own customers and all stakeholders the digital certificates asserting the properties of the output flow.

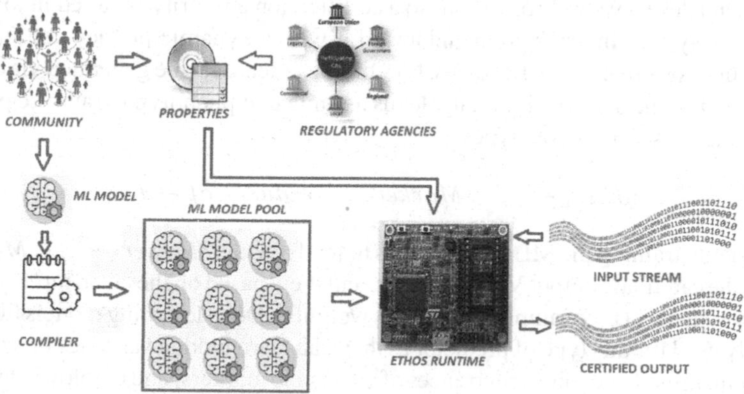

Fig. 4. Actors and eco-system

4 Certifying Differential Privacy

Some crucial and widely studied non-functional properties like differential privacy preservation [42], which are usually enforced a priori by randomizing input distributions D_{in} on the training set, could be obtained via randomization of the ML model output. In the a priori (non-interactive) case, given a privacy budget ε and an *([I],O)* entry of the ML model's training set (*[I]* being a randomized version of the deterministic input I), it is possible to compute the Laplacian probability distribution of *[I]* centered around *I*, that will deliver a probability (1-ε) for an attacker of correctly guessing the output *O'* corresponding to *I* based on *O*. If $\varepsilon = 0$, no randomization is present, and *O* can be obtained with certainty by feeding the model with *I* and reading the output; if $\varepsilon = 1$, the output value *O'* is entirely random. The budget notion can be useful for establishing at run-time (interactively) the value of $x_{S,T}$ in the bail example introduced above; if the budget for $x_{S,T}$ is large, any output distribution (all the ML models in the pool) will satisfy the constraint. If we enforce the "fair" value $x_{S,T} = |S|/|T|$, fewer models in the pool will provide a high score. The budget is used to instantiate the choice strategy of the MAB monitor as discussed in the sample scenario below.

Background. Alice has decided to have her genome sequenced by an online service to explore her family tree. The ancestry service is free but requires Alice's consensus to contribute her genomic data and some of her phenotypic traits to support the research being done at a medical school, which will perform the DNA extraction. Research at the medical school is funded by a software company who runs a software pipeline where user genomics data are first stripped of personal data and then used to train an ML model on the company's premises.

The trained model is deployed as a cloud-based service for hospitals to use. After some time, Alice discovers that the local hospital where she is treated has started to use the ML model developed by the medical school for predicting phenotypic traits. She is worried that the hospital staff – knowing her healthcare information and her traits - might identify her as a contributor to the model's training set. The country where the software resides mandates privacy certification, so a certification authority is called in to monitor the software system in production and award it with a dynamic privacy certificate that will reassure users like Alice. Let us see how this certificate can be generated via dynamic monitoring. For the sake of simplicity, let us assume that phenotypic traits are predicted based on markers within genotypes.

$$Marker_1 - \ldots - Marker_n \rightarrow Value - of - Trait$$

In general, training the ML model means to feed it with $[Marker_1 - \ldots - Marker_n]$, compute the candidate output Value-of-Trait, and then, based on the error, modify the predictor's parameters (for example the model's weights). Deep Learning models like CNNs are widely used for this type of predictor as they can learn hidden variables that represent features emerging from joint occurrences of genotypic markers, and employ convolution, sampling and dropout strategies to reduce the complexity of high-dimensional marker data. For the sake of simplicity, let us consider a simple Voronoi model (also called Nearest-Neighbor), where the Voronoi grid coincides with the training set. When the model is fed with a vector of markers, the model provides the Value-of-Trait of the closest element on the grid, according to a suitable distance function.

The Four Steps. The software house's development environment deploys a model pool of n reduced Voronoi models, each missing some points in the Voronoi grid (*Step 1*). At each step, the MAB monitor chooses the model whose Voronoi grid excludes enough points to ensure that each point's contribution to the sequence of inferences made until that moment lies within budget ε. In the simplest case, say $\varepsilon = 0.2$, the MAB monitor will ensure that the model omitting Alice's point will be used at least 20% of the times. Any observer looking at the inference flow will not be able to assert with certainty Alice's data presence in the training dataset.

The MAB monitor publishes a signed digital certificate with the value of ε (*Step 2*). Composing this certificate with the certificates of other pipeline stages (e.g., about the stripping of personal data) the hospital generates a virtual privacy certificate for the entire system (*Step 3*). The certificate is dynamic: if the monitor cannot deliver the budget ε (for instance due to insufficient pool size), it will immediately invalidate it. The certificate is also virtual, being a machine-readable artefact valid while the software pipeline instance is in operation.

The certificate is periodically re-generated and signed to be checked by users like Alice (*Step 4*).

5 Conclusions

We put forward the notion of a novel certification framework for ML models based on three ideas: *(i)* define the desired behavioral properties of ML models as score functions on the models' output; *(ii)* use statistical monitoring of ML models' behavior at inference time to efficiently compute scores and check that the desired behavioral properties are achieved; *(iii)* compose monitors' outcome within dynamic virtual certificates for composite AI software applications.

Acknowledgements. Research supported, in parts, by EC H2020 Project CONCORDIA GA 830927 and Università degli Studi di Milano under the program "Piano sostegno alla ricerca 2018".

References

1. Mitkov, R.: Anaphora resolution: the state of the art. School of Languages and European Studies, University of Wolverhampton, pp. 1–34 (1999)
2. Lewis, B.: In the game: the interface between Watson and Jeopardy! IBM J. Res. Dev. **56**(34), 171–176 (2012)
3. Dafoe, A.: AI Governance: A Research Agenda. Governance of AI Program, Future of Humanity Institute, University of Oxford, Oxford, UK (2018)
4. Guizzardi, R., Li, F.-L., Borgida, A., Mylopoulos, J.: An ontological interpretation of non-functional requirements. In: Frontiers in Artificial Intelligence and Applications, Proceedings of the 8th International Conference on Formal Ontology in Information Systems (FOIS) (2014)

5. Kesner, R.P., Gilbert, P.E., Wallenstein, G.V.: Testing neural network models of memory with behavioral experiments. Curr. Opin. Neurobiol. **10**(2), 260–265 (2000)
6. Schelter, S., Böse, J.-H., Kirschnick, J., Klein, T., Seufert, S.: Automatically tracking metadata and provenance of machine learning experiments. In: Proceedings of Workshop on ML Systems at NIPS 2017, Long Beach, CA, USA (2017)
7. Ardagna, C.A., Bellandi, V., Bezzi, M., Ceravolo, P., Damiani, E., Hebert, C.: Model-based big data analytics-as-a-service. Take Big Data to the Next Level. IEEE Transactions on Services Computing (Early Access) (2018)
8. Redavid, D., et al.: Semantic support for model based big data analytics-as-a-service (MBDAaaS). In: Proceedings of 12th International Conference on Complex, Intelligent, and Software Intensive Systems (CISIS-2018), pp. 1012–1021, Matsue, Japan (2018)
9. Di Martino, B., Esposito, A., Damiani, E.: Towards AI-powered multiple cloud management. IEEE Internet Comput. **23**(1), 64–71 (2019)
10. Khosravi, P., Liang, Y., Choi, Y., Van den Broeck, G.: What to expect of classifiers? reasoning about logistic regression with missing features. In: Proceedings of the ICML Workshop on Tractable Probabilistic Modeling (TPM), pp. 2716–2724, Macao, China (2019)
11. Biggio, B., Roli, F.: Wild patterns: Ten years after the rise of adversarial machine learning. Pattern Recogn. **84**, 317–331 (2018)
12. Bryson, J., Winfield, A.: Standardizing ethical design for artificial intelligence and autonomous systems. Computer **50**(5), 116–119 (2017)
13. Morley, J., Floridi, L., Kinsey, L., Elhalal, A.: From What to How: An Overview of AI Ethics Tools, Methods and Research to Translate Principles into Practices (2019). arXiv: 1905.06876v1
14. McNamara, D., Soon Ong, C., Williamson, R.C.: Costs and benefits of fair representation learning. In: Proceedings of the 2019 AAAI/ACM Conference on AI, Ethics, and Society, pp. 263–270, ACM, Honolulu, HI, USA (2019)
15. Adel, T., Valera, I., Ghahramani, Z., Weller, A.: One-network adversarial fairness. In: Proceedings of 33rd AAAI Conference on Artificial Intelligence, Honolulu, HI, USA (2019)
16. Raff, E., Sylvester, J., Mills, S.: Fair forests: regularized tree induction to minimize model bias. In: Proceedings of the 2018 AAAI/ACM Conference on AI, Ethics, and Society, pp. 243–250, ACM, New Orleans, LA (2018)
17. Madras, D., Creager, E., Pitassi, T., Zemel, R.: Fairness through causal awareness: learning causal latent-variable models for biased data. In: Proceedings of the ACM Conference on Fairness, Accountability, and Transparency, pp. 349–358, ACM, Atlanta, GA (2019)
18. Rafael, Y., et al.: Four ethical priorities for neurotechnologies and AI. Nat. News **551**(7679), 159 (2017)
19. Morris, J., Lee, G., Parker, K., Bundell, G.A., Lam, C.P.: Software component certification. Computer **34**(9), 30–36 (2001)
20. Damiani, E., Manã, A.: Toward ws-certificate. In: Proceedings of the 2009 ACM Workshop on Secure Web Services, pp. 1–2, ACM, Chicago, IS, USA (2009)
21. Damiani, E., Ardagna, C.A., El Ioini, N.: Open Source Systems Security Certification. Springer, Berlin (2009). https://doi.org/10.1007/978-0-387-77324-7
22. Spanoudakis, G., Damiani, E., Maña, A.: Certifying services in cloud: the case for a hybrid, incremental and multi-layer approach. In: Proceedings of IEEE 14th International Symposium on High-Assurance Systems Engineering (HASE 2012), pp. 175–176, IEEE, Singapore (2012)
23. Anisetti, M., Ardagna, C.A., Damiani, E., Polegri, G.: Test-based security certification of composite services. ACM Trans. Web (TWEB) **13**(1), 3 (2019)
24. Méry, D., Kumar Singh, N.: Trustable formal specification for software certification. In: Proceedings of Symposium on Leveraging Applications of Formal Methods, Verification and Validation, 2010. Lecture Notes in Computer Science, vol. 6416, pp. 312–326 (2010)

25. Denney, E., Pai, G.: Evidence arguments for using formal methods in software certification. In: Proceedings of IEEE Symposium on Software Reliability Engineering (ISSRE), pp. 375–380, IEEE, Pasadena, CA, USA (2013)
26. Armando, A., et al.: The AVISPA tool for the automated validation of internet security protocols and applications. In: Proceedings of CAV 2005: Computer Aided Verification. Lecture Notes in Computer Science, vol. 3576, pp. 281–285 (2005)
27. Clarkson, M.R., Schneider, F.B.: Hyperproperties. J. Comput. Secur. **18**(6), 1157–1210 (2010)
28. Datta, A., Franklin, J., Garg, D., Jia, L., Kaynar, D.: On adversary models and compositional security. IEEE Secur. Priv. **9**(3), 26–32 (2011)
29. Fuchs, A., Gürgens, S.: Preserving confidentiality in component compositions. In: Proceedings of International Conference on Software Composition. Lecture Notes in Computer Science, vol. 8088, pp. 33–48 (2013)
30. Mellado, D., Fernández-Medina, E., Piattini, M.: A common criteria based security requirements engineering process for the development of secure information systems. Comput. Stan. Interfaces **29**(2), 244–253 (2007)
31. Scott, S.L.: Multi-armed bandit experiments in the online service economy. Appl. Stochast. Models Bus. Ind. **31**, 37–49 (2015)
32. Leite, R., Pavel, B., Vanschoren, J,: Selecting classification algorithms with active testing. In: Proceedings of MLDM 2012: Machine Learning and Data Mining in Pattern Recognition. Lecture Notes in Computer Science, vol. 7376, pp. 117–131 (2012)
33. Antos, A., Grover, V., Szepesvári, C.: Active learning in multi-armed bandits. In: Freund, Y., Györfi, L., Turán, G., Zeugmann, T. (eds.) ALT 2008. LNCS (LNAI), vol. 5254, pp. 287–302. Springer, Heidelberg (2008). https://doi.org/10.1007/978-3-540-87987-9_25
34. Menke, J., Martinez, T.R.: Using permutations instead of student's t distribution for p-values in paired-difference algorithm comparisons. In: Proceedings of 2004 IEEE International Joint Conference on Neural Networks, vol. 2, pp. 1331–1335, IEEE, Budapest, Hungary (2004)
35. Vanschoren, J.: Meta-learning: A survey (2018). arXiv:1810.03548
36. Damiani, E., Tettamanzi, A., Liberali, V.: On-line evolution of FPGA-based circuits: a case study on hash functions. In: Proceedings of the First NASA/DoD Workshop on Evolvable Hardware, pp. 26–33, IEEE, Pasadena, CA, USA (1999)
37. Brennet, T., Dieterich, W., Ehret, B.: Evaluating the predictive validity of the COMPAS risk and needs assessment system. Crim. Justice Behav. **36**(1), 21–40 (2008)
38. Lai, T.L., Robbins, H.: Asymptotically efficient adaptive allocation rules. Adv. Appl. Math. **6**(1), 4–22 (1985)
39. Corbett-Davies, S., Pierson, E., Feller, A., Goel, S., Huq, A.: Algorithmic decision making and the cost of fairness. In: Proceedings of the 23rd ACM SIGKDD International Conference on Knowledge Discovery and Data Mining, pp. 797–806, ACM, Halifax, NS, Canada (2017)
40. Scott, S.L.: Applied stochastic models in business and industry. Appl. Stoch. Models Bus. Ind. **26**, 639–658 (2010)
41. Anisetti, M., Ardagna, C.A., Gaudenzi, F., Damiani, E., Diomede, N., Tufarolo, P.: Moon cloud: a cloud platform for ICT security governance. In: Proceedings of IEEE Global Communications Conference (GLOBECOM 2018), pp. 1–7, IEEE, Abu Dhabi, UAE (2018)
42. Dwork, C., McSherry, F., Nissim, K., Smith, A.: Calibrating noise to sensitivity in private data analysis. TCC 2006: Theory of Cryptography. Lecture Notes in Computer Science, vol. 3876, pp. 265–284 (2006)

The Lost Recipes from the Four Schools of Amathus
Invited Talk Extended Abstract

Gunnar W. Klau(✉)

Department of Computer Science, Algorithmic Bioinformatics group, Heinrich Heine University Düsseldorf, Düsseldorf, Germany
gunnar.klau@hhu.de

Abstract. This paper tells the story of the Four Schools of Amathus and the lost recipes for a legendary dish that attracted many people to the ancient royal city. Deciphering the recipes from snippets of the ancient scrolls that have been found recently seems to be an impossible task. Fortunately, the problem bears a strong resemblance to the haplotype phasing problem, which has been studied in more recent times. We point out the similarities between these problems, how they can be formulated as optimization problems and survey different solution strategies.

Keywords: Amathus · Haplotype phasing · Polyploid species

1 The Four Schools of Amathus

Amathus was an ancient royal city on the beautiful island of Cyprus. Its remains are located on the southern coast of the island, close to the present city of Limassol. Amathus is known as one of main centres of worship of Aphrodite, the goddess of love and beauty in Greek mythology. Also, according to one of the many versions of the Ariadne myth, it is the place where pregnant Ariadne was abandoned by Theseus and then died during childbirth. What many people do not know, however, is that the city was also famous for a delicious dish, called Amathubrosia. The story of this dish is related to the Four Schools of Amathus and the subject of many legends.

Here is what happened, according to the legends: Each year in midsummer, during Aphrodisia, the main festival to worship the goddess Aphrodite, the four famous chefs Athena, Barnabas, Charalambos, and Daphne met in the Temple of Aphrodite to create Amathubrosia in honor of the goddess. This attracted people from all over the Mediterranean Sea who came to Amathus to taste their creation. Although the four chefs had slightly different ideas on how to prepare Amathubrosia, they always found a compromise and agreed on a common recipe according to which they prepared the dish.

Funded by the Deutsche Forschungsgemeinschaft (DFG, German Research Foundation) – 395192176.

A. Chatzigeorgiou et al. (Eds.): SOFSEM 2020, LNCS 12011, pp. 16–23, 2020.
https://doi.org/10.1007/978-3-030-38919-2_2

One year, however, the four chefs did not agree anymore and had a terrible argument. What used to be cooperation became competition, and everybody kept their recipe as a secret. Each chef founded a school, which, together, became to be known as the Four Schools of Amathus. See Fig. 1 for an illustration. The four chefs wrote down their recipes for Amathubrosia on scrolls, which were kept in sacred rooms in the corresponding schools. The four schools even added to the fame of Amathus, and more and more people came to visit the Aphrodisia festival each year to find out which of the four Amathubrosia variants they liked best. Preparing these dishes became a quite profitable business, and, as a consequence, many more chefs were needed.

Fig. 1. The founders of the Four Schools of Amathus. From left to right: Athena, Barnabas, Charalambos and Daphne.

To accommodate the increased demand, the schools took on apprentice chefs, who were taught the basics of preparing food by the four famous heads of the schools. Whenever an apprentice chef passed the final cooking exam, they were allowed to enter the secret room of their school where the original recipe of Athena, Barnabas, Charalambos or Daphne was kept. The apprentices were allowed to make a copy for themselves, which they had to keep as a secret as long as they lived. Because they did not have much time to copy the recipe and because the rooms were rather dark, they all made a number of errors when copying the recipes.

Whenever a chef died, it was tradition that his or her copy of the school's recipe was ripped to many small pieces and buried along the chef in a cave in a sacred grove close to Amathus, which was the traditional burying ground of all chefs of the four schools.

Unfortunately, when Amathus was destroyed, the Four Schools of Amathus were destroyed as well and the tradition of preparing the tasty dish was lost. People still try to prepare Amathubrosia these days, but the result lacks the small changes added by the famous four chefs and just tastes boring. Luckily, the cave was rediscovered recently and archaeologists got hold of the bones of the ancient chefs and the snippets of their recipes.

As the snippets are all mixed together and the many copies of recipes from the different schools all bear individual errors made by the chefs who copied the recipe it seems an impossible task to reconstruct the original recipes from the Four Schools of Amathus. How this can be done and how this task is related to a more modern problem is the topic of the remainder of this paper.

2 Deciphering the Scrolls

Now that the knowledge on how to prepare the delicious dish has reappeared, it seems an obvious idea to decipher the four ancient recipes in order to prepare the original dishes. This is, however, not so easy as it may seem. What we have at hand is a large number of snippets, which, as explained above, are a mixture of small fragments of recipes from different cooks of the four different schools. One strategy to get to the original four recipes would be to compare the snippets to the only reference we have today, which is the recipe for the boring nowadays version of Amathubrosia, which can be found in the library of Limassol. We can map all fragments to this recipe and can thus discover locations of the recipe where the four variants differ. This information can be characterized by an $n \times m$ fragment matrix F, where n is the number of positions in the reference recipe where differences occur and m is the number of snippets. Each entry $F_{i,j}$ in the matrix tells us how snippet i looks at position j of the reference recipe. We either have $F_{i,j} =$ '–' if the snippet does not carry any information at all for this position or $F_{i,j} \in \{0, 1, 2, 3\}$, where the differences are numbered arbitrarily but consistently for each position.

In principle, we now seek a four-partition of the snippets, where each partition gives us the original recipe of the corresponding school. The problem, however, is that the snippets not only contain the differences of the recipes of the Four Schools, but also the individual errors each apprentice made when copying from the master scrolls.

Should we ever want to prepare an Amathubrosia dish as delicious as the ones created by Athena, Barnabas, Charalambos, or Daphne, we have to find a way to distinguish between the important differences of the four variants of the dish and the individual errors.

3 Computational Genomics and Read-Based Haplotype Phasing

Fortunately, computational genomics, a seemingly distant area of research, may help us with the recovery of the four ancient recipes. The young field of genomics already has a major impact on individuals and society and will do so much more in the near future. Recent advances in sequencing technology are transforming medical and fundamental research: Large genotype-phenotype studies are now being carried out routinely and yield new insights about the genetic basis of disease and drug response. These advances in medical genomics enable precision-medicine approaches for the treatment of patients, which are becoming more and more widespread and successful. Other fields, such as population genomics, benefit from the possibility to study millions of genetic loci in large populations.

In most species, the genome of individual cells is organized in a number of chromosomes. Each chromosome exists in k copies, which are usually inherited from mother and father. The parameter k is called the *ploidy* of the genome. Many species, for example humans, are *diploid*, that is $k = 2$. However, organisms of higher ploidy exist such as many plant species including important food crops like potato, wheat or maize.

One way to study individual genomes is on the level of *genotypes*. Genotyping refers to determining the alleles inherited from the parents present at a particular genetic locus and can be achieved using various technologies including microarrays and short-read sequencing. Using genotype-level genomics it remains unknown, however, on which of the k chromosome copies such a variant resides, which makes the information passed on to down-stream analyses incomplete. The full sequences of the chromosomal copies are known as haplotypes. In contrast to genotyping, *haplotyping* aims to reconstruct the full sequences of all k haplotypes. Moving from genotypes to haplotypes is known as *phasing*.

Most work for read-based haplotype phasing has been presented for the diploid case ($k = 2$), where the most popular model is the Minimum Error Correction (MEC) model [4]. The survey [3] is a guided tour of computational haplotyping and focuses on the characteristics of problem instances resulting from different present day technologies and a survey of relevant techniques. The following presentation is an extension of the notation used in [3] and [8].

As the key challenge in molecular haplotyping is to distinguish true genetic variability from sequencing errors, the MEC model asks for a minimum cost correction of the sequencing data to allow a conflict-free partition of the reads to k chromosomal copies.

The input for MEC is a *fragment matrix* $F \in \{0, 1, \ldots, t-1, \text{`}-\text{'}\}^{m \times n}$, where the rows $1 \leq i \leq m$ correspond to the fragments and the columns $1 \leq j \leq n$ correspond to the variant positions and $t$ is the maximum number of alleles considered. A non-gap entry $F(i, j) \neq \text{`}-\text{'}$ specifies that read i covers variant j and gives evidence for one of the t possible alleles at this position. Typical genomic analyses consider only a major and a minor allele at each position, that is $t = 2$. In any case, t is bounded by the ploidy k.

Two rows i_1 and i_2 of F are in *conflict* if there is a position j such that $F(i_1, j) \neq \text{`}-\text{'}$ and $F(i_2, j) \neq \text{`}-\text{'}$ but $F(i_1, j) \neq F(i_2, j)$. A set of rows is *conflict-free* if it does not contain conflicting row pairs.

A fragment matrix F is *k-feasible* if a partition $(I^0, \ldots, I^{k-1})$ of its rows exists such that all parts $I^0, \ldots, I^{k-1}$ are conflict-free. Such a partition determines the k haplotypes h^l with $l \in \{0, \ldots, k-1\}$ in the following, natural way:

$$h^l(j) = F(i, j) \text{ for some } i \in I_l.$$

In practice, the entries in a fragment matrix are associated with 'phred-scaled' base qualities $Q \in \mathbb{N}^{m \times n}$ that correspond to the estimated probabilities of $10^{-Q(i,j)/10}$ that entry $F(i,j)$ has been wrongly sequenced. These phred scores serve as costs of flipping entries and allow less confident base calls to be corrected at lower cost compared to high confidence ones. The *error distance* of two fragment matrices F and F' is

$$d_Q(F, F') = \sum_{i=1}^{m} \sum_{j=1}^{n} \begin{cases} 0 & F(i,j) = F'(i,j) \\ Q(i,j) & F(i,j) \neq F'(i,j). \end{cases}$$

We can now state the weighted Minimum Error Correction problem in the k-ploid case formally as follows: Given ploidy k, a fragment matrix F and a quality matrix Q, find a k-feasible fragment matrix F' with minimum error distance $d_Q(F, F')$.

As most existing approaches focus on the important diploid special case, we first summarize the contributions we made for $k = 2$ before we move on the more challenging task to phase polyploid genomes.

Diploid Phasing

Diploid phasing in the MEC model is NP-hard [4] and even hard to approximate within a constant factor [8]. The problem remains hard even in the gapless case, that is, if all non-gap entries in the fragment matrix appear in consecutive order [2]. In the binary case, that is, if no gaps exist at all, it is an open problem whether the problem is hard.

For phasing diploid genomes, we presented WhatsHap [7], which is a dynamic programming algorithm that solves the weighted MEC model for $k = 2$ to provable optimality. The algorithm is linear in the number of variants and thus in the size of the genome, but exponential in the maximum coverage of a position. By enumerating read bipartitions in Gray code order, WhatsHap achieves a runtime of $O(2^c n)$, where c is the maximum coverage across all columns. In particular, the running time of the DP does not depend on the read length, which is beneficial for long-read sequencing data.

Later, Pirola et al. [8] considered a restricted variant of MEC, in which up to k corrections are allowed per position, and presented an FPT algorithm that runs in time $O(c^k L n)$, where L is the maximal number of variants covered by any read.

These DP-based algorithms work well for maximum coverage values up to 25. Instances with higher coverage cannot be solved to optimality in reasonable computing time with these methods. However, a clever read sampling strategy as described in [5] leads to excellent results in practice.

Polyploid Phasing

Phasing diploid genomes has become a fast routine step where solutions based on the MEC model as described above provide excellent solutions, whose quality depends more on the quality of the read data than on the underlying algorithmic approach.

Polyploid phasing, however, presents a bigger challenge. As for diploid phasing most work has been based on the MEC model, in which the higher ploidy of course increases the size of the solution space and makes the solution computationally even more challenging. Clearly, as the diploid MEC is a special case, most theoretical results also hold for the polyploid case, see also [1]. Interestingly, the binary MEC, in which no gaps appear in the input, is NP-hard for $k \geq 3$ [2], but the question whether the problem is approximable within a constant factor is still open. Regarding practical algorithms, the problem is still considered to be largely unsolved. The authors of a recent survey concluded diplomatically that there is "clearly room for improvement in polyploid haplotyping algorithms" [6].

In our current work [9], we identify the MEC model as one of the reasons why current polyploid phasing algorithms do not work well in practice. Polyploid genomes usually exhibit large regions of two or more identical haplotypes. For the MEC model there is no benefit in assigning conflict-free reads to the same haplotype. Instead, the model favors collapsing regions of locally identical haplotypes into one partition and uses the free partitions for noisy reads in order to reduce the MEC score. Consequently, MEC-based approaches for polyploid phasing struggle in such regions.

We therefore propose a new model that differs from the limited MEC paradigm. The key assumption is that each haplotype should be covered by a uniformly divided share of reads. We take the coverage into account within a newly established threading step, in which the haplotypes are threaded through clusters of reads that are likely to belong to the same haplotype or to identical haplotypes. This enables us to detect and properly phase regions where multiple haplotypes coincide. We also introduce cuts within the haplotypes at positions with increased phasing uncertainty and thereby output phased blocks that ensure high accuracy within the fragments. We provide a sensible way to compute these block boundaries at varying, user-defined degrees of strictness. This way, we enable a configurable trade-off between longer blocks that potentially contain errors and shorter but highly accurate blocks. We show that our method returns results that are more accurate than those computed by the state-of-the-art tools, especially in regions of identical haplotypes. The approach scales to gigabase-sized genomes: we can, for example, phase an artificial human tetraploid chromosome 1 in less than 3.5 h on a single core of a standard desktop. See Fig. 2 for an illustration.

The exact model the algorithm is addressing is difficult to describe. What comes close is the following formulation: Given ploidy k, a fragment matrix F and a quality matrix Q, find a k-feasible fragment matrix F' with minimum score

$$\alpha\, d_Q(F, F') + (1 - \alpha) \sum_{j=1}^{n} \sum_{l=1}^{k} w(c_{jl}, c_j, k),$$

where $d_Q(F, F')$ is the MEC error distance from above and $w(c_{jl}, c_j, k)$ is a term that penalizes haplotype l at position j depending on its coverage c_{jl} in

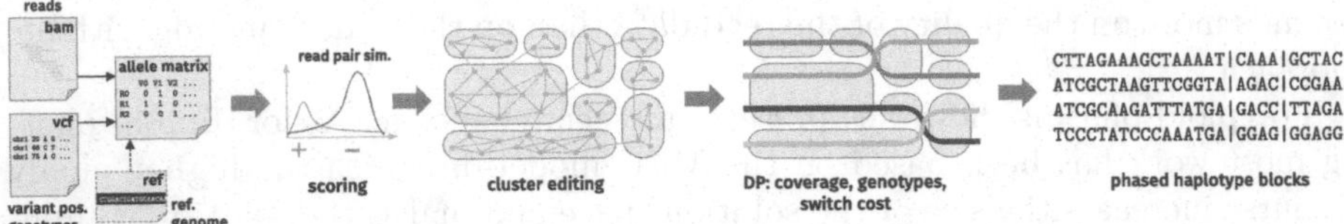

Fig. 2. Overview of the new polyploid phasing model and algorithm. After parsing the input and generating the allele matrix F, we compute a statistical score for each read pair. This score is a first indication whether two reads should rather be assigned to the same haplotype (or identical haplotypes) or to two different haplotypes. We apply cluster editing, a graph-based clustering technique, to a read graph weighted by these scores (grey round shapes). Subsequently, we take the coverage into account by threading k haplotypes (colored lines) through the clusters (here $k = 4$). Here, we can balance out violations of the coverage for each haplotype with costs for switching between haplotypes. We can also incorporate genotype information at this point, if available. The procedure results in k phased haplotypes, which are subdivided into blocks (vertical lines).

relation to the total coverage c_j at this position. Parameter α models the trade-off between these possibly conflicting terms. Note that the coverage term can, for example, be set according to a log likelihood model based on the binomial distribution.

We encourage researchers to study this model from a theoretical perspective, that is, to investigate computational complexity variants and special cases, to establish approximation results and to analyze the fixed parameter tractability of this model with respect to meaningful parameters. We also hope to stimulate further algorithmic work addressing this model, which may lead to practically efficient polyploid phasing tools.

4 Back to Amathus

Obviously, reconstructing the recipes can be cast and solved as a haplotype phasing problem, where $k = 4$, and the number of alleles is $t = 4$. Since it is conceivable that also in the recipes long stretches are similar—after all, the chefs disagreed, but they did not disagree maximally on each detail—a strategy as outlined at the end of Sect. 3 may lead to the four original recipes. Of course, this depends also on the length of the fragments, the number of variants they cover and the total number of fragments. This is not so much different for data in computational genomics. Bon appetit!

Acknowledgments. I wish to thank all the co-authors of the various papers on haplotype phasing mentioned in Sect. 3, especially Sven Schrinner and Tobias Marschall, Jana Ebler and Rebecca Serra Mari for the nice collaboration on the polyploid phasing problem. Thanks to Philipp Spohr for inspirations shaping the Amathus story and to Nguyen Khoa Tran for providing the beautiful drawing of Athena, Barnabas, Charalambos and Daphne.

References

1. Bonizzoni, P., Dondi, R., Klau, G.W., Pirola, Y., Pisanti, N., Zaccaria, S.: On the minimum error correction problem for haplotype assembly in diploid and polyploid genomes. J. Comput. Biol. **23**(9), 718–736 (2016)
2. Cilibrasi, R., van Iersel, L., Kelk, S., Tromp, J.: On the complexity of several haplotyping problems. In: Casadio, R., Myers, G. (eds.) WABI 2005. LNCS, vol. 3692, pp. 128–139. Springer, Heidelberg (2005). https://doi.org/10.1007/11557067_11
3. Klau, G.W., Marschall, T.: A guided tour to computational haplotyping. In: Kari, J., Manea, F., Petre, I. (eds.) CiE 2017. LNCS, vol. 10307, pp. 50–63. Springer, Cham (2017). https://doi.org/10.1007/978-3-319-58741-7_6
4. Lippert, R., Schwartz, R., Lancia, G., Istrail, S.: Algorithmic strategies for the single nucleotide polymorphism haplotype assembly problem. Briefings Bioinf. **3**(1), 23–31 (2002)
5. Martin, M., et al.: WhatsHap: fast and accurate read-based phasing. bioRxiv (2016). https://doi.org/10.1101/085050, https://www.biorxiv.org/content/early/2016/11/14/085050
6. Motazedi, E., Finkers, R., Maliepaard, C., de Ridder, D.: Exploiting next generation sequencing to solve the haplotyping puzzle in polyploids: a simulation study. Briefings Bioinf. **19**(3), 387–403 (2018)
7. Patterson, M., et al.: WhatsHap: weighted haplotype assembly for future-generation sequencing reads. J. Comput. Biol. **22**(6), 498–509 (2015)
8. Pirola, Y., Zaccaria, S., Dondi, R., Klau, G.W., Pisanti, N., Bonizzoni, P.: HapCol: accurate and memory-efficient haplotype assembly from long reads. Bioinformatics **32**(11), 1610–1617 (2015). (Oxford, England)
9. Schrinner, S.D., et al.: Haplotype threading: accurate polyploid phasing from long reads (2019, in preparation)

Sharing Energy for Optimal Edge Performance

Erol Gelenbe[1,2(✉)] and Yunxiao Zhang[2,3]

[1] Institute of Theoretical and Applied Informatics, Polish Academy of Sciences, ul. Baltycka 5, 44100 Gliwice, Poland
gelenbe.erol@orange.fr

[2] Laboratoire I3S, Université Côte d'Azur, 06108 Cedex2, Nice, France
yunxiao.zhang15@imperial.ac.uk

[3] Imperial College, London, UK

Abstract. Using the Energy Packet Network (EPN) model, we show how energy can be shared between heterogenous servers at the edge to minimize the overall average response time of jobs. The system is modeled as a probabilistic network where energy and jobs are being dispatched to the edge servers using G-Networks with a product-form solution for the equilibrium probability distribution of system state. The approach can also be used to design energy dispatching systems when renewable energy is used to improve the sustainability of edge computing.

1 Introduction

Heterogeneous sensors, other digital devices and computer servers or workstations (WS) are being incorporated into the Internet of Things (IoT) [3,7,28,33, 41] to manage cities, services and industry [44] with applications in practically all areas of social activity [1,4,26], creating massive energy requirements that can benefit from energy harvesting from wind, fluid flows, photovoltaic, and electromagnetic fields, with energy stores (ES) such as batteries to buffer the effect of intermittent energy sources [5,36,40]. Harvested energy can contribute to the sustainability of information and computer technology (ICT) [19,39], but it raises new questions. Research is needed to understand how system Quality of Service (QoS) can be maintained in the presence of intermittent harvested energy [6,10,17,30,34,42], including optimal network routing for energy savings [31,35], data transmission for scheduling for energy usage optimization [2], and greater energy efficiency in data centers [32] needed to process the massive data from the IoT.

Recent work on the Energy Packet Network (EPN) paradigm [9,14–16] has proposed a discrete state-space modeling approach to evaluate the QoS and energy consumption in systems where computer jobs, data packets, and energy packets (EPs), interact in complex interconnected information processing and data transmission systems. EPNs were recently applied to backhaul networks operating with renewable energy sources [18]. Other work has suggested hardware schemes for simultaneously forwarding both data packets and energy [37,38].

A. Chatzigeorgiou et al. (Eds.): SOFSEM 2020, LNCS 12011, pp. 24–36, 2020.
https://doi.org/10.1007/978-3-030-38919-2_3

Previous work [25], discussed some optimization algorithms based on qenueing networks for dispatching network packets so as to minimize a composite cost function that combines overall network energy consumption and QoS. In [20], the EPN model has been used to study architectures which interconnect energy prosumer systems, so that energy consumption and leakage, and the response time to service requests, are minimized. In [21] a utility function, which is a linear combination of the throughput and the probability that the system does not run out of energy, maximized. The EPN paradigm has recently generated interest and further work [9,24,29,43] to model and optimize sensor networks and servers that operate with harvested energy.

In this paper we consider servers or workstations (WS), each of which is powered by a battery or energy store (ES) which is charged from a source of intermittent energy such as wind or photovoltaic. We assume that the energy is represented by discretized EPs, where one EP is the amount of energy needed to process one or more jobs depending on the different jobs being considered; this approach generalizes previous work where one EP corresponds to the amount of energy needed to process exactly one job or forward on data packet. EPs can also circulate in the system so that an ES can process transfer them to other WSs. Energy in batteries may be lost through leakage at a rate that depends on the particular EB.

Based on these assumptions, we assume that neither EPs nor jobs may be moved are not moved between WSs, and that the system receives a total fixed power rate, expressed in EPs per second. We are given a fixed distribution for the number of jobs that a single EP can process at any given WS, but this distribution may be different at the different WSs. The problem is then to select the fraction of jobs that we send to each of the WSs so as to minimize the overall average response time W of jobs. The case where we move a fraction D_i of the jobs at node i to some other server j according to a probability matrix $M = [M_{ij}]$ is discussed in [27].

In Sect. 2, we briefly discuss G-Networks, and relate the EPN model to this more general quenueing network model. We present the EPN model parameters in Sect. 3. Then we detail the optimization problem and a numerical example. We present conclusions and suggestions for further work in Sect. 4.

2 EPN and Its G-Network Representation

The EPN system considered is schematically presented in Fig. 1. In the approach taken in this paper, jobs or tasks that need to be executed in the system are modelled as ordinary customers in a queueing network. They arrive to any one of N WSs which are represented as queues. Jobs first arrive to a given WS, call it W_i, at rate λ_i jobs/sec. Each W_i has an energy storage battery denoted E_i, so that there are a total of N ESs. EPs arrive from an external intermittent energy source at rate γ_i EPs/sec to E_i which can be viewed as a "queue of EPs". We denote the number of jobs at WS W_i at time t by $K_i(t)$, while $B_i(t)$ denotes the number of EPs at E_i. EPs at the E_i are expended (locally consumed) or moved in the following manner:

- If $B_i(t) > 0$ then E_i will leak energy at some rate $\delta_i \geq 0$ EPs/sec. Thus when $B_i(t) > 0$, after a time of average value δ_i^{-1}, we will have one less EP at ES i due to energy leakage.
- ES S_i provides EPs at rate w_i when $B_i(t) > 0$. With probability M_{ij} an EP is moved to another other E_j so that $B_i(t^+) = B_i(t) - 1$, and $B_j(t^+) = B_j(t) + 1$. Such transfers may be made to share energy with other ESs which are being depleted more rapidly.
- Or with probability $d_i = 1 - \sum_{j=1}^{N} P_{ij}$ the EP is forwarded to W_i and with probability $1 \geq D_i \geq 0$, one EP is expended to serve a batch of up to b_i jobs at the WS. If $K_i(t) > 0$ then the EP will serve $max[K_i(t), b_i]$ jobs in one step and after service we end up with $K_i(t^+) = K_i(t) - max[K_i(t), b_i]$. Since different jobs may have different energy requirements when running at a given W_i, we assume that b_i (i.e., the number of jobs that are processed with a single EP at W_i), is a random variable with probability distribution $\pi_{is} = \Pr[b_i = s]$, $s = 1, 2, ...$
- With probability $1 - D_i$, if $K_i(t) > 0$ one EP will be used to serve just one job, and then forward that job to another W_j according to the transition probability matrix $M = [M_{ij}]$. As a result we will have $K_i(t^+) = K_i(t) - 1$, $K_j(t^+) = K_j(t) + 1$.
- If an EP arrives to a WS i and $K_i(t) = 0$, then the EP will just be expended to keep the WS in working order, and no jobs will be processed or moved.

2.1 The G-Network Model

The EP is a special case of G-Networks [8,11,12] which are queueing networks that have the remarkable "product form solution" which simplifies their computational structure. An EPN is a multi-class G-Network with Batch Removal [13,23]. This is an open queueing network with v of service stations or WSs. The EPN "jobs" can be computer programs that need to be executed, or data packets that need to be transmitted, and belong to one of C classes. Each ach customer class has distinct arrival rates to the network, and distinct routing probabilities in the network. Customer also belong to three Types, of "positive" and "negative" customers, or "triggers". Other types of customers include "resets" [22] and "adders" [8].

Positive customers are the normal queueing network customers which request and obtain service at the queues, and belong to one of the C classes. At all of the v queues, positive customers have i.i.d. exponential service times of rate $r(1), \ldots, r(v)$ which are identical for all classes of customers. After completing service and leaving a node i, a positive customer of class c can change into a positive customer of class c' at node j with probability $\Pi^+_{c,i,c',j}$, the corresponding transition probability matrix is $\Pi^+ = [\Pi^+_{c,i,c',j}]$, or the positive customer leaves the network with probability $l_{c,i}$, or it changes into a negative customer of class c' and join node j with probability $\Pi^-_{c,i,c',j}$, in which case it will remove, or "instantaneously serve", a batch of positive customers of class c', and the batch is of

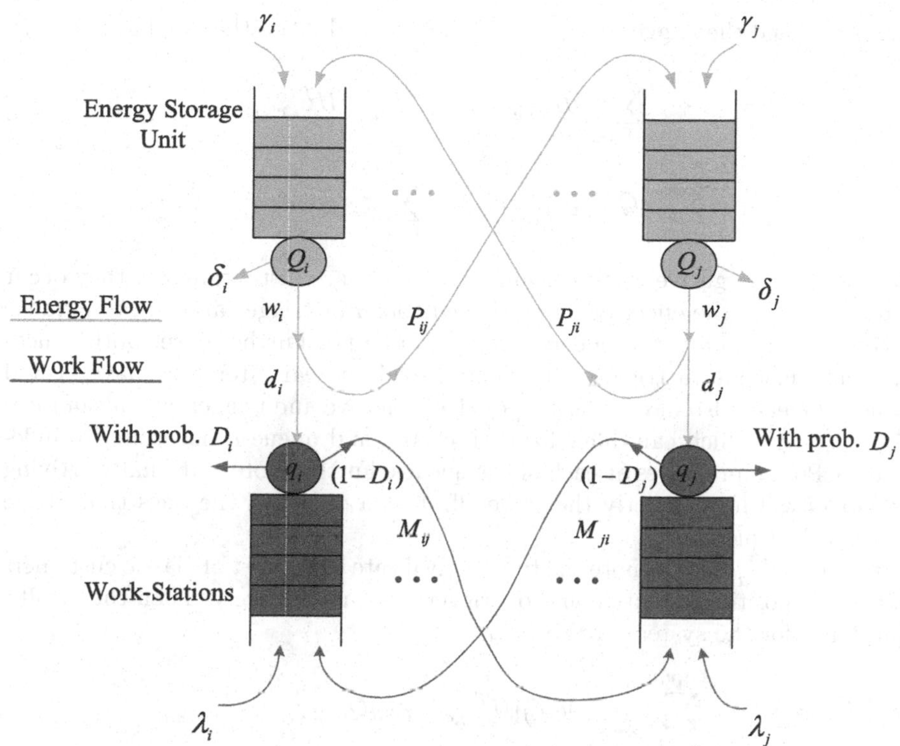

Fig. 1. A EPN system with N WSs and ESs. EPs are accumulated in the ESs denoted E_i, and jobs queue at the WSs denote W_i. The EPs can be forwarded to the corresponding W_i or moved to other ESs. Jobs in the W_i can finish processing locally or they may be forwarded to other WSs for further processing.

maximum size $B_{j,c',j}$ at queue j, where $B_{c',j}$ is a random variable with probability distribution $\pi_{c',j,s} = \Pr[B_{c',j} = s] \geq 0,\ s \geq 1$. If a negative customer of class c at node i arrives to queue j as a class c' customer at time t, and the number positive customers of class c' at j is $K_{c',j}(t)$, then a total of $max\ [K_{c',j}(t), B_{c',j}]$ positive customers of class c' will be instantaneously removed from the queue at j so that $K_{c',j}(t^+) = 0$ if $B_{,c',j} \geq K_{c',j}(t)$, and $K_{c',j}(t^+) = K_{c',j} - B_{c',j}$ if $B_{c',j} < K_{c',j}(t)$, and the negative customer disappears at time t^+. If $K_{c',j}(t) = 0$ then the negative customer disappears and no customer is removed from queue j. The positive customer of class c leaving queue i can become a "trigger" of class c' at queue j with probability $\Pi^T_{c,i,c',j}$, in which case it will move a class c' customer from queue j to queue l, and that customer becomes a class c'' customer at queue l, with probability $Q_{c',j,c'',l} \geq 0$. If queue j does not contain a class c' customer when the trigger arrives to queue j, then no customer is transferred from j to l, and the trigger disappears. These probabilities that we have satisfy:

$$1 = l_{c,i} + \sum_{c'=1,j=1}^{C,v} [\Pi^{+}_{c,i,c',j} + \Pi^{-}_{c,i,c',j} + \Pi^{T}_{c,i,c',j}],$$

$$1 = \sum_{c''=1,l=1}^{C,v} Q_{c',j,c'',l},\ and\ 1 = \sum_{s=1}^{\infty} \pi_{c',j,s},\ for\ all\ c', j.$$

The effect of a negative customer and of a trigger are instantaneous: they occur in zero time; i.e. the effect of a negative customer or trigger arriving to a queue at time t will modify the queue's state at time t^{+}. Furthermore, both a negative customer and a trigger will themselves disappear after they have visited queue j. Queues also have *external* positive, negative and trigger arrivals of rates $\lambda^{+}_{c,i}$, $\lambda^{-}_{c,i}$, $\lambda^{T}_{c,i}$ which can differ for each class c and queue i, according to independent Poison processes at each of the queues. Furthermore, externally arriving customers will have exactly the same effect at a queue as the ones that arrive from another queue.

Let $\Lambda^{+}_{c,i}$, $\Lambda^{-}_{c,i}$, $\Lambda^{T}_{c,i}$ denote the total arrival rate to queue i of class c customers that are of positive, negative and of trigger type, respectively. Then the "traffic equations" for the system are given by:

$$\Lambda^{+}_{c,i} = \lambda^{+}_{c,i} + \sum_{c',j,l}^{|C|,v,v} r(c',j) q_{c',j} [\Pi^{T}_{c',j,c'',l} q_{c'',l} Q_{c'',l,c,i} + \Pi^{+}_{c',j,c,i}],$$

$$\Lambda^{-}_{c,i} = \lambda^{-}_{c,i} + \sum_{c',j}^{|C|,v} r(c',j) q_{c',j} \Pi^{-}_{c',j,c,i},\ \Lambda^{T}_{c,i} \lambda^{T}_{c,i} + \sum_{c',j}^{|C|,v} r(c',j) q_{c',j} \Pi^{T}_{c',j,c,i},$$

$$where\ q_{c,i} = \frac{\Lambda^{+}_{c,i}}{r(c,i) + \Lambda^{T}_{c,i} + \Lambda^{-}_{c,i} \cdot [\frac{1-\sum_{s=1}^{\infty} q^{s}_{c,i} \pi_{c,i,s}}{1-q_{c,i}}]}. \quad (1)$$

In the sequel we will assume that at any queue i only positive, negative customers, and triggers of a single class c_i can arrive. Thus for a specific c_i we have $\Lambda^{T}_{c,i} = \Lambda^{-}_{c,i} = \Lambda^{+}_{c,i} = 0$, if $c \neq c_i$, $\Lambda^{T}_{c_i,i} \geq 0$, $\Lambda^{-}_{c_i,i} \geq 0$, $\Lambda^{+}_{c_i,i} \geq 0$. Also we assume that service rates are the same for all classes of positive customers $r(c',i) = r(c,i) = r(i)$. As a consequence we have:

$$q_{c_i,i} = \frac{\Lambda^{+}_{c_i,i}}{r(i) + \Lambda^{T}_{c_i,i} + \Lambda^{-}_{c_i,i} \cdot [\frac{1-\sum_{s=1}^{\infty} q^{s}_{c_i,i} \pi_{c_i,i,s}}{1-q_{c_i,i}}]}. \quad (2)$$

With these assumptions, the following result follows from previous work [13,23]:

Result. Let $K(t) = (K_1(t), \ldots, K_v(t))$. If the Eq. (1) have an unique solution such that all the $0 < q_{c,i} < 1$, for $1 \leq i \leq v$ and $1 \leq c \leq C$, then denoting by $q_i = q_{c_i,i}$, the following result holds:

$$\lim_{t \to \infty} \Pr[K(t) = (k_1, \ldots, k_v)] = \prod_{i=1}^{v} q_i^{k_i} (1 - q_i). \quad (3)$$

Directly following from the above PFS, we can show that the marginal queue length probability distribution for any queue j is given by:

$$\lim_{t\to\infty} \Pr[K_j(t) = k_j]$$
$$= \sum_{i=1, i\neq j}^{v} \sum_{k_i=1,\ i\neq j}^{v} [\prod_{i=1}^{v} q_i^{k_i}(1-q_i)]$$
$$= q_j^{k_j}(1-q_j). \tag{4}$$

3 The EPN System

The EPN of Fig. 1 can be represented by a G-Network with $v = 2N$ queues, where the WSs are represented by the queues $1, \ldots, N$, while the ESs are represented by the queues $N+1, \ldots, 2N$.

With regard to the notation in Sects. 2 and 2.1, the network has $C = 2$, i.e. two classes of customers where Class 1 refers to the jobs, while Class 2 refers to the EPs, and negative customers and triggers cannot arrive to any of the queues from the outside world, i.e. $\lambda_{c,i}^- = \lambda_{c,i}^T = 0$ for $c = 1,\ 2$ and $i \in \{1, \ldots, 2N\}$. Class 1 customers are "positive customers" representing jobs being served at the WSs with $\lambda_{1,i}^+ = \lambda_i$, and $\lambda_{2,i}^+ = \lambda_{1,i}^- = \lambda_{2,i}^- = 0$ for $i = 1, \ldots, N$. Furthermore jobs at the WSs are only removed, or moved to another WS, under the effect of EPs, i.e. $r(i) = 0$ and $l_{1,i} = l_{2,i} = 0$ for $i = 1, \ldots, N$.

Class 2 customers are EPs acting as positive customers at the storage units or ESs, represented by queues $N+1,\ \ldots\ , 2N$. Hence for $i, j \in \{N+1, \ldots 2N\}$: $\lambda_{2,i}^+ = \gamma_i$, $\lambda_{2,i}^- = 0$, $\lambda_{1,i}^+ = \lambda_{1,i}^- = 0$, and $r(i) = w_i + \delta_i$. Also $\Pi_{2,i,2,j}^+ = P_{ij}$, and $\Pi_{2,i,2,j}^- = 0$; note that $l_{2,i} = \frac{\delta_i}{\delta_i + w_i}$. EPs become negative customers or triggers when they arrive with probability $d_j.\frac{w_j}{\delta_j + w_j}$ to a queue i from a queue $j = N+i$, $i \in \{1, \ldots, N\}$. With probability D_i an EP becomes a negative customer with batch removal, so that the EP is used to process one or more jobs at a WS and the probability distribution of the size of the batch of jobs that can are served is $\pi_{1,i,s} = \Pr[B_{1,i} = s]$, and $\Pi_{2,j,1,i}^- = D_i.d_j.\frac{w_j}{\delta_j + w_j}$, with $j \in \{N+1, \ldots 2N\}$ and $i = j - N$.

With probability $1 - D_i$ an EP becomes a trigger, so that $\Pi_{2,j,1,i}^T = (1 - D_i)d_j.\frac{w_j}{\delta_j + w_j}$, and $q_{1,i,1,m} = M_{im}$, for $j \in \{N+1, \ldots 2N\}$, $i = j - N$, $1 \leq m \leq N$. Note that $\Pi_{2,j,2,i}^T = \Pi_{1,j,2,i}^T = \Pi_{1,j,1,i}^T = 0$ for all $i, j \in \{1, \ldots, 2N\}$, and $\Pi_{2,j,1,i}^T = 0$ if $i \neq j - N$ for $N+1 \leq j \leq 2N$. Also, $P_{1,i,1,j}^+ = (1 - D_i)M_{ij}$, $P_{1,i,1,j}^+ = (1 - D_i)P_{ij}$, $P_{1,i,2,j}^+ = 0$, $P_{2,i,1,j}^+ = 0$, $l_{1,i} = 0$, for $i, j \in \{1, \ldots N\}$. Furthermore $l_{1,i} = 0$, $l_{2i} = 0$ for $i = 1, \ldots, N$, and $l_{1i} = 0$, $l_{2,i} = \frac{\delta_i}{\delta_i + w_i}$ for $i = N+1, \ldots, 2N$. Finally $1 - d_i = \sum_{j=1}^N P_{ij}$ for $i = 1, \ldots, N$, and $\sum_{j=1}^N M_{ij} = 1$ for $i = 1, \ldots, N$.

Regarding to (2) in the G-Network Model, in the EPN model, the two classes have two utilization equations:

$$q_{1,i} = \frac{\Lambda^{+}_{1,i}}{q_{2,i+N}w_i d_i[(1-D_i) + D_i\frac{1-\sum_{s=1}^{\infty} q_{1,s}^{s}\pi_{1,i,s}}{1-q_{1,i}}]}, \tag{5}$$

where $\Lambda^{+}_{1,i} = \lambda_i + \sum_{j=1}^{N} q_{1,j}(1-D_j)d_j w_j M_{ji} q_{2,j+N}$ and

$$q_{2,i+N} = \frac{\gamma_i + \sum_{j=1}^{N} w_j q_{2,j+N} P_{ji}}{w_i + \delta_i}. \tag{6}$$

According to G-Network Theory outlined in the Sect. 2.1, the following expression holds:

$$\lim_{t\to\infty} \Pr[K(t) = (k_{1,1}, \ldots, k_{1,N}, k_{2,N+1}, \ldots, k_{2,2N})] = \prod_{i=1}^{N} q_{1,i}^{k_{1,i}}(1-q_{1,i}) q_{2,i+N}^{k_{2,i+N}}(1-q_{2,i+N}). \tag{7}$$

if (5) and (6) have an unique solution such that all the $0 < q_{c,i} < 1$ for $1 \leq i \leq 2N$ and $1 \leq c \leq 2$. The marginal probability of the queue length for the queue i and class c is

$$\lim_{t\to\infty} \Pr[K_{c,i}(t) = k_{c,i}] = q_{c,i}^{k_{c,i}}(1-q_{c,i}) \tag{8}$$

3.1 Cost Function, Parameters and Optimization

G-networks were proposed to control energy consumption in packet networks [25], and the model was used in [20] to determine the best architecture, distributed or centralized, for storing and dispatching harvested energy. In [21] the EPN model is used under the assumption that one EP is the amount of energy needed to process a job. Here will address two related optimization problems that are outlined below. The objective is to minimize the average response time for jobs that come into the system, where the jobs arrive from the outside world to WS i at a given rate λ_i. Furthermore, the total arrival rate of EPs is fixed at some value γ and each of the ESs has a transfer rate of EPs to the corresponding WS given by w_i and a local energy leakage rate δ_i, for $i = 1, \ldots, N$.

In order to obtain an intuitively appealing result, we will assume that $\pi_{1,i,s} = \frac{(1-u_i)u_i^s}{u(i)}$ where $0 < u_i < 1$ is a real number and $\sum_{s=1}^{\infty}(1-u_i)u_i^{s-1} = 1$.

Consider the case where the EPs cannot moved between ESs so that $M_{ji} = 0$ and $d_i = 1$. Also assume that jobs cannot be moved between WSs, i.e. $D_i = 1$. In this case, assume that the total renewable energy flow into WS i is $\gamma_i = p_i.\gamma$.

The cost function that needs to be minimized represents the overall average job response time:

$$W = \frac{1}{\sum_{i=1}^{N}\lambda_i}\sum_{i=1}^{N}\frac{q_{1,i}}{1-q_{1,i}}. \tag{9}$$

Regarding Eqs. (5) and (6) with the specific restrictions for this case with $d_i = 1$, $D_i = 1$, for $1 \leq i \leq N$, we have:

$$q_{1,i} = \frac{\lambda_i}{q_{2,i} w_i [\frac{1-\sum_{s=1}^{\infty} q_{1,i}^s \cdot \pi_{1,i,s}}{1-q_{1,i}}]}, \tag{10}$$

$$q_{2,i+N} = \frac{\gamma p_i}{w_i + \delta_i}. \tag{11}$$

Our problem is then to choose $p = (p_1, \dots, p_N)$ so as to minimize W for a given value of γ and for given energy leakage rate δ_i at each ES i.

Using Little's Formula we can write:

$$W = \frac{1}{\lambda^+} \sum_{i=1}^{N} \frac{q_{1,i}}{1 - q_{1,i}}, \ where \ \lambda^+ = \sum_{i=1}^{N} \lambda_i. \tag{12}$$

Note that $\Lambda_{1,i}^+ = \lambda_i$ when $D_i = 1$ for all $i = 1, \dots, N$. Substituting $\frac{(1-u_i)u_i^s}{u_i}$ into (10), we have

$$q_{1,i} = \frac{\lambda_i}{q_{2,i+N} w_i} \times [\frac{1 - \sum_{s=1}^{\infty} \frac{(1-u_i)u_i^s}{u_i} q_{1,i}^s}{1 - q_{1,i}}]^{-1} = \frac{\lambda_i}{u_i \lambda_i + q_{2,i+N} w_i}. \tag{13}$$

Substituting (13) into the cost function W, we get:

$$W = \frac{1}{\lambda^+} \sum_{i=1}^{N} \frac{\lambda_i}{\sigma_i \gamma p_i + \lambda_i (u_i - 1)}, \ with \ \sigma_i = \frac{w_i}{w_i + \delta_i}. \tag{14}$$

where σ_i is the energy efficiency with regard to leakage, of $i - th$ ES node.

Choosing the $p_i \geq 0$ so as to minimize W is an optimization problem subject to the constraint $\sum_{i=i}^{N} p_i = 1$. Therefore we apply the method of Lagrange multipliers and choose the Lagrangian

$$\mathcal{L} = W + \beta (\sum_{i=1}^{N} p_i - 1), \tag{15}$$

where the Lagrange multiplier β is a real number. Suppose p^* is a local solution of the optimization problem. Then the necessary Kuhn-Tucker conditions are:

$$\nabla_p \mathcal{L}(p^*, \beta^*) = 0, \ and \ \sum_{i=1}^{N} p_i^* - 1 = 0, \tag{16}$$

from which we derive

$$\frac{\partial W}{\partial p_i} = \frac{-\lambda_i \sigma_i \gamma}{\lambda^+ \left[\sigma_i \gamma p_i + \lambda (u_i - 1)\right]^2} = -\beta. \tag{17}$$

Then rearranging (17), the solution p_i^* is

$$p_i^* = \frac{\lambda_i(1-u_i)}{\sigma_i\gamma} + \sqrt{\frac{\lambda_i}{\lambda^+\sigma_i\gamma\beta}}. \tag{18}$$

Moreover, the second necessary condition

$$\sum_{i=1}^{N}\left(\frac{\lambda_i(1-u_i)}{\sigma_i\gamma} + \sqrt{\frac{\lambda_i}{\lambda^+\sigma_i\gamma\beta}}\right) = 1, \tag{19}$$

also must hold. Solving (18) and (19) simultaneously, we see that the optimal solution must be:

$$p_i^* = \frac{\lambda_i(1-u_i)}{\sigma_i\gamma} + \frac{\sqrt{\frac{\lambda_i}{\sigma_i}}}{\sum_{i=1}^{N}\sqrt{\frac{\lambda_i}{\sigma_i}}}\left(1 - \sum_{i=1}^{N}\frac{\lambda_i(1-u_i)}{\sigma_i\gamma}\right). \tag{20}$$

However, the sufficient condition that there exists an optimum solution p^* also needs to be examined. To guarantee the existence of the strict constrained local minimum, the Hessian $\nabla_{pp}\mathcal{L}$ must be positive definite. Notice that $\nabla_{pp}\mathcal{L}$ is a diagonal matrix with diagonal entries:

$$\frac{\partial^2\mathcal{L}(p^*,\beta^*)}{\partial p_i^2} = \frac{\partial^2 W}{\partial p_i^2} = \frac{2\lambda_i\sigma_i^2\gamma^2}{\lambda^+\left[\sigma_i\gamma p_i^* + \lambda_i(u_i-1)\right]^3}. \tag{21}$$

Thus the sufficient condition holds if the inequality

$$\sigma_i\gamma p_i^* > \lambda_i(1-u_i), \tag{22}$$

is satisfied for all $i = 1,\ldots,N$. Substituting p_i^* into (22), we see that the inequality is equivalent to:

$$\gamma > \sum_{i=1}^{N}\frac{\lambda_i}{\sigma_i}(1-u_i). \tag{23}$$

This condition is physically meaningful since it implies that the total rate of harvested EPs has to be sufficiently large so as to provide enough energy so as to power the WSs despite the energy leakage that also will occur.

3.2 An Example

In order to illustrate the optimal solution, consider a system with three WSs and ESs with parameters given in Table 1.

The sufficient condition (22) allows us to determine the range of p_1, p_2 and p_3 that guarantee that every ES can provide sufficient power to its corresponding WS:

$$0.2933 < p_1 < 1,\ 0.1760 < p_2 < 1,\ 0.0597 < p_3 < 1,$$

Table 1. Parameters for the system with three WSs and ESs

Parameters	Values
γ	150 EPs/sec
$\lambda_1,\ \lambda_2,\ \lambda_3$	$50, 30, 10$ jobs/sec
D_1, D_2, D_3	1, 1, 1
$w_1,\ w_2,\ w_3$	$100, 80, 50$ EPs/sec
$u_1,\ u_2,\ u_3$	$0.2, 0.2, 0.2$
M_{ij} for all i, j	0
M_{ij} for all i, j	0
$\delta_1,\ \delta_2,\ \delta_3$	$10, 8, 6$ EPs/sec
$d_1,\ d_2,\ d_3$	$1, 1, 1$

with the constraint $p_1 + p_2 + p_3 = 1$. The resulting values of W for all (p_1, p_2, p_3) are shown in Figs. 2 and 3 where the x and y axes are p_1 and p_2, and $p_3 = 1 - p_1 - p_2$. From (20) we obtain the optimum operating point which minimizes the total average response time as being $(p_1^*, p_2^*, p_3^*) = (0.5049, 0.3399, 0.1552)$ with the minimum value $W^* = 42.9$ ms.

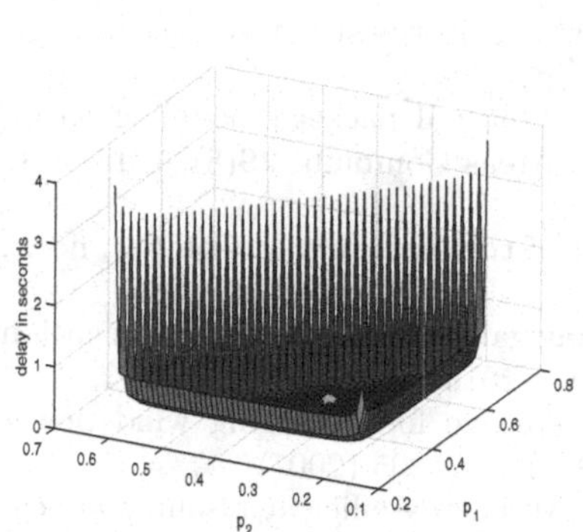

Fig. 2. Average response time with all (p_1, p_2) pairs. The red dot is the optimal solution of Eq. (20). The range of the values p_i is not $[0, 1]$ due to the constraint and the sufficient conditions, and the curve is not convex.

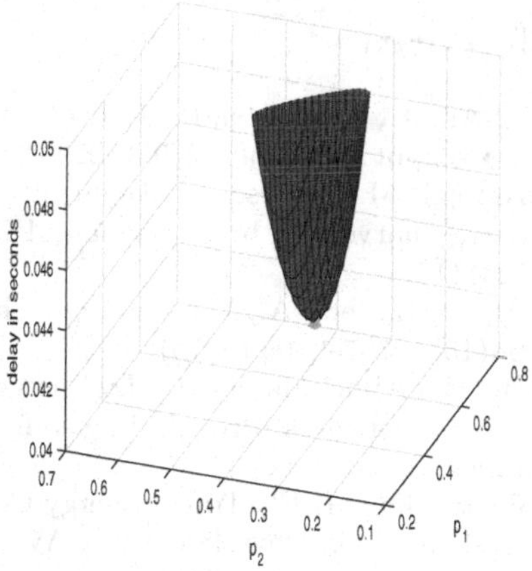

Fig. 3. The neighbourhood of the optimum point at a much smaller scale of W along the z-axis.

4 Conclusions

We have considered an EPN model where jobs and energy packets cannot be transferred to other workstations, so that each workstation executes locally the jobs that it receives, using energy from its own energy storage unit. We have derived a key result where a common flow of energy is distributed optimally among the workstations so that the average response time of jobs can be minimized. The problem has been solved analytically for the geometrically distributed number of jobs processed with one energy packet. In other work [27] the average response time has been minimized when jobs can be moved among WSs according to a given probability transition matrix, but each station decides locally whether to move a job or not. Future work will investigate the minimization of a cost function that combines the average response time of jobs, and the energy wastage through leakage or due to idle workstations which consume energy when they do not process jobs.

Acknowledgements. This research was supported by the European Union's Horizon 2020 research and innovation programme under grant agreement No 780572, through the SKK4ED project which aims to minimize the cost, development time and complexity of low-energy software development processes, by providing tools for automatic optimization of multiple quality requirements, such as technical debt, energy efficiency, dependability and performance.

References

1. Al-Ali, A., Aburukba, R.: Role of Internet of Things in the smart grid technology. J. Comput. Commun. **3**(05), 229 (2015)
2. Antepli, M.A., Uysal-Biyikoglu, E., Erkal, H.: Optimal packet scheduling on an energy harvesting broadcast link. IEEE J. Sel. Areas Commun. **29**(8), 1712–1731 (2011)
3. Atzori, L., Iera, A., Morabito, G.: The internet of things: a survey. Comput. netw. **54**(15), 2787–2805 (2010)
4. Bi, H., Abdelrahman, O.H.: Energy-aware navigation in large-scale evacuation using g-networks. Probab. Eng. Inf. Sci. **1**, 1–13 (2016)
5. Black, M., Strbac, G.: Value of bulk energy storage for managing wind power fluctuations. IEEE Trans. Energy Convers. **22**(1), 197–205 (2007)
6. Coskun, C.C., Davaslioglu, K., Ayanoglu, E.: An energy-efficient resource allocation algorithm with QOS constraints for heterogeneous networks. In: 2015 IEEE on Global Communications Conference (GLOBECOM), pp. 1–7. IEEE, December 2015
7. Da Xu, L., He, W., Li, S.: Internet of things in industries: a survey. IEEE Trans. Ind. Inform. **10**(4), 2233–2243 (2014)
8. Fourneau, J.M., Gelenbe, E.: G-networks with adders. Fut. Internet **9**(3), 34 (2017). https://doi.org/10.3390/fi9030034
9. Fourneau, J., Marin, A., Balsamo, S.: Modeling energy packets networks in the presence of failures. In: 24th IEEE International Symposium on Modeling, Analysis and Simulation of Computer and Telecommunication Systems, MASCOTS 2016, London, United Kingdom, 19–21 September 2016, pp. 144–153 (2016). https://doi.org/10.1109/MASCOTS.2016.44

10. Gatzianas, M., Georgiadis, L., Tassiulas, L.: Control of wireless networks with rechargeable batteries. IEEE Trans. Wireless Commun. **9**(2), 581–593 (2010). https://doi.org/10.1109/TWC.2010.080903
11. Gelenbe, E.: Réseaux neuronaux aléatoires stables. Comptes rendus de l'Académie des sciences. Série 2, Mécanique, Physique, Chimie, Sciences de l'Univers, Sciences de la Terre **310**(3), 177–180 (1990)
12. Gelenbe, E.: G-networks by triggered customer movement. J. Appl. Probab. **30**(3), 742–748 (1993)
13. Gelenbe, E.: G-networks with signals and batch removal. Probab. Eng. Inf. Sci. **7**(3), 335–342 (1993). https://doi.org/10.1017/S0269964800002953
14. Gelenbe, E.: Energy packet networks: ICT based energy allocation and storage. In: Rodrigues, J.J.P.C., Zhou, L., Chen, M., Kailas, A. (eds.) GreeNets 2011. LNICST, vol. 51, pp. 186–195. Springer, Heidelberg (2012). https://doi.org/10.1007/978-3-642-33368-2_16
15. Gelenbe, E.: Energy packet networks: adaptive energy management for the cloud. In: CloudCP 2012 Proceedings of the 2nd International Workshop on Cloud Computing Platforms, p. 1. ACM (2012). https://doi.org/10.1145/2168697.2168698
16. Gelenbe, E.: Energy packet networks: smart electricity storage to meet surges in demand. In: Proceedings of the 5th International ICST Conference on Simulation Tools and Techniques, pp. 1–7. ICST (Institute for Computer Sciences, Social-Informatics and Telecommunications Engineering) (2012)
17. Gelenbe, E.: A sensor node with energy harvesting. ACM SIGMETRICS Perform. Eval. Rev. **42**(2), 37–39 (2014)
18. Gelenbe, E., Abdelrahman, O.H.: An energy packet network model for mobile networks with energy harvesting. Nonlinear Theor. Appl. IEICE **9**(3), 1–15 (2018). https://doi.org/10.1587/nolta.9.1
19. Gelenbe, E., Caseau, Y.: The impact of information technology on energy consumption and carbon emissions. Ubiquity **2015**, 1 (2015)
20. Gelenbe, E., Ceran, E.T.: Central or distributed energy storage for processors with energy harvesting. In: Sustainable Internet and ICT for Sustainability (SustainIT), 2015, pp. 1–3. IEEE (2015)
21. Gelenbe, E., Ceran, E.T.: Energy packet networks with energy harvesting. IEEE Access **4**, 1321–1331 (2016)
22. Gelenbe, E., Fourneau, J.M.: G-networks with resets. Perform. Eval. **49**(1), 179–191 (2002)
23. Gelenbe, E., Labed, A.: G-networks with multiple classes of signals and positive customers. Eur. J. Oper. Res. **108**, 293–305 (1998)
24. Gelenbe, E., Marin, A.: Interconnected wireless sensors with energy harvesting. In: Gribaudo, M., Manini, D., Remke, A. (eds.) ASMTA 2015. LNCS, vol. 9081, pp. 87–99. Springer, Cham (2015). https://doi.org/10.1007/978-3-319-18579-8_7
25. Gelenbe, E., Morfopoulou, C.: A framework for energy-aware routing in packet networks. Comput. J. **54**(6), 850–859 (2010)
26. Gelenbe, E., Wu, F.J.: Future research on cyber-physical emergency management systems. Fut. Internet **5**(3), 336–354 (2013)
27. Gelenbe, E., Zhang, Y.: Performance optimization with energy packets. IEEE Syst. J. **13**(4), 2770–2780 (2019)
28. Gubbi, J., Buyya, R., Marusic, S., Palaniswami, M.: Internet of things (IoT): a vision, architectural elements, and future directions. Fut. Gener. Comput. Syst. **29**(7), 1645–1660 (2013)
29. Kadioglu, Y.M.: Finite capacity energy packet networks. Probab. Eng. Inf. Sci. **31**(4), 477–504 (2017)

30. Kadioglu, Y.M., Gelenbe, E.: Product form solution for cascade networks with intermittent energy. IEEE Systems Journal (2018, accepted for publication)
31. Mao, Z., Koksal, C.E., Shroff, N.B.: Near optimal power and rate control of multi-hop sensor networks with energy replenishment: Basic limitations with finite energy and data storage. IEEE Trans. Autom. Control **57**(4), 815–829 (2012). https://doi.org/10.1109/TAC.2011.2166310
32. Newcombe, L.: Data centre energy efficiency metrics: existing and proposed metrics to provide effective understanding and reporting of data centre energy. BCS: British Computer Society (2008). http://www.bcs.org/upload/pdf/data-centre-energy.pdf
33. Perera, C., Liu, C.H., Jayawardena, S., Chen, M.: A survey on internet of things from industrial market perspective. IEEE Access **2**, 1660–1679 (2014). https://doi.org/10.1109/ACCESS.2015.2389854
34. Rahimi, A., Zorlu, Ö., Muhtaroglu, A., Kulah, H.: Fully self-powered electromagnetic energy harvesting system with highly efficient dual rail output. IEEE Sens. J. **12**(6), 2287–2298 (2012)
35. Sarkar, S., Khouzani, M.H.R., Kar, K.: Optimal routing and scheduling in multihop wireless renewable energy networks. IEEE Trans. Autom. Control **58**(7), 1792–1798 (2013). https://doi.org/10.1109/TAC.2013.2250074
36. Shaikh, F.K., Zeadally, S., Exposito, E.: Enabling technologies for green Internet of Things. IEEE Syst. J. **11**(2), 983–994 (2017). https://doi.org/10.1109/JSYST.2015.2415194
37. Takahashi, R., Azuma, S.I., Tashiro, K., Hikihara, T.: Design and experimental verification of power packet generation system for power packet dispatching system. In: American Control Conference (ACC), 2013, pp. 4368–4373. IEEE (2013)
38. Takahashi, R., Takuno, T., Hikihara, T.: Estimation of power packet transfer properties on indoor power line channel. Energies **5**(7), 2141–2149 (2012)
39. Van Heddeghem, W., Lambert, S., Lannoo, B., Colle, D., Pickavet, M., Demeester, P.: Trends in worldwide ict electricity consumption from 2007 to 2012. Comput. Commun. **50**, 64–76 (2014)
40. Wade, N.S., Taylor, P., Lang, P., Jones, P.: Evaluating the benefits of an electrical energy storage system in a future smart grid. Energy Policy **38**(11), 7180–7188 (2010)
41. Whitmore, A., Agarwal, A., Da Xu, L.: The Internet of Things: a survey of topics and trends. Inf. Syst. Front. **17**(2), 261–274 (2015)
42. Yang, J., Ulukus, S.: Optimal packet scheduling in an energy harvesting communication system. IEEE Trans. Commun. **60**(1), 220–230 (2012)
43. Yin, Y.: Optimum energy for energy packet networks. Probab. Eng. Inf. Sci. **31**(4), 516–539 (2017)
44. Zanella, A., Bui, N., Castellani, A., Vangelista, L., Zorzi, M.: Internet of things for smart cities. IEEE Internet of Things J. **1**(1), 22–32 (2014)

Foundations of Computer Science – Regular Papers

A Characterization of the Context-Free Languages by Stateless Ordered Restart-Delete Automata

Friedrich Otto(✉)

Fachbereich Elektrotechnik/Informatik, Universität Kassel, 34109 Kassel, Germany
f.otto@uni-kassel.de

Abstract. We consider stateless ordered restart-delete automata, which are actually just stateless ordered restarting automata (stl-ORWW-automata) that have an additional delete operation. While the stl-ORWW-automata just accept the regular languages, we show that the context-free languages are characterized by the swift stateless ordered restart-delete automaton, that is, by the stateless ordered restart-delete automaton that can move its window to any position after performing a restart.

Keywords: Restarting automaton · Ordered rewriting · Context-free language

1 Introduction

The restarting automaton was introduced in [5] to model the linguistic technique of analysis by reduction (see, e.g., [7]). Despite its linguistic motivation many classical classes of formal languages have been characterized by various types of restarting automata (for a survey see, e.g., [10]). A particularly simple type of restarting automaton is the *ordered restarting automaton* (ORWW-automaton, for short) that has been introduced in [9] in relation with the processing of picture languages.

An ORWW-automaton consists of a finite-state control, a tape with end-markers, a read-write window of size three, and a (partial) ordering on its tape alphabet. Based on the actual state and window contents, the automaton can move its window one position to the right and change its state, or it can replace the letter in the middle of the window by a smaller letter and restart, or it can accept. During a restart, the window is moved back to the left end of the tape, and the finite-state control is reset to the initial state. In [11], it is shown that deterministic ORWW-automata (det-ORWW-automata) don't need states and that they characterize the class of regular languages. On the other hand, the nondeterministic ORWW-automata yield an abstract family of languages that is incomparable to the (deterministic) context-free languages, the Church-Rosser languages, and the growing context-sensitive languages with

A. Chatzigeorgiou et al. (Eds.): SOFSEM 2020, LNCS 12011, pp. 39–50, 2020.
https://doi.org/10.1007/978-3-030-38919-2_4

respect to inclusion [6]. However, stateless nondeterministic ORWW-automata (stl-ORWW-automata) are just as expressive as det-ORWW-automata, that is, they yield another characterization for the regular languages.

In [12] (see, also [13]), the det-ORWW-automaton was extended by an additional delete/restart operation that allows to delete the symbol from the middle of the window and to restart, obtaining the *deterministic ordered restart-delete automaton* (or det-ORD-automaton, for short). It turned out that these automata don't need states, and that the class of languages they accept properly includes the class of deterministic context-free languages, while it is contained in the intersection of the (unambiguous) context-free languages and the Church-Rosser languages.

Here we turn to the nondeterministic variant of the ordered restart-delete automaton. As this model extends both, the det-ORD-automaton as well as the ORWW-automaton, it is immediate that the resulting language class is quite large. In particular, it contains languages that are not even growing context-sensitive. Therefore, we restrict our attention to the stateless variant of these automata, the *stateless ordered restart-delete automaton* (or stl-ORD-automaton, for short). In fact, we concentrate on a restricted variant called *swift* stl-ORD-automaton that can always perform move-right steps unless its window is already at the right end of its tape. While the stl-ORWW-automaton just accepts the regular languages, we will see that the swift stl-ORD-automaton yields another characterization for the class of context-free languages.

This paper is structured as follows. In Sect. 2, we define the stl-ORD-automaton and we illustrate it by a detailed example. In the next section, we prove that each swift stl-ORD-automaton can be simulated by a pushdown automaton (PDA), showing that all languages accepted by swift stl-ORD-automata are necessarily context-free. Finally, in Sect. 4, we show conversely that each context-free language is accepted by some swift stl-ORD-automaton by providing a simulation of a PDA by a swift stl-ORD-automaton, therewith completing the proof of our main result. In the concluding section, we relate the swift stl-ORD-automaton to the *1-context rewriting systems* of [3] and state a number of open problems.

2 The Stateless Ordered Restart-Delete Automaton

An alphabet Σ is a finite set of letters. For all $n \geq 0$, Σ^n is the set of words over Σ of length n, Σ^+ is the set of non-empty words, and $\Sigma^* = \Sigma^+ \cup \{\lambda\}$, where λ denotes the empty word. For $w \in \Sigma^*$, $|w|$ denotes the length of w. A language over Σ is any subset of Σ^*. Of particular interest are the classes REG, DCFL, CFL, CRL, and GCSL of regular, deterministic context-free, context-free, Church-Rosser [8], and growing context-sensitive languages [2,4]. Further, for any type of automaton X, $\mathcal{L}(\mathsf{X})$ will denote the class of languages that are accepted by automata of that type.

Definition 1. *A* stateless ordered restart-delete automaton *(stl-ORD-automaton) has a flexible tape with endmarkers and a read/write window of size 3. It is* *defined by a 6-tuple* $M = (\Sigma, \Gamma, \triangleright, \triangleleft, \delta, >)$*, where* Σ *is a finite input alphabet,* Γ *is a finite tape alphabet such that* $\Sigma \subseteq \Gamma$*, the symbols* $\triangleright, \triangleleft \notin \Gamma$*, called* sentinels, *serve as markers for the left and right border of the work space, respectively,* $>$ *is a* partial ordering *on* Γ*, and*

$$\delta : (((\Gamma \cup \{\triangleright\}) \cdot \Gamma \cdot (\Gamma \cup \{\triangleleft\})) \cup \{\triangleright\triangleleft\}) \to 2^{\Gamma \cup \{\lambda, \mathsf{MVR}\}} \cup \{\mathsf{Accept}\}$$

is the transition relation *that describes four types of transition steps:*

(1) A move-right step *has the form* $\mathsf{MVR} \in \delta(a_1a_2a_3)$*, where* $a_1 \in \Gamma \cup \{\triangleright\}$ *and* $a_2, a_3 \in \Gamma$*. It causes* M *to shift the window one position to the right.*
(2) A rewrite/restart step *has the form* $b \in \delta(a_1a_2a_3)$*, where* $a_1 \in \Gamma \cup \{\triangleright\}$*,* $a_2, b \in \Gamma$*, and* $a_3 \in \Gamma \cup \{\triangleleft\}$ *such that* $a_2 > b$ *holds. It causes* M *to replace the symbol* a_2 *in the middle of its window by* b *and to restart (see above). Observe that this operation requires that the newly written letter* b *is smaller than the letter* a_2 *being replaced with respect to the partial ordering* $>$*.*
(3) A delete/restart step *has the form* $\lambda \in \delta(a_1a_2a_3)$*, where* $a_1 \in \Gamma \cup \{\triangleright\}$*,* $a_2 \in \Gamma$*, and* $a_3 \in \Gamma \cup \{\triangleleft\}$*. It causes* M *to delete the symbol* a_2 *in the middle of its window and to restart. Through this step, the tape field that contains* a_2 *is also removed, that is, the length of the tape is reduced.*
(4) An accept step *has the form* $\delta(a_1a_2a_3) = \mathsf{Accept}$*, where* $a_1 \in \Gamma \cup \{\triangleright\}$*,* $a_2 \in \Gamma$*, and* $a_3 \in \Gamma \cup \{\triangleleft\}$*. It causes* M *to halt and accept. In addition, we allow an accept step of the form* $\delta(\triangleright\triangleleft) = \mathsf{Accept}$*.*

If $\delta(u)$ is undefined for some word u, then M necessarily halts, when its window contains the word u, and we say that M *rejects* in this situation. Further, the letters in $\Gamma \smallsetminus \Sigma$ are called *auxiliary symbols.*

A *configuration* of a stl-ORD-automaton M is a word $\alpha \in \{\triangleright\} \cdot \Gamma^* \cdot \{\triangleleft\}$ in which a factor of length three (or all of α if $|\alpha| < 3$) is underlined. Here it is understood that α is the current contents of the tape and that the window contains the underlined factor. A *restarting configuration* has the form $\triangleright w \triangleleft$, where the prefix of length three is underlined; if $w \in \Sigma^*$, then this restarting configuration is also called an *initial configuration.* A configuration that is reached by an accept step is an *accepting configuration*, denoted by Accept, and a configuration of the form α such that $\delta(\beta)$ is undefined, where β is the underlined factor of length three, is a *rejecting configuration.* A *halting configuration* is either an accepting or a rejecting configuration. By $\vdash_M$ we denote the *single-step computation relation* that M induces on its set of configurations, and $\vdash_M^*$, its reflexive and transitive closure, is the *computation relation* of M.

Any computation of a stl-ORD-automaton M consists of certain phases. A phase, called a *cycle*, starts in a restarting configuration, the window is moved along the tape by MVR steps until a rewrite/restart or a delete/restart step is performed and thus, a new restarting configuration is reached. If no further rewrite or delete operation is performed, any computation necessarily finishes in a halting configuration – such a phase is called a *tail.* By $\vdash_M^c$ we denote the execution of a complete cycle, and $\vdash_M^{c^*}$ is the reflexive transitive closure of $\vdash_M^c$. It is the reduction relation that M induces on its set of restarting configurations.

An input $w \in \Sigma^*$ is accepted by M, if there exists a computation of M which starts with the initial configuration $\rhd w \lhd$ and ends with an accept step. The language consisting of all words that are accepted by M is denoted by $L(M)$.

As each cycle ends with a rewrite operation, which replaces a symbol a by a symbol b that is strictly smaller than a with respect to the given ordering $>$, or with a delete operation, we see that each computation of M on an input of length n consists of at most $|\Gamma| \cdot n$ many cycles. Each cycle can be simulated by a nondeterministic Turing machine in linear time, and hence, M can be simulated by a nondeterministic Turing machine in time $O(n^2)$.

As each det-ORD-automaton can be simulated by a stateless det-ORD-automaton [12], we obtain the following inclusion.

Proposition 2. $\mathcal{L}(\mathsf{det\text{-}ORD}) \subseteq \mathcal{L}(\mathsf{stl\text{-}ORD})$.

The following example illustrates how stl-ORD-automata work.

Example 3. Let $L = \{ a^n b^n \mid n \geq 0 \} \cup \{ a^n b^{2n} \mid n \geq 0 \}$. It is well-known that L is a context-free language that is not deterministic context-free. Further, this language is not accepted by any ORWW-automaton [6]. However, L is accepted by the stl-ORD-automaton $M = (\Sigma, \Gamma, \rhd, \lhd, \delta, >)$ that is defined by taking $\Sigma = \{a, b\}$ and $\Gamma = \Sigma \cup \{ a_1, e, e_1, f, f_1, f_2\}$, by choosing the partial ordering $>$ such that $a > a_1$, $b > e > e_1$, and $b > f > f_1 > f_2$, and by defining the transition relation δ as follows:

$$
\begin{array}{lll}
(0)\ \delta(xyz) \ni \mathsf{MVR} & \text{for all } x \in \Gamma \cup \{\rhd\} \text{ and } y, z \in \Gamma, & \\
(1)\ \delta(\rhd\lhd) = \mathsf{Accept}, & & \\
(2)\ \delta(ab\lhd) = \{e\}, & (10)\ \delta(a_1ee) \ni e_1, & (18)\ \delta(a_1ff) \ni f_1, \\
(3)\ \delta(bb\lhd) = \{e, f\}, & (11)\ \delta(a_1e\lhd) \ni e_1, & (19)\ \delta(f_1ff) \ni f_2, \\
(4)\ \delta(bbe) \ni e, & (12)\ \delta(aa_1e_1) \ni \lambda, & (20)\ \delta(f_1f\lhd) \ni f_2, \\
(5)\ \delta(abe) \ni e, & (13)\ \delta(\rhd a_1e_1) \ni \lambda, & (21)\ \delta(a_1f_1f_2) \ni \lambda, \\
(6)\ \delta(bbf) \ni f, & (14)\ \delta(ae_1e) \ni \lambda, & (22)\ \delta(aa_1f_2) \ni \lambda, \\
(7)\ \delta(abf) \ni f, & (15)\ \delta(\rhd e_1\lhd) \ni \lambda, & (23)\ \delta(\rhd a_1f_2) \ni \lambda, \\
(8)\ \delta(aae) \ni a_1, & (16)\ \delta(aaf) \ni a_1, & (24)\ \delta(af_2f) \ni \lambda, \\
(9)\ \delta(\rhd ae) \ni a_1, & (17)\ \delta(\rhd af) \ni a_1, & (25)\ \delta(\rhd f_2\lhd) \ni \lambda.
\end{array}
$$

For an input of the form $a^m b^n$, first the factor b^n is rewritten, from right to left, into e^n or into f^n using instructions (2) to (7). In the former case, it is then checked whether $m = n$ by alternatingly rewriting the last letter a into a_1, the first letter e into e_1 and then deleting a_1 and e_1 using instructions (8) to (15). In the latter case, it is checked whether $n = 2m$ by alternatingly rewriting the last letter a into a_1 and the first factor ff into $f_1 f_2$ and then deleting f_1, a_1 and f_2 using instructions (16) to (25). For example, given $w = aabbbb$ as input, M can execute the following computation (Recall that we underline the letters inside the window):

$$
\begin{array}{llllllllll}
\underline{\rhd aa}bbbb\lhd & \vdash^5_{(0)} & \rhd aabb\underline{bb\lhd} & \vdash_{(3)} & \underline{\rhd aa}bbbf\lhd & \vdash^4_{(0)} & \rhd aab\underline{bbf}\lhd & \vdash_{(6)} & \underline{\rhd aa}bbff\lhd \\
 & \vdash^* & \underline{\rhd aa}ffff\lhd & \vdash_{(0)} & \rhd \underline{aaf}fff\lhd & \vdash_{(16)} & \underline{\rhd aa_1}ffff\lhd & \vdash^2_{(0)} & \rhd a\underline{a_1ff}ff\lhd \\
 & \vdash_{(18)} & \underline{\rhd aa_1}f_1fff\lhd & \vdash^3_{(0)} & \rhd aa_1\underline{f_1ff}\lhd & \vdash_{(19)} & \underline{\rhd aa_1}f_1f_2ff\lhd & \vdash^2_{(0)} & \rhd a\underline{a_1f_1f_2}ff\lhd \\
 & \vdash_{(21)} & \underline{\rhd aa_1}f_2ff\lhd & \vdash_{(0)} & \rhd \underline{aa_1f_2}ff\lhd & \vdash_{(22)} & \underline{\rhd af_2}ff\lhd & \vdash_{(0)} & \rhd \underline{af_2f}f\lhd \\
 & \vdash_{(24)} & \underline{\rhd af}f\lhd & \vdash^* & \underline{\rhd f_2\lhd} & \vdash_{(25)} & \underline{\rhd\lhd} & \vdash_{(1)} & \mathsf{Accept}.
\end{array}
$$

It follows that $L(M) = L$.

3 Swift Stl-ORD-Automata Only Accept Context-Free Languages

We want to show that each language that is accepted by a stl-ORD-automaton is necessarily context-free. To prove this result, we would like to simulate the accepting computations of a stl-ORD-automaton by a (nondeterministic) PDA. For this simulation, we would like to use an extension of the one that is given in [12] for simulating a stl-det-ORD-automaton by a PDA. Unfortunately, this approach leads to a serious problem. We will then overcome this problem by restricting our attention to just a subclass of stl-ORD-automata, the *swift* stl-ORD-automata.

We first present a simple example in order to describe the data structure that will be used for the simulation. As the purpose of this example is simply to illustrate the dynamics of the simulation, the language accepted by the given automaton is of no importance.

Example 4. Let M be the stl-ORD-automaton on the input alphabet $\Sigma = \{a_1, a_2, a_3, a_4, a_5\}$, the tape alphabet $\Gamma = \Sigma \cup \{b_1, b_2, b_3, b_4, c_1\}$ and the ordering $a_i > b_i > c_1$ $(1 \leq i \leq 4)$, where the transition relation is given by the following table:

$$
\begin{array}{lll}
\delta(\rhd a_1 a_2) = \{\mathsf{MVR}, b_1\}, & \delta(a_1 a_2 a_3) = \{\mathsf{MVR}\}, & \delta(a_2 a_3 a_4) = \{b_3\}, \\
\delta(a_1 a_2 b_3) = \{b_2\}, & \delta(\rhd a_1 b_2) = \{\mathsf{MVR}\}, & \delta(a_1 b_2 b_3) = \{\lambda\}, \\
\delta(\rhd a_1 b_3) = \{\mathsf{MVR}, b_1\}, & \delta(b_1 b_3 a_4) = \{\mathsf{MVR}\}, & \delta(b_3 a_4 a_5) = \{\mathsf{MVR}, b_4\}, \\
\delta(a_4 a_5 \lhd) = \{\lambda\}, & \delta(b_3 a_4 \lhd) = \{b_4\}, & \delta(b_3 b_4 \lhd) = \{\lambda\}, \\
\delta(\rhd b_1 b_3) = \{\mathsf{MVR}, c_1\}, & \delta(b_1 b_3 b_4) = \{\mathsf{MVR}\}, & \delta(b_1 b_3 \lhd) = \mathsf{Accept}.
\end{array}
$$

Given the word $w = a_1 a_2 a_3 a_4 a_5$ as input, M can execute the following accepting computation:

$$
\begin{array}{llll}
\underline{\rhd a_1 a_2} a_3 a_4 a_5 \lhd & \vdash_M \rhd \underline{a_1 a_2 a_3} a_4 a_5 \lhd & \vdash_M \rhd a_1 \underline{a_2 a_3 a_4} a_5 \lhd & \vdash_M \underline{\rhd a_1 a_2} b_3 a_4 a_5 \lhd \\
 & \vdash_M \rhd \underline{a_1 a_2 b_3} a_4 a_5 \lhd & \vdash_M \underline{\rhd a_1 b_2} b_3 a_4 a_5 \lhd & \vdash_M \rhd \underline{a_1 b_2 b_3} a_4 a_5 \lhd \\
 & \vdash_M \underline{\rhd a_1 b_3} a_4 a_5 \lhd & \vdash_M \underline{\rhd b_1 b_3} a_4 a_5 \lhd & \vdash_M \rhd \underline{b_1 b_3 a_4} a_5 \lhd \\
 & \vdash_M \rhd b_1 \underline{b_3 a_4 a_5} \lhd & \vdash_M \rhd b_1 b_3 \underline{a_4 a_5 \lhd} & \vdash_M \underline{\rhd b_1 b_3} a_4 \lhd \\
 & \vdash_M \rhd \underline{b_1 b_3 a_4} \lhd & \vdash_M \rhd b_1 \underline{b_3 a_4 \lhd} & \vdash_M \underline{\rhd b_1 b_3} b_4 \lhd \\
 & \vdash_M \rhd \underline{b_1 b_3 b_4} \lhd & \vdash_M \rhd b_1 \underline{b_3 b_4 \lhd} & \vdash_M \underline{\rhd b_1 b_3} \lhd \\
 & \vdash_M \rhd \underline{b_1 b_3 \lhd} & \vdash_M \mathsf{Accept}. &
\end{array}
$$

To encode this computation in a compact way, we introduce a 3-tuple of vectors $T_i = (L_i, W_i, R_i)$ for each letter w_i of w, $1 \leq i \leq |w|$, where

- W_i is a sequence $(x_1, x_2, \ldots, x_r)$ over $\Gamma \cup \{\lambda\}$ such that $w_i = x_1 > x_2 > \cdots > x_{r-1}$ and ($x_r \in \Gamma$ and $x_{r-1} > x_r$) or $x_r = \lambda$,
- L_i is a sequence of letters $(y_1, y_2, \ldots, y_{r-1})$ over $\Gamma \cup \{\rhd\}$, and
- R_i is a sequence of letters $(z_1, z_2, \ldots, z_{r-1})$ over $\Gamma \cup \{\lhd\}$ such that $\delta(y_j x_j z_j) = x_{j+1}$ holds for all $j = 1, 2, \ldots, r-1$.

The idea is that W_i encodes the sequence of letters that are produced by M in an accepting computation for a particular field, and L_i and R_i encode the information on the neighboring letters to the left and to the right that are used to perform the corresponding rewrite steps. For example, the triple $(y_1, x_1, z_1) \in (L_i, W_i, R_i)$ means that x_1 is rewritten into x_2, while the left neighboring field contains the letter y_1 and the right neighboring field contains the letter z_1. In particular, if $x_r = \lambda$, then the instruction $\lambda \in \delta(y_{r-1}x_{r-1}z_{r-1})$ was used to delete the letter x_{r-1}. For the above computation, we obtain thus the following sequence of triples:

L_1	W_1	R_1	L_2	W_2	R_2	L_3	W_3	R_3	L_4	W_4	R_4	L_5	W_5	R_5
$\rhd$	a_1	b_3	a_1	a_2	b_3	a_2	a_3	a_4	b_3	a_4	$\lhd$	a_4	a_5	$\lhd$
	b_1		a_1	b_2	b_3		b_3		b_3	b_4	$\lhd$		λ	
				λ						λ				

The triple $(L_2, W_2, R_2) = \left|\begin{array}{ccc} a_1 & a_2 & b_3 \\ a_1 & b_2 & b_3 \\ & \lambda & \end{array}\right|$ expresses the fact that the rewrite operation $b_2 \in \delta(a_1a_2b_3)$ has been applied in the above computation to replace the letter a_2 by b_2, and the delete operation $\lambda \in \delta(a_1b_2b_3)$ has been used to delete the letter b_2. However, from $R_2 = (b_3, b_3)$ we see that these operations have been applied only after the letter a_3 has been rewritten into b_3.

Conversely, if $(L_i, W_i, R_i)_{i=1,2,\ldots,n}$ is sequence of triples that describe an accepting computation of M on input $w = w_1w_2 \cdots w_n$, then we can extract the sequence of rewrite and delete operations of M from this sequence. Indeed, from the above sequence we see that $w = a_1a_2a_3a_4a_5$. Further, from the first triple $(L_1, W_1, R_1) = \left|\begin{array}{ccc} \rhd & a_1 & b_3 \\ & b_1 & \end{array}\right|$, we see that M only executes the rewrite step $b_1 \in \delta(\rhd a_1b_3)$ at position 1. As this is the first position, we know that the left neighboring field contains the left sentinel $\rhd$, that is, with respect to the left, this sequence of reductions is correct.

Now we consider the second triple $(L_2, W_2, R_2) = \left|\begin{array}{ccc} a_1 & a_2 & b_3 \\ a_1 & b_2 & b_3 \\ & \lambda & \end{array}\right|$. Thus, at position 2, M executes the sequence of rewrite steps $b_2 \in \delta(a_1a_2b_3)$ and $\lambda \in \delta(a_1b_2b_3)$. As initially field 1 contains the letter a_1, we see that these steps are correct with respect to the left.

At position 3, M only executes the rewrite step $b_3 \in \delta(a_2a_3a_4)$. As the initial letter at position 2 is a_2, we see that this rewrite step is correct with respect to the left. Now that a_3 has been rewritten into b_3, we see that the rewrite and delete steps at position 2 are correct with respect to the right. Thus, all rewrite and delete steps at position 2 have been verified, and as W_2 ends with λ, we can remove the triple (L_2, W_2, R_2). But then the triple (L_3, W_3, R_3) becomes the right neighbor of (L_1, W_1, R_1), which shows now that the rewrite at position 1 is also correct with respect to the right.

At position 4, M executes the rewrite step $b_4 \in \delta(b_3a_4\lhd)$ and the delete step $\lambda \in \delta(b_3b_4\lhd)$. As this position initially contains the letter a_4, we conclude that the rewrite at position 3 is correct with respect to the right, which then shows that the steps at position 4 are correct with respect to the left.

Finally, at position 5, M only executes the delete step $\lambda \in \delta(a_4a_5\lhd)$, which is correct with respect to the left and with respect to the right. Thus, we can remove the triple (L_5, W_5, R_5), which means that the right sentinel $\lhd$ is now the new right neighbor of position 4. This in turn implies that the rewrite and delete steps at position 4 are correct with respect to the right, and hence, we can remove the triple (L_4, W_4, R_4) as well. Finally, as $\mathsf{Accept} \in \delta(b_1b_3\lhd)$, we see that the above sequence of triples does indeed describe an accepting computation of M on input $a_1a_2a_3a_4a_5$, *provided M can always move to the required position by a sequence of move-right steps.* □

In the case of stl-ORWW-automata, that is, when no letter can be deleted, the required MVR-steps can easily be inferred from the sequence of triples considered above, and so, by checking the transition relation of M, it can be verified whether they are actually possible. However, for stl-ORD-automata, that is, when delete operations are used, this is not at all clear, as particular move-right steps may or may not use letters at positions that are at some point deleted. Therefore, in order to turn our idea into a correct simulation of a stl-ORD-automaton by a PDA, we turn to a restricted class of stl-ORD-automata.

Definition 5. *A stl-ORD-automaton $M = (\Sigma, \Gamma, \rhd, \lhd, \delta, >)$ is called* swift *if* $\mathsf{MVR} \in \delta(a_1a_2a_3)$ *for all $a_1 \in \Gamma \cup \{\rhd\}$ and all $a_2, a_3 \in \Gamma$. By* swift-ORD *we denote the class of all swift stl-ORD-automata.*

Thus, from a restart configuration $\rhd w \lhd$, a swift-ORD-automaton M can move its window to any position on the tape. Thus, a computation of M cannot be blocked by a factor across which M cannot move its window. The stl-ORD-automaton M of Example 3 is actually a swift-ORD-automaton. As M performs its rewrite and delete steps strictly from right to left, it is easily seen that this ability to freely move to any position on the tape does not lead to the acceptance of any words that do not belong to the language $L = \{\, a^nb^n \mid n \geq 0 \,\} \cup \{\, a^nb^{2n} \mid n \geq 0 \,\}$.

Based on the discussion in Example 4, we can now formulate the following theorem.

Theorem 6. $\mathcal{L}(\mathsf{swift\text{-}ORD}) \subseteq \mathsf{CFL}$.

Proof. Let M be a swift-ORD-automaton, where we assume that M only accepts at the right sentinel, and let P be the PDA that proceeds as follows given a word $w \in \Sigma^*$ as input. Each time P reads an input letter $a \in \Sigma$, it guesses a triple of the form (L, W, R) as described in the above example. The first such triple should have either $L = \Lambda$ (that is, no rewrite or delete step is executed at position 1) or $L = (\rhd, \rhd, \ldots, \rhd)$, otherwise, P rejects immediately. Then P verifies that all rewrite steps encoded in this triple are correct with respect to the transition relation of M and stores the triple on its pushdown. On reading the next input letter, P guesses the next triple (L, W, R), verifies that all its lines correspond to transition steps of M, and compares this triple to the topmost triple, say (L', W', R'), on its pushdown. Using these two triples it determines which of the transitions of the triple (L', W', R') are correct with respect to the right and which of the transitions of the triple (L, W, R) are correct with respect to the left. If an incorrect transition is detected, then P rejects immediately, otherwise it marks the transitions verified. If a contradiction is detected, then P halts without acceptance. Otherwise, if all transitions of (L', W', R') have been verified completely, and if W' ends with the entry λ, then the triple (L', W', R') is popped from the pushdown. In the latter case, the triple (L, W, R) is compared to the now topmost triple on the pushdown in order to verify the correctness of further rewrite steps. Finally, once it has been verified that all transitions of (L, W, R) are correct with respect to the left, then this triple is pushed onto the pushdown and the next input letter is read. This process continues until the input has been read completely. As M accepts at the right sentinel, P must also check that the transition relation of M contains the corresponding accept step. If all these tests are positive, then P accepts.

As the sequences of triples describe accepting computations of M in such a way that from a sequence the correctness of the corresponding computation can be checked, we see that P has an accepting computation for input $w \in \Sigma^*$ if and only if $w \in L(M)$. Hence, it follows that $L(P) = L(M)$, which means that $L(M)$ is indeed a context-free language. □

4 Each Context-Free Language Is Accepted by a Swift-ORD-Automaton

Here we prove that each context-free language is accepted by a swift-ORD-automaton.

Theorem 7. $\mathsf{CFL} \subseteq \mathcal{L}(\mathsf{swift\text{-}ORD})$.

Proof. Let $L \subseteq \Sigma^*$ be a context-free language. There is a context-free grammar $G = (V, \Sigma, S, P)$ in quadratic Greibach normal form for the language $L \smallsetminus \{\lambda\}$ [14], that is, each production $(A \to r) \in P$ satisfies the restriction that $r \in \Sigma \cdot (V^2 \cup V \cup \{\lambda\})$. From this grammar, a PDA $A = (Q, \Sigma, \Delta_A, q, S, \delta_A)$ can be obtained such that $N(A) = L$, that is, L is the language accepted by A with empty pushdown. Here we can require that A only has a single state, that it does not execute any λ-transitions, and that, in each step, it replaces the topmost symbol on its pushdown by a word of length at most two, that is, $Q = \{q\}$, $\Delta_A = V$, and $(q, \alpha^R) \in \delta_A(q, a, B)$ iff $(B \to a\alpha) \in P$, where $a \in \Sigma$, $B \in V$, and $\alpha \in (V^2 \cup V \cup \{\lambda\})$. In our encoding below, the bottom (top) of the pushdown will always be on the left (right).

We now construct a swift-ORD-automaton $M = (\Sigma, \Gamma, \triangleright, \triangleleft, \delta, >)$ for L that simulates the PDA A. Essentially, it works as the automaton in Example 3, that is, it determines the transition of A that is to be applied next, it marks the letters that are to be rewritten, and then it replaces (or deletes) the corresponding letters. Notice that the simulation of A can be performed in a non-length-increasing fashion using an appropriate encoding of the pushdown. The swift-ORD-automaton M is defined as follows:

- $\Gamma = \Sigma \cup \{\, a', a'' \mid a \in \Sigma \,\} \cup \{\, [\alpha] \mid \alpha \in \Delta_A \cup \Delta_A^2 \,\} \cup \{[q,\lambda]\} \cup$
  $\{\, [q,\alpha], [q,\alpha]'' \mid \alpha \in \Delta_A \cup \Delta_A^2 \,\} \cup \{\, [q,\alpha]_a \mid \alpha \in \Delta_A \cup \Delta_A^2, a \in \Sigma \,\}$,
- for all $a, b \in \Sigma$, $\alpha \in \Delta_A^+$, and $x \in \Delta_A$,

$$a > a' > a'' > [\alpha x] > [q,\alpha x]'' > [q,\alpha x] > [q,\alpha x]_b > [\alpha] > [q,\lambda],$$

- and the transition relation δ is defined through the following table, where $a, b \in \Sigma$, $c \in \Sigma \cup \{\lhd\}$, $\alpha, \gamma \in \Delta_A^*$, $\alpha_1 \in \Delta_A^+$, $X \in \{\rhd\} \cup \{\, [\alpha] \mid 1 \le |\alpha| \le 2 \,\}$, and $x \in \Delta_A$:

$$\begin{array}{lll}
(0)\ \delta(XYZ) & \ni \mathsf{MVR} & \text{for all } X \in \Gamma \cup \{\rhd\} \text{ and } Y, Z \in \Gamma, \\
(1)\ \delta(\rhd a \lhd) & = \mathsf{Accept} & \text{for all } a \in L \cap (\Sigma \cup \{\lambda\}), \\
(2)\ \delta(\rhd ab) & \ni [q,\alpha], & \text{if } \delta_A(q,a,S) \ni (q,\alpha), \\
(3)\ \delta([q,\alpha x]ac) & \ni a', & \text{if } \delta_A(q,a,x) \ni (q,\gamma), \\
(4)\ \delta(X[q,\alpha x]a') & \ni [q,\alpha x]_a, & \text{if } \delta_A(q,a,x) \ni (q,\gamma), \\
(5)\ \delta([q,\alpha x]_a a'c) & \ni [q,\gamma]'', & \text{if } \delta_A(q,a,x) \ni (q,\gamma), \gamma \neq \lambda, \\
(6)\ \delta(X[q,\alpha x]_a[q,\gamma]'') & \ni [\alpha], & \text{if } \delta_A(q,a,x) \ni (q,\gamma), \gamma \neq \lambda, \text{ and } \alpha \neq \lambda, \\
(7)\ \delta(X[q,x]_a[q,\gamma]'') & \ni \lambda, & \text{if } \delta_A(q,a,x) \ni (q,\gamma), \gamma \neq \lambda, \\
(8)\ \delta(X[q,\gamma]''c) & \ni [q,\gamma], & \\
(9)\ \delta([q,\alpha x]_a a'c) & \ni a'', & \text{if } \delta_A(q,a,x) \ni (q,\lambda), \\
(10)\ \delta(X[q,\alpha x]_a a'') & \ni [q,\alpha], & \text{if } \delta_A(q,a,x) \ni (q,\lambda) \text{ and } \alpha \neq \lambda, \\
(11)\ \delta([q,\alpha]a''c) & \ni \lambda, & \\
(12)\ \delta([\alpha_1][q,x]_a a'') & \ni \lambda, & \text{if } \delta_A(q,a,x) \ni (q,\lambda), \\
(13)\ \delta(X[\alpha_1]a'') & \ni [q,\alpha_1], & \\
(14)\ \delta(\rhd[q,x]_a a'') & \ni [q,\lambda], & \text{if } \delta_A(q,a,x) \ni (q,\lambda), \\
(15)\ \delta(\rhd[q,\lambda]\lhd) & = \mathsf{Accept}. &
\end{array}$$

In order to illustrate this definition, we consider a simple example. Let $A = (\{q\}, \{a,b\}, \{S,B,C\}, q, S, \delta_A)$, where δ_A only contains the following transitions:

$$\begin{aligned}
\delta_A(q,a,S) &= \{(q,BC),(q,B)\}, \\
\delta_A(q,b,B) &= \{(q,\lambda)\}, \\
\delta_A(q,a,C) &= \{(q,BC),(q,B)\}.
\end{aligned}$$

Then the language $N(A)$ that is accepted by A with empty pushdown is the language $\{\, a^n b^n \mid n \ge 1 \,\}$, and, for example, A can execute the following accepting computation:

$$\begin{aligned}
(q, aaabbb, S) \vdash_A (q, aabbb, BC) \vdash_A (q, abbb, BBC) \vdash_A (q, bbb, BBB) \\
\vdash_A \quad (q, bb, BB) \quad \vdash_A \quad (q, b, B) \quad \vdash_A \quad (q, \lambda, \lambda).
\end{aligned}$$

This computation is now simulated by the corresponding swift-ORD-automaton M as follows:

$$\begin{aligned}
\underline{\rhd aa}abbb\lhd \vdash_{(2)} \rhd[q,BC]aabbb\lhd \vdash_{(0)} \rhd\underline{[q,BC]aa}bbb\lhd \\
\vdash_{(3)} \underline{\rhd[q,BC]a'}abbb\lhd \vdash_{(4)} \rhd\underline{[q,BC]_a a'a}bbb\lhd
\end{aligned}$$

$$\begin{array}{llll}
\vdash_{(0)} & \rhd[q,BC]_a a'abbb\lhd & \vdash_{(5)} & \rhd[q,BC]_a[q,BC]''abbb\lhd \\
\vdash_{(6)} & \rhd[B][q,BC]''abbb\lhd & \vdash_{(0)} & \rhd[B][q,BC]''abbb\lhd \\
\vdash_{(8)} & \rhd[B][q,BC]abbb\lhd & \vdash^2_{(0)} & \rhd[B][q,BC]abbb\lhd \\
\vdash_{(3)} & \rhd[B][q,BC]a'bbb\lhd & \vdash_{(0)} & \rhd[B][q,BC]a'bbb\lhd \\
\vdash_{(4)} & \rhd[B][q,BC]_a a'bbb\lhd & \vdash^2_{(0)} & \rhd[B][q,BC]_a a'bbb\lhd \\
\vdash_{(5)} & \rhd[B][q,BC]_a[q,B]''bbb\lhd & \vdash_{(0)} & \rhd[B][q,BC]_a[q,B]''bbb\lhd \\
\vdash_{(6)} & \rhd[B][B][q,B]''bbb\lhd & \vdash^2_{(0)} & \rhd[B][B][q,B]''bbb\lhd \\
\vdash_{(8)} & \rhd[B][B][q,B]bbb\lhd & \vdash^3_{(0)} & \rhd[B][B][q,B]bbb\lhd \\
\vdash_{(3)} & \rhd[B][B][q,B]b'bb\lhd & \vdash^2_{(0)} & \rhd[B][B][q,B]b'bb\lhd \\
\vdash_{(4)} & \rhd[B][B][q,B]_b b'bb\lhd & \vdash^3_{(0)} & \rhd[B][B][q,B]_b b'bb\lhd \\
\vdash_{(9)} & \rhd[B][B][q,B]_b b''bb\lhd & \vdash^2_{(0)} & \rhd[B][B][q,B]_b b''bb\lhd \\
\vdash_{(12)} & \rhd[B][B]b''bb\lhd & \vdash_{(0)} & \rhd[B][B]b''bb\lhd \\
\vdash_{(13)} & \rhd[B][q,B]b''bb\lhd & \vdash^2_{(0)} & \rhd[B][q,B]b''bb\lhd \\
\vdash_{(11)} & \rhd[B][q,B]bb\lhd & \vdash^2_{(0)} & \rhd[B][q,B]bb\lhd \\
\vdash_{(3)} & \rhd[B][q,B]b'b\lhd & \vdash_{(0)} & \rhd[B][q,B]b'b\lhd \\
\vdash_{(4)} & \rhd[B][q,B]_b b'b\lhd & \vdash^2_{(0)} & \rhd[B][q,B]_b b'b\lhd \\
\vdash_{(9)} & \rhd[B][q,B]_b b''b\lhd & \vdash_{(0)} & \rhd[B][q,B]_b b''b\lhd \\
\vdash_{(12)} & \rhd[B]b''b\lhd & \vdash_{(13)} & \rhd[q,B]b''b\lhd \\
\vdash_{(0)} & \rhd[q,B]b''b\lhd & \vdash_{(11)} & \rhd[q,B]b\lhd \\
\vdash_{(0)} & \rhd[q,B]b\lhd & \vdash_{(3)} & \rhd[q,B]b'\lhd \\
\vdash_{(4)} & \rhd[q,B]_b b'\lhd & \vdash_{(0)} & \rhd[q,B]_b b'\lhd \\
\vdash_{(9)} & \rhd[q,B]_b b''\lhd & \vdash_{(14)} & \rhd[q,\lambda]b''\lhd \\
\vdash_{(0)} & \rhd[q,\lambda]b''\lhd & \vdash_{(11)} & \rhd[q,\lambda]\lhd \\
\vdash_{(15)} & \text{Accept.} & &
\end{array}$$

Let $w = a_1a_2\cdots a_n \in \Sigma^+$. Starting from the initial configuration $\rhd w \lhd$, M repeatedly reaches configurations of the form

$$\rhd [\alpha_1][\alpha_2]\cdots[\alpha_{r-1}]\underline{[q,\alpha_r]a_s a_{s+1}}\cdots a_n \lhd,$$

where $\alpha_1\alpha_2\cdots\alpha_r$ encodes the contents of the pushdown of A and $a_s a_{s+1}\cdots a_n$ is the part of the input w that has not yet been read by A. Using the marked letters of the form a'_s, $[q,\alpha_r]_{a_s}$, a''_s, and $[q,\gamma]''$, M can now simulate the next step of A. Thus, by induction on the length of an accepting computation of A, it can be shown that $N(A) \subseteq L(M)$. Further, based on the way in which the marked letters are used, it can be seen that all rewrite and delete steps of M are executed at the border between the prefix ending with $[q,\alpha_r]$ and the suffix starting with a_s. A sequence of move-right steps that takes the window of M further to the right will not lead to any rewrite or delete step, that is, such a sequence will result in an unsuccessful computation. Finally, it can be seen that each accepting computation of M is a simulation of an accepting computation of A. Thus, $L(M) = N(A) = L$ follows. □

Thus, we have derived the following characterization.

Corollary 8. $\mathcal{L}$(swift-ORD) = CFL.

5 Conclusion

By introducing an additional delete/restart operation, we have extended the stl-ORWW-automaton to the stl-ORD-automaton. While the former just accepts the regular languages, we have seen that the latter accepts all context-free languages. In fact, by restricting the stl-ORD-automaton to its swift variant, we obtained a characterization for the class of context-free languages. However, it still remains open whether stl-ORD-automata that are not swift can accept any languages that are not context-free. Further, the descriptional complexity of swift stl-ORD-automata has not yet been studied.

After a restart, a swift stl-ORD-automaton M can move its window to any position on the tape. Hence, M can be described by a string-rewriting system

$$S_M = \{\, a_1a_2a_3 \to a_1ba_3 \mid b \in \delta(a_1a_2a_3)\,\} \cup \{\, a_1a_2a_3 \to a_1a_3 \mid \lambda \in \delta(a_1a_2a_3)\,\}.$$

By using an additional auxiliary symbol #, we can modify M in such a way that instead of simply accepting at some point, it writes the letter #, and then in subsequent cycles, it deletes all other letters. Then $L(M) = \{\, w \in \Sigma^* \mid \rhd w \lhd \vdash_M^* \rhd \# \lhd \,\}$, and so $L(M) = \{\, w \in \Sigma^* \mid \rhd w \lhd \Rightarrow_{S_M}^* \rhd \# \lhd \,\}$, where $\Rightarrow_{S_M}^*$ denotes the rewrite relation that is induced by S_M (see, e.g., [1]).

In [3], P. Černo and F. Mráz study so-called context rewriting systems. For any $k \geq 1$, a *k-context rewriting system* is a system $R = (\Sigma, \Gamma, I)$, where Σ is an input alphabet, Γ is a working alphabet that contains Σ but not the sentinels $\rhd$ and $\lhd$, and I is a finite set of instructions of the form $(x, z \to t, y)$, where $x \in \Gamma^k \cup \rhd \cdot \Gamma^{\leq k-1}$ is called the *left context*, $y \in \Gamma^k \cup \Gamma^{\leq k-1} \cdot \lhd$ is called the *right context*, and $z \to t$, $z, t \in \Gamma^*$, is called a *rule*. A word $w = \rhd uzv \lhd$ can be rewritten into $\rhd utv \lhd$ (denoted as $\rhd uzv \lhd \to_R \rhd utv \lhd$) if and only if there exists an instruction $i = (x, z \to t, y) \in I$ such that x is a suffix of $\rhd u$ and y is a prefix of $v\lhd$.

The *reduction language* associated with R is defined as

$$L^-(R) = \{\, w \in \Sigma^* \mid \rhd w \lhd \to_R^* \rhd\lhd \,\},$$

where $\to_R^*$ is the reflexive and transitive closure of $\to_R$. Observe that the empty word λ always belongs to the language $L^-(R)$. The string-rewriting system S_M obtained from a swift stl-ORD-automaton M can easily be interpreted as a 1-context rewriting system. For each of the resulting instructions $i = (x, z \to t, y)$, we have $|x| = |z| = |y| = 1$ and $|t| \leq 1$. In addition, if $|t| = 1$, then $z > t$ with respect to the partial ordering $>$ on Γ that is provided by M. In [3], P. Černo and F. Mráz introduce the *clearing restarting automaton* which is a context rewriting system such that $|z| \geq 1$ and $t = \lambda$ for each instruction $i = (x, z \to t, y)$. They show that 1-clearing restarting automata only accept a proper subclass of the context-free languages, while already 2-clearing restarting automata accept some non-context-free languages. Furthermore, they show that there are context-free languages that are not accepted by any clearing restarting automaton.

Accordingly, they extend their study to *Δ-clearing restarting automata*. Here Δ is an additional symbol, and such an automaton admits instructions of the form $i = (x, z \to t, y)$, where $z \in \Gamma^+$ and $t = \lambda$ or $t = \Delta$. These automata accept some context-free languages that are not accepted by clearing restarting automata, but it remains open whether all context-free languages can be accepted by them. The 1-context rewriting systems of the form S_M that we obtain from the swift stl-ORD-automata can be interpreted as a generalization of the 1-Δ-clearing restarting automata. Thus, the difference in the rewrite operations of swift stl-ORD-automata and the rewrite instructions of 1-Δ-clearing restarting automata illustrates the gap that is to be bridged in order to show that 1-Δ-clearing restarting automata accept all context-free languages.

References

1. Book, R., Otto, F.: String-Rewriting Systems. Springer, New York (1993). https://doi.org/10.1007/978-1-4613-9771-7_3
2. Buntrock, G., Otto, F.: Growing context-sensitive languages and Church-Rosser languages. Inf. Comput. **141**, 1–36 (1998)
3. Černo, P., Mráz, F.: Clearing restarting automata. Fund. Inform. **104**, 17–54 (2010)
4. Dahlhaus, E., Warmuth, M.: Membership for growing context-sensitive grammars is polynomial. J. Comput. Syst. Sci. **33**, 456–472 (1986)
5. Jančar, P., Mráz, F., Plátek, M., Vogel, J.: Restarting automata. In: Reichel, H. (ed.) FCT 1995. LNCS, vol. 965, pp. 283–292. Springer, Heidelberg (1995). https://doi.org/10.1007/3-540-60249-6_60
6. Kwee, K., Otto, F.: On the effects of nondeterminism on ordered restarting automata. In: Freivalds, R.M., Engels, G., Catania, B. (eds.) SOFSEM 2016. LNCS, vol. 9587, pp. 369–380. Springer, Heidelberg (2016). https://doi.org/10.1007/978-3-662-49192-8_30
7. Lopatková, M., Plátek, M., Sgall, P.: Towards a formal model for functional generative description: analysis by reduction and restarting automata. Prague Bull. Math. Linguist. **87**, 7–26 (2007)
8. McNaughton, R., Narendran, P., Otto, F.: Church-Rosser Thue systems and formal languages. J. ACM **35**, 324–344 (1988)
9. Mráz, F., Otto, F.: Ordered restarting automata for picture languages. In: Geffert, V., Preneel, B., Rovan, B., Štuller, J., Tjoa, A.M. (eds.) SOFSEM 2014. LNCS, vol. 8327, pp. 431–442. Springer, Cham (2014). https://doi.org/10.1007/978-3-319-04298-5_38
10. Otto, F.: Restarting automata. In: Ésik, Z., Martín-Vide, C., Mitrana, V. (eds.) Recent Advances in Formal Languages and Applications, Studies in Computational Intelligence, vol. 25, pp. 269–303. Springer, Berlin (2006)
11. Otto, F.: On the descriptional complexity of deterministic ordered restarting automata. In: Jürgensen, H., Karhumäki, J., Okhotin, A. (eds.) DCFS 2014. LNCS, vol. 8614, pp. 318–329. Springer, Cham (2014). https://doi.org/10.1007/978-3-319-09704-6_28
12. Otto, F.: On deterministic ordered restart-delete automata. In: Hoshi, M., Seki, S. (eds.) DLT 2018. LNCS, vol. 11088, pp. 529–540. Springer, Cham (2018). https://doi.org/10.1007/978-3-319-98654-8_43
13. Otto, F.: On deterministic ordered restart-delete automata. Theor. Comp. Sci. **795**, 257–274 (2019)
14. Rosenkrantz, D.J.: Matrix equations and normal forms for context-free grammars. J. ACM **14**, 501–507 (1967)

A Constructive Arboricity Approximation Scheme

Markus Blumenstock(✉) and Frank Fischer

Institute of Computer Science, Johannes Gutenberg University Mainz, Mainz, Germany
{mablumen,frank.fischer}@uni-mainz.de

Abstract. The arboricity Γ of a graph is the minimum number of forests its edge set can be partitioned into. Previous approximation schemes were nonconstructive, i.e., they approximate the arboricity as a value without computing a corresponding forest partition. This is because they operate on pseudoforest partitions or the dual problem of finding dense subgraphs.

We propose an algorithm for converting a partition of k pseudoforests into a partition of $k+1$ forests in $\mathcal{O}(mk \log k + m \log n)$ time with a data structure by Brodal and Fagerberg that stores graphs of arboricity k. A slightly better bound can be given if perfect hashing is used. When applied to a pseudoforest partition obtained from Kowalik's approximation scheme, our conversion implies a constructive $(1+\epsilon)$-approximation algorithm for the arboricity with runtime $\mathcal{O}(m \log n \log \Gamma \, \epsilon^{-1})$ for every $\epsilon > 0$. For fixed ϵ, the runtime can be reduced to $\mathcal{O}(m \log n)$.

Moreover, our conversion implies a near-exact algorithm that computes a partition into at most $\Gamma + 2$ forests in $\mathcal{O}(m \log n \, \Gamma \log^* \Gamma)$ time. It might also pave the way to faster exact arboricity algorithms.

Keywords: Approximation algorithms · Matroid partitioning

1 Introduction

Given a simple graph $G = (V, E)$ with n vertices and m edges, the arboricity $\Gamma(G)$ is the minimum number of forests on V that the edge set E can be partitioned into. Such a partition can be computed in polynomial time [11,16,17,26], and a linear-time 2-approximation algorithm is known [3,12]. In graphs of bounded arboricity, some NP-hard problems become tractable [2,13], and for several algorithms, it is possible to show better runtime estimates [10,13,19] or approximation factors [5]. There are distributed algorithms that operate directly on forest partitions for the maximal independent set problem [6] and the minimum dominating set problem [22].

An interesting relationship of the arboricity and dense subgraphs becomes apparent by the classic Nash-Williams formula [25]

$$\Gamma(G) = \lceil \gamma(G) \rceil, \text{ where } \gamma(G) := \max_{\substack{(V_H, E_H) \subseteq G \\ |V_H| \geq 2}} \frac{|E_H|}{|V_H| - 1} \tag{1}$$

A. Chatzigeorgiou et al. (Eds.): SOFSEM 2020, LNCS 12011, pp. 51–63, 2020.
https://doi.org/10.1007/978-3-030-38919-2_5

is called the *fractional arboricity*. A recent approximation scheme by Worou and Galtier [28] approximates γ (and hence, Γ) by constructing a subgraph of high density, but it does not construct a forest partition.

A *pseudoforest* is a graph in which each connected component contains at most one cycle. The pseudoarboricity $p(G)$ is defined analogously, and a similar formula holds [26]:

$$p(G) = \lceil d^*(G) \rceil, \text{ where } d^*(G) := \max_{(V_H, E_H) \subseteq G} \frac{|E_H|}{|V_H|} \tag{2}$$

is called the *maximum density*. It is evident from (1) and (2) that Γ and p must be very close.

Theorem 1 ([26]). *For a simple graph G, we have $p(G) \leq \Gamma(G) \leq p(G) + 1$.*

Thus, if we compute a pseudoforest partition approximating p, we directly know an approximation of the *value* Γ. Kowalik's approximation scheme [20] computes a partition of $K \leq \lceil (1+\epsilon)d^* \rceil$ pseudoforests in time $\mathcal{O}(m \log n \log p\, \epsilon^{-1})$. However, the algorithm in [17] for converting a partition of K pseudoforests into a partition of K or $(K+1)$ forests takes $\mathcal{O}(mn \log K)$ time. Kowalik thus raised the question whether a faster (approximate) conversion exists.

The main result of this paper is a fast conversion of k pseudoforests into $k+1$ forests (in particular, a new proof of Theorem 1), which implies a fast constructive approximation scheme for the arboricity: We divide the K pseudoforests obtained from the pseudoarboricity approximation scheme into k-tuples. Each k-tuple is converted into $k+1$ forests. The number k is chosen minimally such that $(k+1)/k \leq 1 + \epsilon$ for the given ϵ.

Our conversion uses the notion of a *surplus graph*: By removing one edge on every cycle in each of the pseudoforests $P_1, \ldots, P_k$ we obtain forests $F_1, \ldots, F_k$ and a surplus set M of edges. The edges in M inherit the index (color) of the pseudoforest they were removed from. We then make $H = (V, M)$ acyclic by applying a sequence of two procedures: The first procedure moves edges from M to the k forests such that each connected component of H has at most one edge of each color. It uses color swap operations in H and a certain union-find data structure [1] for representing the F_i. The second procedure exchanges edges in H with edges in the forests in order to remove all cycles in H. It uses link-cut trees [27] and adjacency queries in $F_1 \cup \cdots \cup F_k$. For the queries, perfect hashing [15] or a data structure for storing graphs of arboricity at most k [9] is used.

2 Paper Outline and Contributions

Section 3 gives a literature review of the arboricity and pseudoarboricity problems. Notation and definitions are introduced in Sect. 4. In Sects. 5, 6 and 7, we show the conversion of k pseudoforests into $k+1$ forests in time $\mathcal{O}(mk \log k + m \log n)$, or alternatively in time $\mathcal{O}(mk + m \log n)$ when a perfect hash function is constructed beforehand in $\mathcal{O}(m)$ expected time. Applying it to a pseudoforest partition obtained from Kowalik's approximation scheme yields the following main theorem.

Theorem 2. *For every $\epsilon > 0$, a simple graph can be partitioned into at most $\lceil(1+\epsilon)\cdot\lceil(1+\epsilon)d^*\rceil\rceil$ forests in time $\mathcal{O}(m\log n\log\Gamma\,\epsilon^{-1})$. Furthermore, if ϵ is fixed, the runtime can be bounded as $\mathcal{O}(m\log n)$.*

The 'furthermore'-part follows from a small modification that terminates the binary search of Kowalik's scheme once the ratio of the upper and lower bound falls below $1+\epsilon$. This eliminates the factor $\log\Gamma$ in the runtime and is described in the arXiv version of this paper.[1] Therein, we add some minor results and correct a few mistakes in the literature. Our conversion also implies a near-exact arboricity algorithm whose runtime scales with Γ, which is shown in Sect. 8:

Theorem 3. *A simple graph can be partitioned into at most $\Gamma+2$ forests in $\mathcal{O}(m\log n\,\Gamma\log^*\Gamma)$ time.*

Here, $\log^*$ denotes the iterated binary logarithm.

3 Related Work

Both the set of forests and the set of pseudoforests on a graph are matroids, thus the arboricity and pseudoarboricity can be computed with Edmonds' matroid partitioning algorithm [11] in polynomial time. Picard and Queyranne [26] reduce the problems to 0–1 fractional programming problems that can be solved with $\mathcal{O}(n)$ and $\mathcal{O}(\log n)$ maximum flow computations, respectively. Gabow and Westermann [17] describe matroid partitioning algorithms specialized to these two matroids. Gabow's algorithm [16], which uses Newton's method for fractional optimization and flow algorithms, is the fastest known exact arboricity algorithm with a runtime of $\mathcal{O}(m^{3/2}\log(n^2/m))$.

To the best of our knowledge, no constructive arboricity algorithm with an approximation factor $1 < c < 2$ is known in general graphs. The linear-time greedy algorithm [3,12] is constructive. It computes an acyclic orientation that minimizes the maximum indegree among all acyclic orientations [8]. This indegree equals the degeneracy of the graph [24] and is at most $2\Gamma - 1$ [12]. An acyclic k-orientation can be converted into a forest k-partition (implicit in [20]). Cyclic orientations cannot be used in this manner directly, so the approach is exhausted.

The approximation scheme of Worou and Galtier [28] computes for $\epsilon > 0$ a $1/(1+\epsilon)$-approximation of the fractional arboricity γ in $\mathcal{O}(m\log^2(n)\log(\frac{m}{n})\,\epsilon^{-2})$ time. It constructs a subgraph that attains this density in the sense of the right-hand side of (1), but apparently no forest partition is computed.

A pseudoforest k-partition can be converted into a k-orientation, and vice versa, in linear time [20]. Hence the pseudoarboricity problem is equivalent to finding an orientation where the maximum indegree (or outdegree) is minimized. Dinitz' algorithm, which has a runtime of $\mathcal{O}(m\min(\sqrt{m},n^{2/3}))$ on unit capacity networks [14], can be employed to find a k-orientation in the same runtime, if

[1] https://arxiv.org/abs/1811.06803.

it exists [4,7,20] (see also the related [18]). Recently, faster flow algorithms for unit capacities were given with runtimes $\tilde{\mathcal{O}}(|E|^{10/7})$ [23] and $\tilde{\mathcal{O}}(|E|\sqrt{|V|})$ [21].

A binary search for the minimum k introduces a factor of $\mathcal{O}(\log p)$ in the runtime. The total runtime of the approach based on Dinitz' algorithm can be reduced to $\mathcal{O}(m \min(\sqrt{m \log p}, (n \log p)^{2/3}))$ using the balanced binary search technique of Gabow and Westermann [17]. Blumenstock [7] improves the first bound to $\mathcal{O}(m^{3/2}\sqrt{\log\log p})$ by employing a pseudoarboricity approximation scheme [20] to shrink the interval for the search beforehand.[2]

Kowalik's approximation scheme [20] works by terminating Dinitz' algorithm early. It computes a $\lceil(1+\epsilon)d^*\rceil$-orientation in time $\mathcal{O}(m \log n \log p\, \epsilon^{-1})$. A partition of k pseudoforests can be converted into a partition of $k+1$ forests, and k if possible, in $\mathcal{O}(mn \log k)$ time. This is implicit in [17]. (We claim in the arXiv version that the runtime bound $\mathcal{O}(m^2/k \log k)$ in [17] is incorrect).

4 Notation and Preliminaries

We consider finite simple graphs $G = (V, E)$, i.e., G is undirected and has no loops. We follow the standard graph-theoretic terminology. For technical reasons we assume $n \geq 2$ and $m \geq n$ for the input graphs. If every vertex in a (sub-)graph has degree zero or one, the edge set is called a *matching*.

In an *orientation* of a simple graph G, every edge of G is present once, directed in one of the two possible directions. If all indegrees in the orientation are at most k, it is called a k-orientation.

An acyclic simple graph is called a *forest*. Its connected components are called *trees*. A tree of n vertices has exactly $n-1$ edges. We denote the disjoint union of sets by $\dot{\cup}$. If E is partitioned as $E = F_1 \dot{\cup} \cdots \dot{\cup} F_k$ where (V, F_i) is a forest for all $i = 1, \ldots, k$, we call $(F_1, \ldots, F_k)$ a forest k-partition. The *arboricity* $\Gamma(G)$ is the smallest integer k such that a forest k-partition of G exists.

If a graph has at most one cycle per connected component, it is called a *pseudoforest*. Its connected components are called *pseudotrees*. A component that is a pseudotree but not a tree is said to be *unicyclic*. We define pseudoforest k-partitions $(P_1, \ldots, P_k)$ and the pseudoarboricity $p(G)$ analogously. A basic property of a unicyclic component is that removing an arbitrary edge on its cycle leaves a tree. Note that connecting two different trees by an edge results in a single tree, but this does not carry over to pseudotrees.

Let (V, P) be a pseudoforest. For every cycle $C \subseteq P$, select one edge $e_C \in C$ arbitrarily. The set M of all these selected edges is a matching. We call this kind of matching M a *P-matching*. The following lemma is obvious.

Lemma 1. *A pseudoforest (V, P) can be partitioned into a forest and a P-matching in linear time.*

[2] We note that this algorithm can be formulated in terms of flows entirely without any knowledge of matroid theory. While not explicitly stated in [7], within the same runtime an 'almost densest subgraph' of density greater $\lceil d^*\rceil - 1$ can be determined using the network of [18] once for parameter $p - 1 = \lceil d^*\rceil - 1$.

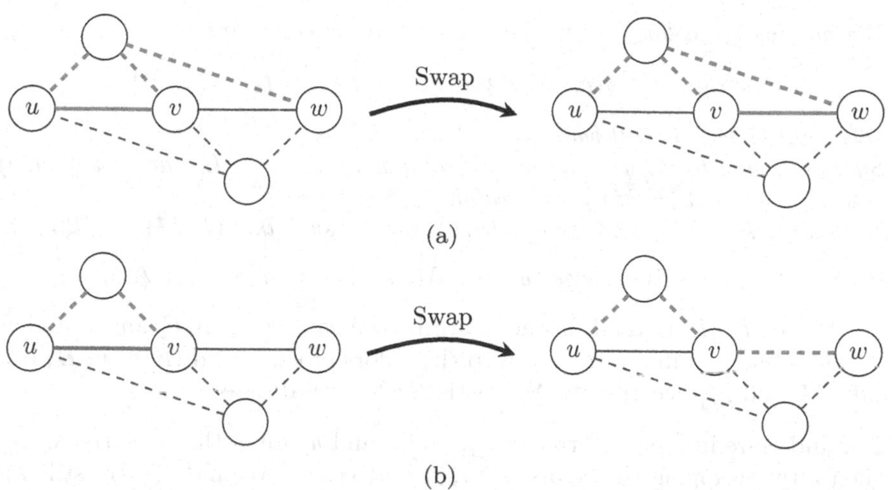

Fig. 1. **(a)** Situation 1. of Lemma 2. The colors i (red, thick) and j (blue, thin) of the edges $(u, v) \in M_i$ and $(v, w) \in M_j$ are swapped. Dotted edges represent the forests F_i and F_j. **(b)** Situation 2. of Lemma 2. v and w are in different trees of forest F_i (red, thick, dotted), hence the edge (v, w) can be inserted into F_i after swapping colors. (Color figure online)

In the following sections, we will describe a conversion of a pseudoforest k-partition into a forest $(k+1)$-partition using P-matchings within a reasonable runtime. In the arXiv version, we show how to obtain linear time for $k \in \{2, 3\}$.

5 The Surplus Graph

We maintain the edges E of the graph as a partition $E = F \dot{\cup} M$, where $F = F_1 \dot{\cup} \cdots \dot{\cup} F_k$ for forests $F_1, \ldots, F_k$ and $M = M_1 \dot{\cup} \cdots \dot{\cup} M_k$ such that $F_i \cup M_i = P_i$ is a pseudoforest and M_i is a P_i-matching for $i = 1, \ldots, k$. We call (F, M) a *valid partition* of the graph. Initially, a valid partition is obtained by applying Lemma 1 to each P_i of a given pseudoforest k-partition. Edges in both F_i and M_i are considered to have color i. The graph $H = (V, M)$ is called the *surplus graph*. Note that any two adjacent edges of H must have different colors.

Turning H into a forest while keeping (F, M) valid yields a constructive proof of Theorem 1. The following lemma provides a *swap operation* that will be used to move edges from H to the forests $F_1, \ldots, F_k$. It is illustrated in Fig. 1.

Lemma 2. *Let (F, M) be a valid partition of a simple graph G, and let $H = (V, M)$ be its surplus graph. Let $(u, v) \in M$ with color i and $(v, w) \in M$ with color $j \neq i$. Then one of the following applies.*

1. *We may swap the colors of (u, v) and (v, w) in H, i.e., modify*

$$M_i \leftarrow M_i \setminus \{(u, v)\} \cup \{(v, w)\} \text{ and } M_j \leftarrow M_j \setminus \{(v, w)\} \cup \{(u, v)\}$$

such that (F, M) is still valid.

2. *We may assign color* j *to* (u,v) *in* H *and insert* (v,w) *into* F_i*, i.e., modify*

$$M_j \leftarrow M_j \setminus \{(v,w)\} \cup \{(u,v)\} \text{ and } F_i \leftarrow F_i \cup \{(v,w)\}$$

such that (F,M) *is still valid.*
3. *Symmetrically to 2., we may assign color* i *to* (v,w) *in* H *and insert* (u,v) *into* F_j *such that* (F,M) *is still valid.*
4. *We may insert* (v,w) *into* F_i *and* (u,v) *into* F_j *such that* (F,M) *is still valid.*

Furthermore, if there is an edge $(w,x) \in M$ *of color* i*, then 2. or 4. applies.*

Proof. Since (F,M) is valid, u and v are in the same tree in F_i and v and w are in the same tree in F_j. Let us swap the colors of (u,v) and (v,w) in H, i.e., modify M_i and M_j accordingly. We distinguish several cases:

1. If u and v are in the same tree of F_j, and v and w are in the same tree of F_i, then after swapping the colors of (u,v) and (v,w), M_i and M_j are still P_i- and P_j-matchings, respectively. This is illustrated in Fig. 1a.
2. If v and w are in different trees in F_i, then $F_i \cup \{(v,w)\}$ is a forest. If u and v are in the same tree in F_j, change the color of (u,v) to j, now no edge in M_i exists whose endpoints are both in the tree of F_i that v is contained in. Since (F,M) had been valid, there is at most one edge in M_i whose endpoints are both in the tree of F_i that w is contained in. Hence after inserting (v,w) into F_i, there is still at most one such edge for the joined tree. This is illustrated in Fig. 1b.
3. If v and w are in the same tree in F_i, and u and v are in different trees in F_j, we have a case that is symmetric to 2.
4. If v and w are in different trees in F_i, and u and v are in different trees in F_j, then we can insert (v,w) into F_i and (u,v) into F_i. (F,M) is easily seen to be valid.

For the 'furthermore'-claim, we observe that if $(w,x) \in M_i$, then v and w must be in different trees of F_i because (F,M) is valid. □

We can implement swap operations with k union-find data structures that keep track of the vertex sets of the connected components in each F_i. An edge $(u,v) \in M$ connects two different trees in F_i if and only if the vertex sets $S_u = \text{find}_i(u)$ and $S_v = \text{find}_i(v)$ returned by the union-find structure of F_i are different. To insert (u,v) with $S_u \neq S_v$ into F_i, $\text{union}_i(S_u, S_v)$ is called, which merges S_u and S_v into a single set.

Since we will be performing $\mathcal{O}(mk)$ find operations, but only $\mathcal{O}(m)$ union operations, we use the union-find data structure of [1] that has a total runtime of $\mathcal{O}(f + n \log n)$ for a sequence of f find and up to $n-1$ union operations. Note that this does not imply a time bound of $\mathcal{O}(f + m' \log n)$ for $m' < m$ union operations dispersed among the k union-find structures. In order to achieve this, we first contract every connected component of an F_i into a 'super-vertex'. These are then handled by the union-find structure instead. A table of size kn is constructed that allows constant-time look-up of the super-vertex containing a $v \in V$ in F_i. We omit the details.

Lemma 3. *Let a path in the surplus graph H be given by edges $(e_1, \dots, e_l)$ where $e_1, e_l \in M_i$ for the same color i. Then we can modify (F, M) such that the cardinality of M decreases while maintaining validity.*

Proof. We can move the color i from e_1 towards e_l in a sequence of swaps using Lemma 2, i.e., swap the colors of e_t and e_{t+1} for $t = 1, \dots$ until one of the cases 2.–4. of Lemma 2 applies. This happens at the latest when e_{l-2} has color i, because then we are in the 'furthermore'-part of Lemma 2. Thus we can move some edge on the path from M to F while maintaining validity. □

An application of Lemma 3 does not necessarily remove an edge of the duplicate color in question, but possibly an edge of some other color along the path. Irrespective of this, we will charge the entire cost including finding the duplicate color to the removed edge. There are at most m removals.

A connected component of the surplus graph H is called *colorful* if every color appears at most once in it. A surplus graph is called colorful if all its connected components are colorful. Note that each component in a colorful surplus graph has at most k edges, a fact that we will later exploit for our runtime analyses.

Lemma 4. *A colorful surplus graph can be obtained in $\mathcal{O}(mk + m \log n)$ time.*

Proof. Obtain an arbitrary surplus graph H in linear time. We can initially build union-find structures for all (contracted) forests $F_1, \dots, F_k$ in total time $\mathcal{O}(nk + m)$ from the connected components of each F_i.

A duplicate color in a connected component of H can be identified by performing a depth-first search in it: Record the colors encountered in the search in a Boolean array of length k. If there is a duplicate color i, we will encounter one such color and recognize it after at most $k + 1$ steps of the DFS (not counting backtracking steps). Otherwise, the search is unsuccessful and the component already is colorful. The number of unsuccessful searches is at most n.

We can now apply Lemma 3 to a path of length at most $k + 1$ from the edge of color i encountered first to the edge of color i encountered second. We charge the costs of the $\mathcal{O}(k)$ find and at most two union operations to some edge that was removed in the swap sequence and start the next search. We perform the searches in each connected component of H until all duplicate colors have been eliminated. Note that components may disconnect during the algorithm. There are at most m successful searches. The total cost is thus $\mathcal{O}(mk + m \log n)$. □

The union-find structures do not store the edges that we insert. Thus, we store them and their colors separately in a list so we can reconstruct the F_i later on.

6 Exchanging Edges on Cycles

In order to remove cycles from a colorful surplus graph H, we want to replace an edge e in H that is on some cycle in a connected component C with an edge from some F_i that goes to a vertex outside of C. This reduces the number of edges that are on at least one cycle in H. After at most m such operations, H

will be a forest. To do so, we will insert e into a carefully chosen F_i and exchange it with an adjacent edge on the resulting cycle. We call this the *cycle exchange.*

First, we store the forests $F_1, \ldots, F_k$ in k link-cut tree data structures [27] in total time $\mathcal{O}(nk + m \log n)$. In these structures, each tree is considered to be a rooted tree (which is stored in a compressed way) with all edges oriented towards the root, and the root of the tree containing vertex u can be accessed via root(u). There is an operation evert(u) that makes u the root of its tree. The operation cut(u) deletes the parent edge of u and thereby splits the tree. There is an operation link(u,v), where u is a root and v is in a different tree than u, that makes u point to v. All these operations can be performed in $\mathcal{O}(\log n)$ amortized (in fact, even worst-case) time [27]. There will be $\mathcal{O}(m)$ such operations in total.

We will also maintain the union-find structures. The reason for this is that the vertex sets of the connected components of an F_i do not change in the cycle exchange. Thus, the union-find structures still work correctly. We will perform $\mathcal{O}(mk)$ find operations, but $\mathcal{O}(m)$ union and link-cut tree operations, so it is advantageous to simultaneously keep the union-find structure for the faster find runtime.

An edge suitable for the cycle exchange always exists if H is cyclic. How to determine this edge fast will be shown in the next section.

Lemma 5. *Let $C = (V_C, E_C)$ be a colorful cyclic connected component of a surplus graph. For any $v \in V_C$, there is a color i such that there is an edge of color i in E_C, and v has no neighbors in F_i that are in V_C.*

Proof. Since C is colorful, exactly $|E_C|$ different colors $c_1, \ldots, c_{|E_C|}$ appear in C. As C is cyclic, we have $|E_C| \geq |V_C|$. If v had a neighbor among the vertices V_C in every F_i, $i = c_1, \ldots, c_{|E_C|}$, then v would have at least $|V_C|$ neighbors among V_C in G, a contradiction. □

Lemma 6. *Let H be a colorful surplus graph. If for a vertex v on a cycle in H a color i as in Lemma 5 can be determined in time $T(k, n, m)$ with $P(k, n, m)$ preprocessing time, then we can obtain an acyclic colorful surplus graph in time $\mathcal{O}(mT(k, n, m) + mk + m \log n + P(k, n, m))$.*

Proof. We start from a colorful surplus graph H. We can determine if a connected component C of H is cyclic in time $\mathcal{O}(k)$ with DFS. If it is, let (u, v) be an edge on this cycle. Determine the color i as in Lemma 5 in time $T(k, n, m)$.

Determine a path from the edge of color i in C to v. As in the proof of Lemma 4, move i towards (u, v) in a sequence of color swaps. If an edge is removed from H by this, we charge the costs including the $\mathcal{O}(k)$ find and at most two union operations to the removed edge. We then start looking for cycles again. There can be at most m such removals in H in total.

If no edge is removed from H, then v is now incident to an edge (u, v) of color i in H. Since (F, M) is valid, we know that inserting (u, v) into F_i would create a cycle. Make u the root of its link-cut tree in F_i by calling evert (u), i.e., the link-cut tree represents the tree where all edges are directed towards u. If (u, v) were to be inserted into this tree, then it would create a cycle passing through u

and v. Call parent (v) to obtain an edge (v, w) on this cycle incident with v. By the choice of i, $w \notin V_C$, i.e., this edge must leave C in H and can hence not be on a cycle. Call cut (v) to remove the edge from the link-cut tree, which breaks it into a tree rooted at u and the subtree rooted at v. Call link (u, v) to insert the edge (u, v) into the link-cut tree. As remarked earlier, the union-find structure still represents the trees of F_i after these changes. The number of edges in H that are on at least one cycle decreases, which may happen at most m times, so the costs for all cycle exchanges amount to $\mathcal{O}(m \log n)$ in total.

$C \setminus \{(u, v)\}$ is joined to another colorful connected component of H via (v, w). Duplicate colors in the resulting component of size $\mathcal{O}(k)$ are detected and removed as in the proof of Lemma 4 in order to keep H colorful. Here, we must charge costs to the decrease of $|M|$ because an edge may re-enter M in cycle exchanges. The costs of $\mathcal{O}(k)$ of an unsuccessful search are charged to the cycle exchange. □

We have now obtained an alternative and algorithmic proof of Theorem 1. In addition, each connected component of H has at most k edges.

7 Finding the Exchange Edge Fast

We next describe two ways of finding the exchange edge with a runtime that does not depend on n. The first approach uses the data structure of Brodal and Fagerberg [9, Section 4], which stores a graph of arboricity at most k and hence can be used for $F_1 \cup \cdots \cup F_k$. It allows querying whether two vertices are adjacent in time $\mathcal{O}(\log k)$, inserting an edge in $\mathcal{O}(\log k)$ amortized time, and deleting an edge in $\mathcal{O}(\log n)$ amortized time. The structure can initially be built for a given graph in $\mathcal{O}(m \log k + n)$ time (with a little effort, $\mathcal{O}(m + n)$ is possible).

The representation used by the data structure is an orientation of the graph such that every vertex has indegree at most $4k$. Every edge is stored only once, namely in the adjacency list/balanced search tree of the vertex it points to. Hence the size of each list/search tree is $\mathcal{O}(k)$. We can store the current color of each edge with it without affecting the runtimes.

Lemma 7. *In the situation of Lemma 5, we can determine the exchange edge in time $T(k, n, m) \in \mathcal{O}(k \log k)$ using the data structure of Brodal and Fagerberg with preprocessing time $P(k, n, m) \in \mathcal{O}(m + n)$. All other operations have the same asymptotic complexity as in Lemma 6.*

Proof. Create the data structure for $F_1 \cup \cdots \cup F_k$ in time $\mathcal{O}(m + n)$. When looking for a cycle in a component $C = (V_C, E_C)$ of the colorful surplus graph (with $|V_C| \leq k + 1$), we use a Boolean array of size k to mark the colors of the component and remember the respective edges. When some vertex v on a cycle has been determined, we test for each $u \in V_C \setminus \{v\}$ whether $(u, v) \in E \setminus M$ in $\mathcal{O}(\log k)$ with the color-augmented Brodal-Fagerberg data structure. If the edge is present in some F_i, then we obtain color i from the data structure and unmark it in the Boolean array. Once all $u \in V_C \setminus \{u\}$ have been tested, search for a color i that is still marked: the attached edge is the one we were looking for, i.e., v has no neighbors in V_C in F_i. All these operations cost $\mathcal{O}(k \log k)$ in total.

During the cycle exchange algorithm in Lemma 6, at most m edges are inserted into the forests $F_1, \ldots, F_k$. An edge is only deleted in a cycle exchange, which happens at most m times. Thus, these cost amount to $\mathcal{O}(m)$ insertions and deletions in the data structure, and each such operation costs $\mathcal{O}(\log n)$ amortized time. Hence the runtime of Lemma 6 can indeed be achieved. □

We now prove the main theorem.

Proof (Theorem 2). We can obtain a partition into $K \leq \lceil (1+\epsilon)d^* \rceil$ pseudoforests in time $\mathcal{O}(m \log n \log p\, \epsilon^{-1})$ with Kowalik's approximation scheme [20]. For fixed ϵ, the factor $\log p$ can be removed (see the arXiv version).

We can assume that $\epsilon \geq 1/K$. Let $k \leq K$ be the smallest integer such that $(k+1)/k \leq 1+\epsilon$. Note that $k \in \mathcal{O}(\epsilon^{-1})$ and $\log k \in \mathcal{O}(\log n)$. Divide the K pseudoforests evenly into k-tuples of pseudoforests, if possible, otherwise $l \leq k-1$ pseudoforests remain. Convert each k-tuple into $k+1$ forests and the remaining l pseudoforests into $l+1$ forests with Lemmas 6 and 7. □

The second approach uses perfect hashing: For the set E of m edges from the universe $V \times V$, we construct a perfect hash function and maintain the set $E \setminus M = F$ in a hash table and store the current color information of each edge with it. The perfect hashing scheme of Fredman et al. [15] allows worst-case constant runtimes for querying, insertion, and deletion. Constructing the perfect hash function is possible in $\mathcal{O}(m)$ expected time (deterministic construction is possible in $\mathcal{O}(n^2 m)$). The following lemma is proved analogously to Lemma 7.

Lemma 8. *In the situation of Lemma 5, we can determine the exchange edge in $\mathcal{O}(k)$ time with $\mathcal{O}(m)$ expected time for preprocessing using perfect hashing. All other operations have the same asymptotic complexity as in Lemma 6.*

While Lemma 8 has the downside of being randomized, it may be useful for the development of an exact randomized algorithm for arboricity.

8 (Near-)Exact Arboricity Algorithms

Gabow's exact arboricity algorithm has a runtime of $\mathcal{O}(m^{3/2} \log(n^2/m))$ [16]. Note that $p \leq \Gamma \in \mathcal{O}(\sqrt{m})$ [7,10,17]. Hence, even in the worst case we can convert p pseudoforests into $p+1$ forests in time $\mathcal{O}(m^{3/2})$ with Lemmas 6 and 8 after the perfect hash function has been constructed. Since algorithms for pseudoarboricity are known that run in time $\mathcal{O}(m^{3/2}\sqrt{\log \log p})$ [7] and even $\mathcal{O}(m^{3/2})$ with recent flow algorithms [23], we would obtain a faster exact (randomized) algorithm if we can insert all edges of the constructed $(k+1)$-th forest into $F_1, \ldots, F_k$ fast enough if this is feasible. By (1), an infeasibility certificate would be a set $S \subseteq V$ for which $F_i[S]$ is a tree for every $i = 1, \ldots, k$, and an additional edge whose end vertices are both in S.

We are able to give a near-exact algorithm whose runtime scales with Γ.

Proof (Theorem 3). Compute a 2-approximation of d^* in linear time (see Sect. 3) in order to set $\epsilon \simeq 1/d^*$. Compute a partition into at most $\lceil(1+1/d^*)d^*\rceil = p+1$ pseudoforests in $\mathcal{O}(m \log n\, p \log p)$ time using Kowalik's approximation scheme. The runtime bound can be improved to $\mathcal{O}(m \log n\, p \log^* p)$ using the iterated interval shrinking technique of [7].

Using Lemmas 6 and 7, we convert this partition into a partition of at most $p+2 \leq \Gamma+2$ forests in $\mathcal{O}(mp\log p + m\log n)$ time. The claim follows. □

9 Conclusion and Outlook

We presented a fast conversion of k pseudoforests into $k+1$ forests. This implies a constructive approximation scheme for the arboricity. It remains to investigated how a constant number of forests can be inserted into a forest k-partition fast (if feasible), say with a runtime of $\mathcal{O}(mk + m\log n)$. Our conversions would then imply an exact randomized algorithm with runtime $\mathcal{O}(m^{3/2})$, being slightly faster than Gabow's, and an exact algorithm with runtime $\mathcal{O}(m \log n\, \Gamma \log^* \Gamma)$.

A related open question is whether Kowalik's approximation scheme for pseudoarboricity can be used to determine a $1/(1+\epsilon)$-approximation to the densest subgraph (by (2), it approximates the value d^*). As it has inversely linear dependence on ϵ, it would be preferable to the approximation scheme in [28] (for the slightly different measure γ) that has an inversely quadratic dependence.

Acknowledgements. The authors thank Łukasz Kowalik for discussions and Ernst Althaus for simplifying the algorithm that eliminates duplicate colors.

References

1. Aho, A.V., Hopcroft, J.E., Ullman, J.D.: The Design and Analysis of Computer Algorithms. Addison-Wesley, Reading (1974)
2. Alon, N., Gutner, S.: Linear time algorithms for finding a dominating set of fixed size in degenerated graphs. Algorithmica **54**(4), 544 (2008). https://doi.org/10.1007/s00453-008-9204-0
3. Arikati, S.R., Maheshwari, A., Zaroliagis, C.D.: Efficient computation of implicit representations of sparse graphs. Discret. Appl. Math. **78**(1–3), 1–16 (1997). https://doi.org/10.1016/S0166-218X(97)00007-3
4. Asahiro, Y., Miyano, E., Ono, H., Zenmyo, K.: Graph orientation algorithms to minimize the maximum outdegree. Int. J. Found. Comput. Sci. **18**(02), 197–215 (2007). https://doi.org/10.1142/S0129054107004644
5. Bansal, N., Umboh, S.W.: Tight approximation bounds for dominating set on graphs of bounded arboricity. Inf. Process. Lett. **122**, 21–24 (2017). https://doi.org/10.1016/j.ipl.2017.01.011
6. Barenboim, L., Elkin, M.: Sublogarithmic distributed MIS algorithm for sparse graphs using Nash-Williams decomposition. Distrib. Comput. **22**(5), 363–379 (2010). https://doi.org/10.1007/s00446-009-0088-2
7. Blumenstock, M.: Fast algorithms for pseudoarboricity. In: Proceedings of the Eighteenth Workshop on Algorithm Engineering and Experiments, ALENEX 2016, Arlington, Virginia, USA, 10 January 2016, pp. 113–126 (2016). https://doi.org/10.1137/1.9781611974317.10

8. Borradaile, G., Iglesias, J., Migler, T., Ochoa, A., Wilfong, G., Zhang, L.: Egalitarian graph orientations. J. Graph Algorithms Appl. **21**(4), 687–708 (2017). https://doi.org/10.7155/jgaa.00435
9. Brodal, G.S., Fagerberg, R.: Dynamic representations of sparse graphs. In: Dehne, F., Sack, J.-R., Gupta, A., Tamassia, R. (eds.) WADS 1999. LNCS, vol. 1663, pp. 342–351. Springer, Heidelberg (1999). https://doi.org/10.1007/3-540-48447-7_34
10. Chiba, N., Nishizeki, T.: Arboricity and subgraph listing algorithms. SIAM J. Comput. **14**(1), 210–223 (1985). https://doi.org/10.1137/0214017
11. Edmonds, J.: Minimum partition of a matroid into independent subsets. J. Res. Nat. Bur. Stand. Sect. B **69B**, 67–72 (1965)
12. Eppstein, D.: Arboricity and bipartite subgraph listing algorithms. Inf. Process. Lett. **51**(4), 207–211 (1994). https://doi.org/10.1016/0020-0190(94)90121-X
13. Eppstein, D., Löffler, M., Strash, D.: Listing all maximal cliques inlarge sparse real-world graphs. ACM J. Exp. Algorithmics **18** (2013). https://doi.org/10.1145/2543629
14. Even, S., Tarjan, R.E.: Network flow and testing graph connectivity. SIAM J. Comput. **4**(4), 507–518 (1975). https://doi.org/10.1137/0204043
15. Fredman, M.L., Komlós, J., Szemerédi, E.: Storing a sparse table with $o(1)$ worst case access time. J. ACM **31**(3), 538–544 (1984). https://doi.org/10.1145/828.1884
16. Gabow, H.N.: Algorithms for graphic polymatroids and parametric $\bar{s}$-sets. J. Algorithms **26**(1), 48–86 (1998). https://doi.org/10.1006/jagm.1997.0904
17. Gabow, H.N., Westermann, H.H.: Forests, frames, and games: algorithms for matroid sums and applications. Algorithmica **7**(1–6), 465–497 (1992). https://doi.org/10.1007/BF01758774
18. Goldberg, A.V.: Finding a maximum density subgraph. Technical report, University of California at Berkeley, Berkeley, CA, USA (1984)
19. Golovach, P.A., Villanger, Y.: Parameterized complexity for domination problems on degenerate graphs. In: Broersma, H., Erlebach, T., Friedetzky, T., Paulusma, D. (eds.) WG 2008. LNCS, vol. 5344, pp. 195–205. Springer, Heidelberg (2008). https://doi.org/10.1007/978-3-540-92248-3_18
20. Kowalik, Ł.: Approximation scheme for lowest outdegree orientation and graph density measures. In: Asano, T. (ed.) ISAAC 2006. LNCS, vol. 4288, pp. 557–566. Springer, Heidelberg (2006). https://doi.org/10.1007/11940128_56
21. Lee, Y.T., Sidford, A.: Path finding methods for linear programming: solving linear programs in $\tilde{O}(\sqrt{rank})$ iterations and faster algorithms for maximum flow. In: 2014 IEEE 55th Annual Symposium on Foundations of Computer Science, pp. 424–433, October 2014. https://doi.org/10.1109/FOCS.2014.52
22. Lenzen, C., Wattenhofer, R.: Minimum dominating set approximation in graphs of bounded arboricity. In: Lynch, N.A., Shvartsman, A.A. (eds.) DISC 2010. LNCS, vol. 6343, pp. 510–524. Springer, Heidelberg (2010). https://doi.org/10.1007/978-3-642-15763-9_48
23. Mądry, A.: Navigating central path with electrical flows: from flows to matchings, and back. In: 54th Annual IEEE Symposium on Foundations of Computer Science, FOCS 2013, Berkeley, CA, USA, 26–29 October 2013, pp. 253–262. IEEE Computer Society (2013). https://doi.org/10.1109/FOCS.2013.35
24. Matula, D.W., Beck, L.L.: Smallest-last ordering and clustering and graph coloring algorithms. J. ACM **30**(3), 417–427 (1983). https://doi.org/10.1145/2402.322385
25. Nash-Williams, C.S.J.A.: Decomposition of finite graphs into forests. J. London Math. Soc. **39**(1), 12 (1964). https://doi.org/10.1112/jlms/s1-39.1.12

26. Picard, J.C., Queyranne, M.: A network flow solution to some nonlinear 0-1 programming problems, with applications to graph theory. Networks **12**(2), 141–159 (1982). https://doi.org/10.1002/net.3230120206
27. Sleator, D.D., Tarjan, R.E.: A data structure for dynamic trees. J. Comput. Syst. Sci. **26**(3), 362–391 (1983). https://doi.org/10.1016/0022-0000(83)90006-5
28. Worou, B.M.T., Galtier, J.: Fast approximation for computing the fractional arboricity and extraction of communities of a graph. Discret. Appl. Math. **213**, 179–195 (2016). https://doi.org/10.1016/j.dam.2014.10.023

A Game of Cops and Robbers on Graphs with Periodic Edge-Connectivity

Thomas Erlebach and Jakob T. Spooner(✉)

School of Informatics, University of Leicester, Leicester, England
{te17,jts21}@leicester.ac.uk

Abstract. This paper considers a game in which a single cop and a single robber take turns moving along the edges of a given graph G. If there exists a strategy for the cop which enables it to be positioned at the same vertex as the robber eventually, then G is called cop-win, and robber-win otherwise. In contrast to previous work, we study this classical combinatorial game on edge-periodic graphs. These are graphs with an infinite lifetime comprised of discrete time steps such that each edge e is assigned a bit pattern of length l_e, with a 1 in the i-th position of the pattern indicating the presence of edge e in the i-th step of each consecutive block of l_e steps. Utilising the known framework of reachability games, we obtain an $O(\mathsf{LCM}(L) \cdot n^3)$ time algorithm to decide if a given n-vertex edge-periodic graph G^τ is cop-win or robber-win as well as compute a strategy for the winning player (here, L is the set of all edge pattern lengths l_e, and $\mathsf{LCM}(L)$ denotes the least common multiple of the set L). For the special case of edge-periodic cycles, we prove an upper bound of $2 \cdot l \cdot \mathsf{LCM}(L)$ on the minimum length required of any edge-periodic cycle to ensure that it is robber-win, where $l = 1$ if $\mathsf{LCM}(L) \geq 2 \cdot \max L$, and $l = 2$ otherwise. Furthermore, we provide constructions of edge-periodic cycles that are cop-win and have length $1.5 \cdot \mathsf{LCM}(L)$ in the $l = 1$ case and length $3 \cdot \mathsf{LCM}(L)$ in the $l = 2$ case.

1 Introduction

Pursuit-evasion games are games played between two teams of players, who take turns moving within the confines of some abstract arena. Typically, one team – the *pursuers* – are tasked with catching the members of the other team – the *evaders* – whose task it is to evade capture indefinitely. The study of such games has led to their application in a number of real-world scenarios, one widely-studied example of which would be their application to the problem of guiding robots through real-world environments [8]. From a theoretical standpoint, other variants of the game have been studied for their intrinsic links to important graph parameters; for example, in one particular variant in which each pursuer can, in a single turn, move to an arbitrary vertex of the given graph G, it is well known that establishing the minimum number of pursuers it takes to catch one evader also establishes the treewidth of G [20].

A. Chatzigeorgiou et al. (Eds.): SOFSEM 2020, LNCS 12011, pp. 64–75, 2020.
https://doi.org/10.1007/978-3-030-38919-2_6

The variant most closely resembled by the one considered in this paper was first studied separately by Quilliot [18], and by Nowakowski and Winkler [15], as the discrete *Cops and Robbers* game: One cop (pursuer) and one robber (evader) take turns moving across an edge (or remaining at their current vertex) in a given graph G, with the cop aiming to catch the robber, and the robber attempting to avoid capture. (By 'catching the robber' we mean that the cop occupies the same vertex as the robber.) In this paper, we consider a variant of this game where the game arena is an *edge-periodic* graph [7]. We call this game *Edge-Periodic Cops and Robbers*, or EPCR for short. Such graphs can be thought of as traditional static graphs equipped with an additional function mapping each edge e to a *pattern* of length l_e that dictates in which time steps e is present within each consecutive period of l_e steps. Formal definitions of edge-periodic graphs, which can be seen as a subclass of temporal graphs [14], and EPCR are given in Sect. 2. As far as we are aware, pursuit-evasion games have not yet been studied in the context of temporal graphs.

Paper Outline and Our Results. The remainder of this section discusses related work. Section 2 gives preliminaries. In Sect. 3, we consider the problem of deciding, given an edge-periodic graph G^τ, whether a game of edge-periodic cops and robbers played on G^τ is won by the cop or won by the robber. We exploit the connection (which was previously noted, e.g., in [11]) between the game of cops and robbers and *reachability games* to solve the one cop, one robber variant of cops and robbers on edge-periodic graphs. Our algorithm runs in polynomial-time whenever the lowest common multiple of the lengths of each edges appearance-pattern is n and $\max L$; we remark, however, that the algorithms has exponential running-time in the worst-case (more in Sect. 5). In Sect. 4, we consider edge-periodic graphs whose underlying graph is a cycle. We prove an upper bound of $2 \cdot l \cdot \mathsf{LCM}(L)$ on the length required of any such cycle C^τ in order to guarantee that it is robber-win, where $l = 1$ if $\mathsf{LCM}(L) \geq 2 \cdot \max L$, and $l = 2$ otherwise. Here, L is the set of the lengths of the bit patterns assigned to the edges of the cycle, and $\mathsf{LCM}(L)$ their least common multiple. We also give lower bound constructions showing that there exist cop-win edge-periodic cycles of length $\frac{3}{2} \cdot \mathsf{LCM}(L)$ and $3 \cdot \mathsf{LCM}(L)$ in the $l = 1$ and $l = 2$ case, respectively. Section 5 concludes the paper.

Related Work. The introduction of pursuit-evasion type combinatorial games is most often attributed to Parsons, who studied a problem in which a team of rescuers search for a lost spelunker in a circular cave system [16]. By representing the cave as a cycle graph, he showed that one rescuer is not enough to guarantee that the spelunker is found, but that two are. In a similar vein, the *Cop and Robber* problem, in which one cop attempts to catch a robber in a given graph G, was introduced independently by Quilliot [18], and by Nowakowski and Winkler [15]. Their papers characterise precisely those graphs for which one cop is enough to guarantee that the robber is caught. Aigner and Fromme [1] considered a generalised variant of the game, in which k cops attempt to catch a single robber; their paper introduced the notion of the *cop-number* of a graph, i.e., the minimum number of cops required to guarantee that the robber is caught.

Reductions from the standard game of cops and robbers to a game played on a directed graph, and algorithms that can decide, for a given graph, whether cop or robber wins, were given in [2,4,10]. Kehagias and Konstantinidis [11] note a connection between these approaches and reachability games. Reachability games are a well-studied class of 2-player *token-pushing* games, in which two players push a token along the edges of a directed graph in turn – one with the aim to push the token to some vertex belonging to a prespecified subset of the graph's vertex set, and the other with the aim to ensure the token never reaches such a vertex [9]. The winner of a reachability game played on a given directed graph G can be established in polynomial time [3,9]. For more information regarding cops and robbers/pursuit-evasion games, as well as their connection to reachability games, we refer the reader to [3,5,8,9,11,12,17].

In this paper, we consider the game of cops and robbers within the context of *temporal graphs*. Temporal graphs are a relatively new object of interest, and incorporate an aspect of time-variance into the combinatorial structure of traditional static graphs [14]. One previously considered way of viewing a temporal graph $\mathcal{G}$ is as a sequence of L subgraphs of a given *underlying graph* G (where L is the *lifetime* of the graph) [13], with each subgraph indexed by the time steps $t \in [L]$. For problems within this model, it is often natural to assume that each subgraph G_t in all time steps $t \in [L]$ is connected [13]. The edge-periodic graphs considered in this paper differ in that this connectivity assumption is dropped – similar graphs were introduced in [7]. For further related work on temporal graphs, we refer the reader to, e.g., [6,13,14].

2 Graph Model and Game Rules

For any positive integer k we write $[k]$ for the set $\{0, 1, \ldots, k-1\}$.

Definition 1 (Edge-periodic graph G^τ). *An edge-periodic graph $G^\tau = (V, E, \tau)$ is a temporal graph with underlying (directed or undirected) graph $G = (V, E)$ and infinite lifetime, and an additional function $\tau : E \to \{0,1\}^*$ that maps each edge $e \in E$ to a pattern $\tau(e) = b_e(0)b_e(1)\cdots b_e(l_e - 1)$ of length $l_e > 0$. Each $\tau(e)$ consists of l_e Boolean values, such that e is present in a time step $t \geq 0$ if and only if $b_e(t \bmod l_e) = 1$. We can assume that for any edge $e \in E(G^\tau)$, $b_e(i) = 1$ for at least one $i \in [l_e]$, so that every edge e is present at least once in any period of l_e time steps.*

For a given temporal graph $G^\tau = (V, E, \tau)$, we refer to the length l_e of the bit pattern assigned to edge e as the *period* of e. Furthermore, we use $L = \{l_e : e \in E\}$ to denote the set of all edge periods and $\mathsf{LCM}(L)$ to denote the least common multiple of the elements in L. When the set L is clear from the context, we omit it from the notation, writing LCM in place of $\mathsf{LCM}(L)$.

We consider a game of cops and robbers identical in its rule set to the one introduced in [18] and [15] (in particular, the variant with 1 cop and 1 robber), but with edge-periodic graphs as the game arenas. We call the resulting game *edge-periodic cop(s) and robber(s)*, or EPCR for short. In this paper we only consider the undirected case, but all results translate to the directed case easily.

Rules of EPCR. Initially, the two players (cop C and robber R) each select a start vertex on a given edge-periodic graph G^τ. C chooses first, followed by R, whose choice is made in full knowledge of C's choice. After the start vertices have been chosen, in each time step $t \geq 0$, players take alternating turns moving over an edge in the graph that is incident to their current vertex or choosing to remain at their current vertex, following the convention that in any particular time step, C moves first, in full knowledge of R's position, followed by R; again, R's move is made with full knowledge of the move that C just made. Whenever C or R are situated at a vertex $v \in V(G^\tau)$ during some time step t and it is their turn to make a move, they may only traverse those edges $\{v, u\}$ with $b_{\{v,u\}}(t \bmod l_{\{v,u\}}) = 1$. The game terminates only when, at the end of either player's move, C and R are situated at the same vertex in G^τ. If there exists a strategy for C that ensures that the game terminates, we say that G^τ is *cop-win*. Otherwise, there must exist a strategy for R that enables infinite evasion of C; in this case we call G^τ *robber-win*.

3 Determining the Winner of a Game of EPCR

In this section, we prove the following theorem:

Theorem 1. *Let G^τ be an edge-periodic graph with n nodes, and let $L = \{l_e : e \in E(G^\tau)\}$. Then, it can be decided in $O(\mathsf{LCM} \cdot n^3)$ time whether G^τ is cop-win or robber-win. A winning strategy for the winning player can be computed in the same time bound.*

The proof mainly uses a transformation from a given edge-periodic graph G^τ to a finite directed graph G'. The transformation is such that the playing of an instance of EPCR on G^τ is essentially equivalent to the playing of a *reachability game* on G'. For this, we need a way of translating a particular state of an instance of EPCR played on G^τ to a corresponding state in the reachability game played on G'. The following definition introduces the notion of a *position* that represents the current state in a game of EPCR on an edge-periodic graph G^τ.

Definition 2 (Position in G^τ). *A* position *of a game of EPCR played on an edge-periodic graph G^τ is a 4-tuple $P = (c_P, r_P, s_P, t_P)$, where $c_P \in V(G^\tau)$ is C's current vertex, $r_P \in V(G^\tau)$ is R's current vertex, $s_P \in \{C, R\}$ is the player whose turn it is to move next, and t_P is the current time step.*

We call any position P such that $c_P = r_P$ a *terminating position*, since this indicates that both players are situated on the same vertex and hence C has won. Next, we formally introduce *reachability games* [9]:

Definition 3 (Reachability game G'). *A reachability game is a directed graph G', given as a 3-tuple*

$$G' = (V_0 \cup V_1, E', F),$$

where $V_0 \cup V_1$ is a partition of the node set V' (also referred to as the state set); $E' \subseteq V' \times V'$ is a set of directed edges; and $F \subseteq V'$ is a set of final states.

The game is played by two opposing players, Player 0 and Player 1; V_0 and V_1 are the (disjoint) sets of Player 0/Player 1 owned nodes, respectively. A token is placed at some initial vertex v_0 at the start of the game. Depending on whether $v_0 \in V_0$ or $v_0 \in V_1$, the corresponding player then selects one of the outgoing edges of v_0 and pushes the token along that edge. When the token arrives at the next vertex, the player who owns that vertex then selects an outgoing edge and pushes the token along it. This process continues, and such a sequence of moves constitutes a *play* of the reachability game on G'. Formally, a play $\phi = v_0, v_1, \ldots$ is a (possibly infinite) sequence of vertices in V', such that $(v_i, v_{i+1}) \in E'$ for all $i \geq 0$. We say that a play ϕ is *won* by Player 0 if there exists some i such that $v_i \in F$. Otherwise, ϕ is of infinite length and for no i is $v_i \in F$, and ϕ is won by Player 1.

3.1 Transformation

We now detail our transformation from a given edge-periodic graph G^τ to a reachability game $G' = (V', E', F)$.

State Set V'. We define the state set (i.e., vertex set) of the directed graph G' to be a set of 4-tuples, each corresponding to a position in the game of EPCR on G^τ as follows:

$$V' = \{(c, r, s, t) : c, r \in V(G^\tau), s \in \{\mathsf{C}, \mathsf{R}\}, t \in [\mathsf{LCM}]\}.$$

Let $V_0 = \{(c, r, s, t) \in V' : s = \mathsf{C}\}$ and $V_1 = \{(c, r, s, t) \in V' : s = \mathsf{R}\}$ be the sets of Player 0 (or C) owned nodes, and Player 1 (or R) owned nodes, respectively. We can restrict the range of t to [LCM] without losing information because all edge periods divide LCM and hence the set of edges present at any time t is the same as the set of edges present at time t mod LCM.

Edge Set E'. In order to construct the edge set $E' \subseteq (V_0 \times V_1) \cup (V_1 \times V_0)$ we include the edge (S, S') for $S = (c, r, s, t)$ and $S' = (c', r', s', t')$ in E' if and only if the following conditions are satisfied:

$$\begin{aligned}
(1)\ s = \mathsf{C} \implies & (c = c' \vee (\{c, c'\} \in E(G^\tau) \text{ and } b_{\{c,c'\}}(t \bmod l_{\{c,c'\}}) = 1)) \\
& \wedge (r = r') \wedge (t' = t) \wedge (s' = \mathsf{R}), \\
(2)\ s = \mathsf{R} \implies & (r = r' \vee (\{r, r'\} \in E(G^\tau) \text{ and } b_{\{r,r'\}}(t \bmod l_{\{r,r'\}}) = 1)) \\
& \wedge (c = c') \wedge (t' = (t+1) \bmod \mathsf{LCM}) \wedge (s' = \mathsf{C}).
\end{aligned}$$

Condition (1) ensures that C can only stay at a vertex or move over an adjacent edge that is present in every time step t'' with t'' mod LCM $= t$, and that the next state will be a state in the same time step where R has to move. Condition (2) is the analogous condition for R, but the next state will be in the following time step (modulo LCM) and C will have to move next.

Set of Final States F. Let $F = \{(c, r, s, t) \in V' : c = r\}$, so that the set of final states consists of all states that correspond to a position in G^τ where C is positioned on the same vertex as R (i.e., where C has won the game).

3.2 Proof of Theorem 1

We first introduce the elements of the theory of reachability games that are required for the proof of Theorem 1, starting with the definition of the *attractor set*:

Definition 4 (Attractor set $Attr(F)$ **[3]).** *The sequence* $(Attr_i(F))_{i\geq 0}$ *is recursively defined as follows:*

$$\begin{aligned} Attr_0(F) &= F \\ Attr_{i+1}(F) &= Attr_i(F) \cup \{v \in V_0 \mid \exists (v,u) \in E' : u \in Attr_i(F)\} \cup \\ &\qquad \{v \in V_1 \mid \forall (v,u) \in E' : u \in Attr_i(F)\} \end{aligned}$$

The sets $Attr_i(F)$ *are a sequence of subsets of* V' *that is monotone with respect to set-inclusion. Let*

$$Attr(F) = \bigcup_{i\geq 0} Attr_i(F).$$

Since G' *is finite,* $Attr(F)$ *is the least fixed point of the sequence* $(Attr_i(F))_{i\geq 0}$.

Intuitively, the states in $Attr(F)$ are the states from which Player 0 can win the game. For $x \in \{0,1\}$, a *memoryless* strategy of Player x is a partial function $\sigma_x : V_x \to V'$ that specifies for each state in V_x (except states in $V_x \cap F$) the state to which Player x pushes the token from that state. The strategy is called memoryless because the move a player selects only depends on the current state, not on the history of the game. A winning strategy of Player 0 from any state in $Attr(F)$ consists of selecting for each state u in $(Attr_{i+1}(F) \setminus Attr_i(F)) \cap V_0$, for any $i \geq 0$, an arbitrary outgoing edge leading to a state in $Attr_i(F)$. The states in $Attr_i(F)$, for any $i \geq 0$, have the property that Player 0 wins the game after at most i further moves (in total for both players) when following that strategy. Similarly, $V' \setminus Attr(F)$ is the set of states from which Player 1 can win the game. The winning strategy for Player 1 from any such state consists of selecting for each state u in $V_1 \setminus Attr(F)$ an arbitrary outgoing edge leading to a state that is not in $Attr(F)$. These winning strategies are memoryless.

Theorem 2 (Berwanger [3], Grädel et al. [9]). *In a given reachability game* $G' = (V', E', F)$, *Player 0 has a winning strategy from any state* $S \in Attr(F)$, *and Player 1 has a winning strategy from any state* $S \in V' \setminus Attr(F)$. *There exists an algorithm which computes the set* $Attr(F)$ *and a memoryless winning strategy for the winning player in time* $O(|V'| + |E'|)$.

Our transformation produces, from a given edge-periodic graph G^τ, a directed graph $G' = (V', E', F)$ such that there is a correspondence between positions in the game of EPCR on G^τ and states in V'. Let $Attr(F)$ be the attractor set for G'. Winning strategies for G' translate directly into winning strategies for EPCR on G^τ from any winning position by moving according to the outgoing edges chosen by the winning strategy in G'. Using the notation S_P to refer to the state in V' that corresponds to the position P in the game of EPCR on G^τ, Theorem 2 then implies the following:

Lemma 1. *C can force a win from a position P if and only if the state $S_P \in V'$ satisfies $S_P \in Attr(F)$. Starting from a position P such that $S_P \notin Attr(F)$, R can force the sequence of moves to never reach any state $S \in F$, and, as such, the EPCR game can be won by R.*

Lemma 2. *An edge-periodic graph G^τ is cop-win if and only if there exists a vertex $v \in V(G^\tau)$ such that $(v, r, C, 0) \in Attr(F)$ for all $r \in V(G^\tau)$.*

Proof. ($\Rightarrow$) Assume not, so that G^τ is cop-win but there exists no vertex $v \in V(G^\tau)$ such that $(v, r, \mathsf{C}, 0) \in Attr(F)$ for all $r \in V(G^\tau)$. Then for every start vertex c that C can choose, there exists at least one vertex u such that the state $(c, u, \mathsf{C}, 0) \notin Attr(F)$. Let R choose such a vertex u as its start vertex. Since R chooses u in full knowledge of C's choice of c, it follows that R can force the equivalent reachability game on G' to begin from a state $S_{(c,u,\mathsf{C},0)} \notin Attr(F)$, hence winning the reachability game regardless of C's choice of c. This is a contradiction since, by assumption, G^τ is cop-win.

($\Leftarrow$) If C chooses a vertex v with the stated property as its initial vertex, the resulting position P will correspond to a state $S_P \in Attr(F)$ no matter which vertex R chooses as its initial vertex, and by Lemma 1 C has a winning strategy. □

Proof (of Theorem 1). Since $n = |V(G^\tau)|$, our transformation produces, given an edge-periodic graph G^τ, a directed graph $G' = (V', E', F)$, such that $|V'| = O(\mathsf{LCM} \cdot n^2)$. This is because V' contains tuples (c, r, s, t) for n choices of c, n choices of r, two choices of s, and LCM choices of t. Next, note that each state $S_P \in V'$ has at most n outgoing edges because the player whose turn it is can only stay at its vertex or move to one of at most $n - 1$ neighbouring vertices. It follows that $|E'| = O(\mathsf{LCM} \cdot n^3)$. Furthermore, the transformation can be done in $O(|V'| + |E'|) = O(\mathsf{LCM} \cdot n^3)$ time.

By Theorem 2, the attractor set $Attr(F)$ of G' can be computed in time $O(\mathsf{LCM} \cdot n^3)$. By Lemma 2, we can then determine whether G^τ is cop-win by checking if there exists at least one vertex $c \in V(G^\tau)$ such that $(c, r, \mathsf{C}, 0) \in Attr(F)$ for all $r \in V(G^\tau)$: if such a c exists, G^τ is cop-win, otherwise it is robber-win. This check can be done in $O(n^2)$ time.

By Theorem 2, we also obtain a memoryless winning strategy σ_0 for Player 0 from all states in $Attr(F)$, and a memoryless winning strategy σ_1 for Player 1 from all states in $V' \setminus Attr(F)$, in $O(|V'| + |E'|)$ time. If G^τ is cop-win, we obtain a winning-strategy for C by letting C choose as its initial vertex any vertex satisfying the condition of Lemma 2 and then behave in line with σ_0: When it is $\mathsf{C}'s$ turn in a current position $P = (c_P, r_P, \mathsf{C}, t_P)$, C constructs from it the state S_P, looks up the state $\sigma_0(S_P) = (c', r', \mathsf{R}, t_P \bmod \mathsf{LCM})$, and moves to c' (or stays at c_P if $c_P = c'$). Similarly, if G^τ is robber-win, we obtain a winning-strategy for R by letting R choose its initial vertex r (in response to C's choice of its initial vertex c) in such a way that $S_P \notin Attr(F)$ for $P = (c, r, \mathsf{C}, 0)$ and then behave in line with σ_1. □

We remark that, as long as LCM is polynomial in n and $\max L$, the winner of EPCR on a given graph G^τ can be determined in polynomial time. In particular, if the periods l_e are bounded by some constant for all $e \in E(G^\tau)$, the winner can be determined in $O(n^3)$ time.

Finally, we note that Theorem 1 can be generalised to a setting with k cops at the expense of increasing the algorithm's running time to $O(\mathsf{LCM} \cdot k \cdot n^{k+2})$. The idea is to fix an arbitrary ordering of the cops and create $k+1$ layers of states during every time step $t \in [\mathsf{LCM}]$ (one for each of the k cops' moves, followed finally by the robber's move). By allowing the players to play their moves in each time step in this serialised fashion the resulting game graph requires $O(\mathsf{LCM} \cdot k)$ layers with n^{k+1} states in each, with at most n edges leading from every state to states in the following layer.

4 An Upper Bound on the Length Required to Ensure an Edge-Periodic Cycle Is Robber-Win

In this section, we consider edge-periodic cycles, a restricted subclass of edge-periodic graphs where the underlying graph is a cycle. We are interested in how long (in terms of number of edges) the cycle needs to be to ensure that the robber can escape the cop indefinitely. First, we show that any edge-periodic infinite path for which the set L of its edge periods is finite is robber-win. After this, we show how the strategy for such infinite paths can be adapted to the cycle case. Let the given edge periodic cycle be $C^\tau = (V, E, \tau)$, and let $L = \{l_e : e \in E\}$ denote the set of edge periods. In the remainder of this section, we write LCM as short-hand for LCM.

We first consider infinite paths, which will later allow us to handle the case in which the cop chases the robber around the cycle in a fixed direction.

Lemma 3. *Let P be an infinite edge-periodic path, $L = \{l_e : e \in E(P)\}$, and assume that $|L|$ is finite. Then, starting from any time step t, there exists a winning strategy for R from any vertex with distance at least $2 \cdot \mathsf{LCM}$ from C's start vertex if $\mathsf{LCM} = \max L$, and with distance at least LCM otherwise.*

Proof. First, notice that since we assume that $|L|$ is finite, so must be LCM. Let C's vertex at the start of time step t be $c_t \in P$. Denote R's initial vertex by r_t, and assume without loss of generality that r_t is a vertex in P that lies to the right of c_t. Assume from now onward that P is a path starting at c_t and extending infinitely to the right, and that C moves right whenever possible (it is clear that this is the best strategy for capturing R).

Consider the set L and its constituent elements. Either (1) there exists $x \in L$ such that $\max L$ is not a multiple of x – then $\mathsf{LCM} \geq 2 \cdot \max L$, since it cannot be the case that $\mathsf{LCM} = j \cdot \max L$ for any $j < 2$; or (2) for every $x \in L$, $\max L = x \cdot i$ for some integer $i \geq 1$; then $\mathsf{LCM} = \max L$. With this in mind, define $B = \mathsf{LCM}$ if (1) holds and $B = 2 \cdot \mathsf{LCM}$ if (2) holds. Now, let us define the *strips* S_i $(i \geq 1)$ to be finite subpaths of P, such that for all edges $e \in S_i$, e is first traversed by C in some time step $t_e \in [t + (i-1)B, t + iB - 1]$ (assuming that C moves right whenever it can). Note that $B \geq 2 \cdot \max L$ and hence each S_i must contain at least two edges. By convention, we call the leftmost and rightmost edges (vertices) of any S_i its *first* and *last* edges (vertices), respectively. Note also that the last vertex of S_i and the first vertex of S_{i+1} are one and the same, for all $i \geq 1$.

Note that the first vertex of S_2 is at most B edges away from c_t. By the condition of the lemma, R is located at least B edges away from c_t. For the remainder of the analysis, we assume that R is located at the first vertex of S_2 and moves right whenever possible. If R can escape C indefinitely under this assumption, it is clear that R can also do so if it starts further to the right.

We now demonstrate that R wins the game. Note that the set of edges that are present in each step repeats every B time steps as LCM divides B. Thus, we have that C and R traverse strips in a synchronised fashion: For any $i \geq 1$, during the interval $[t + (i-1)B, t + iB - 1]$ of time steps, C traverses S_i and R traverses S_{i+1}. The only possibility for C to catch R would be for C to reach the last vertex of S_i before R leaves the first vertex of S_{i+1}. However, C reaches the last vertex of S_i in a time step $t' = t + iB - j$ for some $1 \leq j \leq \max L$, as the last edge of S_i is available at least once in $\max L$ consecutive time steps. On the other hand, R leaves the first vertex of S_{i+1} in a time step $t'' = t + (i-1)B + j'$ for some $0 \leq j' < \max L$. As $B \geq 2 \max L$, it follows that $t' > t''$, showing that C cannot catch R. □

Theorem 3. *Let $C^\tau = (V, E, \tau)$ be an edge-periodic cycle on n vertices and $L = \{l_e : e \in E\}$. If $n \geq 2 \cdot l \cdot \mathsf{LCM}(L)$, then C^τ is robber-win (where $l = 1$ if $\mathsf{LCM}(L) \geq 2 \cdot \max L$, and $l = 2$ otherwise).*

Proof. For any $t \geq 0$, we let c_t and r_t denote the vertex at which C and R are positioned at the start of time step t, respectively. Consider now some edge $e \in E(C^\tau)$ and classify its vertices as a 'left' and 'right' vertex arbitrarily; let the left vertex of each edge be the right vertex of the following edge in the cycle. We proceed by specifying a strategy for R. Initially, let C choose c_0; R chooses r_0 to be the vertex antipodal to c_0 in C^τ. (If n is odd then R selects r_0 to be either of the two vertices that are furthest away from c_0; we will refer to both these vertices as antipodal to c_0, and treat vertices in all steps $t \geq 0$ in the same way.) We now distinguish between two modes of play, *Hide* and *Escape*, and specify R's strategy in each of them.

Hide Mode: A *Hide period* begins in step 0 and in any step $t \geq 2$ such that c_t and r_t are antipodal, but c_{t-1} and r_{t-1} were not. Note that any game in which R follows our strategy begins in a Hide period. The Hide period beginning at step t is the interval $[t, t + x]$ such that $c_{t'}$ and $r_{t'}$ are antipodal for all $t' \in [t, t + x]$, but c_{t+x+1} and r_{t+x+1} are not. (If no such step $t + x + 1$ exists, the Hide period is $[t, \infty)$.) Any Hide period (except if it is of the form $[t, \infty)$) is followed directly by an Escape period, which will start in step $t + x + 1$.

R's **Hide Strategy:** If the game is in a Hide period during step t, R observes C's choice of c_{t+1} and tries to move to (or stay at) a vertex antipodal to it. Clearly, R cannot be caught in any step belonging to a Hide period, as regardless of whether $\mathsf{LCM} = \max L$ or $\mathsf{LCM} \geq 2 \cdot \max L$, we have that $n \geq 4 \cdot \max L \geq 4$. As a result, antipodal vertices in C^τ are at least distance 2 from one another.

Escape Mode: An *Escape period* always begins in a step t such that step $t-1$ was the last step of some Hide period. As such, an Escape period is an interval $[t, t+x]$ such that $c_{t'}$ and $r_{t'}$ are not antipodal for any $t' \in [t, t+x]$, but c_{t+x+1} and r_{t+x+1} are. The last step of the Escape period is then $t+x$, and the first step of the next Hide period is $t+x+1$. If there is no step $t+x+1$ in which c_{t+x+1} and r_{t+x+1} are antipodal, the Escape period is $[t, \infty)$.

R's **Escape Strategy:** Assume that some Escape period starts in step t. Then, at the start of step $t-1$, c_{t-1} and r_{t-1} were antipodal to one another, and during step $t-1$, we had a situation in which C was able to move towards R in some direction, but the edge incident to r_{t-1} leading in the same direction was not present. Now, recall that if $l=2$, so that $\mathsf{LCM} = \max L$, then $n \geq 4 \cdot \mathsf{LCM}$; and if $l=1$, so that $\mathsf{LCM} \geq 2 \cdot \max L$, then $n \geq 2 \cdot \mathsf{LCM}$. Therefore, since c_{t-1} and r_{t-1} are antipodal in C^τ, if $l=2$ holds we have that the distance between them is at least $2 \cdot \mathsf{LCM}$ and if $l=1$ holds, the distance between them is at least LCM. Observe now that we are able to view any edge-periodic cycle of finite length as an infinite path whose edge patterns repeat infinitely often. We can thus view the Escape period as an instance of the game on an infinite edge-periodic path starting at time step $t-1$, to which Lemma 3 applies. Hence, R can evade C until the Escape period ends (or indefinitely, in case the Escape period never ends).

Since every step t belongs to either a Hide period or an Escape period, we have shown that C can never catch R, and the proof is complete. □

We now give lower bounds on the length required of a strictly edge-periodic cycle to ensure that it is robber-win.

Theorem 4. *There exists an edge-periodic cycle of length $3 \cdot \mathsf{LCM}$ with edge pattern lengths in the set L that is both cop-win and satisfies $\mathsf{LCM} = \max L$.*

Proof. Let $M > 1$ be an integer and consider an edge-periodic cycle C with $3M$ edges and with edge pattern lengths in $L = \{1, M\}$. Let two consecutive edges have patterns 0...01 of length M, and all $3M-2$ remaining edges have pattern 1. We refer to the subpath of C consisting of the two edges with period M as the M-path, and the subpath with edges labelled with 1 as the 1-path.

We now specify a strategy for C and show that it is in fact a winning strategy: Let C position itself initially at either of the two vertices belonging to the 1-path that are distance $M-1$ from one extreme point of the M-path, and $2M-1$ from the opposite extreme point (where distance is taken to mean the length of the path to that extreme point that avoids the edges of the M-path). Call that chosen vertex c_0, and notice that it splits the 1-path into two subpaths that intersect only in c_0 – one of length $M-1$ which we will call P^-, the other of length $2M-1$ which we call P^+. If R chooses its initial position to be some vertex lying on P^-, then C can move along all edges of P^- in the first $M-1$ steps and catch R. If R chooses its initial position as some vertex lying on P^+, then in the first $2M-1$ steps C can traverse all edges of P^+. The only way for R to leave P^+ without encountering C is via the M-path. R can traverse only one edge of the M-path (in the M-th step) and will be stuck at the middle vertex of the M-path until the $2M$-th step (i.e., until time step $2M-1$). In step $2M-1$, C will be positioned at the vertex that lies on both the M-path and P^+. C will move first and catch R. It remains to be shown that R will be caught if it chooses the middle vertex of the M-path as its start vertex: here, R will not be able to move until step $M-1$, so C can traverse all edges of P^- in the first $M-1$ steps and then, in step $M-1$, catch R before R can make its move. □

A small amount of modification to the construction in the proof of Theorem 4 yields the following lower bound for the case when $\mathsf{LCM} \geq 2 \cdot \max L$:

Theorem 5. *There exists an edge-periodic cycle of length* $1.5 \cdot \mathsf{LCM}$ *with edge periods in the set L that is both cop-win and satisfies* $\mathsf{LCM} \geq 2 \cdot \max L$.

Proof. Perform the construction from the proof of Theorem 4, taking $M > 1$ to be odd. Again let C select one of the vertices that has distance $M-1$ and $2M-1$ from opposite ends of the M-path as its start vertex, calling that vertex x. Consider the strategy from the proof of Theorem 4 and observe that there are two edges that C may cross in the second step. Select either one of these edges and replace its pattern of 1 with the pattern 01 (with period 2). C can now follow the strategy in the proof of Theorem 4 – this works since the edge with pattern 01 has been selected so that it is present whenever C's strategy crosses that edge. Since M is odd, we have that $\mathsf{LCM} = 2M$. Since the constructed cycle has length $3M$, the theorem follows. □

5 Conclusion

We have introduced a cops and robbers game on edge-periodic graphs and shown that there exists an algorithm with running time $O(\mathsf{LCM} \cdot n^3)$ that decides whether the cop or robber wins and computes a winning strategy for the winning player. The running-time of the algorithm is polynomial if $\mathsf{LCM}(L)$ is polynomial in n and $\max L$. A natural open question is: What is the complexity of deciding whether cop or robber wins when the least common multiple of the edge periods is exponential in the size of the input? We note that $\mathsf{LCM}(\{1,...,n\}) = \mathsf{e}^{\phi(n)}$, where $\phi(n) \in \Theta(n)$ is Chebyshev's function [19], and thus there are edge-periodic graphs where the running-time of our algorithm is exponential. It would be interesting to establish whether there exists a better algorithm or whether the problem is *NP*-hard for this case. More generally, one could also examine the cops and robbers game within the context of other temporal graph models. It would also be interesting to reduce the gap between our upper and lower bounds on the minimum length required of an edge-periodic cycle to be guaranteed to be robber-win.

Acknowledgements. The authors would like to thank Maciej Gazda for helpful discussions regarding reachability games, as well as an anonymous reviewer for a suggestion leading to the running-time for the variant with k cops mentioned at the end of Sect. 3.

References

1. Aigner, M., Fromme, M.: A game of cops and robbers. Discret. Appl. Math. **8**(1), 1–12 (1984). https://doi.org/10.1016/0166-218X(84)90073-8
2. Berarducci, A., Intrigila, B.: On the cop number of a graph. Adv. Appl. Math. **14**(4), 389–403 (1993). https://doi.org/10.1006/aama.1993.1019
3. Berwanger, D.: Graph games with perfect information. arXiv:1407.1647 (2013)
4. Bonato, A., MacGillivray, G.: A general framework for discrete-time pursuit games (2015). Unpublished manuscript
5. Bonato, A., Nowakowski, R.: The Game of Cops and Robbers on Graphs, Student Mathematical Library, vol. 61. American Mathematical Society, Providence (2011). https://doi.org/10.1090/stml/061
6. Casteigts, A.: A Journey Through Dynamic Networks (with Excursions). Habilitation à diriger des recherches, University of Bordeaux, June 2018. https://tel.archives-ouvertes.fr/tel-01883384
7. Casteigts, A., Flocchini, P., Quattrociocchi, W., Santoro, N.: Time-varying graphs and dynamic networks. In: Frey, H., Li, X., Ruehrup, S. (eds.) ADHOC-NOW 2011. LNCS, vol. 6811, pp. 346–359. Springer, Heidelberg (2011). https://doi.org/10.1007/978-3-642-22450-8_27
8. Chung, T.H., Hollinger, G.A., Isler, V.: Search and pursuit-evasion in mobile robotics. Auton. Robot. **31**(4), 299–316 (2011). https://doi.org/10.1007/s10514-011-9241-4
9. Grädel, E., Thomas, W., Wilke, T. (eds.): Automata Logics, and Infinite Games: A Guide to Current Research. Springer, New York (2002). https://doi.org/10.1007/3-540-36387-4
10. Hahn, G., MacGillivray, G.: A note on k-cop, l-robber games on graphs. Discret. Math. **306**(19), 2492–2497 (2006). https://doi.org/10.1016/j.disc.2005.12.038. Creation and Recreation: A Tribute to the Memory of Claude Berge
11. Kehagias, A., Konstantinidis, G.: Cops and robbers, game theory and Zermelo's early results. arXiv:1407.1647 (2014)
12. Kehagias, A., Mitsche, D., Pralat, P.: The role of visibility in pursuit/evasion games. Robotics **4**, 371–399 (2014)
13. Michail, O.: An introduction to temporal graphs: an algorithmic perspective. Internet Math. **12**(4), 239–280 (2016). https://doi.org/10.1080/15427951.2016.1177801
14. Michail, O., Spirakis, P.G.: Elements of the theory of dynamic networks. Commun. ACM **61**(2), 72–72 (2018). https://doi.org/10.1145/3156693
15. Nowakowski, R., Winkler, P.: Vertex-to-vertex pursuit in a graph. Discret. Math. **43**(2), 235–239 (1983). https://doi.org/10.1016/0012-365X(83)90160-7
16. Parsons, T.D.: Pursuit-evasion in a graph. In: Alavi, Y., Lick, D.R. (eds.) Theory and Applications of Graphs, pp. 426–441. Springer, Heidelberg (1978). https://doi.org/10.1007/BFb0070400
17. Patsko, V., Kumkov, S., Turova, V.: Pursuit-evasion games. In: Basar, T., Zaccour, G. (eds.) Handbook of Dynamic Game Theory, pp. 1–87. Springer, Heidelberg (2017). https://doi.org/10.1007/978-3-319-27335-8-30-1
18. Quilliot, A.: Jeux et pointes fixes sur les graphes. Ph.D. thesis, University of Paris VI (1978)
19. Rankin, B.A.: Ramanujan: Twelve lectures on subjects suggested by his life and work. Math. Gaz. **45**(352), 166 (1961). https://doi.org/10.1017/S0025557200044892
20. Seymour, P., Thomas, R.: Graph searching and a min-max theorem for tree-width. J. Comb. Theory Ser. B **58**(1), 22–33 (1993). https://doi.org/10.1006/jctb.1993.1027

Approximating Shortest Connected Graph Transformation for Trees

Nicolas Bousquet[1] and Alice Joffard[2](✉)

[1] Univ. Grenoble Alpes, CNRS, Laboratoire G-SCOP, Grenoble-INP, Grenoble, France
nicolas.bousquet@grenoble-inp.fr
[2] LIRIS, Université Claude Bernard, Lyon, France
alice.joffard@liris.cnrs.fr

Abstract. Let G, H be two connected graphs with the same degree sequence. The aim of this paper is to find a transformation from G to H via a sequence of flips maintaining connectivity. A flip of G is an operation consisting in replacing two existing edges uv, xy of G by ux and vy.

Taylor showed that there always exists a sequence of flips that transforms G into H maintaining connectivity. Bousquet and Mary proved that there exists a 4-approximation algorithm of a shortest transformation. In this paper, we show that there exists a 2.5-approximation algorithm running in polynomial time. We also discuss the tightness of the lower bound and show that, in order to drastically improve the approximation ratio, we need to improve the best known lower bounds.

1 Introduction

Sorting by reversals problem. The problem of sorting by reversals has been widely studied in the last twenty years in genomics. The reversal of a sequence of DNA is a common mutation of a genome, that can lead to major evolutionary events. It consists, given a DNA sequence that can be represented as a labelled path $x_1, \ldots, x_n$ on n vertices, in turning around a part of it. More formally, a reversal is a transformation that, given two integers $1 \leq i < j \leq n$, transforms the path $x_1, \ldots, x_n$ into $x_1, \ldots, x_{i-1}, x_j, x_{j-1}, \ldots, x_i, x_{j+1}, \ldots, x_n$. It is easy to prove that, given two paths on the same vertex set (and with the same leaves), there exists a sequence of reversals that transforms the first into the second. Biologists want to find the minimum number of reversals needed to transform a genome (i.e. a path) into another in order to compute the evolutionary distance between different species.

An input of the Sorting by Reversals problem consists of two paths P, P' with the same vertex set (and the same leaves) and an integer k. The output is positive if and only if there exists a sequence of at most k reversals that transforms P into P'. Capraca proved that the Sorting by Reversals problem is NP-complete [4]. Kececioglu and Sankoff first proposed an algorithm

This work was supported by ANR project GrR (ANR-18-CE40-0032).

A. Chatzigeorgiou et al. (Eds.): SOFSEM 2020, LNCS 12011, pp. 76–87, 2020.
https://doi.org/10.1007/978-3-030-38919-2_7

that computes a sequence of reversals of size at most twice the length of an optimal solution in polynomial time [10]. Then, Christie improved it into a 3/2-approximation algorithm [5]. The best polynomial time algorithm known so far is a 1.375-approximation due to Berman et al. [2].

A reversal can be equivalently defined as follows: given a path P and two edges ab and cd, a reversal consists in the deletion of the edges ab and cd and the addition of ac and bd that keeps the connectivity of the graph. Indeed, when we transform $x_1, \ldots, x_n$ into $x_1, \ldots, x_{i-1}, x_j, x_{j-1}, \ldots, x_i, x_{j+1}, \ldots, x_n$, we have deleted the edges $x_{i-1}x_i$ and x_jx_{j+1} and have created the edges $x_{i-1}x_j$ and x_ix_{j+1}. In this paper, we study the generalization of the SORTING BY REVERSALS problem for trees and general graphs that has also been extensively studied in the last decades.

SHORTEST CONNECTED GRAPH TRANSFORMATION *problem.* Let $G = (V, E)$ be a graph where V denotes the set of vertices and E the set of edges. For basic definitions on graphs, the reader is referred to [6]. All along the paper, the graphs are loop-free but may admit multiple edges. A *tree* is a connected graph which does not contain any cycle (a multi-edge being considered as a cycle).

The *degree sequence* of a graph G is the sequence of the degrees of its vertices in non-increasing order. Given a non-increasing sequence of integers $S = \{d_1, \ldots, d_n\}$, a graph $G = (V, E)$ whose vertices are labeled as $V = \{v_1, \ldots, v_n\}$ *realizes* S if $d(v_i) = d_i$ for all $i \leq n$. Senior [12] gave necessary and sufficient conditions to guarantee that, given a sequence of integers $S = \{d_1, \ldots, d_n\}$, there exists a connected multigraph realizing S. Hakimi [7] then proposed a polynomial time algorithm that outputs a connected (multi)graph realizing S if such a graph exists or returns no otherwise.

A *flip* σ (also called *swap* or *switch* in the literature) on two edges ab and cd consists in deleting the edges ab and cd and creating the edges ac and bd (or ad and bc)[1]. The flip operation that transforms the edges ab and cd into the edges ac and bd is denoted $(ab, cd) \to (ac, bd)$. When the target edges are not important we will simply say that we flip the edges ab and cd.

Let $S = \{d_1, \ldots, d_n\}$ be a non-increasing sequence and let G and H be two graphs on n vertices $v_1, \ldots, v_n$ realizing S. The graph G can be *transformed* into H if there is a sequence $(\sigma_1, \ldots, \sigma_k)$ of flips that transforms G into H. Note that since flips do not modify the degree sequence, all the intermediate graphs also realize S. Let $\mathcal{G}(S)$ be the graph whose vertices are the loop-free multigraphs realizing S and where two vertices G and H of $\mathcal{G}(S)$ are adjacent if G can be transformed into H via a single flip. Since the flip operation is reversible, the graph $\mathcal{G}(S)$ is an undirected graph called the *reconfiguration graph of* S. Note that there exists a sequence of flips between any pair of graphs realizing S if and only if the graph $\mathcal{G}(S)$ is connected. Hakimi [8] proved that, for any non-increasing sequence S, if the graph $\mathcal{G}(S)$ is not empty then it is connected.

One can wonder if the reconfiguration graph is still connected when we restrict to graphs with stronger properties. For a graph property Π, let us denote

[1] In the case of multigraphs, we simply decrease by one the multiplicities of ab and cd and increase by one the ones of ac and bd.

by $\mathcal{G}(S,\Pi)$ the subgraph of $\mathcal{G}(S)$ induced by the graphs realizing S that have the property Π. If we respectively denote by $\mathscr{C}$ and $\mathscr{S}$ the property of being connected and simple, Taylor proved in [13] that $\mathcal{G}(S,\mathscr{C})$, $\mathcal{G}(S,\mathscr{S})$ and $\mathcal{G}(S,\mathscr{C}\wedge\mathscr{S})$ are connected (where $\wedge$ stands for "and"). Let G, H be two graphs of $\mathcal{G}(S,\Pi)$. A sequence of flips *transforms* G *into* H *in* $\mathcal{G}(S,\Pi)$ if the sequence of flips transforms G into H and all the intermediate graphs also have the property Π. In other words, a sequence of flips that transforms G into H in $\mathcal{G}(S,\Pi)$ is a path between G and H in $\mathcal{G}(S,\Pi)$. Since [13] ensures that $\mathcal{G}(S,\Pi)$ is connected, one can ask what is the minimum length of such a transformation between G and H. This problem is known to be NP-hard, see e.g. [4]. In this paper we will study the following problem:

SHORTEST CONNECTED GRAPH TRANSFORMATION

Input: Two connected multigraphs G, H with the same degree sequence.
Output: The minimum number of flips needed to transform G into H in $\mathcal{G}(S,\mathscr{C})$.

Note that SHORTEST CONNECTED GRAPH TRANSFORMATIONis a generalization of SORTING BY REVERSALS since, when the degree sequence consists of $n-2$ vertices of degree 2 and two vertices of degree 1, we simply want to find a sequence of reversals of minimum length between two paths. Bousquet and Mary [3] proposed a 4-approximation algorithm for SHORTEST CONNECTED GRAPH TRANSFORMATION. Our main result is the following:

Theorem 1. SHORTEST CONNECTED GRAPH TRANSFORMATION *admits a* 2.5-*approximation algorithm.*

Section 3 is devoted to the proof of Theorem 1. In order to prove it, we will mainly focus on the SHORTEST TREE TRANSFORMATION problem which is the same as SHORTEST CONNECTED GRAPH TRANSFORMATIONexcept that the input consists of trees with the same degree sequence. Informally speaking, it is due to the fact that if an edge of the symmetric difference appears in some cycle, then we can reduce the size of the symmetric difference in one flip, as observed in [3].

When we desire to give some explicit bound on the quality of a solution, we need to compare it with the length of an optimal transformation. When we do not want to keep connectivity, Will [14] gives an explicit formula of the number of steps in a minimum transformation. When we want to keep connectivity, no such formula is known. Our 5/2-approximation algorithm is obtained by comparing it to the formula of Will (which is a lower bound when we want to keep connectivity). In Sect. 4, we discuss the tightness of this lower bound. We exhibit two graphs G and H such that the length of a shortest transformation between G and H is at least 1.5 times larger than the bound given by [14], and even twice longer under some assumptions on the set of possible flips. In order to prove this result, we generalize some notions introduced for sorting by reversals in [5] to general graphs.

This example ensures that if we want to find an approximation algorithm with a ratio better than 1.5, we might have to improve the algorithm, but overall, we need to improve the lower bound. The formal point and the two graphs G and H can be found in Sect. 4.

Related Works

Mass Spectrometry. Mass spectrometry is a technique used by chemists in order to obtain the formula of a molecule. It provides the mass-to-charge (m/z) ratio spectrum of the molecule from which we can deduce how many atoms of each element the molecule has. With this formula, we would like to find out the nature of the molecule, i.e. the bonds between the different atoms. But the existence of structural isomers points out that there could exist several solutions for this problem. Thus, we would like to find all of them. Since the valence of each atom is known, this problem actually consists in finding all the connected loop-free multigraphs whose degree sequence is the sequence of the valences of those atoms. The reconfiguration problem we are studying here can be a tool for an enumeration algorithm consisting in visiting the reconfiguration graph.

Flips and Reconfiguration. The SHORTEST CONNECTED GRAPH TRANSFORMATIONproblem belongs to the class of reconfiguration problems that received a considerable attention in the last few years. Reconfiguration problems consist, given two solutions of the same problem, in transforming the first solution into the second via a sequence of "elementary" transformations (such as flips) maintaining some properties all along. For more information on reconfiguration problems, the reader is referred for instance to [11].

2 Preliminaries

2.1 Symmetric Difference

Unless specified otherwise, we consider unoriented loop-free multigraphs. Let $G = (V(G), E(G))$ be a graph where $V(G)$ is the set of vertices of G and $E(G)$ is its set of edges. The *intersection* of two graphs G and H on the same set of vertices V is the graph $G \cap H$ with vertex set V, and such that $e \in E(G \cap H)$, with multiplicity m, if the minimum multiplicity of e in both graphs is m. Their *union*, $G \cup H$, has vertex set V, and $e \in E(G \cup H)$, with multiplicity m, if and only if the maximum multiplicity of e in G and H is m. Finally, the *difference* $G - H$ has vertex set V and $e \in E(G - H)$ with multiplicity m if and only if the difference between its multiplicities in G and H is $m > 0$. The *symmetric difference* of G and H is $\Delta(G, H) = (G - H) \cup (H - G)$. We denote by $\delta(G, H)$ the number of edges of $\Delta(G, H)$.

Let G, H be two graphs with the same degree sequence. An edge e of G is *good* if it is in $G \cap H$ and is *bad* otherwise. Note that since G and H have the same degree sequence, the graph $\Delta(G, H)$ has even degree on each vertex and the number of edges of G incident to v is equal to the number of edges of H incident to v.

Each flip removes at most 4 edges of the symmetric difference. Therefore, the length of a transformation from G to H is at least $\delta(G,H)/4$. In fact, it is possible to obtain a slightly better bound on the length of the transformation. A cycle C in $\Delta(G,H)$ is *alternating* if edges of G and H alternate in C. Since the number of edges of G incident to v is equal to the number of edge of H incident to v in $\Delta(G,H)$, the graph $\Delta(G,H)$ can be partitioned into a collection of alternating cycles. We denote by $mnc(G,H)$ the maximal number of cycles in a partition $\mathcal{C}$ of $\Delta(G,H)$ into alternating cycles. Will [14] proved the following:

Theorem 2 (Will [14]). *Let G, H be two graphs with the same degree sequence. A shortest sequence of flips that transforms G into H (that does not necessarily maintain the connectivity of the intermediate graphs) has length exactly $\frac{\delta(G,H)}{2} - mnc(G,H)$.*

Note that Theorem 2 indeed provides a lower bound for a transformation of Shortest Connected Graph Transformation.

2.2 Basic Facts Concerning Flips

Let $G = (V,E)$ be an unoriented graph and $v \in V(G)$. The set $N_G(v)$ of *neighbours* of v in G is the set of vertices u such that $uv \in E(G)$. Let D be a directed graph and $v \in V(D)$. The set $N_D^-(v)$ of *in-neighbours* of v in D is the set of vertices u such that uv is an arc of D, and the set $N_D^+(v)$ of *out-neighbours* of v in D is the set of vertices u such that vu is an arc of D. When G and D are obvious from the context we will simply write $N(v), N^-(v), N^+(v)$.

The *inverse* σ^{-1} of a flip σ is the flip such that $\sigma \circ \sigma^{-1} = id$, i.e. applying σ and then σ^{-1} leaves the initial graph. The *opposite* $-\sigma$ of a flip σ is the unique other flip that can be applied to the two edges of σ. If we consider a flip $\sigma = (ab, cd) \to (ac, bd)$, then $\sigma^{-1} = (ac, bd) \to (ab, cd)$ and $-\sigma = (ab, cd) \to (ad, bc)$. Note that $-\sigma$ is a flip deleting the same edges as σ while σ^{-1} cancels the flip σ. When we transform a graph G into another graph H, we can flip the edges of G or the edges of H. Indeed, applying the sequence of flips $(\sigma_1, \dots, \sigma_i)$ to transform G into a graph K, and the sequence of flips $(\tau_1, \dots, \tau_j)$ to transform H into K is equivalent to applying the sequence $(\sigma_1, \dots, \sigma_i, \tau_j^{-1}, \dots, \tau_1^{-1})$ to transform G into H.

Let $G = (V,E)$ be a connected graph and let H be a graph with the degree sequence of G. A flip is *good* if it flips bad edges and creates at least one good edge. It is *bad* otherwise. A *connected flip* is a flip such that its resulting graph is connected. Otherwise, it is *disconnected.* A *path* from $a \in V$ to $b \in V$ is a sequence of vertices $(v_1, \dots, v_k)$ such that $a = v_1$, $b = v_k$, for every integer $i \in [k-1]$, $v_i v_{i+1} \in E(G)$ and there is no repetition of vertices. Similarly, a path from e to f with $e, f \in E(G)$ is a path from an endpoint of e to an endpoint of f that does not contain the other endpoint of e and of f. A path between x and y (vertices or edges) is a path from x to y or a path from y to x. The *content* of a path is its set of vertices. We say that an edge e *belongs to* (or *is on*) a path P if both endpoints of e appear consecutively in P. The *intersection* $P_1 \cap P_2$ of two paths P_1 and P_2 is the intersection of their contents. The vertices of a sequence $(v_1, \ldots, v_k)$ are *aligned* in G if there exists a path P which is the concatenation of $k-1$ paths $P_1 P_2 \ldots P_{k-1}$ where P_i is a path from v_i to v_{i+1} for $i \in [k-1]$. Note that we might have $v_i = v_{i+1}$ and then $P_i = v_i$.

Note that, for every connected graph G, if $ab, cd \in E(G), ab \neq cd$, then (a,b,c,d), (a,b,d,c), (b,a,c,d), or (b,a,d,c) are aligned. Moreover, if G is a tree, exactly one of them is aligned. Let G be a connected graph and $a,b,c,d \in V(G)$ such that (a,b,c,d) are aligned. The *in-area* of the two edges ab and cd is the connected component of $G \setminus \{ab, cd\}$ containing the vertices b and c. The other components are called *out-areas*. The following lemma links the connectivity of a flip and the alignment of its vertices:

Lemma 1. *($\star$) Let G be a connected graph and $ab, cd \in E(G)$ where a,b,c and d are pairwise distinct vertices of G. If (a,b,c,d) or (b,a,d,c) are aligned in G, then the flip $(ab,cd) \to (ac,bd)$ is connected. If G is a tree, then it is also a necessary condition.*

The proofs of all the statements marked with a $\star$ are not included in this extended abstract. Lemma 1 ensures that, for trees, exactly one of the two flips σ and $-\sigma$ is connected.

Let e and f be two vertex-disjoint edges of a tree T, and let σ_2 be a flip in T that does not flip e nor f. The flip σ_2 *depends on e and f* if applying the connected flip on e and f changes the connectivity of σ_2. By abuse of notation, for any two flips σ_1 and σ_2 on pairwise disjoint edges, σ_2 *depends on* σ_1 if σ_2 depends on the edges of σ_1. The flip σ_1 *sees* σ_2 if exactly one of the edges of σ_2 is on the path linking the two edges of σ_1 in G.

The following lemma links the dependency of two flips and the position of their edges in a tree:

Lemma 2. *($\star$) Let T be a tree and σ_1 and σ_2 be two flips on T, whose edges are pairwise distinct. The three following points are equivalent:*

1. *σ_2 depends on σ_1,*
2. *σ_1 depends on σ_2,*
3. *σ_2 sees σ_1 and σ_1 sees σ_2.*

We now give two consequences of applying a connected flip.

Lemma 3. *($\star$) Let T be a tree and σ_1 and σ_2 be two flips on T with pairwise disjoint edges, where σ_1 is connected. Let T' be the tree obtained after applying σ_1 to T. The flip σ_1^{-1} sees σ_2 in T' if and only if σ_1 sees σ_2 in T. And σ_2 sees σ_1^{-1} in T' if and only if σ_2 sees σ_1 in T.*

Lemma 4. *($\star$) Let T be a tree and σ_1, σ_2 and σ_3 be three flips on T whose edges are pairwise disjoint and such that σ_1 sees σ_2, σ_2 sees σ_3, and σ_2 is connected. Let T' be the tree obtained by applying the flip σ_2 to T. The flip σ_1 sees σ_3 in T if and only if σ_1 does not see σ_3 in T'.*

3 Upper Bound

Let us recall a result of [3].

Lemma 5. *Let G, H be two connected graphs with the same degree sequence. There exists a sequence of at most two flips that decreases $\delta(G,H)$ by at least 2. Moreover, if there is an alternating C_4 in $\Delta(G,H)$, it can be removed in at most 2 steps, without modifying the rest of the graph.*

Lemma 5 immediately implies the following:

Corollary 1. Shortest Connected Graph Transformation *admits a polynomial time 4-approximation algorithm.*

The goal of the rest of this section is to improve the approximation ratio. The crucial lemma is the following:

Lemma 6. *Let G, H be two trees with the same degree sequence. There exists a sequence of at most 3 flips that decreases $\delta(G,H)$ by at least 4. Moreover, this sequence only flips bad edges.*

Proof. Let G' be the graph whose vertices are the connected components of $G \cap H$ and where two vertices S_1 and S_2 of G' are incident if there exists an edge in G between a vertex of S_1 and a vertex of S_2. In other words, G' is obtained from G by contracting every connected component of $G \cap H$ into a single vertex. Note that the edges of G' are the edges of $G - H$. Moreover, as G is a tree, G' also is. We can similarly define H'. Note that G' and H' have the same degree sequence.

Let S_1 be a leaf of G' and S_2 be its parent in G'. Let us show that S_2 is not a leaf of G'. Indeed, otherwise G' would be reduced to a single edge. In particular, $E(G-H)$ would contain only one edge. Since the degree sequence of $G-H$ and $H-G$ are the same, the edge of $H-G$ would have to be the same, a contradiction. Thus, we can assume that S_2 is not a leaf. Let u_1u_2 be the edge of $G-H$ between $u_1 \in S_1$ and $u_2 \in S_2$. Since $G-H$ and $H-G$ have the same degree sequence and S_1 is a leaf of G', there exists a unique vertex v_1 such that $u_1v_1 \in E(H-G)$. Moreover there exists a vertex v_2 such that $u_2v_2 \in E(H-G)$.

Let us first assume that $v_1 = v_2$. Then there exists a vertex w distinct from u_1 and u_2 such that $v_1w \in E(G-H)$ since v_1 has degree at least 2 in $H-G$. Since S_1 is a leaf of G', $w \notin S_1$ and either (u_1, u_2, v_1, w) or (u_1, u_2, w, v_1) are aligned in G. If (u_1, u_2, v_1, w) are aligned then the flip $(u_1u_2, v_1w) \to (u_1v_1, u_2w)$ in G is connected and creates the edge u_1v_1. If (u_1, u_2, w, v_1) are aligned then $(u_1u_2, v_1w) \to (u_1w, u_2v_1)$ is connected and creates the edge $u_2v_1 = u_2v_2$. In both cases, we reduce the size of the symmetric difference by at least 2 in one flip, and we can conclude with Lemma 5.

From now on, we assume that $v_1 \neq v_2$. We focus on the alignment of u_1, v_1, u_2 and v_2 in H. Since S_1 is a leaf of G', it is also a leaf of H'. Thus, v_1 is on the path from u_1 to u_2 and either (u_1, v_1, u_2, v_2) or (u_1, v_1, v_2, u_2) are aligned. If (u_1, v_1, u_2, v_2) are aligned then Lemma 1 ensures that $(u_1v_1, u_2v_2) \to (u_1u_2, v_1v_2)$

is connected in H and reduces the size of the symmetric difference by at least 2. We can conclude with Lemma 5. Thus, we can assume that (u_1, v_1, v_2, u_2) are aligned in H (see Fig. 1 for an illustration).

Let us first remark that if u_2 has degree at least 2 in $H - G$ (or equivalently in $G - H$), then we are done. Indeed, if there exists $w \neq v_2$ such that $u_2w \in E(H-G)$ then, since (u_1, v_1, v_2, u_2) are aligned, (u_1, v_1, u_2, w) have to be aligned. Indeed, v_2u_2 is the only edge of $H - G$ on the path from v_1 to u_2 incident to u_2. Thus the flip $(u_1v_1, u_2w) \rightarrow (u_1u_2, v_1w)$ is connected in H. Since it reduces $\delta(G, H)$ by at least 2, we can conclude with Lemma 5.

From now on, we will assume that u_2 has degree 1 in $H - G$. Let H_3 (resp. H_4) be the connected component of v_1 and v_2 (resp. u_2) in $H \setminus \{u_1v_1, u_2v_2\}$, which exists since (u_1, v_1, v_2, u_2) are aligned. Note that the third component of $H \setminus \{u_1v_1, u_2v_2\}$ is reduced to S_1. By definition, H_3 is the in-area of u_1v_1 and u_2v_2.

We now show that there exists an edge $u_3u_4 \in E(G - H)$, with $u_3 \in H_3$, $u_4 \in H_4$, and such that the connected component S_4 of $G \cap H$ containing u_4 is not a leaf of G'. Indeed, since G is connected, there exists a path P from v_1 to u_2 in G. Since u_1u_2 is the only edge of $G - H$ that has an endpoint in S_1, this path does not contain any vertex of S_1. Thus, it necessarily contains an edge u_3u_4 between a vertex u_3 of H_3 and a vertex u_4 of H_4. Since H_3 and H_4 are anticomplete in $G \cap H$, $u_3u_4 \in E(G - H)$. Moreover, the connected component S_4 of $G \cap H$ containing u_4 is not a leaf of G', as it is either S_2 which is not a leaf, or P has to leave S_4 at some point with an edge of $G - H$ since P ends in $u_2 \in S_2$.

Since u_3 and u_4 have the same degree in $G - H$ and $H - G$, there exist v_3, v_4 such that $u_3v_3, u_4v_4 \in E(H - G)$. Moreover, since S_4 is not a leaf of G' (and thus of H'), there exists an edge of $H - G$ between a vertex $u_5 \in S_4$ and a vertex $v_5 \in V \setminus S_4$ where $u_5v_5 \neq u_4v_4$.

Let us prove that u_3, v_3, u_4 and v_4 are pairwise distinct. By definition, we have $u_3 \neq v_3$, $u_4 \neq v_4$ and $u_3 \neq u_4$. Moreover, since $u_3u_4 \in E(G - H)$, $u_3 \neq v_4$ and $u_4 \neq v_3$. Thus, the only vertices that can be identical are v_3 and v_4. If $v_3 = v_4$, since $u_3 \in H_3$, $u_4 \in H_4$, and v_2u_2 is the only edge of $H - G$ from H_3 to H_4, then either $v_3 = v_4 = v_2$ or $v_3 = v_4 = u_2$. In the first case, $u_4 = u_2$ since v_2u_2 is the only edge of $H - G$ from H_3 to H_4. Thus, u_2 is the endpoint of both u_1u_2 and u_2u_3 in $G - H$. In the second case, u_2 is the endpoint of both u_2u_3 and u_2u_4 in $H - G$. Thus, in both cases, u_2 has degree at least 2 in $H - G$, a contradiction.

We now focus on the alignment of u_3, u_4, v_3 and v_4 in H. If (v_3, u_3, v_4, u_4) or (u_3, v_3, u_4, v_4) are aligned, then the flip $(u_3v_3, u_4v_4) \rightarrow (u_3u_4, v_3v_4)$ is connected in H and reduces the size of the symmetric difference by at least 2, since $u_3u_4 \in E(G - H)$. Note that the flip is well-defined since all the vertices are distinct. Thus, we can conclude with Lemma 5. Therefore, we can assume that (u_3, v_3, v_4, u_4) or (v_3, u_3, u_4, v_4) are aligned in H.

We give, in each case, a sequence of three flips that decreases the size of the symmetric difference by at least 4. Due to space restriction, the proof that those flips can be applied and maintain the connectivity are not included in this extended abstract.

Case 1. (u_3, v_3, v_4, u_4) are aligned. (See Fig. 1 for an illustration).
We successively apply the flips $\sigma_1 : (u_2v_2, u_5v_5) \to (u_2v_5, u_5v_2)$, $\sigma_2 : (u_3x, u_4v_4) \to (u_3u_4, xv_4)$ where $x = u_5$ if $u_3 = v_2$ and $v_3 = u_2$, and $x = v_3$ otherwise, and $\sigma_3 : (u_1v_1, u_2v_5) \to (u_1u_2, v_1v_5)$ in H. Since $u_1u_2, u_3u_4 \in E(G-H)$, this sequence of flips indeed reduces $\delta(G, H)$ by at least 4.

Case 2. (v_3, u_3, u_4, v_4) are aligned.
We apply $\sigma_1 : (u_2v_2, u_4v_4) \to (u_2v_4, u_4v_2)$, $\sigma_2 : (u_3v_3, u_4v_2) \to (u_3u_4, v_2v_3)$ then $\sigma_3 : (u_1v_1, u_2v_4) \to (u_1u_2, v_1v_4)$ to H. Again, $u_1u_2, u_3u_4 \in E(G-H)$ and it reduces $\delta(G, H)$ by at least 4.

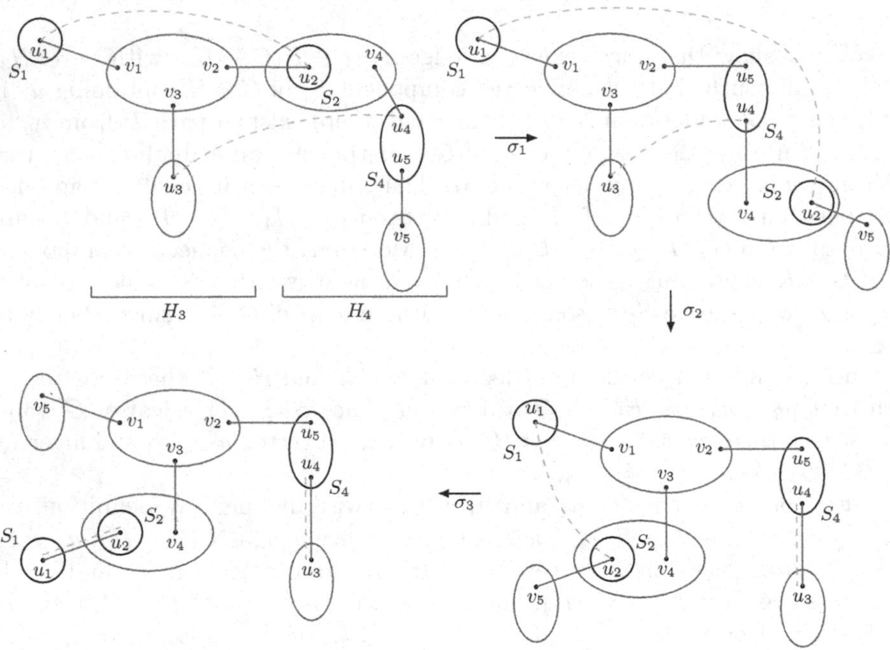

Fig. 1. The three flips $\sigma_1 : (u_2v_2, u_5v_5) \to (u_2v_5, u_5v_2)$, $\sigma_2 : (u_3v_3, u_4v_4) \to (u_3u_4, v_3v_4)$ and $\sigma_3 : (u_1v_1, u_2v_5) \to (u_1u_2, v_1v_5)$ applied to the graph H where (u_3, v_3, v_4, u_4) are aligned. The blue full edges are in $E(H-G)$ and the red dashed edges are in $E(G-H)$. (Color figure online)

Therefore, in all the cases, we have found a sequence of three flips whose edges are in the symmetric difference and that reduce $\delta(G, H)$ by at least 4. Moreover, the proof immediately provides a polynomial time algorithm to find such a sequence. □

Note that Lemma 6 allows to obtain a 3-approximation algorithm for Shortest Connected Graph Transformation. Indeed, as shown in the proof of

Lemma 1 in [3], as long as there exists an edge of the symmetric difference in a cycle of G, one can reduce the size of the symmetric difference by 2 in one step. Afterwards, we can assume that the remaining graphs $G - H$ and $H - G$ are trees. By Lemma 6, in three flips, the symmetric difference of the optimal solution decreases by at most 12 while our algorithm decreases it by at least 4. (Note that free to try all the flips, finding these flips is indeed polynomial). But we can actually improve the approximation ratio. The idea consists in treating differently short cycles. A *short* cycle is a C_4, a *long* cycle is a cycle of length at least 6. We now give the main result of this section.

Theorem 3. SHORTEST CONNECTED GRAPH TRANSFORMATION *admits a 5/2-approximation algorithm running in polynomial time. It becomes a 9/4-approximation algorithm if $\Delta(G,H)$ does not contain any short cycle.*

Proof. Let $\mathcal{C}$ be an optimal partition of $\Delta(G,H)$ into alternating cycles, i.e. a partition with $mnc(G,H)$ cycles. Let c be the number of short cycles in $\mathcal{C}$. Bereg and Ito [1] provide a polynomial time algorithm to find a partition of $\Delta(G,H)$ into alternating cycles having at least $\frac{c}{2}$ short cycles. Lemma 5 ensures that we can remove their $2c$ edges from the symmetric difference in at most c flips. If an edge of the symmetric difference is in a cycle of G or H, then in one step we can reduce the symmetric difference by 2 [3]. Otherwise, by Lemma 6, we can remove the remaining $\delta(G,H) - 2c$ edges using at most $\frac{3(\delta(G,H)-2c)}{4}$ flips in polynomial time. Therefore, we can transform G into H with at most $c + \frac{3(\delta(G,H)-2c)}{4}$ flips.

Let us now provide a lower bound on the length of a shortest transformation from G to H. By definition, $\mathcal{C}$ contains c short cycles. Theorem 2 ensures that we need at least c steps to remove the short cycles, plus $\ell - 1$ flips to remove each cycle of length 2ℓ. Therefore, we need at least $\frac{\delta(G,H)-4c}{3}$ flips to remove the $\delta(G,H) - 4c$ remaining edges from the symmetric difference.

The ratio between the upper bound and the lower bound is

$$f(c) := \frac{c + \frac{3\delta(G,H)-6c}{4}}{c + \frac{\delta(G,H)-4c}{3}} = \frac{3(3\delta(G,H) - 2c)}{4(\delta(G,H) - c)}.$$

The function f being increasing and since the number of short cycles in $\mathcal{C}$ cannot exceed $\frac{\delta(G,H)}{4}$, we have $f(c) \leq f(\frac{\delta(G,H)}{4}) = \frac{5}{2}$. It gives a $\frac{5}{2}$-approximation in polynomial time. Moreover, when there is no alternating short cycle in $\Delta(G,H)$, $c = 0$. Since $f(0) = \frac{9}{4}$, we obtain a $\frac{9}{4}$-approximation. □

4 Discussion on the Tightness of the Lower Bound

In this section, we discuss the quality of the lower bound of Theorem 2. We first prove that if we only flip bad edges of the same cycle of the symmetric difference then the length of a shortest transformation can be almost twice longer than the one given by the lower bound of Theorem 2. In order to prove it, we generalize several techniques and results of Christie [5], proved for the SORTING BY REVERSALS problem.

Note that the result of Hannenhalli and Pevzner [9] actually proves that in the case of paths, when the symmetric difference only contains vertex-disjoint short cycles, it is not necessarily optimal to only flip edges of the same cycle. However, studying this restriction gives us a better understanding of the general problem.

We also prove that, if we only flip bad edges (which are not necessarily in the same cycle of the symmetric difference), then the length of a shortest transformation can be almost 3/2 times longer than the one given by the lower bound. Note that all the existing approximation algorithms for Sorting by Reversals and Shortest Connected Graph TransformationOnly flip bad edges. But again no formal proof guarantees that there always exists a shortest transformation where we only flip bad edges.

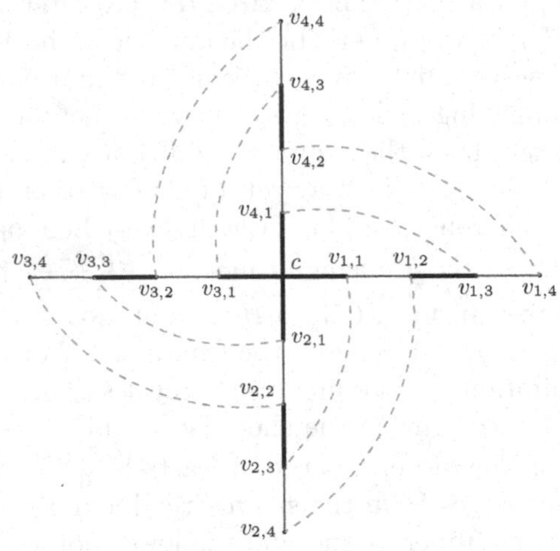

Fig. 2. The graphs G_4 and H_4. The black thick edges are in $E(G_4 \cap H_4)$, the blue thin edges are in $E(G_4 - H_4)$ and the red dashed edges are in $E(H_4 - G_4)$. (Color figure online)

Both results (whose proofs are not included in this extended abstract) are obtained with the same graphs G_k and H_k represented in Fig. 2 for $k = 4$. For any $k \geq 2$, let $G_k = (V_k, E(G_k))$ and $H_k = (V_k, E(H_k))$ be the graphs with $V_k = \{v_{i,j}, 1 \leq i \leq k, 1 \leq j \leq 4\} \cup \{c\}$, $E(G_k) = \bigcup_{i \in [k]} \{cv_{i,1}, v_{i,1}v_{i,2}, v_{i,2}v_{i,3}, v_{i,3}v_{i,4}\}$, and $E(H_k) = \bigcup_{i \in [k]} \{cv_{i,1}, v_{i,1}v_{i+1,3}, v_{i,2}v_{i,3}, v_{i,2}v_{i+1,4}\}$, where the additions are defined modulo k. One can easily check that, in this construction, both G_k and H_k are the subdivisions of a star where each branch has 4 vertices. Note that $\Delta(G, H)$ is the disjoint union of k short cycles. Moreover, the partition of $\Delta(G, H)$ into alternating cycles is unique.

The proof of Lemma 5 ensures that there is a transformation from G_k to H_k in at most $2k$ steps where we only flip bad edges in the same cycle of the symmetric difference. So our first result is tight with our assumptions. We conjecture that the length of a shortest transformation from G_k to H_k is at least $2k-1$ without any assumption on the set of possible flips.

References

1. Bereg, S., Ito, H.: Transforming graphs with the same graphic sequence. J. Inf. Process. **25**, 627–633 (2017)
2. Berman, P., Hannenhalli, S., Karpinski, M.: 1.375-approximation algorithm for sorting by reversals. In: Möhring, R., Raman, R. (eds.) ESA 2002. LNCS, vol. 2461, pp. 200–210. Springer, Heidelberg (2002). https://doi.org/10.1007/3-540-45749-6_21
3. Bousquet, N., Mary, A.: Reconfiguration of graphs with connectivity constraints. In: Epstein, L., Erlebach, T. (eds.) WAOA 2018. LNCS, vol. 11312, pp. 295–309. Springer, Cham (2018). https://doi.org/10.1007/978-3-030-04693-4_18
4. Caprara, A.: Sorting by reversals is difficult. In: Proceedings of the First Annual International Conference on Computational Molecular Biology, RECOMB 1997, pp. 75–83. ACM (1997)
5. Christie, D.A.: A 3/2-approximation algorithm for sorting by reversals. In: Proceedings of the Ninth Annual ACM-SIAM Symposium on Discrete Algorithms, SODA 1998, pp. 244–252 (1998)
6. Diestel, R.: Graph Theory. Graduate Texts in Mathematics, vol. 173, 3rd edn. Springer, Heidelberg (2005). https://doi.org/10.1007/978-3-662-53622-3
7. Hakimi, S.L.: On realizability of a set of integers as degrees of the vertices of a linear graph. i. J. Soc. Ind. Appl. Math. **10**(3), 496–506 (1962)
8. Hakimi, S.L.: On realizability of a set of integers as degrees of the vertices of a linear graph ii. Uniqueness. J. Soc. Ind. Appl. Math. **11**(1), 135–147 (1963)
9. Hannenhalli, S., Pevzner, P.A.: Transforming cabbage into turnip: polynomial algorithm for sorting signed permutations by reversals. J. ACM (JACM) **46**(1), 1–27 (1999)
10. Kececioglu, J.D., Sankoff, D.: Exact and approximation algorithms for sorting by reversals, with application to genome rearrangement. Algorithmica **13**(1/2), 180–210 (1995)
11. Nishimura, N.: Introduction to reconfiguration (2017, preprint)
12. Senior, J.: Partitions and their representative graphs. Am. J. Math. **73**(3), 663–689 (1951)
13. Taylor, R.: Contrained switchings in graphs. In: McAvaney, K.L. (ed.) Combinatorial Mathematics VIII. LNM, vol. 884, pp. 314–336. Springer, Heidelberg (1981). https://doi.org/10.1007/BFb0091828
14. Will, T.G.: Switching distance between graphs with the same degrees. SIAM J. Discrete Math. **12**(3), 298–306 (1999)

Approximating Weighted Completion Time for Order Scheduling with Setup Times

Alexander Mäcker, Friedhelm Meyer auf der Heide, and Simon Pukrop(✉)

Heinz Nixdorf Institute and Computer Science Department, Paderborn University, Fürstenallee 11, 33102 Paderborn, Germany
{amaecker,fmadh,simonjp}@mail.uni-paderborn.de

Abstract. Consider a scheduling problem in which jobs need to be processed on a single machine. Each job has a weight and is composed of several operations belonging to different families. The machine needs to perform a setup between the processing of operations of different families. A job is completed when its latest operation completes and the goal is to minimize the total weighted completion time of all jobs.

We study this problem from the perspective of approximability and provide constant factor approximations as well as an inapproximability result. Prior to this work, only the NP-hardness of the unweighted case and the polynomial solvability of a certain special case were known.

Keywords: Order scheduling · Multioperation jobs · Total completion time · Approximation · Setup times

1 Introduction

Many models for scheduling problems assume jobs to be atomic. There are, however, numerous natural situations where it is more suitable to model jobs as compositions and consider the problem as an *order scheduling* formulation: In this case a job is called order and is assumed to be composed of a set of operations, which are requests for products. A job is considered to be finished as soon as all its operations are finished and a natural objective is the minimization of the sum of completion times of all jobs.

Another important aspect in such scenarios can be the consideration of setup times that may occur due to the change of tools on a machine, the reconfiguration of hardware, cleaning activities or any other preparation work [1–3]. We model this aspect by assuming the set of operations to be partitioned into several families. The machine needs to perform a setup whenever it switches from

This work was partially supported by the German Research Foundation (DFG) within the Collaborative Research Centre "On-The-Fly Computing" under the project number 160364472—SFB 901/3.

A. Chatzigeorgiou et al. (Eds.): SOFSEM 2020, LNCS 12011, pp. 88–100, 2020.
https://doi.org/10.1007/978-3-030-38919-2_8

processing an operation belonging to one family to an operation of a different family. No setup is required between operations of the same family.

This kind of order scheduling (with setups) has several applications, which have been reported in the literature and we name a few of them here: It can model situations in a food manufacturing environment [9]. Several base ingredients need to be produced on a single machine and then assembled, setup times effectively occur due to cleaning activities between producing different base ingredients. Another example [9] are customer orders where each order requests several products, which need to be produced by a single machine, and an order can be shipped to the customer only as soon as all products have been produced. Finally, our primary motivation for considering multioperation jobs comes from its applicability within our project on "On-The-Fly Computing" [11]. The main idea is that IT-services are (automatically) composed of several small, elementary services that together provide the desired functionality. Setup times occur due to the reconfiguration of hardware or for the provisioning of data that is required, depending on the type of elementary service to be executed. Due to space constraints, some of the proofs in this work have been moved to the full version of this paper [17].

1.1 Contribution and Results

We consider the aforementioned problem, which is formally introduced in Sect. 2 and which in the survey [16] by Leung et al. was termed *fully flexible case of order scheduling with arbitrary setup times*, for the case of a single machine. Because it is known that the problem is NP-hard as mentioned in Sect. 3 where we summarize relevant related work, we study the problem with respect to its approximability. The key ingredient of our approach is based on the following idea. We define a simplified variant of the considered problem, in which we only require that, before any operation of a given family is processed, a setup for this family is performed once at *some* (arbitrary) earlier time. Solutions to this simplified variant already carry a lot of information for solving the original problem. We show that they can be transformed into $(1+\sqrt{2})$-approximate solutions for our original problem in polynomial time in Sect. 4. We then provide an algorithm that solves the simplified variant optimally leading to a $(1+\sqrt{2})$-approximation for the original problem in Sect. 5. The runtime of the approach, however, is $O(\text{poly}(n) \cdot K!)$, where K denotes the number of families, and n is the number of operations. Thus, it is only polynomial for a constant number of families, which turns out to be no coincidence as we also observe that solving the simplified variant optimally for non-constant K is NP-hard. We show how an algorithm by Hall et al. [10] can be combined with our approach from Sect. 4 to obtain a runtime of $O(\text{poly}(n, K))$ while worsening the approximation factor to $2(1+\sqrt{2})$ in Sect. 6. We complement this result by a hardness result for approximations with a factor less than 2 assuming a certain variant of the Unique Games Conjecture.

2 Model

We consider a scheduling problem in which a set $\mathcal{J} = \{j_1, \dots, j_n\}$ of n jobs needs to be processed by a single machine. Each job j has a weight $w(j) \in \mathbb{R}_{\geq 0}$ and consists of a set of operations $j = \{o_1^j, o_2^j, \dots\}$. Each operation o_i^j is characterized by a processing time $p(o_i^j) \in \mathbb{R}_{\geq 0}$ and belongs to a family $f(o_i^j) \in \mathcal{F} = \{f_1, \dots, f_K\}$. If the schedule starts with an operation of family f and whenever the machine switches from processing operations of one family f' to an operation of another family f, a setup taking $s(f) \in \mathbb{R}_{\geq 0}$ time needs to take place first. Given this setting, the goal is to compute a schedule that minimizes the weighted sum of job completion times, where a job is considered to be completed as soon as all its operations are completed. More formally, a schedule is implicitly given by a permutation π on $\bigcup_{i=1}^n j_i$ and the completion time of an operation o is given by the accumulated setup times and processing times of jobs preceding operation o. That is, for $\pi = (o_1, o_2, \dots)$ the completion time of operation o_i is given by $C_{o_i}^\pi = \sum_{k=1}^i p(o_k) + \sum_{k=1}^i I(f(o_{k-1}), f(o_k))s(f(o_k))$, where I is an indicator being 0 if and only if its arguments are the same and 1 otherwise. Then, the completion time of a job j is given by $C_j^\pi = \max_{o \in j} C_o^\pi$ and the goal is to minimize the total weighted completion time given by $C^\pi = \sum_{j \in \mathcal{J}} w(j) C_j^\pi$.

Using the classical three-field notation for scheduling problems and following Gerodimos et al. [9], we denote the problem by $1|s_f, \textit{assembly}|\sum w_j C_j$. We study this problem in terms of its approximability. A polynomial-time algorithm $\mathcal{A}$ has an approximation factor of α if, on any instance, $C^\pi \leq \alpha \cdot C^{\text{OPT}}$, where C^π and C^{OPT} denotes the total weighted completion time of $\mathcal{A}$ and an optimal solution, respectively.

3 Related Work

The problem $1|s_f, \textit{assembly}|\sum w_j C_j$, and the more general version with multiple machines, are also known as order scheduling. More precisely, it was termed order scheduling in the flexible case with setup times in the survey [16] by Leung et al. As previously mentioned, it is known that this problem is already NP-hard for the unweighted case and a single machine as proven by Ng et al. [19]. Besides this hardness result, only one single positive result is known. Due to Gerodimos et al. [9], a special case can be solved optimally in time $O(n^{K+1})$. This special case requires that the jobs can be renamed so that job j_{i+1} contains, for each operation $o \in j_i$, an operation o' such that $f(o) = f(o')$ and $p(o') > p(o)$. A related positive result is due to Divakaran and Saks [8]. They designed a 2-approximation algorithm for our problem in case all jobs consist of a single operation. Monma and Potts worked on algorithms for the same model with a variety of objective functions. One result is an optimal algorithm for total weighted completion time with the constraint that the number of families is constant [18]. Their approach however relies on the fact that there is a trivial order inside each family, and the problem only arises in interleaving the families. Since we are dealing with multioperation jobs we cannot assume such an order.

Taking a broader perspective of the problem, it can be seen as a generalization of the classical problem of minimizing the total (weighted) completion time when there are no setups and all jobs are atomic (i.e., we only have single-operation jobs). It is well-known that sequencing all jobs in the order of non-decreasing processing times (shortest processing time ordering, SPT) minimizes the total completion time on a single machine [15]. In case jobs have weights and the objective is to minimize the total weighted completion time, a popular result is due to Smith [20]. He showed that weighted shortest processing time (WSPT), that is, sort the jobs non-decreasingly with respect to their ratio of processing time and weight, is optimal for this objective. Besides these two results, the problem has been studied quite a lot and in different variants with respect to the number of machines, potential precedences between jobs, the existence of release times and even more. For a single machine it was shown by Lenstra and Kan [15] and independently by Lawler [14] that adding precedences among jobs to the (unweighted) problem makes it NP-hard. In their paper, Hall et al. [10] analyzed algorithms based on different linear programming formulations and obtained constant factor approximations for several variants including the minimization of the total weighted completion time on a single machine with precedences. Particularly, they obtained a 2-approximation for this problem, which we will later make use of for our approximation algorithm for non-constant K. Actually, the factor 2 they achieve is essentially optimal, as Bansal and Khot [4] were able to show that a $(2-\varepsilon)$-approximation is impossible for any $\varepsilon > 0$ assuming a stronger version of the Unique Games Conjecture.

More loosely related is a model due to Correa et al. [7] in which jobs can be split into arbitrary parts (that can be processed in parallel) and where each part requires a setup time to start working on it. They proposed a constant factor approximation for weighted total completion time on parallel machines. Recently, some approximation results for the minimization of the makespan for single operation jobs have been achieved for different machine environments with setup times [6,12,13]. Finally, scheduling with setup times in general is a large field of research, primarily with respect to heuristics and exact algorithms, and the interested reader is referred to the three exhaustive surveys due to Allahverdi et al. [1–3].

4 The One-Time Setup Problem

In this section, we introduce a relaxation of $1|s_f, \mathit{assembly}|\sum w_j C_j$ and show how solutions to this relaxation can be transformed into solutions to the original problem by losing a small constant factor. The *one-time setup* problem ($1|\mathit{ot}\text{-}s_f$, $\mathit{assembly}|\sum w_j C_j$) is a relaxation of $1|s_f, \mathit{assembly}|\sum w_j C_j$ in which setups are not required on each change to operations of a different family. Instead we only require that, for any family f, a setup for f is performed once at some time before any operation belonging to f is processed. Formally, we introduce a new (setup) operation o_f^s for each family f with $p(o_f^s) = s(f)$, $w(o_f^s) = 0$ and a precedence relation between o_f^s and each operation belonging to f that ensures that o_f^s is processed before the respective operations. A schedule π is then implicitly given by a permutation on all operations (those belonging to jobs as well as those representing setups). We only consider those permutations which adhere to the precedence constraints. The completion time of an operation o_i under schedule $\pi = (o_1, o_2, \ldots)$ is given by $C_{o_i}^{\pi} = \sum_{k=1}^{i} p(o_k)$. The remaining definitions such as the completion time of a job and total weighted completion time remain unchanged. Note that this problem is indeed a relaxation of our original problem in the sense that the total weighted completion time cannot increase when only requiring one-time setups.

Before we turn to our approach to transform solutions to $1|ot\text{-}s_f,\ assembly|\sum w_j C_j$ into feasible solutions for $1|s_f,\ assembly|\sum w_j C_j$, we first give a simple observation. It shows that we can, intuitively speaking, glue all operations of a job together and focus on determining the order of such glued jobs. Formally, given a schedule π, a job is *glued* if all of j's operations are processed consecutively without other operations in between. We have the following lemma.

Lemma 1. *Every schedule can be transformed into one in which all jobs are glued without increasing the total weighted completion time.*

Due to the previous result, we assume in the rest of the paper that each job j in an instance of $1|ot\text{-}s_f,\ assembly|\sum w_j C_j$ only consists of a single operation o^j. This operation has a processing time $p(o^j) = \sum_{o \in j} p(o) =: p(j)$ and the precedence relation is extended so that each setup operation with a precedence to some $o \in j$ now has a precedence to o^j.

4.1 Transforming One-Time Setup Solutions

In this section, we present our algorithm Transform to transform a solution π for the one-time setup problem $1|ot\text{-}s_f,\ assembly|\sum w_j C_j$ into a solution π_{out} for our original problem $1|s_f,\ assembly|\sum w_j C_j$. Initially, π_{out} is the sequence of operations as implied by the solution π after splitting the glued jobs into its original operations again and not including setup operations. A *batch* is a (maximal) subsequence of consecutive (non-setup) operations of the same family. Intuitively, the schedule implied by π_{out} probably already has a useful order for the jobs but is missing a good batching of operations of the same family. This would lead to way too many setups to obtain a good schedule. Therefore, the main idea of Transform is to keep the general ordering of the schedule but to make sure that each setup is "worth it", i.e., that each batch is sufficiently large to justify a setup. We will achieve that by filling up each batch with operations of the same family scheduled later until the next operation would increase the length of the batch too much. More precisely, let $B_1, B_2, \ldots$ be the batches in π_{out} (in this order). We iterate over the batches B_i in the order of increasing i and for each batch B_i of family f do the following: Move as many operations of family f from the closest batches $B_{i'}, i' > i$, to B_i as possible while ensuring that $p(B_i) < \beta \cdot s(f)$, where β (we call it the *pull factor*) is some fixed constant and $p(B_i) := \sum_{j \in B_i} p(j)$. If a batch gets empty before being considered, it is removed from π_{out} (and hence, *not* considered in later iterations). We show the following theorem on the quality of TRANSFORM.

Theorem 1. *If* $C^{\pi} \leq c \cdot C^{\text{OPT}}$, *then* $C^{\pi_{out}} \leq (1+\beta) \cdot c \cdot C^{\text{OPT}}$, *for any* $\beta \geq \sqrt{2}$.

Proof. We only need to show that $C^{\pi_{\text{out}}} \leq (1+\beta) \cdot C^{\pi}$. For the analysis we will compare the completion time of each operation o in π to the one in π_{out} (in their respective cost model). We denote by $\pi(\ldots o)$ the schedule π up to and including operation o and by $f \in \pi(\ldots o)$ that some operation in $\pi(\ldots o)$ is of family f. We have

$$C_o^{\pi} = \sum_{f \in \mathcal{F} | f \in \pi(\ldots o)} \underbrace{\left(p(o_f^s) + \sum_{o' \in \bigcup_{j \in \mathcal{J}} j | o' \in \pi(\ldots o) \wedge f(o') = f} p(o') \right)}_{=:(C_o^{\pi})_f}.$$

We will now analyze the contribution $(C_o^{\pi_{\text{out}}})_f$ of some family f to the completion time of o in π_{out}. We have

$$(C_o^{\pi_{\text{out}}})_f \leq (C_o^{\pi})_f \underbrace{-s(f)}_{\text{removed } o_f^s} + \underbrace{\beta \cdot s(f)}_{\text{added operations}} + \underbrace{\left(\left\lceil \frac{(C_o^{\pi})_f - s(f)}{\frac{\beta}{2} s(f)} \right\rceil \cdot s(f) \right)}_{\text{cost of setups}},$$

due to the following reasoning. The first three summands describe the contribution of class f's jobs to the completion time of o in π_{out}. Compared to $(C_o^{\pi})_f$, we move operations of length at most $\beta \cdot s(f)$ belonging to family f in front of o (recall that empty batches are removed in the process of TRANSFORM; only the last batch of some family f before o pulls operations from behind o in front of o) and we do not consider the one-time setup operation. The last summand represents the contribution due to setups for family f. We need to do at most $\left\lceil \frac{(C_o^{\pi})_f - s(f)}{\frac{\beta}{2} s(f)} \right\rceil$ many setups for operations of family f that contribute to the completion time of o in π_{out}. This is true because of the following reasoning. From our construction we know that for two batches of the same family, with no other batches of the same family in between, the processing time of those batches combined has to be at least $\beta \cdot s(f)$, otherwise they would have been combined. If there is an odd number of batches we cannot say anything about the last batch, except that it has a nonzero processing time. This factor is captured by the rounding. Therefore we obtain

$$\begin{aligned}
(C_o^{\pi_{\text{out}}})_f &\le (C_o^{\pi})_f + (\beta-1)s(f) + \left(\left\lceil \frac{(C_o^{\pi})_f - s(f)}{\frac{\beta}{2}s(f)} \right\rceil \cdot s(f)\right) \\
&\le (C_o^{\pi})_f + \beta \cdot s(f) + \left(\frac{(C_o^{\pi})_f}{\frac{\beta}{2}s(f)} \cdot s(f)\right) - \frac{2}{\beta}s(f) \\
&\le (C_o^{\pi})_f + (\beta - \frac{2}{\beta}) \cdot s(f) + \frac{2(C_o^{\pi})_f}{\beta} \\
&\le (1+\frac{2}{\beta})(C_o^{\pi})_f + (\beta - \frac{2}{\beta}) \cdot s(f) \overset{\beta \ge \sqrt{2}}{<} (1+\beta)(C_o^{\pi})_f,
\end{aligned}$$

where the last inequality holds because a family f can only contribute to the completion time of o in π_{out} if it contributed to the completion of o in π and in this case $(C_o^{\pi})_f \ge s(f)$ by definition. (If o itself got moved to the front there might be a family that contributed to C_o^{π} but does not to $C_o^{\pi_{\text{out}}}$.) Since each operation's completion time in π_{out} is at most $(1+\beta)$ times as big as in π, we know that for each job $j \in \mathcal{J}$, $C_j^{\pi_{\text{out}}} \le (1+\beta) \cdot C_j^{\pi}$. □

Actually, one can show that there are instances in which $C^{\pi_{\text{out}}} \ge (1+\beta) \cdot C^{\pi}$ and therefore, that the analysis of TRANSFORM is indeed tight (cf. full version [17]). However, it is also worth mentioning that these instances are rather "artificial" as the jobs' processing times (and even their sum) are negligible while setup operations essentially dominate the completion times. In less nastily constructed instances, we would expect that even for moderate values $\beta > \sqrt{2}$, $(1+\frac{2}{\beta})(C_o^{\pi})_f$ significantly dominates $(\beta - \frac{2}{\beta}) \cdot s(f)$ for most of the operations o as $(C_o^{\pi})_f$ grows the later o is scheduled while $s(f)$ stays constant. This would then lead to $C^{\pi_{\text{out}}} \approx 1 + \frac{2}{\beta}$. This observation is also discussed and supported by our simulations (cf. full version [17]).

5 Approximations for Constant Number of Families

In this section, we study approximations for the problem $1|s_f,\ assembly|\sum w_j C_j$ and a fixed number of families K. The general idea is to first solve the $1|ot\text{-}s_f,\ assembly|\sum w_j C_j$ problem optimally and then to use the TRANSFORM algorithm as described in the previous section, leading to $(1+\sqrt{2})$-approximate solutions for instances of $1|s_f,\ assembly|\sum w_j C_j$. To solve the problem $1|ot\text{-}s_f,\ assembly|\sum w_j C_j$ optimally, we describe a two-step algorithm and two possible approaches for its second step. The first one is a direct application of a known approach by Lawler [14]. We also propose a new, alternative approach, which is much simpler as it is specifically tailored to our problem.

To solve $1|ot\text{-}s_f,\ assembly|\sum w_j C_j$ optimally, in the first step, we exhaustively enumerate all possible permutations of setups. In the second step, we then find, for each permutation, the optimal schedule under the assumption that the order of setup operations is fixed according to the permutation. After we have performed both of these steps, we can simply select the best result, which is the optimal solution to the $1|ot\text{-}s_f,\ assembly|\sum w_jC_j$ problem.

5.1 Series Parallel Digraph and Lawler's Algorithm

Lawler [14] proposed an algorithm that optimally solves $1|prec|\sum w_jC_j$ in polynomial time under the condition that the precedences can be described by a series parallel digraph. To solve $1|ot\text{-}s_f,\ assembly|\sum w_jC_j$, the general idea is to modify the precedence graph of a given $1|ot\text{-}s_f,\ assembly|\sum w_jC_j$ instance so that it becomes series parallel and then apply Lawler's algorithm. We create a series parallel digraph that represents both the jobs reliance on setups as well as the predetermined order of setup operations as follows. Given a permutation $\tau = (o^s_{f'_1}, o^s_{f'_2}, \dots)$ of setup operations, we create a precedence chain of nodes $o^s_{f'_1} \to o^s_{f'_2} \to \dots$. Then for each operation o^j, we add an edge from $o^s_{f'_i}$ to o^j such that i is the smallest value for which all operations in j belong to a family in $\{f'_1, f'_2, \dots, f'_i\}$. Intuitively, since we have fixed the order of setups for each operation, we can easily see which setup operation is the last one that is necessary to process the operation. We do not care about the other precedences because they became redundant after fixing the setup order.

Having done this we have a $1|prec|\sum w_jC_j$ problem with a series parallel digraph that is equivalent to the $1|ot\text{-}s_f,\ assembly|\sum w_jC_j$ problem. At this point we can use the result by Lawler [14] to solve this in polynomial time.

5.2 Simple Local Search Algorithm

In this section, we propose a simple algorithm to solve $1|ot\text{-}s_f,\ assembly|$ $\sum w_jC_j$ optimally in polynomial time given that K is fixed. Since our algorithm is tailored to this specific problem, it is a lot simpler and works with less overhead which justifies introducing it here alongside the aforementioned solution.

We first show that we can assume optimal schedules to fulfill a natural generalization of the weighted SPT-order to the setting with setup times. We define this notion for our original problem as follows: A schedule π is in *generalized weighted SPT-order* if the following is true: For every j_i, j_k with $\frac{p(j_i)}{w(j_i)} < \frac{p(j_k)}{w(j_k)}$, j_i is scheduled before j_k or j_k is scheduled at a position where j_i cannot be scheduled (because precedences would be violated). If $\frac{p(j_i)}{w(j_i)} = \frac{p(j_k)}{w(j_k)}$, $j_\ominus$ is scheduled before $j_\oplus$ or $j_\oplus$ is scheduled at a position where $j_\ominus$ cannot be scheduled, where $\ominus = \min\{i,k\}$ and $\oplus = \max\{i,k\}$.

Lemma 2. *Any schedule π with total weighted completion time C^{OPT} can be transformed into one in generalized weighted SPT-order without increasing the total weighted completion time.*

Proof. If there are two jobs j_i, j_k with $\frac{p(j_i)}{w(j_i)} < \frac{p(j_k)}{w(j_k)}$ that do not fulfill the desired property, π cannot be optimal due to the following reasoning. Let J be the set of operations scheduled after j_i and before j_k. Let $p(J) = \sum_{o \in J} p(o)$ and $w(J) = \sum_{o \in J} w(o)$ denote the processing time of all jobs and setups and all weights in J, respectively. We show that moving j_k directly behind j_i (move$_1$) or moving j_i directly before j_k (move$_2$) reduces the total weighted completion time of π.

The change of the total weighted completion time due to move$_1$ is given by $\Delta_1 = -w(J)p(j_k) + p(J)w(j_k) - w(j_i)p(j_k) + w(j_k)p(j_i)$. If $\Delta_1 < 0$, move$_1$ decreases the total weighted completion time and we are done.

Otherwise, if $\Delta_1 \geq 0$, we show that move$_2$ leads to a decrease. Since $\frac{p(j_i)}{w(j_i)} < \frac{p(j_k)}{w(j_k)}$ we know that $-w(j_i)p(j_k) + w(j_k)p(j_i) < 0$. Therefore, $-w(J)p(j_k) + p(J)w(j_k) > 0$. The change in total weighted completion time Δ_2 of move$_2$ is given by $\Delta_2 = -(-w(J)p(j_i) + p(J)w(j_i)) - w(j_i)p(j_k) + w(j_k)p(j_i)$. Since $\frac{p(j_i)}{w(j_i)} < \frac{p(j_k)}{w(j_k)}$ there exist $x, y \in \mathbb{R}^+$ with $p(j_i) = x \cdot p(j_k)$ and $w(j_i) = x \cdot w(j_k) + y$. Plugging those in we get

$$\begin{aligned}
\Delta_2 &= -(-w(J)p(j_i) + p(J)w(j_i)) - w(j_i)p(j_k) + w(j_k)p(j_i) \\
&= -(-w(J)xp(j_k) + p(J)xw(j_k) + p(J)y) - w(j_i)p(j_k) + w(j_k)p(j_i) \\
&= \underbrace{-x \cdot (-w(J)p(j_k) + p(J)w(j_k))}_{<0} \underbrace{-p(J)y}_{<0} \underbrace{-w(j_i)p(j_k) + w(j_k)p(j_i)}_{<0} < 0.
\end{aligned}$$

Therefore, in both cases we get a contradiction to the optimality of π and hence, no such jobs j_i and j_k can exist.

It remains to argue about pairs of jobs j_i and j_k such that $\frac{p(j_i)}{w(j_i)} = \frac{p(j_k)}{w(j_k)}$ (case works analogously to the previous one, refer to the full version [17]). Repeated application of this process leads to a schedule with the desired properties. □

Due to the previous lemma, we will restrict ourselves to schedules that are in generalized weighted SPT-order. We call the (possibly empty) sequence of jobs between two consecutive setup operations in a schedule a *block*. We therefore particularly require that in any schedule we consider, the jobs within a block are ordered according to the weighted SPT-order.

We execute a local search algorithm started on the initial schedule π_τ^{init} given by the input setup operation order τ followed by all jobs in weighted SPT-order (ties are broken in favor of jobs with lower index). An optimal schedule is then computed by iteratively improving this schedule by a local search algorithm. Given a schedule π, a *move* of job j is given by the block into which j is placed subject to the constraint that the resulting schedule remains feasible. Note that due to our assumption that we only consider schedules in generalized weighted SPT-order, a schedule π and a move of a job j uniquely determine a new feasible schedule. A move of job j is called a GREEDY move if it improves the total weighted completion time and among all moves of j, no other move leads to a larger improvement. Among all greedy moves for job j we call the one that places j closest to the beginning of π GREEDY$^+$ move. Our local search algorithm iteratively applies, in weighted SPT-order, one single GREEDY$^+$ move for each job. For ease of presentation, we assume in the following that we have guessed the permutation τ of setup operations correctly and that in the following the initial schedule in all considerations is always assumed to be π_τ^{init}.

Lemma 3. *Each schedule* π *in generalized weighted SPT-order can be reached by applying, in weighted SPT-order, a single move for each job. Additionally, each intermediate schedule is in generalized weighted SPT-order.*

Due to the previous lemma, from now on we assume the following. A sequence $\langle\sigma_1,\ldots,\sigma_i\rangle$ of moves defines the schedule obtained by applying the moves $\sigma_1,\ldots,\sigma_i$ (in this order) to the respective first i jobs in weighted SPT-order to the initial schedule π_τ^{init}. Our next step is to show that an optimal schedule can be found by GREEDY moves.

Lemma 4. *Suppose there is a sequence* $\langle\sigma_1,\ldots,\sigma_n\rangle$ *of moves such that the resulting schedule* π *has total weighted completion time* C^{OPT}. *Then all moves are* GREEDY *moves.*

The next corollary follows by the previous three lemmas.

Corollary 1. *There is an optimal schedule that can be reached by applying, in weighted SPT-order, a single* GREEDY *move per job.*

Using similar arguments as in the proof of the previous lemma, we can finally show that our tie breaker (by which GREEDY and GREEDY$^+$ moves differ) does not do any harm when searching for an optimal solution.

Lemma 5. *Applying, in weighted SPT-order, a* GREEDY$^+$ *move for each job, leads to an optimal schedule.*

By the previous lemma, we have the final theorem of this section.

Theorem 2. *The local search algorithm computes optimal solutions for the one-time setup problem in time* $O(n \cdot \log(n) \cdot K!)$. *In combination with the* TRANSFORM *algorithm from Sect. 4, this yields an approximation algorithm with approximation factor* $1+\sqrt{2}$ *for our original problem.*

6 Arbitrary Number of Families

In the previous section, we have seen that $1|s_f,\ \mathit{assembly}|\sum w_jC_j$ can be solved in time $O(n\cdot\log(n)\cdot K!)$, which is polynomial for a fixed number K of families. At this point, one might ask whether there are approximation algorithms running in time poly(n, K), and whether the non-polynomial dependence on K is inherent to $1|\mathit{ot}\text{-}s_f,\ \mathit{assembly}|\sum w_jC_j$. The latter is indeed true because Woeginger has shown in his paper [21] that different special cases of the $1|\mathit{prec}|\sum w_jC_j$ model, including one being equivalent to our (glued) $1|\mathit{ot}\text{-}s_f,\ \mathit{assembly}|\sum w_jC_j$ with the restriction that all job weights are 1, are equally hard to approximate. Therefore, optimally solving the one-time setup problem is indeed NP-hard for non-constant K. On the positive side, we show how $1|s_f,\ \mathit{assembly}|\sum w_jC_j$

problems can be approximated in time poly(n, K) in Sect. 6.1. This approach, however, worsens the approximation by a factor of 2 from $(1+\sqrt{2})$ to $2(1+\sqrt{2})$. Lastly we show that $1|s_f,\ assembly|\sum w_j C_j$ is inapproximable within factor $2-\varepsilon$, assuming a version of the Unique Games Conjecture, by applying results from Woeginger [21] and Bansal and Khot [4].

6.1 Approximation Algorithm

The general idea of our approximation algorithm is the same as for the case of a constant K: We first solve $1|ot\text{-}s_f,\ assembly|\sum w_j C_j$ and then use TRANSFORM from Sect. 4.1 to obtain a feasible schedule for our original $1|s_f,\ assembly|\sum w_j C_j$ problem. Recall that $1|ot\text{-}s_f,\ assembly|\sum w_j C_j$ is a special case of $1|prec|\sum w_j C_j$. As this problem has been studied a lot, there are different approximation algorithms in the literature and, for example, [5,10] provide 2-approximation algorithms. Therefore, we conclude with the following theorem.

Theorem 3. *$1|s_f,\ assembly|\sum w_j C_j$ can be approximated with an approximation factor of $2(1+\sqrt{2})$ in polynomial time.*

6.2 Lower Bound on the Approximability

Theorem 4. *Assuming a stronger version of the Unique Games Conjecture [4], $1|s_f,\ assembly|\sum w_j C_j$ is inapproximable within $2-\varepsilon$ for any $\varepsilon > 0$.*

Proof. Woeginger [21] showed that the general $1|prec|\sum w_j C_j$ and some special cases of the problem have the same approximability threshold. Bansal and Khot [4] could prove that, assuming a stronger version of the Unique Games Conjecture, $1|prec|\sum w_j C_j$, and therefore also the special cases in [21], are inapproximable within $2-\varepsilon$ for any $\varepsilon > 0$. The special case we are interested in was defined by Woeginger as: *[the] special case where every job has either $p_j = 0$ and $w_j = 1$, or $p_j = 1$ and $w_j = 0$, and where the existence of a precedence constraint $J_i \to J_j$ implies that $p_i = 1$ and $w_i = 0$, and that $p_j = 0$ and $w_j = 1$* [21]. It is easy to see that an α-approximation for $1|s_f,\ assembly|\sum w_j C_j$ also yields an α-approximation for the stated special case by transforming an instance of the special case in the following way: For every job i with $w_i = 0$ add a family f_i with $s(f_i) = 1$. For every job l with $w_l = 1$ add a job j_l with $w_{j_l} = 1$. For every precedence $J_i \to J_l$ add an operation $o_{i,l}$ to j_l with $f(o_{i,l}) = f_i$ and $p(o_{i,l}) = 0$. It is easy to see that the optimal solutions of both problems have the same weight. In both representations the difficult part is to decide the order of weight 0 jobs or setups, respectively. All jobs or operations with processing time 0 can be scheduled as early as possible in an optimal solution.

Therefore we can conclude that $1|s_f,\ assembly|\sum w_j C_j$ has at least an equally high appoximability threshold as $1|prec|\sum w_j C_j$. □

7 Future Work

It might be interesting whether there is a better algorithm for transforming solutions for the one-time setup problem to their respective original problem. One could also try to improve the approximation factor by designing algorithms that directly solve our original problem without the detour via the one-time setup problem. Another interesting direction for the future is the question whether our lower bound can be increased. For the special case with a constant number of families, the question whether that problem is already NP-hard also remains open.

References

1. Allahverdi, A.: The third comprehensive survey on scheduling problems with setup times/costs. Eur. J. Oper. Res. **246**(2), 345–378 (2015)
2. Allahverdi, A., Gupta, J.N., Aldowaisan, T.: A review of scheduling research involving setup considerations. Omega **27**(2), 219–239 (1999)
3. Allahverdi, A., Ng, C.T., Cheng, T.C.E., Kovalyov, M.Y.: A survey of scheduling problems with setup times or costs. Eur. J. Oper. Res. **187**(3), 985–1032 (2008)
4. Bansal, N., Khot, S.: Optimal long code test with one free bit. In: Proceedings of the 50th Annual IEEE Symposium on Foundations of Computer Science (FOCS), pp. 453–462. IEEE (2009)
5. Chekuri, C., Motwani, R.: Precedence constrained scheduling to minimize sum of weighted completion times on a single machine. Discrete Appl. Math. **98**(1–2), 29–38 (1999)
6. Correa, J.R., et al.: Strong LP formulations for scheduling splittable jobs on unrelated machines. Math. Program. **154**(1–2), 305–328 (2015)
7. Correa, J.R., Verdugo, V., Verschae, J.: Splitting versus setup trade-offs for scheduling to minimize weighted completion time. Oper. Res. Lett. **44**(4), 469–473 (2016)
8. Divakaran, S., Saks, M.E.: Approximation algorithms for problems in scheduling with set-ups. Discrete Appl. Math. **156**(5), 719–729 (2008)
9. Gerodimos, A.E., Glass, C.A., Potts, C.N., Tautenhahn, T.: Scheduling multi-operation jobs on a single machine. Ann. OR **92**, 87–105 (1999)
10. Hall, L.A., Schulz, A.S., Shmoys, D.B., Wein, J.: Scheduling to minimize average completion time: off-line and on-line approximation algorithms. Math. Oper. Res. **22**(3), 513–544 (1997)
11. Happe, M., Meyer auf der Heide, F., Kling, P., Platzner, M., Plessl, C.: On-the-fly computing: a novel paradigm for individualized IT services. In: Proceedings of the 16th IEEE International Symposium on Object/Component/Service-Oriented Real-Time Distributed Computing (ISORC), pp. 1–10. IEEE Computer Society (2013)
12. Jansen, K., Klein, K., Maack, M., Rau, M.: Empowering the configuration-IP-new PTAS results for scheduling with setups times. In: Proceedings of the 10th Innovations in Theoretical Computer Science Conference (ITCS). LIPIcs, vol. 124, pp. 1–19. Schloss Dagstuhl - Leibniz-Zentrum fuer Informatik (2019)
13. Jansen, K., Maack, M., Mäcker, A.: Scheduling on (un-)related machines with setup times. In: Proceedings of the 2019 IEEE International Parallel and Distributed Processing Symposium (IPDPS), pp. 145–154. IEEE Computer Society (2019)

14. Lawler, E.L.: Sequencing jobs to minimize total weighted completion time subject to precedence constraints. Ann. Discrete Math. **2**, 75–90 (1978)
15. Lenstra, J.K., Kan, A.H.G.R.: Complexity of scheduling under precedence constraints. Oper. Res. **26**(1), 22–35 (1978)
16. Leung, J.Y., Li, H., Pinedo, M.: Order scheduling models: an overview. In: Kendall, G., Burke, E.K., Petrovic, S., Gendreau, M. (eds.) Multidisciplinary Scheduling: Theory and Applications, pp. 37–53. Springer, Boston (2005). https://doi.org/10.1007/0-387-27744-7_3
17. Mäcker, A., Meyer auf der Heide, F., Pukrop, S.: Approximating weighted completion time for order scheduling with setup times. arXiv e-prints arXiv:1910.08360, October 2019
18. Monma, C.L., Potts, C.N.: On the complexity of scheduling with batch setup times. Oper. Res. **37**(5), 798–804 (1989)
19. Ng, C.T., Cheng, T.C.E., Yuan, J.J.: Strong NP-hardness of the single machine multi-operation jobs total completion time scheduling problem. Inf. Process. Lett. **82**(4), 187–191 (2002)
20. Smith, W.E.: Various optimizers for single-stage production. Naval Res. Logistics Q. **3**(1–2), 59–66 (1956)
21. Woeginger, G.J.: On the approximability of average completion time scheduling under precedence constraints. Discrete Appl. Math. **131**(1), 237–252 (2003)

Bounds for the Number of Tests in Non-adaptive Randomized Algorithms for Group Testing

Nader H. Bshouty[1], George Haddad[2], and Catherine A. Haddad-Zaknoon[1](✉)

[1] Technion, Haifa, Israel
{bshouty,catherine}@cs.technion.ac.il
[2] The Orthodox Arab College, Grade 11, Haifa, Israel
haddadgeorge9@gmail.com

Abstract. We study the group testing problem with non-adaptive randomized algorithms. Several models have been discussed in the literature to determine how to randomly choose the tests. For a model $\mathcal{M}$, let $m_{\mathcal{M}}(n,d)$ be the minimum number of tests required to detect at most d defectives within n items, with success probability at least $1-\delta$, for some constant δ. In this paper, we study the measures

$$c_{\mathcal{M}}(d) = \lim_{n\to\infty} \frac{m_{\mathcal{M}}(n,d)}{\ln n} \text{ and } c_{\mathcal{M}} = \lim_{d\to\infty} \frac{c_{\mathcal{M}}(d)}{d}.$$

In the literature, the analyses of such models only give upper bounds for $c_{\mathcal{M}}(d)$ and $c_{\mathcal{M}}$, and for some of them, the bounds are not tight. We give new analyses that yield tight bounds for $c_{\mathcal{M}}(d)$ and $c_{\mathcal{M}}$ for all the known models $\mathcal{M}$.

Keywords: Group testing · Randomized algorithms · Non-adaptive algorithms

1 Introduction

Group testing is a strategy to identify d *defective* items from a pile of n elements by testing groups of items rather than testing each one individually. A group test is identified by a subset of items. The test response is *positive* if it includes at least one defective item, and *negative* otherwise. The problem of group testing is the task of identifying all the d items with a minimum number of group tests.

Formally, let $S=[n]:=\{1,2,\ldots,n\}$ be the set of the n *items* and let $I\subseteq S$ be the set of *defective items*. Suppose that we know that the number of defective items, $|I|$, is bounded by some integer d. A *test* is a set $J\subset S$. The answer to the test is $T(I,J)=1$ if $I\cap J\neq\emptyset$ and 0 otherwise. The problem is to find the defective items with a minimum number of tests.

Although the group testing scheme was originally introduced as a potential solution for an economical mass blood testing during WWII [9], many researchers

A. Chatzigeorgiou et al. (Eds.): SOFSEM 2020, LNCS 12011, pp. 101–112, 2020.
https://doi.org/10.1007/978-3-030-38919-2_9

have suggested applying this approach in a variety of practical problems. Du and Hwang [11], for example, outline a wide range of applications in DNA screening that involve group testing. On the other hand, Wolf [26] presents an applicable group testing generalization to the random access communications problem. For a brief history and other applications, the reader is referred to [2,6,7,10,11,14, 15,20,21,23].

Generally, the algorithm operates in *stages* or *rounds*. In each round, the tests are defined in advance and are tested in a single parallel step. Tests in some round might depend on the answers of the previous rounds. An algorithm that includes only one stage is called a *non-adaptive algorithm*, while a multi-stage algorithm is called an *adaptive algorithm*.

Since tests might be time consuming, in most practical applications, performing the tests simultaneously is highly required. Therefore, non-adaptive algorithms are extremely desirable in practice. It is well known, however, that any non-adaptive *deterministic* algorithm must do at least $\Omega(d^2 \log n/ \log d)$ tests [1,13,22,24]. This is $O(d/ \log d)$ times more than the number of tests of the folklore non-adaptive randomized algorithm that does only $O(d \log n)$ tests. Due to their reduced number of tests, randomized non-adaptive algorithms for group testing have drawn the attention of researchers for the past few decades, and many algorithms have been proposed [3,4,11,12,16,17].

The set of tests in any non-adaptive deterministic (resp. randomized) algorithm can be identified with a binary (resp. random) $m \times n$ test matrix M (also called pool design). Each row in M corresponds to an assignment $a = (a_1, \cdots, a_n) \in \{0,1\}^n$ where $a_i = 1$ if and only if $i \in J$ or equivalently, the ith item participates in the test defined by the subset J. For random algorithms, the following models are studied in the literature for constructing an $m \times n$ random test matrix M.

1. *Random incidence design* (RID algorithms). The entries in M are chosen randomly and independently to be 0 with probability p and 1 with probability $1 - p$.
2. *Random r-size design* (RrSD algorithms). The rows in M are chosen randomly and independently from the set of all vectors $\{0,1\}^n$ of weight r.
3. *Random s-set design* (RsSD algorithms). The columns in M are chosen randomly and independently from the set of all vectors $\{0,1\}^m$ of weight s.
4. *Uniform Transversal Design with alphabet of size q* (UTDq algorithms) A design matrix M is called *transversal* if the rows of M can be divided into disjoint families, where each family is a partition of all items [11]. A well known method for constructing transversal designs is using a *q-ary matrix*. A q-ary matrix M' is a matrix over the alphabet $\Sigma = \{1, \ldots, q\}$, for some fixed $2 \le q \in [n]$. Transforming a q-ary matrix M' to a binary matrix M is as follows. Each row r in M' is translated to q binary rows in M. For each $\sigma \in \Sigma$, replace each entry that is equal to σ by 1 and the others convert to 0. Therefore, if the matrix M' is of dimension $m' \times n$, then M is an $m \times n$ binary matrix where $m = qm'$. We say that a $q-$ary matrix M' (UTDq algorithms) is a *uniform random* matrix if its entries are chosen randomly and independently

to be any symbol of the alphabet with probability $1/q$. We say that M is a *q- transversal random matrix*, if there is a uniform random q−ary matrix M' such that M is derived from M' according to the previous procedure.

One advantage of RID and RrSD algorithms over RsSD and UTDq algorithms is that, in parallel machines, the tests can be generated by different processors (or laboratories) without any communication between them. In those models all the machines use the same distribution, draw a sample and perform the test. Those algorithms are called *strong non-adaptive* in the sense that the rows of the matrix M can also be non-adaptively generated in one parallel step.

For a model $\mathcal{M}$, let $m_{\mathcal{M}}(n,d)$ be the minimum number of tests of n items with at most d defective items that is required in order to ensure success (finding the defective items) with probability at least $1-\delta$, for some constant δ. In this paper, we study the constant $c_{\mathcal{M}}$ for each model $\mathcal{M}$, where $c_{\mathcal{M}}$ is defined as follows:

$$c_{\mathcal{M}}(d) = \lim_{n\to\infty} \frac{m_{\mathcal{M}}(n,d)}{\ln n} \text{ and } c_{\mathcal{M}} = \lim_{d\to\infty} \frac{c_{\mathcal{M}}(d)}{d}.$$

To the best of our knowledge, there has been little discussion about any non-trivial lower bound on the number of tests required in a non-adaptive randomized algorithm for group testing [8]. Moreover, the analyses of the previous models known in the literature give only upper bounds for $c_{\mathcal{M}}(d)$ and $c_{\mathcal{M}}$, and some of these bounds are not tight. For some models, the used techniques do not even lead to an upper bound, and other relaxed measures are examined, such as the expected number of non-defective items that are eliminated after each test.

The objective of this paper is to establish lower and upper bounds on the number of tests required for a non-adaptive randomized algorithm to identify d defectives among n items with a success probability at least $1-\delta$. We develop new techniques that give tight bounds for $c_{\mathcal{M}}(d)$ and $c_{\mathcal{M}}$ over the models: $\mathcal{M} =$RID, RrSD, RsSD and UTDq.

1.1 Old and New Results

Let M be an $m \times n$ test matrix. Let $I \subseteq S = [n]$, $|I| \le d$ be the set of defective items. Let $T(I,M)$ denote the vector of answers to the tests (rows of M), that is, $T(I,M) := \vee_{i\in I} M^{(i)}$ where $\vee$ is bit-wise "or" and $M^{(i)}$ is the ith column of M. A matrix M is called (n,I)*-separable* if for every $J \subseteq [n]$, $|J| \le d$ and $J \ne I$, we have $T(J,M) \ne T(I,M)$. That is, the only set J of up to d items that is consistent with the answers of the tests $\vee_{i\in I} M^{(i)}$ is I.

While the separability property is obviously sufficient to guarantee identifying the defectives successfully, unfortunately, the analysis of such property seems to be very involved [5]. Therefore, a more relaxed property is required. A matrix M is called (n,I)*-disjunct* with respect to some subset $I \subset [n], |I| \le d$, if for each $i \notin I$, there is a test that contains it but does not contain any of the defective items. Formally, for any $i \notin I$, there is a row $t \in [m]$ such that $M_{t,i} = 1$ and for all $j \in I$, $M_{t,j} = 0$. Since no defective item participates in such test, the response of the oracle on it will be negative (0) and hence, is a witness for the fact that the item i is not defective.

If the test matrix M is (n, I)-disjunct, the decoding algorithm reveals the defective items according to the following procedure. It starts with a set $X = S$. After making all the tests defined by M, for every negative answer of a row a in M, it removes from X all the items i where $a_i = 1$. Since M is (n, I)-disjunct, all the non-defective items are guaranteed to have a test that eliminates them from X. Therefore, X will eventually contain the defective items only. This can be done in linear time in the size of M.

The folklore non-adaptive randomized algorithm randomly chooses M such that, for any set of at most d defective items I, with probability at least $1 - \delta$, M is (n, I)-disjunct, and then applies the previous algorithm to identify the defective items. This is why the property of *disjunction* is well studied in the literature [3,4,11,16–19]. It is well known (and very easy to see) that if M is (n, I)-disjunct then M is (n, I)-separable.

Let δ be some constant. For a model $\mathcal{M}$, let $m^D_{\mathcal{M}}(n, d)$ be the minimum number of tests that is required in order to ensure that for any set of at most d defective items I, with probability at least $1-\delta$, the test matrix is (n, I)-disjunct. We define,

$$c^D_{\mathcal{M}}(d) = \lim_{n\to\infty} \frac{m^D_{\mathcal{M}}(n, d)}{\ln n} \text{ and } c^D_{\mathcal{M}} = \lim_{d\to\infty} \frac{c^D_{\mathcal{M}}(d)}{d}.$$

Since an (n, I)-disjunct matrix is (n, I)-separable, for every model $\mathcal{M}$, we have

$$c_{\mathcal{M}}(d) \le c^D_{\mathcal{M}}(d). \tag{1}$$

Consider $m_{\mathcal{M}}(n, d+1)$ random tests in the model $\mathcal{M}$ and their corresponding matrix M. Let I be any set of size d. Then, with probability at least $1 - \delta$, M is (n, I)-separable and therefore for any $j \notin I$ we have $\vee_{i\in I\cup\{j\}} M^{(i)} \ne \vee_{i\in I} M^{(i)}$. Notice that $\vee_{i\in I\cup\{j\}} M^{(i)} \ne \vee_{i\in I} M^{(i)}$ implies that there is a row a of M that satisfies $a_i = 0$ for all $i \in I$ and $a_j = 1$. Therefore, with probability at least $1-\delta$, M is (n, I)-disjunct. Thus, $m^D_{\mathcal{M}}(n, d) \le m_{\mathcal{M}}(n, d + 1)$ and

$$c_{\mathcal{M}}(d + 1) \ge c^D_{\mathcal{M}}(d). \tag{2}$$

Since $c_{\mathcal{M}}(d), c^D_{\mathcal{M}}(d) = O(d)$ and from (1) and (2) it follows that $c_{\mathcal{M}} = c^D_{\mathcal{M}}$. The best lower bound for $c_{\mathcal{M}}$ is $c_{\mathcal{M}} \ge 1/\ln 2$ and $c_{\mathcal{M}}(d), c^D_{\mathcal{M}}(d) \ge d/\ln 2$. This bound follows from the trivial information-theoretic lower bound $d \log n = (d/\ln 2)\ln n$.

Sebö, [25], studies the RID model for the simple case when the number of defective items is *exactly* d. He shows that the best probability for the random test matrix (i.e., that gives a minimum number of tests) is $p = 1 - 1/2^{1/d}$ and, in this case, the number of tests meets the information-theoretic lower bound.

The general case is studied in [3–5,11,12,16,17]. The technique used in most of the studies relies on obtaining the probability that maximizes the expected number of items eliminated in one test. In [3], it is shown that for the RID model this probability is $p = 1 - 1/(d+1)$. Moreover, it is shown that using this probability $c^D_{\rm RID}(d) \le ed + (e-1)/2 + O(1/d)$ and therefore $c_{\rm RID} = c^D_{\rm RID} \le e$. [3,5]. In their work, Bshouty et. al. [5] study the separability property and show that $c_{\rm RID}(d) \le ed - (e+1)/2 + O(1/d)$.

In this paper we give lower and upper bounds on the number of tests required by any non-adaptive randomized group testing algorithm when tests are chosen according to the RID, RrSD, RsSD and UTDq models. For random designs selected according to the RID model, we show that

$$c^D_{\rm RID}(d) = ed, \qquad \text{and therefore,} \qquad c_{\rm RID} = e = 2.718.$$

The optimal probability that derives this result is $p = e^{-1/d}$. This, in particular, shows that finding the probability that maximizes the expected number of items that are eliminated in one test does not necessarily give the probability that minimizes the total number of tests. Considering the RrSD model, we prove that

$$c^D_{\rm RrSD}(d) = c^D_{\rm RID}(d) = ed, \qquad \text{and} \qquad c_{\rm RrSD} = c_{\rm RID} = e = 2.718.$$

Moreover, for the RsSD model, we show that

$$c^D_{\rm RsSD}(d) = \frac{1}{(\ln 2)\max_{0<\alpha\le 1}\left(H(\alpha) - \beta H\left(\frac{\alpha}{\beta}\right)\right)} \qquad \text{and} \qquad c_{\rm RsSD} = \frac{1}{(\ln 2)^2} = 2.081$$

where $\beta = 1 - (1-\alpha)^d$. Regarding the UTDq model, we prove that

$$c^D_{\rm UTDq}(d) = \min_q \frac{q}{-\ln P_{q,d}} \qquad \text{and} \qquad c_{\rm UTDq} = \frac{1}{(\ln 2)^2},$$

where

$$P_{q,d} = \left(\prod_{i=1}^{d}\left(\frac{i}{q}\right)^{R_{q,d,i}}\right)^{1/q^d},$$

and $R_{q,d,i}$ is the number of strings in $[q]^d$ that contains exactly i symbols.

In addition, for small d, Table 1 outlines the values of $c^D_{\mathcal{M}}(d)/d$ across the above four models.

2 The RID Model

In this section, we study the Random Incidence Design (RID algorithms). We recall that in this model, the entries in M are chosen randomly and independently to be 0 with probability p, and 1 with probability $1-p$. We prove:

Theorem 1. *We have,* $c^D_{\rm RID}(d) = ed$ *and* $c_{\rm RID} = e$.

In Lemma 1, by choosing $p = e^{-\frac{1}{d}}$, we develop an upper bound on the number of queries required for the group testing problem when M is designed according to the RID model. For this choice of p, we establish, in Lemma 2, a lower bound that shows that this choice of p gives the minimum number of tests in this model.

As in the introduction, let the set I denote defective items set. We say that a row i in M is a *good* row if for every $j \in I$, we have $M_{i,j} = 0$.

Table 1. Leading constant of $d\ln(n)$ for small d and $n \to \infty$ for RID, RrSD, RsSD, and UTDq models.

d	RID	RrSD	RsSD	UTDq
2	2.718	2.718	1.95	2.417
3	2.718	2.718	1.96	2.31
4	2.718	2.718	1.992	2.225
5	2.718	2.718	2.01	2.221
6	2.718	2.718	2.02	2.198
7	2.718	2.718	2.03	2.182
8	2.718	2.718	2.04	2.17
9	2.718	2.718	2.044	2.16
10	2.718	2.718	2.05	2.152
$\to\infty$	$e = 2.718$	$e = 2.718$	$\frac{1}{\ln^2 2} = 2.081$	$\frac{1}{\ln^2 2} = 2.081$

Lemma 1. *Let M be an $m \times n$ RID matrix with $p = e^{-\frac{1}{d}}$ where*

$$m - \sqrt{2em\ln\frac{2}{\delta}} = ed\ln\frac{2n}{\delta}. \tag{3}$$

Then, for any I of size at most d, with probability at least $1-\delta$, M is an (n, I)-disjunct matrix. In particular,

$$m = ed\ln\frac{2n}{\delta} + \Theta\left(\sqrt{d\ln\frac{n}{\delta}\ln\frac{1}{\delta}}\right), \qquad c^D_{\mathrm{RID}}(d) \le ed \quad \text{and} \quad c_{\mathrm{RID}} \le e.$$

Proof. For any $1 \le i \le m$, let X_i be a random variable that is equal to 1 if the row i in M is a good row and 0 otherwise. The probability that a row i in M is a good row is $(e^{-1/d})^d = e^{-1}$. Let $X = X_1 + \cdots + X_m$, be the number of the good rows in M. Then, $\mathbf{E}[X_i] = \Pr[X_i = 1] = e^{-1}$ and $\mu := \mathbf{E}[X] = e^{-1}m$. Let $m' = ed\ln(2n/\delta)$. Let A be the event indicating that the number of the good rows in M is less than $m'p^d = m'/e$. Let $T \subseteq [m]$ be the set of the good row indexes in M. That is, for each row $t \in T$, $M_{t,j} = 0$ for all $j \in I$. Let B be the event indicating that there is a column $j \notin I$ in M such that $M_{t,j} = 0$ for all $t \in T$, and therefore, M is not (n, I)-disjunct. Then we can say that,

$$\Pr[B|\overline{A}] \le (n-d)p^{m'p^d} \le ne^{-\frac{m'}{ed}} = \frac{\delta}{2}. \tag{4}$$

Using Chernoff bound, for m as specified in (3) and since $\mu = e^{-1}m$ we can conclude,

$$\Pr[A] = \Pr\left[X < \frac{m'}{e}\right] = \Pr\left[X < \left(1 - \left(1 - \frac{m'}{m}\right)\right)\mu\right] \le e^{-\frac{(1-\frac{m'}{m})^2 m}{2e}} = \frac{\delta}{2}. \tag{5}$$

Using (4) and (5) we get,

$$\Pr[B] = \Pr[B|A]\Pr[A] + \Pr[B|\overline{A}]\Pr[\overline{A}] \le \Pr[A] + \Pr[B|\overline{A}] \le \delta. \square$$

Lemma 2. *Let M be an $m \times n$ RID matrix with probability $0 \le p \le 1$. If*

$$m = ed\ln n - e^3\sqrt{3m},$$

then with probability at least $1/3$, M is not (n, I)-disjunct.

In particular, to have success probability at least $1/3$, we must have

$$m \ge ed\ln n - \Theta(\sqrt{d\ln n}).$$

Therefore,

$$c_{\rm RID}^{D}(d) \ge ed \quad \textit{and} \quad c_{\rm RID} \ge e.$$

Proof. Consider $c = d\ln(1/p)$. Then $p = e^{-c/d}$. Let $m' = ed\ln n$ and $w = e^3\sqrt{3m}$. Then $m = m' - w$. Let X be a random variable that is equal to the number of good rows in M. Then $\mu := \mathbf{E}[X] = e^{-c}m$. Let F be the event that M is not (n, I)-disjunct. The probability that M is (n, I)-disjunct is the probability that in every column that corresponds to a non-defective item, not all the entries of the good rows are zero. Therefore, given that $X = x$, it is easy to see that $\Pr[\overline{F}|X = x] = (1 - p^x)^{n-|I|}$. We distinguish between two cases: $c \ge 6$ and $c \le 6$.

Case I. $c \le 6$. By Chernoff bound, and since $ce^{1-c} \le 1$ for every $c \le 6$,

$$\begin{aligned}
\Pr[F] &\ge \Pr[F \wedge X \le e^{-c}m'] = \Pr[F \mid X \le e^{-c}m'] \cdot \Pr[X \le e^{-c}m'] \\
&\ge \left(1 - \left(1 - p^{e^{-c}m'}\right)^{n-d}\right)\left(1 - \Pr\left[X \ge e^{-c}m'\right]\right) \\
&\ge \left(1 - \left(1 - n^{-ce^{1-c}}\right)^{n-d}\right)\left(1 - \Pr\left[X \ge e^{-c}m\left(1 + \frac{w}{m}\right)\right]\right) \\
&\ge \left(1 - \left(1 - n^{-ce^{1-c}}\right)^{n-d}\right)\left(1 - e^{-\frac{e^{-c}w^2}{3m}}\right) \\
&\ge \left(1 - \left(1 - \frac{1}{n}\right)^{n-d}\right)\left(1 - e^{-1}\right) = \left(1 - e^{-1+o(1)}\right)\left(1 - e^{-1}\right) > \frac{1}{3},
\end{aligned}$$

where the last inequality is correct due to the fact that $d = o(n)$.

Case II. $c \ge 6$. By Markov bound, since $m \le ed\ln n$ and $2ce^{1-c} < 0.1$, then for $c \ge 6$ and $d = o(n)$ we have,

$$\begin{aligned}
\Pr[F] &\ge \Pr[F \wedge X \le 2e^{-c}m] = \Pr[F \mid X \le 2e^{-c}m] \cdot (1 - \Pr[X \ge 2e^{-c}m]) \\
&\ge \left(1 - \left(1 - p^{2e^{-c}m}\right)^{n-d}\right) \cdot \frac{1}{2} \ge \frac{1}{2}\left(1 - \left(1 - p^{2e^{-c}ed\ln n}\right)^{n-d}\right) \\
&\ge \frac{1}{2} \cdot \left(1 - \left(1 - n^{-2ce^{1-c}}\right)^{n-d}\right) \ge \frac{1}{2} \cdot \left(1 - \left(1 - \frac{1}{n^{0.1}}\right)^{n-d}\right) > \frac{1}{3}. \square
\end{aligned}$$

3 The RrSD Model

In this section we study the Random r-Size Design (RrSD algorithms). As defined previously, in this model, the rows in M are chosen randomly and independently from the set of all vectors $\{0,1\}^n$ of weight r.

Theorem 2. *We have,* $c^D_{\mathrm{RrSD}}(d) = ed$ *and* $c_{\mathrm{RrSD}} = e$.

In Lemma 3, we give an upper bound using $r = (1 - e^{-\frac{1}{d}})n$. In Lemma 4, we establish a lower bound that shows that this choice of r gives the minimum number of tests for this model. The proof of both lemmas is very similar to the upper and lower bound proofs of the RID model, and are given in details in the full version of the paper.

Lemma 3. *Let M be an $m \times n$ RrSD matrix with $r = (1-p)(n-d+1)$, where $p = e^{-\frac{1}{d}}$ and*

$$m - \sqrt{2em \ln \frac{2}{\delta}} = ed \ln \frac{2n}{\delta}.$$

Then, for any I of size at most d, with probability at least $1-\delta$, M is an (n, I)-disjunct matrix. In particular,

$$m = ed \ln \frac{2n}{\delta} + \Theta\left(\sqrt{d \ln \frac{n}{\delta} \ln \frac{1}{\delta}}\right), \quad c^D_{\mathrm{RrSD}}(d) \le ed \quad \textit{and} \quad c_{\mathrm{RrSD}} \le e.$$

Lemma 4. *Let M be an $m \times n$ RrSD matrix where $0 \le r \le n$ and $d < n^{1/2}/\ln^3 n$. If*

$$m = ed\ln(n/e) - e^3\sqrt{3m},$$

then, with probability at least $1/3$, M is not (n, I)-disjunct. In particular, to have success probability at least $1/3$, we must have

$$m \ge ed \ln n - \Theta(d + \sqrt{d \ln n}).$$

And therefore,

$$c^D_{\mathrm{RrSD}}(d) \ge ed \quad \textit{and} \quad c_{\mathrm{RrSD}} \ge e.$$

4 The RsSD Model

In this section we study the Random s-Set Design (RsSD algorithms). We recall that in this model, the columns of M are chosen randomly and independently from the set of all vectors $\{0,1\}^m$ of weight s.

We prove,

Theorem 3. *We have,*

$$c^D_{\mathrm{RsSD}}(d) = \frac{1}{(\ln 2) \max_{0<x\le d}\left(H(\alpha) - \beta H\left(\frac{\alpha}{\beta}\right)\right)} \quad \textit{and} \quad c_{\mathrm{RsSD}} = \frac{1}{(\ln 2)^2}$$

where $\alpha = x/d$ and $\beta = 1-(1-\alpha)^d$.

An upper bound for this model is established in Lemma 5 using $s = \ln 2 \cdot m/d$. Moreover, in Lemma 6, a lower bound is developed indicating that this choice of s gives the minimum number of tests for this model. For detailed proof of both lemmas, the reader is referred to the full version of the paper.

Lemma 5. *Let M be an $m \times n$ RsSD matrix with $\alpha = \frac{s}{m} = \frac{x}{d}$ and $\beta = 1-(1-\alpha)^d$ where $0 < x \le d$ is any real number. Let*

$$m' = \frac{\ln n + \ln \frac{3}{\delta} + \frac{1}{2}\ln\left(\frac{\beta(1-\alpha)}{\beta-\alpha}\right)}{(\ln 2)\left(H(\alpha) - \beta H\left(\frac{\alpha}{\beta}\right)\right)},$$

and $\lambda = \frac{2}{\sqrt{\delta(1-\alpha)^d m}} < 1$. Then, for $m = (1+\lambda)m'$, with probability at least $1-\delta$, M is an (n, I)-disjunct matrix. In particular

$$c^D_{\text{RsSD}}(d) \le \frac{1}{(\ln 2)\max_{0<x\le d}\left(H(\alpha) - \beta H\left(\frac{\alpha}{\beta}\right)\right)} \quad \text{and} \quad c_{\text{RsSD}} \le \frac{1}{(\ln 2)^2}.$$

The lower bound for this model is given in the following lemma.

Lemma 6. *Let M be an $m \times n$ RsSD matrix with $\alpha = s/m = \frac{x}{d}$ for any real number $0 < x \le d$, and let $\beta = 1-(1+\lambda)(1-\alpha)^d$, where $\lambda < 1/10$ is any small constant. If*

$$m = \frac{\ln n + \ln 2 + \frac{1}{2}\ln\left(\frac{\beta(1-\alpha)}{\beta-\alpha}\right)}{\left(H(\alpha) - \beta H\left(\frac{\alpha}{\beta}\right)\right)\ln 2},$$

then, with probability at least $3\lambda/16$, M is not (n, I)-disjunct. In particular

$$c^D_{\text{RsSD}}(d) \ge \frac{1}{(\ln 2)\max_{0<x\le d}\left(H(\alpha) - \beta H\left(\frac{\alpha}{\beta}\right)\right)} \quad \text{and} \quad c_{\text{RsSD}} \ge \frac{1}{(\ln 2)^2}.$$

5 Random Uniform Transversal Design Model

In this section we develop bounds for the UTDq design. For ease of the analysis, we assume that $d \le q$. In the full paper, we show that all the results are also true for any $q \ge 1$. For $d \le q$, we define

$$P_{q,d} = \left(\prod_{i=1}^{d}\left(\frac{i}{q}\right)^{R_{q,d,i}}\right)^{1/q^d},$$

where $R_{q,d,i}$ is the number of strings in $[q]^d$ that contains exactly i symbols. It is easy to see that

$$R_{q,d,i} = \binom{q}{i} N_{d,i},$$

where $N_{d,i}$ is the number of strings of length d over the alphabet $\Sigma_i := \{1, 2, \ldots, i\}$ that contains all the symbols in Σ_i.

In this section we prove

Theorem 4. *We have*

$$c^D_{\mathrm{UTDq}}(d) = \min_q \frac{q}{-\ln P_{q,d}} \quad \textit{and} \quad c_{\mathrm{UTDq}} = \frac{1}{(\ln 2)^2}.$$

For any set $K \subseteq [n]$, let $S_{i,K}(M) = \{M_{i,t} | t \in K\}$ be the set of the symbols that appear in the entries that correspond to the row i and the columns of K. Throughout this section, we will assume, w.l.o.g., that the set of the defective item is $I = [d]$. The following lemmas are proved in the full version of the paper. We start with:

Lemma 7. *Let M be an $m \times n$ q-transversal random matrix. The probability that M is not $(n, [d])$-disjunct is*

$$1 - \left(1 - \prod_{i=1}^{m'} \frac{|S_{i,[d]}(M')|}{q}\right)^{n-d}.$$

The following lemma provides an upper bound on $c^D_{\mathrm{UTDq}}(d)$.

Lemma 8. *Let M be an $m \times n$ q-transversal random matrix where*

$$m = \frac{1}{(1-\lambda)} \frac{q \ln(2n/\delta)}{-\ln P_{q,d}} = \frac{q \ln(2n/\delta)}{-\ln P_{q,d}} + o(\ln n), \tag{6}$$

and $\lambda = \sqrt{(2q^{d+1}/m)\ln(2q^d/\delta)}$. Then, with probability at least $1 - \delta$, M is $(n, [d])$−disjunct. Therefore,

$$c^D_{\mathrm{UTDq}}(d) \le \min_q \frac{q}{-\ln P_{q,d}}.$$

We now prove a lower bound on $c^D_{\mathrm{UTDq}}(d)$.

Lemma 9. *Let M be an $m \times n$ q-transversal random matrix. Let*

$$m = \frac{1}{(1+\lambda)} \frac{q \ln(8(n-d))}{-\ln P_{q,d}} = \frac{q \ln n}{-\ln P_{q,d}} + o(\ln n) \tag{7}$$

where $\lambda = \sqrt{(3q^{d+1}/m)\ln(16q^d)}$. Then, with probability at least $3/4$, M is not $(n, [d])$−disjunct. Therefore,

$$c^D_{\mathrm{UTDq}}(d) \ge \min_q \frac{q}{-\ln P_{q,d}}.$$

It is not clear, however, how to compute the $c^D_{\mathrm{UTDq}}(d)/d$ when $d \to \infty$ in order to get c_{UTDq}. The following two lemmas give a different analysis that enables us to approximate $c^D_{\mathrm{UTDq}}(d)$, and then to compute c_{UTDq}. The proofs of both lemmas are given in details in the full version of the paper.

Lemma 10. *Let M be an $m \times n$ q-transversal random matrix where*

$$m = \frac{q\ln(n/\delta)}{-\ln\left(1-\left(1-\frac{1}{q}\right)^d\right)}. \tag{8}$$

Then, with probability at least $1-\delta$, M is an $(n,[d])$-disjunct matrix. In particular,

$$c^D_{\mathrm{UTDq}}(d) \le \min_q \frac{q}{-\ln\left(1-\left(1-\frac{1}{q}\right)^d\right)} \quad \textit{and} \quad c_{\mathrm{UTDq}} \le \frac{1}{(\ln 2)^2}.$$

In the following Lemma 11, we prove a tight lower bound for c_{UTDq}.

Lemma 11. *Let be an $m \times n$ q-transversal random matrix where*

$$m = \frac{q\ln((n-d)/2)}{-\ln\left(1-\left(1-\frac{1}{q}\right)^d - O(q^{-1/3})\right)}.$$

Then, with probability at least $1/4$, the matrix M is not $(n,[d])$-disjunct.
In particular, we have

$$c_{\mathrm{UTDq}} \ge \frac{1}{\ln^2(2)}.$$

References

1. Rykov, V.V., D'yachkov, A.G.: Bounds on the length of disjunctive codes. Probl. Peredachi Inf. **18**, 7–13 (1982)
2. Angluin, D.: Queries and concept learning. Mach. Learn. **2**(4), 319–342 (1987)
3. Balding, D.J., Bruno, W.J., Torney, D.C., Knill, E.: A comparative survey of non-adaptive pooling designs. In: Speed, T., Waterman, M.S. (eds.) Genetic Mapping and DNA Sequencing. IMA, vol. 81, pp. 133–154. Springer, New York (1996). https://doi.org/10.1007/978-1-4612-0751-1_8
4. Bruno, W.J., et al.: Efficient pooling designs for library screening. Genomics **26**(1), 21–30 (1995)
5. Bshouty, N.H., Diab, N., Kawar, S.R., Shahla, R.J.: Non-adaptive randomized algorithm for group testing. In International Conference on Algorithmic Learning Theory, ALT 2017, 15–17 October 2017, Kyoto University, Kyoto, Japan, pp. 109–128 (2017)
6. Cicalese, F.: Group testing. Fault-Tolerant Search Algorithms. MTCSAES, pp. 139–173. Springer, Heidelberg (2013). https://doi.org/10.1007/978-3-642-17327-1_7
7. Cormode, G., Muthukrishnan, S.: What's hot and what's not: tracking most frequent items dynamically. ACM Trans. Database Syst. **30**(1), 249–278 (2005)
8. Damaschke, P., Muhammad, A.S.: Randomized group testing both query-optimal and minimal adaptive. In: Bieliková, M., Friedrich, G., Gottlob, G., Katzenbeisser, S., Turán, G. (eds.) SOFSEM 2012. LNCS, vol. 7147, pp. 214–225. Springer, Heidelberg (2012). https://doi.org/10.1007/978-3-642-27660-6_18

9. Dorfman, R.: The detection of defective members of large populations. Ann. Math. Stat. **14**(4), 436–440 (1943)
10. Du, D.-Z., Hwang, F.K.: Combinatorial Group Testing and Its Applications. World Scientfic Publishing, Singapore (1993)
11. Du, D.-Z., Hwang, F.K.: Pooling Designs and Nonadaptive Group Testing: Important Tools for DNA Sequencing. World Scientfic Publishing, Singapore (2006)
12. Erdös, P., Rényi, A.: On two problems of information theory, pp. 241–254 (1963). Publications of the Mathematical Institute of the Hungarian Academy of Sciences
13. Füredi, Z.: On r-cover-free families. J. Comb. Theory Ser. A **73**(1), 172–173 (1996)
14. Hong, E.S., Ladner, R.E.: Group testing for image compression. IEEE Trans. Image Process. **11**(8), 901–911 (2002)
15. Hwang, F.K.: A method for detecting all defective members in a population by group testing. J. Am. Stat. Assoc. **67**(339), 605–608 (1972)
16. Hwang, F.K.: Random k-set pool designs with distinct columns. Probab. Eng. Inf. Sci. **14**(1), 49–56 (2000)
17. Hwang, F.K., Liu, Y.C.: The expected numbers of unresolved positive clones for various random pool designs. Probab. Eng. Inf. Sci. **15**(1), 57–68 (2001)
18. Hwang, F.K., Liu, Y.C.: A general approach to compute the probabilities of unresolved clones in random pooling designs. Probab. Eng. Inf. Sci. **18**(2), 161–183 (2004)
19. Hwang, F.K., Liu, Y.: Random pooling designs under various structures. J. Comb. Optim. **7**, 339–352 (2003)
20. Kautz, W., Singleton, R.: Nonrandom binary superimposed codes. IEEE Trans. Inf. Theory **10**(4), 363–377 (1964)
21. Macula, A.J., Popyack, L.J.: A group testing method for finding patterns in data. Discrete Appl. Math. **144**(1), 149–157 (2004). Discrete Mathematics and Data Mining
22. Porat, E., Rothschild, A.: Explicit nonadaptive combinatorial group testing schemes. IEEE Trans. Inf. Theory **57**(12), 7982–7989 (2011)
23. Ngo, H.Q., Du, D.-Z.: A survey on combinatorial group testing algorithms with applications to DNA library screening. Discrete Math. Theor. Comput. Sci. **55**, 171–182 (2000). DIMACS Series
24. Ruszinkó, M.: On the upper bound of the size of the r-cover-free families. J. Comb. Theory Ser. A **66**(2), 302–310 (1994)
25. Sebö, A.: On two random search problems. J. Stat. Plan. Inference **11**(1), 23–31 (1985)
26. Wolf, J.: Born again group testing: multiaccess communications. IEEE Trans. Inf. Theory **31**(2), 185–191 (1985)

Burning Two Worlds
Algorithms for Burning Dense and Tree-Like Graphs

Shahin Kamali, Avery Miller(✉), and Kenny Zhang

University of Manitoba, Winnipeg, MB, Canada
{shahin.kamali,avery.miller}@umanitoba.ca,
zhangyt3@myumanitoba.ca

Abstract. Graph burning is a model for the spread of social influence in networks. The objective is to measure how quickly a fire (e.g., a piece of fake news) can be spread in a network. The burning process takes place in discrete rounds. In each round, a new fire breaks out at a selected vertex and burns it. Meanwhile, the old fires extend to their adjacent vertices and burn them. A *burning schedule* selects where the new fire breaks out in each round, and the *burning problem* asks for a schedule that burns all vertices in a minimum number of rounds, termed the *burning number* of the graph. The burning problem is known to be NP-hard even when the graph is a tree or a disjoint set of paths. For connected graphs, it has been conjectured [3] that burning takes at most $\lceil \sqrt{n}\, \rceil$ rounds.

In this paper, we approach the algorithmic study of graph burning from two directions. First, we consider connected n-vertex graphs with minimum degree δ. We present an algorithm that burns any such graph in at most $\sqrt{\frac{24n}{\delta+1}}$ rounds. In particular, for graphs with $\delta \in \Theta(n)$, all vertices are burned in a constant number of rounds. More interestingly, even when δ is a constant that is independent of n, our algorithm answers the graph-burning conjecture in the affirmative by burning the graph in at most $\lceil \sqrt{n} \rceil$ rounds. Then, we consider burning connected graphs with bounded pathlength or treelength. This includes many graph families, e.g., interval graphs (pathlength 1) and chordal graphs (treelength 1). We show that any connected graph with pathlength pl and diameter d can be burned in $\lceil \sqrt{d-1} \rceil + pl$ rounds. Our algorithm ensures an approximation ratio of $1 + o(1)$ for graphs of bounded pathlength. We also give an algorithm that achieves an approximation ratio of $2 + o(1)$ for burning connected graphs of bounded treelength. Our approximation factors are better than the best known approximation factor of 3 for burning general graphs.

Keywords: Graph algorithms · Approximation algorithms · Graph burning problem · Social contagion · Pathlength · Treelength

1 Introduction

With the recent rapid growth of social networks, numerous approaches have been proposed to study social influence in these networks [7,11,16,17]. These studies

A. Chatzigeorgiou et al. (Eds.): SOFSEM 2020, LNCS 12011, pp. 113–124, 2020.
https://doi.org/10.1007/978-3-030-38919-2_10

focus on how fast a contagion can spread in a network. A contagion can be an emotional state or a piece of data such a political opinion, a piece of fake news, or gossip. Interestingly, the spread of a contagion does not require point-to-point communication. For example, an experimental study on Facebook suggests that users can experience different emotional states after being exposed to other users' posts, e.g., without direct communication and without their awareness [17].

Given the fact that a contagion is distributed without the active involvement and awareness of users, one can argue that it is merely defined by the structure of the underlying network [3]. A graph's *burning number* has been suggested as a parameter that measures how prone a social network is to the spread of a contagion, which is modeled via a set of fires. Given an undirected and unweighted graph that models a social network, the fires spread in the network in synchronous rounds in the following way. In round 1, a fire is initiated at a vertex; a vertex at which a fire is started is called an *activator*. In each round that follows, two events take place. First, all existing fires spread to their neighboring vertices, e.g., in round 2, the neighboring vertices of the first activator will be burned (i.e., they are now on fire). Second, a new fire can be started elsewhere in the network: a new vertex is selected as an activator at which a new fire is initiated. This continues until the first round in which all vertices are on fire, at which time we say the burning 'completes'. The choice of activators affects how quickly the burning process completes. A *burning schedule* specifies a *burning sequence* of vertices: the i'th vertex in the sequence is the activator in round i (Fig. 1).

The *burning number* of a graph G, denoted by $bn(G)$, is the minimum number of rounds required to complete the burning of G. The graph burning problem asks for a burning schedule that completes in $bn(G)$ rounds. Unfortunately, this problem is NP-hard even for simple graphs such as trees or disjoint sets of paths [1]. So, the focus of this paper is on algorithms that provide close-to-optimal solutions, that is, algorithms that burn graphs in a small (but not necessarily an optimal) number of rounds.

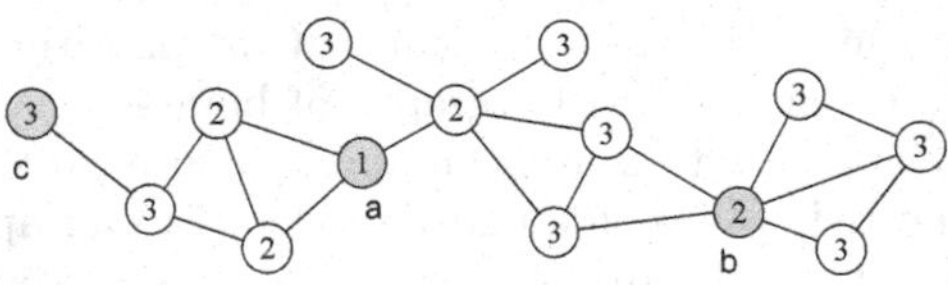

Fig. 1. Burning a graph using burning schedule $\langle a, b, c\rangle$. The number at each vertex x indicates the round at which the fire starts at x. The burning completes in 3 rounds.

Previous Work

The graph burning problem was introduced by Bonato et al. [3,4] as a way to model the spread of a contagion in social networks. Bonato et al. [3] proved that the burning number of any connected graph is at most $2\lceil\sqrt{n}\rceil - 1$, and conjectured that it is always at most $\lceil\sqrt{n}\rceil$. Land and Lu improved the upper bound to $\frac{\sqrt{6}}{2}\sqrt{n}$ [18]. The conjecture, known as graph burning conjecture, is still open but verified for basic graph families [6,9]. Bessy et al. [1] showed that the burning problem is NP-complete, and it remains NP-hard for simple graph families such as graphs with maximum degree three, spider graphs, and path forests. Recently, several heuristics were experimentally studied [28]. Bonato and Kamali [5] studied approximation algorithms for the problem. Using a simple algorithm inspired by the k-center problem (see, for example, [27]), they showed that there is a polynomial time algorithm that burns any graph G in at most $3bn(G)$ rounds. They also provided a 2-approximation algorithm for trees and a polynomial time approximate scheme (PTAS) for path-forests. A line of research has been focused on characterizing the burning number for different graph families. This includes grid graphs [2,23], Cartesian products and the strong products of graphs [22,23], binomial random graphs [22], random geometric graphs [22], spider graphs [1,6,9], path-forests [1,5,6], generalized Petersen graphs [26], and Theta graphs [20].

Our Contributions

In this paper, we approach the algorithmic study of graph burning from two directions. In Sect. 2, we consider dense connected graphs, i.e., graphs with lower bound δ on the minimum degree. We provide an algorithm that burns such graphs on n vertices in at most $\sqrt{\frac{24n}{\delta+1}}$ rounds. In particular, for dense graphs with $\delta \in \Theta(n)$, all vertices are burned in a constant number of rounds. More interestingly, even when δ is a sufficiently large constant that is independent of the graph size, our algorithm answers the graph-burning conjecture of Bonato et al. [3] in the affirmative by burning the graph in at most $\lceil\sqrt{n}\,\rceil$ rounds.

In Sect. 3, we provide parameterized algorithms for burning connected graphs with small pathlength and treelength. A graph has pathlength at most pl (respectively treelength at most tl), if there is a Robertson-Seymour path decomposition (respectively tree decomposition) of G such that the distance between any two vertices in the same bag of the decomposition is at most pl (respectively tl). Intuitively speaking, these are graphs that can be transformed into a path (respectively tree) by contracting groups of vertices that are all at close to each other. A formal definition can be found in Sect. 3. Graphs with small pathlength or treelength span several well-known families of graphs. For example, a graph is an interval graph if and only if its pathlength is at most 1 [13,19], and a chordal graph if and only if its treelength is at most 1 [12,19].

We provide algorithms that burn connected graphs of bounded pathlength and treelength. First, we observe that if the diameter is bounded by a constant, an optimal burning schedule can be computed in polynomial time using an exhaustive approach. So, we focus on a more interesting *asymptotic setting* where the diameter of the graph is asymptotically large. We show that any connected graph G of diameter d and pathlength at most pl can be burned in at most $\lceil\sqrt{d-1}\rceil + pl \leq \lceil\sqrt{n}\rceil + pl$ rounds. Since $\lceil\sqrt{d}\rceil$ is a lower bound for the burning number, our algorithm achieves $1 + o(1)$ approximation factor for connected graphs of bounded pathlength. In particular, our algorithm achieves an nearly-optimal solution for burning connected interval graphs. We also present an approximation algorithm for burning connected graphs of small treelength. For a graph with treelength at most tl, our algorithm has an approximation factor of at most $2 + (4tl + 1)/d$, which is $2 + o(1)$ for graphs of bounded treelength (e.g., chordal graphs). Our approximation factors are improvements over the best known approximation ratio of 3 for arbitrary graphs [5]. Due to space constraints, the full proofs appear in [15].

2 Dense Graphs

For any graph G, the *degree* of a vertex is the number of edges incident to v. In this section, we present an algorithm that constructs a burning schedule whose length is parameterized by the minimum degree of the graph, which is defined as the minimum vertex degree taken over all of its vertices. As expected, increasing the minimum degree of the graph will decrease the number of rounds needed to burn all of the vertices, and our result sheds light on the nature of this tradeoff. An interesting consequence of our algorithm is that we make progress towards resolving the conjecture from [3] that every connected graph on n vertices can be burned in at most $\lceil \sqrt{n} \rceil$ rounds. We prove that the conjecture holds for all graphs with minimum degree at least 23.

To describe and analyze our algorithm, we denote by $d(v, w)$ the length of the shortest path between v and w in G, i.e., the distance between v and w. We denote by $N_r(v)$ the set of vertices whose distance from v is at most r. For any vertex $v \in G$, let $ecc(v)$ denote the eccentricity of v, i.e., the maximum distance between v and any other vertex of G. Let $rad(G)$ denote the radius of G, i.e., the minimum eccentricity taken over all vertices in G.

The algorithms works as follows: for a well-chosen even integer $2r$ (to be specified later), our algorithm picks a maximal set of vertices such that the distance between every pair is greater than $2r$. This can be done efficiently in a greedy manner: pick any vertex v, add v to A, remove $N_{2r}(v)$ from G, and repeat the above until G is empty.

To analyze the algorithm, we start by finding an upper bound on $|A|$ with respect to r. This bound will rely on the following fact that a lower bound on the degree implies a lower bound on the size of $N_r(v)$ for any $v \in G$.

Proposition 1. *Consider any connected graph G that has minimum degree δ. For any vertex $v \in G$ and any $r \in \{1, \ldots, ecc(v)\}$, we have that $N_r(v) \geq \lfloor \frac{r+2}{3} \rfloor (\delta + 1)$.*

Proof. Let v be an arbitrary vertex in G. Define $L_i = \{w \mid d(v, w) = i\}$, i.e., L_i is the set of vertices whose distance from v in G is exactly i. Define $S_0 = L_0 \cup L_1$, i.e., S_0 consists of v and its neighbors. Note that $|S_0| \geq \delta + 1$ since v has degree at least δ. Next, for each $j \in \{1, \ldots, \lfloor \frac{r+2}{3} \rfloor - 1\}$, define $S_j = L_{3j-1} \cup L_{3j} \cup L_{3j+1}$. Note that, for all $j \in \{1, \ldots, \lfloor \frac{r+2}{3} \rfloor - 1\}$, we have $3j + 1 \leq 3(\lfloor \frac{r+2}{3} \rfloor - 1) + 1 \leq r \leq ecc(v)$.

The fact that $3j+1 \le r$ means that all vertices in $L_{3j-1} \cup L_{3j} \cup L_{3j+1}$ are within distance r from v, i.e., $S_j \in N_r(v)$ for each $j \in \{1, \ldots, \lfloor \frac{r+2}{3} \rfloor - 1\}$. Moreover, the fact that $3j+1 \le ecc(v)$ means that each of L_{3j-1}, L_{3j}, and L_{3j+1} is non-empty. In particular, we can pick an arbitrary vertex in L_{3j}, which by assumption has at least δ neighbors, and each of these neighbors must be in one of L_{3j-1}, L_{3j}, or L_{3j+1} (i.e., in S_j), which implies that $|S_j| \ge \delta + 1$ for all $j \in \{1, \ldots, \lfloor \frac{r+2}{3} \rfloor - 1\}$. Finally, by construction, $S_j \cap S_{j'} = \emptyset$ for any two distinct $j, j' \in \{0, \ldots, \lfloor \frac{r+2}{3} \rfloor - 1\}$. So $|N_r(v)| \ge \sum_{j=0}^{\lfloor \frac{r+2}{3} \rfloor - 1} |S_j| \ge \lfloor \frac{r+2}{3} \rfloor (\delta + 1)$. □

We apply the preceding lower bound to $N_r(v)$ for each $v \in A$, and then use the fact that these neighborhoods are disjoint to find an upper bound on $|A|$.

Lemma 1. *Consider any n-vertex connected graph G with minimum degree δ. Suppose that A is a subset of the vertices of G such that, for some $r \in \{1, \ldots, rad(G)\}$, the distance between each pair of vertices in A is greater than $2r$. Then $|A| \le \frac{3n}{r(\delta+1)}$.*

Proof. Denote by $v_1, \ldots, v_{|A|}$ the vertices in A. Since the distance between each pair of these vertices is greater than $2r$, the sets $N_r(v_1), \ldots, N_r(v_{|A|})$ are disjoint, so $n \ge \sum_{i=1}^{|A|} |N_r(v_i)|$. As $r \le rad(G)$, it follows that $r \le ecc(v_i)$ for each $i \in \{1, \ldots, |A|\}$, so, by Proposition 1, we know that $|N_r(v_i)| \ge \lfloor \frac{r+2}{3} \rfloor (\delta+1) \ge \frac{r(\delta+1)}{3}$. Therefore, $n \ge \sum_{i=1}^{|A|} |N_r(v_i)| \ge |A| \frac{r(\delta+1)}{3}$, which implies the desired result. □

As A is a maximal set of vertices with pairwise distance greater than $2r$, each vertex in G is within distance $2r$ from some vertex in A. Burning one activator from A in each of the first $|A|$ rounds, the fire then spreads and burns all vertices within an additional $2r$ rounds. We find the value for r such that $|A| + 2r$ is minimized, which leads to the following bound on burning time.

Theorem 1. *For any connected graph G on n vertices that has minimum degree δ, our algorithm produces a burning sequence that burns G within $\left\lceil \sqrt{\frac{24n}{\delta+1}} \right\rceil$ rounds.*

Corollary 1. *For any connected graph G on n vertices with minimum degree $\delta \ge 23$, the burning number is at most $\lceil \sqrt{n} \rceil$.*

3 Graphs of Small Pathlength or Treelength

In this section, we provide efficient algorithms for burning connected graphs of small pathlength or treelength. Our algorithms achieve good approximation ratios when the diameter of the input graph is asymptotically large. For graphs of small diameter, the problem can be optimally solved using brute force, as given by the following theorem.

Theorem 2. *The burning problem can be optimally solved in polynomial time if the diameter of the input graph is bounded by a constant.*

3.1 Preliminaries

The concepts of *path decomposition* and *tree decomposition* [14,24] were initially intended to measure, via the pathwidth and treewidth parameters, how close a graph is to a path and a tree, respectively. Pathlength and treelength are related parameters that are also based on the same definition of path decomposition.

Definition 1 (Decompositions, treelength, pathlength).

- *A* tree decomposition $\mathcal{T}$ *of a graph G is a tree whose vertex set is a finite set of* bags $\{B_i \mid 1 \leq i \leq \xi \in \mathbb{N}\}$, *where: each bag is a subset of the vertices of G; for every edge* $\{v, w\}$, *at least one bag contains both v and w; and, for every vertex v of G, the set of bags containing v forms a connected subtree of* $\mathcal{T}$. *When* $\mathcal{T}$ *is a path, then the decomposition is called a* path decomposition *of G.*
- *A* rooted tree decomposition *is a tree decomposition with a designated root bag, and parent/child relationships between bags are defined in the usual way. For any bag B in a rooted decomposition* $\mathcal{T}$, *we denote by* $\mathcal{T}_B$ *the subtree of the decomposition rooted at B.*
- *The* length *of a decomposition is the maximum distance between two vertices in the same bag, i.e.,* $\max_{1 \leq i \leq \xi}\{d(x, y) \mid x, y \in B_i\}$. *The* treelength of G, *denoted by tl, is defined to be the minimum length taken over all tree decompositions of G. The* pathlength of G, *denoted by pl, is defined to be the minimum length taken over all path decompositions of G.*

Figure 2 illustrates the concepts of pathlength and treelength. We always refer to vertices of $\mathcal{T}$ as bags to distinguish them from vertices of G. We also assume that the input graph is connected. For any graph G, the pathlength of G cannot be smaller than its treelength, that is, the family of graphs with bounded treelength includes graphs with bounded pathlength as a sub-family. It is known that a graph has pathlength 1 if and only if it is an interval graph [13], and treelength 1 if and only if it is a chordal graph [12]. So, the path/tree decomposition of these graphs can be computed in linear time using the algorithm of Booth and Lueke [8] for interval graphs, and a lexicographic breadth-first search [25] for chordal graphs. However, we cannot extend these algorithms to larger values of pathlength or treelength: it is known that the problem of determining whether a given graph has treelength at most k is NP-hard for any $k \geq 2$ [21]. On the positive side, there are algorithms with approximation factor 2 for computing pathlength [19], and approximation factor 3 for computing treelength [10]. Given these results, it is safe to assume a path/tree decomposition of a given graph is provided together with the graph (otherwise, we use these algorithms to achieve decompositions that are a constant factor away from the optimal decomposition).

In the remainder of the paper, every rooted tree decomposition $\mathcal{T}$ is assumed to be *trimmed*, in the sense that if there is a bag B such that the set of all vertices in bags of $\mathcal{T}_B$ is a subset of B's parent bag, we remove the entire subtree rooted at B from $\mathcal{T}$. This does not change the length of the decomposition.

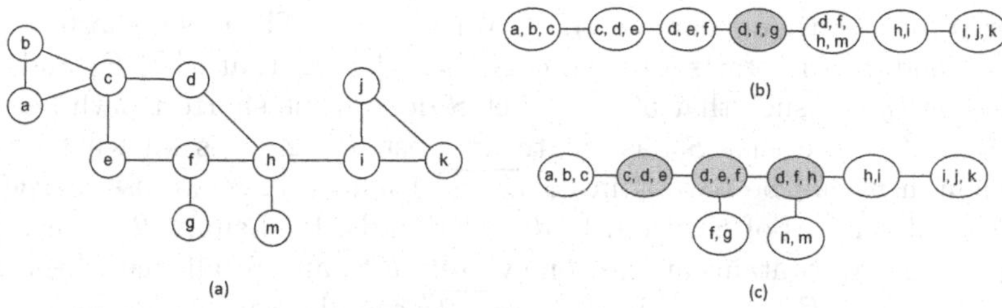

Fig. 2. (a) A graph G (b) A path decomposition of G with pathlength 3. Note that the distance between d and g in the highlighted bag is 3. (c) A tree decomposition of G with treelength 2. Each highlighted bag contains a pair of vertices with distance 2.

Observation 1. *For any connected graph G and any rooted tree decomposition $\mathcal{T}'$, there is a trimmed rooted tree decomposition $\mathcal{T}$ of G of same length as $\mathcal{T}'$.*

The following result establishes a useful structural property about rooted tree decompositions that will be used in the remainder of the paper.

Lemma 2. *Consider any rooted tree decomposition or any path decomposition $\mathcal{T} = \{B_i \mid 1 \leq i \leq \xi \in \mathbb{N}\}$ of a connected graph G. Let u and v be any two distinct vertices in G, let B_u be any bag of $\mathcal{T}$ that contains u, and let B_v be any bag of $\mathcal{T}$ that contains v. If P is a shortest path between u and v in G, then each bag in the shortest path between B_u and B_v in $\mathcal{T}$ contains a vertex of P.*

3.2 Burning Graphs of Small Pathlength

The following theorem shows that a graph of bounded pathlength can be burned in a nearly-optimal number of rounds.

Theorem 3. *Given a connected graph G of diameter $d \geq 1$ and a path decomposition of G with pathlength pl, it is possible to burn G in $\lceil \sqrt{d-1}\, \rceil + pl$ rounds.*

Proof. Consider a path decomposition $\mathcal{T} = \{B_i \mid 1 \leq i \leq \xi \in N\}$ of G such that the bags are indexed in increasing order from one leaf to the other. Further, assume that $\mathcal{T}$ has the following form: the first bag B_1 contains a vertex x that is absent in B_2, and the last bag B_ξ contains a vertex y that is absent in bag $B_{\xi-1}$. For any path decomposition of G that is not of this form, at least one of $B_1 \subseteq B_2$ or $B_\xi \subseteq B_{\xi-1}$ holds. If $B_1 \subseteq B_2$, we can remove B_1 to get another path decomposition of G, and if $B_\xi \subseteq B_{\xi-1}$, we can remove B_ξ to get another path decomposition of G. If $\mathcal{T}$ consists of one bag B_1, then the diameter of G is pl, so G can be burned within pl rounds by choosing any vertices of G as activators. So we proceed under the assumption that $\xi \geq 2$.

Since $\mathcal{T}$ is a valid path decomposition, each neighbor x' of x must appear together with x in at least one bag, and this must be B_1: as we assumed that x is in B_1 and not B_2, we know that x does not appear in any bag B_i with $i \geq 2$ as the bags containing x must form a connected subgraph of $\mathcal{T}$. Similarly, each neighbor y' of y must appear together with y in B_ξ. Thus, the shortest path between x and y in G starts with an edge $\{x, x'\}$ such that $x' \in B_1$ and ends with an edge $\{y', y\}$ such that $y' \in B_\xi$. Let S denote the shortest path between x' and y' in G; note that S has length at most $d-2$. It is known that any path of length m can be burned in $\lceil\sqrt{m+1}\rceil$ rounds [4], so we use a schedule that burns all vertices of S within $\lceil\sqrt{d-1}\rceil$ rounds. By Lemma 2, each bag in $\{B_2, B_3, \ldots, B_{\xi-1}\}$ contains at least one vertex of S, and recall that $x' \in S$ is in B_1 and $y' \in S$ is in B_ξ. So, within the $\lceil\sqrt{d-1}\rceil$ rounds that it takes to burn the vertices of S, at least one vertex in each bag of the decomposition is burned. In the *pl* rounds that follow, all vertices will be burned since the distance between any two vertices in each bag is at most *pl*. □

The study of pathlength is relatively new, and its relationship with other graph families is not fully discovered yet. Regardless, we can still use Theorem 3 to state the following two corollaries about grids and interval graphs.

Corollary 2. *Consider a grid graph G of size $n = n_1 \times n_2$ and $n_1 \leq n_2$. It is possible to burn G in $\sqrt{n} + o(\sqrt{n})$ rounds.*

Corollary 3. *Any connected interval graph G of diameter d and size n can be burned within $\lceil\sqrt{d}\,\rceil + 1 \leq \lceil\sqrt{n}\,\rceil + 1$ rounds.*

Finally, we show that the algorithm used to prove Theorem 3 guarantees a $1 + o(1)$-approximation factor.

Corollary 4. *Given any connected graph G of bounded pathlength, there is an algorithm with approximation factor $1 + o(1)$.*

Proof. First, if the diameter d of G is bounded by a constant, use Theorem 2 to optimally burn G. Next, assume G has asymptotically large diameter. Given a path decomposition, we apply Theorem 3 to burn G in $\lceil\sqrt{d-1}\,\rceil + pl$ rounds. An optimal burning schedule requires at least $\lceil\sqrt{d+1}\,\rceil$ rounds to burn G [4]. So, our algorithm achieves an approximation ratio of $\frac{\lceil\sqrt{d-1}\rceil + pl}{\lceil\sqrt{d+1}\rceil} < 1 + pl/\sqrt{d}$, which is $1 + o(1)$ as *pl* is bounded by a constant and d is asymptotically large. □

3.3 Burning Graphs of Small Treelength

In this section, we consider the burning problem in connected graphs of bounded treelength. This class includes trees, as trees have treelength 1 since they are chordal. For trees, there is a known algorithm with approximation factor 2 [5]. Our algorithm can be seen as an extension to all graphs of bounded treelength.

We first define a procedure named `BurnGuess` which takes as input a graph G and a positive integer g. A rooted tree decomposition $\mathcal{T}$ of G is given, and we pick an arbitrary vertex in the root bag called the "origin" vertex, denoted by o. The output of `BurnGuess` is either: (I) `no-schedule`, indicating that there does not exist a schedule such that the burning process completes within g rounds, or, (II) a schedule such that all vertices are burned within $2g + 4tl + 1$ rounds.

Procedure BurnGuess works by marking the vertices of G in iterations. Initially, no vertex is marked. At the beginning of each iteration $i \geq 1$, an arbitrary unmarked vertex at maximum distance in G from the origin o is selected and called *terminal* t_i. Let B_i be a bag of $\mathcal{T}$ with minimum depth (distance from the root) that contains t_i. We traverse $\mathcal{T}$ starting from B_i towards the root of $\mathcal{T}$ until we find a bag B'_i such that all vertices in B'_i are at distance at least g from t_i in G. If there is no such B'_i, the root of $\mathcal{T}$ is chosen as B'_i. We select an arbitrary vertex in B'_i as the ith activator and denote it by c_i. After selecting c_i, all vertices in G that are within distance $(2g-i+1)+4tl$ from c_i are marked, and iteration i ends. The above process continues until all vertices in $\mathcal{T}$ are marked or when the number of iterations exceeds $g+1$. Figure 3 illustrates the algorithm.

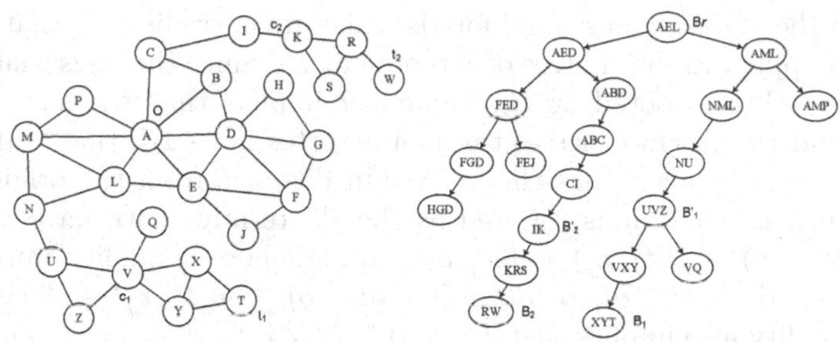

Fig. 3. An illustration of BurnGuess with $g = 2$ on chordal graph G (left). On the right, a tree decomposition of length $tl = 1$ is rooted at bag B_r; vertex $A \in B_r$ is chosen as origin o. In iteration $i = 1$, the furthest unmarked vertex from A is vertex T; so $t_1 = T$ and $B_1 = \{X, Y, T\}$. From B_1, the first ancestor bag in which all vertices have distance at least $g = 2$ from T in G is $B'_1 = \{U, V, Z\}$. A vertex from B'_1 is selected as an activator, say, $c_1 = V$. All vertices at distance $(2g - i + 1) + 4tl = 8$ from V in G are marked. Only W remains unmarked, so it is selected as t_2 in the next iteration.

We now establish the following lemma that provides an upper bound on the burning time of any burning schedule returned by BurnGuess. The idea is that an activator c_i chosen in iteration i will be burned in round i, and all nodes marked in iteration i will burn within the next $(2g - i + 1) + 4tl$ rounds.

Lemma 3. *If BurnGuess on input G returns a burning schedule A, then the burning process corresponding to schedule A completes within $2g+4tl+1$ rounds.*

Our next goal is to establish a lower bound for the burning number of graph G in the case that BurnGuess returns no-schedule on input G. To this end, we first provide the following technical lemma.

Lemma 4. *For any connected graph G, after each iteration i of BurnGuess on input G, each vertex in B'_i is within distance $g + 2tl - 1$ of t_i in G.*

Lemma 5. *If the BurnGuess procedure returns no-schedule for inputs G, g, then there is no burning schedule such that the corresponding burning process burns all vertices of G in fewer than g rounds.*

Proof. From the definition of `BurnGuess`, the value `no-schedule` is returned when $g+1$ iterations have been completed and there exists an unmarked vertex in G. For any iteration i, Lemma 4 ensures that the distance between c_i and t_i is at most $g+2tl-1$. In iteration i, all vertices within distance $(2g-i+1)+4tl \geq g+2tl-1$ from c_i in G are marked, so vertex t_i is marked by the end of iteration i. As a result, the $g+1$ iterations involve $g+1$ different terminal vertices $t_1, \ldots, t_{g+1}$.

Let S be the set consisting of the terminal vertices $t_1, \ldots, t_{g+1}$, excluding the terminal t_k whose corresponding activator c_k is located in the root bag B_r of $\mathcal{T}$, if such a terminal exists. Note that there are at least g vertices in S. The following claim gives a useful fact about the bags that contain terminals from S.

Claim 1: For all $t_i, t_j \in S$ with $i < j$, terminal t_j appears in a bag of $\mathcal{T} \setminus \mathcal{T}_{B'_i}$.

To prove the claim, we assume, for the sake of contradiction, that all bags containing t_j appear in the subtree of $\mathcal{T}$ rooted at B'_i. Since the bags that contain t_j are in the subtree rooted at B'_i, Lemma 2 implies that the shortest path between t_j and the origin o passes through a vertex $x_j \in B'_i$, that is $d(t_j, o) = d(t_j, x_j) + d(x_j, o)$. Since t_i has the maximum distance from the origin among unmarked vertices when it is selected as the ith terminal, we have $d(t_i, o) \geq d(t_j, o)$. So $d(t_i, o) \geq d(t_j, x_j) + d(x_j, o)$. The triangle inequality implies that $d(t_i, c_i) + d(c_i, o) \geq d(t_i, o)$, so $d(t_i, c_i) + d(c_i, o) \geq d(t_j, x_j) + d(x_j, o)$. The triangle inequality also implies that $d(t_j, x_j) \geq d(t_j, c_i) - d(x_j, c_i)$ and $d(x_j, o) \geq d(c_i, o) - d(c_i, x_j)$, so $d(t_i, c_i) + d(c_i, o) \geq (d(t_j, c_i) - d(x_j, c_i)) + (d(c_i, o) - d(c_i, x_j))$. Simplifying this inequality, we get $d(t_i, c_i) \geq d(t_j, c_i) - 2d(c_i, x_j)$. As c_i and x_j are both in B'_i, we get $d(c_i, x_j) \leq tl$, so $d(t_i, c_i) \geq d(t_j, c_i) - 2tl$. Lemma 4 implies that $d(t_i, c_i) \leq g + 2tl - 1$, so $g + 2tl - 1 \geq d(t_j, c_i) - 2tl$, and it follows that $d(t_j, c_i) \leq g + 4tl - 1$. However, this means that t_j is marked in iteration i of `BurnGuess`, which contradicts that t_j was chosen as a terminal from unmarked vertices in iteration $j > i$. This completes the proof of Claim 1.

Using Claim 1, we prove that any two terminals in S are far apart in G.

Claim 2: For any two terminals $t_i, t_j \in S$, $d(t_i, t_j) \geq 2g$ in G.

To prove the claim, assume, without loss of generality, that $i < j$. As B'_i is an ancestor of B_i, Claim 1 implies that the shortest path in $\mathcal{T}$ between bags B_i and B_j passes through B'_i. By Lemma 2, it follows that B'_i contains a vertex x on the shortest path between t_i and t_j in G. As x is on a shortest path, we can write $d(t_i, t_j) = d(t_i, x) + d(t_j, x)$. By the triangle inequality, $d(t_i, x) + d(t_j, x)$ is at least $(d(t_i, c_i) - d(x, c_i)) + (d(t_j, c_i) - d(x, c_i))$, which simplifies to $d(t_i, c_i) + d(t_j, c_i) - 2d(x, c_i)$. Since x and c_i both appear in B'_i, it follows that $d(x, c_i) \leq tl$, so $d(t_i, c_i) + d(t_j, c_i) - 2d(x, c_i) \geq d(t_i, c_i) + d(t_j, c_i) - 2tl$. By definition, all vertices in B'_i, e.g., c_i, are at distance at least g from t_i, e.g., $d(t_i, c_i) \geq g$, so $d(t_i, c_i) + d(t_j, c_i) - 2tl \geq g + d(t_j, c_i) - 2tl$. Since terminals are chosen from unmarked vertices, and t_j is chosen in iteration $j > i$, it follows that t_j was unmarked at the end of iteration i, which means $d(t_j, c_i) > (2g - i + 1) + 4tl \geq g + 4tl$. Thus, $g + d(t_j, c_i) - 2tl > g + (g + 4tl) - 2tl > 2g$. Hence, we have shown that $d(t_i, t_j) > 2g$, which completes the proof of Claim 2.

To complete the proof of the lemma, assume, for the sake of contradiction, that there is a burning schedule A that burns G in $t \leq g-1$ rounds. It follows that A specifies at most $t \leq g-1$ activators. In a burning process that lasts t rounds, a vertex v is burned only if v is within distance $g-1$ from at least one activator. So, if every vertex of G is burned, then each terminal is within distance $g-1$ from at least one activator. However, no two terminals $t_i, t_j \in S$ are within distance $g-1$ from the same activator, since, by Claim 2, $d(t_i, t_j)$ is at least $2g$. This implies that A contains at least $|S| = g$ activators, a contradiction. □

Theorem 4. *Given a connected graph G of bounded treelength, and a tree decomposition $\mathcal{T}$ of G with bounded length tl, there is a polynomial-time algorithm for burning G that achieves an approximation factor of $2 + o(1)$.*

Proof. If the diameter d of G is bounded, use Theorem 2 to burn it optimally in polynomial time. Otherwise, repeatedly apply `BurnGuess` to find the smallest parameter g^* for which the algorithm returns a schedule A. By Lemma 3, schedule A burns the graph within $2g^* + 4tl + 1$ rounds. Meanwhile, by Lemma 5, no schedule can burn all vertices of G within $g^* - 1$ rounds. The approximation ratio of the algorithm will be $\frac{2g^*+4tl+1}{g^*} = 2 + (4tl+1)/g^*$, which is $2 + o(1)$ as the diameter of G, and hence g^*, is asymptotically large.

With respect to the time complexity, we can find g^* using binary search on the range $[1, d]$, i.e., calling `BurnGuess` $O(\log d)$ times. Each call to `BurnGuess` with parameter g has at most $g + 1 \leq d + 1$ iterations. In each iteration i, the distance calculations can be performed using Dijkstra's algorithm in $O(n^2)$ time, and marking all vertices within distance $(2g-i+1)+4tl$ from c_i can be done in a Breadth-First manner in $O(n^2)$ time. So, the complexity of `BurnGuess` is $O(dn^2)$, and the algorithm's overall time complexity is $O(n^2 d \log d) \in O(n^3 \log n)$. □

References

1. Bessy, S., Bonato, A., Janssen, J.C.M., Rautenbach, D., Roshanbin, E.: Burning a graph is hard. Discrete Appl. Math. **232**, 73–87 (2017)
2. Bonato, A., Gunderson, K., Shaw, A.: Burning the plane: densities of the infinite cartesian grid. Preprint (2018)
3. Bonato, A., Janssen, J.C.M., Roshanbin, E.: Burning a graph as a model of social contagion. In: Workshop of Workshop on Algorithms and Models for the Web Graph, pp. 13–22 (2014)
4. Bonato, A., Janssen, J.C.M., Roshanbin, E.: How to burn a graph. Internet Math. **12**(1–2), 85–100 (2016)
5. Bonato, A., Kamali, S.: Approximation algorithms for graph burning. In: Theory and Applications of Models of Computation Conference (TAMC), pp. 74–92 (2019)
6. Bonato, A., Lidbetter, T.: Bounds on the burning numbers of spiders and path-forests. ArXiv e-prints, July 2017
7. Bond, R.M., et al.: A 61-million-person experiment in social influence and political mobilization. Nature **489**(7415), 295–298 (2012)
8. Booth, K.S., Lueker, G.S.: Testing for the consecutive ones property, interval graphs, and graph planarity using pq-tree algorithms. J. Comput. Syst. Sci. **13**(3), 335–379 (1976)

9. Das, S., Dev, S.R., Sadhukhan, A., Sahoo, U., Sen, S.: Burning spiders. In: Panda, B.S., Goswami, P.P. (eds.) CALDAM 2018. LNCS, vol. 10743, pp. 155–163. Springer, Cham (2018). https://doi.org/10.1007/978-3-319-74180-2_13
10. Dourisboure, Y., Gavoille, C.: Tree-decompositions with bags of small diameter. Discrete Math. **307**(16), 2008–2029 (2007)
11. Fajardo, D., Gardner, L.M.: Inferring contagion patterns in social contact networks with limited infection data. Netw. Spat. Econ. **13**(4), 399–426 (2013)
12. Gavril, F.: The intersection graphs of subtrees in trees are exactly the chordal graphs. J. Comb. Theory Ser. B **16**(1), 47–56 (1974)
13. Gilmore, P.C., Hoffman, A.J.: A characterization of comparability graphs and of interval graphs. Can. J. Math. **16**, 539–548 (1964)
14. Halin, R.: S-functions for graphs. J. Geom. **8**(1–2), 171–186 (1976)
15. Kamali, S., Miller, A., Zhang, K.: Burning two worlds: Algorithms for burning dense and tree-like graphs. CoRR abs/1909.00530 (2019). http://arxiv.org/abs/1909.00530
16. Kramer, A.D.I.: The spread of emotion via facebook. In: CHI Conference on Human Factors in Computing Systems, (CHI), pp. 767–770 (2012)
17. Kramer, A.D.I., Guillory, J.E., Hancock, J.T.: Experimental evidence of massive-scale emotional contagion through social networks. In: Proceedings of the National Academy of Sciences, pp. 8788–8790 (2014)
18. Land, M.R., Lu, L.: An upper bound on the burning number of graphs. In: Proceedings of Workshop on Algorithms and Models for the Web Graph, pp. 1–8 (2016)
19. Leitert, A.: Tree-Breadth of Graphs with Variants and Applications. Ph.D. thesis, Kent State University, College of Arts and Sciences, Department of Computer Science (2017)
20. Liu, H., Zhang, R., Hu, X.: Burning number of theta graphs. Appl. Math. Comput. **361**, 246–257 (2019)
21. Lokshtanov, D.: On the complexity of computing treelength. Discrete Appl. Math. **158**(7), 820–827 (2010). third Workshop on GraphClasses, Optimization, and Width Parameters Eugene, Oregon, USA, October 2007
22. Mitsche, D., Pralat, P., Roshanbin, E.: Burning graphs: a probabilistic perspective. Graphs and Combinatorics **33**(2), 449–471 (2017)
23. Mitsche, D., Pralat, P., Roshanbin, E.: Burning number of graph products. Theor. Comput. Sci. **746**, 124–135 (2018)
24. Robertson, N., Seymour, P.D.: Graph minors iii planar tree-width. J. Comb. Theory Ser. B **36**(1), 49–64 (1984)
25. Rose, D.J., Tarjan, R.E., Lueker, G.S.: Algorithmic aspects of vertex elimination on graphs. SIAM J. Comput. **5**(2), 266–283 (1976)
26. Sim, K.A., Tan, T.S., Wong, K.B.: On the burning number of generalized Petersen graphs. Bull. Malays. Math. Sci. Soc. **6**, 1–14 (2017)
27. Vazirani, V.V.: Approximation Algorithms. Springer, Heidelberg (2003). https://doi.org/10.1007/978-3-662-04565-7
28. Šimon, M., Huraj, L., Dirgova Luptáková, I., Pospichal, J.: Heuristics for spreading alarm throughout a network. Appl. Sci. **9**(16), 3269 (2019). https://doi.org/10.3390/app9163269

Faster STR-EC-LCS Computation

Kohei Yamada[1(✉)], Yuto Nakashima[1], Shunsuke Inenaga[1,2], Hideo Bannai[1], and Masayuki Takeda[1]

[1] Department of Informatics, Kyushu University, Fukuoka, Japan
{kohei.yamada,yuto.nakashima,inenaga,bannai,takeda}@inf.kyushu-u.ac.jp
[2] PRESTO, Japan Science and Technology Agency, Kawaguchi, Japan

Abstract. The longest common subsequence (LCS) problem is a central problem in stringology that finds the longest common subsequence of given two strings A and B. More recently, a set of four constrained LCS problems (called generalized constrained LCS problem) were proposed by Chen and Chao [J. Comb. Optim, 2011]. In this paper, we consider the substring-excluding constrained LCS (STR-EC-LCS) problem. A string Z is said to be *an STR-EC-LCS of two given strings* A and B *excluding* P if, Z is one of the longest common subsequences of A and B that does not contain P as a substring. Wang et al. proposed a dynamic programming solution which computes an STR-EC-LCS in $O(mnr)$ time and space where $m = |A|, n = |B|, r = |P|$ [Inf. Process. Lett., 2013]. In this paper, we show a new solution for the STR-EC-LCS problem. Our algorithm computes an STR-EC-LCS in $O(n|\Sigma| + (L+1)(m-L+1)r)$ time where $|\Sigma| \leq \min\{m, n\}$ denotes the set of distinct characters occurring in both A and B, and L is the length of the STR-EC-LCS. This algorithm is faster than the $O(mnr)$-time algorithm for short/long STR-EC-LCS (namely, $L \in O(1)$ or $m-L \in O(1)$), and is at least as efficient as the $O(mnr)$-time algorithm for all cases.

1 Introduction

The *longest common subsequence* (*LCS*) problem of finding an LCS of given two strings, is a classical and important problem in Theoretical Computer Science. Given two strings A and B of respective lengths m and n, it is well known that the LCS of A and B can be computed by a standard dynamic programming technique [13]. Since LCS is one of the most fundamental similarity measures for string comparison, there are a number of studies on faster computation of LCS and its applications [2,3,11,14]. It is also known that there is a conditional lower bound which states that the LCS of two strings of length n each cannot be computed in $O(n^{2-\epsilon})$ time for any constant $\epsilon > 0$, unless the famous popular Strong Exponential Time Hypothesis (SETH) fails [1]. Thus, it is highly likely that one needs to use almost quadratic time for computing LCS in the worst case. Still, it is possible to design algorithms for computing LCS whose running time depends on other parameters. One of such algorithms was proposed by Nakatsu et al. [10], which finds an LCS of given two strings A and B in $O(n(m-l))$

A. Chatzigeorgiou et al. (Eds.): SOFSEM 2020, LNCS 12011, pp. 125–135, 2020.
https://doi.org/10.1007/978-3-030-38919-2_11

time and space, where l is the length of the LCS of the two given strings. This algorithm is efficient when l is large, namely, A and B are very similar.

Of a variety of extensions to LCS that have been extensively studied, this paper focuses on a class of problems called the *constrained LCS* problems, first considered by Tsai [12]. We are given strings A, B and constraint string P of length r, and the CLCS problem is to find a longest subsequence common to A and B, such that the subsequence has P as a subsequence. He also presented a dynamic programming algorithm which solves the problem in $O(m^2n^2r)$ time and space. The motivation for introducing constraints is to reflect some a-priori knowledge (e.g., biological knowledge) to the solutions. Later, the *generalized constrained LCS* (*GC-LCS*) problems were introduced by Chen et al. [4]. GC-LCS consists of four variants of the constrained LCS problem, which are respectively called *SEQ-IC-LCS*, *SEQ-EC-LCS*, *STR-IC-LCS*, and *STR-EC-LCS*. For given strings A, B and P, the problem is to find a longest subsequence common to A and B such that the subsequence includes/excludes/includes/excludes P as a subsequence/subsequence/substring/substring, respectively for SEQ-IC-LCS/SEQ-EC-LCS/STR-IC-LCS/STR-EC-LCS. We remark that CLCS is the same as SEQ-IC-LCS. The best known results for these problems were proposed in [4–6,15].

The quadratic bound for STR-IC-LCS seems to be very difficult to improve, since STR-IC-LCS is a special case of LCS (recall the afore-mentioned conditional lower bound for LCS). Since the other three variants require cubic time, it is important to discover more efficient solutions for these problems. There exist faster dynamic programming solutions for SEQ-IC-LCS and STR-IC-LCS which are based on run-length encodings [8,9]. However, no faster solutions to STR-EC-LCS than the one with $O(mnr)$ running time [15] are known to date.

In this paper, we revisit the STR-EC-LCS problem. More formally, we say that a string Z is *an STR-EC-LCS of two given strings* A and B *excluding* P if, Z is one of the longest common subsequences of A and B that does not contain P as a substring. We show a new dynamic programming solution for the STR-EC-LCS problem which runs in $O(n|\Sigma| + (L+1)(m-L+1)r)$ time and space, where Σ is the set of distinct characters occurring in both A and B, and L is the length of the solution. Note that $|\Sigma| \leq \min\{m, n\}$ always holds. Our algorithm is built on Nakatsu et al.s' method for the (original) LCS problem [10]. Assume w.l.o.g. that $m \leq n$. When the length of STR-EC-LCS is quite short or long (namely, $L \in O(1)$ or $m - L \in O(1)$), our algorithm runs only in $O(n|\Sigma| + mr) = O((n+r)m) = O(nm)$ time and space, since $r \leq n$. Even in the worst case where $L \in \Theta(m)$ and $m - L \in \Theta(m)$, which happens when $L = cm$ for any constant $0 < c < 1$, our algorithm is still as efficient as $O(mnr)$ since $|\Sigma| \leq \min\{m, n\}$.

This paper is organized as follows; we will give notations which we use in this paper in Sect. 2, we will propose our dynamic programming solution for the STR-EC-LCS problem in Sect. 3, finally, we will explain our algorithm for the STR-EC-LCS in Sect. 4.

2 Preliminaries

2.1 Strings

Let Σ be an integer *alphabet*. An element of Σ^* is called a *string*. The length of a string w is denoted by $|w|$. The empty string ε is a string of length 0. For a string $w = xyz$, x, y and z are called a *prefix*, *substring*, and *suffix* of w, respectively. The i-th character of a string w is denoted by $w[i]$, where $1 \leq i \leq |w|$. For a string w and two integers $1 \leq i \leq j \leq |w|$, let $w[i..j]$ denote the substring of w that begins at position i and ends at position j. For convenience, let $w[i..j] = \varepsilon$ when $i > j$.

A string Z is a *subsequence* of A if Z can be obtained from A by removing zero or more characters. In this paper, we consider common subsequences of two strings A and B of respective lengths m and n. For this sake, we can perform a standard preprocessing on A and B that removes every character that occurs only in either A or B, because such a character is never contained in any common subsequences of A and B. Assuming $n \geq m$, this preprocessing can be done in $O(n \log n)$ time with $O(n)$ space for general ordered alphabets, and in $O(n)$ time and space for integer alphabets of polynomial size in n (c.f. [7]). In what follows, we consider the latter case of integer alphabets, and assume that A and B have been preprocessed as above. In the sequel, let Σ denote the set of distinct characters that occur in both A and B. Note that $|\Sigma| \leq \min\{m, n\} = m$ holds.

2.2 STR-EC-LCS

Let A, B and P be strings. A string Z is said to be *an STR-EC-LCS of two given strings* A and B *excluding* P if, Z is one of the longest common subsequences of A and B that does not contain P as a substring. For instance, `bcaac`, `bcaba`, `acaac`, `acaba`, `abaac` and `ababa` are STR-EC-LCS of $A =$ `abcabac` and $B =$ `acbcaacbaa` excluding $P =$ `abc`. Although `abcaba` and `abcaac` are longest common subsequences of A and B, they are not STR-EC-LCS of the same strings (since they have P as a substring).

In Sect. 3, we revisit the STR-EC-LCS problem defined as follows.

Problem 1 (STR-EC-LCS problem [4]*).* Given strings A, B, and P, compute an STR-EC-LCS (and/or its length) of given strings.

In the rest of the paper, m, n, and r respectively denote the length of A, B and P. It is easy to see that STR-EC-LCS problem is the same as LCS problem when $r > \min\{m, n\}$. We assume that $r \leq m \leq n$ without loss of generality.

3 Dynamic Programming Solution for the STR-EC-LCS Problem

Our aim of this section is to show our dynamic programming solution for the STR-EC-LCS problem. We first give short descriptions of a dynamic programming solution for the LCS problem proposed by Nakatsu et al. [10], and a

s \ i	0	1	2	3	4	5	6	7
0	0	0	0	0	0	0	0	0
1	*	2	2	1	1	1	1	1
2	*	*	3	3	2	2	2	2
3	*	*	*	4	4	4	3	3
4	*	*	*	*	7	6	6	4
5	*	*	*	*	*	*	7	7
6	*	*	*	*	*	*	*	*
7	*	*	*	*	*	*	*	*

Fig. 1. This is an example for table e of given strings $A = \mathtt{aabacab}$ and $B = \mathtt{baabbcaa}$. For the sake of visibility, the value $n + 1 = 9$ is replaced by asterisk ($*$). The last row in the table which has a value smaller than $n + 1$ is 5; that is, the length of an LCS of A and B is 5.

dynamic programming solution for the STR-EC-LCS problem proposed by Wang et al. [15].

3.1 Solution for LCS by Nakatsu et al.

Nakatsu et al. proposed a dynamic programming solution for computing an LCS of given strings A and B. Here, we give a slightly modified description of their solution in order to describe our algorithm. For any $0 \leq i, s \leq m$, let $e(i, s)$ be the length of the shortest prefix $B[1..e(i, s)]$ of B such that the length of the longest common subsequence of $A[1..i]$ and $B[1..e(i, s)]$ is s. For convenience, $e(i, s) = n + 1$ if no such prefix exists or if $s > i$ holds. The values $e(i, s)$ will be computed using dynamic programming, where i represents the column number, and s represents the row number. Let $\tilde{s}$ be the largest value such that $e(i, s) < n + 1$ for some i, i.e, $\tilde{s}$ is the last row in the table of e, which has a value smaller than $n + 1$. We can see that the length of the longest common subsequence of A and B is $\tilde{s}$. We give an example in Fig. 1.

Now we explain how to compute e efficiently. Assume that $e(i - 1, s)$ and $e(i - 1, s - 1)$ have already been computed. We consider $e(i, s)$. It is easy to see that $e(i, s) \leq e(i - 1, s)$. If $e(i, s) < e(i - 1, s)$, an LCS of $A[1..i]$ and $B[1..e(i, s)]$ must use the character $A[i]$ as the last character. Then, we can see that $e(i, s)$ is the index of the leftmost occurrence of $A[i]$ in $B[e(i - 1, s - 1) + 1..n]$. Let $j_{i,s}$ be the the index of the leftmost occurrence of $A[i]$ in $B[e(i - 1, s - 1) + 1..n]$. From these facts, the following recurrence formula holds for e:

$$e(i, s) = \min\{e(i - 1, s), j_{i,s}\}.$$

If we add more information, we can backtrack on the table in order to compute an LCS (as a string), and not just its length.

3.2 Solution for STR-EC-LCS by Wang et al.

Wang et al. proposed a dynamic programming solution for STR-EC-LCS problem of given strings A, B and P. Here, we describe a key idea of their solution.

Definition 1. *For any string S, $\sigma(S)$ is the length of the longest prefix of P which is a suffix of S.*

By using this notation, they considered a table f defined as follows: let $f(i, j, k)$ be the length of the longest common subsequence Z of $A[1..i]$ and $B[1..j]$ such that Z does not have P as a substring and $\sigma(Z) = k$. They also showed a recurrence formula for f. By the definition of f, the length of an STR-EC-LCS is $\max\{f(m, n, t) \mid 0 \leq t < r\}$.

3.3 Our Solution for STR-EC-LCS

Our solution is based on the idea of Sect. 3.1. We maintain occurrences of a prefix of P as a suffix of a common subsequence by using the idea of Sect. 3.2.

For convenience, we introduce the following notation.

Definition 2. *A string Z is said to satisfy* Property(i, s, k) *if*

- *Z is a subsequence of $A[1..i]$,*
- *Z does not have P as a substring,*
- *$|Z| = s$, and*
- *$\sigma(Z) = k$.*

Thanks to the above notation, we can simply introduce our table d for computing STR-EC-LCS as follows. Let d be a 3-dimensional table where $d(i, s, k)$ is the length of the shortest prefix $B[1..d(i, s, k)]$ of B such that there exists a subsequence which satisfies Property(i, s, k) (if no such subsequence exists, then $d(i, s, k) = n + 1$ for convenience).

We can obtain the following observation about the length of an STR-EC-LCS by the definition of d.

Observation 1. *Let $\tilde{s}$ be the largest $1 \leq s \leq m$ such that $d(i, s, k) < n + 1$ for some i and k. $\tilde{s}$ is the length of an STR-EC-LCS by the definition of d.*

We give an example of a table in Fig. 2.

The next lemma shows a recurrence formula for d. We use this lemma for computing the length of a STR-EC-LCS.

Lemma 1.

$$d(i, s, k) = \min(\{d(i-1, s, k)\} \cup \{j_t \mid 0 \leq t < r\})$$

holds, where j_t is the smallest position j in $B[d(i-1, s-1, t) + 1..n]$ such that $A[i] = B[j]$, and there exists a string Z which satisfies Property$(i-1, s-1, t)$ *and $\sigma(ZA[i]) = k$ (if no such Z exists for t, then $j_t = n + 1$).*

$k = 0$

s\i	0	1	2	3	4	5	6	7
0	0	0	0	0	0	0	0	0
1	*	*	*	1	1	1	1	1
2	*	*	*	4	4	4	4	4
3	*	*	*	*	*	6	6	4
4	*	*	*	*	*	*	*	*
5	*	*	*	*	*	*	*	*
6	*	*	*	*	*	*	*	*
7	*	*	*	*	*	*	*	*

$k = 1$

s\i	0	1	2	3	4	5	6	7
0	*	*	*	*	*	*	*	*
1	*	2	2	2	2	2	2	2
2	*	*	*	*	2	2	2	2
3	*	*	*	*	7	7	7	7
4	*	*	*	*	*	*	7	7
5	*	*	*	*	*	*	*	*
6	*	*	*	*	*	*	*	*
7	*	*	*	*	*	*	*	*

$k = 2$

s\i	0	1	2	3	4	5	6	7
0	*	*	*	*	*	*	*	*
1	*	*	*	*	*	*	*	*
2	*	*	3	3	3	3	3	3
3	*	*	*	*	*	*	3	3
4	*	*	*	*	*	*	8	8
5	*	*	*	*	*	*	*	*
6	*	*	*	*	*	*	*	*
7	*	*	*	*	*	*	*	*

Fig. 2. This is our table d for given strings $A = \mathtt{aabacab}$, $B = \mathtt{baabbcaa}$, and $P = \mathtt{aab}$. In this figure, the value $n + 1 = 9$ is replaced by asterisk ($*$) for convenience. The lowest row which has a value smaller than $n + 1 = 9$ is $\tilde{s} = 4$. Thus, the length of a STR-EC-LCS is 4.

Proof. We show the following inequations to prove this lemma;

1. $d(i, s, k) \le \min(\{d(i-1, s, k)\} \cup \{j_t \mid 0 \le t < r\})$,
2. $d(i, s, k) \ge \min(\{d(i-1, s, k)\} \cup \{j_t \mid 0 \le t < r\})$.

We start from proving the first inequation. By the definition of d, $d(i, s, k) \le d(i-1, s, k)$ always holds. If $\{j_t \mid 0 \le t < r\} = \emptyset$, then the first inequation holds. We assume that $\{j_t \mid 0 \le t < r\} \ne \emptyset$, and j_{t_1} is in the set $(0 \le t_1 < r)$. Then, there exists a subsequence Z_1 of $B[1..d(i-1, s-1, t_1)]$ which satisfies $\mathsf{Property}(i-1, s-1, t_1)$. Since $A[i] = B[j_{t_1}]$ and $j_{t_1} > d(i-1, s-1, t_1)$, $Z_1 A[i]$ is a subsequence of $B[1..j_{t_1}]$ that satisfies $\mathsf{Property}(i, s, k)$ and $\sigma(Z_1 A[i]) = k$. This implies that $d(i, s, k) \le j_{t_1}$. Thus, the first inequation holds.

Suppose that the second inequation does not hold, namely,

$$d(i, s, k) < \min(\{d(i-1, s, k)\} \cup \{j_t \mid 0 \le t < r\}) \tag{1}$$

holds. If $d(i, s, k) = n+1$, then the above inequation does not hold. Now we consider the case $d(i, s, k) < n+1$. By the definition of d, there exists a subsequence Z_2 of $B[1..d(i, s, k)]$ that satisfies $\mathsf{Property}(i, s, k)$. Let $Z_2' = Z_2[1..|Z_2|-1]$. Then, Z_2' is a length $s-1$ subsequence of $A[1..i-1]$ which does not have P as a substring. Since Z_2' satisfies $\mathsf{Property}(i-1, s-1, \sigma(Z_2'))$, $d(i-1, s-1, \sigma(Z_2')) < d(i, s, k)$ holds. Moreover, $\sigma(Z_2' B[d(i, s, k)]) = k$ holds. If $A[i] = B[d(i, s, k)]$, then, $j_{\sigma(Z_2')} \le d(i, s, k)$ holds. This fact contradicts Inequation (1). Now we can assume that $A[i] \ne B[d(i, s, k)]$. This implies that Z_2 is a common subsequence of $A[1..i]$ and $B[1..d(i, s, k) - 1]$, or a common subsequence of $A[1..i-1]$ and $B[1..d(i, s, k)]$. The first case implies a contradiction by the definition of d. The second case implies that $d(i, s, k) = d(i-1, s, k)$, a contradiction. Thus, $d(i, s, k) \ge \min(\{d(i-1, s, k)\} \cup \{j_t \mid 0 \le t < r\})$ holds. □

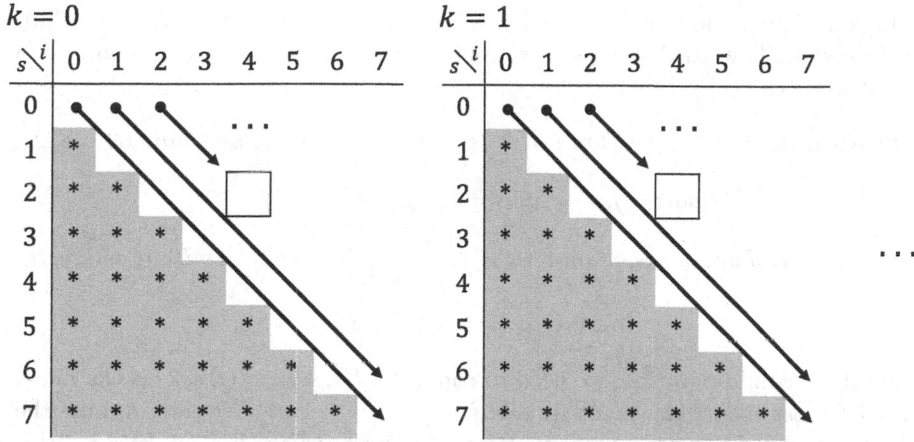

Fig. 3. This figure shows the order of computation for table d. For each table (i.e., for each k), we do not need to compute the lower left part (satisfying $s > i$). We start from computing values on the leftmost arrow for each table. In each step (i, s), we compute $d(i, s, k)$ for all tables (for instance, squared values in the figure will be computed in the same step).

4 Algorithm

In this section, we show how to compute STR-EC-LCS by using Lemma 1. We mainly explain our algorithm to compute the length of an STR-EC-LCS (we will explain how to compute an STR-EC-LCS at the end of this section).

To use Lemma 1, we need $d(i-1, s, k)$ and $d(i-1, s-1, t)$ for all $0 \leq t < r$ for computing $d(i, s, k)$. We compute our table for every diagonal line from upper left to lower right in left-to-right order. In each step of our algorithm, we will fix $0 \leq i, s \leq m$ (we use (i, s) to denote the step for fixed i and s). Then we compute $d(i, s, k)$ for any $0 \leq k < r$ in the step. We can see from a simple observation that $d(i, s, k) = n + 1$ holds for any input strings if $i < s$ (since no STR-EC-LCS of length s exists). Thus, we do not compute $d(i, s, k)$ explicitly such that $i < s$. We also describe this strategy in Fig. 3.

Now we consider how to compute $d(i, s, k)$ for any $0 \leq k < r$. Let $Z(i, s, k)$ be a subsequence of $B[1..d(i-1, s-1, k)]$ satisfying $\mathsf{Property}(i-1, s-1, k)$. Due to Lemma 1, string $Z(i, s, k)A[i]$ is a witness for value $d(i, s, \sigma(Z(i, s, k)A[i]))$ if a (leftmost) position j in $B[d(i-1, s-1, k)+1..n]$ such that $A[i] = B[j]$ exists. For any i, s, k, let $J(i, s, k)$ denote the position j described above. Thus, we can compute $d(i, s, k)$ for any k in step (i, s) as follows.

1. Set $d(i-1, s, k)$ as the initial value for $d(i, s, k)$ for each k.
2. Compute $J(i, s, k)$ and $\sigma(Z(i, s, k)A[i])$ for each k.
3. If $J(i, s, k) < d(i, s, \sigma(Z(i, s, k)A[i]))$, then update $d(i, s, \sigma(Z(i, s, k)A[i]))$ to $J(i, s, k)$.

Lemma 1 and the above discussion ensure the correctness of this algorithm. Next we show how to do these operations efficiently. We use the following two data structures.

Definition 3. *For any position j in B (i.e., $j \in [1, n]$) and any character $\alpha \in \Sigma$,*

$$\mathsf{next}_B(j, \alpha) = \min\{q \mid B[q] = \alpha, q \geq j\}.$$

Definition 4. *For any position t in P (i.e., $t \in [0, r-1]$) and any character $\alpha \in \Sigma$,*

$$\mathsf{next}_\sigma(t, \alpha) = \sigma(P[1..t]\alpha).$$

At the second operation, we need to compute $J(i, s, k)$. $J(i, s, k)$ is the index of the leftmost occurrence of $A[i]$ in $B[d(i-1, s-1, k)+1..n]$. We can compute the occurrence by using next_B, namely, $J(i, s, k) = \mathsf{next}_B(d(i-1, s-1, k)+1, A[i])$.

Moreover, we need to compute $\sigma(Z(i, s, k)A[i])$. We know that $\sigma(Z(i, s, k)) = k$, namely, $Z(i, s, k)$ has $P[1..k]$ as a suffix. By the definition of $\sigma(\cdot)$, $\sigma(S)+1 \geq \sigma(S\alpha)$ holds for any string S and $\alpha \in \Sigma$. This implies that $\sigma(Z(i, s, k)A[i]) = \sigma(P[1..t]A[i])$. Thus, we can compute $\sigma(Z(i, s, k)A[i])$ by using $\mathsf{next}_\sigma(\cdot)$, namely, $\sigma(Z(i, s, k)A[i]) = \sigma(P[1..t]A[i]) = \mathsf{next}_\sigma(t, A[i])$.

We can easily compute next_B in linear time and space (we give a pseudo-code in Algorithm 1). next_σ was introduced in [15] (as table λ). They also showed that this table can be computed in linear time and space (we give a pseudo-code in Algorithm 2).

Algorithm 1: Construction for next_B

Input: String B of length n, Alphabet Σ
Output: next_B

foreach *character* $\alpha \in \Sigma$ **do** $\mathsf{next}_B(n, \alpha) = n + 1$;
for $j = n - 1$ **to** 0 **do**
  foreach $\alpha \in \Sigma$ **do**
    if $\alpha = B[j + 1]$ **then** $\mathsf{next}_B(j, \alpha) = j + 1$;
    else $\mathsf{next}_B(j, \alpha) = \mathsf{next}_B(j + 1, \alpha)$;
return next_B

We have finished describing how to compute d. This algorithm computes $O(m^2r)$ values (i.e., the size of the table d). We can see that every operation can be done in constant time. Thus, this algorithm takes $O(n|\Sigma| + m^2r)$ time and space. This complexity is similar to Wang et al.s' result (algorithm described in Sect. 3.2). We can modify our algorithm to compute d more efficiently by using the following two observations.

Observation 2. *Assume that we have already computed table d until the i-th diagonal line (i.e., the diagonal line which has $d(i, 0, \cdot)$). Let s' be the lowest row*

Algorithm 2: Construction for next_σ

Input: String P of length r, Alphabet Σ
Output: next_σ

$kmp(0) \leftarrow -1$;
$kmp(1) \leftarrow 0$;
$k \leftarrow 0$;
for $i = 2$ **to** r **do**
  while $k \geq 0$ *and* $P[k+1] \neq P[i]$ **do** $k \leftarrow kmp(k)$;
  $k \leftarrow k + 1$;
  $kmp(i) \leftarrow k$;
$\mathsf{next}_\sigma(0, P[1]) \leftarrow 1$;
foreach $\alpha \in \Sigma - \{P[1]\}$ **do**
  $\mathsf{next}_\sigma(0, \alpha) \leftarrow 0$;
for $k = 1$ **to** $r - 1$ **do**
  foreach $\alpha \in \Sigma$ **do if** $\alpha = P[k+1]$ **then** $\mathsf{next}_\sigma(k, \alpha) \leftarrow k + 1$;
  else $\mathsf{next}_\sigma(k, \alpha) \leftarrow \mathsf{next}_\sigma(\mathrm{kmp}(k), \alpha)$;
return next_σ

which has a value smaller than $n+1$. Then, we do not need to compute the last $s'+1$ diagonal lines since these diagonal lines do not make better candidates for STR-EC-LCS.

Observation 3. *If $d(i, s, k) = n + 1$ for all k, then $d(i + 1, s + 1, k) = \ldots = d(i + (m - i), s + (m - i), k) = n + 1$ holds for any k.*

Thanks to the above observations, the number of values which we need to compute is $O((L+1)(m-L+1)r)$ where L is the length of STR-EC-LCS (see also Fig. 4).

Finally, we discuss how to store d. We consider computing the i-th diagonal line (i.e., $d(i, 0, k), \ldots, d(i + (m - i), m - i, k)$). Suppose that $d(i, 0, k), \ldots, d(i + t - 1, t - 1, k)$ have already been computed. Then, we store these values by using an array of size $2^{\lceil \log t \rceil}$. If the array filled with values for the line (i.e., $d(i + 2^{\lceil \log t \rceil} - 1, 2^{\lceil \log t \rceil} - 1, k) < n + 1$ for some k), we make new array of size $2^{\lceil \log t \rceil + 1}$ for values $d(i, 0, k), \ldots, d(i + 2^{\lceil \log t \rceil + 1} - 1, 2^{\lceil \log t \rceil + 1} - 1, k)$ on the line. By Observation 3, we will compute at most $L + 2$ values for each line, the total length of arrays for each line is $O(L)$, where L is the length of an STR-EC-LCS. Therefore, we can compute the length of an STR-EC-LCS in $O(n|\Sigma| + (L + 1)(m - L + 1)r)$ time and space.

Computing an STR-EC-LCS. If we want to compute an STR-EC-LCS, we store a pair (s', k') for every $d(i, s, k)$. The pair (s', k') represents that $d(i, s, k)$ was given by $d(i - 1, s', k')$. By using these information, we can compute an STR-EC-LCS from right to left. We show an example in Fig. 5.

Since we can store (s', k') in constant time and space for each $d(i, s, k)$, and compute an STR-EC-LCS in $O(m)$ time, we can get the following main result.

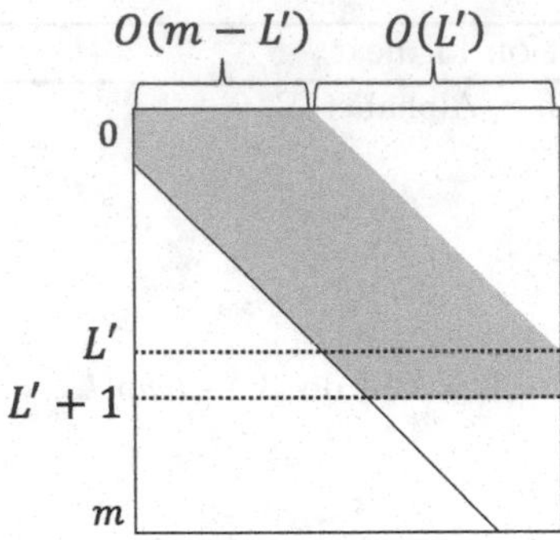

Fig. 4. This is a table for some k. Due to Observations 2 and 3, we do not need to compute values in white part (there might exist positions which do not need their values). The maximum number of values which we need to compute (namely, the total area of the r gray parts) is $O((L+1)(m-L+1)r)$.

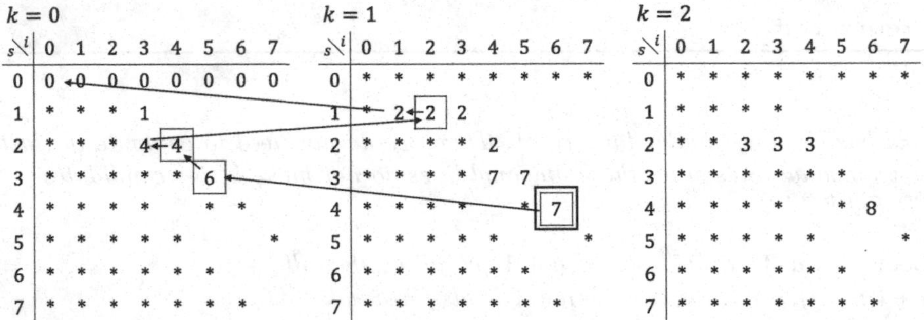

Fig. 5. In this figure, an arrow represents additional information for backtracking. For instance, $d(6,4,1) = 7$ was given by $d(5,3,0) = 6$ while computing d. We can get an STR-EC-LCS `abca` of $A =$ `aabacab`, $B =$ `baabbcaa`, and $P =$ `aab`.

Theorem 1. *For given strings A, B and P, we can compute an STR-EC-LCS in $O(n|\Sigma| + (L+1)(m-L+1)r)$ time and space where m, n, r and L are the length of A, B, P and the STR-EC-LCS, respectively.*

Acknowledgments. This work was supported by JSPS KAKENHI Grant Numbers JP18K18002 (YN), JP17H01697 (SI), JP16H02783 (HB), JP18H04098 (MT), and by JST PRESTO Grant Number JPMJPR1922 (SI).

References

1. Abboud, A., Backurs, A., Williams, V.V.: Tight hardness results for LCS and other sequence similarity measures. FOCS **2015**, 59–78 (2015)
2. Ahsan, S.B., Aziz, S.P., Rahman, M.S.: Longest common subsequence problem for run-length-encoded strings. In: 2012 15th International Conference on Computer and Information Technology (ICCIT), pp. 36–41, December 2012

3. Bunke, H., Csirik, J.: An improved algorithm for computing the edit distance of run-length coded strings. Inf. Process. Lett. **54**(2), 93–96 (1995). http://www.sciencedirect.com/science/article/pii/002001909500005W
4. Chen, Y.C., Chao, K.M.: On the generalized constrained longest common subsequence problems. J. Comb. Optim. **21**(3), 383–392 (2011). https://doi.org/10.1007/s10878-009-9262-5
5. Chin, F.Y., Santis, A.D., Ferrara, A.L., Ho, N., Kim, S.: A simple algorithm for the constrained sequence problems. Inf. Process. Lett. **90**(4), 175–179 (2004). http://www.sciencedirect.com/science/article/pii/S0020019004000614
6. Deorowicz, S.: Quadratic-time algorithm for a string constrained lcs problem. Inf. Process. Lett. **112**(11), 423–426 (2012). http://www.sciencedirect.com/science/article/pii/S0020019012000567
7. Inenaga, S., Hyyrö, H.: A hardness result and new algorithm for the longest common palindromic subsequence problem. Inf. Process. Lett. **129**, 11–15 (2018)
8. Kuboi, K., Fujishige, Y., Inenaga, S., Bannai, H., Takeda, M.: Faster STR-IC-LCS computation via RLE. In: Kärkkäinen, J., Radoszewski, J., Rytter, W. (eds.) 28th Annual Symposium on Combinatorial Pattern Matching, CPM 2017, 4–6 July 2017, Warsaw, Poland. LIPIcs, vol. 78, pp. 20:1–20:12. Schloss Dagstuhl - Leibniz-Zentrum fuer Informatik (2017). https://doi.org/10.4230/LIPIcs.CPM.2017.20
9. Liu, J.J., Wang, Y.L., Chiu, Y.S.: Constrained longest common subsequences with run-length-encoded strings. Comput. J. **58**(5), 1074–1084 (2014). https://doi.org/10.1093/comjnl/bxu012
10. Nakatsu, N., Kambayashi, Y., Yajima, S.: A longest common subsequence algorithm suitable for similar text strings. Acta Inf. **18**, 171–179 (1982). https://doi.org/10.1007/BF00264437
11. Stern, H., Shmueli, M., Berman, S.: Most discriminating segment - longest common subsequence (MDSLCS) algorithm for dynamic hand gesture classification. Pattern Recogn. Lett. **34**(15), 1980–1989 (2013). http://www.sciencedirect.com/science/article/pii/S0167865513000512, smart Approaches for Human Action Recognition
12. Tsai, Y.T.: The constrained longest common subsequence problem. Inf. Process. Lett. **88**(4), 173–176 (2003). http://www.sciencedirect.com/science/article/pii/S002001900300406X
13. Wagner, R.A., Fischer, M.J.: The string-to-string correction problem. J. ACM **21**(1), 168–173 (1974). https://doi.org/10.1145/321796.321811
14. Wang, C., Zhang, D.: A novel compression tool for efficient storage of genome resequencing data. Nucleic Acids Res. **39**(7), e45 (2011). https://doi.org/10.1093/nar/gkr009
15. Wang, L., Wang, X., Wu, Y., Zhu, D.: A dynamic programming solution to a generalized LCS problem. Inf. Process. Lett. **113**(19–21), 723–728 (2013). https://doi.org/10.1016/j.ipl.2013.07.005

Kernels of Sub-classes of Context-Free Languages

Martin Kutrib(✉)

Institut für Informatik, Universität Giessen, Arndtstr. 2, 35392 Giessen, Germany
kutrib@informatik.uni-giessen.de

Abstract. While the closure of a language family $\mathscr{L}$ under certain language operations is the least family of languages which contains all members of $\mathscr{L}$ and is closed under all of the operations, a kernel of $\mathscr{L}$ is a greatest family of languages which is a subfamily of $\mathscr{L}$ and is closed under all of the operations. Here we investigate properties of kernels of general language families and operations defined thereon as well as kernels of (deterministic) (linear) context-free languages with a focus on Boolean operations. While the closures of language families usually are unique, this uniqueness is not obvious for kernels. We consider properties of language families and operations that yield unique and non-unique, that is a set, of kernels. For the latter case, the question whether the union of all kernels coincides with the language family, or whether there are languages that do not belong to any kernel is addressed. Furthermore, the intersection of all kernels with respect to certain operations is studied in order to identify sets of languages that belong to all of these kernels.

1 Introduction

Classical and well-developed concepts to represent (formal) languages are, for example, grammars, language equations, or accepting automata. Similarly, families of languages can be represented in several ways. For example, a language family can be defined to be the family of all languages represented by a certain type of grammar, automaton model, language equation, or by applying appropriate operations on other language families. From a practical point of view, there is often a considerable interest in language families that are robust with respect to language operations, that is, the families are preferably closed under the operations, and/or in language families that admit efficient recognizers. A good example are context-free languages, that are one of the most important and most developed area of formal language theory. However, the family is not closed under the two Boolean operations complementation and intersection. Moreover, the known upper bound on the time complexity for context-free language recognition still exceeds $O(n^2)$. As an approach to characterize language families having strong closure properties and efficient recognizers but decrease the expressive capacity only slightly, closures of sub-classes of the context-free languages have been investigated.

A. Chatzigeorgiou et al. (Eds.): SOFSEM 2020, LNCS 12011, pp. 136–147, 2020.
https://doi.org/10.1007/978-3-030-38919-2_12

The Boolean closure of the linear context-free languages offers a significant increase in expressive capacity compared with the linear context-free languages itself. In addition, it preserves the attractively efficient recognition algorithm taking $O(n^2)$ time and $O(n)$ space [11]. In [12], a characterization of deterministic real-time one-way cellular automata by so-called linear conjunctive grammars has been shown. Linear conjunctive grammars are basically linear context-free grammars augmented with an explicit intersection operation, where the number of intersections is, in some sense, not bounded as in a Boolean formula. The systematic investigation of the Boolean closures of arbitrary and deterministic context-free languages started in [14–16], in particular, motivated by the question "How much more powerful is nondeterminism than determinism?" The closure of deterministic languages under the regular operations is studied in [1], while the regular closure of the linear context-free languages is considered in [10].

Here we are interested in language families with strong closure properties obtained by looking into a given family instead of closing and, thus, extending the family. To this end, we study the notion of kernels of language families. Basically, a kernel of some family $\mathscr{L}$ with respect to some language operations defined on $\mathscr{L}$ is a greatest sub-family of $\mathscr{L}$ that is closed under the operations. For example, the family of linear context-free languages is not closed under complementation. Its complementation kernel consists of all linear context-free languages whose complement is also linear context free. This kernel is also known as the family of strongly linear context-free languages that is considered in [8] with respect to its expressive capacity and closure properties. Another question that motivates the concept is as follows. Given a language such that also its complement belongs to the same family, the description of which of both is more economic [8]? For example, it is known that a nondeterministic finite automaton can require 2^n states to accept the complement of a language accepted by an n-state nondeterministic finite automaton [9]. So, a representation of the complement by the n-state automaton together with a bit that says that actually the complement of the language accepted is meant is much more economic from the descriptional complexity point of view. A machine characterization of the complementation kernel of the context-free languages in terms of self-verifying pushdown automata is obtained in [2].

Another well-understood kernel is the family of recursive languages. It is the complementation kernel of the recursively enumerable languages.

The paper is organized as follows. After presenting the basic definitions and notions in the next section, Sect. 3 deals with the uniqueness of kernels. The underlying results are as general as possible while clarifying examples often deal with sub-classes of context-free languages. The question whether any language of a family belongs to some kernel based on given operations is dealt with in Sect. 4. More precisely, we are interested in the question whether the union of all kernels coincides with the language family. The intersection of all of these kernels and its related questions are considered in Sect. 5. Finally, we discuss some interesting untouched problems and questions for further research in Sect. 6.

2 Preliminaries

We write Σ^* for the set of all words over a finite alphabet Σ. The *empty word* is denoted by λ, and we set $\Sigma^+ = \Sigma^* \setminus \{\lambda\}$. The *reversal* of a word w is denoted by w^R, and for the *length* of w we write $|w|$. Set *inclusion* is denoted by $\subseteq$ and *strict set inclusion* by $\subset$.

A subset of Σ^* is called a *(formal) language* over Σ. A *language operation* is an operation whose finite number of parameters are languages, and whose result is a language. For example, the *complement* of a language is defined with respect to the underlying alphabet Σ. For a language $L \subseteq \Sigma^*$, the *complement* $\overline{L}$ of L is $\{ w \in \Sigma^* \mid w \notin L \}$. For all $k \geq 1$, a $k$ary language operation $\circ$ is said to be *idempotent* if $\circ(L, L, \ldots, L) = L$, for all L in the domain of $\circ$. For easier writing, here we call even a unary language operation $\circ$ with the property $\circ(L) = L$ idempotent (so we do *not* require $\circ(\circ(L)) = \circ(L)$).

Let Ω be an infinite enumerable set of letters. The set $\mathscr{L}$ is a *family of languages* over Ω if for each $L \in \mathscr{L}$ there is a finite subset $\Sigma \subset \Omega$ such that $L \subseteq \Sigma^*$. In the sequel we tacitly omit Ω when it is understood. For a family of languages $\mathscr{L}$, the family of complements CO-$\mathscr{L}$ is defined to be $\{ \overline{L} \mid L \in \mathscr{L} \}$.

Let $\mathscr{L}$ be a family of languages and $op_1, op_2, \ldots, op_k$, $k \geq 1$, be a finite number of operations defined on $\mathscr{L}$.

1. Then $\Gamma_{op_1,op_2,\ldots,op_k}(\mathscr{L})$ denotes the $(op_1, op_2, \ldots, op_k)$ *closure of* $\mathscr{L}$. That is, the *least family of languages which contains all members of* $\mathscr{L}$ *and is closed under* $op_1, op_2, \ldots, op_k$. In other words, there exists no language family $\mathscr{L}'$ that is closed under $op_1, op_2, \ldots, op_k$ such that $\mathscr{L} \subseteq \mathscr{L}' \subset \Gamma_{op_1,op_2,\ldots,op_k}(\mathscr{L})$.
2. By $\gamma_{op_1,op_2,\ldots,op_k}(\mathscr{L})$ we denote the set of $(op_1, op_2, \ldots, op_k)$ *kernels of* $\mathscr{L}$. That is, the set of *greatest families of languages which are subfamilies of* $\mathscr{L}$ *and are closed under* $op_1, op_2, \ldots, op_k$. In other words, for all kernels $\kappa \in \gamma_{op_1,op_2,\ldots,op_k}(\mathscr{L})$ there exists no language family $\mathscr{L}'$ that is closed under $op_1, op_2, \ldots, op_k$ such that $\kappa \subset \mathscr{L}' \subseteq \mathscr{L}$.

In particular, we consider the operations complementation ($\sim$), union ($\cup$), and intersection ($\cap$), which are called *Boolean operations.* Accordingly, we write Γ_{BOOL} for $\Gamma_{\sim,\cup,\cap}$ and γ_{BOOL} for$\gamma_{\sim,\cup,\cap}$.

Since special attention is paid to sub-classes of context-free languages, we recall briefly the notion of a context-free grammar and refer to the literature, for example to [7], for detailed definitions of the characterizing automata models.

A *context-free grammar* is a system $G = \langle N, T, S, P \rangle$, where N and T are the disjoint alphabets of nonterminals and terminals, $S \in N$ is the axiom, and P is the finite set of productions of the form $A \to u$, where $A \in N$ and $u \in (N \cup T)^*$. A context-free grammar is said to be *linear* if and only if for all productions the right-hand side u contains at most one nonterminal, that is, $u \in (T^*NT^*) \cup T^*$. A linear grammar is said to be *left-linear* if and only if a nonterminal may only appear as leftmost symbol at the right-hand side of the productions, that is, $u \in (NT^*) \cup T^*$.

The language *generated* by G is the set $\{w \in T^* \mid S \Rightarrow^* w\}$, where $\Rightarrow^*$ denotes the reflexive, transitive closure of the derivation relation $\Rightarrow$.

The families of languages that can be generated by context-free, linear, and left-linear grammars are called context-free (CFL), linear (LIN), and regular (REG) languages. The automaton model for the recognition of context-free languages is the nondeterministic pushdown automaton. Its deterministic variant characterizes the deterministic context-free languages (DCFL). As for DCFL there is an automaton model for linear languages. Restricting a pushdown automaton such that it may switch from increasing the height of its pushdown to decreasing it only once, thus performing only one turn, leads to the definition of one-turn pushdown automata [5]. It is known that nondeterministic one-turn pushdown automata characterize the linear languages and deterministic one-turn pushdown automata define the deterministic linear languages (DLIN).

3 Uniqueness of Kernels

While the closures of language families under all of the usually considered operations are unique language families, this uniqueness is not obvious for kernels. In fact, it does not always hold. On the other hand, if the kernels are based on unary operations then they are unique, that is, the corresponding set of kernels γ is a singleton.

Proposition 1. *Let $\mathscr{L}$ be a family of languages and $\circ$ be a unary operation defined on $\mathscr{L}$. Then the set $\gamma_\circ(\mathscr{L})$ is a singleton.*

Proof. For any language L from $\mathscr{L}$, the application of $\circ$, that is $\circ(L)$, either does belong to $\mathscr{L}$ or not. Now we consider the iterated application of $\circ$ to $L \in \mathscr{L}$ and define $\circ^1 = \circ$ and, for $1 \leq i$,

$$\circ^{i+1}(L) = \begin{cases} \circ(\circ^i(L)) & \text{if } \circ^i(L) \in \mathscr{L} \\ \text{undefined} & \text{else} \end{cases}.$$

So, the iterated application of $\circ$ to languages from $\mathscr{L}$ induces a finite or infinite sequence of (not necessarily different) languages.

If this sequence is finite for some $L \in \mathscr{L}$ then language L does not belong to any $\circ$ kernel of $\mathscr{L}$, since otherwise the kernel would not be closed under $\circ$.

If this sequence is infinite then language L does belong to all $\circ$ kernels of $\mathscr{L}$. If not, all languages $L, \circ^1(L), \circ^2(L), \dots$ could be added to the kernel without affecting its closure under $\circ$ or its containment in $\mathscr{L}$, a contradiction to the maximality of the kernel.

We conclude that any language from $\mathscr{L}$ either belongs to all $\circ$ kernels or to none $\circ$ kernel. So, the kernel is uniquely determined. □

In general, the uniqueness is lost for *k*ary operations if $k \geq 2$.

Theorem 2. *Let $\mathscr{L}$ be a family of languages, $k \geq 2$, and $\circ$ be a kary idempotent operation defined on $\mathscr{L}$. Then the set $\gamma_\circ(\mathscr{L})$ includes more than one kernel if and only if $\mathscr{L}$ is not closed under $\circ$.*

Proof. If $\mathscr{L}$ is closed under $\circ$, it is its own $\circ$ kernel and, thus, $\gamma_\circ(\mathscr{L})$ is a singleton.

Now assume that $\mathscr{L}$ is not closed under $\circ$ and let $L_1, L_2, \ldots, L_k \in \mathscr{L}$ be witnesses for the non-closure. That is, $\circ(L_1, L_2, \ldots, L_k) \notin \mathscr{L}$. First, we argue that any of the witness languages, say L_i, belongs to a $\circ$ kernel of $\mathscr{L}$. To this end, it suffices to consider the set $\{L_i\}$ which is a subset of $\mathscr{L}$. Since $\circ$ is idempotent the set $\{L_i\}$ is closed under $\circ$. So, either it is a kernel or it is a subset of some kernel.

Now it remains to be concluded that not all of the languages $L_1, L_2, \ldots, L_k$ can belong to the same kernel, since this would violate the closure under $\circ$. So, there are at least two different kernels in $\gamma_\circ(\mathscr{L})$. $\square$

So far, we obtained that the $\circ$ kernel of some language family is unique if $\circ$ is a unary operation or if the family is closed under $\circ$, and that there are more than one kernels if $\circ$ is a $k$ary idempotent operation, for $k \geq 2$, and $\mathscr{L}$ is not closed under $\circ$. The following examples reveal that a finite as well as an infinite number of kernels may exist.

Example 3. Let $\mathscr{L}$ be defined as union of CFL with $\{L_{\text{expo}}\}$, where L_{expo} is the non-context-free unary language $\{ a^{2^n} \mid n \geq 0 \}$. Family $\mathscr{L}$ is not closed under the idempotent operation union since, for example, $L_{\text{expo}} \cup \{aaa\}$ is not context free and, thus, does not belong to $\mathscr{L}$. By Theorem 2, $\gamma_\cup(\mathscr{L})$ includes more than one kernel. In particular, CFL is included in $\gamma_\cup(\mathscr{L})$, since CFL is closed under union. This is the only union kernel of $\mathscr{L}$ that does not include L_{expo}.

On the other hand, there must exist a kernel in $\gamma_\cup(\mathscr{L})$ having $\{L_{\text{expo}}\}$ as subset, since $\{L_{\text{expo}}\}$ is closed under union and a subset of $\mathscr{L}$. We show that there is exactly one union kernel of $\mathscr{L}$ that includes L_{expo}.

Let $U = \{ L \mid L \text{ is finite subset of } L_{\text{expo}} \}$ be the set of finite languages whose words belong to L_{expo}, and let $R = \{ L \in \text{CFL} \mid (L \cup L_{\text{expo}}) \cap a^* \in \text{REG} \}$ be the set of context-free languages whose unary words from a^* form a regular language when joint with L_{expo}. We claim that $\kappa = U \cup R \cup \{L_{\text{expo}}\}$ is the sole union kernel of $\mathscr{L}$ that includes L_{expo}.

Clearly, we have the inclusion $\kappa \subset \mathscr{L}$. To show that κ is closed under union, let $u, u' \in U$ and $r, r' \in R$. We obtain $u \cup L_{\text{expo}} = L_{\text{expo}} \in \kappa$, $u \cup u' \in U \subset \kappa$, and $u \cup r \in \text{CFL}$, $(u \cup r \cup L_{\text{expo}}) \cap a^* = (r \cup L_{\text{expo}}) \cap a^*$ and, thus $u \cup r \in R \subset \kappa$. Further, we have $r \cup L_{\text{expo}} \cup L_{\text{expo}} = r \cup L_{\text{expo}}$ and, therefore, $r \cup L_{\text{expo}} \in R \subset \kappa$, and $(r \cup r' \cup L_{\text{expo}}) \cap a^* = \big((r \cup L_{\text{expo}}) \cap a^*\big) \cup \big((r' \cup L_{\text{expo}}) \cap a^*\big) \in \text{REG}$ and, thus $r \cup r' \in R \subset \kappa$. We conclude that κ is closed under union.

Finally, it remains to be shown that none of the languages $\mathscr{L} \setminus \kappa$ can belong to any union kernel of $\mathscr{L}$ that includes L_{expo}. This implies that κ is maximal and therefore, in fact, a kernel, and that it is the unique.

So, let $L \in \mathscr{L} \setminus \kappa$. If L includes at least one word that is not of the form a^*, the union $L \cup L_{\text{expo}}$ is not equal to L_{expo}. Since L does not belong to R, we have that $(L \cup L_{\text{expo}}) \cap a^*$ is unary but not regular. So, it is not context free either. Since context-free languages are closed under intersection with regular languages, $L \cup L_{\text{expo}}$ is not context free. It follows that no union kernel of $\mathscr{L}$ that includes L_{expo} includes L.

Next, assume that all words in L are of the form a^*. Since L does not belong to R we now from the previous case that $(L \cup L_{\text{expo}}) \cap a^* = L \cup L_{\text{expo}}$ is not context free. So, if L belongs to the kernel, $L \cup L_{\text{expo}}$ has to be equal to L_{expo}. This implies $L \subseteq L_{\text{expo}}$. Since $L \notin \kappa$ the inclusion is proper: $L \subset L_{\text{expo}}$. Since any infinite subset of L_{expo} is not context free and any finite subset does belong to $U \in \kappa$, we obtain the contradiction that L cannot belong to $\mathscr{L} \setminus \kappa$.

So, we have shown that the set $\gamma_{\cup}(\mathscr{L})$ consists of exactly two kernels, one includes L_{expo} and the other does not. ■

Example 4. The family DLIN is not closed under intersection. We consider the number of kernels in $\gamma_{\cap}(\text{DLIN})$. To this end, for $k \geq 2$, define language $L_k = \{ a^n(\$a^*)^{k-2}\$a^n(\$a^*)^* \mid n \geq 0 \}$ that belongs to DLIN. However, for all $2 \leq i < j$, the intersection $L_i \cap L_j$ is language

$$\{ a^n(\$a^*)^{i-2}\$a^n(\$a^*)^{j-i-1}\$a^n(\$a^*)^* \mid n \geq 0 \}$$

which is not even context free. We conclude that for $2 \leq i, j$ the languages L_i and L_j do not belong to the same kernel if $i \neq j$. On the other hand, for $2 \leq k$, there must exist a kernel in $\gamma_{\cap}(\text{DLIN})$ having $\{L_k\}$ as subset, since it is closed under intersection and a subset of DLIN. So, the set $\gamma_{\cap}(\text{DLIN})$ includes infinitely many kernels. ■

4 Union of Kernels

Next we turn to the question whether any language of a family belongs to some kernel based on given operations. Or are there languages that do not belong to any of such kernels. More precisely, we are interested in the question whether the union of all kernels coincides with the language family.

Theorem 5. *Let $\mathscr{L}$ be a family of languages and $op_1, op_2, \ldots, op_k$, $k \geq 1$, be a finite number of idempotent operations defined on $\mathscr{L}$. Then*

$$\{ L \mid L \in \kappa \ \textit{for some} \ \kappa \in \gamma_{op_1,op_2,\ldots,op_k}(\mathscr{L}) \} = \mathscr{L}.$$

Proof. The inclusion in $\mathscr{L}$ is trivial. So, it remains to be shown that any language from $\mathscr{L}$ does belong to some $(op_1, op_2, \ldots, op_k)$ kernel of $\mathscr{L}$.

To this end, let $L \in \mathscr{L}$ be an arbitrary language from the family. We consider the set $\nu = \{L\}$. Since it contains only one language and all operations $op_1, op_2, \ldots, op_k$ are idempotent, it is closed under $op_1, op_2, \ldots, op_k$. So, either ν is itself a $(op_1, op_2, \ldots, op_k)$ kernel of $\mathscr{L}$, or there exist a kernel in $\gamma_{op_1,op_2,\ldots,op_k}(\mathscr{L})$ having ν as subset. □

Example 6. Consider the families DLIN, LIN, DCFL, as well as CFL and the idempotent operations union and intersection. Theorem 5 says that any language from one of the families belongs to some $(\cup, \cap)$ kernel of that family. That is,

$$\{ L \mid L \in \kappa \ \text{for some} \ \kappa \in \gamma_{\cup,\cap}(\mathscr{L}) \} = \mathscr{L},$$

for $\mathscr{L} \in \{\text{DLIN}, \text{LIN}, \text{DCFL}, \text{CFL}\}$. ■

Theorem 5 reveals in particular that idempotent operations do not prevent languages from belonging to a kernel. Let us discuss the role played by the requirement that the operations have to be idempotent. If a unary operation is idempotent, *any* language family is closed under this operation (in fact, the operation is the identity). However, if at least one unary operation under which the family is not closed is in the list, the situation changes.

Proposition 7. *Let $\mathscr{L}$ be a family of languages not closed under the unary operation $\circ$, and $op_1, op_2, \ldots, op_k$, $k \geq 0$, be a finite number of further operations defined on $\mathscr{L}$. Then $\{ L \mid L \in \kappa$ for some $\kappa \in \gamma_{\circ,op_1,op_2,\ldots,op_k}(\mathscr{L}) \} \subset \mathscr{L}$.*

Proof. The inclusion claimed is trivial. So, it remains to be shown that the inclusion is strict.

Since $\mathscr{L}$ is not closed under $\circ$, there is a language $L \in \mathscr{L}$ such that $\circ(L) \notin \mathscr{L}$. So, L cannot belong to any $(\circ, op_1, op_2, \ldots, op_k)$ kernel of $\mathscr{L}$, since the containment would violate the closure of the kernel under $\circ$. □

Example 8. It is well-known that the family CFL is not closed under complementation. Applying Proposition 7 shows that not all context-free languages belong to some Boolean kernel. That is, $\{ L \mid L \in \kappa$ for some $\kappa \in \gamma_{\text{BOOL}}(\text{CFL}) \} \subset$ CFL. ■

In general, the condition of Proposition 7, namely that the family $\mathscr{L}$ has not to be closed under the unary operation, cannot be relaxed. The following proposition shows this fact. It is in contrast to Example 8.

Proposition 9. *Any deterministic context-free language belongs to some kernel $\kappa \in \gamma_{BOOL}(\text{DCFL})$.*

Proof. Let $L \in$ DCFL be some language over the alphabet Σ. We consider the set $\nu = \{L, \overline{L}, \Sigma^*, \emptyset\}$ which is clearly closed under complementation, union, and intersection.

Since DCFL is closed under complementation and includes the regular languages Σ^* and $\emptyset$, either ν is itself a Boolean kernel of DCFL, or there exists a kernel in $\gamma_{\text{BOOL}}(\text{DCFL})$ having ν, and thus $\{L\}$, as subset. □

In order to continue the discussion of the requirement that the operations have to be idempotent, we present a further example considering the binary non-idempotent operation of marked concatenation.

Example 10. The family LIN is not closed under the binary non-idempotent operation of marked concatenation ($\bullet$). In fact, it has been shown in [6] that the marked concatenation of two linear context-free languages is linear context free if and only if at least one of the languages is regular.

We consider $\gamma_\bullet(\text{LIN})$. Since the family REG is closed under marked concatenation, there must be some $\kappa \in \gamma_\bullet(\text{LIN})$ such that REG $\subseteq \kappa$. On the other hand, let $L \in \text{LIN} \setminus \text{REG}$ be an arbitrary linear context-free language that is not regular. Then L cannot belong to any kernel in $\gamma_\bullet(\text{LIN})$ since $L \bullet L$ is not linear context free due to [6]. Therefore, REG is the sole marked concatenation kernel of LIN. That is, $\gamma_{\bullet}(\text{LIN}) = \{\text{REG}\}$ and, thus, the marked concatenation kernel of LIN is unique. Moreover, $\{\, L \mid L \in \kappa \text{ for some } \kappa \in \gamma_{\bullet}(\text{LIN}) \,\} \subset \text{LIN}$. ■

It is worth mentioning that literally Example 10 also applies to the family DLIN.

5 Intersection of Kernels

We now turn to the question which languages belong to all kernels based on given operations. So, we consider the intersection of all of these kernels.

Proposition 11. *Let $\mathscr{L} \in \{\text{CFL}, \text{LIN}, \text{DCFL}, \text{DLIN}\}$. All intersection kernels and union kernels of $\mathscr{L}$ include* REG.

Proof. In contrast to the assertion assume that there is a kernel $\nu \in \gamma_{\cap}(\mathscr{L})$ such that $\text{REG} \not\subseteq \nu$.

In order to obtain a contradiction we show that ν is strictly included in a kernel from $\gamma_{\cap}(\mathscr{L})$ and, thus, cannot be an intersection kernel of $\mathscr{L}$ at all. To this end, we join ν with REG and build the intersection closure of the union. That is, we consider $\kappa = \Gamma_{\cap}(\nu \cup \text{REG})$.

Any language $L \in \kappa$ has a representation of the form K, R, or $K \cap R$, where $K \in \nu$ and $R \in \text{REG}$. Since $\mathscr{L}$ includes the regular languages and is closed under intersection with regular languages, language L belongs to $\mathscr{L}$. So, we have $\Gamma_{\cap}(\nu \cup \text{REG}) \subseteq \mathscr{L}$. This shows the assertion for intersection kernels.

Since $\mathscr{L}$ is closed under union with regular languages as well, the argumentation for union kernels follows by replacing intersection with union. □

Of particular interest are the languages that belong to *all* Boolean kernels.

Theorem 12. *Let $\mathscr{L} \supseteq \mathscr{T}$ be two families of languages. If $\mathscr{L}$ is closed under union and under intersection with languages from $\mathscr{T}$, and $\mathscr{T}$ is closed under the Boolean operations then $\mathscr{T} \subseteq \kappa$ for all $\kappa \in \gamma_{BOOL}(\mathscr{L})$.*

Proof. In contrast to the assertion assume that there is a kernel $\nu \in \gamma_{\text{BOOL}}(\mathscr{L})$ such that $\mathscr{T} \not\subseteq \nu$.

In order to obtain a contradiction we show that ν is strictly included in a kernel from $\gamma_{\text{BOOL}}(\mathscr{L})$ and, thus, cannot be a Boolean kernel of $\mathscr{L}$ at all. To this end, we join ν with $\mathscr{T}$ and build the Boolean closure of the union. That is, we consider $\kappa = \Gamma_{\text{BOOL}}(\nu \cup \mathscr{T})$. We show that κ is included in $\mathscr{L}$.

Let $L \in \kappa$. Then, for some $m, l_1, l_2, \ldots, l_m \geq 0$, language L has a representation $\bigcup_{1 \leq i \leq m} \bigcap_{1 \leq j \leq l_i} L_{i,j}$ such that $L_{i,j} \in (\nu \cup \mathscr{T})$ or $L_{i,j} \in \text{CO-}(\nu \cup \mathscr{T})$. Since ν as well as $\mathscr{T}$ are closed under complementation, we have $(\nu \cup \mathscr{T}) = \text{CO-}(\nu \cup \mathscr{T})$, and may safely assume that $L_{i,j} \in (\nu \cup \mathscr{T})$.

Now, for $1 \leq i \leq m$, let $L_i = L_{i,1} \cap L_{i,2} \cap \cdots \cap L_{i,l_i}$. Since ν as well as $\mathscr{T}$ are closed under intersection, we have $L_i = K_i \cap T_i$ or $L_i = K_i$ or $L_i = T_i$, for some

$K_i \in \nu$ and $T_i \in \mathscr{T}$. Moreover, since ν and $\mathscr{T}$ are sub-families of $\mathscr{L}$, and $\mathscr{L}$ is closed under intersection with languages from $\mathscr{T}$, language L_i belongs to $\mathscr{L}$.

Finally, $L = \bigcup_{1 \le i \le m} L_i$ and the closure of $\mathscr{L}$ under union implies that L belongs to $\mathscr{L}$. Therefore, κ is included in $\mathscr{L}$. □

Corollary 13. *Let $\mathscr{L} \supseteq \mathscr{T}$ be two families of languages. If $\mathscr{L}$ is closed under intersection and under union with languages from $\mathscr{T}$, and $\mathscr{T}$ is closed under the Boolean operations then $\mathscr{T} \subseteq \kappa$ for all $\kappa \in \gamma_{BOOL}(\mathscr{L})$.*

Proof. The corollary can be shown almost literally as Theorem 12, where the representation of language $L \in \kappa$ is given as $\bigcap_{1 \le i \le m} \bigcup_{1 \le j \le l_i} L_{i,j}$, and by interchanging union and intersection in the reasoning. □

Example 14. The families CFL and LIN are closed under union and under intersection with regular languages. The family of regular languages is closed under the Boolean operations. So, by applying Theorem 12 we obtain that *all* Boolean kernels of CFL and LIN include REG.

Moreover, applying Corollary 13 shows that *all* Boolean kernels of CO-CFL and CO-LIN include REG. ■

Since any intersection, union, and complementation kernel of CFL, LIN, CO-CFL, and CO-LIN includes a Boolean kernel which, in turn, includes REG, all of these kernels include REG as well. Moreover, for all unary operations $\circ$ under which the family of regular languages is closed, the unique $\circ$ kernel of CFL, LIN, CO-CFL, and CO-LIN includes REG (see Proposition 1). This immediately raises the question whether these kernels are characterized by REG. Or are there certain non-regular languages that belong to *all* kernels of a certain type. Example 10 shows that REG is the sole marked concatenation kernel of LIN and, thus, characterizes the kernel. However, in the following we turn to show that there are non-regular languages belonging to the intersection of all Boolean kernels of CFL, LIN, CO-CFL, and CO-LIN.

To this end, we recall the notion of semilinear languages. Consider, for some fixed positive integer m, the vectors in $\mathbb{N}^m$. A set of the form

$$\{ v_0 + x_1 v_1 + x_2 v_2 + \cdots + x_k v_k \mid x_i \ge 0, 1 \le i \le k \},$$

where $v_0, v_1, \ldots, v_k \in \mathbb{N}^m$, is said to be *linear.* A *semilinear* set is a finite union of linear sets. It is known that the family of semilinear subsets of $\mathbb{N}^m$ is closed under union, intersection, and complementation [3]. For an alphabet $\Sigma = \{a_1, a_2, \ldots, a_m\}$ the *Parikh mapping* $\Psi\colon \Sigma^* \to \mathbb{N}^m$ is defined by $\Psi(w) = (|w|_{a_1}, |w|_{a_2}, \ldots, |w|_{a_m})$, where $|w|_{a_i}$ denotes the number of occurrences of a_i in the word w. In [13] a fundamental result concerning the distribution of symbols in the words of a context-free language has been shown. It says that for any context-free language L, the Parikh image $\Psi(L) = \{ \Psi(w) \mid w \in L \}$ is semilinear.

In the following we consider semilinear languages that are subsets of a^*b^*, where the number of b's depends linearly on the number of a's. The dependency is given by linear functions $\varphi\colon \mathbb{N} \to \mathbb{N}$ with $\varphi(n) = c_1 \cdot n + c_0$, for some $c_0, c_1 \geq 0$. So, we define $L_\varphi = \{\, a^n b^{\varphi(n)} \mid n \geq 0 \,\}$. Note that there are functions φ such that L_φ is context free but not regular (for example $\varphi(n) = n$, $\varphi(n) = 2n$, etc.), or L_φ is regular (for example $\varphi(n)$ is constant). However, the linearity of φ implies that L_φ is a semilinear language, where $\Psi(L_\varphi) = \left\{ \binom{0}{c_0} + x\binom{1}{c_1} \,\middle|\, x \geq 0 \right\}$.

Theorem 15. *Let $\varphi\colon \mathbb{N} \to \mathbb{N}$ be a linear function. For an arbitrary context-free language L, the intersection $L \cap L_\varphi$ belongs to* DLIN.

Proof. We consider the Parikh image

$$S = \Psi(L \cap L_\varphi) = \Psi((L \cap a^*b^*) \cap L_\varphi) = \Psi(L \cap a^*b^*) \cap \Psi(L_\varphi).$$

The set S is semilinear since $L \cap a^*b^*$ is context free and, thus, semilinear [13], language L_φ is semilinear, and semilinear sets are closed under intersection [3].

Let $\pi_1\colon \mathbb{N}^2 \to \mathbb{N}$ be the canonical projection on the first factor. Then $\pi_1(S)$ is semilinear. So, the language $U = \{\, a^n \mid n \geq 0, a^n b^{\varphi(n)} \in L \,\} = \Psi^{-1}(\pi_1(S))$ is regular since it is unary and semilinear.

Now, let M be a deterministic finite automaton accepting U. From M one can easily construct a deterministic one-turn pushdown automaton accepting $\{\, a^n b^{\varphi(n)} \mid n \geq 0, a^n \in U \,\} = L \cap L_\varphi$. So, the theorem follows. □

Example 16. Let $\varphi\colon \mathbb{N} \to \mathbb{N}$ be a linear function. Then, for all families $\mathscr{L}$ from $\{\mathrm{CFL}, \mathrm{LIN}, \mathrm{DCFL}, \mathrm{DLIN}\}$, all intersection kernels of $\mathscr{L}$ include all, even non-regular, languages L_φ.

Similar as above we obtain a contradiction when we assume that there is an intersection kernel $\nu \in \gamma_\cap(\mathscr{L})$ such that there is $L_\varphi \notin \nu$.

Consider $\kappa = \Gamma_\cap(\nu \cup \{L_\varphi\})$. Each language $L \in \kappa$ has a representation as K, L_φ, or $K \cap L_\varphi$, where $K \in \nu$.

Since L_φ belongs to $\mathscr{L}$, $K \cap L_\varphi \in \mathrm{DLIN} \subseteq \mathscr{L}$ by Theorem 15, and $\nu \subseteq \mathscr{L}$, the closure $\Gamma_\cap(\nu \cup \{L_\varphi\})$ is included in $\mathscr{L}$, which gives a contradiction to the maximality of ν. ■

The situation changes when in Theorem 15 the language L_φ is replaced by its complement $\overline{L_\varphi}$. It is an immediate observation that in this case the determinism is not generally achieved. However, we can show that the property of being context free or linear context free can be preserved. To this end, we first provide the next lemma.

It has already been shown in [4] that a language $L \subseteq a^*b^*$ is context free if and only if it is semilinear. We turn to strengthen this result to linear context-free languages. Basically, it shows that there are no non-linear context-free languages $L \subseteq a^*b^*$ at all.

Proposition 17. *A language $L \subseteq a^*b^*$ is linear context free if and only if it is semilinear.*

Proof. If language L is linear context free, it is semilinear. So, it is sufficient to show the converse. To this end, let $L \subseteq a^*b^*$ be semilinear. A semilinear subset S of $\mathbb{N}^2$ determines uniquely a language $\Psi^{-1}(S)$ whose words are of the form a^*b^*, that is $L = \Psi^{-1}(\Psi(L))$. Now let the Parikh image $\Psi(L)$ be given by a finite union of sets of the form

$$\left\{ \begin{pmatrix} u_0 \\ v_0 \end{pmatrix} + x_1 \begin{pmatrix} u_1 \\ v_1 \end{pmatrix} + x_2 \begin{pmatrix} u_2 \\ v_2 \end{pmatrix} + \cdots + x_k \begin{pmatrix} u_k \\ v_k \end{pmatrix} \middle| \, x_i \geq 0, 1 \leq i \leq k \right\},$$

where $u_0, v_0, u_1, v_1, \ldots, u_k, v_k \in \mathbb{N}$.

For each of these sets, say set S', we construct a linear context-free grammar that generates $\Psi^{-1}(S')$. Since the family of linear context-free languages is closed under union, this shows the lemma.

The linear context-free grammar for S' is $G = \langle N, T, A, P \rangle$, where $N = \{A\}$, $T = \{a, b\}$, and $P = \{\, A \to a^{u_i} A b^{v_i} \mid 1 \leq i \leq k \,\} \cup \{A \to a^{u_0} b^{v_0}\}$. □

Theorem 18. *Let* $\varphi \colon \mathbb{N} \to \mathbb{N}$ *be a linear function,* $\mathscr{L} \in \{\text{CFL}, \text{LIN}\}$, *and* $L \in \mathscr{L}$ *be arbitrary. Then the intersection* $L \cap \overline{L_\varphi}$ *belongs to* $\mathscr{L}$.

Proof. The intersection $L \cap \overline{L_\varphi}$ consists of all words from L that are not of the form a^*b^*, and all words from L of the form a^*b^* where the number of b's is different from φ applied to the number of a's. So, we have the representation $L \cap \overline{L_\varphi} = (L \setminus a^*b^*) \cup ((L \cap a^*b^*) \setminus L_\varphi)$.

Since $\mathscr{L}$ is closed under set difference with regular languages, $L \setminus a^*b^*$ belongs to $\mathscr{L}$. Since $\mathscr{L}$ is closed under intersection with regular languages, $L \cap a^*b^*$ belongs to $\mathscr{L}$ and, thus, is semilinear. Further, L_φ is semilinear. The family of semilinear languages is closed under set difference [3]. Therefore, $(L \cap a^*b^*) \setminus L_\varphi$ is a semilinear language which, in turn, is linear context free by Proposition 17 and, thus, belongs to $\mathscr{L}$ as well.

Since $\mathscr{L}$ is closed under union, the intersection $L \cap \overline{L_\varphi}$ belongs to $\mathscr{L}$. □

Now we are prepared to show that there are non-regular languages belonging to the intersection of all Boolean kernels of CFL, LIN, CO-CFL, and CO-LIN.

Theorem 19. *Let* $\varphi \colon \mathbb{N} \to \mathbb{N}$ *be a linear function. Then, for all families* $\mathscr{L}$ *from* $\{\text{CFL}, \text{LIN}, \text{CO-CFL}, \text{CO-LIN}\}$, *all Boolean kernels of* $\mathscr{L}$ *include all, even non-regular, languages* L_φ.

6 Untouched Questions

We have started to study the properties of kernels of general language families and operations defined thereon systematically as well as kernels of (deterministic) (linear) context-free languages with a focus on Boolean operations.

Since only less is known about kernels a bunch of questions and problems remain open or untouched. Exemplarily, we mention four of them: (1) The non-trivial closure properties of kernels themselves are of natural interest. (2) Are there hierarchies of kernels? (3) A machine characterization of the complementation kernel of the context-free languages in terms of self-verifying pushdown automata is known [2]. Basically, the characterization is given by a machine for the underlying language family, where the acceptance condition is modified. Are there machine characterizations of other kernels?

Acknowledgment. The author would like to thank Henning Fernau for fruitful discussions at an early stage of the paper.

References

1. Bertsch, E., Nederhof, M.J.: Regular closure of deterministic languages. SIAM J. Comput. **29**, 81–102 (1999)
2. Fernau, H., Kutrib, M., Wendlandt, M.: Self-verifying pushdown automata. In: Non-Classical Models of Automata and Applications (NCMA 2017), vol. 329, pp. 103–117. Austrian Computer Society, Vienna (2017). books@ocg.at
3. Ginsburg, S.: The Mathematical Theory of Context-Free Languages. McGraw Hill, New York (1966)
4. Ginsburg, S., Spanier, E.H.: Bounded ALGOL-like languages. Trans. Am. Math. Soc. **113**, 333–368 (1964)
5. Ginsburg, S., Spanier, E.H.: Finite-turn pushdown automata. SIAM J. Contr. **4**, 429–453 (1966)
6. Greibach, S.A.: The unsolvability of the recognition of linear context-free languages. J. ACM **13**, 582–587 (1966)
7. Harrison, M.A.: Introduction to Formal Language Theory. Addison-Wesley, Reading (1978)
8. Ilie, L., Păun, G., Rozenberg, G., Salomaa, A.: On strongly context-free languages. Discrete Appl. Math. **103**, 158–165 (2000)
9. Jirásková, G.: State complexity of some operations on binary regular languages. Theoret. Comput. Sci. **330**, 287–298 (2005)
10. Kutrib, M., Malcher, A.: Finite turns and the regular closure of linear context-free languages. Discrete Appl. Math. **155**, 2152–2164 (2007)
11. Kutrib, M., Malcher, A., Wotschke, D.: The Boolean closure of linear context-free languages. Acta Inform. **45**, 177–191 (2008)
12. Okhotin, A.: Automaton representation of linear conjunctive languages. In: Ito, M., Toyama, M. (eds.) DLT 2002. LNCS, vol. 2450, pp. 393–404. Springer, Heidelberg (2003). https://doi.org/10.1007/3-540-45005-X_35
13. Parikh, R.J.: On context-free languages. J. ACM **13**, 570–581 (1966)
14. Wotschke, D.: Nondeterminism and Boolean operations in PDA's. J. Comput. Syst. Sci. **16**, 456–461 (1978)
15. Wotschke, D.: The Boolean closures of the deterministic and nondeterministic context-free languages. In: Brauer, W. (ed.) GI 1973. LNCS, vol. 1, pp. 113–121. Springer, Heidelberg (1973). https://doi.org/10.1007/3-540-06473-7_11
16. Wotschke, D.: Degree-languages: a new concept of acceptance. J. Comput. Syst. Sci. **14**(2), 187–209 (1977)

Minimal Unique Substrings and Minimal Absent Words in a Sliding Window

Takuya Mieno[1(✉)], Yuki Kuhara[1], Tooru Akagi[1], Yuta Fujishige[1,2], Yuto Nakashima[1], Shunsuke Inenaga[1,3], Hideo Bannai[1], and Masayuki Takeda[1]

[1] Department of Informatics, Kyushu University, Fukuoka, Japan
{takuya.mieno,yuki.kuhara,toru.akagi,yuta.fujishige,yuto.nakashima, inenaga,bannai,takeda}@inf.kyushu-u.ac.jp

[2] Japan Society for Promotion of Science, Tokyo, Japan

[3] PRESTO, Japan Science and Technology Agency, Kawaguchi, Japan

Abstract. A substring u of a string T is called a minimal unique substring (MUS) of T if u occurs exactly once in T and any proper substring of u occurs at least twice in T. A string w is called a minimal absent word (MAW) of T if w does not occur in T and any proper substring of w occurs in T. In this paper, we study the problems of computing MUSs and MAWs in a sliding window over a given string T. We first show how the set of MUSs can change in a sliding window over T, and present an $O(n \log \sigma)$-time and $O(d)$-space algorithm to compute MUSs in a sliding window of width d over T, where σ is the maximum number of distinct characters in every window. We then give tight upper and lower bounds on the maximum number of changes in the set of MAWs in a sliding window over T. Our bounds improve on the previous results in Crochemore et al. (2017).

1 Introduction

Processing massive string data is a classical and important task in theoretical computer science, with a variety of applications such as data compression, bioinformatics, and text data mining. It is natural and common to assume that such a massive string is given in an *online* fashion, one character at a time from left to right, and that the memory usage is limited to some pre-determined space. This is a so-called *sliding window model*, where the task is to process all substrings $T[i..i+d-1]$ of pre-fixed length d in a string T of length n in an incremental fashion, for increasing $i = 1, \ldots, n-d+1$. Usually the window size d is set to be much smaller than the string length n, and thus the challenge here is to design efficient algorithms that processes all such substrings using only $O(d)$ working space. A typical application to the sliding window model is data compression; examples are the famous Lempel-Ziv 77 (the original version) [16] and PPM [2].

In this paper, we study the following classes of strings in the sliding window model: *Minimal Unique Substrings* (*MUSs*) and *Minimal Absent Words*

A. Chatzigeorgiou et al. (Eds.): SOFSEM 2020, LNCS 12011, pp. 148–160, 2020.
https://doi.org/10.1007/978-3-030-38919-2_13

(*MAWs*). MUSs have been heavily utilized for solving the *Shortest Unique Substring* (*SUS*) problem [6,8,11,14], and MAWs have applications to data compression based on *anti-dictionaries* [3,10]. However, despite the fact that there is a common application field to MUSs and MAWs such as bioinformatics [1,5,11,13], to our knowledge, these two objects were considered to be quite different and were studied separately. This paper is the first that brings a light to their similarities by observing that a string w is a MUS (resp. MAW) of a string S if the number of occurrences of w in S is one (resp. zero), and the number of occurrences of any proper substring of w is at least two (resp. at least one).

We begin with combinatorial results on MUSs in a sliding window. Namely, we show that the number of MUSs that are added or deleted by one slide of the window is always $O(1)$ (Sect. 3). We then present the first efficient algorithm that maintains the set of MUSs for a sliding window of length d over a string of length n in a total of $O(n \log \sigma)$ time and $O(d)$ working space (Sect. 4). Our main algorithmic tool is the suffix tree for a sliding window that requires $O(d)$ space and can be maintained in $O(n \log \sigma)$ time [7,12]. Our algorithm for computing MUSs in a sliding window is built on our combinatorial results, and it keeps track of three different loci over the suffix tree, all of which can be maintained in $O(\log \sigma)$ amortized time per each sliding step.

MAWs in a sliding window have already been studied by Crochemore et al. [4]. They studied the number of MAWs to be added/deleted when the current window is shifted, and we improve some of these results (Sect. 5): For any string T over an alphabet of size σ, let $\mathsf{MAW}(T[i..j])$ be the set of all MAWs in the substring $T[i..j]$. Crochemore et al. [4] showed that $|\mathsf{MAW}(T[i..i+d]) \setminus \mathsf{MAW}(T[i..i+d-1])| \leq (s_i - s_\alpha)(\sigma - 1) + \sigma + 1$ and $|\mathsf{MAW}(T[i-1..i+d-1]) \setminus \mathsf{MAW}(T[i..i+d-1])| \leq (p_i - p_\beta)(\sigma - 1) + \sigma + 1$, where s_i, s_α, p_i, and p_β are the lengths of the longest repeating suffix of $T[i..i+d-1]$, of the longest suffix of $T[i..i+d-1]$ having an internal occurrence immediately followed by $\alpha = T[i+d]$, of the longest repeating prefix of $T[i..i+d-1]$, and of the longest prefix of $T[i..i+d-1]$ having an internal occurrence immediately preceded by $\beta = T[i-1]$. Since both $s_i - s_\alpha$ and $p_i - p_\beta$ are in $\Theta(d)$ in the worst case, it leads to an $O(\sigma d)$ upper bound. We improve this by showing that both $|\mathsf{MAW}(T[i..i+d]) \setminus \mathsf{MAW}(T[i..i+d-1])|$ and $|\mathsf{MAW}(T[i-1..i+d-1]) \setminus \mathsf{MAW}(T[i..i+d-1])|$ are at most $d+\sigma'+1$, where σ' is the number of distinct characters in $T[i..i+d-1]$. Since $\sigma' \leq d$, this leads to an improved $O(d)$ upper bound. We also show that this is tight. Crochemore et al. [4] also showed that $\sum_{i=1}^{n-d} |\mathsf{MAW}(T[i..i+d-1]) \triangle \mathsf{MAW}(T[i+1..i+d])| \in O(\sigma n)$. We give an improved upper bound $O(\min\{\sigma, d\} n)$ and show that this is tight.

All proofs omitted due to lack of space can be found in a full version [9].

2 Preliminaries

Strings. Let Σ be an alphabet. An element of Σ is called a character. An element of Σ^* is called a string. The length of a string T is denoted by $|T|$. The empty string ε is the string of length 0. If $T = xyz$, then x, y, and z are called a *prefix*, *substring*, and *suffix* of T, respectively. They are called a *proper prefix*, *proper substring*, and *proper suffix* of T if $x \neq T$, $y \neq T$, and $z \neq T$, respectively. If a string b is a prefix of T and is a suffix of T, b is called a *border* of T. For any $1 \leq i \leq |T|$, the i-th character of T is denoted by $T[i]$. For any $1 \leq i \leq j \leq |T|$, $T[i..j]$ denote the substring of T starting at i and ending at j. For convenience, $T[i'..j'] = \varepsilon$ for $i' > j'$. For any $1 \leq i \leq |T|$, let $T[..i] = T[1..i]$ and $T[i..] = T[i..|T|]$. For a string w, the set of beginning positions of occurrences of w in T is denoted by $occ_T(w) = \{i \mid T[i..i+|w|-1] = w\}$. Let $\#occ_T(w) = |occ_T(w)|$. For convenience, let $\#occ_T(\varepsilon) = |T|+1$. In what follows, we consider an arbitrarily fixed string T of length $n \geq 1$ over an alphabet Σ of size $\sigma \geq 2$.

Minimal Unique Substrings and Minimal Absent Words. Any string w is said to be *absent* from T if $\#occ_T(w) = 0$, and *present* in T if $\#occ_T(w) \geq 1$. For any substring w of T, w is called *unique* in T if $\#occ_T(w) = 1$, *quasi-unique* in T if $1 \leq \#occ_T(w) \leq 2$, and *repeating* in T if $\#occ_T(w) \geq 2$. A unique substring w of T is called *a minimal unique substring* of T if any proper substring of w is repeating in T. Since a unique substring w of T has exactly one occurrence in T, it can be identified with a unique interval $[s,t]$ such that $1 \leq s \leq t \leq n$ and $w = T[s..t]$. We denote by $\mathsf{MUS}(T) = \{[s,t] \mid T[s..t]$ is a MUS of $T\}$ the set of intervals corresponding to the MUSs of T. From the definition of MUSs, it is clear that $[s,t] \in \mathsf{MUS}(T)$ if (a) $T[s..t]$ is unique in T, (b) $T[s+1..t]$ is repeating in T, and (c) $T[s..t-1]$ is repeating in T.

An absent string w from T is called a *minimal absent word* of T if any proper substring of w is present in T. We denote by $\mathsf{MAW}(T)$ the set of all MAWs of T. From the definition of MAWs, it is clear that $w \in \mathsf{MAW}(T)$ if (A) w is absent from T, (B) $w[2..]$ is present in T, and (C) $w[..|w|-1]$ is present in T.

This paper deals with the problems of computing MUSs/MAWs in a sliding window of fixed length d over a given string T, formalized as follows:

Input: String T of length n and positive integer d ($< n$).
Output for the MUS problem: $\mathsf{MUS}(T[i..i+d-1])$ for all $1 \leq i \leq n-d+1$.
Output for the MAW problem: $\mathsf{MAW}(T[i..i+d-1])$ for all $1 \leq i \leq n-d+1$.

Suffix Trees. The *suffix tree* of a string T, denoted $STree_T$, is a *compacted trie* that represents all suffixes of T. We consider a version of suffix trees a.k.a. *Ukkonen trees* [15]: Namely, $STree_T$ is a rooted tree such that (1) each edge is labeled by a non-empty substring of T, (2) each internal node has at least two children, (3) the out-going edges of each node begin with mutually distinct characters, (4) the suffixes of T that are unique in T are represented by paths from the root to the leaves, and the other suffixes of T that are repeating in T are represented by paths from the root that end either on internal nodes or on edges. To simplify the description of our algorithm, we assume that there is an auxiliary node $\perp$ which is the parent of only the root node. The out-going edge of $\perp$ is labeled with Σ; This means that we can go down from $\perp$ by reading any character in Σ.

For each node v in $STree_T$, $par(v)$ denotes the parent of v, $str(v)$ denotes the path string from the root to v, $depth(v)$ denotes the *string depth* of v (i.e. $depth(v) = |str(v)|$), and $subtree(v)$ denotes the subtree of $STree_T$ rooted at v. For each leaf ℓ in $STree_T$, $start(\ell)$ denotes the starting position of $str(\ell)$ in T. For each non-empty substring w of T, $hed(w) = v$ denotes the *highest explicit descendant* where w is a prefix of $str(v)$ and $depth(par(v)) < |w| \leq depth(v)$. For each substring w of T, $locus(w) = (u, h)$ represents the locus in $STree_T$ where the path that spells out w from the root terminates, such that $u = hed(w)$ and $h = depth(u) - |w| \geq 0$. We say that a substring w of T with $locus(w) = (u, h)$ is represented by an *explicit node* if $h = 0$, and by an *implicit node* if $h \geq 1$. We remark that in the Ukkonen tree $STree_T$ of a string T, some repeating suffixes may be represented by implicit nodes. An implicit node which represents a suffix of T is called an *implicit suffix node.* For any non-empty substring w that is represented by an explicit node v, the *suffix link* of v is a reversed edge from v to the explicit node that represents $w[2..]$. The suffix link of the root that represents ε points to $\perp$.

3 Combinatorial Results on MUSs in a Sliding Window

Throughout this section, we consider positions i, j ($1 \leq i \leq j \leq n$) such that $T[i..j]$ denotes the sliding window for the i-th position over the input string T. The following arguments hold for *any* values of i and j, and hence, they will be useful for sliding windows of any length d.

Let $lrs_{i,j}$ be the longest repeating suffix of $T[i..j]$, $sqs_{i,j}$ be the shortest quasi-unique suffix of $T[i..j]$, and $sqp_{i,j}$ be the shortest quasi-unique prefix of $T[i..j]$. Note that $lrs_{i,j}$ can be the empty string, and that both $sqs_{i,j}$ and $sqp_{i,j}$ are always non-empty strings.

The next lemmas are useful for analyzing combinatorial properties on MUSs and for designing an efficient algorithm for computing them in a sliding window.

Lemma 1. *The following three statements are equivalent: (1)* $|lrs_{i,j}| \geq |sqs_{i,j}|$; *(2)* $\#occ_{T[i..j]}(lrs_{i,j}) = 2$; *(3)* $\#occ_{T[i..j]}(sqs_{i,j}) = 2$.

Lemma 2. $|lrs_{i,j+1}| \leq |lrs_{i,j}| + 1$.

3.1 Changes to MUSs When Appending a Character to the Right

In this subsection, we consider an operation that slides the right-end of the current window $T[i..j]$ with one character by appending the next character $T[j+1]$ to $T[i..j]$. We use the following observation.

Observation 1. For each non-empty substring s of $T[i..j]$, $\#occ_{T[i..j+1]}(s) \leq \#occ_{T[i..j]}(s) + 1$. Also, $\#occ_{T[i..j+1]}(s) = \#occ_{T[i..j]}(s) + 1$ if and only if s is a suffix of $T[i..j+1]$.

MUSs to Be Deleted When Appending a Character to the Right. Due to Observation 1, we obtain Lemma 3 which describes MUSs to be deleted when a new character $T[j+1]$ is appended to the current window $T[i..j]$.

Lemma 3. *For any* $[s,t]$ *with* $i \leq s < t \leq j$, $[s,t] \in \mathsf{MUS}(T[i..j])$ *and* $[s,t] \notin \mathsf{MUS}(T[i..j+1])$ *if and only if* $T[s..t] = sqs_{i,j+1}$ *and* $\#occ_{T[i..j+1]}(sqs_{i,j+1}) = 2$.

Proof. ($\Rightarrow$) Let $w = T[s..t]$. Since $[s,t] \in \mathsf{MUS}(T[i..j])$ and $[s,t] \notin \mathsf{MUS}(T[i..j+1])$, $\#occ_{T[i..j]}(w) = 1$ and $\#occ_{T[i..j+1]}(w) \geq 2$. It follows from Observation 1 that $\#occ_{T[i..j+1]}(w) = 2$ and w is a suffix of $T[i..j+1]$. If we assume that w is a proper suffix of $sqs_{i,j+1}$, then $\#occ_{T[i..j+1]}(w) \geq 3$ by the definition of $sqs_{i,j+1}$, but this contradicts with $\#occ_{T[i..j+1]}(w) = 2$. If we assume that $sqs_{i,j+1}$ is a proper suffix of w, then $\#occ_{T[i..j]}(sqs_{i,j+1}) \geq \#occ_{T[i..j]}(T[s+1..t]) \geq 2$. Also, $\#occ_{T[i..j+1]}(sqs_{i,j+1}) = \#occ_{T[i..j]}(sqs_{i,j+1}) + 1 \geq 3$ by Observation 1, but this contradicts with the definition of $sqs_{i,j+1}$. Therefore, we obtain $w = sqs_{i,j+1}$. Moreover, $\#occ_{T[i..j+1]}(sqs_{i,j+1}) = 2$ since $w = sqs_{i,j+1}$ is a substring of $T[i..j]$.
($\Leftarrow$) Since $w = T[s..t]$ is a suffix of $T[i..j+1]$ and $\#occ_{T[i..j+1]}(w) = 2$, w is unique in $T[i..j]$. By the definition of sqs_{j+1}, a proper suffix $w[2..] = T[s+1..t]$ of $w = sqs_{i,j+1}$ occurs at least three times in $T[i..j+1]$, i.e. $T[s+1..t]$ is repeating in $T[i..j]$. Also, a prefix $w[..|w|-1] = T[s..t-1]$ of $w = sqs_{i,j+1}$ is clearly repeating in $T[i..j]$. Therefore, $w = T[s..t]$ is a MUS of $T[i..j]$ and is not a MUS of $T[i..j+1]$. □

By Lemma 3, at most one MUS can be deleted when appending $T[j+1]$ to the current window $T[i..j]$, and such a deleted MUS must be $sqs_{i,j+1}$.

MUSs to Be Added When Appending a Character to the Right. First, we consider a MUS to be added when appending $T[j+1]$ to $T[i..j]$, which is a suffix of $T[i..j+1]$. The next observation follows from the definition of $lrs_{i,j}$:

Observation 2. If $[s,j] \in \mathsf{MUS}(T[i..j])$, then $s = j - |lrs_{i,j}|$. Namely, if there is a MUS of $T[i..j]$ that is a suffix of $T[i..j]$, then it must be the suffix of $T[i..j]$ that is exactly one character longer than $lrs_{i,j}$.

Lemma 4. $[j+1-k, j+1] \in \mathsf{MUS}(T[i..j+1])$ *if and only if* $T[j+1-k..j+1] = \alpha^{k+1}$ *or* $k \leq |lrs_{i,j}|$, *where* $k = |lrs_{i,j+1}|$ *and* $\alpha = T[j+1]$.

Proof. ($\Rightarrow$) Assume on the contrary that $T[j+1-k..j+1] \neq \alpha^{k+1}$ and $k > |lrs_{i,j}|$. By the assumptions and Lemma 2, $|lrs_{i,j}| = k-1$, and thus, $T[j-|lrs_{i,j}|..j] = T[j+1-k..j]$. Since $T[j+1-k..j+1]$ is a MUS of $T[i..j+1]$, $T[j+1-k..j] = T[j-|lrs_{i,j}|..j]$ occurs at least twice in $T[i..j+1]$. On the other hand, $T[j-|lrs_{i,j}|..j]$ is unique in $T[i..j]$ by the definition of $lrs_{i,j}$, hence $T[j-|lrs_{i,j}|..j]$ occurs in $T[i..j+1]$ as a suffix of $T[i..j+1]$. Consequently, we have $T[j-|lrs_{i,j}|..j] = T[j+1-|lrs_{i,j}|..j+1]$, i.e. $T[j-k..j+1] = T[j+1-k..j+1] = \alpha^{k+1}$ with $\alpha = T[j+1]$, a contradiction.
($\Leftarrow$) By definition, $T[j+2-k..j+1] = lrs_{i,j+1}$ is repeating in $T[i..j+1]$ and $T[j+1-k..j+1]$ is unique in $T[i..j+1]$. Now it suffices to show $T[j+1-k..j]$ is

repeating in $T[i..j+1]$. If $T[j+1-k..j+1] = \alpha^{k+1}$, then clearly $T[j+1-k..j] = \alpha^k$ is repeating in $T[i..j+1]$. If $k \le |lrs_{i,j}|$, then $T[j+1-k..j]$ is a suffix of $T[j+1-|lrs_{i,j}|..j]$. Thus $\#occ_{T[i..j+1]}(T[j+1-k..j]) \ge \#occ_{T[i..j]}(T[j+1-k..j]) \ge \#occ_{T[i..j]}(T[j+1-|lrs_{i,j}|..j]) \ge 2$. □

Next, we consider MUSs to be added when appending $T[j+1]$ to $T[i..j]$, which are *not* suffixes of $T[i..j+1]$.

Lemma 5. *For each* $[s,t] \in \mathsf{MUS}(T[i..j+1])$ *with* $t \ne j+1$, *if* $[s,t] \notin \mathsf{MUS}(T[i..j])$ *then* $\#occ_{T[i..j+1]}(sqs_{i,j+1}) = 2$ *and* $sqs_{i,j+1}$ *is a proper substring of* $T[s..t]$.

Proof. Since $t \ne j+1$, $T[s..t]$ is not a suffix of $T[i..j+1]$. Moreover, since $[s,t] \in \mathsf{MUS}(T[i..j+1])$, $T[s..t]$ is unique in $T[i..j]$. Since $T[s..t]$ is not a MUS of $T[i..j]$, there exists a MUS u of $T[i..j]$ which is a proper substring of $T[s..t]$. Assume on the contrary that $\#occ_{T[i..j+1]}(sqs_{i,j+1}) = 1$ or $u \ne sqs_{i,j+1}$. Then, it follows from Lemma 3 that u is a MUS of $T[i..j+1]$. However, this contradicts with $[s,t] \in \mathsf{MUS}(T[i..j+1])$. Therefore, $\#occ_{T[i..j+1]}(sqs_{i,j+1}) = 2$ and $u = sqs_{i,j+1}$ is a proper substring of $T[s..t]$. □

Namely, a MUS which is not a suffix is added by appending one character only if there is a MUS to be deleted by the same operation. Moreover, such added MUSs must contain the deleted MUS.

Lemma 6. *If* $\#occ_{T[i..j+1]}(sqs_{i,j+1}) = 2$, *then there are three integers* p_l, p_s, q *such that* $i \le p_l \le p_s \le q < j+1$ *and* $T[p_s..q] = sqs_{i,j+1}$ *and* $T[p_l..q] = lrs_{i,j+1}$. *Also, the following propositions hold:*

(a) If there is no MUS of $T[i..j]$ *ending at* $q+1$, *then* $[p_s, q+1] \in \mathsf{MUS}(T[i..j+1])$.
(b) If there is no MUS of $T[i..j]$ *starting at* $p_l - 1$ *and* $p_l \ge i+1$, *then* $[p_l - 1, q] \in \mathsf{MUS}(T[i..j+1])$.

Now we have the main result of this subsection:

Theorem 1. *For any* $1 \le i \le j < n$, $|\mathsf{MUS}(T[i..j+1]) \triangle \mathsf{MUS}(T[i..j])| \le 4$ *and* $-1 \le |\mathsf{MUS}(T[i..j+1])| - |\mathsf{MUS}(T[i..j])| \le 2$. *Furthermore, these bounds are tight for any* σ, i, j *with* $\sigma \ge 3$, $1 \le i \le j < n$, *and* $j - i + 1 \ge 5$.

3.2 Changes to MUSs When Deleting the Leftmost Character

In this subsection, we consider an operation that deletes the leftmost character $T[i-1]$ from $T[i-1..j]$. Basically, we can use symmetric arguments to the previous subsection where we considered appending a character to the right of the window. We omit the details here in the case of deleting the leftmost character, but all necessary observations and lemmas are available in the full version of this paper [9].

The main result of this subsection is the following:

Theorem 2. *For any* $1 < i \leq j \leq n$, $|\mathsf{MUS}(T[i-1..j]) \triangle \mathsf{MUS}(T[i..j])| \leq 4$ *and* $-1 \leq |\mathsf{MUS}(T[i-1..j])| - |\mathsf{MUS}(T[i..j])| \leq 2$. *Furthermore, these bounds are tight for any* σ, i, j *with* $\sigma \geq 3$, $1 < i \leq j \leq n$, *and* $j - i + 1 \geq 5$.

The next corollary is immediate from Theorems 1 and 2.

Corollary 1. *Given a positive integer* $d < n$. *For every* i *with* $1 \leq i \leq n - d$, $|\mathsf{MUS}(T[i..i+d-1]) \triangle \mathsf{MUS}(T[i+1..i+d])| \in O(1)$.

4 Algorithm for Computing MUSs in a Sliding Window

This section presents our algorithm for computing MUSs in a sliding window.

4.1 Updating a Suffix Tree and Three Loci in a Suffix Tree

First, we introduce some additional notions. Since we use Ukkonen's algorithm [15] for updating the suffix tree when a new character $T[j+1]$ is appended to the right end of the window $T[i..j]$, we maintain the locus for $lrs_{i,j}$ as in [15]. Also, in order to compute the changes of MUSs, we can use $sqs_{i,j}$. (c.f. Lemmas 3 and 6). Thus, we also maintain the locus for $sqs_{i,j}$.

The locus for $lrs_{i,j}$ (resp. $sqs_{i,j}$) in $STree_{T[i..j]}$ is called the *primary active point* (resp. the *secondary active point*) and is denoted by $\mathsf{pp}_{i,j}$ (resp. $\mathsf{sp}_{i,j}$). Additionally, in order to maintain $\mathsf{sp}_{i,j}$ efficiently, we also maintain the locus for the longest suffix of $T[i..j]$ which occurs at least three times in $T[i..j]$. We call this locus the *tertiary active point* that is denoted by $\mathsf{tp}_{i,j}$.

Appending One Character. When $T[i..j]$ is the empty string (the base case, where $i = 1$ and $j = 0$), we set all the three active points $(root, 0)$. Then we increase j, and the suffix tree grows in an online manner until $j = d$ using Ukkonen's algorithm. Then, for each $j > d$, we also increase i each time j increases, so that the sliding window is shifted to the right, by using sliding window algorithm for the suffix tree [7,12].

When $T[j+1]$ is appended to the right end of $T[i..j]$, we first update the suffix tree to $STree_{T[i,,j+1]}$ and compute $\mathsf{pp}_{i,j+1}$. Since $\mathsf{pp}_{i,j+1}$ coincides with the *active point*, $\mathsf{pp}_{i,j+1}$ can be found in amortized $O(\log \sigma)$ time [7,12,15].

After updating the suffix tree, we can compute $\mathsf{tp}_{i,j+1}$ and $\mathsf{sp}_{i,j+1}$ as follows:

1. Traverse character $T[j+1]$ from $\mathsf{tp}_{i,j}$, and set $w \leftarrow str(\mathsf{tp}_{i,j})T[i+1]$ which is the suffix of $T[i..j+1]$ that is one character longer than $\mathsf{tp}_{i,j}$. Then, w corresponds to a candidate for $\mathsf{tp}_{i,j+1}$.
2. While $\#occ_{T[i..j+1]}(w) < 3$, set $w \leftarrow w[2..]$ and search for the locus for w by using suffix links in $STree_{T[i..j+1]}$. This w is a new candidate for $\mathsf{tp}_{i,j+1}$.
3. After breaking the while-loop, obtain $\mathsf{tp}_{i,j+1} = locus(w)$ since w is the longest suffix of $T[i..j+1]$ which occurs more than twice in $T[i..j+1]$.
4. Also, $\mathsf{sp}_{i,j+1}$ equals the locus which is the very previous candidate for $\mathsf{tp}_{i,j+1}$.

As is described in the above algorithm, we can locate $\mathsf{tp}_{i,j+1}$ using suffix link, in as similar manner to the active point $\mathsf{pp}_{i,j+1}$. Thus, the cost for locating $\mathsf{tp}_{i,j+1}$ for each increasing j is amortized $O(\log \sigma)$ time, again by a similar argument to the active point ($\mathsf{pp}_{i,j+1}$). What remains is, for each candidate w for $\mathsf{tp}_{i,j+1}$, how to quickly determine whether $\#occ_{T[i..j+1]}(w) < 3$ or not. In what follows, we show that it can be checked in $O(1)$ time for each candidate.

Observation 3. For each suffix s of a string $T[i..j+1]$, let $locus(s) = (u, h)$.

Case 1. If u is an internal node, s occurs at least three times in $T[i..j+1]$.
Case 2. If u is a leaf and $h = 0$, s occurs exactly once in $T[i..j+1]$.
Case 3. If u is a leaf and $h \neq 0$,
 Case 3.1. if there is a suffix s' of $T[i..j+1]$ with $hed(s') = hed(s)$ which is longer than s, s occurs at least three times in $T[i..j+1]$.
 Case 3.2. otherwise, s occurs exactly twice in $T[i..j+1]$.

For any suffix s of $T[i..j+1]$, if we are given $locus(s) = (u, h)$, then we can obviously determine in constant time whether s occurs at least three times in $T[i..j+1]$ or not, except Case 3. The next lemma allows us to determine it in constant time in Case 3.

Lemma 7. *Suppose the locus $\mathsf{pp}_{i,j+1}$ in $STree_{T[i..j+1]}$ is already computed. Given a leaf ℓ of $STree_{T[i..j+1]}$, it can be determined in $O(1)$ time whether there is an implicit suffix node on the edge $(par(\ell), \ell)$ and if so, the locus of the lowest implicit suffix node on $(par(\ell), \ell)$ can be computed in $O(1)$ time.*

Deleting the Leftmost Character. When the leftmost character $T[i-1]$ is deleted from $T[i-1..j]$, we first update the suffix tree and compute $\mathsf{pp}_{i,j}$ by using the sliding window algorithm for the suffix tree [7,12]. Each pair of position pointers for the edge-labels of the suffix tree can be maintained in amortized $O(1)$ time so that these pointers always refer to positions within the current sliding window, by a simple *batch update* technique (see [12] for details). After that, we compute $\mathsf{tp}_{i,j}$ and $\mathsf{sp}_{i,j}$ in a similar way to the case of appending a new character shown previously.

It follows from the above arguments in this subsection that we can update the suffix tree and the three active points in amortized $O(\log \sigma)$ time, each time the window is shifted by one character.

4.2 Computing $sqp_{i-1,j}$

In order to compute the changes of MUSs when the leftmost character $T[i-1]$ is deleted from $T[i-1, j]$, we can use $sqp_{i-1,j}$ before updating the suffix tree. Thus, we present an efficient algorithm for computing $sqp_{i-1,j}$. First, we consider the following cases, where ℓ is the leaf corresponding to $T[i-1..j]$:

Case A. $hed(lrs_{i-1,j}) = \ell$.
Case B. $hed(lrs_{i-1,j}) \neq \ell$ and $subtree(par(\ell))$ has more than two leaves.
Case C. $hed(lrs_{i-1,j}) \neq \ell$ and $subtree(par(\ell))$ has exactly two leaves.

For Case A, the next lemma holds:

Lemma 8. *Given $STree_{T[i-1..j]}$ and $\mathsf{pp}_{i-1,j}$. Let ℓ be the leaf corresponding to $T[i-1..j]$. If $\mathsf{pp}_{i-1,j}$ is on the edge $(par(\ell), \ell)$, the following propositions hold:*

(a) $occ_{T[i-1..j]}(sqp_{i-1,j}) = \{i-1, j-|lrs_{i-1,j}|+1\}$.

(b) If there is exactly one implicit suffix node on $(par(\ell), \ell)$, $sqp_{i-1,j} = T[i-1..i-1+depth(par(\ell))]$.

(c) If there are more than one implicit suffix node on $(par(\ell), \ell)$, then $|lrs_{i-1,j}| > \lfloor (j-i+2)/2 \rfloor$ and $sqp_{i-1,j} = T[i-1..j-2h+1]$, where $\mathsf{pp}_{i-1,j} = (\ell, h)$.

Proof. Let $\mathsf{pp}_{i-1,j} = (\ell, h)$ and $L = |lrs_{i-1,j}|$.

(a) Since $\mathsf{pp}_{i-1,j}$ is on the edge $(par(\ell), \ell)$, $sqp_{i-1,j}$ is a prefix of $lrs_{i-1,j}$, and $\#occ_{T[i-1..j]}(lrs_{i-1,j}) = \#occ_{T[i-1..j]}(sqp_{i-1,j}) = 2$. Therefore, we obtain that $occ_{T[i-1..j]}(sqp_{i-1,j}) = occ_{T[i-1..j]}(lrs_{i-1,j}) = \{i-1, j-L+1\}$.

(b) In this case, it is clear that $sqp_{i-1,j} = T[i-1..i-1+depth(par(\ell))]$.

(c) Let (ℓ, h') be the locus of the implicit suffix node which is the lowest on the edge $(par(\ell), \ell)$ except $\mathsf{pp}_{i-1,j}$. Also, let x be the string corresponding to the locus (ℓ, h'). In this case, x occurs exactly three times in $T[i-1..j]$. Also, x is the longest border of $lrs_{i-1,j}$. Assume on the contrary that $L \le \lfloor (j-i+2)/2 \rfloor$. Then, two occurrences of $lrs_{i-1,j}$ in $T[i-1..j]$ are not overlapping, and thus $\#occ_{T[i-1..j]}(x) \ge 2 \times \#occ_{T[i-1..j]}(lrs_{i-1,j}) = 4$, it is a contradiction. Therefore, $L > \lfloor (j-i+2)/2 \rfloor$.
Next, we consider a relation between h and h'. By the definition, $h = |T[i-1..j]| - L = j-i+2-L$. Since $L > \lfloor (j-i+2)/2 \rfloor$, x matches the intersection of two occurrences of $lrs_{i-1,j}$, i.e. $x = T[j-L+1..i+L-2]$. Thus, $h' = |T[i-1..j]| - |x| = j-i+2-(2L-j+i-2) = 2(j-i+2-L) = 2h$. Therefore $sqp_{i-1,j} = T[i-1..j-h'+1] = T[i-1..j-2h+1]$. □

In Case B, it is clear that $sqp_{i-1,j} = T[i-1..i-1+depth(p)]$ since $str(p)$ occurs at least three times in $T[i-1..j]$.

For Case C, the next lemma holds:

Lemma 9. *Given $STree_{T[i-1..j]}$ and $\mathsf{pp}_{i-1,j}$. Let ℓ be the leaf corresponding to $T[i-1..j]$, $p = par(\ell)$, and $q = par(p)$. If $subtree(p)$ has exactly two leaves and there are no implicit suffix nodes on any edges in $subtree(p)$, then it can be determined in $O(1)$ time whether there is an implicit suffix node on (q,p). If such an implicit node exists, then the locus of the lowest implicit suffix node on (q,p) can be computed in $O(1)$ time.*

We can design an algorithm for computing $sqp_{i-1,j}$ by using the above lemmas, as follows. Let ℓ be the leaf corresponding to $T[i-1..j]$, $p = par(\ell)$ and $q = par(p)$.

In Case A. $sqp_{i-1,j}$ is computed by Lemma 8.

In Case B. $sqp_{i-1,j} = T[i-1..i-1+depth(p)]$ and $\#occ_{T[i-1..j]}(sqp_{i-1,j}) = 1$.

In Case C. We divide this case into some subcases by the existence of an implicit suffix node on edges (p, ℓ') and (q,p) where ℓ' is the sibling of ℓ. We first determine the existence of an implicit suffix node on (p, ℓ') (by Lemma 7).

- If there is an implicit suffix node on (p, ℓ'), then $sqp_{i-1,j} = T[i-1..i-1+depth(p)]$ and $\#occ_{T[i-1..j]}(sqp_{i-1,j}) = 1$.
- If there is no implicit suffix node on both (p, ℓ) and (p, ℓ'), we can determine in constant time the existence of an implicit suffix node on (q, p) (by Lemma 9). If there is an implicit suffix node on (q, p), $sqp_{i-1,j} = T[i-1..depth(p)-h+1]$ and $occ_{T[i-1..j]}(sqp_{i-1,j}) = \{i-1, start(\ell')\}$. Otherwise, $sqp_{i-1,j} = T[i-1..depth(q)+1]$ and $occ_{T[i.-1.j]}(sqp_{i-1,j}) = \{i-1, start(\ell')\}$.

It follows from the above arguments in this subsection that $sqp_{i-1,j}$ can be computed in $O(1)$ time by using the suffix tree and the (primary) active point.

4.3 Detecting MUSs to Be Added/Deleted

By using the afore-mentioned lemmas in this section, we can design an efficient algorithm for detecting MUSs to be added/deleted. The details of our algorithm can be found in the full version [9].

The main result of this section is the following:

Theorem 3. *We can maintain the set of MUSs in a sliding window of length d on a string T of length n over an alphabet of size σ, in a total of $O(n \log \sigma)$ time and $O(d)$ working space.*

Corollary 2. *There exists an online algorithm to compute all MUSs in a string T of length n over an alphabet of size σ in a total of $O(n \log \sigma)$ time with $O(n)$ working space.*

5 Combinatorial Results on MAWs in a Sliding Window

5.1 Changes to MAWs When Appending Character to the Right

We consider the number of changes of MAWs when appending $T[j+1]$ to $T[i..j]$.

For the number of deleted MAWs, the next lemma is known:

Lemma 10 ([4]). *For any $1 \leq i \leq j < n$, $|\mathsf{MAW}(T[i..j]) \backslash \mathsf{MAW}(T[i..j+1])| = 1$.*

Next, we consider the number of added MAWs. We classify each MAW w in $\mathsf{MAW}(T[i..j+1]) \setminus \mathsf{MAW}(T[i..j])$ to the following three types[1]. Let σ' be the number of distinct characters occurring in $T[i..j]$.

Type 1. $w[2..]$ and $w[..|w|-1]$ are both absent from $T[i..j]$.
Type 2. $w[2..]$ is present in $T[i..j]$ and $w[..|w|-1]$ is absent from $T[i..j]$.
Type 3. $w[2..]$ is absent from $T[i..j]$ and $w[..|w|-1]$ is present in $T[i..j]$.

We denote by $\mathcal{M}_1$, $\mathcal{M}_2$, and $\mathcal{M}_3$ the set of MAWs of Type 1, Type 2 and Type 3, respectively. The next lemma holds:

[1] At least one of $w[2..]$ and $w[..|w|-1]$ is absent from $T[i..j]$, because $w \notin \mathsf{MAW}(T[i..j])$.

Lemma 11. *For any* $1 \le i \le j < n$, $|\mathsf{MAW}(T[i..j+1]) \setminus \mathsf{MAW}(T[i..j])| \le \sigma' + d$, *where* $d = j - i + 1$.

Proof. In [4], it is shown that $|\mathcal{M}_1| \le 1$. It is also shown in [4] that the last characters of all MAWs in $\mathcal{M}_2$ are all different. Furthermore, by the definition of $\mathcal{M}_2$, the last character of each MAW in $\mathcal{M}_2$ occurs in $T[i..j]$. Thus, $|\mathcal{M}_2| \le \sigma'$. In the rest of the proof, we show that the number of MAWs of Type 3 is at most $d-1$. We show that there is an injection $f : \mathcal{M}_3 \to [i, j-1]$ that maps each MAW $w \in \mathcal{M}_3$ to the ending position of the leftmost occurrence of $w[..|w|-1]$ in $T[i..j]$. By the definition of $\mathcal{M}_3$, w is absent from $T[i..j+1]$ and $w[|w|] = T[j+1]$ for each $w \in \mathcal{M}_3$, and thus, no occurrence of $w[..|w|-1]$ in $T[i..j]$ ends at position j. Hence, the range of f does not contain the position j, i.e. it is $[i..j-1]$. Next, for the sake of contradiction, we assume that f is not an injection, i.e. there are two distinct MAWs $w_1, w_2 \in \mathcal{M}_3$ such that $f(w_1) = f(w_2)$. W.l.o.g., assume $|w_1| \ge |w_2|$. Since $w_1[|w_1|] = w_2[|w_2|] = T[j+1]$ and $f(w_1) = f(w_2)$, w_2 is a suffix of w_1. If $|w_1| = |w_2|$, then $w_1 = w_2$ and it contradicts with $w_1 \ne w_2$. If $|w_1| > |w_2|$, then w_2 is a proper suffix of w_1, and it contradicts with the fact that w_2 is absent from $T[i..j+1]$. Therefore, f is an injection and $|\mathcal{M}_3| \le j - 1 - i + 1 = d - 1$. □

The next lemma follows from Lemmas 10 and 11.

Lemma 12. *For any* $1 \le i \le j < n$, $|\mathsf{MAW}(T[i..j+1]) \triangle \mathsf{MAW}(T[i..j])| \le \sigma' + d + 1$, *where* $d = j - i + 1$. *The upper bound is tight when* $\sigma \ge 3$ *and* $\sigma' + 1 \le \sigma$.

5.2 Changes to MAWs When Deleting the Leftmost Character

Next, we analyze the number of changes of MAWs when deleting the leftmost character from a string. By a symmetric argument to Lemma 12, we obtain the next lemma:

Lemma 13. *For any* $1 < i \le j \le n$, $|\mathsf{MAW}(T[i..j]) \triangle \mathsf{MAW}(T[i-1..j])| \le \sigma' + d + 1$ *where* $d = j - i + 1$ *and* σ' *is the number of distinct characters occurs in* $T[i..j]$. *Also, the upper bound is tight when* $\sigma \ge 3$, *and* $\sigma' + 1 \le \sigma$.

Finally, by combining Lemmas 12 and 13, we obtain the next corollary:

Corollary 3. *Let* d *be the window length. For a string* T *of length* $n > d$ *and each integer* i *with* $1 \le i \le n - d$, $|\mathsf{MAW}(T[i..i+d-1]) \triangle \mathsf{MAW}(T[i+1..i+d])| \in O(d)$. *Also, there exists a string* T' *which satisfies* $|\mathsf{MAW}(T'[j..j+d-1]) \triangle \mathsf{MAW}(T'[j+1..j+d])| \in \Omega(d)$ *for some* j *with* $1 \le j \le |T'| - d$.

5.3 Total Changes of MAWs When Sliding the Window on a String

In this subsection, we consider the total number of changes of MAWs when sliding the window of length d from the beginning of T to the end of T. We denote the total number of changes of MAWs by $\mathcal{S}(T, d) = \sum_{i=1}^{n-d} |\mathsf{MAW}(T[i..i+d-1]) \triangle \mathsf{MAW}(T[i+1..i+d])|$. The following lemma is known:

Lemma 14 ([4]). *For a string T of length $n > d$ over an alphabet Σ of size σ, $\mathcal{S}(T, d) \in O(\sigma n)$.*

The aim of this subsection is to give a more rigorous bound for $\mathcal{S}(T, d)$. We first show that the above bound is tight under some conditions.

Lemma 15. *The upper bound of Lemma 14 is tight when $\sigma \leq d$ and $n - d \in \Omega(n)$.*

Next, we consider the case where $\sigma \geq d + 1$.

Lemma 16. *For a string T of length $n > d$ over an alphabet Σ of size σ, $\mathcal{S}(T, d) \in O(d(n - d))$, and this upper bound is tight when $\sigma \geq d + 1$.*

The main result of this section follows from the above lemmas:

Theorem 4. *For a string T of length $n > d$ over an alphabet Σ of size σ, $\mathcal{S}(T, d) \in O(\min\{d, \sigma\}n)$. This upper bound is tight when $n - d \in \Omega(n)$.*

We remark that $n - d \in \Omega(n)$ covers most interesting cases for the window length d, since the value of d can range from $O(1)$ to cn for any $0 < c < 1$.

References

1. Chairungsee, S., Crochemore, M.: Using minimal absent words to build phylogeny. Theor. Comput. Sci. **450**, 109–116 (2012)
2. Cleary, J.G., Witten, I.H.: Data compression using adaptive coding and partial string matching. IEEE Trans. Commun. **32**(4), 396–402 (1984)
3. Crochemore, M., Mignosi, F., Restivo, A., Salemi, S.: Data compression using antidictionaries. Proc. IEEE **88**(11), 1756–1768 (2000)
4. Crochemore, M., Héliou, A., Kucherov, G., Mouchard, L., Pissis, S.P., Ramusat, Y.: Minimal absent words in a sliding window and applications to on-line pattern matching. In: Klasing, R., Zeitoun, M. (eds.) FCT 2017. LNCS, vol. 10472, pp. 164–176. Springer, Heidelberg (2017). https://doi.org/10.1007/978-3-662-55751-8_14
5. Gräf, S.: Optimized design and assessment of whole genome tiling arrays. Bioinformatics **23**(13), i195–i204 (2007)
6. Hu, X., Pei, J., Tao, Y.: Shortest unique queries on strings. In: Moura, E., Crochemore, M. (eds.) SPIRE 2014. LNCS, vol. 8799, pp. 161–172. Springer, Cham (2014). https://doi.org/10.1007/978-3-319-11918-2_16
7. Larsson, N.J.: Extended application of suffix trees to data compression. In: DCC 1996, pp. 190–199 (1996)
8. Mieno, T., Inenaga, S., Bannai, H., Takeda, M.: Shortest unique substring queries on run-length encoded strings. In: MFCS 2016, pp. 69:1–69:11 (2016)
9. Mieno, T., et al.: Minimal unique substrings and minimal absent words in a sliding window. CoRR abs/1909.02804 (2019)
10. Ota, T., Fukae, H., Morita, H.: Dynamic construction of an antidictionary with linear complexity. Theor. Comput. Sci. **526**, 108–119 (2014)
11. Pei, J., Wu, W.C., Yeh, M.: On shortest unique substring queries. In: ICDE 2013, pp. 937–948 (2013)

12. Senft, M.: Suffix tree for a sliding window: an overview. In: WDS, vol. 5, pp. 41–46 (2005)
13. Silva, R.M., Pratas, D., Castro, L., Pinho, A.J., Ferreira, P.J.S.G.: Three minimal sequences found in Ebola virus genomes and absent from human DNA. Bioinformatics **31**(15), 2421–2425 (2015)
14. Tsuruta, K., Inenaga, S., Bannai, H., Takeda, M.: Shortest unique substrings queries in optimal time. In: Geffert, V., Preneel, B., Rovan, B., Štuller, J., Tjoa, A.M. (eds.) SOFSEM 2014. LNCS, vol. 8327, pp. 503–513. Springer, Cham (2014). https://doi.org/10.1007/978-3-319-04298-5_44
15. Ukkonen, E.: On-line construction of suffix trees. Algorithmica **14**(3), 249–260 (1995)
16. Ziv, J., Lempel, A.: A universal algorithm for sequential data compression. IEEE Trans. Inf. Theory **IT–23**(3), 337–349 (1977)

On Synthesis of Specifications with Arithmetic

Rachel Faran(✉) and Orna Kupferman

The Hebrew University of Jerusalem, Jerusalem, Israel
{rachelmi,orna}@cs.huji.ac.il

Abstract. *Variable automata with arithmetic* enable the specification of reactive systems with variables over an infinite domain of numeric values and whose operation involves arithmetic manipulation of these values [9]. We study the *synthesis problem* for such specifications. While the problem is in general undecidable, we define a fragment, namely *semantically deterministic* variable automata with arithmetic, for which the problem is decidable. Essentially, an automaton is semantically deterministic if the restrictions on the possible assignments to the variables that are accumulated along its runs resolve its nondeterministic choices. We show that semantically deterministic automata can specify many interesting behaviors – many more than deterministic ones, and that the synthesis problem for them can be reduced to a solution of a two-player game. For automata with simple guards, the game has a finite state space, and the synthesis problem can be solved in time polynomial in the automaton and exponential in the number of its variables.

1 Introduction

Synthesis is the automated construction of systems from their specifications [5,18]. The specification is typically given by a temporal-logic formula or an automaton, and it distinguishes between outputs, generated by the system, and inputs, generated by its environment. The system should *realize* the specification, namely satisfy it against all possible environments. Since its introduction, synthesis has been one of the most studied problems in formal methods, with extensive research on wider settings, heuristics, and applications [1].

Until recently, all studies of the synthesis problem considered *finite-state* transducers that realize specifications given by temporal-logic formulas over a finite set of Boolean propositions or by finite-state automata over finite alphabets. Many real-life systems, however, have an infinite state space. One class of infinite-state systems, motivating this work, consists of systems in which the control is finite and the source of infinity is the domain of the variables in the systems. This includes, for example, software with integer parameters [3], datalog systems with infinite data domain [22], and many more [4,6]. Lifting automata-based methods to the setting of such systems requires the introduction of automata with *infinite alphabets*. The latter include *registers* [21], *pebbles* [16], *variables* [10], and *data* [2] automata. These formalisms refer to the infinite values by comparing them to each other. Thus, the exact value is abstracted: one can specify, for example, that each value received as input is generated as an output at least once during the next 10 transitions, but cannot specify, for example, that if a

A. Chatzigeorgiou et al. (Eds.): SOFSEM 2020, LNCS 12011, pp. 161–173, 2020.
https://doi.org/10.1007/978-3-030-38919-2_14

value $x \in \mathbb{Q}$ is received as input then the next 10 transitions include only outputs of values in $[x-5, x+5]$.

In [9], we introduced automata with arithmetic, which do support specifications as the latter. Here, we consider *nondeterministic looping word automata with arithmetic* (NLWAs, for short), which define languages over alphabets of the form $\Sigma \times \mathbb{Q}$, for some finite set Σ. Each NLWA has a finite set X of variables over $\mathbb{Q}$. The transitions of an NLWA are labeled by both letters from Σ and guards involving values in $\mathbb{Q}$, variables in X, and the symbol $\star$, which refers to the $\mathbb{Q}$-value of the letter read. A word w is accepted by an NLWA $\mathcal{A}$ if there is an assignment to the variables that appear in $\mathcal{A}$ such that there is an infinite run of $\mathcal{A}$ on w. In particular, all the guards along the run are satisfied. The *looping* acceptance condition is a special case of the Büchi condition. It captures *safety properties*, and we use it here in order to circumvent technical challenges that are irrelevant for the challenge of handling arithmetic. It is shown in [9] that many decision problems on NLWAs are decidable, essentially by replacing queries about reachability via runs by queries about the satisfaction of guards accumulated during runs.

Recent work on synthesis shows that while the synthesis problem for register automata is undecidable [7], the *register-bounded* synthesis problem is decidable [8, 12, 13]. There, the number of registers of the system and/or the environment is bounded, which enables an abstraction of the exact values stored in the registers. The abstraction, however, strongly depends on the fact that the only operation that register automata apply to the stored values is comparison. As discussed above, such comparisons cannot handle specifications that refer to the values, in particular ones with arithmetic.

We study the synthesis problem for NLWAs and point to a decidable fragment. In the setting of finite alphabets, the specifications are over an alphabet $2^{I \cup O}$, for finite sets I and O of input and output signals. The synthesis problem for specifications given by deterministic automata is solvable by a reduction to a two-player game [1]. The positions of the game are the states of the automaton. In each round of the game, the environment player provides an input in 2^I, the system player responds with an output in 2^O, and the game transits to the corresponding successor state. The system wins the game if, no matter which inputs the environment provides, the run generated along the interaction between the players is accepting. When the automaton is nondeterministic, the system responds not only with an output, but also with a transition that should be taken. This is problematic, as this choice of a transition should accommodate all possible future choices of the environment. In particular, if different future choices of the environment induce computations that are all in the language of the automaton yet require different nondeterministic choices, the system cannot win.

The need to work with deterministic automata is a known barrier for synthesis in practice: algorithms involve complicated determinization constructions [19] or acrobatics for circumventing determinization [15]. In the case of automata with arithmetic, the challenge is bigger: First, even when the automata are deterministic, the strategies of the players depend on values in $\mathbb{Q}$, thus the game has infinitely many configurations. Second, determinism significantly reduces the expressive power of NLWAs. Indeed, it requires that for every state q and letter $\sigma \in \Sigma$, at most one guard in all the σ-transitions from q is satisfiable. For example, we cannot have two σ-transitions from q, one guarded by $x > 8$

and the second by $x \leq 5$. We suggest to distinguish between three levels of determinism in an NLWA $\mathcal{A}$. In addition to the standard definition, by which $\mathcal{A}$ is deterministic (DLWA) if every state has at most one satisfiable transition for every input, we say that $\mathcal{A}$ is *deterministic per assignment* (DPA-NLWA) if for every assignment to the variables in X, there is at most one run of $\mathcal{A}$ on every word. Then, $\mathcal{A}$ is *semantically deterministic* (SD-NLWA) if the restrictions on the possible assignments to the variables in X that are accumulated along the run resolve its nondeterministic choices. In the example above, if every run that leads to the state q is such that at most one of $\gamma \wedge (x > 8)$ or $\gamma \wedge (x \leq 5)$ is satisfiable, where γ is the conjunction of guards accumulated during the run, then there is no real choice to make when σ is read in state q.

We show that while DPA-NLWAs are as expressive as NLWAs, they are not useful in synthesis. Indeed, the assignment to the variables in X in the different runs on $\mathcal{A}$ is not known in advance to the system and the environment, and may be different for the different computations they generate. On the other hand, SD-NLWAs can be soundly used in the synthesis game. Intuitively, as has been the case with *good for games* (GFG) automata [11, 14], SD-NLWAs can resolve their nondeterministic choices in a way that only depends on the past. While in GFG automata, resolving of nondeterminism concerns the limit behavior of the run, in SD-NLWAs it also concerns the restrictions on the possible assignment to the variables in X. Consequently, while GFG automata with an acceptance condition γ (e.g., Büchi) are as expressive as deterministic γ automata [14, 17], SD-NLWAs are strictly more expressive than DLWAs. Moreover, we argue that natural NLWAs are semantically deterministic. Indeed, in a typical specification, one first assigns values into the variables in the specification and then resolves guards that depend on the assignments.

We solve the synthesis problem for SD-NLWAs by reducing it to a two-player game. While semantic determinism handles the nondeterminism, there is also the challenge of the infinite variable domain. In order to obtain a finite game, we show that when the guards of the SD-NLWA are simple, namely each term refers to at most one variable or to $\star$, then we can exploit the density of $\mathbb{Q}$ and abstract the infinitely many values in $\mathbb{Q}$ by finitely many partitions of $\mathbb{Q}$ and orders on X induced by the guards in the SD-NLWAs. The transducers induced by a winning strategy use the same set X of variables, and are also semantically deterministic, in the sense that guards over the input values are used to resolve nondeterminism. The game, and hence also the synthesis problem, can be solved in time polynomial in the SD-NLWA and exponential in X.

Due to lack of space, some examples and proofs are omitted, and can be found in the full version, in the authors' URLs.

2 Preliminaries

For a finite set X of variables, the set of *terms over* X, denoted Θ_X, is defined inductively as follows.

- m, x, and $\star$, for $m \in \mathbb{Q}$, $x \in X$, and the symbol $\star$.
- $t_1 + t_2$ and $t_1 - t_2$, for $t_1, t_2 \in \Theta_X$.

A term is *simple* if it contains at most one element in $X \cup \{\star\}$. For a number $k \in \mathbb{Q}$ and an assignment $f : X \to \mathbb{Q}$, let $f^k : X \cup \{\star\} \to \mathbb{Q}$ be an extension of f where for all

$x \in X$, we have $f^k(x) = f(x)$, and for the symbol $\star$ we have $f^k(\star) = k$. Given k and f, we can extend f^k to terms over X in the expected way; for example, $f^k : \Theta_X \to \mathbb{Q}$ is such that $f^k(t_1 + t_2) = f^k(t_1) + f^k(t_2)$.

The set of *guards over* X, denoted $\mathcal{G}_X$, is defined inductively as follows.

- $t_1 < t_2$ and $t_1 = t_2$, for $t_1, t_2 \in \Theta_X$,
- $\neg\gamma_1$ and $\gamma_1 \wedge \gamma_2$, for $\gamma_1, \gamma_2 \in \mathcal{G}_X$.

For $k \in \mathbb{Q}$, an assignment $f : X \to \mathbb{Q}$, and a guard $\gamma \in \mathcal{G}_X$, we define when k *satisfies* γ *under* f, denoted $k \models_f \gamma$, by induction on the structure of γ as follows.

- For two terms $t_1, t_2 \in \Theta_X$, we have that $k \models_f (t_1 < t_2)$ iff $f^k(t_1) < f^k(t_2)$, and $k \models_f (t_1 = t_2)$ iff $f^k(t_1) = f^k(t_2)$.
- For guards $\gamma_1, \gamma_2 \in \mathcal{G}_X$, we have that $k \models_f \neg\gamma$ if $k \not\models_f \gamma$, and $k \models_f \gamma_1 \wedge \gamma_2$ if $k \models_f \gamma_1$ and $k \models_f \gamma_2$.

Using Boolean operations, we can compare terms also by the $\leq, \geq$, and $>$ relations. We refer to guards of the form $t_1 \sim t_2$, for $\sim \in \{\leq, \geq, =, <, >\}$ as *atomic guards*. As it is useless to have $\star$ in both t_1 and t_2, we assume that when an atomic guard includes $\star$, then $\star$ appears only in t_1. A guard is *simple* if all its terms are simple. For example, $x_1 \leq x_2 + 5$ and $\star = x_1$ are simple, whereas $x_1 + x_2 \leq 5$ is not.

We are interested in languages of infinite words in $(\Sigma \times \mathbb{Q})^\omega$. A *nondeterministic looping word automaton with arithmetic* (NLWA, for short) is a tuple $\mathcal{A} = \langle \Sigma, X, Q, Q_0, \Delta \rangle$, where Σ is an alphabet, X is a set of variables, Q is a set of states, $Q_0 \subseteq Q$ is a set of initial states, and $\Delta \subseteq Q \times \Sigma \times \mathcal{G}_X \times Q$ is a transition relation. Thus, each transition is labeled by both a letter in Σ and a guard in $\mathcal{G}_X$. A *run* of $\mathcal{A}$ on an infinite word $\langle \sigma_0, k_0 \rangle, \langle \sigma_1, k_1 \rangle, \ldots$ over $\Sigma \times \mathbb{Q}$ is a sequence of states $q_0, q_1, \ldots$, where $q_0 \in Q_0$, and there is an assignment $f : X \to \mathbb{Q}$ such that for every position $i \geq 0$, there is a transition $\langle q_i, \sigma_i, \gamma, q_{i+1} \rangle \in \Delta$ such that $k_i \models_f \gamma$. Note that the assignment f is fixed throughout the run, and that all the runs are infinite. The *language* of $\mathcal{A}$, denoted $L(\mathcal{A})$, is the set of all words $w \in (\Sigma \times \mathbb{Q})^\omega$ such that there is a run of $\mathcal{A}$ on w. For example, the NLWA below accepts all the words in which all the letters agree on the $\mathbb{Q}$-component, possibly only until the Σ-component is *start* and the $\mathbb{Q}$-component is increased by 1.

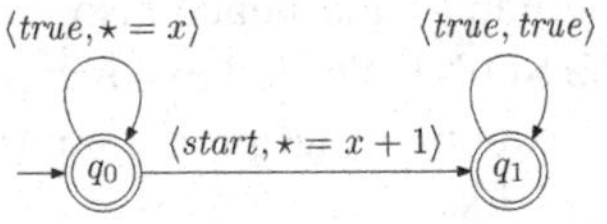

$\langle true, \star = x \rangle$ $\langle true, true \rangle$

$\langle start, \star = x + 1 \rangle$

q_0 q_1

We turn to define determinism for NLWAs. Consider an NLWA $\mathcal{A} = \langle \Sigma, X, Q, Q_0, \Delta \rangle$. We say that $\mathcal{A}$ is *deterministic per assignment* (DPA-NLWA, for short) if $|Q_0| = 1$, and for every function $f : X \to \mathbb{Q}$, state $q \in Q$, and letter $\langle \sigma, k \rangle \in \Sigma \times \mathbb{Q}$, there is at most one transition $\langle q, \sigma, \gamma, q' \rangle \in \Delta$ such that $k \models_f \gamma$. In other words, $\mathcal{A}$ is deterministic per assignment if for every assignment to X, there is at most one run of $\mathcal{A}$ on every word in $(\Sigma \times \mathbb{Q})^\omega$. We say that $\mathcal{A}$ is *deterministic* (DLWA) if the choice of the transition does not depend on the assignment to X. Formally, we have that $\mathcal{A}$ is deterministic if $|Q_0| = 1$, and for every state $q \in Q$ and letter $\langle \sigma, k \rangle \in \Sigma \times \mathbb{Q}$, there is at most one transition $\langle q, \sigma, \gamma, q' \rangle \in \Delta$ such that there exists $f : X \to \mathbb{Q}$ for which $k \models_f \gamma$.

Finally, an NLWA is *semantically deterministic* (SD-NLWA, for short) if for every state q and every run r that reaches q, the restrictions on the variable values accumulated throughout r resolve the nondeterminism in q. Formally, for a finite word $w = \langle \sigma_0, k_0 \rangle, \ldots \langle \sigma_t, k_t \rangle \in (\Sigma \times \mathbb{Q})^*$, let $r = q_0, q_1, q_2, \ldots, q_{t+1}$ be a run on w, and let γ_i be the guard that labels the σ_i-transition from q_i to q_{i+1}. We denote $\gamma^r = \bigwedge_{1 \leq i \leq t} \gamma_i$. We say that $\mathcal{A}$ is semantically deterministic if $|Q_0| = 1$, and for every state $q \in Q$, every run r from q_0 to q, and every letter $\langle \sigma, k \rangle \in \Sigma \times \mathbb{Q}$, there is at most one transition $\langle q, \sigma, \gamma, q' \rangle \in \Delta$ such that there exists $f : X \to \mathbb{Q}$ for which $k \models_f \gamma \wedge \gamma^r$.

In the context of open systems, we define automata over $\Sigma = 2^{I \cup O}$ for finite sets I and O of input and output signals, respectively. Also, rather than a single value in $\mathbb{Q}$, we let each position in the computation include two such values – input and output. Thus, a computation is $\pi = \langle \sigma_0, k_0^I, k_0^O \rangle, \langle \sigma_1, k_1^I, k_1^O \rangle, \cdots \in (2^{I \cup O} \times \mathbb{Q} \times \mathbb{Q})^\omega$. In order to indicate whether a guard refers to the input or output value, we use a variant of NLWAs, termed NLWAs^{IO}, in which the $\star$ in the guards is parameterized by the letters I and O. Thus, the atomic guards in an NLWA^{IO} with variables in X include $\star_I \sim t$ and $\star_O \sim t$, for a term $t \in \Theta_X$ and $\sim \in \{\leq, \geq, =, <, >\}$. Then, for a pair of numbers $\langle k^I, k^O \rangle \in \mathbb{Q} \times \mathbb{Q}$, we have that $\langle k^I, k^O \rangle$ satisfies $\star_I \sim t$ iff $k^I \sim t$. Likewise, $\langle k^I, k^O \rangle$ satisfies $\star_O \sim t$ iff $k^O \sim t$. The semantics of NLWAs^{IO} is similar to that of NLWAs, except that for $\gamma \in \mathcal{G}_X$, an assignment $f : X \to \mathbb{Q}$ and a position $j \geq 0$, we have that $(\pi, j) \models_f \gamma$ iff $\langle k_j^I, k_j^O \rangle \models_f \gamma$. Note that the notions of determinism, determinism per assignment and semantic determinism can be easily extended to NLWAs^{IO}.

For finite sets I and O of input and output signals, respectively, a finite-state I/O *transducer over* $\mathbb{Q}$ is $\mathcal{T} = \langle I, O, S, s_0, \mathcal{G}^I, \mathcal{G}^O, \rho, \tau \rangle$, where S is a finite set of states, $s_0 \in S$ is an initial state, $\mathcal{G}^I, \mathcal{G}^O \subset \mathcal{G}_X$ are finite sets of atomic guards that may include $\star_I$ and $\star_O$, respectively, $\rho : S \times 2^I \times \mathcal{G}^I \to S$ is a transition function, and $\tau : S \to 2^O \times \mathcal{G}^O$ is a labelling function on the states. Note that $\mathcal{T}$ abstracts the concrete input and output values, and partitions the infinitely many values according to satisfaction of guards in $\mathcal{G}^I$ and $\mathcal{G}^O$.

Intuitively, $\mathcal{T}$ models the interaction of an environment that generates at each moment in time a letter in 2^I and a value in $\mathbb{Q}$ with a system that responds with a letters in 2^O and a value in $\mathbb{Q}$. Let $\mathcal{F} = \{F : F \subseteq \mathbb{Q}^X\}$, and consider an input word $w = \langle i_0, k_0^I \rangle \cdot \langle i_1, k_1^I \rangle \cdots \in (2^I \times \mathbb{Q})^\omega$. A *run* of $\mathcal{T}$ on w is a sequence in $(S \times 2^O \times \mathbb{Q} \times \mathcal{F})^\omega$, where the S-components describe the states visited along the run, the 2^O- and $\mathbb{Q}$-components describe the outputs, and the $\mathcal{F}$-components describe the set of possible assignments to X that are consistent with the restrictions accumulated during the run – restrictions imposed by both the guards along the transitions and their combination with the values of the inputs and the assignments guards in the states and their combination with the values of the outputs. Accordingly, $F_0 = \mathbb{Q}^X$, and for all $j \geq 0$, we have that $s_{j+1} = \rho(s_j, i_j, \gamma_j^I)$ for γ_j^I such that $k_j^I \models_f \gamma_j^I$ for some $f \in F_j$, $\tau(s_j) = \langle o_j, \gamma_j^O \rangle$ for γ_j^O such that $k_j^I \models_f \gamma_j^I$ for some $f \in F_j \cap \{f \in \mathbb{Q}^X : k_j^I \models_f \gamma_j^I\}$, and $F_{j+1} = F_j \cap \{f \in \mathbb{Q}^X : k_j^I \models_f \gamma_j^I\} \cap \{f \in \mathbb{Q}^X : k_j^O \models_f \gamma_j^O\}$. We require ρ to be *receptive* and *deterministic*, in the sense that for every input word $w = \langle i_0, k_0^I \rangle \cdot \langle i_1, k_1^I \rangle \cdots \in (2^I \times \mathbb{Q})^\omega$ and $j \geq 0$, there is exactly one state $s_{j+1} = \rho(s_j, i_j, \gamma_j^I)$ such that $k_j^I \models_f \gamma_j^I$ for some $f \in F_j$. An *output* of $\mathcal{T}$ on w is $\langle o_1, k_1^O \rangle \cdot \langle o_2, k_2^O \rangle \cdots \in (2^O \times \mathbb{Q})^\omega$ such that there is a run $\langle s_0, o_0, k_0^O, F_0\rangle, \langle s_1, o_1, k_1^O, F_1\rangle, \langle s_2, o_2, k_2^O, F_2\rangle \ldots$ of $\mathcal{T}$ on w. Note that the first output assignment is that of s_1, thus $\tau(s_0)$ is ignored. This reflects the fact that the environment initiates the interaction. A *computation of* $\mathcal{T}$ *on* w is then $\mathcal{T}(w) = \langle i_0 \cup o_1, k_0^I, k_1^O\rangle, \langle i_1 \cup o_2, k_1^I, k_2^O\rangle, \ldots \in (2^{I \cup O} \times \mathbb{Q} \times \mathbb{Q})^\omega$.

For an NLWA^{IO} $\mathcal{A}$, we say that $\mathcal{T}$ *realizes* $L(\mathcal{A})$ if for every input word $w \in (2^I \times \mathbb{Q})^\omega$, all the computations of $\mathcal{T}$ on w are in $L(\mathcal{A})$. The *synthesis* problem for NLWA^{IO} is then to decide, given an NLWA^{IO} $\mathcal{A}$, whether $L(\mathcal{A})$ is realizable, and if so, to return an I/O-transducer that realizes it.

3 Different Levels of Nondeterminism in NLWAs

In this section we study the different levels of nondeterminism in NLWAs. We start with the expressive power of DPA-NLWAs and SD-NLWAs, with respect to NLWAs, DLWAs, and each other, and continue to the problem of deciding the nondeterminism level of a given automaton.

We first prove that determinism per assignment does not restrict the expressive power of NLWAs. Thus, DPA-NLWAs are as expressive as NLWA. The proof is constructive and the idea is based on an elaboration of the subset construction. There, given a nondeterministic automaton $\mathcal{A}$ with state space Q, a deterministic equivalent automaton $\mathcal{A}'$ has state space 2^Q, and the transitions are defined so that $\mathcal{A}'$ visits a state $S \in 2^Q$ after reading a prefix w if S is the set of states that $\mathcal{A}$ may reach in at least one run after reading w. Since the path traversed by $\mathcal{A}$ when it reads w is not important (recall we consider looping automata), the subset construction maintains all the information needed. In the case of NLWAs, the paths traversed are important – they induce restrictions on possible assignments to X. Accordingly, an adoption of the subset construction to DPA-NLWAs involves a duplication of the set of variables – one copy for each transition. Then, the state space of the equivalent DPA-NLWA consists of subsets of Q along with an indication, for each $x \in X$, which copies of x should be assigned the same value. The detailed construction appears in the full version.

Theorem 1. *DPA-NLWAs are as expressive as NLWAs.*

Theorem 1 is quite surprising, but is of no real help in the context of synthesis. Indeed (see formal proof in Lemma 1), determinization per assignment is not useful when we run the automaton simultaneously on all the computations of a transducer, as different computations may be accepted with different assignments. Accordingly, we turn to focus on semantically-deterministic NLWAs. As we show in Lemma 2, this model of determinism is useful for solving the synthesis problem.

Theorem 2. *SD-NLWAs are strictly less expressive than NLWAs.*

Proof. The NLWA $\mathcal{A}$ in the left of Fig. 1 accepts all the words in which the projection on the Σ-component is in $a^\omega + a^* \cdot b^\omega$, where in the second case the first b comes after two letters that agree on their $\mathbb{Q}$-component. In the full version, we show that $\mathcal{A}$ does not have an equivalent SD-NLWA. Intuitively, it follows from the fact that the nondeterministic choice between a^ω and $a^* \cdot b^\omega$ should be taken before x is assigned a value. □

Theorem 3. *SD-NLWAs are strictly more expressive than DLWAs.*

Proof. Consider the NLWA $\mathcal{A}$ in the right of Fig. 1. It is easy to see that $\mathcal{A}$ is an SD-NLWA. Indeed, when a run reaches q_1, the variable x is already assigned a value, thus the nondeterminism in q_1 can be resolved. In the full version, we show that $\mathcal{A}$ does not have an equivalent DLWA. □

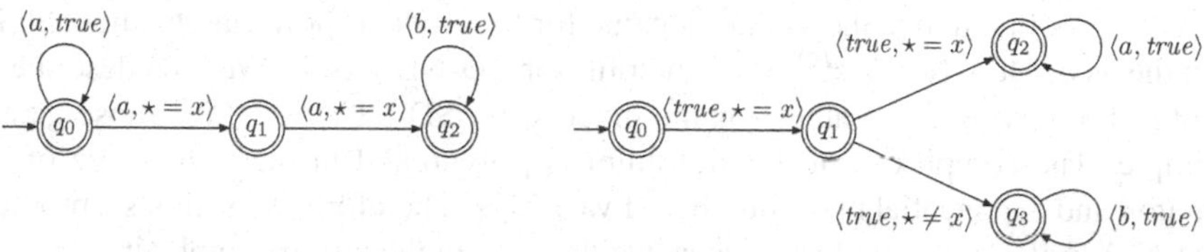

Fig. 1. An NLWA with no equivalent SD-NLWA, and an SD-NLWA with no equivalent DLWA.

We turn to discuss the problem of deciding the type of a given automaton. Note that we consider the syntactic questions, namely whether the given automaton is deterministic per assignment or semantically deterministic, and not the semantic one, namely whether it has an equivalent DPA-NLWA or SD-NLWA.

Theorem 4. *The problems of deciding whether a given NLWA is a DPA-NLWA or is an SD-NLWA are co-NP-complete.*

Proof. Given an NLWA $\mathcal{A} = \langle \Sigma, X, Q, Q_0, \Delta \rangle$, a nondeterministic Turing machine can decide in polynomial time that $\mathcal{A}$ is not a DPA-NLWA by guessing a reachable state $q \in Q$, a letter $\langle \sigma, k \rangle \in \Sigma \times \mathbb{Q}$, and two transitions $\langle q, \sigma, \gamma_1, q' \rangle, \langle q, \sigma, \gamma_2, q'' \rangle$ such that $k \models_f \gamma_1 \wedge \gamma_2$ for some $f : X \to \mathbb{Q}$. Deciding that $\mathcal{A}$ is not an SD-NLWA can be done by guessing, in addition, two assignments $f_1, f_2 : X \to \mathbb{Q}$, a finite word $w \in (\Sigma \times \mathbb{Q})^*$, and a run r on it from the initial state q_0 to q, where all the guards throughout r are satisfiable by the corresponding letters in w under both f_1 and f_2. Then, the letter $\langle \sigma, k \rangle$ and the transitions $\langle q, \sigma, \gamma_1, q' \rangle, \langle q, \sigma, \gamma_2, q'' \rangle$ should be such that $k \models_{f_1} \gamma_1$ and $k \models_{f_2} \gamma_2$. In the full version, we show that these guesses are polynomial, which implies, by [20], the desired upper bounds.

For the lower bounds, consider a Boolean formula φ over a set $X = \{x_1, \ldots, x_n\}$ of variables and the NLWA $\mathcal{A}$ described in Fig. 2. Note that the only state in $\mathcal{A}$ in which nondeterminism may appear is q_n, and that φ is satisfiable iff both of the edges that leave q_n are satisfiable. Thus, $\mathcal{A}$ is a DPA-NLWA iff φ is not satisfiable. Since every path from q_0 to q_n induces an assignment $f : X \to \{0, 1\}$, we also have that $\mathcal{A}$ is an SD-NLWA iff φ is not satisfiable, and the lower bounds follow. □

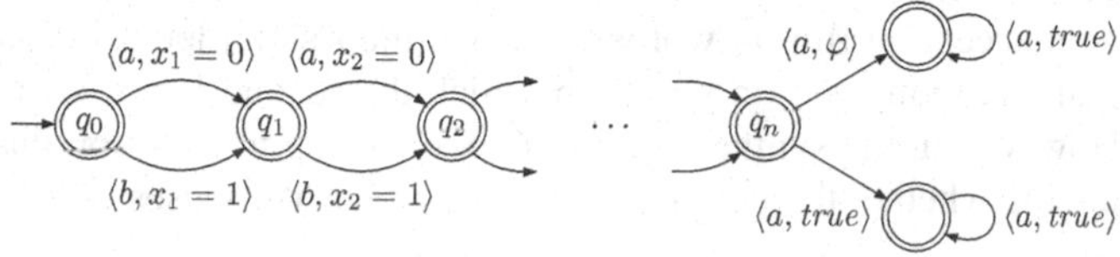

Fig. 2. The NLWA $\mathcal{A}$ is an SD-NLWA and a DPA-NLWA iff φ is not satisfiable.

4 Synthesis

Recall that in synthesis, the goal is to decide, given an automaton $\mathcal{A}$ over $2^{I \cup O}$, whether $L(\mathcal{A})$ is realizable, and if so, to return an I/O-transducer that realizes it. As described in Sect. 1, the synthesis problem is reduced to deciding the winner in a two-player game that is played over $\mathcal{A}$. For finite-state deterministic automata, the game is finite and can be decided in polynomial time. In this section we study the synthesis problem for NLWAs^{IO}. We first define the synthesis game for them, then show that it is not helpful for synthesis of DPA-NLWAs^{IO}, but is helpful for SD-NLWAs^{IO}. We then describe an algorithm for solving the synthesis game induced by SD-NLWAs^{IO} all whose guards are simple. The complexity of the algorithm is polynomial in the state space of the automaton and exponential in the number of variables. Therefore, the synthesis problem for SD-NLWAs^{IO} with simple guards is decidable with the above complexity.

4.1 The Synthesis Game

Consider a NLWA^{IO} $\mathcal{A} = \langle 2^{I \cup O}, X, Q, q_0, \Delta\rangle$. The players in the *synthesis game* $G^{\mathcal{A}}$ are OR (the system) and AND (the environment), its possible locations are $Q \cup \{\bot\}$, and its initial location is q_0. Let q_j be the location of the game at the start of the j-th round. The j-th round of a play consists of two parts: first, AND chooses a letter $\langle i_j, k_j^I\rangle \in 2^I \times \mathbb{Q}$. Then, OR chooses a letter $\langle o_j, k_j^O\rangle \in 2^O \times \mathbb{Q}$ and a state q_{j+1} such that there is a transition $\langle q_j, i_j \cup o_j, \gamma_j, q_{j+1}\rangle \in \Delta$ for some $\gamma_j \in \mathcal{G}_X$, and there exists $f : X \to \mathbb{Q}$ such that $\langle k_t^I, k_t^O\rangle \models_f \gamma_t$ for all $0 \leq t \leq j$. If no such state exists, or if $q_j = \bot$, then OR chooses $q_{j+1} = \bot$. The successive location of the game is q_{j+1}. That is, in every round, every player chooses in its turn a letter, and Player OR chooses a transition that respects all the choices made so far. If no such transition exists, then the game moves to the location $\bot$. If q_{j+1} is $\bot$, then AND wins. Otherwise, the game continues forever and OR wins. Indeed, then, the word $\langle \sigma_0, k_0^I, k_0^O\rangle, \langle \sigma_1, k_1^I, k_1^O\rangle, \dots \in (2^{I \cup O} \times \mathbb{Q} \times \mathbb{Q})^\omega$ that AND and OR generate during an infinite play is accepted by $\mathcal{A}$.

Note that a position of $G^{\mathcal{A}}$ includes more information than its location. Namely, it maintains the guards on the transitions and the numbers that were chosen by Players AND and OR during previous rounds. However, this amounts an unbounded information. Thus, a graph that describes the positions of $G^{\mathcal{A}}$ has an infinite state space. Below we show how to represent this graph symbolically, for NLWAs^{IO} with simple guards. Essentially, rather than maintaining the values accumulated so far, the graph only maintains restrictions they induce. Before we solve the game, we examine its usefulness.

Lemma 1. *There is a realizable DPA-NLWA^{IO} $\mathcal{A}$, such that* OR *does not win $G^{\mathcal{A}}$.*

Proof. Consider the DPA-NLWA $\mathcal{A}$ in Fig. 3, over a single variable x and $I = \{i\}$. It is easy to see that $\mathcal{A}$ is DPA-NLWA and that $L(\mathcal{A}) = (2^I \times \mathbb{Q})^\omega$. Indeed, the run $q_0 \cdot q_1 \cdot {q_3}^\omega$ is an accepting run on words in which the second letter is i, and the run $q_0 \cdot q_2 \cdot {q_4}^\omega$ is an accepting run on words in which the second letter is $\neg i$. Hence, $\mathcal{A}$ is realizable. However, in the synthesis game $G^{\mathcal{A}}$, choosing a transition that leaves q_0 amounts to guessing whether the next input is i or $\neg i$, thus OR looses $G^{\mathcal{A}}$. □

Lemma 2. *For every SD-NLWA^{IO} $\mathcal{A}$, we have that* OR *wins $G^{\mathcal{A}}$ iff $\mathcal{A}$ is realizable.*

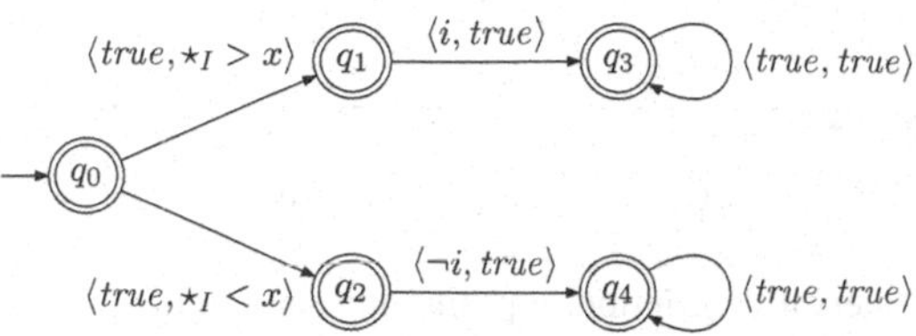

Fig. 3. A realizable DPA-NLWA^{IO} for which OR looses the synthesis game.

Proof. Consider an SD-NLWA^{IO} $\mathcal{A}$. It is easy to see that a winning strategy for OR in $G^{\mathcal{A}}$ induces a transducer that realizes $\mathcal{A}$. We prove the second direction. Assume that $\mathcal{A}$ is realizable. Thus, there is a function $f : (2^I \times \mathbb{Q})^* \to (2^O \times \mathbb{Q})$ such that for every $w_I = \langle i_0, k_0^I\rangle \cdot \langle i_1, k_1^I\rangle \cdots \in (2^I \times \mathbb{Q})^\omega$, we have that $w_I \oplus f(w_I) \in L(\mathcal{A})$, where $f(w_I) = f(\langle i_0, k_0^I\rangle) \cdot f(\langle i_0, k_0^I\rangle \cdot \langle i_1, k_1^I\rangle) \cdots \in (2^O \times \mathbb{Q})^\omega$ and $w_I \oplus f(w_I)$ is the infinite word over $2^{I \cup O} \times \mathbb{Q} \times \mathbb{Q}$ combined from w_I and $f(w_I)$. We describe a winning strategy for OR in $G^{\mathcal{A}}$. For $j \geq 0$, let $w_I^j = \langle i_1, k_1^I\rangle, \ldots, \langle i_j, k_j^I\rangle$ be the sequence of letters that AND chose until the j-th round. In the j-th round, if the location of the game is q, then a winning strategy for OR is to move to the only state q' such that $q' \in \delta(q, w_I^j \oplus f(w_I^j))$. Since $\mathcal{A}$ is semantically deterministic, there is exactly one such q', and by the definition of f, this strategy is winning for OR. □

4.2 Solving the Synthesis Game

In this section we solve the synthesis game for SD-NLWAs^{IO} with simple guards. As stated in Lemma 2, semantic determinization guarantees that the game captures the synthesis problem, and we are left with the challenge of handling the infinite state space. We first need some definitions and notations.

For a set of variables X, we say that a set of brackets $B_X \subset \{\#_x : x \in X, \# \in \{(,), [,]\}\}$ is *legal* if it includes exactly one right and one left bracket for every variable. For example, for $X = \{x, y\}$, the set $\{(_x,]_x, [_y,]_y\}$ is legal, and the set $\{(_x,]_x, [_y\}$ is not. An *interval set* is $N \cup \{-\infty, \infty\} \cup B_X$, where $N \subset \mathbb{Q}$ is a set of numbers and B_X is a legal set of brackets. Consider an interval set $\mathcal{I} = N \cup \{-\infty, \infty\} \cup B_X$, and let $\preceq$ be a total order relation on the elements of $\mathcal{I}$. We say that $\langle \mathcal{I}, \preceq\rangle$ is an *interval description* (ID, for short) if the following hold.

- $k_1 \preceq k_2$ iff $k_1 \leq k_2$ for all $k_1, k_2 \in N \cup \{-\infty, \infty\}$,
- $-\infty \preceq b$ and $\infty \succeq b$ for all $b \in B_X$. If $b \preceq -\infty$, then $b \in \{(_x\colon x \in X\}$, and if $b \succeq \infty$, then $b \in \{)_x : x \in X\}$,
- $b_1 \preceq b_2$ for all $x \in X$ and $b_1 \in \{(_x, [_x\}, b_2 \in \{)_x,]_x\}$. If $b_2 \preceq b_1$, then $b_1 = [_x$ and $b_2 =]_x$.

As an example, consider $\mathcal{I} = \{0, 1\} \cup \{-\infty, \infty\} \cup \{(_x,]_x, [_y,)_y\}$, and an order relation $\preceq$ such that $-\infty = (_x \prec [_y \prec 0 \prec]_x \prec 1 =)_y \prec \infty$. Then, $\langle \mathcal{I}, \preceq\rangle$ is the ID illustrated in Fig. 4a. Intuitively, an ID describes intervals of possible values for variables in X, relatively to each other and to the numbers in N. The tuple $\langle \mathcal{I}, \preceq\rangle$ indicates that $-\infty < x \leq k_1$ for some $0 < k_1 < 1$ and that $k_2 \leq y < 1$, for some $k_2 < 0$.

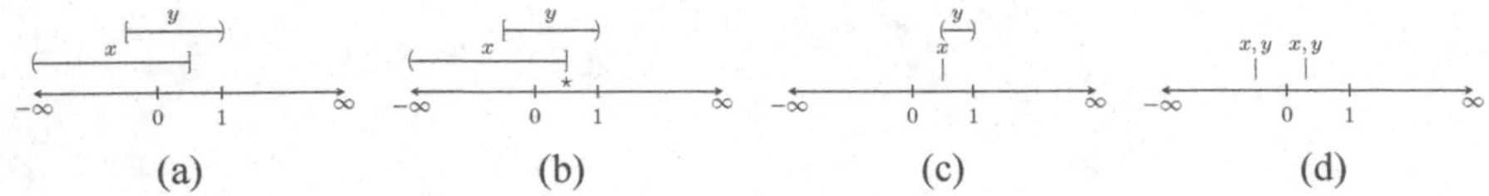

Fig. 4. Example of an interval description, a positioning description, and their updates. (Color figure online)

A *positioning description* (PD, for short) is a pair $\langle \mathcal{I}_\star, \preceq \rangle$, where $\mathcal{I}_\star = \mathcal{I} \cup \{\star\}$, and $\preceq$ is a total order relation on $\mathcal{I}_\star$ that satisfies all the conditions as for IDs, and in addition, $\star \succ -\infty$ and $\star \prec \infty$. Let $\preceq'$ be the order relation obtained from $\preceq$ by reducing $\preceq$ to the elements in $\mathcal{I}$. We then say that $\langle \mathcal{I}_\star, \preceq \rangle$ is obtained from $\langle \mathcal{I}, \preceq' \rangle$. Note that different PDs can be obtained from a single ID. Intuitively, a PD abstracts a choice of a number by denoting it with $\star$ and describing its position with respect to the intervals of possible values for the variables.

Given a PD $\langle \mathcal{I}_\star, \preceq \rangle$ and a simple guard $\gamma \in \mathcal{G}_X$, one can compute the set $update(\langle \mathcal{I}_\star, \preceq \rangle, \gamma)$ of all IDs that combine the restrictions in γ with the interval-restrictions in $\langle \mathcal{I}_\star, \preceq \rangle$. As an example, consider the PD $\langle \mathcal{I}_\star, \preceq \rangle$ that is illustrated in Fig. 4b, namely $-\infty = ({}_x \prec [{}_y \prec 0 \prec]_x = \star \prec 1 =)_y \prec \infty$. Figure 4c illustrates the (single) ID in $update(\langle \mathcal{I}_\star, \preceq \rangle, (\star = x) \wedge (\star < y))$, namely $-\infty \prec 0 \prec [{}_x =]_x = ({}_y \prec 1 =)_y \prec \infty$. Figure 4d illustrates, in red and in blue, two IDs in $update(\langle \mathcal{I}_\star, \preceq \rangle, (x = y))$. Note that the guard may contradict the restrictions in the positioning vector. For example, $update(\langle \mathcal{I}_\star, \preceq \rangle, (\star < x)) = \emptyset$.

Note that we define the updating only for simple guards, that is, ones in which each term refers to at most one member in $X \cup \{\star\}$. For example, we do not aim to express the restrictions imposed by $\langle \mathcal{I}_\star, \preceq \rangle$ and the guard $\star \geq x - y + 1$ via an intervals vector. In addition, as we demonstrated above, there might be several ways to combine restrictions in a positioning vector with a guard. This multiplicity may derive either from the fact that the guard does not refer to $\star$ (as in the example above), or from disjunctions in the guard. Finally, note that one may handle atomic guards of the form $\star \sim x + m$ and $y \sim x + m$, for $\sim \in \{\leq, \geq, =, <, >\}$, $x, y \in X$, and $m \in \mathbb{Q}$, by adding $\{\#_{x+m} : \# \in \{(,), [,]\}\}$ to the brackets set. This requires a prior knowledge of which atomic guards we may have to handle. However, as we show below, when we solve synthesis, the information about possible guards is available.

Theorem 5. *Let $\mathcal{A} = \langle \Sigma, X, Q, Q_0, \Delta \rangle$ be a simple NLWA^{IO}. The synthesis game $G^{\mathcal{A}}$ is solvable in time polynomial in $|Q|$ and exponential in $|X|$.*

Proof. Given an NLWA^{IO} $\mathcal{A}$, we model $G^{\mathcal{A}}$ by an and-or graph $\langle V_{\text{AND}}, V_{\text{OR}}, E \rangle$. Every vertex in this graph abstracts a position of $G^{\mathcal{A}}$, indicating whose is the current turn of the play, the location of the game, the letters and numbers AND and OR choose in the current round, and OR's commitments regarding the values of the variables, induced by previous transitions he chose. These commitments are expressed via IDs, and the $\mathbb{Q}$-choices of the players are expressed via PDs. We maintain a single PD with two $\star$-s, abbreviated with I and O to indicate whose choice they are. We extend the notion of PDs accordingly, so that they include both $\star_I$ and $\star_O$.

Formally, let $\mathcal{A}$ be an NLWA^{IO} over a set of variables X, and let $N \subset \mathbb{Q}$ be the set of constants that appear in the guards of $\mathcal{A}$. We denote by $\hat{\mathcal{I}}$ the set of all interval sets $N \cup \{-\infty, \infty\} \cup B_X$ for legal B_X, and define $P = \{\langle \mathcal{I}, \preceq \rangle : \mathcal{I} \in \hat{\mathcal{I}}$ and $\langle \mathcal{I}, \preceq \rangle$ is an ID$\}$. In addition, let $P_{\star_I}$ and $P_{\star_I,\star_O}$ be the set of all PDs obtained from IDs in P by adding to them the symbols $\star_I$ and $\star_I, \star_O$, respectively.

We define $V_{\text{AND}} = Q \times P$ and $V_{\text{OR}} = (Q \times 2^I \times P_{\star_I}) \cup (Q \times 2^{I \cup O} \times P_{\star_I,\star_O})$. Let $V = V_{\text{AND}} \cup V_{\text{OR}}$. Then, $E \subseteq V \times V$, is the transition relation that consists of the following transitions, for every $q \in Q$, $i \in 2^I$ and $o \in 2^O$:

- $(\langle q, \langle \mathcal{I}, \preceq \rangle \rangle, \langle q, i, \langle \mathcal{I}_{\star_I}, \preceq' \rangle \rangle)$, where $\langle \mathcal{I}_{\star_I}, \preceq' \rangle \in P_{\star_I}$ is a PD obtained from the ID $\langle \mathcal{I}, \preceq \rangle \in P$,
- $(\langle q, i, \langle \mathcal{I}_{\star_I}, \preceq \rangle \rangle, \langle q, i \cup o, \langle \mathcal{I}_{\star_I,\star_O}, \preceq' \rangle \rangle)$, where $\langle \mathcal{I}_{\star_I,\star_O}, \preceq' \rangle \in P_{\star_I,\star_O}$ is a PD obtained from the PD $\langle \mathcal{I}_{\star_I}, \preceq \rangle \in P_{\star_I}$, and
- $(\langle q, i \cup o, \langle \mathcal{I}_{\star_I,\star_O}, \preceq \rangle \rangle, \langle q', \langle \mathcal{I}, \preceq' \rangle \rangle)$, where there is a transition $\langle q, i \cup o, \gamma, q' \rangle$ in $\mathcal{A}$, and $\langle \mathcal{I}, \preceq' \rangle \in \mathit{update}(\langle \mathcal{I}_{\star_I,\star_O}, \preceq \rangle, \gamma)$.

Note that vertices in $Q \times 2^{I \cup O} \times P_{\star_I,\star_O}$ might not have a successor, for example, if q does not have an $(i \cup o)$-successor, or if $\mathit{update}(\langle \mathcal{I}_{\star_I,\star_O}, \preceq \rangle, \gamma) = \emptyset$.

Finally, we define $v_0 = \langle q_0, \langle \mathcal{I}^0, \preceq_0 \rangle \rangle$, where $\mathcal{I}^0 = N \cup \{-\infty, \infty\} \cup \{(_x : x \in X\} \cup \{)_x : x \in X\}$, and $\preceq_0$ is the only possible order relation on the elements in $\mathcal{I}^0$.

The players generate a play over $\langle V_{\text{AND}}, V_{\text{OR}}, E \rangle$, where in every round, if the current position is $v \in V_j$, for $j \in \{\text{AND}, \text{OR}\}$, then player j chooses a successor v' of v, and the play proceeds to position v'. Recall that $V = V_{\text{OR}} \cup V_{\text{AND}}$. A *strategy* for a player $j \in \{\text{AND}, \text{OR}\}$ is a function $f_j : V^* \times V_j \rightarrow V$ such that for every $u \in V^*$ and $v \in V_j$, we have that $\langle v, f_j(u, v) \rangle \in E$. Thus, a strategy for Player j maps the history of the game, when it ends in a position v owned by player j, to a successor of v. Two strategies $f_{\text{AND}}, f_{\text{OR}}$ and the initial position v_0 induce a play $\pi = v_0, v_1, v_2 \cdots \in V^\omega$, where for every $i \geq 0$, if $v_i \in V_j$ for $j \in \{\text{AND}, \text{OR}\}$, then $v_{i+1} = f_j((v_0, \ldots, v_{i-1}), v_i)$. We say that π is the *outcome* of $f_{\text{OR}}, f_{\text{AND}}$ and v_0, and denote $\pi = \mathit{outcome}(f_{\text{OR}}, f_{\text{AND}}, v_0)$. A play π is *winning* for OR if it is infinite. A position $v \in V$ is winning for OR if there exists a strategy f_{OR} such that for every strategy f_{AND}, we have that $\mathit{outcome}(f_{\text{OR}}, f_{\text{AND}}, v)$ is winning for OR. Then, deciding who wins $G^{\mathcal{A}}$ reduces to deciding whether the vertex that abstracts the initial position of the game is winning for AND.

It is not hard to see that the game $\langle V_{\text{AND}}, V_{\text{OR}}, E \rangle$ models the synthesis game. A winning strategy for OR in this graph induces a transducer that realizes $L(\mathcal{A})$. Indeed, a PD essentially abstracts the $\mathbb{Q}$-choices of the environment and the system. The former can be described using guards over X and $\star_I$ that label the transitions that leave the according state in the transducer. The later induces a guard over X and $\star_O$ that labels the according state in the transducer. Finally, since the graph is finite and every infinite play is winning, the problem of deciding whether v_0 is winning for Player OR reduces to the problem of reachability in and-or graphs, which can be solved in time polynomial in the size of the graph. It is not hard to see that the size of the graph is polynomial in $|Q|$ and in $|P \cup P_{\star_I} \cup P_{\star_I,\star_O}|$, and that the later is exponential in $|X|$. Therefore, one can decide who wins $G^{\mathcal{A}}$ in time polynomial in $|Q|$ and exponential in $|X|$. □

Together with Lemma 2, Theorem 5 implies the following.

Corollary 1. *The synthesis problem for a simple SD-NLWA^{IO} over a set X of variables with a set Q of states can be is solved in time polynomial in $|Q|$ and exponential in $|X|$.*

Note that while for model checking, a framework that handles $\mathbb{Q}$ is more general than one that handles $\mathbb{N}$, for synthesis this is not the case. That is, there are specifications that are realizable over $\mathbb{Q}$, but not over $\mathbb{N}$. For example, a specification in which the system has to choose a number between two numbers given by the environment. In particular, the abstraction in our synthesis algorithm exploits the density of $\mathbb{Q}$. We leave the question of solving synthesis for NLWAs over $\Sigma \times \mathbb{N}$ open.

References

1. Bloem, R., Chatterjee, K., Jobstmann, B.: Graph games and reactive synthesis. In: Clarke, E., Henzinger, T., Veith, H., Bloem, R. (eds.) Handbook of Model Checking, pp. 921–962. Springer, Cham (2018). https://doi.org/10.1007/978-3-319-10575-8_27
2. Bojańczyk, M., Muscholl, A., Schwentick, T., Segoufin, L.: Two-variable logic on data trees and XML reasoning. J. ACM **56**(3), 1–48 (2009)
3. Bouajjani, A., Habermehl, P., Mayr, R.R.: Automatic verification of recursive procedures with one integer parameter. TCS **295**, 85–106 (2003)
4. Ceri, S., Fraternali, P., Bongio, A., Brambilla, M., Comai, S., Matera, M.: Designing Data-Intensive Web Applications. Morgan Kaufmann Publishers Inc., San Francisco (2002)
5. Church, A.: Logic, arithmetics, and automata. In: Proceedings of the International Congress of Mathematicians, 1962, pp. 23–35. Institut Mittag-Leffler (1963)
6. Delzanno, G., Sangnier, A., Traverso, R.: Parameterized verification of broadcast networks of register automata. In: Abdulla, P.A., Potapov, I. (eds.) RP 2013. LNCS, vol. 8169, pp. 109–121. Springer, Heidelberg (2013). https://doi.org/10.1007/978-3-642-41036-9_11
7. Ehlers, R., Seshia, S.A., Kress-Gazit, H.: Synthesis with identifiers. In: McMillan, K.L., Rival, X. (eds.) VMCAI 2014. LNCS, vol. 8318, pp. 415–433. Springer, Heidelberg (2014). https://doi.org/10.1007/978-3-642-54013-4_23
8. Exibard, L., Filiot, E., Reynier, P.-A.: Synthesis of data word transducers. In: Proceedings of the 30th CONCUR (2019)
9. Faran, R., Kupferman, O.: LTL with arithmetic and its applications in reasoning about hierarchical systems. In: Proceedings of the 22nd LPAR. EPiC, vol. 57, pp. 343–362 (2018)
10. Grumberg, O., Kupferman, O., Sheinvald, S.: An automata-theoretic approach to reasoning about parameterized systems and specifications. In: Van Hung, D., Ogawa, M. (eds.) ATVA 2013. LNCS, vol. 8172, pp. 397–411. Springer, Cham (2013). https://doi.org/10.1007/978-3-319-02444-8_28
11. Henzinger, T.A., Piterman, N.: Solving games without determinization. In: Ésik, Z. (ed.) CSL 2006. LNCS, vol. 4207, pp. 395–410. Springer, Heidelberg (2006). https://doi.org/10.1007/11874683_26
12. Khalimov, A., Kupferman, O.: Register bounded synthesis. In: Proceedings of the 30th CONCUR (2019)
13. Khalimov, A., Maderbacher, B., Bloem, R.: Bounded synthesis of register transducers. In: Lahiri, S.K., Wang, C. (eds.) ATVA 2018. LNCS, vol. 11138, pp. 494–510. Springer, Cham (2018). https://doi.org/10.1007/978-3-030-01090-4_29
14. Kupferman, O., Safra, S., Vardi, M.Y.: Relating word and tree automata. Ann. Pure Appl. Logic **138**(1–3), 126–146 (2006)
15. Kupferman, O., Vardi, M.Y.: Safraless decision procedures. In: Proceedings of the 46th FoCS, pp. 531–540 (2005)

16. Neven, F., Schwentick, T., Vianu, V.: Towards regular languages over infinite alphabets. In: Sgall, J., Pultr, A., Kolman, P. (eds.) MFCS 2001. LNCS, vol. 2136, pp. 560–572. Springer, Heidelberg (2001). https://doi.org/10.1007/3-540-44683-4_49
17. Niwiński, D., Walukiewicz, I.: Relating hierarchies of word and tree automata. In: Morvan, M., Meinel, C., Krob, D. (eds.) STACS 1998. LNCS, vol. 1373, pp. 320–331. Springer, Heidelberg (1998). https://doi.org/10.1007/BFb0028571
18. Pnueli, A., Rosner, R.: On the synthesis of a reactive module. In: Proceedings of the 16th POPL, pp. 179–190 (1989)
19. Safra, S.: On the complexity of ω-automata. In: Proceedings of the 29th FoCS, pp. 319–327 (1988)
20. Schrijver, A.: Theory of Linear and Integer Programming. Wiley-Interscience Series in Discrete Mathematics and Optimization. Wiley, Hoboken (1999)
21. Shemesh, Y., Francez, N.: Finite-state unification automata and relational languages. Inf. Comput. **114**, 192–213 (1994)
22. Vianu, V.: Automatic verification of database-driven systems: a new frontier. In: ICDT 2009, pp. 1–13 (2009)

On the Average State Complexity of Partial Derivative Transducers

Stavros Konstantinidis[1], António Machiavelo[2], Nelma Moreira[3], and Rogério Reis[3](✉)

[1] Saint Mary's University, Halifax, NS, Canada
s.konstantinidis@smu.ca

[2] CMUP & DM, Faculdade de Ciências da Universidade do Porto, Porto, Portugal
ajmachia@fc.up.pt

[3] CMUP & DCC, Faculdade de Ciências da Universidade do Porto, Porto, Portugal
{nam,rvr}@dcc.fc.up.pt

Abstract. $2D$ regular expressions represent rational relations over two alphabets. In this paper we study the average state complexity of partial derivative standard transducers ($\mathcal{T}_{\mathrm{PD}}$) that can be defined for (general) $2D$ expressions where basic terms are pairs of ordinary regular expressions ($1D$). While in the worst case the number of states of $\mathcal{T}_{\mathrm{PD}}$ can be $O(n^2)$, where n is the size of the expression, asymptotically and on average that value is bounded from above by $O(n^{\frac{3}{2}})$. Moreover, asymptotically and on average the alphabetic size of a $2D$ expression is half of the size of that expression. All results are obtained in the framework of analytic combinatorics considering generating functions of parametrised combinatorial classes defined implicitly by algebraic curves. In particular, we generalise the methods developed in previous work to a broad class of analytic functions.

1 Introduction

We consider $2D$ expressions that represent rational relations over two alphabets. Expressions and transducers with labels over finitely generated monoids were studied by Konstantinidis et al. [11,12], and also by Demaille [7]. Partial derivative methods have become a standard method to manipulate several kinds of expressions [1,2,5–7,14], not only because they are in general more succinct than other equivalent constructions, but for some operators they are easier to define (e.g. for intersection [2]). For regular languages, the average complexity of partial derivative automata ($\mathcal{A}_{\mathrm{PD}}$), considering different sets of operations, has been studied [2,3,5]. Using the framework of analytic combinatorics, for ordinary ($1D$) regular expressions of (tree-)size n (with concatenation, union and Kleene star) it was shown that, asymptotically and on average, the number of

This work was partially supported by NSERC, Canada and CMUP (UID/MAT/00144/2019), which is funded by FCT, FEDER, and PT2020.

A. Chatzigeorgiou et al. (Eds.): SOFSEM 2020, LNCS 12011, pp. 174–186, 2020.
https://doi.org/10.1007/978-3-030-38919-2_15

states of $\mathcal{A}_{\text{PD}}$ is $\frac{1}{4}n$, (being the worst-case $O(n^2)$) while for expressions with intersection of (tree-)size n that number is upper bounded by $(1.056 + o(1))^n$ (being the worst-case $O(2^n)$) [2–4]. In this paper we consider general $2D$ expressions where basic terms are pairs of $1D$ regular expressions. We define a partial derivative standard transducer construction ($\mathcal{T}_{\text{PD}}$) from these expressions, and study its average state complexity. The analytic combinatorial methods used for ordinary $1D$ regular expressions could not be applied for $2D$ expressions. In particular, to get explicit expressions for the generating functions involved would be unmanageable. So, generating functions implicitly defined by algebraic curves must be used, and in previous work it was shown how to get the required information for the asymptotic estimates with an indirect use of the existence of Puiseux expansions at singularities [6]. In this paper, as the involved algebraic curves are more intricate, we needed to refine the methods described in the literature, and use Puiseux expansions together with the Newton's polygon technique to find the estimates for the asymptotic behaviours of parametrised families of combinatorial classes. This new, more refined, method is introduced in Sect. 4. Section 2 reviews the partial derivative construction for ordinary $1D$ regular expressions. In Sect. 3 we define $2D$ expressions, and present the corresponding construction of partial derivative transducers ($\mathcal{T}_{\text{PD}}$). Section 5 presents the average complexity results obtained using the framework of Sect. 4. We show that for general $2D$ expressions, while in the worst case the number of states of $\mathcal{T}_{\text{PD}}$ can be $O(n^2)$, where n is the size of the expression, asymptotically and on average, that value is bounded from above by $O(n^{\frac{3}{2}})$. Restricting to pairs of $1D$ expressions, the previous bound is already reached, showing that these kind of expressions are responsible for the increasing of complexity. Furthermore, the same bounds apply to sums or concatenations of pairs of $1D$ expressions, i.e., regular relations.

2 Preliminares

A *nondeterministic finite automaton* (NFA) is a five-tuple $A = \langle Q, \Sigma, \delta, I, F\rangle$ where Q is a finite set of states, Σ is a finite alphabet, $I \subseteq Q$ is the set of initial states, $F \subseteq Q$ is the set of final states, and $\delta : Q \times \Sigma \to 2^Q$ is the transition function. The size of an NFA is its number of states. The transition function can be extended to words and to sets of states in the natural way. When $I = \{q_0\}$, we use $I = q_0$. The *language accepted* by A is $\mathcal{L}(A) = \{w \in \Sigma^\star \mid \delta(I, w) \cap F \neq \emptyset\}$. Given an alphabet Σ, the set RE of ($1D$) *regular expressions*, $\mathbf{r}$, over Σ contains $\emptyset$ and the expressions defined by the following grammar:

$$\mathbf{r} := \varepsilon \mid \sigma \in \Sigma \mid (\mathbf{r} + \mathbf{r}) \mid (\mathbf{r} \cdot \mathbf{r}) \mid (\mathbf{r}^\star), \tag{1}$$

where the operator $\cdot$ (concatenation) and the outermost parentheses are often omitted. The *language* associated to $\mathbf{r}$ is denoted by $\mathcal{L}(\mathbf{r})$ and defined as usual (with ε representing the empty word). Two expressions $\mathbf{r}_1$ and $\mathbf{r}_2$ are *equivalent*, $\mathbf{r}_1 \sim \mathbf{r}_2$ if $\mathcal{L}(\mathbf{r}_1) = \mathcal{L}(\mathbf{r}_2)$. If $S \subseteq$ RE, $\mathcal{L}(S) = \cup_{\mathbf{r} \in S}\mathcal{L}(\mathbf{r})$. The *(tree-)size* $|\mathbf{r}|$ of $\mathbf{r} \in$ RE is the number of symbols in $\mathbf{r}$ (disregarding parentheses). The *alphabetic size* $|\mathbf{r}|_\Sigma$ is the number of letters occurring in $\mathbf{r}$. We define the *constant part* of $\mathbf{r}$, $\mathsf{c}(\mathbf{r})$, by $\mathsf{c}(\mathbf{r}) = \varepsilon$ if $\varepsilon \in \mathcal{L}(\mathbf{r})$, and $\mathsf{c}(\mathbf{r}) = \emptyset$ otherwise. This function is extended to sets of expressions by $\mathsf{c}(S) = \varepsilon$ if and only if exists $\mathbf{r} \in S$ such that $\mathsf{c}(\mathbf{r}) = \varepsilon$. In the case of a singleton $\{s\}$ we write it simply as s. Given $L \subseteq \Sigma^\star$ and $\sigma \in \Sigma$, let $\sigma^{-1}L = \{\, w \mid \sigma w \in L\}$. This notion can be extended to words and languages. The partial derivative automaton of a regular expression was introduced independently by Mirkin [15] and Antimirov [1]. For a regular expression $\mathbf{r} \in \mathrm{RE}$, let the *linear form* of $\mathbf{r}$, $\mathsf{n} : \mathrm{RE} \to 2^{\Sigma\times\mathrm{RE}}$, be inductively defined by

$$\begin{aligned} \mathsf{n}(\emptyset) = \mathsf{n}(\varepsilon) &= \emptyset, & \mathsf{n}(\mathbf{r}+\mathbf{r}') &= \mathsf{n}(\mathbf{r}) \cup \mathsf{n}(\mathbf{r}'),\\ \mathsf{n}(\sigma) &= \{(\sigma,\varepsilon)\}, & \mathsf{n}(\mathbf{r}\mathbf{r}') &= \mathsf{n}(\mathbf{r})\mathbf{r}' \cup \mathsf{c}(\mathbf{r})\mathsf{n}(\mathbf{r}'),\\ & & \mathsf{n}(\mathbf{r}^\star) &= \mathsf{n}(\mathbf{r})\mathbf{r}^\star, \end{aligned} \tag{2}$$

where for any $S \subseteq \Sigma \times \mathrm{RE}$, we define $S\emptyset = \emptyset S = \emptyset$, $S\varepsilon = \varepsilon S = S$, and $S\mathbf{r}' = \{\,(\sigma,\mathbf{r}\mathbf{r}') \mid (\sigma,\mathbf{r}) \in S \wedge \mathbf{r} \neq \varepsilon\,\} \cup \{\,(\sigma,\mathbf{r}') \mid \exists(\sigma,\varepsilon) \in S\,\}$ if $\mathbf{r}' \neq \emptyset, \varepsilon$ (and analogously for $\mathbf{r}'S$).

Proposition 1 ([1]). *For all* $\boldsymbol{r} \in \mathrm{RE}$, $\boldsymbol{r} \sim \bigcup_{(\sigma,\boldsymbol{r}')\in \mathsf{n}(\boldsymbol{r})} \sigma\boldsymbol{r}' \cup \mathsf{c}(\boldsymbol{r})$.

For a regular expression $\mathbf{r} \in \mathrm{RE}$ and a symbol $\sigma \in \Sigma$, *the set of partial derivatives of* $\mathbf{r}$ *w.r.t.* σ is defined by $\partial_\sigma(\mathbf{r}) = \{\,\mathbf{r}' \mid (\sigma,\mathbf{r}') \in \mathsf{n}(\mathbf{r})\,\}$. We have $\mathcal{L}(\partial_\sigma(\mathbf{r})) = \sigma^{-1}\mathcal{L}(\mathbf{r})$. Partial derivatives can be extended w.r.t words and set of partial derivatives of an expression $\mathbf{r}$ can be defined by iterating the linear form. Let $\pi_0(\mathbf{r}) = \downarrow_2 (\mathsf{n}(\mathbf{r}))$, where $\downarrow_2 (s,t) = t$ is the standard second projection on pairs of objects and naturally extended to sets of pairs. Iteratively applying the operator π_0 we have, $\pi_i(\mathbf{r}) = \pi_0(\pi_{i-1}(\mathbf{r}))$, for $i \in \mathbb{N}$, and $\pi(\mathbf{r}) = \bigcup_{i\in\mathbb{N}_0} \pi_i(\mathbf{r})$. The set $\mathsf{PD}(\mathbf{r}) = \pi(\mathbf{r}) \cup \{\mathbf{r}\}$ is the *set of partial derivatives* of $\mathbf{r}$ and $\pi(\mathbf{r})$ is the *support*[1].

Proposition 2 ([15]). *The support* $\pi(\boldsymbol{r})$ *is inductively defined by*

$$\begin{aligned} \pi(\emptyset) &= \emptyset, & \pi(\boldsymbol{r}+\boldsymbol{r}') &= \pi(\boldsymbol{r}) \cup \pi(\boldsymbol{r}'),\\ \pi(\varepsilon) &= \emptyset, & \pi(\boldsymbol{r}\boldsymbol{r}') &= \pi(\boldsymbol{r})\boldsymbol{r}' \cup \pi(\boldsymbol{r}'),\\ \pi(\sigma) &= \{\varepsilon\}, & \pi(\boldsymbol{r}^\star) &= \pi(\boldsymbol{r})\boldsymbol{r}^\star, \end{aligned}$$

where, for any $S \subseteq \mathrm{RE}$, *we define* $S\emptyset = \emptyset S = \emptyset$, $S\varepsilon = \varepsilon S = S$, *and* $S\boldsymbol{r}' = \{\,\boldsymbol{r}\boldsymbol{r}' \mid \boldsymbol{r} \in S \wedge \boldsymbol{r} \neq \varepsilon\,\} \cup \{\,\boldsymbol{r}' \mid \exists \varepsilon \in S\,\}$ *if* $\boldsymbol{r}' \neq \emptyset, \varepsilon$ *(and analogously for* $\boldsymbol{r}'S$*).*

Proposition 3 ([1,15]). $|\pi(\boldsymbol{r})| \leq |\boldsymbol{r}|_\Sigma$ *and* $|\mathsf{PD}(\boldsymbol{r})| \leq |\boldsymbol{r}|_\Sigma + 1$.

The *partial derivative automaton* of $\mathbf{r}$ is $\mathcal{A}_{\mathrm{PD}}(\mathbf{r}) = \langle \mathsf{PD}(\mathbf{r}), \Sigma, \delta_{\mathrm{PD}}, \mathbf{r}, F\rangle$, where $F = \{\,\mathbf{r}_1 \in \mathsf{PD}(\mathbf{r}) \mid \mathsf{c}(\mathbf{r}_1) = \varepsilon\,\}$, and $\delta_{\mathrm{PD}} = \{\,(\mathbf{r}_1,\sigma,\mathbf{r}') \mid \mathbf{r}_1 \in \mathsf{PD}(\mathbf{r}) \wedge (\sigma,\mathbf{r}') \in \mathsf{n}(\mathbf{r}_1)\,\}$.

Proposition 4 ([1,15]). *For all* $\boldsymbol{r} \in \mathrm{RE}$, $\mathcal{L}(\mathcal{A}_{PD}(\boldsymbol{r})) = \mathcal{L}(\boldsymbol{r})$.

[1] Extending partial derivatives w.r.t. words, one could also define $\mathsf{PD}(r) = \bigcup_{w\in\Sigma^\star} \partial_w(\mathbf{r})$.

3 2*D* Expressions

Let Σ and Δ be two alphabets. A relation R is any subset of $\Sigma^\star \times \Delta^\star$. The concatenation of two relations R and S is the relation $RS = \{(u_1u_2, v_1v_2) \mid (u_1, v_1) \in R \wedge (u_2, v_2) \in S\}$. The Kleene closure of the relation R is the relation $R^\star = \bigcup_{n\geq 0} R^n$. The monoid $\Sigma^\star \times \Delta^\star$ has the identity $(\varepsilon, \varepsilon)$, and the following set of generators $\{(\sigma, \varepsilon), (\varepsilon, \tau) \mid \sigma \in \Sigma \wedge \tau \in \Delta\}$ with the set of equations

$$\{\,(\sigma,\varepsilon)(\varepsilon,\tau) \doteq (\sigma,\tau), (\varepsilon,\tau)(\sigma,\varepsilon) \doteq (\sigma,\tau) \mid \sigma \in \Sigma \wedge \tau \in \Delta\,\}. \tag{3}$$

For a relation $R \subseteq \Sigma^\star \times \Delta^\star$, the quotient of R by a symbol is defined as before, but one needs to take into account the above equations. For instance, for $\sigma \in \Sigma$ and $\tau \in \Delta$, $(\sigma,\varepsilon)^{-1}R = \{\,(\varepsilon,\tau)w \mid (\sigma,\tau)w \in R\}$ and $(\varepsilon,\tau)^{-1}R = \{\,(\sigma,\varepsilon)w \mid (\sigma,\tau)w \in R\}$. The set of *rational relations* is the smallest set of relations that contains the finite relations and is closed under union, concatenation and Kleene closure. Rational relations are accepted by transducers. A *finite transducer in standard-form* (SFT) over two alphabets Σ and Δ is defined as an NFA, except that the transition function is $\delta : Q \times (\Sigma_\varepsilon \times \Delta_\varepsilon) \to 2^Q$, where for a set X, $X_\varepsilon = X \cup \{\varepsilon\}$. The relation realised by an SFT t is denoted by $\mathcal{R}(t)$. In this section we consider 2*D* expressions that represent rational relations. The notions of linear form, of partial derivative and of partial derivative transducers are extend to 2*D* expressions. In Sect. 5 we study the average state complexity of these transducers. Recently, Demaille [7] defined derivative automata for multitape weighted regular expressions. The expressions and transducers studied in this paper are restrictions of those models to two tapes and the Boolean semiring.

To represent rational relations one could just consider 1*D* expressions where basic terms are the generators of $\Sigma^\star \times \Delta^\star$. Those expressions are called *standard* 2*D* regular expressions (S2D-RE) and are a particular case of the ones considered in [11]. For *standard* 2*D* regular expressions, and using the same methods, it can be shown that the asymptotic bounds for partial derivative transducers are the same as for partial derivative automata (for 1*D* expressions) [3].

A *(general)* 2*D regular expression* (2D-RE) over Σ and Δ, where Σ is the *input alphabet* and Δ the *output alphabet*, is an expression that is either $\emptyset$, or can be defined by the following grammar

$$\mathbf{g} := \mathbf{r}/\mathbf{r}' \mid (\mathbf{g}+\mathbf{g}) \mid (\mathbf{g}\cdot\mathbf{g}) \mid (\mathbf{g}^\star), \tag{4}$$

where $\mathbf{r} \in$ RE over Σ and $\mathbf{r}' \in$ RE over Δ. The *relation* $\mathcal{R}(\mathbf{g}) \subseteq \Sigma^\star \times \Delta^\star$ realised by a 2D-RE $\mathbf{g}$ is defined inductively as follows $\mathcal{R}(\mathbf{r}/\mathbf{r}') = \mathcal{L}(\mathbf{r}) \times \mathcal{L}(\mathbf{r}')$, $\mathcal{R}(\mathbf{g}\cdot\mathbf{g}') = \mathcal{R}(\mathbf{g})\mathcal{R}(\mathbf{g}')$, and $\mathcal{R}(\mathbf{g}^\star) = (\mathcal{R}(\mathbf{g}))^\star$. Two expressions $\mathbf{g}, \mathbf{g}'$ are *equivalent*, $\mathbf{g} \sim \mathbf{g}'$, if $\mathcal{R}(\mathbf{g}) = \mathcal{R}(\mathbf{g}')$. A relation is *rational* if and only if it is represented by a 2D-RE[2]. The *constant part* of a 2D-RE expression $\mathbf{g}$ is given by c : 2D-RE $\longrightarrow \{\emptyset, \varepsilon/\varepsilon\}$ such that $\mathsf{c}(\mathbf{g}) = \varepsilon/\varepsilon$ if $(\varepsilon, \varepsilon) \in \mathcal{R}(\mathbf{g})$, and $\mathsf{c}(\mathbf{g}) = \emptyset$, otherwise. For $S \subseteq$ 2D-RE or $S \subseteq (\Sigma_\varepsilon \times \Delta_\varepsilon) \times$ 2D-RE and an expression $\mathbf{g}$, we adopt the same conventions as for 1*D* expressions regarding $\mathbf{g}S$ and $S\mathbf{g}$. In particular, we let $(\varepsilon/\varepsilon)S = S(\varepsilon/\varepsilon) = S$ (and also $S\varepsilon = \varepsilon S = S$).

[2] This follows from the definition above.

For the *linear form* of an expression $\mathbf{g} \in$ 2D-RE, $\mathsf{n} : \text{2D-RE} \to 2^{(\Sigma_\varepsilon \times \Delta_\varepsilon)\times \text{2D-RE}}$, one only needs to extend the definition for expressions of the form $\mathbf{r}_1/\mathbf{r}_2$, being the remaining cases as in Eq. (2), considering expressions $\mathbf{g} \in$ 2D-RE. We note that one possibility was to consider $\mathsf{n}(\mathbf{r}_1/\mathbf{r}_2) = \{(\mathbf{r}_1/\mathbf{r}_2, \varepsilon/\varepsilon)\}$ (see [11]), but then one could not construct directly an SFT. Here we define

$$\mathsf{n}(\mathbf{r}_1/\mathbf{r}_2) = (\mathsf{n}(\mathbf{r}_1)||\mathsf{n}(\mathbf{r}_2)) \cup \mathsf{c}(\mathbf{r}_2)(\mathsf{n}(\mathbf{r}_1)||\{(\varepsilon,\varepsilon)\}) \cup \mathsf{c}(\mathbf{r}_1)(\{(\varepsilon,\varepsilon)\}||\mathsf{n}(\mathbf{r}_2)), \tag{5}$$

where for $N \subseteq \Sigma_\varepsilon \times \text{RE}$ and $M \subseteq \Delta_\varepsilon \times \text{RE}$,

$$N||M = \{\, ((\gamma,\gamma'), \mathbf{r}/\mathbf{r}') \mid (\gamma,\mathbf{r}) \in N \wedge (\gamma',\mathbf{r}') \in M \,\}.$$

The correctness of the previous definition is given by the following proposition.

Proposition 5. *For all* $r_1, r_2 \in$ RE, $r_1/r_2 \sim \bigcup_{((\gamma,\gamma'),g')\in \mathsf{n}(r_1/r_2)} (\gamma/\gamma')g' \cup \mathsf{c}(r_1/r_2)$.

Then, we have

Proposition 6. *For all* $g \in$ 2D-RE, $g \sim \bigcup_{((\gamma,\gamma'),g')\in \mathsf{n}(g)} (\gamma/\gamma')g' \cup \mathsf{c}(g)$.

As before, one can obtain the *support* of an expression $\mathbf{g}$, $\pi(\mathbf{g})$, by iterating the linear form. Only the base case differs from the ones in Proposition 2.

Proposition 7. *For all* $r_1, r_2 \in$ RE,

$$\pi(r_1/r_2) \subseteq \pi(r_1)||\pi(r_2) \cup \pi(r_1)||\{\varepsilon\} \cup \{\varepsilon\}||\pi(r_2),$$

where for $S, T \subseteq$ RE, $S||T = \{\, r/r' \mid r \in S \wedge r' \in T \,\}$.

Note that the inclusion in Proposition 7 can be strict, as $\pi(ab/abc) = \{b/bc, \varepsilon/c, \varepsilon/\varepsilon\}$, $\pi(ab) = \{\varepsilon, b\}$, $\pi(abc) = \{\varepsilon, bc, c\}$ and $\mathsf{c}(ab) = \mathsf{c}(abc) = \emptyset$. Proposition 7 and Proposition 2 ensure that for every $\mathbf{g} \in$ 2D-RE, the support $\pi(\mathbf{g})$ is finite and in the worst-case of size $O(n^2)$, where n is the size of $\mathbf{g}$. The quadratic blow-up is achieved if one considers $\mathbf{r}_n = (a^\star)^n$, $n \geq 1$, and the 2D-RE $\mathbf{r}_n/\mathbf{r}_n$.

Corollary 8. *For all* $g \in$ 2D-RE, $|\pi(g)| \leq (|g|_{\Sigma\cup\Delta})^2$, *where* $|g|_{\Sigma\cup\Delta}$ *is the alphabetic size of* g.

The *partial derivative transducer* of $\mathbf{g}$ is $\mathcal{T}_{\text{PD}}(\mathbf{g}) = \langle \pi(\mathbf{g}) \cup \{\mathbf{g}\}, \Sigma, \Delta, \delta_{\text{PD}}, \mathbf{g}, F\rangle$, where $F = \{\, \mathbf{g}_1 \in \pi(\mathbf{g}) \cup \{\mathbf{g}\} \mid \mathsf{c}(\mathbf{g}_1) = \varepsilon/\varepsilon \,\}$, and $\delta_{\text{PD}} = \{\, (\mathbf{g}_1, (\gamma,\gamma'), \mathbf{g}') \mid \mathbf{g}_1 \in \pi(\mathbf{g}) \cup \{\mathbf{g}\} \wedge ((\gamma,\gamma'), \mathbf{g}') \in \mathsf{n}(\mathbf{g}_1) \,\}$.

Proposition 9. *For all* $g \in$ 2D-RE, $\mathcal{R}(\mathcal{T}_{PD}(g)) = \mathcal{R}(g)$.

An upper bound of the number of states of $\mathcal{T}_{\text{PD}}(\mathbf{g})$ is obtained if one assumes that

$$\pi(\mathbf{r}_1/\mathbf{r}_2) = \pi(\mathbf{r}_1)||\pi(\mathbf{r}_2) \cup \pi(\mathbf{r}_1)||\{\varepsilon\} \cup \{\varepsilon\}||\pi(\mathbf{r}_2)$$

always holds, and as usual $\pi(\mathbf{g} + \mathbf{g}') = \pi(\mathbf{g}) \cup \pi(\mathbf{g}')$, $\pi(\mathbf{g}\mathbf{g}') = \pi(\mathbf{g})\mathbf{g}' \cup \pi(\mathbf{g}')$, and $\pi(\mathbf{g}^\star) = \pi(\mathbf{g})\mathbf{g}^\star$. These equalities are used in Sect. 5 to obtain an upper bound for the average case size of partial derivative transducers. In the next section we set up the analytic combinatorics framework that allows to obtain those estimates.

4 The Analytic Combinatorics Framework

Given some measure of the objects of a combinatorial class, $\mathcal{A}$, for each $n \in \mathbb{N}_0$ let a_n be the sum of the values of this measure for all objects of size n. Let $A(z) = \sum_n a_n z^n$ be the corresponding generating function (*cf.* [8]). We will use the notation $[z^n]A(z)$ for a_n. The generating function $A(z)$ can be seen as a complex analytic function, and the study of its behaviour around its dominant singularity ρ, when unique, gives us access to the asymptotic form of its coefficients. In particular, if $A(z)$ is analytic in some indented disc neighbourhood of ρ, then one has the following [4,8]:

Theorem 10. *The coefficients of the series expansion of the complex function*

$$f(z) = (1 - z)^{\alpha},$$

where $\alpha \in \mathbb{C} \setminus \mathbb{N}_0$, *have the following asymptotic approximation:*

$$[z^n]f(z) = \frac{n^{-\alpha-1}}{\Gamma(-\alpha)} + o\left(n^{-\alpha-1}\right).$$

Here Γ *is Euler's gamma function.*

The combinatorial classes that we deal with in the present paper give rise to generating functions implicitly defined by algebraic curves that are quite a bit more convoluted than those previously described in the literature. We, therefore, needed to refine the method to pursue these calculations, and we will expound that, in some detail, here. Generically, from an unambiguous generating grammar, one obtains a set of polynomial equations involving the generating functions for the objects corresponding to the variables of the grammar, in particular the one whose coefficients we want to asymptotically estimate. Computing a Gröbner basis for the ideal generated by those polynomials, one gets an algebraic equation for that generating function $w = w(z)$, i.e., an equation of the form

$$G(z, w) = 0,$$

where $G(z, w)$ is a polynomial in $\mathbb{Z}[z][w]$ of which $w(z)$ is a root.

Since $w(z)$ is the generating function of a combinatorial class, thus a series with non-negative integer coefficients which is not a polynomial, it must have, by Pringsheim's Theorem (*cf.* [8], Thm IV.6), a real positive singularity, ρ, smaller than or equal to 1. In all that follows we will assume that there is no other singularity with that norm, which is the case of all generating functions dealt with in this paper, as we will see. At this singularity, ρ, two cases may occur:

Case I: $\lim_{z\to\rho} w(z) = a$, where a is a positive real number.
Case II: $\lim_{z\to\rho} w(z) = +\infty$.

In the first case the curve defined by G has a shape similar to the one depicted in Fig. 1, on the left, and

$$\frac{\partial G}{\partial w}(\rho, a) = 0. \tag{6}$$

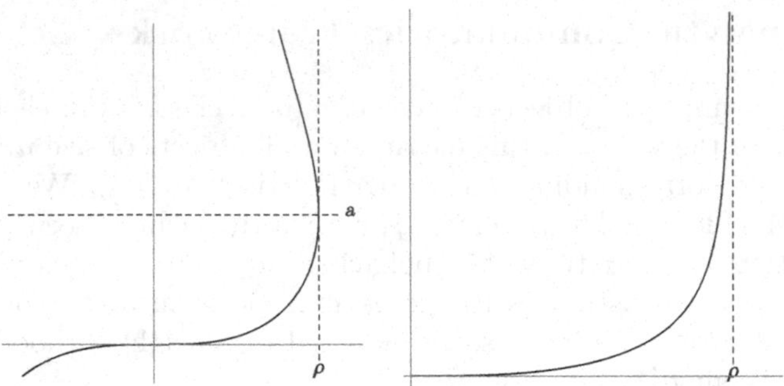

Fig. 1. Generic shape of $G(z, w)$ near its dominant singularity (cases I and II).

This, together with the fact that $G(\rho, a) = 0$, shows that ρ is a root of the resultant, $\text{res}_w(G(z,w), \frac{\partial G}{\partial w}(z,w))$, of $G(z,w)$ and $\frac{\partial G}{\partial w}(z,w)$ with respect to w (*cf.* [13, p. 204]). With the help of a numerical solver and drawing the relevant part of the algebraic curve $G(z,w) = 0$, one can, by an elimination process, find out the minimum polynomial, in $\mathbb{Q}[z]$, of ρ. We will denote this polynomial by $m(z)$. Using now the $\text{res}_z(G(z,w), \frac{\partial}{\partial w}G(z,w))$ one can get, in a similar fashion, an irreducible polynomial that has a as a root.

In Case II, the irreducible polynomial for ρ is a factor of the leading coefficient of $G(z,w)$, seen as a polynomial in w (*cf* [10], Th. 12.2.1).

In Case I, after making the change of variable $s = 1 - z/\rho$, one knows that $w = w(s)$ has a Puiseux series expansion at the singularity $s = 0$, i.e., there exists a slit neighbourhood of that point in which $w(s)$ has a representation as a power series with fractional powers (*cf.* [10], Chap. 12). In particular, w must have the form

$$w(s) = a - g(s)s^{\alpha}, \tag{7}$$

for some $\alpha \in \mathbb{Q}^+$, the first positive exponent of that expansion, and where $g(s)$ is such that $g(s) = b + h(s)s^{\beta}$, $h(0) \neq 0$, $\beta \in \mathbb{Q}^+$, and $b \in \mathbb{R}^{\star}$.

The value of α can be obtained by looking at the Taylor expansion of $G(z,w)$ at (ρ, a),

$$G(z,w) = \sum_{i,j \geq 0} \frac{1}{i!j!} \frac{\partial^{i+j} G}{\partial z^i w^j}\bigg|_{\substack{z=\rho \\ w=a}} (z-\rho)^i (w-a)^j.$$

Noticing that $z = \rho - \rho s$, and using Eq. (7), one has

$$G(\rho - \rho s, a - g(s)s^{\alpha}) = \sum_{i,j \geq 0} \frac{(-1)^{i+j}}{i!j!} \frac{\partial^{i+j} G}{\partial z^i w^j}\bigg|_{\substack{z=\rho \\ w=a}} \rho^i g(s)^j s^{i+j\alpha}. \tag{8}$$

Using that $G(z, w(z)) = 0$, $G(\rho, a) = 0$, and (6), and dividing it through by s^{α}, one gets

$$0 = \sum_{\substack{i,j \geq 0 \\ (i,j) \notin \{(0,0),(0,1)\}}} \frac{(-1)^{i+j}}{i!j!} \frac{\partial^{i+j} G}{\partial z^i w^j}\bigg|_{\substack{z=\rho \\ w=a}} \rho^i g(s)^j s^{i+(j-1)\alpha}. \tag{9}$$

One can now compute $p_{ij}(z) = \text{res}_w \left(G(z,w), \dfrac{\partial^{i+j} G}{\partial z^i w^j} \right)$, and $\gcd(p_{ij}(z), m(z))$ to see which derivatives are non-zero at ρ. Then, one can use the Newton's polygon technique to find α [9,17,18]. The points of Newton polygon that lead to the value of α correspond to the terms of (9) with the lowest exponent, that must cancel out together. This conduces, after setting $s = 0$, to a polynomial equation for the value b defined in the sentence containing (7). One then uses this value in Theorem 10 to get the desired asymptotic approximation. In conclusion, for the case where $\lim_{z\to\rho} w(z) = a$, one has

$$[z^n]w(z) \sim \frac{-b}{\Gamma(-\alpha)} \rho^{-n} n^{-\alpha-1}. \tag{10}$$

In Case II, the one where $\lim_{z\to\rho} w(z) = +\infty$, making $v = \frac{1}{w}$ one concludes as above that $v = cs^\alpha - g(s)s^{\alpha+\beta}$, for some $0 < \alpha < 1$, $\beta > 0$, and for some Puiseux series $g(s)$, with non-negative exponents. Denoting by m the degree of G relative to w, the polynomial satisfied by v is then

$$H(z,v) = v^m G\left(z, \frac{1}{v}\right), \tag{11}$$

which is the reciprocal polynomial of $G(z,w)$ with respect to the variable w. In this case the equation that corresponds to Eq. (8) is:

$$H(\rho - \rho s, cs^\alpha - g(s)s^{\alpha+\beta}) = \sum_{i,j\geq 0} \frac{(-1)^i}{i!j!} \frac{\partial^{i+j} H}{\partial z^i w^j}\bigg|_{\substack{z=\rho \\ w=0}} \rho^i (c - g(s)^\beta)^j s^{i+j\alpha}. \tag{12}$$

Using the same procedure as above, one computes ρ, and then the value of c. Since

$$\begin{aligned} w = \frac{1}{cs^\alpha - g(s)s^{\alpha+\beta}} &= \frac{1}{c} s^{-\alpha} \frac{1}{1 - \frac{g(s)}{c} s^\beta} \\ &= \frac{1}{c} s^{-\alpha} \left(1 + \frac{g(s)}{c} s^\beta + \frac{g(s)^2}{c^2} s^{2\beta} + \cdots \right), \end{aligned}$$

one sees, using again Theorem 10, that

$$[z^n]w(z) \sim \frac{1}{c\,\Gamma(\alpha)} \rho^{-n} n^{\alpha-1}. \tag{13}$$

Summing up, we have the following.

Theorem 11. *With the notations and in the conditions above described, one has*

$$[z^n]w(z) \sim \begin{cases} \frac{-b}{\Gamma(-\alpha)} \rho^{-n} n^{-\alpha-1}, & \text{if } \lim_{z\to\rho} w(z) = a, \\ \frac{1}{c\,\Gamma(\alpha)} \rho^{-n} n^{\alpha-1}, & \text{if } \lim_{z\to\rho} w(z) = +\infty, \end{cases}$$

where b, c, ρ and α can be computed as above described.

5 Average Descriptional Complexity Results

Using the framework just described, we obtain asymptotic estimates for an upper bound of the average state complexity of partial derivative transducer for $2D$ expressions of size $n \geq 0$. Those estimates depend on the size of the alphabets Σ and Δ, which we assume both to be equal to some integer $k > 0$. Moreover we denote by RE_k the set of $1D$ expressions over an alphabet of size k.

5.1 Average State Complexity of $\mathcal{T}_{\mathrm{PD}}$ for 2D-RE

The generating function $G_k(z)$ associated with $\mathbf{g} \in$ 2D-RE is the following[3], where $R_k(z)$ is the generating function of regular expressions $\mathbf{r} \in \mathrm{RE}_k$ [4].

$$G_k(z) = zR_k(z)^2 + zG_k(z) + 2zG_k(z)^2, \tag{14}$$
$$R_k(z) = (k+1)z + zR_k(z) + 2zR_k(z)^2. \tag{15}$$

Considering Proposition 2, let $p(\mathbf{r})$ be the size of the support of an expression $\mathbf{r} \in \mathrm{RE}_k$ which is defined by $p(\varepsilon) = 0$, $p(\sigma) = 1$, $p(\mathbf{s}+\mathbf{s}') = p(\mathbf{s}\cdot\mathbf{s}') = p(\mathbf{s})+p(\mathbf{s}')$, and $p(\mathbf{s}^\star) = p(\mathbf{s})$. An upper bound for the size of the support $\pi(\mathbf{g})$, $q(\mathbf{g})$, is defined by $q(\mathbf{r}/\mathbf{r}') = p(\mathbf{r})p(\mathbf{r}') + p(\mathbf{r}) + p(\mathbf{r}')$, $q(\mathbf{g}+\mathbf{g}') = q(\mathbf{g}\cdot\mathbf{g}') = q(\mathbf{g}) + q(\mathbf{g}')$, and $q(\mathbf{g}^\star) = q(\mathbf{g})$. Thus, the generating function $Q_k(z) = \sum_{\mathbf{g}} q(\mathbf{g})z^{|\mathbf{g}|}$ for $\pi(\mathbf{g})$ satisfies the following equation,

$$Q_k(z) = zQ_k(z) + 4zQ_k(z)G_k(z) + 2zP_k(z)R_k(z) + zP_k(z)^2, \tag{16}$$

where $P_k(z)$ is the generating function for the support of regular expressions in RE_k, which satisfy

$$P_k(z) = kz + zP_k(z) + 4zR_k(z)P_k(z). \tag{17}$$

From Eqs. (15), (17), (14) and (16), using Gröbner basis, one obtains algebraic equations for $G_k(z)$ and $Q_k(z)$:

$$\mathcal{C}_G(z,w) = 16z^3w^4 + 16(z^3 - z^2)w^3 - g_2(z)w^2 + g_1(z)w + (1+k)^2z^3 = 0, \tag{18}$$

where $g_2(z) = 2z((1+4k)z^2 + 6z - 3)$ and $g_1(z) = (1-z)((3+4k)z^2 + 2z - 1)$ and

$$\mathcal{C}_Q(z,w) = p(z)^4q_4(z)w^4 - k^2z^2p(z)^2q_2(z)w^2 + k^4z^8q_0(z)^2 = 0, \tag{19}$$

where

$$\begin{aligned}
p(z) &= (8k+7)z^2 + 2z - 1\\
q_4(z) &= (16k^2 + 40k + 23)z^4 - 4(4k+3)z^3 + (8k+2)z^2 + 4z - 1\\
q_2(z) &= (200k^3 + 544k^2 + 474k + 133)z^6 - (48k^2 + 24k - 10)z^5\\
&\quad + (24k^2 - 44k - 41)z^4 + 28(2k+1)z^3 + (3 - 14k)z^2 - 6z + 1\\
q_0(z) &= (25k^2 + 37k + 14)z^2 + (6k+4)z - (3k+2).
\end{aligned}$$

[3] I.e. $[z^n]G_k(z)$ gives the number of expressions $\mathbf{g}$ of size n.

For $G_k(z)$, we conclude to be in Case I. The irreducible polynomial that implicitly defines the singularity ρ_k of $G_k(z)$ is, computed using the resultant $\text{res}_w(\mathcal{C}_G(z,w), \frac{\partial \mathcal{C}_G}{\partial w}(z,w))$. In this case we obtain two candidates for the minimal polynomial $m_G(z)$ of the singularity ρ_k, each one having only one root in $]0,1[$. Using a computer algebra system, one can show that those roots are only equal for $k = -1$. This implies, by continuity (in k), that they always keep their relative position, for all $k > -1$. Now, $\text{res}_z(\mathcal{C}_G(z,w), \frac{\partial \mathcal{C}_G}{\partial w}(z,w))$ factors into three irreducible polynomials, one of which has a_k as a root. These three polynomials have, among them, four positive roots, which a computer algebra system can find, as a function of k. Then, one can check which pairs (ρ'_k, a'_k), where ρ'_k is a candidate for ρ_k, and a'_k a candidate for a_k, belong to the curve $\mathcal{C}_G$, and their relative location (Fig. 2). By a simple topological argument, one then can conclude that $m_G(z) = (8k+7)z^2 + 2z - 1$, $\rho_k = \frac{1}{1+\sqrt{8k+8}}$, and $a_k = \frac{\sqrt{2}-1}{2}\sqrt{k+1}$. One then checks that $\dfrac{\partial \mathcal{C}_G}{\partial z}(\rho_k, a_k)$ and $\dfrac{\partial^2 \mathcal{C}_G}{\partial w^2}(\rho_k, a_k)$ are both non-zero, for all k, which entails that $\alpha = \frac{1}{2}$. The value for b_k can then be computed, and $b_k \sim \sqrt{\frac{k}{2}}$. As for $Q_k(z)$, one sees that Case II applies, and that the minimal polynomial is either $p(z)$ or $q_4(z)$. It turns out that each of these polynomials has exactly one positive real root, ρ_k and ζ_k. One can then check that these roots coincide only for $k = -1$, and so that one of them is always bigger than the other for all positive values of k, namely ρ_k. One then can check that the curve $\mathcal{C}_Q$ crosses the vertical line $z = \zeta_k$ exactly once above the z-axis, which makes clear that the singularity for $Q_k(z)$ is ρ_k, thus the same as for $G_k(z)$. In this case, the Newton polygon analysis shows that $\alpha = 1$ and that the polynomial satisfied by c, as explained after (9), and noticing that here we make use of inversion explained in (11), is given by

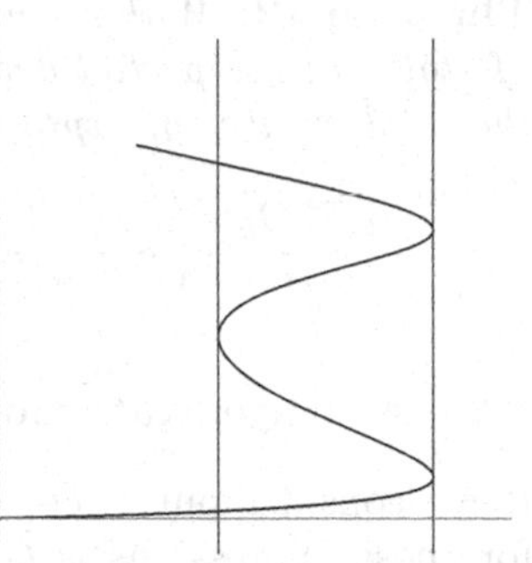
Fig. 2. Possible values for (ρ_k, a_k).

$$\left.\frac{\partial^4 H}{\partial v^4}\right|_{\substack{z=\rho\\v=0}} c^4 + 6\left.\frac{\partial^4 H}{\partial z^2 \partial v^2}\right|_{\substack{z=\rho\\v=0}} \rho^2 c^2 + \left.\frac{\partial^4 H}{\partial z^4}\right|_{\substack{z=\rho\\v=0}} \rho^4 = 0.$$

This is a quadratic equation in c^2, whose discriminant can be seen to be zero. One gets

$$c_k^2 = -3\rho_k^2 \left(\left.\frac{\partial^4 H}{\partial z^2 \partial v^2}\right|_{\substack{z=\rho\\v=0}} \right) \Big/ \left(\left.\frac{\partial^4 H}{\partial v^4}\right|_{\substack{z=\rho\\v=0}} \right). \tag{20}$$

From all this, it follows that

Theorem 12. *With the notations above introduced, the ratio of the total number of states in the partial derivative transducer $\mathcal{T}_{PD}(\boldsymbol{g})$ of expressions of size n to the total number of expressions of the same size is given by*

$$\frac{[z^n]\,Q_k(z)}{[z^n]\,G_k(z)} \sim \frac{-\Gamma(-\frac{1}{2})}{b_k c_k} n^{\frac{3}{2}}, \text{ for all } k, \text{ and } \lim_{k\to\infty} \frac{-\Gamma(-\frac{1}{2})}{b_k c_k} = \frac{\sqrt{\pi}}{8\sqrt{2}}.$$

5.2 Average State Complexity of $\mathcal{T}_{PD}$ for Pairs of *REs*

If we consider only $2D$-expressions of the form $\mathbf{r}/\mathbf{r}'$, the generating function for these expressions is $G'_k(z) = 2zR_k(z)$ and for the support π is, following Proposition 7, $Q'_k(z) = 2zP_k(z)R_k(z) + zP_k^2(z)$. From these, one can deduce the following algebraic equations for $G'_k(z)$ and $Q'_k(z)$:

$$\mathcal{C}_{G'}(z,w) = w^2 + (z-1)w + 2(k+1)z^2 = 0, \tag{21}$$

and

$$\mathcal{C}_{Q'}(z,w) = p(z)^2 w^2 + kzg'_1(z)w + k^2z^4 g'_0(z) = 0, \tag{22}$$

where $p(z)$ is as above, and

$$g'_1(z) = (80k^2 + 126k + 49)z^4 + 4(9k+7)z^3 - 2(9k+5)z^2 - 4z + 1,$$
$$g'_0(z) = (25k^2 + 37k + 14)z^2 + (6k+4)z - 3k - 2.$$

Let us first deal with $G'_k(z)$. We easily conclude that we are in Case I. The irreducible polynomial that implicitly defines the singularity ρ_k of $G'_k(z)$ is computed using $\mathrm{res}_w(\mathcal{C}_{G'}(z,w), \frac{\partial \mathcal{C}_{G'}}{\partial w}(z,w))$. In this case we obtain a single candidate for the minimal polynomial, $m_{G'}(z)$, of the singularity, ρ_k, namely

$$m_{G'}(z) = (8k+7)z^2 + 2z - 1,$$

and thus $\rho_k = \frac{1}{1+\sqrt{8k+8}}$. One has

$$\mathrm{res}_w(\mathcal{C}_{G'}(z,w), \frac{\partial \mathcal{C}_{G'}}{\partial w}(z,w)) = (7+8k)w^2 - 8(1+k)w + 2(1+k),$$

from which one gets $a_k = \frac{4(1+k)-\sqrt{2(1+k)}}{7+8k}$, where $a_k = G'_k(\rho_k)$.

Using now the Newton's polygon method, one gets that $\alpha = \frac{1}{2}$, and

$$b_k = \sqrt{\frac{2\rho_k \frac{\partial \mathcal{C}_{G'}}{\partial z}(\rho_k, a_k)}{\frac{\partial^2 \mathcal{C}_{G'}}{\partial w^2}(\rho_k, a_k)}} \sim \frac{1}{\sqrt{2}}.$$

As for Q'_k, one sees that one is in Case II, and that the dominant singularity is the same as for G'_k. Using the methods expounded above, one gets that $\alpha = 1$, and that c_k is a zero of the equation

$$\frac{\partial^2 H}{\partial v^2}(\rho_k, 0)c_k^2 - 2\rho_k \frac{\partial^2 H}{\partial z \partial v}(\rho_k, 0)c_k + \rho_k^2 \frac{\partial^2 H}{\partial z^2}(\rho_k, 0) = 0,$$

where $H(z,v) = v^2 G_{Q'}(z, \frac{1}{v})$. It turns out that this equation has a single solution, namely $c_k = \frac{4}{k^2}\left(8 + 8k + (9+8k)\sqrt{2+2k}\right) \sim 32\sqrt{\frac{2}{k}}$. Therefore, in this case an upper bound of the average state complexity of $\mathcal{T}_{\rm PD}(\mathbf{r}/\mathbf{r}')$ is,

Theorem 13. *With the notations above introduced, one has*

$$\frac{[z^n]\,Q_k'(z)}{[z^n]\,G_k'(z)} \sim \frac{-\Gamma(-\frac{1}{2})}{b_k c_k} n^{\frac{3}{2}} \sim \frac{\sqrt{\pi}}{16}\sqrt{k}\,n^{\frac{3}{2}}.$$

6 Conclusions

We defined partial derivative transducers for $2D$ regular expressions over pairs of $1D$ regular expressions. For studying the average state complexity, and given the intricacy of the resulting generating functions, we refine known methods. In Sect. 5, we conclude that for $2D$ expressions of size n, both general and restricted, asymptotically and on average, the state complexity of the partial derivative transducers is bounded from above by $O(n^{\frac{3}{2}})$. For ordinary $1D$ regular expressions, the number of letters in an expression is, asymptotically and on average, $\frac{1}{2}n$ [3, 16]. The same holds for general $2D$ expressions.

References

1. Antimirov, V.M.: Partial derivatives of regular expressions and finite automaton constructions. Theoret. Comput. Sci. **155**(2), 291–319 (1996)
2. Bastos, R., Broda, S., Machiavelo, A., Moreira, N., Reis, R.: On the average complexity of partial derivative automata for semi-extended expressions. J. Autom. Lang. Comb. **22**(1–3), 5–28 (2017)
3. Broda, S., Machiavelo, A., Moreira, N., Reis, R.: On the average state complexity of partial derivative automata: an analytic combinatorics approach. Int. J. Found. Comput. Sci. **22**(7), 1593–1606 (2011)
4. Broda, S., Machiavelo, A., Moreira, N., Reis, R.: A Hitchhiker's Guide to descriptional complexity through analytic combinatorics. Theoret. Comput. Sci. **528**, 85–100 (2014)
5. Broda, S., Machiavelo, A., Moreira, N., Reis, R.: Automata for regular expressions with shuffle. Inf. Comput. **259**(2), 162–173 (2018)
6. Broda, S., Machiavelo, A., Moreira, N., Reis, R.: On average behaviour of regular expressions in strong star normal form. Int. J. Found. Comput. Sci. **30**(6–7), 899–920 (2019)
7. Demaille, A.: Derived-term automata of multitape expressions with composition. Sci. Ann. Comput. Sci. **27**(2), 137–176 (2017)
8. Flajolet, P., Sedgewick, R.: Analytic Combinatorics. Cambridge University Press, Cambridge (2008)
9. Ghys, É.: A Singular Mathematical Promenade. ENS Éditions, Lyon (2017)
10. Hille, E.: Analytic Function Theory, vol. 2. Blaisdell Publishing Company, New York (1962)

11. Konstantinidis, S., Moreira, N., Pires, J., Reis, R.: Partial derivatives of regular expressions over alphabet-invariant and user-defined labels. In: Hospodár, M., Jirásková, G. (eds.) CIAA 2019. LNCS, vol. 11601, pp. 184–196. Springer, Cham (2019). https://doi.org/10.1007/978-3-030-23679-3_15
12. Konstantinidis, S., Moreira, N., Reis, R., Young, J.: Regular expressions and transducers over alphabet-invariant and user-defined labels. In: Câmpeanu, C. (ed.) CIAA 2018. LNCS, vol. 10977, pp. 4–27. Springer, Cham (2018). https://doi.org/10.1007/978-3-319-94812-6_2
13. Lang, S.: Algebra. Graduate Texts in Mathematics, vol. 211, 3rd edn. Springer, New York (2002). https://doi.org/10.1007/978-1-4613-0041-0
14. Lombardy, S., Sakarovitch, J.: Derivatives of rational expressions with multiplicity. Theor. Comput. Sci. **332**(1–3), 141–177 (2005)
15. Mirkin, B.G.: An algorithm for constructing a base in a language of regular expressions. Eng. Cybern. **5**, 51–57 (1966)
16. Nicaud, C.: On the average size of Glushkov's automata. In: Dediu, A.H., Ionescu, A.M., Martín-Vide, C. (eds.) LATA 2009. LNCS, vol. 5457, pp. 626–637. Springer, Heidelberg (2009). https://doi.org/10.1007/978-3-642-00982-2_53
17. Walker, R.J.: Algebraic Curves. Princeton University Press, Princeton (1950)
18. Wall, C.T.C.: Singular Points of Plane Curves. No. 63 in London Mathematical Society Student Texts. Cambridge University Press, Cambridge (2004)

On the Difference Between Finite-State and Pushdown Depth

Liam Jordon(✉) and Philippe Moser(✉)

Computer Science Department, National University of Ireland Maynooth, Maynooth, Co Kildare, Ireland
liam.jordon@mu.ie, pmoser@cs.nuim.ie

Abstract. This paper expands upon existing and introduces new formulations of Bennett's logical depth. A new notion based on pushdown compressors is developed. A pushdown deep sequence is constructed. The separation of (previously published) finite-state based and pushdown based depth is shown. The previously published finite state depth notion is extended to an almost everywhere (a.e.) version. An a.e. finite-state deep sequence is shown to exist along with a sequence that is infinitely often (i.o.) but not a.e. finite-state deep. For both finite-state and pushdown, easy and random sequences with respect to each notion are shown to be non-deep, and that a slow growth law holds for pushdown depth.

Keywords: Algorithmic information theory · Kolmogorov complexity · Bennett's logical depth

1 Introduction

In a seminal paper [2], Bennett introduced a new method to measure the *useful* information contained in a piece of data; called logical depth. Logical depth is different from classical information theory in the following sense. Consider a random binary sequence. According to classical information theory, such a random sequence contains a large amount of information because it cannot be significantly compressed while logical depth says that this information is not of much value. Contrast this with a 10-day weather forecast; from the classical information point of view it contains little information (namely no more than the differential equations from which it was originally simulated), but it contains useful information according to logical depth.

Logical depth helps to formalise the difference between complex and non-complex structures. Deep structures can be thought of as structures that contain an underlying patterns which are extremely difficult to find. Given more and more time and resources, an algorithm could spot these patterns and exploit them (such as to compress a sequence).

L. Jordon—Supported by a postgraduate scholarship from the Irish Research Council, Government of Ireland.

A. Chatzigeorgiou et al. (Eds.): SOFSEM 2020, LNCS 12011, pp. 187–198, 2020.
https://doi.org/10.1007/978-3-030-38919-2_16

Bennett's original notion is based on Kolmogorov complexity [2], and interacts nicely with fundamental notions of computability theory as shown in [12].[1] Due to the uncomputability of Kolmogorov complexity, several researchers have attempted to adapt Bennett's notion to lower complexity levels, aka feasible depth. Most of these notions are centered around polynomial time computations [1,10,11] and finite-state machines [6]. Due to the intrinsic limitations of polynomial time (resp. finite state) algorithms, none can match all the nice properties of Bennett's original, i.e. each feasible notion studied represents a trade-off between advantages and limitations.

Similarly to randomness, there is no absolute notion of logical depth, and all variants mentioned above can be seen as variations of a same theme [11], based on the compression framework. However most notions satisfy some basic properties that could be seen as fundamental. These are:

- Random sequences are not deep (for the appropriate randomness notion).
- Computable sequences are not deep (for the appropriate computability notion).
- A slow growth law: deep sequences cannot be *quickly* computed from shallow ones.
- Some deep sequence exists.

In this paper, we continue the study of depth at the finite-state level. In summary, we construct a new depth notion (called ILPDC-depth), based on information lossless pushdown compressors (see [9] for definitions and a comparison with other compressors). We show our notion satisfies the fundamental depth properties mentioned above. We compare ILPDC-depth to finite-state depth [6], and show the two notions are different. This is somehow surprising as pushdown machines are strictly more capable than finite state machines. This shows that although pushdown machines are strictly stronger than finite state machines, stronger does not necessarily mean better.

We also extend the finite-state notion of [6], by introducing an a.e. version (the original [6] is an i.o. version), and show the two notions differ.

Let us explain our results in more details. In the first part of this paper we introduce a notion of pushdown depth. As observed in [11], most depth notions can be expressed in the compression framework, i.e. fix a compressor type T (e.g. finite state, polynomial time, etc.). A sequence S is T-deep, if for every compressor C of type T, there exists a compressor C' of type T (think of C' as being more powerful than C) such that C' compresses almost every prefix of S, better than C. The meaning of "better" will vary with the corresponding depth notion (and actually has consequences on the computational power of

[1] We acknowledge that logical depth was originally defined as depending on both computational complexity and Kolmogorov complexity, which is a descriptional complexity. The new notions in this paper are focused purely on a descriptional complexity lengths, specifically the ratio between the length of the input and the length of the output to restricted classes of transducers. However we continue to call these depth notions to be consistent with previous literature in [6,11].

the sequence, as shown in [10,12]), but for most notions, bounds considered are $O(1), O(\log n)$ and $O(n)$. From the work in [6], it seems linear bounds are appropriate at the finite state level, and thus we also use linear bounds.

We say a sequence is T-deep if for every compressor C of type T, there exists a compressor C' of type T such that on almost every prefix of S (with length denoted n), C' compresses it at least by αn more bits than C, for some constant α. We define ILPDC-depth by setting type T to be information lossless pushdown compressors (ILPDC). Intuitively, an ILPDC is a pushdown transducer such that when run the transducer on input x, the output y and final state q uniquely determines the input x, hence the name information lossless. Contrary to finite state information lossless transducers, it is not known whether this automatically yields a pushdown decompressor (in the finite state model, the finite state decompressor comes for free [7,8]).

We show ILPDC-depth satisfies all fundamental depth properties highlighted above, i.e. both random and easy sequences are not deep, ILPDC-depth satisfies a slow growth law, and there exists a PD-deep sequence.

Next we compare ILPDC-depth to finite state depth [6] (called i.o. FS-depth), and show the two notions are different: we prove there exists a sequence S which is i.o. FS-deep but not ILPDC-deep.

Most notions of depth, measure the compression difference on almost all prefixes of the sequences. Notable exceptions include the original finite state notion [6] (see [12] for further i.o. notions), where the difference is only required be large on infinitely many prefixes of the sequence. Such depth notions are called i.o. depth. In the second part of this paper, we extend the original i.o. FS-depth of [6], to an almost everywhere notion, called a.e. FS-depth. We show there exists an a.e. FS-depth sequence. We also show a.e. FS-depth is a stronger requirement than i.o. FS-depth, by constructing a sequence that is i.o. FS-deep but not a.e. FS-deep.

Due to lack of space, most proofs are omitted. A final journal version of this paper is in preparation.

2 Preliminaries

$\mathbb{N}$ denotes the set of all non-negative integers. A *finite binary* string is an element of $\{0,1\}^*$. A *binary sequence* is an element of $\{0,1\}^\omega$. The length of a string x is denoted by $|x|$. λ denotes the empty string (the string of length 0). For all $n \in \mathbb{N}$, $\{0,1\}^n$ denotes the set of binary strings of length n. For a string (or sequence) S and $i, j \in \mathbb{N}$, $S[i \dots j]$ denotes the ith through j^{th} bits of S with the convention that if $i > j$ then $S[i \dots j] = \lambda$. $S \upharpoonright j$ denotes $S[0 \dots j-1]$, the first j bits of S. For a string x and a string (or sequence) y, xy denotes the string (or sequence) composed of x concatenated with y. For a string x and $n \in \mathbb{N}$, x^n denotes x concatenated with itself n times. For strings $x, y, z \in \{0,1\}^*$, if $w = xyz$, we say y is a substring of w. For a string x, and a string (or sequence) y, we say x is a prefix of y, written as $x \preceq y$, if $x = y[0 \dots |x|-1]$. In particular we occasionally write $x \prec y$ if x is a prefix of y and $|x| < |y|$. The *lexicographic* ordering of $\{0,1\}^*$ is defined by saying for two strings x, y, x is less than y if either $|x| < |y|$ or else $|x| = |y|$ with $x[n] = 0$ and $y[n] = 1$ for the least n such that $x[n] \neq y[n]$. For a string x, x^{-1} denotes x written in reverse. By intervals of $\mathbb{N}$ we mean closed intervals of $\mathbb{N}$ in the normal sense. All logarithms are taken to be in base 2.

3 Models of Computation

3.1 Finite-State Transducers

We use the standard finite-state transducer model.

Definition 1. *A* finite-state transducer (FST) *is a 4-tuple* $T = (Q, q_0, \delta, \nu)$, *where*

- Q *is a nonempty, finite set of* states,
- $q_0 \in Q$ *is the* initial state.
- $\delta : Q \times \{0,1\} \to Q$ *is the* transition function,
- $\nu : Q \times \{0,1\}^* \to \{0,1\}^*$ *is the* output function,

For all $x \in \{0,1\}^*$ and $b \in \{0,1\}$, the *extended transition function* $\widehat{\delta} : \{0,1\}^* \to Q$, and the transducer output $T : \{0,1\}^* \to \{0,1\}^*$ is defined by the usual recursion.

An FST is *information lossless (IL)* if the function $x \mapsto (T(x), \widehat{\delta}(x))$ is 1–1; i.e. the output and final state of T on input x uniquely identify x. We call an FST that is IL an ILFST. By the identity FST, we mean the ILFST I_{FS} that on every input $x \in \{0,1\}^*$, $I_{\mathrm{FS}}(x) = x$. We write FST to denote the set of all FSTs.

A map $f : \{0,1\}^\omega \to \{0,1\}^\omega$ is said to be *FS computable* (*ILFS computable*) if there is an FST (ILFST) T such that for all $S \in \{0,1\}^\omega$, $\lim\limits_{n\to\infty} |T(S \upharpoonright n)| = \infty$ and for all $n \in \mathbb{N}$, $T(S \upharpoonright n) \preceq f(S)$. In this case we say $T(S) = f(S)$.

It is well known [7,8] that any function computed by an ILFST can be inverted to be approximately computed by another ILFST.

Theorem 1. *For any* ILFST T, *there exists an* ILFST T^{-1} *and a constant* $c \in \mathbb{N}$ *such that for all* $x \in \{0,1\}^*$, $x \upharpoonright (|x| - c) \preceq T^{-1}(T(x)) \preceq x$.

Corollary 1. *For any* ILFST T, *there exists an* ILFST T^{-1} *such that for all* $S \in \{0,1\}^\omega$, $T^{-1}(T(S)) = S$.

3.2 Pushdown Compressors

The model of pushdown compressors we use is the same pushdown compressor model as used in [9]. Note that to keep the model feasible, there is a bound on how long the compressor can empty its stack before it needs to output a symbol.

A *pushdown compressor* (PDC) is an 8-tuple $C = (Q, \Sigma, \Gamma, \delta, \nu, q_0, z_0, c)$ where

1. Q is a non-empty finite set of *states*,
2. Σ is the finite input alphabet,
3. Γ is the finite stack alphabet,
4. $\delta : Q \times (\Sigma \cup \{\lambda\}) \times \Gamma \to Q \times \Gamma^*$ is the *transition function*,
5. $\nu : Q \times (\Sigma \cup \{\lambda\}) \times \Gamma \to \Sigma^*$ is the *output function*,
6. $q_0 \in Q$ is the start state,
7. $z_0 \in \Gamma$ is the special bottom of stack symbol,
8. $c \in \mathbb{N}$ is an upper bound on the number of λ-rules per input bit.

We fix $\Sigma = \{0, 1\}$ and $\Gamma = \{0, 1, z_0\}$. We assume every state in Q is reachable from q_0. We write $\delta = (\delta_Q, \delta_{\Gamma^*})$. The transition function δ accepts λ as an input in addition to $\{0, 1\}$. This means C has the option of altering its stack while not reading an input character. We call this a *λ-rule*. In this case $\delta(q, \lambda, a) = (q', \lambda)$, that is, we pop the top symbol from the top of the stack. To enforce determinism we require that at least one of the following hold for all $q \in Q$ and $a \in \Gamma$:

1. $\delta(q, \lambda, a) = \perp$
2. $\delta(q, b, a) = \perp$ for all $b \in \{0, 1\}$.

δ is restricted so that z_0 cannot be popped off of the stack. That is, for every $q \in Q, b \in \{0, 1\} \cup \{\lambda\}$, either $\delta(q, b, z_0) = \perp$, or $\delta(q, b, z_0) = (q', vz_0)$ where $q' \in Q$ and $v \in \Gamma^*$.

The extended transition function $\delta^* : Q \times \Sigma^* \times \Gamma^+ \to Q \times \Gamma^*$ is defined recursively as usual.

δ^* is abbreviated to δ, and $\delta(q_0, w, z_0)$ to $\delta(w)$. The *output* from state q on input $w \in \{0, 1\}^*$ with $z \in \Gamma^*$ on the top of the stack is defined by the recursion $\nu(q, \lambda, z) = \lambda$,

$$\nu(q, wb, z) = \nu(q, w, z)\nu(\delta_Q(q, w, z), b, \delta_{\Gamma^*}(q, w, z)).$$

The *output* of the compressor C on input $w \in \{0, 1\}^*$ is the string $C(w) = \nu(q_0, q, z_0)$. For a string xy, we write $\bar{\nu}(y)$ as shorthand for $|C(xy)| - |C(y)|$, i.e. the output of C on y after already reading x. It should be clear from context what x is each time this notation is used.

A PDC is said to be *information lossless* (IL) if the function

$$w \mapsto (C(w), \delta_Q(w))$$

is 1–1. A PDC that is IL is called an ILPDC. We write (IL)PDC to be the set of all (IL)PDCs. By the identity PDC I_{PD} we mean the ILPDC that on any input $x \in \{0, 1\}^*$, I_{PD} outputs x without using its stack.

4 Pushdown Depth

Lemma 1 demonstrates the existence of strings that an ILPDC compresses poorly on and is used in proofs throughout this section.

Lemma 1. *Let $d, m \in \mathbb{N}$. Then for all $C \in ILPDC$ with at most d states and for all $x \in \{0,1\}^*$, there exists a string y of length m such that*

$$|C(xy)| - |C(x)| \geq m - \log(d) - 1.$$

The following general depth definition says that S is a.e. T-deep if for every compressor of type T there is a (better) compressor of type T such that the difference of compression, on almost every prefix of S, exceeds some linear bound. More precisely,

Definition 2. *Let S be a sequence. Fix a compressor type* T. *S is a.e.* T*-deep (resp. i.o.* T*-deep) if*

$$(\forall C \in T)(\exists \alpha > 0)(\exists C' \in \mathrm{T})(Qn \in \mathbb{N})\ [|C(S \upharpoonright n)| - |C'(S \upharpoonright n)| \geq \alpha n],$$

where Q is $\forall^\infty$ (resp. $\exists^\infty$).

Observe if S is a.e. T-deep, it is also i.o. T-deep.

To measure how well a compressor compresses a sequence, we use the following compression ratios.

Definition 3. *Let $S \in \{0,1\}^\omega$. Let T be a family of compressor types.*

1. *The* best-case compression ratio *of type T of S is defined as*

$$\rho_T(S) = \inf\{\liminf_{n\to\infty} \frac{|C(S \upharpoonright n)|}{n} : C \in T\}.$$

2. *The* worst-case compression ratio *of type T of S is defined as*

$$R_T(S) = \inf\{\limsup_{n\to\infty} \frac{|C(S \upharpoonright n)|}{n} : C \in T\}.$$

We define pushdown depth to be a.e. ILPDC-depth.

The following results show that pushdown depth satisfies the basic depth properties, in the sense that both easy and random sequences cannot be deep.

Theorem 2. *Let $S \in \{0,1\}^\omega$.*

1. *If $\rho_{\mathrm{ILPDC}}(S) = 1$, then S is not a.e.* ILPDC*-deep.*
2. *If $R_{\mathrm{ILPDC}}(S) = 0$, then S is not a.e.* ILPDC*-deep.*

The following result shows that pushdown depth satisfies a slow growth law.

Theorem 3 (Slow Growth Law). *Let S be any sequence, let $f : \{0,1\}^\omega \to \{0,1\}^\omega$ be ILFS computable and let $S' = f(S)$. If S' is a.e. ILPDC-deep then S is a.e. ILPDC-deep.*

Remark 1. Theorems 2 and 3 also hold true for i.o. ILPDC-depth.

The following result constructs a pushdown deep sequence S. The sequence is a sequence of blocks, where each block is devoted to some pair of compressors C, C'. On such a block, C compresses poorly, while C' compresses very well. This is achieved by having C' simulate C to find strings it cannot compress. The first found string describes the next bit of the block, and so C' is able to compress the block. In blocks not devoted to him, C' simply simulates C. This ensures that C' never compresses S worse than C, and on blocks devoted to it, C' compresses much better. C' detects whether the current block is devoted to him by signal flags interleaved throughout the sequence. To keep C' IL, as soon as C' makes a wrong prediction, it simply outputs what C does from then on onward. To guarantee an a.e. result, blocks devoted to the same pair repeat every constant number of blocks.

A full construction of C' is omitted for space.

Theorem 4. *There exists an a.e. ILPDC-deep sequence.*

5 Finite-State Depth

The finite-depth in [6] is based on finite-state decompression. However, before we begin examining depth, we first must choose a binary representation of all finite-state transducers.

Definition 4. *A* binary representation of finite-state transducers σ_T *is a partially computable map* $\sigma_T : \{0,1\}^* \to \text{FST}$, *such that for every* FST T, *there exists some* $x \in \{0,1\}^*$ *such that* $\sigma_T(x)$ *fully describes* T. *We say* $|T|_{\sigma_T} = \min\{|x| : \sigma_T(x) = T\}$.

For a binary representation of FSTs σ_T, for all $k \in \mathbb{N}$, define

$$\text{FST}^{\leq k}_{\sigma_T} = \{T \in \text{FST} : |T|_{\sigma_T} \leq k\}.$$

For all $k \in \mathbb{N}$ and $x \in \{0,1\}^*$, the *k-finite-state decompression complexity* of x with respect to binary representation σ_T is defined as

$$D^k_{\sigma_T}(x) = \min_{\pi \in \{0,1\}^*} \Big\{ |\pi| \, : T \in \text{FST}^{\leq k}_{\sigma_T} \ \& \ T(\pi) = x \Big\}.$$

Here π is the shortest program that gives x as an output when inputted into an FST of size k or less with respect to the binary representation σ_T. T can be thought of as the FST that can decompress π to reproduce x.

For the purpose of this paper, we fix the following binary representation of finite-state transducers σ_T. Let $T = (Q, q_0, \delta, \nu)$ be an FST. We define the function the function $\Delta : Q \times \{0,1\} \to Q \times \{0,1\}^*$, where $\Delta(q,b) = (\delta(q,b), \nu(q,b))$. This function Δ completely describes the state transitions and outputs of T. In [3], different encoding schemes are presented to represent each transducer via an encoding of this function Δ.

The binary representation σ_T we fix in this paper is as follows. For a transducer T, if $Q = \{q_1, q_2, \ldots, q_n\}$ and $q_0 = q_i$, for $1 \leq i \leq n$, we encode T by the string

$$d(\text{bin}(i))01\rho$$

where $d(\text{bin}(i))$ is the binary encoding of i which acts as a pointer to the start state of T but with every bit doubled and ρ is an encoding of Δ as seen in [3].

We fix this binary representation σ_T as it is needed to prove Lemma 2 which in turn is needed for Theorem 7, Theorem 8 and Theorem 9. However, we later show that if a sequence S is deep with respect to one depth notion, it is deep with respect to every depth notion. Henceforth, we will drop the σ_T notation and instead write $|T|$ for $|T|_{\sigma_T}$, $\text{FST}^{\leq k}$ for $\text{FST}^{\leq k}_{\sigma_T}$ and $D^k_{\text{FS}}(x)$ instead of $D^k_{\sigma_T}(x)$. All other definitions and results hold and can be proved regardless of the binary representation being used.

To measure the randomness density of a sequence, the following notions are useful. For any sequence S,

1. The *finite-state dimension* of S [5] is defined to be

$$\dim_{\text{FS}}(S) = \lim_{k\to\infty} \liminf_{n\to\infty} \frac{D^k_{\text{FS}}(S \upharpoonright n)}{n},$$

2. The *strong finite-state dimension* of S is defined to be

$$\text{Dim}_{\text{FS}}(S) = \lim_{k\to\infty} \limsup_{n\to\infty} \frac{D^k_{\text{FS}}(S \upharpoonright n)}{n}.$$

In [6] a notion[2] of depth based on finite-state transducers is introduced called *i.o. finite-state depth.*

Definition 5. *A sequence S is infinitely often (i.o.) finite-state deep if*

$$(\forall k \in \mathbb{N})(\exists \alpha > 0)(\exists k' \in \mathbb{N})(\exists^\infty n \in \mathbb{N})\, D^k_{\text{FS}}(S \upharpoonright n) - D^{k'}_{\text{FS}}(S \upharpoonright n) \geq \alpha n,$$

and $\text{Dim}_{\text{FS}}(S) \neq 0$.

We introduce an a.e. version of the original finite-state notion [6] called *almost everywhere (a.e) finite-state depth.*

Definition 6. *A sequence S is almost everywhere (a.e.) finite-state deep if*

$$(\forall k \in \mathbb{N})(\exists \alpha > 0)(\exists k' \in \mathbb{N})(\forall^\infty n \in \mathbb{N})\, D^k_{\text{FS}}(S \upharpoonright n) - D^{k'}_{\text{FS}}(S \upharpoonright n) \geq \alpha n,$$

and $\text{Dim}_{\text{FS}}(S) \neq 0$.

Remark 2. The condition that $\text{Dim}_{\text{FS}}(S) \neq 0$ is required as otherwise 0^ω would be considered deep.

---

[2] Actually two notions were introduced, which differ only by the order of quantifiers.

The following result shows that sequences that appear random to finite-state transducers, cannot be finite-state deep. Further study of sequences that appear random to finite-state transducers can be found in [4].

Theorem 5. *Let* $S \in \{0,1\}^\omega$.
If $\dim_{\mathrm{FS}}(S) = 1$, *then* S *is not a.e. finite-state deep.*

Remark 3. We originally hoped to include a version of a slow growth law for a.e. FS-depth. However an adequate notion nor proof has not been found as of yet.

The following theorem demonstrates that if a sequence S is a.e. FS-deep when the size of finite-state transducers are viewed with respect to one binary representation, then it is a.e. FS-deep regardless of what binary representation is used.

Theorem 6. *Let* π_T *be a binary representation of FSTs. Let* S *be an a.e. FS-deep sequence when the size of the finite-state transducers are viewed with respect to a binary representation* π_T. *Then* S *is a.e. FS-deep when the size of the finite-state transducers are viewed with respect to any other binary representation.*

To prove the existence of an a.e. FS-deep sequence we need he following two lemmas.

Lemma 2. *For our fixed binary representation* σ_T, *we have that for* $k \geq 4$, $\forall n \in \mathbb{N}, \forall x, y, z \in \{0,1\}^*$,

$$D^k_{\mathrm{FS}}(xy^n z) \geq D^{3k}_{\mathrm{FS}}(x) + nD^{3k}_{\mathrm{FS}}(y) + D^{3k}_{\mathrm{FS}}(z).$$

Lemma 3. $\forall \epsilon > 0, \forall k \in \mathbb{N}, \exists k' \in \mathbb{N}, \forall x, y, \in \{0,1\}^*$, *whenever* $D^k_{\mathrm{FS}}(x)$ *is sufficiently large*

$$D^{k'}_{\mathrm{FS}}(xy) \leq (1+\epsilon)D^k_{\mathrm{FS}}(x) + D^k_{\mathrm{FS}}(y) + 2.$$

In the following result we construct an a.e. finite-state deep sequence. The sequence is constructed in consecutive blocks, each block is devoted to some pair k, k'. On such a block, transducers of size k do poorly, while some larger transducer of size k' does very well. The key difference with the proof in [6], is that blocks devoted to the same pair k, k' repeat every constant number of blocks. This ensures an a.e. finite-state deep sequence, as opposed to a mere i.o. finite-state deep sequence.

Theorem 7. *There exists an a.e. finite-state deep sequence.*

Remark 4. If $S \in \{0,1\}^\omega$ is a.e. finite-state deep then it is i.o. finite-state deep.

The following result shows that being i.o. FS-deep is a weaker requirement than being a.e. FS-deep.

Theorem 8. *There exists a sequence* S *that is i.o. FS-deep but not a.e. FS-deep.*

5.1 Separation from ILPDC-depth

We next demonstrate a difference between i.o finite state depth [6] to our a.e. pushdown depth notion by constructing a sequence that is i.o. finite-state deep but not a.e. pushdown deep.

Theorem 9. *There exists a sequence S such that S is i.o. finite-state deep, but S is not a.e. ILPDC deep.*

Proof. Fix some $\epsilon > 0$ small. Split $\mathbb{N}$ into intervals $I_1, I_2, I_3, \ldots$ such that $|I_1| = 2^a$ for the smallest constant a such that $2^a > \frac{1}{\epsilon}$, and $|I_j| = 2^{|I_1|+\cdots+|I_{j-1}|}$ for $j \geq 2$. Define $m_j = \min(I_j)$ and $M_j = \max(I_j)$. We construct the sequence $S = S_1 S_2 \cdots$ in stages, with $S_j \in \{0,1\}^{|I_j|}$ for all $j \in \mathbb{N}$. So $S_j = S[m_j \ldots M_j]$.

For S_j, if j is even, S_j is devoted to some FST description bound length $k \in \mathbb{N}$ (what occurs for j odd is discussed later in the proof.) Specifically for each k, k is devoted to every substring S_j where for all $n \geq 0$, $j = 2^k + n2^{k+1}$. $k = 1$ is first devoted to S_2 and every 4th substring after that. $k = 2$ is devoted to S_4 and every 8th interval after that, and so on.

Consider description length k. Let r_k be a string of length $|I_{2^k}|$ such that r_k is $3k$-FS random in the sense that $D^{3k}_{\mathrm{FS}}(r_k) \geq |r_k| - 4k$. Such a string exists as there are at most $|\mathrm{FST}^{\leq 3k}| \cdot 2^{|r_k|-4k} < 2^{|r_k|}$ strings contradicting this. If S_j is devoted to k, we set $S_j = r_k^{\frac{|I_j|}{|r_k|}}$.

First we show S is i.o. FS-deep by examining prefixes of the form $S_1 S_2 \cdots S_j$, for j even. Let $k \geq 4$ and suppose k is devoted to S_j. Then by Lemma 2

$$D^k_{\mathrm{FS}}(S_1 S_2 \cdots S_j) \geq D^{3k}_{\mathrm{FS}}(S_1 S_2 \cdots S_{j-1}) + \frac{|S_j|}{|r_k|} D^{3k}_{\mathrm{FS}}(r_k) \geq \frac{|S_j|}{|r_k|}(|r_k| - 4k).$$

For all $r \in \{0,1\}^*$, define the single state FST $T_r = (\{q_0\}, q_0, \delta, \nu)$, where for $b \in \{0,1\}$, $\nu(q_0, b) = r$. Let k' be large enough so that $I_{\mathrm{FS}}, T_{r_k} \in \mathrm{FST}^{\leq k'}$. Hence $D^{k'}_{\mathrm{FS}}(S_j) \leq \frac{|S_j|}{|r_k|}$ and $D^{k'}_{\mathrm{FS}}(S_1 \cdots S_{j-1}) \leq |S_1 \cdots S_{j-1}|$. Let $\hat{k}$ be from Lemma 3 such that

$$D^{\hat{k}}_{\mathrm{FS}}(S_1 S_2 \cdots S_j) \leq 2D^{k'}_{\mathrm{FS}}(S_1 \cdots S_{j-1}) + D^{k'}_{\mathrm{FS}}(S_j) + 2 \leq 2|S_1 S_2 \cdots S_{j-1}| + \frac{|S_j|}{|r_k|} + 2.$$

Therefore for infinitely many prefixes of the form $S_1 S_2 \cdots S_j$ where S_j is devoted to k

$$\begin{aligned}
D^k_{\mathrm{FS}}(S_1 S_2 \cdots S_j) - D^{\hat{k}}_{\mathrm{FS}}(S_1 S_2 \cdots S_j) &\geq \frac{|S_j|}{|r_k|}(|r_k| - 4k - 1) \\
&\quad - 2|S_1 S_2 \cdots S_j| - 2 \\
&= |S_j|(1 - \frac{4k}{|r_k|} - \frac{1}{|r_k|}) - 2\log|S_j| - 2 \\
&\geq |S_j|(1 - \delta) \qquad (\delta > 17\epsilon) \\
&\geq |S_1 S_2 \cdots S_j|(1 - \alpha) \qquad (\alpha > \delta)
\end{aligned}$$

as $\frac{4k}{|r_k|}$ is maximum for $k = 4$.

For $0 \leq k \leq 3$, the above result follows from the fact that $D^0_{\rm FS}(S \restriction m) \geq \ldots \geq D^3_{\rm FS}(S \restriction m) \geq D^4_{\rm FS}(S \restriction m)$, when we take $S \restriction m = S_1 \cdots S_j$ to be such that S_j is devoted to $k = 4$.

Furthermore S has a non-zero finite-state strong dimension as for $k \geq 4$, there exists infinitely many prefixes of the form $S \restriction m = S_1 \cdots S_j$ such that

$$D^k_{\rm FS}(S \restriction m) \geq |S_j|(1 - \frac{4k}{|r_k|}) \geq |S_j|(1 - 16\epsilon) = (m - \log|S_j|)(1 - 8\epsilon) \geq m(1 - \hat{\epsilon}),$$

where $\hat{\epsilon} > 16\epsilon$ for j large. Therefore S is i.o. finite-state deep.

Next we show S is not a.e. ILPDC-deep.

Let $C_1, C_2, \ldots$ be an enumeration of all ILPDCs such that for a pair of ILPDCs C_p, C_q, if $p \leq q$, C_q has at least as many states as C_p. As the number of machines with k states is bigger than k, we can say that C_k has at most k states.

Henceforth we assume j is odd and examine S_j. Each odd j can be written in the form $2^k - 1 + n2^{k+1}$. Then for ILPDC C_k, C_k is devoted to every interval of the form $2^k - 1 + n2^{k+1}$, $n \geq 0$.

If S_j is devoted to C_k, we set S_j to be a string y of length $|I_j|$ from Lemma 1 that satisfies

$$|C_k(S_1 \cdots S_j)| - |C_k(S_1 \cdots S_{j-1})| \geq |S_j| - \log k - 1.$$

Say $S \restriction m = S_1 \cdots S_j$. Hence we have that for $\beta_1, \beta_2 > 0$, for j large

$$\begin{aligned} |C_k(S \restriction m)| &\geq |S_j| - \log k - 1 > |S_j|(1 - \beta_1) \\ &= (m - O(\log m))(1 - \beta_1) = m(1 - \beta_2). \end{aligned}$$

Hence for all k, for j large and for infinitely many prefixes of the form $S_1 \cdots S_j$ we have

$$|I_{PD}(S_1 \cdots S_j)| - |C_k(S_1 \cdots S_j)| < |S_1 \cdots S_j| - |S_1 \cdots S_j|(1 - \beta_2) = |S_1 \cdots S_j|\beta_2.$$

As β_1, β_2 can be made arbitrarily small, S is not a.e. ILPDC-deep.

□

6 Final Remarks

We introduced pushdown depth and showed our notion is a well behaved depth notion that satisfies all basic depth properties. We showed that is different from i.o. finite-state depth [6]. This gives more weight to the thesis that there is no perfect depth notion, but rather a "best for the job" notion.

It would be interesting to see whether a converse can be proven, i.e. a sequence that is ILPDC-deep but not i.o. finite state deep.

Acknowledgements. The authors would like to thank the anonymous referees for their useful comments, specifically to explore how the chosen binary representations of FSTs affects FS-depth.

References

1. Antunes, L., Fortnow, L., van Melkebeek, D., Vinodchandran, N.: Computational depth: concept and applications. Theoret. Comput. Sci. **354**, 391–404 (2006)
2. Bennett, C.H.: Logical depth and physical complexity. In: Bennett, C. (ed.) The Universal Turing Machine, A Half-Century Survey, pp. 227–257. Oxford University Press, New York (1988)
3. Calude, C.S., Salomaa, K., Roblot, T.K.: Finite state complexity. Theoret. Comput. Sci. **412**(41), 5668–5677 (2011)
4. Calude, C.S., Staiger, L., Stephan, F.: Finite state incompressible infinite sequences. Inf. Comput. **247**, 23–36 (2016)
5. Dai, J., Lathrop, J., Lutz, J., Mayordomo, E.: Finite-state dimension. Theoret. Comput. Sci. **310**, 1–33 (2004)
6. Doty, D., Moser, P.: Feasible depth. In: Cooper, S.B., Löwe, B., Sorbi, A. (eds.) CiE 2007. LNCS, vol. 4497, pp. 228–237. Springer, Heidelberg (2007). https://doi.org/10.1007/978-3-540-73001-9_24
7. Huffman, D.A.: Canonical forms for information-lossless finite-state logical machines. IRE Trans. Circ. Theory CT-6 (Special Supplement) **5**(5), 41–59 (1959)
8. Kohavi, Z.: Switching and Finite Automata Theory, 2nd edn. McGraw-Hill, New York (1978)
9. Mayordomo, E., Moser, P., Perifel, S.: Polylog space compression, pushdown compression, and Lempel-Ziv are incomparable. Theory Comput. Syst. **48**(4), 731–766 (2011)
10. Moser, P.: Polynomial depth, highness and lowness for E. Inf. Comput. (2019, accepted)
11. Moser, P.: On the polynomial depth of various sets of random strings. Theor. Comput. Sci. **477**, 96–108 (2013)
12. Moser, P., Stephan, F.: Depth, highness and DNR degrees. Discrete Math. Theor. Comput. Sci. **19**(4) (2017)

Online Scheduling with Machine Cost and a Quadratic Objective Function

J. Csirik[1(✉)], Gy. Dósa[2], and D. Kószó[1]

[1] Department of Informatics, University of Szeged, Árpád tér 2, Szeged 6720, Hungary
jcsirik@gmail.com

[2] Department of Mathematics, University of Pannonia, Egyetem u. 10, Veszprém 8200, Hungary

Abstract. We will consider a quadratic variant of online scheduling with machine cost. Here, we have a sequence of independent jobs with positive sizes. Jobs come one by one and we have to assign them irrevocably to a machine without any knowledge about additional jobs that may follow later on. Owing to this, the algorithm has no machine at first. When a job arrives, we have the option to purchase a new machine and the cost of purchasing a machine is a fixed constant. In previous studies, the objective was to minimize the sum of the makespan and the cost of the purchased machines. Now, we minimize the sum of squares of loads of the machines and the cost paid to purchase them and we will prove that 4/3 is a general lower bound. After this, we will present a 4/3-competitive algorithm with a detailed competitive analysis.

Keywords: Scheduling · Online algorithms · Analysis of algorithms

1 Introduction

Online scheduling with machine cost is a kind of decision making. In some disciplines it plays a key role. Decision making requires allocating resources to activities which can appear in various forms. To make the best decision, we will optimize one or more performance measures.

In this paper, we will consider a quadratic variant. The model we use was first mentioned in [8]. Let us suppose we have a sequence of independent jobs. They come one by one and each of them has a positive size. We will assign them irrevocably to a machine without prior knowledge about other jobs that may come later on. We have no machine at first. Then, when a job arrives, we have the option to buy a new machine. The cost of purchasing a machine is a fixed constant. In previous studies, the objective was to minimize the sum of the makespan and the cost of purchased machines. Now, in our case, the objective is to minimize the sum of squares of loads of the machines and the total sum needed to purchase all the machines. Formally: Let A be an algorithm and J be an input. Then the *machines of A with respect to J*, denoted by $M_{A,J}$, is a

A. Chatzigeorgiou et al. (Eds.): SOFSEM 2020, LNCS 12011, pp. 199–210, 2020.
https://doi.org/10.1007/978-3-030-38919-2_17

linearly ordered set of machines, used by the algorithm to schedule input J. The *total cost of A on J* is defined by

$$A(J) = \sum_{m \in M_{A,J}} \mathrm{ld}(m)^2 + |M_{A,J}|,$$

where $|B|$ denotes the cardinality of a finite set B, i.e., the number of elements of B and $\mathrm{ld}(m)$ is the sum of job sizes of machine m.

In this way, we would like to achieve a uniform loading of the machines. The quadratic cost function was first introduced for single machine problems in [11] and [10].

We will evaluate the quality of an online algorithm by using a competitive analysis. Here, the standard is the *optimal offline algorithm*. In our case the value of the optimum is well defined but the optimal solution is not unique so we may have different optimal solutions with different number of machines. Now, let us introduce some notations. We will denote an *online algorithm* by A, and one of the *optimal offline* by OPT. Let J be a sequence of jobs. Next, let $A(J)$ be *the total cost of an online algorithm A on a given sequence J*. Similarly, let $OPT(J)$ denote the *optimal offline cost*. We will call *an A online algorithm C-competitive* if $A(J) \leq C \cdot OPT(J)$ for all J.

In [8], it was proved that the competitiveness of each online algorithm is at least $4/3$ with the original objective function. Moreover, a $(1 + \sqrt{5}/2) \approx 1.618$-competitive algorithm is given. In [1], an improved algorithm was presented with a competitive ratio of $(2\sqrt{6}+3)/5 \approx 1.5798$. In [4], it was shown that $\sqrt{2} - \varepsilon$ is a lower bound of the problem. A $(2 + \sqrt{7}/3) \approx 1.5486$-competitive algorithm was also introduced. In addition, it was shown that by applying the lower bounds on the optimal objective value introduced earlier, no algorithm can be proven to be C-competitive with any $C \leq 1.5$. Also, some other variants of the problem were studied in [2,5,6,9]. In [7], the original model was extended with a more general machine cost function. In [3], another possible modification of the model was considered.

The structure of the paper is as follows. In Sect. 2, we will introduce the notations used in this study. In Sect. 3, we will present a general lower bound of 4/3. Lastly, in Sect. 4, we will give a 4/3-competitive algorithm and we will also prove its competitiveness.

2 Preliminaries

Here, we will use the following notations. We shall denote the ith job by j_i and its size by p_i (also known as the processing time). And here when we speak about a job size, we will use size and processing time interchangeably.

Let q be $\sqrt{2}/2 \approx 0.707$. We will consider three different types of jobs. We will call a job *small* if $p_i \leq q$, *medium* if $q < p_i \leq 2q$, and *big* if $2q < p_i$. We will denote the total load by $P(= P(J)) = \sum_{j_i \in J} p_i$ and the total load of all small jobs by $P_s(= P_s(J)) = \sum_{p_i \text{ is small}} p_i$. Note that $P \geq P_s$.

In our algorithm, we will use two types of machines. The first type is called SM, which can receive only small and medium jobs, and its maximum possible load is $2q$. The second type is called B, which can process only big jobs (*big machines*). In the proof we further divide the SM machines into those that process only small jobs called *small machines* (S) and the remaining machines from SM are called *medium machines* (M).

3 Lower Bound

Proposition 1. *Consider two machines with loads $l_1 \geq l_2$. If we reschedule any job with size $x < l_2$ from the second machine to the first one, then the cost will grow.*

Proof. Evidently, $l_1^2 + l_2^2 < (l_1 + x)^2 + (l_2 - x)^2 = l_1^2 + l_2^2 + 2x(l_1 - l_2 + x)$, as $2x(l_1 - l_2 + x) > 0$ since $x > 0$ and $l_1 \geq l_2$. □

Proposition 2. *$2P$ is a lower bound of the cost of the optimal schedule.*

Proof. We will suppose that OPT can distribute P equally among the m machines. In this case $f(m) = m \cdot (P/m)^2 + m$ gives the total cost of OPT. This function has its minimum at $m = P$. So, if we replace m by P, then we will get the optimal value of $2P$. If we cannot distribute the loads equally on the machines, because of the previous proposition the cost will be larger. □

Lemma 1. *An online algorithm which never purchases a second machine is not constant competitive.*

Proof. We prove our statement by contradiction. Let A be a C-competitive online algorithm such that it uses one machine for each input I. Let J be an input having k jobs, each of size 1. Then the optimum will use one machine for each job and so it has a cost of $2k$. Algorithm A will use only one machine and so it will have a cost of $1 + k^2$. But then $A(J)/OPT(J) = k/2 + 1/(2k)$, which is larger than C if k is large enough. This leads to contradiction. □

Proposition 3. *Let J be a finite sequence of arbitrarily small ε jobs, having an even k number of jobs. Then OPT purchases at least two machines if $\sqrt{2} \leq P$.*

Proof. To prove our statement, we have to check whether the cost of having two machines is smaller than having only one. If k is even, this means that

$$2 \cdot (P/2)^2 + 2 \leq P^2 + 1, \tag{1}$$

which is exactly valid if $\sqrt{2} \leq P$. □

Theorem 1. *No online algorithm has a competitive ratio smaller than 4/3.*

Proof. Let A be an online algorithm and J be a finite sequence of arbitrarily small ε jobs. The sequence terminates depending on the situation where A purchases the second machine. If at the moment of purchasing the second machine the number of jobs in J is even then we stop. If the number of jobs is odd at this moment then the input will get one more small job. The best possible algorithm A will schedule this last job to the second machine. (Clearly, A will purchase a second machine since A is constant competitive and because of Lemma 1). We now have an even number of jobs in our input so by Proposition 3, OPT purchases at least two machines if $\sqrt{2} \leq P$. We will only describe in detail the case where the second machine of A has one small job - the other case can be handled similarly.

Thus, we consider the following two cases with respect to P.

1. If $P \leq \sqrt{2}$, then we have

$$\lim_{\varepsilon \to 0+} \frac{A(J)}{OPT(J)} = \lim_{\varepsilon \to 0+} \frac{(P-\varepsilon)^2 + \varepsilon^2 + 2}{P^2+1} = \frac{P^2+2}{P^2+1} \geq 4/3 \ .$$

2. If $\sqrt{2} < P$, then we have

$$\begin{aligned} \lim_{\varepsilon \to 0+} \frac{A(J)}{OPT(J)} &\geq \lim_{\varepsilon \to 0+} \frac{(P-\varepsilon)^2 + \varepsilon^2 + 2}{2 \cdot (\frac{P}{2})^2 + 2} \\ &= \frac{2 \cdot P^2 + 4}{P^2 + 4} \geq 4/3 \ . \end{aligned}$$

□

Lemma 2. *Let* r, s, t_1 *and* t_2 *be positive values with* $4/3 < r/s$, $t_1 \leq (4/3)t_2 < r$ *and* $s > t_2$. *Then* $4/3 < (r - t_1)/(s - t_2)$.

Proof. We will prove this by contradiction. Let us suppose that

$$(r - t_1)/(s - t_2) \leq 4/3.$$

Then from the conditions

$$(r - (4/3)t_2)/(s - t_2) \leq (r - t_1)/(s - t_2)$$

or equivalently $3r - 4t_2 \leq 4s - 4t_2$, i.e. $r/s \leq 4/3$, we arrive at a contradiction. □

4 Algorithm

4.1 Description

We will denote the following algorithm by ALG. Algorithm ALG applies the bin packing algorithm First Fit (FF for short) as a slave algorithm. In the bin packing problem we are given items with positive sizes and unit capacity bins; and we would like to pack the items into as few bins as possible, but the bin capacity cannot be exceeded. Hence, in any bin the total size of items is at most one unit. The FF algorithm packs the items one by one, and the next item is always packed into the first bin it fits. If it does not fit into any bin, we open a new bin for it and pack the item into this new bin.

Algorithm ALG

1. If a small or a medium job arrives, then we will use an SM machine (its maximum possible load is $2q$). We will apply the FirstFit algorithm to decide which machine gets the job. (If there is no SM machine that has enough free space, then we will purchase a new one.)
2. If a big job comes, then we will schedule this to a B machine. We will not schedule any other job to this machine.

We will suppose that ALG uses a small, b medium and c big machines.

In the rest of the paper, we will use the following relaxed problem to estimate $OPT(J)$. We will permit preemption for every small job, but not for any medium or big job. Preemption means that the execution of a job can be divided into non-overlapping time slots, and these parts can be executed by different machines.

We will denote the optimal solution of the relaxed problem by $OPT_R(J)$ for every J and we will call it the *relaxed optimum*. We know that $OPT_R(J) \leq OPT(J)$.

Theorem 2. *$ALG(J)/OPT(J) \leq 4/3$ for every input J.*

Proof. It is enough to prove that $ALG(J)/OPT_R(J) \leq 4/3$ for every input J. First, let us suppose that the opposite is true. Take the case $ALG(J)/OPT_R(J) > 4/3$ for an input J. Input J may contain small, medium and big jobs. If input J contains a big job, and we can leave out one of the big jobs so that for the remaining J $ALG(J)/OPT_R(J) > 4/3$ is still valid, then we will leave out this job and we will repeat this step. It is possible that not all big jobs can be removed. Now, we may suppose that J is the minimal counterexample in the sense of not having a removable big job.

The general flow of our proof will be the following. In the next subsection we will prove some properties of the relaxed optimum and the ALG algorithm. Next, we will give some reduction steps which can be used to modify the minimal counterexample so that its structure is simpler, but it remains a counterexample. In the last subsection we will show that the final reduced example cannot be a counterexample and so the proof is completed.

4.2 Properties of the Relaxed Optimum and Algorithm ALG

First we will prove some properties of the relaxed optimum.

Lemma 3. *Consider the (relaxed) optimal scheduling of a minimal counterexample J. In this case, there is no big job which uses a machine on its own.*

Proof. Suppose a big job X with size x uses a machine alone. This job is also alone in ALG. Let $J' = J \setminus \{X\}$. It follows from Lemma 2 that

$$\frac{ALG(J')}{OPT(J')} = \frac{ALG(J) - 1 - x^2}{OPT(J) - 1 - x^2} > 4/3,$$

which contradicts the fact that J is a minimal counterexample. □

Lemma 4. *Consider the (relaxed) optimal schedule of J. In this case, there is no machine whose jobs can be distributed into two sets and the load of each set is greater than q.*

Proof. Suppose the load of a machine can be distributed into sets S_1 and S_2 with loads $x_1, x_2 > q$. The total cost of this machine is $1 + (x_1 + x_2)^2$. If we schedule S_1 to a new machine and S_2 to a second new machine, then the total cost of the two machines is $2+x_1^2+x_2^2$, which is obviously less than $1+(x_1+x_2)^2$, and hence it is a contradiction. □

The following corollary is a consequence of Lemma 4.

Corollary 1. *In the case of a relaxed optimal schedule:*

(i) Any two big or medium jobs are scheduled to two different machines;
(ii) If a machine processes a big or a medium job, then the rest load is at most q, which can come from only small jobs;
(iii) If a machine processes only small jobs, then its total load is at most $2q$.

Proof. Item (i) and item (ii) are immediate consequences of Lemma 4. To prove item (iii) we can use the preemption possibilities of small jobs: if the total load is larger then $2q$ then we can split this up into two parts where each of them is larger than q and then we can use Lemma 4. □

Proposition 4. *In the case of a relaxed optimal schedule there are no two machines, each with a load less than q.*

Proof. Suppose we have two machines with loads $x_1, x_2 < q$. Then the cost of these two machines is $2 + x_1^2 + x_2^2$, which is obviously more than $1 + (x_1 + x_2)^2$ and hence there is a contradiction. □

Next we will prove some properties of *ALG*.

Lemma 5. *Consider any three small machines of ALG, each having at least two jobs. In this case, the total load of the three small machines is greater than or equal to $4q$ and at most one of the machines can have a load $<(4/3)q$.*

Proof. Consider the last machine of the three. We will suppose that this machine has a load of $<(4/3)q$, and let this load be $(4/3)q - 2x$ for some $0 < x < (2/3)q$. Then, the size of the smallest job is at most $(2/3)q - x$ on this machine. Next, the load of each of the first two machines is at least $(4/3)q + x$ since the smallest job of the third machine does not fit into the first two machines as we apply First Fit packing and the load of these machines cannot be larger then $2q$. So in this case the total load is bigger than $4q$ and only the third machine has a load of $<(4/3)q$.

After this we assume that the load of the third machine is at least $(4/3)q$.

Now we suppose that the load of the second machine is $<(4/3)q$. Let this load be $(4/3)q - 2x$ for some $0 < x < (2/3)q$. Then, the size of the smallest job is at most $(2/3)q - x$ on this machine and this does not fit into the first machine, so the first machine has a load of at least $(4/3)q + x$. The load of the third machine is at least $(4/3)q + 4x$, because there are two jobs and they do not fit into the second machine. So the total load of the three machines is $>4q$ and only the second machine has a load of $<(4/3)q$.

After this we assume that the load of the second and the third machine is at least $(4/3)q$.

If the load of the first machine is less than $(4/3)q$, then let this load be $(4/3)q - 2x$ for some $0 < x < (2/3)q$. Then the load of the second and the third machine is at least $(4/3)q + 4x$, because there are two jobs and they do not fit into the second machine. So the total load of the three machines is $>4q$ and only the first machine has a load of $<(4/3)q$. □

Lemma 6. *Take the schedule of ALG. If $a \geq 2$, then $P_s \geq 2q + (4q/3) \cdot (a-2)$.*

Proof. If $a = 2$, then the Lemma is clearly true as we start the second small machine when the load of the first small machine and the next small job together is larger than $2q$.

Let us suppose that $a = 3$. If any of the three small machines has at least two jobs, the assertion follows from Lemma 5. Suppose there is a machine with only one job. There can be only one such machine as the total load of any two small machines is bigger than $2q$, but the size of any small job is at most q. This is clearly the third (latest) machine. Now let us take the first two machines of three. If one of them has a load $\geq(4/3)q$, then we are done because the other and the third machine have altogether a load of $>2q$. If the first two machines have a load $<(4/3)q$ then the second machine has a load of $<(4/3)q$, and let this load be $(4/3)q - 2x$ for some $0 < x < (2/3)q$. Then, the size of the smallest job is at most $(2/3)q - x$ on this machine. Next, the load of the first machine is at least $(4/3)q + x$ since the smallest job of the second machine does not fit into the first machine as we use First Fit packing and the schedule of these machines cannot be larger then $2q$. So the first machine has a load of $>(4/3)q$ and we are done.

Now we will suppose that $a \geq 4$. If every small machine has at least two jobs then the total load is at least $(4/3)q \cdot a$, which is more than what we need in the lemma. Otherwise there is a small machine with one job, but the load of any other small machine is greater than $(4/3)q$. Now the total load of the small machine with one job and any other small machine is bigger than $2q$, and the load of any other small machine is bigger than $(4/3)q$. Hence, we are done. □

4.3 Modifying the Two Schedules

We will suppose that $ALG(J)/OPT_R(J) > 4/3$ for a fixed minimal counterexample J. We will reduce and modify some jobs in J. We shall rename ALG to A_0 and OPT_R to O_0. Clearly, $A_0/O_0 > 4/3$. We also know that every machine has at most one medium job or a big job: in the relaxed optimal packing because of Corollary 1; and in the ALG because of the scheduling rule.

Now we will modify A_0 and O_0 in several simple steps. After each step their ratio will remain larger than 4/3.

Step 1: We apply this reduction step only if there is at least one big job, otherwise we go directly to Step 2. Therefore suppose there is a big job. We will reduce the size of every big job to $2q = \sqrt{2}$ in both A_0 and O_0. We will denote the new (reduced) schedules by A_1 and O_1. We know that $A_1 \leq A_0$ and $O_1 \leq O_0$. We also note that $A_0 - A_1 \leq O_0 - O_1$. To show this, let us consider a big job with size $2q+x$ with $x > 0$. In O_0, the machine processing this particular job has a load $2q+x+y$ with $y > 0$ because a big machine of O_0 contains small job(s)s as well. After reducing the job size the load of this particular machine will be $2q + y$. In A_1, the new load is $2q$ instead of $2q + x$. The difference of squares in A_1 is $4qx + x^2$. In O_1, $(2q+x+y)^2$ becomes $(2q+y)^2$ with difference $2(2q + y)x + x^2$. Then we get from Lemma 2 that $A_1/O_1 > 4/3$.

Step 2: We will decrease the total cost of O_1 by using $2P$ instead of the actual cost of O_1. Since $2P \leq O_1$ by Proposition 2, clearly $A_1/(2P) > 4/3$. Thus let $O_2 = 2P$. We do not change in this step A_1, hence we let $A_2 = A_1$.

Step 3: In A_2, we will reduce the load of every medium machine to q by decreasing the size of the medium job to exactly q and deleting the small job(s) here if they exist. We will call it A_3. Due to Lemma 2, $A_3/O_3 > 4/3$ where $O_3 = 2P$. To show this, let us consider a medium machine. Its total load is $(q + x)$ with $0 < x \leq q$. After the reduction, it is only q. So the difference in cost is $2qx + x^2$. $2P$ is then decreased by $2x$. To use Lemma 2 we need $2qx + x^2 \leq (4/3) \cdot 2x$, i.e. $2q + x \leq 8/3$, which is true, because $x \leq q$. After this step on all medium machines in A_3, we will have one medium size job of size q.

Step 4: Now, we will change the loads of small machines of A_3. Let the new schedule be A_4. Next, $A_4 \geq A_3$, but P is not changed. Let us consider two small machines of A_3. We will move some loads from one to the other, until one machine has a load $2q$ or the other one has a load of 0. We notice that here we can use a relaxation regarding small jobs, as it will only increase the cost of *ALG*. The total cost will then increase. Now, in A_4, except at most one, every small machine has a load of exactly $2q$ or 0. If we have the small machine with a load different from 0 and $2q$, then we will denote its load by x. Note that we keep the machines with load 0, because the cost of purchasing these machines is included in the total cost. According to the lower bound of Lemma 6,

- if $a = 3k + 2$, then we have at least $2k + 1$ machines with load of $2q$,
- if $a = 3k + 3$, then we have at least $2k + 1$ machines with load of $2q$. If there are exactly $2k + 1$ machines with this load, then at least one further machine has load of at least $(4/3)q$,
- if $a = 3k + 1, k \geq 1$, then we have at least $2k$ machines with load of $2q$. If there are exactly $2k$ machines with this load, then at least one other machine has a load of at least $(2/3)q$.

Step 5: In the last step, we will change the load of those small machines whose load is greater than the lower bound of Lemma 6. From A_4, we will get A_5 and from O_4 we will get O_5.

This means that

- if $a = 3k + 2$ then we keep $2k + 1$ machines with load of $2q$ and we set the size of all other jobs to zero;

- if $a = 3k + 3$ then we keep $2k + 1$ machines with load of $2q$ and one machine with load of q and we set all other sizes to zero;
- if $a = 3k+1, k \geq 1$ then we keep $2k$ machines with load of $2q$ and one machine with load of $(2/3)q$ and we set all other sizes to zero.

First we note that it may happen that more machines will have a full load of $2q$ than we need in the proof (whose total load is provided by Lemma 6). In this case we will delete the jobs of these small machines as follows.

- Some load of $0 \leq x \leq 2q$ is reduced to zero. In this case the cost of A_4 will decrease by x^2 and the lower bound of the optimal algorithm will decrease by $2x$.
- The load is $(4/3)q \leq x \leq 2q$ and it is decreased to q, or the load is $(2/3)q \leq x \leq 2q$ and it is decreased to $(2/3)q$.

It is easy to see that in both cases Lemma 2 can be applied and $A_5/O_5 > 4/3$ still holds.

4.4 Competitivness

So we reduced and modified our minimal counterexample. We proved that the input after Step 5 is still a counterexample. Now we will prove that it cannot be a counterexample.

After the reduction we have $a \geq 0$ small machines. Among them, there a_1 such machines where the load of a machine is exactly $2q$, and a_2 machines with load 0. Moreover, we have at most one additional small machine; and if it exists then its load is denoted by x ($2q \geq x \geq 0$), and according to the subcase above, here $x = q$ or $x = (2/3)q$.

We also know that in the A_5 schedule the load of the medium machines is exactly q, and the load of the big machines is exactly $2q$ (where $q = 1/\sqrt{2}$). To get the contradiction, we have to prove the following inequality:

$$a + b + c + a_1 \cdot (2q)^2 + x^2 + b \cdot q^2 + c \cdot (2q)^2 \leq \frac{4}{3} \cdot 2(a_1 \cdot 2q + x + b \cdot q + c \cdot 2q),$$

where on the right hand side we used the $2P$ lower bound of the optimum value from Proposition 2, which is actually the same as O_5. This inequality (using $q^2 = 1/2$) leads after simplification to

$$a + 2a_1 + x^2 + \frac{3}{2}(b + 2c) \leq \frac{4\sqrt{2}}{3}(2a_1) + \frac{8}{3} \cdot x + \frac{4\sqrt{2}}{3}(b + 2c). \quad (2)$$

Here the coefficient of $(b + 2c)$ is $\frac{4\sqrt{2}}{3} \approx 1.8856$ on the right hand side, while it is (only) 1.5 on the left hand side. This means that if the inequality is valid for an input with certain $(b + 2c)$ value, then it is also valid for the modified input where the small machines are the same (after the reduction) but the value of $(b+2c)$ is bigger. Hence we will consider our main inequality (2) only if $b+2c \leq 1$. If the inequality is valid even for $b + 2c = 0$, then we are done (as it is also valid for bigger values of $b + 2c$). Otherwise we will consider the case where $b+2c = 1$.

Now we will consider three cases according to the remainder of a divided by three. Several small cases will remain, and these remaining cases will be considered so that instead of the lower bound of the optimum value (i.e. $2P$) sometimes it will be easier to compare the objective value of the algorithm with the objective value of the optimum.

Case 1, $a = 3k+2$; moreover $k \geq 1$ or $b+2c \geq 1$. In this case according to the reduction Step 5 we get that $a_1 = 2k+1$ so among the small machines there are $2k+1$ machines with load $2q$ and there are $k+1$ machines with 0 load (here $x = 0$). The total load of the small machines (after the reduction) is $(2k+1) \cdot 2q$. In the case $b = c = 0$, (2) looks like

$$3k + 2 + 2(2k+1) \leq \frac{8\sqrt{2}}{3} \cdot (2k+1)$$

$$7k + 4 \leq \frac{16\sqrt{2}}{3}k + \frac{8\sqrt{2}}{3}.$$

If $k = 1$, the inequality looks like $11 \leq 8\sqrt{2} \approx 11.314$, and since the coefficient of k on the right hand side is $\frac{16\sqrt{2}}{3} \approx 7.5425 > 7$, the inequality holds for any $k \geq 1$. If $k = 0$ then we may assume that $b + 2c = 1$, so (as we saw above) it suffices to show that $4 + 3/2 \leq \frac{8\sqrt{2}}{3} + \frac{4\sqrt{2}}{3} = 5.6569$. As this holds once again, we are done.

Case 2, $a = 3k+3$, where $k \geq 0$. We suppose that $b = c = 0$. In this case after the reduction we have $a_1 = 2k+1$ and $x = q$. Then (2) looks like the following:

$$(3k+3) + 2(2k+1) + \frac{1}{2} \leq \frac{4\sqrt{2}}{3}(4k+2) + (4/3)(\sqrt{2}),$$

which is

$$7k + 5.5 \leq \frac{16\sqrt{2}}{3}k + 4\sqrt{2}$$

and for $k = 0$ it means $5.5 < 4\sqrt{2} = 5.656$, which is true.

Case 3, $a = 3k+1$, where $k \geq 1$. We suppose that $b = c = 0$. After the reduction Step 5 we have $2k$ machines with load $2q$, one machine with load $\frac{2}{3}q$, and k machines with load 0. We need to show that

$$(3k+1) + 2 \cdot (2k) + \frac{2}{9} \leq \frac{4\sqrt{2}}{3}(4k) + \frac{8}{9} \cdot \sqrt{2}$$

i.e.

$$7k + 11/9 \leq \frac{16\sqrt{2}}{3}k + \frac{8\sqrt{2}}{9},$$

where it is enough to examine the case $k = 1$. Here, the inequality looks like $8.222 \approx 7 + 11/9 \leq \frac{16\sqrt{2}}{3} + \frac{8\sqrt{2}}{9} = 8.799$, which is true.

At this point we have seen that the algorithm is at most (4/3)-competitive in the above cases. Below, we will continue with the cases not yet covered. First, let us see what cases have already been investigated, and what remain:

	Covered	Remain
Case 1:	$a = 3k + 2$; where $k \geq 1$ or $b + 2c > 0$	$k = 0$ and $b = c = 0$
Case 2:	$a = 3k + 3$, where $k \geq 0$	–
Case 3:	$a = 3k + 1$, where $k \geq 1$	$k = 0$

We realize that two cases remain. One possibility is that there are two small machines, and no other machine (first row in the table, Case R1 below). The only other possibility is that there is exactly one small machine, and there are possibly several medium and/or big machines (last row in the table, Case R2 below). If there is no medium and no big machine the schedule is optimal; so we can assume that there is also at least one machine which is not small.

Case R1, there are two small machines and no other machine. Let us consider the moment when the first job is assigned to the second (small) machine by the algorithm. At that moment the sum of the loads of the machines is bigger than $2q$. After applying the reduction let the loads of the two machines be $2q$ and x, respectively. Here $0 < x \leq 2q = \sqrt{2}$. The objective value of the algorithm is $2 + x^2 + (2q)^2 = 4 + x^2$.

Suppose that in the optimal solution the jobs are assigned to one machine. Then $OPT = 1 + (x + 2q)^2 \geq 1 + 2 + x^2 = 3 + x^2$, and we are done. If they are assigned to two machines then $OPT \geq 2 + \frac{(2q+x)^2}{2} = 2 + \frac{2+4qx+x^2}{2} = 3 + x^2/2 + 2qx \geq 3 + x^2$ and we are done again. If they are assigned to three machines, then $OPT \geq 3 + \frac{(2q+x)^2}{3}$, thus $\frac{4}{3}OPT \geq 4 + \frac{4}{9}(2+4qx+x^2) > 4 + \frac{4}{9}(2x \cdot x + x^2) > 4 + x^2$. It is easy to see that the optimal solution will not use four or more machines.

Case R2, $a = 1$ and $b + c > 0$. After simple calculation we get that our main inequality (2) is valid if $b + 2c \geq 3$. Thus it remains for us to consider the case where $b + 2c \leq 2$. Within this we will distinguish three subcases, and compare the objective value of the algorithm to the optimum value (instead of its lower bound) as follows.

Subcase R2.1. $a = 1, b = 0, c = 1$. Because the reduction, the size of the big job is $y = \sqrt{2}$. The total size of the small jobs is x for some $0 < x \leq 2q = \sqrt{2}$. The objective value of the algorithm is $2 + x^2 + y^2 = 4 + x^2$. Suppose that in the optimal solution the jobs are assigned to one machine. Then $OPT = 1 + (x + y)^2 \geq 1 + x^2 + y^2 = 3 + x^2$, and we are done. If they are assigned to two machines then the schedule of the algorithm is optimal.

Subcase R2.2. $a = 1, b = 2, c = 0$. In this case the reduction should be applied in a different way. Note that the total load of any two machines is more than $2q$ using the algorithmic rule. Hence let us perform the reduction so that we decrease the load of some machine, and at the same time increase the load of one other machine. During this time, the load of some machine will reach $2q$. After this we perform another reduction to make the loads of the two other machines as unbalanced as possible. The next two cases can happen after the reduction.

a, The loads are 0, $2q$ and $q + x$ with some $0 < x \leq q$.
b, The loads are x, $2q$ and $2q$ with some $0 < x \leq 2q$.

Note that in the optimal solution the two medium jobs are assigned to different machines. We make the calculations in these cases one by one.

Case a: The objective value of the algorithm is $3 + 2 + x^2 = 5 + x^2 \leq 5.5$. Let us see the optimal value. If the jobs are assigned to two machines then $OPT \geq 2 + \frac{(3q+x)^2}{2} = 2 + \frac{(9/2+6qx+x^2)}{2} = 17/4 + x^2/2 + 3qx$, thus $\frac{4}{3}OPT \geq 17/3+\frac{2}{3}x^2+4qx > 5+x^2$. If they are assigned to three machines, then $\frac{4}{3}OPT \geq \frac{4}{3}(3 + \frac{9/2}{3}) = 6$. Optimum will certainly not use four or more machines.

Case b: The objective value of the algorithm is $3 + 2 + 2 + x^2 = 7 + x^2 \leq 9$. Let us see the optimal value. If the jobs are assigned to two machines then $OPT \geq 2+\frac{(4q+x)^2}{2} = 2+\frac{(8+8qx+x^2)}{2} = 6+x^2/2+4qx$, similarly as before, we are done. If they are assigned to three machines, then $\frac{4}{3}OPT \geq \frac{4}{3}(3 + \frac{8+8qx+x^2}{3}) = \frac{4}{9}x^2+\frac{32}{9}qx+\frac{68}{9}$, we are done. If they are assigned to four machines, then $\frac{4}{3}OPT \geq \frac{4}{3}(4 + \frac{8+8qx+x^2}{4}) = \frac{1}{3}x^2 + \frac{8}{3}qx + 8$ which is enough. Optimum will certainly not use five or more machines.

Subcase R2.3. $a = 1$, $b = 1, c = 0$. In this case the reduction is similar to that of performed in case R1, namely we decrease the load of the small machine and increase the load of the medium machine. The load of the medium machine will grow to reach $2q$. Let the load of the small machine be x for some $0 < x \leq 2q = \sqrt{2}$. Then we have that the value of the objective is $4 + x^2$, and from this point the proof is the same.

References

1. Dósa, Gy., He, Y.: Better online algorithms for scheduling with machine cost. SIAM J. Comput. **33**(5), 1035–1051 (2004)
2. Dósa, Gy., He, Y.: Scheduling with machine cost and rejection. J. Comb. Optim. **12**(4), 337–350 (2006)
3. Dósa, Gy., Imreh, Cs.: The generalization of scheduling with machine cost. Theoret. Comput. Sci. **510**, 102–110 (2013)
4. Dósa, Gy., Tan, Z.: New upper and lower bounds for online scheduling with machine cost. Discrete Optim. **7**(3), 125–135 (2010)
5. Han, S., Jiang, Y., Hu, J.: Online algorithms for scheduling with machine activation cost on two uniform machines. J. Zhejiang Univ.-Sci. A **8**(1), 127–133 (2007)
6. He, Y., Cai, S.: Semi-online scheduling with machine cost. J. Comput. Sci. Technol. **17**(6), 781–787 (2002)
7. Imreh, Cs.: Online scheduling with general machine cost functions. Discrete Appl. Math. **157**(9), 2070–2077 (2009)
8. Imreh, C., Noga, J.: Scheduling with machine cost. In: Hochbaum, D.S., Jansen, K., Rolim, J.D.P., Sinclair, A. (eds.) APPROX/RANDOM -1999. LNCS, vol. 1671, pp. 168–176. Springer, Heidelberg (1999). https://doi.org/10.1007/978-3-540-48413-4_18
9. Nagy-György, J., Imreh, Cs.: Online scheduling with machine cost and rejection. Discrete Appl. Math. **155**(18), 2546–2554 (2007)
10. Szwarc, W., Mukhopadhyay, S.K.: Minimizing a quadratic cost function of waiting times in single-machine scheduling. J. Oper. Res. Soc. **46**(6), 753–761 (1995)
11. Townsend, W.: The single machine problem with quadratic penalty function of completion times: a branch-and-bound solution. Manage. Sci. **24**(5), 530–534 (1978)

Parallel Duel-and-Sweep Algorithm for the Order-Preserving Pattern Matching

Davaajav Jargalsaikhan(✉), Diptarama Hendrian, Ryo Yoshinaka, and Ayumi Shinohara

Graduate School of Information Sciences, Tohoku University, Sendai, Japan
davaajav@ecei.tohoku.ac.jp, {diptarama,ryoshinaka,ayumis}@tohoku.ac.jp

Abstract. Given a text and a pattern over an alphabet, the classic exact matching problem searches for all occurrences of pattern P in text T. Unlike the exact matching problem, *order-preserving pattern matching* considers the relative order of elements, rather than their exact values. In this paper, we propose the first parallel algorithm for the OPPM problem. Our algorithm is based on the "duel-and-sweep" algorithm. For a pattern of length m and a text of length n, our algorithm runs in $O(\log^3 m)$ time and $O(n \log^3 m)$ work on the Priority CRCW PRAM.

Keywords: String matching · Order-preserving pattern matching · Parallel algorithm

1 Introduction

Given a text and a pattern, the exact matching problem searches for all occurrence positions of the pattern in the text. Unlike the exact matching problem, the *order-preserving pattern matching* (OPPM) problem [5,6] considers the relative order of elements, rather than their real values. For instance, for exact matching $(12, 35, 5) \neq (25, 30, 21)$. However, for OPPM, $(12, 35, 5)$ matches $(25, 30, 21)$, since the relative order of the elements is same. Namely, the first element is the second smallest, the second element is the largest and the third element is the smallest among $(12, 35, 5)$, $(25, 30, 21)$, respectively. Order-preserving matching has gained much interest in recent years, due to its applicability in problems where the relative order matters, such as share prices in stock markets, weather data or musical notes.

There are several serial OPPM algorithms proposed in recent years. Kubica et al. [6] and Kim et al. [5] independently proposed a solution for the OPPM problem based on the KMP algorithm. Cho et al. [1] brought forward another algorithm based on the Horspool algorithm that uses q-grams. Jargalsaikhan et al. [4] proposed a duel-and-sweep algorithm for the OPPM problem.

Actually, duel-and-sweep is a technique developed for *parallel* pattern matching algorithms. The first duel-and-sweep algorithm was proposed by Vishkin [8]

This research was partially supported by JSPS KAKENHI Grant Numbers JP15H05-706 and JP19K20208.

A. Chatzigeorgiou et al. (Eds.): SOFSEM 2020, LNCS 12011, pp. 211–222, 2020.
https://doi.org/10.1007/978-3-030-38919-2_18

as a parallel algorithm for exact pattern matching. The goal of this paper is to propose a parallel algorithm for the OPPM problem based on those algorithms by Jargalsaikhan et al. [4] and Vishkin [8].

Adapting an exact matching algorithm to OPPM is no trivial task. The difficulty of OPPM mainly comes from the fact that we cannot determine the isomorphism by comparing the symbols in the text and the pattern on each position independently; instead, we have to consider their respective relative orders in the pattern and in the text. For instance, consider strings S_1, S_2, T_1, T_2 of equal length. Suppose that S_1 matches T_1 and S_2 matches T_2. In exact matching, the concatenation of S_1 and S_2 will match that of T_1 and T_2. In OPPM, the two concatenations will not necessarily match each other.

We choose the Priority Concurrent Read Concurrent Write Parallel Random-Access Machines (P-CRCW PRAM) [3] to model the parallel algorithm. Given the text of length n and the pattern of length m, our algorithm runs in $O(\log^3 m)$ time using $O(n \log^3 m)$ work on the P-CRCW PRAM. To the best of our knowledge, this is the first parallel algorithm for solving the OPPM problem.

2 Preliminaries

We use Σ to denote an alphabet of integer symbols such that the comparison of any two symbols can be done in constant time. Σ^* denotes the set of strings over Σ. For a string $S \in \Sigma^*$, we denote the i-th element of S by $S[i]$ and the substring of S that starts at the location i and ends at j by $S[i:j]$. We say that two strings S and S' of equal length are *order-isomorphic*, written $S \approx S'$, if

$$S[i] \leq S[j] \iff S'[i] \leq S'[j] \quad \text{for all } 1 \leq i, j \leq |S|.$$

For instance, $(12, 35, 5) \approx (25, 30, 21) \not\approx (11, 13, 20)$. If $S \not\approx S'$, then, there must exist a pair $\langle i, j \rangle$ of locations ($i < j$) such that the condition above does not hold. We call such $\langle i, j \rangle$ a *mismatch location pair* for S and S'. We say that a mismatch location pair $\langle i, j \rangle$ is *tight* if $S[1:j-1] \approx S'[1:j-1]$ and $S[1:j] \not\approx S'[1:j]$.

Given a text and a pattern, in the OPPM problem, we find all positions of substrings of the text that are order-isomorphic to the pattern.

Definition 1 (OPPM problem).

Input: *A text* $T \in \Sigma^*$ *of length* n *and a pattern* $P \in \Sigma^*$ *of length* $m \leq n$.
Output: *All positions* j *of the text such that* $T[j:j+m-1] \approx P$.

In the remainder of this paper, we fix a text T of length n and a pattern P of length m. For an integer x with $1 \leq x \leq n-m+1$, a *candidate* is the substring of T starting from x of length m, i.e., $T_x = T[x:x+m-1]$. When a candidate T_x is order-isomorphic to the pattern, we call x an *occurrence* of the pattern inside the text.

Definition 2 (Block-based period). *Given a string* S *of length* n, *an integer* p *is called a* block-based period *of* S, *if*

$$S[1:p] \approx S[kp+1:kp+p] \text{ for } k \in \{1, \ldots, \lfloor n/p \rfloor - 1\} \text{ and,}$$
$$S[1:r] \approx S[n-r+1:n] \text{ for } r = n \bmod p.$$

String S of length n is *block-periodic*, if there exists a block-based period $p \geq 2$ of S such that $n \geq 2p$. Otherwise, it is *block-aperiodic.*

Definition 3 (Border-based period). *Given a string S of length n, an integer p is called a* border-based period *of S if $S[1:n-p] \approx S[p+1:n]$.*

If p is a border-based period of S, then p is also a block-based period of S [7]. The reverse does not necessarily hold true. For instance, $S = (13, 7, 10, 21, 14, 18, 22, 15, 20, 28, 11, 25)$ has a block-based period 3, since $S[1:3] \approx S[4:6] \approx S[7:9] \approx S[10:12]$. However, S does not have a border-based period 3, since $S[1:9] \not\approx S[4:12]$, for which $\langle 1, 5\rangle$ and $\langle 3, 5\rangle$ are tight mismatch location pairs.

Lemma 1. *If $p > 0$ is a border-based period of S, then all multiples of p are also border-based periods of S.*

We use $Lmax_P$ and $Lmin_P$ arrays for the pattern defined by Kubica et al. [6]. Using $Lmax_P$ and $Lmin_P$, the order-isomorphism check between P and some other string S can be performed in $O(m)$ time in serial.

$$Lmax_P[i] = j \ (j < i) \ \text{ if } \ P[j] = \max_{k<i}\{P[k] \mid P[k] \leq P[i]\},$$
$$Lmin_P[i] = j \ (j < i) \ \text{ if } \ P[j] = \min_{k<i}\{P[k] \mid P[k] \geq P[i]\}.$$

If there are more than one such j, any of them can be taken. If there is no such j then we denote it as $Lmin_P[i] = 0$ and $Lmax_P[i] = 0$.

In this paper, we choose the P-CRCW PRAM to model the parallel algorithm. The P-CRCW PRAM allows simultaneous reads and writes into the same memory location, and on the occasion of multiple writes, the processor with the smallest index succeeds. We also assume that $n = 2\,\text{m}$. Larger texts can be cut into overlapping pieces of length $2\,\text{m}$ and processed independently.

3 Parallel Duel-and-Sweep Algorithm for the OPPM Problem

First, we will give the general description of the duel-and-sweep algorithm for OPPM. The descriptions given in the beginning of this section are applicable to both serial and parallel versions of the algorithm. Then, in the following subsections we will explain our parallel algorithm.

The duel-and-sweep algorithm screens all candidates in two stages, called the *dueling* and *sweeping* stages. The dueling stage prunes the candidates, until the remaining candidates are pairwise *consistent.* The sweeping stage "sweeps" through the remaining candidates to determine pattern occurrences. Taking advantage of the fact that the remaining candidates are pairwise consistent, the sweeping stage can be done efficiently.

Definition 4 (Candidate consistency for OPPM). *Two candidates T_x and T_{x+a} are* consistent, *for $0 \leq a < m/2$ if $P[a+1:m] \approx P[1:m-a]$, and $a \geq m/2$. Otherwise we say that T_x and T_{x+a} are not consistent.*

Algorithm 1. Dueling between x and y w.r.t. S, assuming that $x < y$

```
1 Function Dueling(x, y, S)                 // returns the surviving candidate
2     ⟨w1, w2⟩ = W[y − x];
3     if ⟨w1, w2⟩ = ⟨0, 0⟩ then return 0;
4     if P[w1] = P[w2] then
5         if S[y + w1] = S[y + w2] then return y else return x;
6     if P[w1] < P[w2] then
7         if S[y + w1] < S[y + w2] then return y else return x;
8     if P[w1] > P[w2] then
9         if S[y + w1] > S[y + w2] then return y else return x;
```

The candidate consistency comes from the following observation. Consider two overlapping candidates T_x and T_{x+a}, whose overlapping region is $T[x+a: x+m-1]$. If $T_x \approx P$ and $T_{x+a} \approx P$, we can conclude that $T[x+a:x+m-1] \approx P[a+1:m] \approx P[1:m-a]$. In other words, $P[a+1:m] \approx P[1:m-a]$ is a necessary condition for $T_x \approx P \wedge T_{x+a} \approx P$.

To speed up the dueling stage, the pattern is preprocessed, so that the order-isomorphism checks of type $P[a+1:m] \approx P[1:m-a]$ can be performed using constant work. Specifically, the algorithm constructs the *witness table* $W[0 : m/2-1]$, where $W[a] = \langle 0,0\rangle$ if $P[a+1:m] \approx P[1:m-a]$, and otherwise $W[a] \neq \langle 0,0\rangle$ and $W[a]$ stores a *witness* for offset a. (Hereinafter, we will refer to $\langle 0,0\rangle$ as a *zero*.)

Definition 5 (Witness for OPPM [4]**).** *If $P[a+1:m] \not\approx P[1:m-a]$, there must exist at least one mismatch location pair. We call such a mismatch location pair a* witness *for offset a.*

We denote by $\mathcal{W}_P(a)$ the set of all witnesses for the offset a. Obviously, $\mathcal{W}_P(a) = \emptyset$ for $a = 0, m-1$, and m. Note that $\mathcal{W}_P(a)$ is empty, iff a is a border-based period of P. We say that a witness $\langle w_1, w_2\rangle \in \mathcal{W}_P(a)$ is *tight*, if $\langle w_1, w_2\rangle$ is a tight mismatch location pair for $P[1:m-a] \not\approx P[a+1:m]$.

Next, we will define the process called *dueling* between offsets x and $x+a$ w.r.t. the text. The procedure for dueling is described in Algorithm 1. Suppose that $\langle w_1, w_2\rangle \in \mathcal{W}_P(a)$ satisfies $P[w_1] < P[w_2]$ and $P[w_1+a] \geq P[w_2+a]$ (Line 6 of Algorithm 1). If $T[x+a+w_1] < T[x+a+w_2]$, it yields $T_x \not\approx P$ since $P[w_1+a] \geq P[w_2+a]$. Otherwise, $T_{x+a+1} \not\approx P$ since $P[w_1] < P[w_2]$. Thus in any case, we can safely eliminate either candidate T_{x+1} or T_{x+a+1} without looking into other locations. We can perform this process similarly for the other cases.

Dueling can also be performed between two offsets w.r.t. the pattern. Suppose that $\langle w_1, w_2\rangle \in \mathcal{W}_P(a)$. Consider overlapping P on itself with offsets x and $x+a$. If $x+a+w_2 \leq m$, in other words, if the witness pair lies within the overlap region, we can say the following.

(1) If the offset x survives the duel, then $\langle w_1, w_2\rangle \in \mathcal{W}_P(x+a)$.
(2) If the offset $x+a$ survives the duel, then $\langle w_1+a, w_2+a\rangle \in \mathcal{W}_P(x)$.

3.1 Pattern Preprocessing

The goal of the pattern preprocessing is to construct the witness table. The preprocessing in the serial OPPM [4] uses the Z-array [2] modified for OPPM. If Z_S is the modified Z-array for a string S, $Z_S[i]$ stores the length of the longest substring of S that starts at position i and is order-isomorphic to some prefix of S. To obtain the linear construction time, the algorithm uses information in $Z_S[1:i-1]$ to obtain $Z_S[i]$. The prefix-dependent approach cannot be applied to a parallel algorithm.

Vishkin's exact matching parallel algorithm [8] is based on the following idea. To build the witness table it suffices to find the smallest block-based period of the pattern and locate witnesses for offsets less than the smallest block-based period. Assuming that p is the smallest block-based period of P, when $i \bmod p = 0$, $\mathcal{W}_{exact}(i) = \emptyset$. Vishkin's algorithm sets $W_{exact}[i] = W_{exact}[i \bmod p]$ for all offsets $i \in \{1, \ldots, m/2-1\}$.

For OPPM, when the pattern is block-aperiodic, we follow Vishkin's algorithm. When the pattern is block-periodic, we need different ideas. Suppose that p is the smallest block-based period for OPPM. Border-based period is a multiple of the smallest block-based period p, but not every multiple of p is a border-based period of P. Consider the example string $S = (13, 7, 10, 21, 14, 18, 22, 15, 20, 28, 11, 25)$. S has block-based periods $3, 6, 9$, but it only has a single border-based period 9. $\mathcal{W}_P(9) = \emptyset$, but $\mathcal{W}_P(3) \neq \emptyset$ and $\mathcal{W}_P(6) \neq \emptyset$.

Naively checking all multiples of the smallest block-based period will take $O(m^2)$ work on the P-CRCW PRAM, assuming that $Lmax_P$ and $Lmin_P$ arrays has been already computed. Our preprocessing algorithm uses $O(\log^2 m)$ time and $O(m \log^2 m)$ work on the P-CRCW PRAM. Our preprocessing algorithm calls the modified Vishkin's algorithm (Algorithm 5) as a subroutine. The rest of the subsection will describe our preprocessing algorithm.

First, we construct $Lmax_P$, $Lmin_P$ arrays. It is straightforward to parallelize Kubica et al. [6] algorithm for constructing $Lmax_P$ and $Lmin_P$ in serial, using existing sorting algorithms. $Lmax_P$ and $Lmin_P$ arrays can be computed in $O(\log m)$ time and $O(m \log m)$ work on the P-CRCW PRAM.

Order-isomorphism check between the pattern P and a string S can be performed in $O(1)$ time and $O(m)$ work on the P-CRCW PRAM, given $Lmax_P$ and $Lmin_P$ by Lemma 2.

Lemma 2. *For a string of length m, assume that $S[1:b-1] \approx P[1:b-1]$. Given $Lmax_P$ and $Lmin_P$ arrays,* `CheckOrderIsomorphism`S, b, e *in Algorithm 2 returns a tight mismatch location pair if $S[1:e] \not\approx P[1:e]$, and otherwise returns $\langle 0, 0 \rangle$, in $O(1)$ time and $O(m)$ work on the P-CRCW PRAM.*

Next, we show how to construct the witness table on the P-CRCW PRAM. The procedure for constructing witness table is shown in Algorithm 3. All entries of W are initialized to $\langle 0, 0 \rangle$, and each entry is updated at most once. We will define the following properties.

- *witness certainty property (WCP) for* $W[b:e]$. For $i \in \{b, \ldots, e\}$ $W[i] \neq \langle 0, 0 \rangle$ implies $W[i] \in \mathcal{W}_P(i)$.

Algorithm 2. Checks order-isomorphism of S and $P[b\!:\!e]$

```
Function CheckOrderIsomorphism(S, b, e)
    ⟨m1, m2⟩ = ⟨0, 0⟩;
    for i ∈ {b, ..., e} do in parallel
        i_min = Lmin_P[i]; i_max = Lmax_P[i];
        if i_max = 0 and i_min = 0 then
            continue;
        if i_max = 0 and S[i_min] ≤ S[i] then ⟨m1, m2⟩ = ⟨i_min, i⟩;
        else if i_min = 0 and S[i_max] ≥ S[i] then ⟨m1, m2⟩ = ⟨i_max, i⟩;
        else if (P[i_min] = P[i] and S[i_min] ≠ S[i])
                or (P[i_min] > P[i] and S[i_min] ≤ S[i]) then ⟨m1, m2⟩ = ⟨i_min, i⟩;
        else if P[i_max] < P[i] and S[i_max] ≥ S[i] then ⟨m1, m2⟩ = ⟨i_max, i⟩;
    return ⟨m1, m2⟩;
```

Algorithm 3. Preprocessing for pattern P

```
Initialize all values in W[0 : m/2 − 1] to ⟨0, 0⟩;
e = m/2 − 1; p = 1;
while there are some zeros in W[1 : e] do
    W[p] = ⟨w1, w2⟩ = CheckOrderIsomorphism(P[p + 1 : m], 1, m − p);
    if ⟨w1, w2⟩ = ⟨0, 0⟩ then return;
    r = (⌊(m − w2)/p⌋ + 1) · p;
    if r ≤ e then
        RightSubroutine(r, e, p);                    // Algorithm 4
        e = r − p;
    LeftSubroutine(e, p);                            // Algorithm 5
    p = min{i | i ≥ 1, W[i] = ⟨0, 0⟩};
```

- *zero certainty property (ZCP) for* $W[b:e]$. For $i \in \{b, \dots, e\}$ $W[i] = \langle 0, 0\rangle$ implies $\mathcal{W}_P(i) = \emptyset$.

The preprocessing algorithm constructs the witness table from right to left. The suffix of W that satisfies the ZCP becomes longer with each iteration. Once the suffix satisfies the ZCP, we do not need to look into it any further. The prefix satisfies the WCP, and the zeros are located equal distance apart from each other. Suppose that $W[0\!:\!e]$ is the prefix, the algorithm needs to satisfy the ZCP of. The prefix and the suffix of $W[0\!:\!e]$ are processed independently, using Algorithm 5 (`LeftSubroutine`) and Algorithm 4 (`RightSubroutine`) respectively. Putting it formally, we have Lemma 3.

Lemma 3. *After each iteration of the while loop of Algorithm 3,* $W[e+1 : m/2-1]$ *satisfies the ZCP and the WCP,* $W[0\!:\!e]$ *satisfies the WCP, and for* $i \in \{0, \dots, e\}$, $W[i] = \langle 0, 0\rangle$ *iff* $i \bmod p = 0$.

The reason behind dividing $W[0 : e]$ into two comes from the following observation.

Algorithm 4. Algorithm for processing $W[r:e]$

```
Function RightSubroutine(r, e, p0)
    if CheckOrderIsomorphism(P[r+1:m], 1, m−r) = ⟨0,0⟩ then return;
    if CheckOrderIsomorphism(P[e+1:m], 1, m−e) ≠ ⟨0,0⟩ then b = e;
    else
        lo = 0; hi = e/p0;
        while lo < hi do
            mid = lo + ⌊(hi − lo)/2⌋;
            if CheckOrderIsomorphism(P[mid·p0 + 1:m], 1, m − mid·p0) = ⟨0,0⟩
             then hi = mid; else lo = mid + 1;
        b = (lo − 1)·p0;
    ⟨w1, w2⟩ = CheckOrderIsomorphism(P[b+1:m], 1, m−b);
    for i ∈ {r, ..., b} do in parallel
        if W[i] = ⟨0,0⟩ then W[i] = ⟨w1 + (b−i), w2 + (b−i)⟩;
```

Lemma 4. *Suppose that* $\mathcal{W}_P(p)$ *is not empty and let* $\langle w_1, w_2 \rangle$ *be a tight witness for the offset* p. *If* qp *and* $q'p$ *are two offsets such that* $m < qp + w_2 < q'p + w_2$, *then* $P[(q'-q)p+1:m-qp] \approx P[1:m-q'p]$.

Consider overlapping the pattern on itself with offsets qp and $q'p$ as indicated by Lemma 4. The overlap regions of the copies are $P[1:m-qp]$ and $P[1:m-q'p]$, respectively. Lemma 4 claims that the overlap regions of $P[1:m-qp]$ and $P[1:m-q'p]$ are order-isomorphic. As a consequence of Lemma 4, we cannot perform a duel between qp and $q'p$ for the pattern.

At the beginning of an iteration, all zeros in $W[0:e]$ are located p distance apart from each other. We want to update as many zeros in $W[1:e]$ as possible. Let $\langle w_1, w_2 \rangle$ be a tight witness for the offset p. Let e' be the largest multiple of p such that $0 < e' \le e$ and $e' + w_2 \le m$. By Lemma 4, we cannot perform duels for the pattern between offsets that are greater than e'. Thus, we divide $W[0:e]$ into two parts $W[1:e']$ and $W[e'+1:e]$, and process them using different algorithms `LeftSubroutine` and `RightSubroutine`. `LeftSubroutine` relies on the duels for the pattern to verify zeros in $W[0:e']$, while `RightSubroutine` uses Lemma 4 to satisfy the ZCP of $W[e'+1:e]$.

First, we will explain `RightSubroutine`(r, e, p), where $r = e' + p$. Consider overlapping the pattern on itself with offset r. If $P[1:m-r] \approx P[r+1:m]$, then for all offsets $i \in \{r, \cdots, e\}$, $\mathcal{W}_P(i) = \emptyset$ by Lemma 4. Also, if for some offset $j \in \{r, \ldots, e\}$, $\mathcal{W}_P(j) \neq \emptyset$, then for all offsets $i \in \{r, \ldots, j\}$, $\mathcal{W}_P(i) \neq \emptyset$. Based on this observation, `RightSubroutine` uses binary search to find an offset $b \in \{r, \ldots, e\}$ such that for $i \in \{r, \ldots, b\}$ $\mathcal{W}_P(i) \neq \emptyset$, and for $i \in \{b+p, \ldots, e\}$, $\mathcal{W}_P(i) = \emptyset$.

Lemma 5. *Assume that* $W[i] = \langle 0,0 \rangle$ *for any* $i \in \{2p_0, \ldots, e\}$ *satisfying* $i \bmod p_0 = 0$. *After running* `RightSubroutine`(r, e, p_0), $W[i] = \langle 0,0 \rangle$ *if* $\mathcal{W}_P(i) = \emptyset$, *and* $W[i] \in \mathcal{W}_P(i)$ *otherwise for any* $i \in \{r, \ldots, e\}$.

Algorithm 5. Processing for $W[0\!:\!e]$.

```
Function LeftSubroutine(e, p0)
    k = 0;
    while k ≤ ⌊log e⌋ do
        l = 2^k × p0; r = min{e, 2l − 1};
        if there is a zero in W[l:r] then
            p = location of the unique zero in W[l:r];
            W[p] = CheckOrderIsomorphism(P[p + 1:4l], 1, 4l − p + 1);
            if W[p] = ⟨0, 0⟩ then
                c = CertaintySatisfiedUntil(p, k, e);
                while k ≤ c − 1 do
                    SatisfySparsity(2l, e); k = k + 1;
                k = c − 1;
        else SatisfySparsity(2l, e); k = k + 1 ;
```

Algorithm 6. Updating $W[0\!:\!e]$ to satisfy l-sparsity

```
Function SatisfySparsity(l, e)
    for each l-block B of W[0:e] do in parallel
        if there are two zeros in B then
            Let j1 < j2 be the location of two zeros in B; a = j2 − j1;
            surv = Dueling(j1, j2, P);
            ⟨w1, w2⟩ = W[a];
            if surv = j1 then W[j2] = ⟨w1, w2⟩;
            if surv = j2 then W[j1] = ⟨w1 + a, w2 + a⟩;
```

Now, we will explain `LeftSubroutine(`e, p_0`)`. For the sake of simplicity, assume that $p_0 = 1$ and $e = m/2 - 1$. Now, consider partitioning $W[0\!:\!m/2-1]$ into blocks of size 2^k. We will call each block a 2^k*-block*, with the last 2^k-block possibly being shorter than 2^k: $W[i \cdot 2^k : \min\{m/2-1, (i+1) \times 2^k - 1\}]$ for $i = 0, \ldots, \lfloor \frac{m}{2^{k+1}} \rfloor$.

k-sparsity property. Each 2^k-block of $W[0\!:\!m/2-1]$ contains at most one zero. Note that since $W[0] = \langle 0, 0 \rangle$, $W[1\!:\!2^k - 1]$ contains no zeros.
k-lookahead property. For all $i \in \{1, \ldots, 2^k - 1\}$, if $W[i] = \langle w_1, w_2 \rangle$, then $i + w_2 \le 2^{k+1}$.

The general policy for `LeftSubroutine` is to increase the length of the non-zero substring $W[1\!:\!2^k - 1]$ and have zeros in the witness table to be increasingly sparser with each round. The loop invariant of `LeftSubroutine` is as follows.

Lemma 6. *After the round k of* `LeftSubroutine`*, the k-sparsity and the k-lookahead properties hold.*

After running `LeftSubroutine`, zeros in $W[0\!:\!e]$ become sparser. Putting it formally, we have the following.

Algorithm 7. Finds integer $c \leq \log e$ such that p is the block-based period of $P[0:2^c]$ but not of $P[0:2^{c+1}]$.

```
Function CertaintySatisfiedUntil(p, k, e)
    c = k + 2;
    while c ≤ ⌊log e⌋ + 1 do
        r = min{e, 2^{c+1} − 1};
        for q ∈ {2^c, ..., r} s.t. q mod p = 0 do in parallel
            ⟨w_1, w_2⟩ = CheckOrderIsomorphism(P[q + 1 : r + 1], 1, r − p + 1);
            if ⟨w_1, w_2⟩ ≠ ⟨0, 0⟩ then
                for j ∈ {p, ..., 2^c − 1} s.t. j mod p = 0 do in parallel
                    if W[j] = ⟨0, 0⟩ then
                        W[j] = ⟨w_1 + q − j, w_2 + q − j⟩;
                Break from the while loop;
        c = c + 1;
    for i ∈ {p, ..., 2^c − 2^{k+2} − 1} do in parallel
        if W[i] = ⟨0, 0⟩ and i mod p ≠ 0 then W[i] = W[i mod p];
    for the last four 2^k-blocks B of W[0 : 2^c − 1] do in parallel
        If there is a zero in B, let j' be the location of zero in B;
        W[j'] = CheckOrderIsomorphism(P[j' + 1 : r + 1], 1, r − j' + 1);
    return c;
```

Lemma 7. *Assume that for $i \in \{0, \ldots, e\}$, $W[i] = \langle 0, 0\rangle$ iff $i \bmod p_0 = 0$. If $\mathcal{W}_P(p_0)$ is not empty,* $\texttt{LeftSubroutine}(e, p_0)$ *finds some $p = q \cdot p_0$ where $q > 1$ and updates W in such a way that for $i \in \{0, \ldots, e\}$, $W[i] = \langle 0, 0\rangle$ iff $i \bmod p = 0$.*

Theorem 1. *The pattern preprocessing runs in $O(\log^2 m)$ time and $O(m \log^2 m)$ work on the P-CRCW PRAM.*

3.2 Pattern Searching

We define a Boolean array $C[1:m+1]$ and initialize every entry of C to *True*. During the dueling stage, candidates duel with others, until the surviving candidates are pairwise consistent. If T_i remains after the dueling stage, it is indicated by $C[i] =$ *True*. The sweeping stage prunes the surviving candidates from the dueling stage until all remaining candidates are order-isomorphic to the pattern. After the sweeping stage, $C[i] =$ *True* iff $T_i \approx P$. Entries of C are updated at most once during the dueling and sweeping stages. The dueling stage is described in Algorithm 8. The sweeping stage is described in Algorithm 9.

When the pattern is aperiodic, we can use the same approach as Vishkin's parallel exact matching algorithm [8]. When the pattern is periodic, we cannot follow Vishkin's approach used in his parallel exact matching algorithm. Suppose that the pattern is periodic with period p, then we can express it as $P = u^k v$, where $u = P[1:p]$. Vishkin's algorithm finds all occurrences of $u^2 v$, then counts continuous occurrences of u^2v to find all occurrences of P. For OPPM we need different ideas.

Dueling Stage. Suppose that the pattern is block-aperiodic. Analogously to W, we define the k-sparsity property for C: for each 2^k-block of C there is at most one location j such that $C[j] = \mathit{True}$. The procedure for satisfying the k-sparsity is same as in the pattern preprocessing except that duels are done against T. During the round k, the algorithm satisfies the k-sparsity of C. Since we have found witnesses for offsets $i \leq m/2 - 1$, duels can be performed within a $m/2$-block. Furthermore, since the pattern is block-aperiodic, $W[i] \neq \langle 0,0 \rangle$ for $i \in \{1, \ldots, m/2 - 1\}$. Thus, for any two offsets within a $m/2$-block, we can perform duels. After the round $k = \log m - 1$, there are only three locations $1 \leq a \leq m/2$, $m/2+1 \leq b \leq m$ and $m+1$ such that $C[a] = C[b] = C[m+1] = \mathit{True}$. The algorithm checks naively whether T_a, T_b and T_{m+1} are pairwise-consistent.

When the pattern is block-periodic, let $p = \min\{i \mid i \geq 1,\ W[i] = \langle 0,0 \rangle\}$ and $k = \lfloor \log p \rfloor$. The algorithm satisfies the k-sparsity of C first, then start to merge neighboring 2^k-blocks so that the surviving candidates belonging to the same 2^{k+1}-block are pairwise consistent. For each 2^{k+1}-block the surviving candidates are pairwise-consistent.

Consider two neighboring 2^k-blocks $C[q \cdot 2^k + 1 : (q+1) \cdot 2^k]$ and $C[(q+1) \cdot 2^k + 1 : (q+2) \cdot 2^k]$ that are about to be merged. Let $\mathcal{T}_1 = \{i \mid q \cdot 2^k + 1 \leq i \leq (q+1) \cdot 2^k,\ C[i] = \mathit{True}\}$ and $\mathcal{T}_2 = \{i \mid (q+1) \cdot 2^k + 1 \leq i \leq (q+2) \cdot 2^k,\ C[i] = \mathit{True}\}$. If $i_1, j_1 \in \mathcal{T}_1$, then T_{i_1} is consistent with T_{j_1}. Similarly, if $i_2, j_2 \in \mathcal{T}_2$, then T_{i_2} is consistent with T_{j_2}. Suppose that $i_1 < j_1$, $i_2 < j_2$. If T_{i_1} is consistent with T_{j_1} and T_{j_1} is consistent with T_{j_2}, then T_{j_1} is consistent with T_{i_2}. Using the following lemma we can merge the two 2^k-blocks.

Lemma 8. *Suppose that j_1 is the largest element in $\mathcal{T}_1$ and j_2 is the smallest element in $\mathcal{T}_2$ such that T_{j_1} and T_{j_2} are consistent. Let $\mathcal{T}$ be the set consisting of elements i from $\mathcal{T}_1$ such that $i \leq j_1$ and elements i from $\mathcal{T}_2$ such that $i \geq j_2$. Then, if $i, j \in \mathcal{T}$ then T_i and T_j are consistent.*

The while loop invariant of Algorithm 8 is as follows. After each round k, for every 2^k-block of C, if i and j are two locations in the 2^k-block such that $C[i] = C[j] = \mathit{True}$, then T_i is consistent with T_j.

Sweeping Stage. The sweeping stage updates C until $C[i] = \mathit{True}$ iff T_i is an pattern occurrence. Let $\mathcal{T} = \{i \mid 1 \leq i \leq m+1,\ C[i] = \mathit{True}\}$. After the dueling stage for $i, j \in \mathcal{T}$, T_i is consistent with T_j. Now suppose one processor is attached to each entry of T. Then using Lemma 9 the sweeping stage can be performed in $O(\log m)$ time and $O(n \log m)$ work on the P-CRCW PRAM.

Lemma 9. *Let $\mathcal{T}$ be a set of locations such that if $i, j \in \mathcal{T}$ then T_i is consistent with T_j. For any $x \in \mathcal{T}$, let $\mathcal{T}_{<x} = \{i \in \mathcal{T} \mid i < x\}$ and $\mathcal{T}_{>x} = \{i \in \mathcal{T} \mid i > x\}$.*

(1) If $T_x \approx P$, then for $i \in \mathcal{T}_{>x}$, $P[(x-i)+1 : m] \approx T_i[1 : m - (x-i)]$.

Algorithm 8. Algorithm for the dueling stage

```
Initialize array C of length m + 1 to True;
Let p be the smallest integer s.t. W[p + 1] = ⟨0, 0⟩;
Satisfy ⌊log p⌋-sparsity of C in several iterations;
Round k = ⌊log p⌋ + 1;
while k ≤ log m do
    for every 2^k-block B of C do in parallel
        Let s and e be the starting and ending indexes of B;
        l1 = s; r1 = l1 + 2^(k−1) − 1;
        l2 = r1 + 1; r2 = e;
        while l1 < r1 and l2 < r2 do
            while l2 < r2 do
                m1 = l1 + ⌊(r1 − l1)/2⌋; m2 = l2 + ⌊(r2 − l2)/2⌋;
                if T_m1 is consistent with T_m2 then r2 = m2;
                else
                    if T_m1 survives the duel then
                        C[i] = False for all i ≤ m2 in the second half of B;
                        l2 = m2;
                    else
                        C[i] = False for all i ≥ m1 in the first half of B;
                        r1 = m1;
            r2 = e;
            l1 = l1 + ⌊(r1 − l1)/2⌋;
    k = k + 1;
```

(2) If $T_x \not\approx P$, let $\langle m_1, m_2 \rangle$ be a tight mismatch locations. Then for $i \in \mathcal{T}_{<x}$ such that $i + m \geq x + m_2$, $T_i \not\approx P$ and for $i \in \mathcal{T}_{>x}$, $P[(x-i)+1 : m_2 - 1] \approx T_i[1 : (m_2 - 1) - (x - i)]$.

Theorem 2. *The pattern searching runs in $O(\log^3 m)$ time and $O(n \log^3 m)$ work on the P-CRCW PRAM.*

4 Discussion

We have proposed a parallel algorithm for the OPPM problem which based on the duel-and-sweep paradigm [8]. This is the first parallel algorithm for the OPPM problem. Given the text of length n and the pattern of length m, our algorithm runs in $O(\log^3 m)$ depth using $O(m \log^3 m)$ work on the P-CRCW PRAM. If the pattern is block-aperiodic, the preprocessing runs in $O(\log m)$ time and $O(m \log m)$ and the pattern searching runs in $O(\log m)$ time and $O(n)$ work. If the pattern is block-aperiodic, both preprocessing and pattern searching algorithms are work-optimal, that is the work required on the P-CRCW PRAM is same as the time complexity of the fastest serial algorithms [4–6].

Algorithm 9. Algorithm for the sweeping stage

```
Function SweepRecursive(r, T)
    if T = ∅ then return;
    Let x ∈ T such that roughly half of elements T is less than x;
    T_1 = {i ∈ T | i < x}, T_2 = {i ∈ T | i > x};
    if r = x_0 then
        ⟨m_1, m_2⟩ = CheckOrderIsomorphism(T[x : x + m − 1], 1, m);
    else
        ⟨m_1, m_2⟩ = CheckOrderIsomorphism(T[r : x + m − 1], r − x + 1, m);
    if ⟨m_1, m_2⟩ = ⟨0, 0⟩ then
        do in parallel
            SweepRecursive(x_0, T_1); SweepRecursive(x + m, T_2);
    else
        C[x] = False;
        for i ∈ {i ∈ T_1 | i + m ≥ x + m_2} do in parallel
            C[i] = False; Remove i from T_1;
        do in parallel
            SweepRecursive(x_0, T_1); SweepRecursive(x + m_2, T_2);
Function SweepingStage()
    T = {i | 1 ≤ i ≤ m + 1, C[i] = True};
    x_0 = min{T};
    SweepRecursive(x_0, T);
```

References

1. Cho, S., Na, J.C., Park, K., Sim, J.S.: A fast algorithm for order-preserving pattern matching. Inf. Process. Lett. **115**(2), 397–402 (2015)
2. Hasan, M.M., Islam, A.S., Rahman, M.S., Rahman, M.S.: Order preserving pattern matching revisited. Pattern Recogn. Lett. **55**, 15–21 (2015)
3. JáJá, J.: An Introduction to Parallel Algorithms, vol. 17. Addison-Wesley, Reading (1992)
4. Jargalsaikhan, D., Diptarama, Ueki, Y., Yoshinaka, R., Shinohara, A.: Duel and sweep algorithm for order-preserving pattern matching. In: Tjoa, A., Bellatreche, L., Biffl, S., van Leeuwen, J., Wiedermann, J. (eds.) SOFSEM 2018. LNCS, vol. 10706, pp. 624–635. Edizioni della Normale, Cham (2018). https://doi.org/10.1007/978-3-319-73117-9_44
5. Kim, J., et al.: Order-preserving matching. Theoret. Comput. Sci. **525**, 68–79 (2014)
6. Kubica, M., Kulczyński, T., Radoszewski, J., Rytter, W., Waleń, T.: A linear time algorithm for consecutive permutation pattern matching. Inf. Process. Lett. **113**(12), 430–433 (2013)
7. Matsuoka, Y., Aoki, T., Inenaga, S., Bannai, H., Takeda, M.: Generalized pattern matching and periodicity under substring consistent equivalence relations. Theoret. Comput. Sci. **656**, 225–233 (2016)
8. Vishkin, U.: Optimal parallel pattern matching in strings. In: Brauer, W. (ed.) ICALP 1985. LNCS, vol. 194, pp. 497–508. Springer, Heidelberg (1985). https://doi.org/10.1007/BFb0015775

Parameterized Complexity of Synthesizing b-Bounded (m, n)-T-Systems

Ronny Tredup(✉)

Universität Rostock, Institut für Informatik, Theoretische Informatik, Albert-Einstein-Straße 22, 18059 Rostock, Germany
ronny.tredup@uni-rostock.de

Abstract. Let $b \in \mathbb{N}^+$. Synthesis of pure b-bounded (m, n)-T-systems ((m,n)-Synthesis, for short) consists in deciding whether there exists for an input (A, m, n) of transition system A and integers $m, n \in \mathbb{N}$ a pure b-bounded Petri net N as follows: N's reachability graph is isomorphic to A, and each of N's places has at most m incoming and at most n outgoing transitions. In the event of a positive decision, N should be constructed. The problem is known to be NP-complete, and (m,n)-Synthesis parameterized by $m + n$ is in XP [14]. In this paper, we enhance our understanding of (m,n)-Synthesis from the viewpoint of parameterized complexity by showing that it is $W[1]$-hard when parameterized by $m + n$.

1 Introduction

Petri net *synthesis* consists in deciding whether there is a Petri net (PN, for short) that implements a given behavioral specification and in constructing such a net if it exists. Valid synthesis methods yield implementations that are correct by design. The possibility of finding effective or even efficient synthesis algorithms crucially depends on the specification and the searched net. This has been subject of research for many years: It is undecidable whether there is a P/T net implementing a pushdown- or a HMSC-language or whether there is a (pure) bounded P/T net implementing a modal transition systems (MTS, for short) [9,11]. If the specification is a deterministic pushdown-language or -graph, and the search net is a P/T-net, synthesis is decidable [4]. It is also decidable whether there is a b-bounded Petri net that implements an MTS [12]. If the specification is a transition system (TS, for short), and the searched net is a 1-bounded PN, synthesis is NP-complete [2], even if the TS is strongly restricted [15,16]. The synthesis of b-bounded PNs from TSs is NP-complete, even if the searched net is strongly restricted [13,14]. If the bound b is not fixed in advance, the synthesis of bounded PN from TSs is polynomial [1]. If the PN is additionally to be choice-free or a marked graph, even better procedures exist [5,7].

In this paper, we investigate an instance of PN synthesis that is called (m, n)-Synthesis. It consists in deciding whether there exists for an input (A, m, n) of TS A and integers $m, n \in \mathbb{N}$ a pure b-bounded Petri net N as follows: N's reachability graph is isomorphic to A, and each of N's places has at most m incoming

A. Chatzigeorgiou et al. (Eds.): SOFSEM 2020, LNCS 12011, pp. 223–235, 2020.
https://doi.org/10.1007/978-3-030-38919-2_19

and at most n outgoing transitions. The b-bounded (m,n)-T-systems generalize the notion of (weighted) T-systems [6,10] and adapt it to b-bounded PN. In [14], we have shown that (m,n)-SYNTHESIS is NP-complete. We have also argued that (m,n)-SYNTHESIS parameterized by $m+n$ belongs to the complexity class XP. Thus, the question arises whether this parameterization makes the problem fixed parameter tractable. In this paper, we answer this question negatively and show that (m,n)-SYNTHESIS parameterized by $m+n$ is $W[1]$-hard. The proof presents a parameterized reduction from REGULAR INDEPENDENT SET, which restricts the canonical $W[1]$-hard problem to regular graphs [8], to (m,n)-SYNTHESIS. This paper is organized as follows. Section 2 introduces necessary preliminary notions, Sect. 3 presents the $W[1]$-hardness result and Sect. 4 closes the paper.

2 Preliminaries

We assume that the reader is familiar with the concepts relating to fixed-parameter tractability, the standard notions relating to graphs and REGULAR INDEPENDENT SET, the canonical $W[1]$-hard problem restricted to regular graphs. Due to space restrictions, we omit some formal definitions and some proofs. See [8] for the definitions of relevant notions in parameterized complexity theory. In the remainder of this paper, if not stated explicitly otherwise, then $b \in \mathbb{N}^+$ is assumed to be arbitrary but fixed.

Transition Systems. A *transition system* (TS, for short) $A = (S, E, \delta, \iota)$ consists of a finite disjoint set S of states, E of events, a partial *transition function* $\delta : S \times E \rightarrow S$ and an *initial state* $\iota \in S$. A TS A is interpreted as edge-labeled directed graph, and every triple $\delta(s,e) = s'$ is considered an e-labeled edge $s \xrightarrow{e} s'$, called *transition*. An event e *occurs* at state s, denoted by $s \xrightarrow{e}$, if $\delta(s,e) = s'$ for some state s'. This notation is extended to words $w' = we$, $w \in E^*, e \in E$, by inductively defining $s \xrightarrow{\varepsilon} s$ for all $s \in S$ and $s \xrightarrow{w'} s''$ if and only if there is a state $s' \in S$ satisfying $s \xrightarrow{w} s'$ and $s' \xrightarrow{e} s''$. If $w \in E^*$, then $s \xrightarrow{w}$ denotes that there is a state $s' \in S$ such that $s \xrightarrow{w} s'$. If $e \in E$, then by $s_i \xrightarrow{(e)^b} s_{i+b}$ we denote that there are distinct states $s_i, s_{i+1}, \dots, s_{i+b-1}, s_{i+b} \in S$ such that $s_i \xrightarrow{a} s_{i+1} \dots s_{i+b-1} \xrightarrow{a} s_{i+b}$. We assume all TSs to be *reachable*: $\forall s \in S, \exists w \in E^* : s_0 \xrightarrow{w} s$.

***b*-Bounded Petri Nets.** A *b-bounded Petri net* (b-net, for short) $N = (P, T, f, M_0)$ consists of finite and disjoint sets of *places* P and *transitions* T, a (total) *flow function* $f : P \times T \rightarrow \{0, \dots, b\}^2$ and an *initial marking* $M_0 : P \rightarrow \{0, \dots, b\}$. If $f(p,t) = (m,n)$, then $f^-(p,t) = m$ and $f^+(p,t) = n$ define the *consuming* and the *producing* effect of t on p, respectively. The *preset* of a place p is defined by ${}^\bullet p = \{t \in T \mid f^+(p,t) > 0\}$ (transitions producing on p) and its *postset* is defined by $p^\bullet = \{t \in T \mid f^-(p,t) > 0\}$ (transitions consuming from p). Accordingly, the *preset* of a transition t is defined by ${}^\bullet t = \{p \in P \mid f^-(p,t) > 0\}$ (places from which t consumes) and its *postset* by

$t^\bullet = \{p \in P \mid f^+(p,t) > 0\}$ (places on which t produces). A b-net N is *pure* if $\forall (p,t) \in P \times T : f^-(p,t) = 0$ or $f^+(p,t) = 0$, that is, $\forall p \in P : {}^\bullet p \cap p^\bullet = \emptyset$. Let $m, n \in \mathbb{N}$. A b-net N is an *(m, n)-T-system* if $\forall p \in P : |{}^\bullet p| \leq m, |p^\bullet| \leq n$.
The firing rule of b-nets defines their behavior: A transition $t \in T$ can *fire* or *occur* in a marking $M : P \to \{0, \dots, b\}$, denoted by $M \xrightarrow{t}$, if $M(p) \geq f^-(p,t)$ and $M(p) - f^-(p,t) + f^+(p,t) \leq b$ for all places $p \in P$. The firing of t in marking M leads to the marking M' if $M'(p) = M(p) - f^-(p,t) + f^+(p,t)$ for all $p \in P$. This is denoted by $M \xrightarrow{t} M'$. Again, this notation extends to sequences $\sigma \in T^*$, and the *reachability set* $RS(N) = \{M \mid \exists \sigma \in T^* : M_0 \xrightarrow{\sigma} M\}$ contains N's reachable markings. The firing rule preserves N's *b-boundedness* by definition: $M(p) \leq b$ for all $p \in P$ and all $M \in RS(N)$. The *reachability graph* of N is the TS $A_N = (RS(N), T, \delta, M_0)$, such that for all $M, M' \in RS(N)$ and all $t \in T$ we define $\delta(M,t) = M'$ if and only if $M \xrightarrow{t} M'$.

b-Bounded Regions. To find a b-net N implementing a TS A, we want to synthesize N's components purely from the input A. Since A and A_N are to be isomorphic, A's events correspond to N's transitions. However, the notion of a *place* is not known for TSs. A *b-bounded region* R (region, for short) of a TS $A = (S, E, \delta, s_0)$ is a pair $R = (sp, sg)$ of *support* $sp : S \to \{0, \dots, b\}$ and *signature* $sg : E \to \{0, \dots, b\}^2$ such that for every edge $s \xrightarrow{e} s'$ of A holds $sp(s) \geq sg^-(e)$ and $sp(s') = sp(s) - sg^-(e) + sg^+(e)$. If $sg(e) = (m, n)$, then $sg^-(e) = m$ and $sg^+(e) = n$ define e's consuming and producing effect (concerning R), respectively.

A region (sp, sg) models a place p and the corresponding part of the flow function f: $sg^+(e)$ models $f^+(e)$, $sg^-(e)$ models $f^-(e)$ and $sp(s)$ models $M(p)$ in the marking $M \in RS(N)$ corresponding to $s \in S(A)$. The *preset* of R is defined by the *producing events* ${}^\bullet R = \{e \in E \mid sg^+(e) > 0\}$ and its *postset* by the *consuming events* $R^\bullet = \{e \in E \mid sg^-(e) > 0\}$. If $sg(e) = (0, 0)$, then e is called *neutral*. The region R is *pure* if ${}^\bullet R \cap R^\bullet = \emptyset$. Let $\mathcal{R}$ be a set of regions of A, and let $e \in E$. By ${}^\bullet e_{\mathcal{R}} = \{(sp, sg) \in \mathcal{R} \mid sg^-(e) > 0\}$ and $e^\bullet_{\mathcal{R}} = \{(sp, sg) \in \mathcal{R} \mid sg^+(e) > 0\}$ we define the *preset* and *postset* of e (concerning $\mathcal{R}$), respectively. The set $\mathcal{R}$ defines the *synthesized b-net* $N_A^{\mathcal{R}} = (\mathcal{R}, E, f, M_0)$ with flow function $f((sp, sg), e) = sg(e)$ and initial marking $M_0((sp, sg)) = sp(s_0)$ for all $(sp, sg) \in \mathcal{R}, e \in E$. We emphasize again that a *region* R of $\mathcal{R}$ is a *place* of $N_A^{\mathcal{R}}$ with the preset ${}^\bullet R$ and the postset $R^\bullet$; every *event* $e \in E$ is a *transition* of $N_A^{\mathcal{R}}$ with preset ${}^\bullet e = {}^\bullet e_{\mathcal{R}}$ and postset $e^\bullet = e_{\mathcal{R}}{}^\bullet$. It is well known that $A_{N_A^{\mathcal{R}}}$ and A are isomorphic if and only if $\mathcal{R}$'s regions solve certain separation atoms [3], to be introduced next.

A pair (s, s') of distinct states of A defines a *state separation atom* (SSP atom, for short). A region $R = (sp, sg)$ *solves* (s, s') if $sp(s) \neq sp(s')$. The region R is to ensure that $N_A^{\mathcal{R}}$ contains at least one place R such that $M(R) \neq M'(R)$ for the markings M and M' corresponding to s and s', respectively. If there is a b-region that solves (s, s'), then s and s' are called *b-solvable* (solvable, for short). If every SSP atom of A is solvable, then A has the *b-state separation property* (SSP for short). If $e \in E$ and $s \in S$ such that e does not occur at s ($\neg s \xrightarrow{e}$), then the pair (e, s) is an *event state separation atom* (ESSP atom, for short). A b-region $R = (sp, sg)$ *solves* (e, s) if $sg^-(e) > sp(s)$ or $sp(s) - sg^-(e) + sg^+(e) > b$. The meaning of R is to ensure that there is at least one place R in $N_A^{\mathcal{R}}$ such that $\neg M \xrightarrow{e}$ for the marking M corresponding to s. If there is a region that solves (e, s), then e and s are called *b-solvable*; we also say e is solvable at s. If every ESSP atom of A is b-solvable, then A has the *b-event state separation property* (ESSP, for short).

A set $\mathcal{R}$ of regions of A is called *b-admissible* if for every of A's (E)SSP atoms there is a region R in $\mathcal{R}$ that solves it. The following lemma, borrowed from [3, p. 163], summarizes the connection between b-admissible sets of A and synthesis:

Lemma 1 ([3]). *A b-net N has a reachability graph isomorphic to a given TS A if and only if there is a b-admissible set $\mathcal{R}$ of A such that $N = N_A^{\mathcal{R}}$.*

We say a b-net N *solves* A if A_N and A are isomorphic. By Lemma 1, searching for a restricted b-net reduces to finding a b-admissible set of accordingly restricted regions. The following example illustrates this fact.

Example 1. Let $m, n \in \mathbb{N}$, A be a TS and $\mathcal{R}$ be a b-admissible set of pure regions of A. If every region $R \in \mathcal{R}$ satisfies $|{}^\bullet R| \leq m$ and $|R^\bullet| \leq n$, then $N_A^{\mathcal{R}}$ is a pure (m, n)-T-system solving A. In particular, if $b = 2$, then the TS $A = s_0 \xrightarrow{e_1} s_1 \xrightarrow{e_2} s_2$ has the following pure regions:

i	$sp_i(s_1)$	$sp_i(s_2)$	$sp_i(s_3)$	$sg_i(e_1)$	$sg_i(e_2)$	i	$sp_i(s_1)$	$sp_i(s_2)$	$sp_i(s_3)$	$sg_i(e_1)$	$sg_i(e_2)$
1	2	0	0	$(2, 0)$	$(0, 0)$	3	0	2	0	$(0, 2)$	$(2, 0)$
2	0	2	0	$(0, 2)$	$(2, 0)$	4	0	1	2	$(0, 1)$	$(0, 1)$

The set $\mathcal{R} = \{(sp_i, sg_i) \mid 1 \leq i \leq 4\}$ is 2-admissible. Since ${}^\bullet(sp_4, sg_4) = \{e_1, e_2\}$, the solving 2-net $N_A^{\mathcal{R}}$ is not a $(1, 1)$-T-system. However, the set $\mathcal{R}' = \{(sp_i, sig_i) \mid 1 \leq i \leq 3\}$ is 2-admissible, and $N_A^{\mathcal{R}'}$ is a $(1, 1)$-T-system solving A.

3 $W[1]$-Hardness Parameterized by $m + n$

This section is dedicated to the proof of our main result:

Theorem 1. (m, n)-SYNTHESIS *parameterized by $m + n$ is $W[1]$-hard.*

The proof of Theorem 1 consists of a parameterized reduction of REGULAR INDEPENDENT SET to (m, n)-SYNTHESIS. Let (G, k) be an instance of REGULAR INDEPENDENT SET. That is, $G = (V(G), E(G))$ is a graph with set of nodes $V(G) = \{v_1, \ldots, v_n\}$, set of edges $E(G) = \{a_1, \ldots, a_m\}$, and there is an integer $r \in \mathbb{N}$ such that for every node $v \in V(G)$ holds $|\{e \in E(G) \mid v \in e\}| = r$, and k is a positive integer. We reduce (G, k) to an instance $(A, 2rk + 20, 2rk + 20)$ of (m, n)-SYNTHESIS, parameterized by $m + n$, such that G has a k-independent set if and only if A is solvable by a $(2rk + 20, 2rk + 20)$-T-system.

To represent G, the TS A has for every edge $a_i = \{v_{i,1}, v_{i,2}\}$, $i \in \{1, \ldots, m\}$, the following gadget G_i, which uses $a_i, v_{i,1}$ and $v_{i,2}$ as events:

$$\begin{array}{l} g_{i,1} \xrightarrow{\delta_i^1} g_{i,2} \xrightarrow{\zeta_{i,1}^1} g_{i,3} \xrightarrow{(v_{i,1})^b} g_{i,b+3} \xrightarrow{\zeta_{i,2}^1} g_{i,b+4} \xrightarrow{(v_{i,2})^b} g_{i,2b+4} \xrightarrow{\zeta_{i,3}^1} g_{i,2b+5} \xrightarrow{a_i} g_{i,2b+6} \\ \qquad\quad \alpha_1 \downarrow \\ \qquad\quad g_{i,2b+7} \end{array}$$

Let $i \in \{1, \ldots, m\}$. The proof of the *if*-direction bases on the idea to ensure that if A is solvable, then there is a pure region $R = (sp, sg)$ that satisfies the following conditions. Firstly, $sg(\alpha_1) = (b, 0)$, which implies $sp(g_{i,2}) = b$. Secondly, the producing effect of the node events is zero, that is, $sg^+(v_{i,1}) = sg^+(v_{i,2}) = 0$. Thirdly, the ζ-events are neutral, that is, $sg(\zeta_{i,1}^1) = sg(\zeta_{i,2}^1) = sg(\zeta_{i,3}^1) = (0,0)$. As a result, the support value of $g_{i,2b+5}$ is given by $sp(g_{i,2b+5}) = b - b \cdot (sg^-(v_{i,1}) + sg^-(v_{i,2}))$. Moreover, if $sp(g_{i,2b+5}) < b$, then there is *exactly* one $e \in \{v_{i,1}, v_{i,2}\}$ such that $sig^-(e) > 0$. Otherwise we would have the contradiction $sp(g_{i,2b+5}) < 0$. Furthermore, the region R ensures that there are exactly rk edge events with a positive producing effect. That is, there are exactly rk indices $i_1, \ldots, i_{rk} \in \{1, \ldots, m\}$ such that $sg^+(a_{i_j}) > 0$ for all $j \in \{1, \ldots, rk\}$. Since R is pure, this implies $sg^-(a_{i_j}) = 0$ for all $j \in \{1, \ldots, rk\}$. Moreover, by $g_{i_j,2b+5} \xrightarrow{a_{i_j}} g_{i_j,2b+6}$, we obtain $sup(g_{i_j,2b+6}) = sup(g_{i_j,2b+5}) + sig^+(a_{i_j})$. This requires $sp(g_{i_j,2b+5}) < b$, and exactly one of $v_{i_j,1}$ and $v_{i_j,2}$ has a positive consuming effect. The region R ensures that there are *exactly* k node events $v_{\ell_1}, \ldots, v_{\ell_k}$ with a positive consuming effect. Recall, for every node $v \in V(G)$ holds $|\{e \in E(G) \mid v \in e\}| = r$. Thus, if $v_{\ell_1}, \ldots, v_{\ell_k}$ are not independent, then the number of edges which are adjacent to a node of $v_{\ell_1}, \ldots, v_{\ell_k}$ is at most $rk - 1$. Since rk edge events have a positive producing effect, and each of it needs a consuming node, this is a contradiction. Consequently, the set $I = \{v \in V(G) \mid sg^-(v) > 0\}$ defines a k-independent set of G.

For the *only-if*-direction we show that if G has a k-independent set then there is a b-admissible set of regions $\mathcal{R}$ such that $|{}^\bullet R|, |R^\bullet| \leq 2rk + 20$ for all $R \in \mathcal{R}$. The major challenge here is to keep the number of consuming and producing events of solving regions smaller than the parameter. To do so, we exploit G's regularity and the δ- and ζ-events. In what follows, we prove the following lemma:

Lemma 2. *1. If A is solvable, then there is a region (sp, sg) such that the following conditions are true:*

(a) $sg(\alpha_1) = (b, 0)$ and $sg(\zeta_{i,1}^1) = \cdots = sg(\zeta_{i,3}^1) = (0,0)$ for all $i \in \{1, \ldots, m\}$.

(b) If $e \in \{a_1, \ldots, a_m\}$ then $sg^-(e) = 0$ and there are exactly rk events $a_{i_1}, \ldots, a_{i_{rk}} \in \{a_1, \ldots, a_m\}$ with $sg^+(a_{i_j}) > 0$, where $j \in \{1, \ldots, rk\}$.

(c) If $e \in \{v_1, \ldots, v_n\}$ then $sg^+(e) = 0$. Furthermore, there are exactly k events $v_{i_1}, \ldots, v_{i_k} \in \{v_1, \ldots, v_n\}$ with $sg^-(v_{i_j}) > 0$ for all $j \in \{1, \ldots, k\}$.

2. If G has an independent set of size k then there is a b-admissible set $\mathcal{R}$ of A such that $|{}^\bullet R|, |R^\bullet| \leq 2rk + 20$ for all $R \in \mathcal{R}$.

3.1 The Proof of Lemma 2.1

This section introduces the gadgets that ensure Lemma 2.1. For now, we refrain from explaining in which way they are actually conjunct to build A. This conjunction is postponed to Sect. 3.2, which is dedicated to Lemma 2.2.. We let events $e \in E(A)$ occur b times in row to restrict their possible signature in advance:

Lemma 3. *Let A be a TS, and let $e \in E(A)$ be an event that occurs b times in a row: $s_1 \xrightarrow{(e)^b} s_{b+1} \in A$. For any pure region (sp, sg) of A with $sg^+(e) \neq sg^-(e)$ holds either $sg(e) = (1,0)$, $sp(s_1) = b$ and $sp(s_{b+1}) = 0$ or $sg(e) = (0,1)$, $sp(s_1) = 0$ and $sp(s_{b+1}) = b$.*

Proof. The claim follows by $b \geq sp(s_{b+1}) = sp(s_1) + b \cdot (sg^+(e) - sg^-(e)) \geq 0$.

The TS A has for $i \in \{1,2,3\}$ the following w-maker gadget X_i:

$$x_{i,1} \xrightarrow{\delta_i^{10}} x_{i,2} \xrightarrow{(\alpha)^b} x_{i,b+2} \xrightarrow{\zeta} x_{i,b+3} \xrightarrow{w_i} x_{i,b+4} \xrightarrow{(\alpha)^b} x_{i,2b+4}$$

If A is solvable, then there is a region $R = (sp, sg)$ that solves the atom $(\alpha, x_{1,b+3})$, that is, $sg^-(\alpha) > sp(x_{1,b+3})$ or $sp(x_{1,b+3}) - sg^-(\alpha) + sg^+(\alpha) > b$. This implies $sg(\alpha) \neq (0,0)$. Thus, by Lemma 3, we have $sg(\alpha) \in \{(1,0),(0,1)\}$. Since our arguments are symmetrically true for the case $sg(\alpha) = (0,1)$, we assume $sg(\alpha) = (1,0)$ and show that this implies a k-independent set of G.

Since R solves $(\alpha, x_{i,b+3})$, by $sg(\alpha) = (1,0)$, we conclude $sg^-(\alpha) > sp(x_{i,b+3}) = 0$. Moreover, by Lemma 3, we obtain $sp(x_{i,b+2}) = 0$ and $sp(x_{i,b+4}) = b$ for all $i \in \{1,2,3\}$. Furthermore, by $sp(x_{1,b+2}) = sp(x_{1,b+3}) = 0$, we get $sg(\zeta) = (0,0)$. By $sp(x_{i,b+2}) = 0$, this implies $sp(x_{i,b+3}) = 0$ for all $i \in \{2,3\}$. Finally, by $sp(x_{i,b+3}) = 0$ and $sp(x_{i,b+4}) = b$ for all $i \in \{1,2,3\}$, we get the three producing w-events w_1, w_2, w_3: $sg(w_1) = sg(w_2) = sg(w_3) = (0,b)$.

The TS A has for $i \in \{1,\dots,9\}$ a so called α-maker Y_i that uses w_1 and w_2 to manipulate the support of some states and provides the consuming α-event α_i:

$$y_{i,1} \xrightarrow{\delta_i^{11}} y_{i,2} \xrightarrow{w_1} y_{i,3} \xrightarrow{\alpha_i} y_{i,4} \xrightarrow{w_2} y_{i,5}$$

By $sg(w_1) = sg(w_2) = (0,b)$, we have $sp(y_{i,3}) = b$ and $sp(y_{i,4}) = 0$. This implies $sg(\alpha_i) = (b,0)$ for $i \in \{1,\dots,9\}$. The events $\alpha_1,\dots,\alpha_9$ are applied to manipulate the support of some states. For example, by $sg(\alpha_1) = (b,0)$ and $g_{i,2} \xrightarrow{\alpha_1}$, we have $sp(g_{i,2}) = b$ for all $i \in \{1,\dots,m\}$ as discussed before. The following β-makers also exemplify the functionality of the α-events.

The TS A has for every $i \in \{1,\dots,5\}$ the following β-maker Z_i that uses the events α_7 and α_8 to provide the producing β-event β_i:

$$z_{i,1} \xrightarrow{\delta_i^{12}} z_{i,2} \xrightarrow{\alpha_7} z_{i,3} \xrightarrow{\beta_i} z_{i,4} \xrightarrow{\alpha_8} z_{i,5}$$

In particular, by $sg(\alpha_7) = sg(\alpha_8) = (b,0)$, we get $sp(z_{i,3}) = 0$ and $sp(z_{i,4}) = b$. This implies $sg(\beta_i) = (0,b)$ for all $i \in \{1,\dots,5\}$. Just like the α-events, the β-events serve to manipulate the support of some states.

In the remainder of this section, we first introduce the gadgets ensuring that $R = (sp, sg)$ selects exactly rk edge events $a_{i_1}, \dots, a_{i_{rk}}$ such that $sig^+(a_{i_j}) > 0$ for all $j \in \{1, \dots, rk\}$. Secondly, we introduce the gadgets that ensure that there are exactly k node events $v_{\ell_1}, \dots, v_{\ell_k}$ such that $sig^-(v_{\ell_j}) > 0$ for all $j \in \{1, \dots, k\}$. Similar to the already presented gadgets $G_1, \dots, G_m$, these gadgets apply ζ-events, that is, elements of the set $Z = \{\zeta^i_{j,\ell} \mid i, j, \ell \in \mathbb{N}\}$. For the region R, corresponding to Lemma 2.1, these events have to be neutral. For the proof of Lemma 2.2. they allow solving regions with small preset- and postset-cardinality. If $\zeta^i_{j,\ell} \in Z \cap E(A)$, that is, $\zeta^i_{j,\ell}$ actually occurs in A, then A has the following ζ-makers $\bigominus^i_{j,\ell}$ (left) and $\bigoplus^i_{j,\ell}$ (right). These gadgets ensure $\zeta^i_{j,\ell}$'s neutrality:

$$\ominus^i_{j,\ell,1} \xrightarrow{\delta^{13}_{i,j,\ell}} \ominus^i_{j,\ell,2}, \quad \ominus^i_{j,\ell,2} \xrightarrow{\zeta^i_{j,\ell}} \ominus^i_{j,\ell,4}, \quad \ominus^i_{j,\ell,2} \xrightarrow{w_3} \ominus^i_{j,\ell,3}$$

$$\oplus^i_{j,\ell,1} \xrightarrow{\delta^{14}_{i,j,\ell}} \oplus^i_{j,\ell,2}, \quad \oplus^i_{j,\ell,2} \xrightarrow{\zeta^i_{j,\ell}} \oplus^i_{j,\ell,4}, \quad \oplus^i_{j,\ell,2} \xrightarrow{\alpha_9} \oplus^i_{j,\ell,3}$$

By $sg(w_3) = (0, b)$ and $sg(\alpha_9) = (b, 0)$, we get $sp(\ominus^i_{j,\ell,2}) = 0$ and $sp(\oplus^i_{j,\ell,2}) = b$. Moreover, by $0 = sp(\ominus^i_{j,\ell,2}) \geq sg^-(\zeta^i_{j,\ell})$, we obtain $sg^-(\zeta^i_{j,\ell}) = 0$. Finally, by $b \geq sp(\oplus^i_{j,\ell,4}) = sp(\oplus^i_{j,\ell,2}) - sg^-(\zeta^i_{j,\ell}) + sg^+(\zeta^i_{j,\ell})$, implying $b \geq b + sg^+(\zeta^i_{j,\ell})$, we get $sg^+(\zeta^i_{j,\ell}) = 0$.

So far, we have introduced A's gadgets that yield us the α-, β- and ζ-events with the following behavior: If $s \xrightarrow{\alpha}$, then $sp(s) = b$; if $s \xrightarrow{\beta}$, then $sp(s) = 0$; if $s \xrightarrow{\zeta} s'$, then $sp(s) = sp(s')$. These events are applied in the subsequently introduced gadgets, which collaborate to provide the announced behavior of A.

The TS A has for every edge event a_i, $i \in \{1, \dots, m\}$, exactly rk *edge copies* (e-copies, for short) $a_i^1, \dots, a_i^{rk}$. These copies are used to enable the announced selection of rk edge events $a_{i_1}, \dots, a_{i_{rk}}$. To achieve this goal, it is necessary that edge events do not consume and e-copies do not produce. The TS A has for every $i \in \{1, \dots, m\}$ an *edge noCon* C_i. This gadget ensures that a_i does *not consume*. Moreover, for all $i \in \{1, \dots, m\}$ and all $j \in \{1, \dots, rk\}$ it has an *e-copy noPro* $D_{i,j}$. This gadget guarantees that a_i^j does *not produce*.

$$c_{i,1} \xrightarrow{\delta^2_i} c_{i,2} \xrightarrow{\zeta^2_{i,1}} c_{i,3} \xrightarrow{a_i} c_{i,4}, \quad c_{i,2} \xrightarrow{\beta_1} c_{i,5}$$

The edge noCon C_i.

$$d_{i,j,1} \xrightarrow{\delta^3_{i,j}} d_{i,j,2} \xrightarrow{\zeta^3_{i,j}} d_{i,j,3} \xrightarrow{a_i^j} d_{i,j,4}, \quad d_{i,j,2} \xrightarrow{\alpha_2} d_{i,j,5}$$

The e-copy noPro $D_{i,j}$.

By $sg(\beta_1) = (0, b)$ and $sg(\zeta^2_{i,1}) = (0, 0)$, we have $sp(c_{i,3}) = 0$. Since $sp(c_{i,3}) \geq sg^-(a_i)$, this implies $sg^-(a_i) = 0$. Similarly, by $sg(\alpha_2) = (b, 0)$ and $sg(\zeta^3_{i,j}) = (0, 0)$, we obtain $sp(d_{i,j,3}) = b$. The region R is pure. Thus, if $sg^+(a_i^j) > 0$ then $sg^-(a_i^j) = 0$. This implies $sp(d_{i,j,4}) = b + sg^+(a_i^j) > b$, a contradiction. Hence, $sg^+(a_i^j) = 0$ is true.

The region R selects for every $j \in \{1, \dots, rk\}$ exactly one $i \in \{1, \dots, m\}$ such that the e-copy a_i^j has a positive consuming effect, that is, $sig^-(a_i^j) > 0$.

The other e-copies remain neutral. To achieve this, the TS A uses for every $j \in \{1, \ldots, rk\}$ the *edge selector* F_j. The gadget F_j applies the events $a_1^j, \ldots, a_m^j$, that is, the j-th copy of every edge event $a_1, \ldots, a_m$. On F_j, every a_i^j occurs b times consecutively. Separated by ζ-events, these occurrences $(a_1^j)^b, \ldots, (a_m^j)^b$ are placed in a sequence. We abridge $\ell = (m-1)(b+1)$ and define F_j:

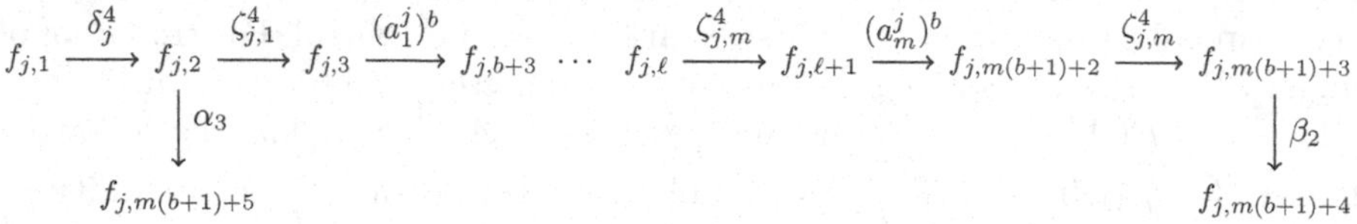

$$
\begin{array}{ccccccccccccccc}
f_{j,1} & \xrightarrow{\delta_j^4} & f_{j,2} & \xrightarrow{\zeta_{j,1}^4} & f_{j,3} & \xrightarrow{(a_1^j)^b} & f_{j,b+3} & \cdots & f_{j,\ell} & \xrightarrow{\zeta_{j,m}^4} & f_{j,\ell+1} & \xrightarrow{(a_m^j)^b} & f_{j,m(b+1)+2} & \xrightarrow{\zeta_{j,m}^4} & f_{j,m(b+1)+3} \\
 & & \downarrow{\scriptstyle \alpha_3} & & & & & & & & & & & & \downarrow{\scriptstyle \beta_2} \\
 & & f_{j,m(b+1)+5} & & & & & & & & & & & & f_{j,m(b+1)+4}
\end{array}
$$

By $sg(\alpha_3) = (b, 0)$, we have $sp(f_{j,2}) = b$ and, by $sg(\beta_2) = (0, b)$, we have $sp(f_{j,m(b+1)+3}) = 0$. The ζ-events are neutral, and $sg^+(a_i^j) = 0$ for all $i \in \{1, \ldots, m\}$. Thus, we obtain $0 = \sum_{i=1}^{m} b \cdot sg^-(a_i^j) < b$. Consequently, there is an $i \in \{1, \ldots, m\}$ such that $sg^-(a_i^j) = 1$, and $sg^-(a_{i'}^j) = 0$ for all $i' \in \{1, \ldots, m\} \setminus \{i\}$. The following edge connectors complete the set of A's gadgets that allow the selection of rk edges $a_{i_1}, \ldots, a_{i_{rk}}$.

The TS A has for all $i \in \{1, \ldots, m\}$ a so called *edge connector* H_i whose purpose is twofold. On the one hand, it ensures that the edge selectors never choose two consuming copies of the same edge event, that is, if $j \neq j'$, $sg^-(a_i^j) > 0$ and $sg^-(a_{i'}^{j'}) > 0$, then $i \neq i'$. On the other hand, $sg^+(a_i) > 0$ if and only if there is a $j \in \{1, \ldots, rk\}$ such that $sg^-(a_i^j) > 0$. Since $F_1, \ldots, F_{rk}$ select rk consuming edge copies, this picks out exactly rk edges $a_{i_1}, \ldots, a_{i_{rk}}$ with a positive producing effect. The gadget H_i applies the event a_i and its rk copies. Separated by ζ-events, two sequences of a_i's copies $a_i^1, \ldots, a_i^{rk}$, each of if it occurring b times consecutively, embrace the event a_i. For readability, we abridge $\ell = (b+1)rk + 2$ and define H_i as follows:

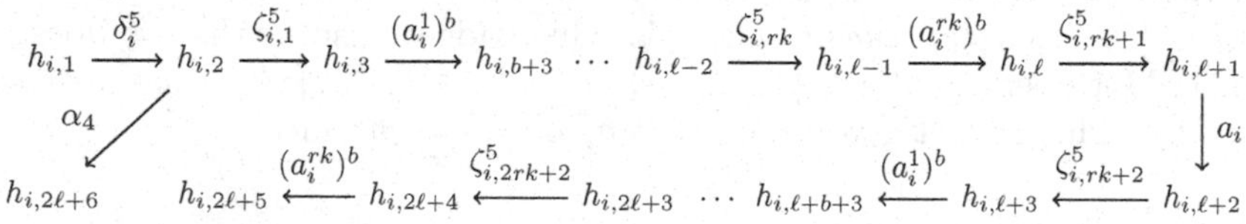

$$
\begin{array}{ccccccccccccccc}
h_{i,1} & \xrightarrow{\delta_i^5} & h_{i,2} & \xrightarrow{\zeta_{i,1}^5} & h_{i,3} & \xrightarrow{(a_i^1)^b} & h_{i,b+3} & \cdots & h_{i,\ell-2} & \xrightarrow{\zeta_{i,rk}^5} & h_{i,\ell-1} & \xrightarrow{(a_i^{rk})^b} & h_{i,\ell} & \xrightarrow{\zeta_{i,rk+1}^5} & h_{i,\ell+1} \\
 & {\scriptstyle \alpha_4} \swarrow & & & & & & & & & & & & & \downarrow{\scriptstyle a_i} \\
h_{i,2\ell+6} & & h_{i,2\ell+5} & \xleftarrow{(a_i^{rk})^b} & h_{i,2\ell+4} & \xleftarrow{\zeta_{i,2rk+2}^5} & h_{i,2\ell+3} & \cdots & h_{i,\ell+b+3} & \xleftarrow{(a_i^1)^b} & h_{i,\ell+3} & \xleftarrow{\zeta_{i,rk+2}^5} & h_{i,\ell+2} & &
\end{array}
$$

By $sg(\alpha_4) = (b, 0)$ it is $sp(h_{i,2}) = b$. The ζ-events are neutral, and $sg^+(a_i^j) = 0$ for all $j \in \{1, \ldots, rk\}$. Thus, it is $sp(h_{(b+1)rk+3}) = b - \sum_{j=1}^{rk} b \cdot sg^-(a_i^j)$, and, by $sp(h_{(b+1)rk+3}) \geq 0$, there is at most one $j \in \{1, \ldots, rk\}$ such that $sg^-(a_i^j) > 0$. Consequently, two copies of the same edge event are never selected by the edge selectors. By Lemma 3, if $sg^-(a_i^j) > 0$, then $sg^-(a_i^j) = 1$. This implies $sp(h_{(b+1)rk+3}) = 0$. Furthermore, a_i^j occurs again b times in a row "after" the occurrence of a_i at $h_{i,(rk+j-1)(b+1)+5}$. This implies $sp(h_{i,(rk+j-1)(b+1)+5}) = b$. Since no edge copy produces, $sg(a_i) = (0, b)$ is immediately implied. Conversely, if $sg^+(a_i) > 0$, then $sp(h_{(b+1)rk+3}) < b$. Thus, by $sp(h_{(b+1)rk+3}) = b - \sum_{j=1}^{rk} b \cdot sg^-(a_i^j)$, there is a consuming copy of a_i. Consequently, $sg^+(a_i) > 0$ if and only if $sg(a_i) = (0, b)$ and there is exactly one $j \in \{1, \ldots, rk\}$ such that $sg^-(a_i^j) = 1$.

So far we have argued that there are exactly rk distinct indices $i_1,\dots,i_{rk} \in \{1,\dots,m\}$ such that $sg(a_{i_1}) = \dots = sg(a_{i_{rk}}) = (0,b)$. Moreover, $sg(a_i) = (0,0)$ for all $i \in \{1,\dots,m\} \setminus \{i_1,\dots,i_{rk}\}$. It remains to argue that these rk "edges" $a_{i_1},\dots,a_{i_{rk}}$ are covered by exactly k "nodes". To achieve this goal, the TS A uses gadgets that work symmetrically to the ones used for the selection of the edges. So called *node noPros* ensure that the node events $v_1,\dots,v_n$ do *not produce*. Moreover, the TS A applies for all $i \in \{1,\dots,n\}$ k *node-copies* $v_i^1,\dots,v_i^k$ and uses *n-copy noCons* to prevent them from consuming. Furthermore, *node selectors* force exactly k node copies to have a producing signature. The *node connectors* ensure that two copies of the same node are never selected and connects a producing node copy v_i^j with its, then consuming, node event v_i. Finally, exactly k nodes $v_{\ell_1},\dots,v_{\ell_k}$ consume. Since these gadgets work symmetrically to the ones for the edges, we only briefly prove their functionality.

The TS A has for $i \in \{1,\dots,n\}$ the so called *node noPro* P_i (left hand side) and for $i \in \{1,\dots,n\}$ and $j \in \{1,\dots,k\}$ the so called *n-copy antiCon* $Q_{i,j}$ (right hand side) which are defined as follows:

$$p_{i,1} \xrightarrow{\delta_i^6} p_{i,2} \xrightarrow{\zeta_{i,1}^6} p_{i,3} \xrightarrow{v_i} p_{i,4}, \qquad p_{i,2} \xrightarrow{\alpha_5} p_{i,5}$$

$$q_{i,j,1} \xrightarrow{\delta_{i,j}^7} q_{i,j,2} \xrightarrow{\zeta_{i,j}^7} q_{i,j,3} \xrightarrow{v_i^j} q_{i,j,4}, \qquad q_{i,j,2} \xrightarrow{\beta_3} q_{i,j,5}$$

By $sg(\alpha_5) = (b,0)$ and $sg(\zeta_{i,1}^6) = (0,0)$, we get $sp(p_{i,3}) = b$ which implies $sg^+(v_i) = 0$. Moreover, by $sg(\beta_3) = (0,b)$ and $sg(\zeta_{i,j}^7) = (0,0)$, we get $sp(q_{i,j,3}) = 0$ which implies $sg^-(v_i^j) = 0$.

The TS A has for every $j \in \{1,\dots,k\}$ a *node selector* T_j. On T_j, separated by ζ-events, the j-th copy of every node event $v_1,\dots,v_n$ occurs b times in a row. We abridge $\ell = (n-1)(b+1)+2$ and define T_j as follows:

$$t_{j,1} \xrightarrow{\delta_j^8} t_{j,2} \xrightarrow{\zeta_{j,1}^8} t_{j,3} \xrightarrow{(v_1^j)^b} t_{j,b+3} \cdots t_{j,\ell} \xrightarrow{\zeta_{j,n}^8} t_{j,\ell+1} \xrightarrow{(v_n^j)^b} t_{j,n(b+1)+2} \xrightarrow{\zeta_{j,n+1}^8} t_{j,n(b+1)+3}$$

$$t_{j,2} \xrightarrow{\beta_4} t_{j,n(b+1)+5}, \qquad t_{j,n(b+1)+3} \xrightarrow{\alpha_6} t_{j,n(b+1)+4}$$

By $sg(\beta_4) = (0,b)$, $sg(\alpha_6) = (b,0)$, the neutrality of the ζ-events and $sg^-(v_1^j) = \dots = sg^-(v_n^j) = 0$, we have $b = sp(t_{j,n(b+1)+3}) = \sum_{i=1}^n b \cdot sg^+(v_i^j) > sp(t_{j,3}) = 0$. Thus, there is exactly one producing j-th copy produces and others are neutral.

Finally, the TS A has for every $i \in \{1,\dots,n\}$ a so called *node connector* U_i that, among others, applies the β-event β_5, the k copies $v_i^1,\dots,v_i^k$ of v_i and the event v_i. We abridge $\ell = (k-1)(b+1)+2$ and define U_i as follows:

$$u_{i,1} \xrightarrow{\delta_i^9} u_{i,2} \xrightarrow{\zeta_{i,1}^9} u_{i,3} \xrightarrow{(v_i^1)^b} u_{i,b+3} \cdots u_{i,\ell} \xrightarrow{\zeta_{i,k}^9} u_{i,\ell+1} \xrightarrow{(v_i^k)^b} u_{i,k(b+1)+2} \xrightarrow{\zeta_{i,k+1}^9} u_{i,k(b+1)+3}$$

$$u_{i,2} \xrightarrow{\beta_5} u_{i,k(b+1)+5}, \qquad u_{i,k(b+1)+3} \xrightarrow{v_i} u_{i,k(b+1)+4}$$

By $sg(\beta_5) = (0, b)$, the neutrality of the ζ-events and $sg^-(v_i^j) = 0$ for all $j \in \{1, \dots, k\}$, it holds $b \geq sp(u_{j,k(b+1)+3}) = \sum_{j=1}^{k} b \cdot sg^+(v_i^j)$. Thus, at most one node-copy v_i^j, $j \in \{0, \dots, k\}$, of v_i is not neutral. In particular, two copies of the same node are never selected by the node selectors. Moreover, if $sg^-(v_i) > 0$, then $sp(u_{j,k(b+1)+3}) > 0$. Consequently, if v_i consumes, then there is a producing copy v_i^j. Since there are at most k producing node copies, there are at most k consuming nodes $v_{\ell_1}, \dots, v_{\ell_k}$. Thus, the rk producing events $a_{i_1}, \dots, a_{i_{rk}}$ are "covered" by exactly k consuming events $v_{\ell_1}, \dots, v_{\ell_k}$. Altogether, this proves that $I = \{v \in V(G) \mid sg^-(v) > 0\}$ defines an independent set of size k of G.

3.2 The Proof of Lemma 2.2

Table 1. The gadgets of A and their corresponding γ-events.

Gadget	G_i	C_i	$D_{i,j}$	F_j	H_i	P_i	$Q_{i,j}$	T_j	U_i	X_i	Y_i	Z_i	$\bigoplus^i_{j,\ell}$	$\bigominus^i_{j,\ell}$
γ-event	γ_i^1	γ_i^2	$\gamma_{i,j}^3$	γ_j^4	γ_i^5	γ_i^6	$\gamma_{i,j}^7$	γ_j^8	γ_i^9	γ_i^{10}	γ_i^{11}	γ_i^{12}	$\gamma_{i,j,\ell}^{13}$	$\gamma_{i,j,\ell}^{14}$

The reduction merges the introduced gadgets to a directed labelled binary tree with initial state $\iota = g_{1,1}$. The resulting TS A consists of 14 blocks, cf. Figure 1. The TS A has for each of its gadgets a γ-event in accordance to Table 1. Using these events, the joining connects the "initial states" of the gadgets as follows:

$$g_{1,1} \xrightarrow{\gamma_1^1} \cdots \xrightarrow{\gamma_{m-1}^1} g_{m,1} \xrightarrow{\gamma_m^1} c_{1,1} \xrightarrow{\gamma_1^2} \cdots \xrightarrow{\gamma_{m-1}^2} c_{m,1} \xrightarrow{\gamma_m^2} d_{1,1,1} \xrightarrow{\gamma_{1,1}^3} \cdots \xrightarrow{\gamma_{1,rk-1}^3} d_{1,rk,1} \xrightarrow{\gamma_{1,rk}^3} d_{2,1,1}$$

$$d_{2,1,1} \xrightarrow{\gamma_{2,1}^3} \cdots \xrightarrow{\gamma_{m,rk-1}^3} d_{m,rk,1} \xrightarrow{\gamma_{m,rk}^3} f_{1,1} \xrightarrow{\gamma_{1,1}^4} \cdots \xrightarrow{\gamma_{n,k-1}^9} \ominus_{n,k,1}^9 \xrightarrow{\gamma_{n,k}^9} \ominus_{n,k+1,1}^9$$

The γ-events $\gamma_{i,j,\ell}^h$, where indices that are 0 are omitted, occur "lexicographically" ordered by $hij\ell$ in accordance to the canonical order on the natural numbers. This defines also an order on the gadgets and makes the conjunction unambiguous.

Due to space restrictions, most of the proof of Lemma 2.2. is omitted. However, the following lemma states the solvability of α and $v_1, \dots, v_n$ and exemplifies in which way A allows regions that respect the parameter.

Lemma 4. *If (G, k) is a yes-instance of* Regular Independent Set *then the events α and $v_1, \dots, v_n$ are solvable by regions that respect the parameter $4rk + 40$.*

Proof. For the sake of space restrictions, we implicitly define solving regions $R_i = (sp_i, sg_i)$ by $sp_i(\iota)$ and sg_i, to be seen in Table 2: The ι-column shows $sp(\iota)$. The event sets occur in the column in accordance to the signature of their elements.

For example, $sp_1(e) = (b, 0)$ for $e \in \{\alpha_1, \ldots, \alpha_9\}$. Moreover, if $e \in E(A)$ does not occur in any presented set corresponding to R_i, then $sg_i(e) = (0, 0)$. In particular, all signatures get along with $(b, 0), (1, 0), (0, 0), (0, 1), (0, b)$. By $sp_i(s') = sp_i(s) - sg_i^-(e) + sp_i^+(e)$ for all $s \xrightarrow{e} s' \in A$, this defines R_i completely.

Solving α: Let $r \in \mathbb{N}^+$ such that every node of G has degree r, and let $I = \{v_{\ell_1}, \ldots, v_{\ell_k}\}$ be a k-independent set of G. The nodes $v_{\ell_1}, \ldots, v_{\ell_k}$ are independent, and each of it has exactly r adjacent edges. Thus, there are exactly rk edges $a_{i_1}, \ldots, a_{i_{rk}} \in E(G)$ such that for all $a \in E(G)$ the following is true. If $a \in E_I = \{a_{i_1}, \ldots, a_{i_{rk}}\}$, then $|a \cap I| = 1$ and otherwise $|a \cap I| = 0$. Using I and E_I, we define region R_1 in accordance to Table 2. If we follow the arguments for the proof of Lemma 2.1, then it is easy to see that R_1 is well defined and solves α at $x_{i,b+2}, x_{i,b+3}$ and $x_{i,2b+4}$ for all $i \in \{1, 2, 3\}$. Moreover, $|{}^\bullet R_1| \leq k(r+1) + 13$ and $|R_1^\bullet| \leq k(r+1) + 11$, cf. Table 2. Thus, the region R_1 respects the parameter. Notice that the latter is possible by grouping "similar" gadgets into blocks. For example, if the node noPros alternated with the node selectors $(P_1, T_1, \ldots, P_n, T_n)$, then the number of consuming and producing γ-events would depend on $|V(G)|$ and would not respect the parameter. The region R_2 of Table 2 solves α at the remaining states of A and respects the parameter.

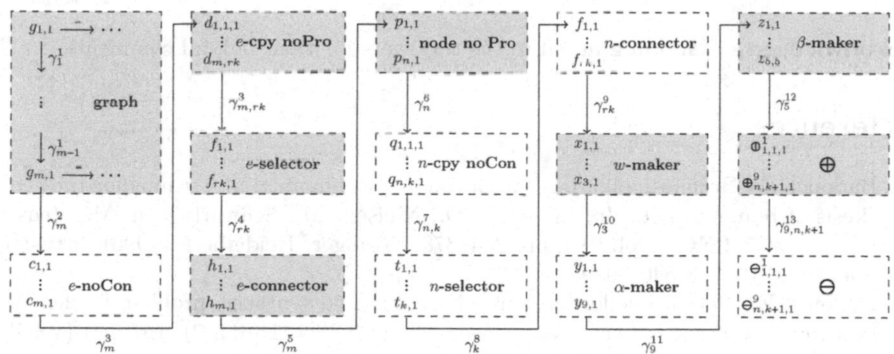

Fig. 1. The gadgets' conjunction to finally build A, consisting of "blocks" in accordance to similar "gadget-types". The red colored areas mark the gadgets whose initial states are mapped to b by R_1 (Table 2) solving $(\alpha, x_{i,b+2}), (\alpha, x_{i,b+3}), (\alpha, x_{i,2b+4})$ for all $i \in \{1, 2, 3\}$. (Color figure online)

Solving v_i, $i \in \{1, \ldots, n\}$: The Region R_3 solves v_i at all states except the sinks of the affected ζ-events. The region R_4 solves v_i at these remaining sinks. The event v_i occurs in $G_{i_1}, \ldots, G_{i_r}$, P_i and U_i. Thus, $|{}^\bullet R_3| \leq r + 2$, $|{}^\bullet R_4| \leq r$, $|R_3^\bullet|$ and $|R_4^\bullet| \leq 1$.

Table 2. Implicitly defined regions of A that solve α and v_i for all $i \in \{1, \ldots, n\}$.

R	ι	$(b, 0)$	$(1, 0)$	$(0, 1)$	$(0, b)$
R_1	b	$\{\alpha_1, \ldots, \alpha_9\}$, $\{\gamma_m^2, \gamma_n^6, \gamma_3^{10}, \gamma_{9,n,k+1}^{13}\}$	$\{\alpha\} \cup I$, $\{a_{i_\ell}^\ell \mid 1 \leq \ell \leq rk\}$	$\{v_{\ell_i}^i \mid 1 \leq i \leq k\}$	$\{w_1, w_2, w_3, \beta_1, \ldots, \beta_5\}$, E_I, $\{\gamma_m^3, \gamma_{rk}^9, \gamma_9^{11}\}$
R_2	0		$\{\alpha\}$		$\{\zeta, \delta_1^{10}, \delta_2^{10}, \delta_3^{10}\}$
R_3	0		$\{v_i\}$		$\{e \in E(A) \mid \xrightarrow{e} s \xrightarrow{v_i}\}$
R_4	0		$\{v_i\}$		$\{\delta_{i_1}^1, \ldots, \delta_{i_r}^1, \delta_i^6, \delta_i^9\}$

4 Conclusion

In this paper, we enhance our understanding of synthesizing (m, n)-T-systems from the viewpoint of parameterized complexity. Although (m, n)-SYNTHESIS parameterized by $m + n$ belongs to XP, we show that there is little hope that this parameterization puts the problem into FPT. Future work might consider the *occupancy number* o_N of a searched net N a parameter. Let $N = (P, T, f, M_0)$ be a pure b-net, and let RS be the set of N's reachable markings. The *occupancy number* o_p of a place $p \in P$ is defined by $o_p = \{M \in RS \mid M(p) > 0\}$, and $o_N = max\{o_p \mid p \in P\}$ defines the *occupancy number* of N. At first glance, this parameter seems promising, at least SYNTHESIS parameterized by o_N is in XP.

Acknowledgements. I'm grateful to the reviewers for their helpful comments.

References

1. Badouel, E., Bernardinello, L., Darondeau, P.: Polynomial algorithms for the synthesis of bounded nets. In: Mosses, P.D., Nielsen, M., Schwartzbach, M.I. (eds.) CAAP 1995. LNCS, vol. 915, pp. 364–378. Springer, Heidelberg (1995). https://doi.org/10.1007/3-540-59293-8_207
2. Badouel, E., Bernardinello, L., Darondeau, P.: The synthesis problem forelementary net systems is NP-complete. Theor. Comput. Sci. **186**(1–2), 107–134 (1997). https://doi.org/10.1016/S0304-3975(96)00219-8
3. Badouel, E., Bernardinello, L., Darondeau, P.: Petri Net Synthesis. TTCSAES. Springer, Heidelberg (2015). https://doi.org/10.1007/978-3-662-47967-4
4. Badouel, E., Darondeau, P.: The synthesis of petri nets from path-automatic specifications. Inf. Comput. **193**(2), 117–135 (2004). https://doi.org/10.1016/j.ic.2004.04.004
5. Best, E., Devillers, R.: Characterisation of the state spaces of live and bounded marked graph petri nets. In: Dediu, A.-H., Martín-Vide, C., Sierra-Rodríguez, J.-L., Truthe, B. (eds.) LATA 2014. LNCS, vol. 8370, pp. 161–172. Springer, Cham (2014). https://doi.org/10.1007/978-3-319-04921-2_13
6. Best, E., Devillers, R.R.: State space axioms for T-systems. Acta Informatica **52**(2–3), 133–152 (2015). https://doi.org/10.1007/s00236-015-0219-0
7. Best, E., Devillers, R.R.: Synthesis of bounded choice-free petri nets. In: CONCUR. LIPIcs, vol. 42, pp. 128–141. Schloss Dagstuhl - Leibniz-Zentrum fuer Informatik (2015). https://doi.org/10.4230/LIPIcs.CONCUR.2015.128
8. Cygan, M., et al.: Parameterized Algorithms. Springer, Cham (2015). https://doi.org/10.1007/978-3-319-21275-3

9. Darondeau, P.: Unbounded petri net synthesis. In: Desel, J., Reisig, W., Rozenberg, G. (eds.) ACPN 2003. LNCS, vol. 3098, pp. 413–438. Springer, Heidelberg (2004). https://doi.org/10.1007/978-3-540-27755-2_11
10. Devillers, R., Hujsa, T.: Analysis and synthesis of weighted marked graph petri nets. In: Khomenko, V., Roux, O.H. (eds.) PETRI NETS 2018. LNCS, vol. 10877, pp. 19–39. Springer, Cham (2018). https://doi.org/10.1007/978-3-319-91268-4_2
11. Schlachter, U.: Bounded petri net synthesis from modal transition systems is undecidable. In: CONCUR. LIPIcs, vol. 59, pp. 15:1–15:14. Schloss Dagstuhl - Leibniz-Zentrum fuer Informatik (2016). https://doi.org/10.4230/LIPIcs.CONCUR.2016.15
12. Schlachter, U., Wimmel, H.: k-bounded petri net synthesis from modal transition systems. In: CONCUR. LIPIcs, vol. 85, pp. 6:1–6:15. Schloss Dagstuhl - Leibniz-Zentrum fuer Informatik (2017). https://doi.org/10.4230/LIPIcs.CONCUR.2017.6
13. Tredup, R.: Hardness results for the synthesis of b-bounded petri nets. In: Donatelli, S., Haar, S. (eds.) PETRI NETS 2019. LNCS, vol. 11522, pp. 127–147. Springer, Cham (2019). https://doi.org/10.1007/978-3-030-21571-2_9
14. Tredup, R.: Synthesis of structurally restricted b-bounded petri nets: complexity results. In: Filiot, E., Jungers, R., Potapov, I. (eds.) RP 2019. LNCS, vol. 11674, pp. 202–217. Springer, Cham (2019). https://doi.org/10.1007/978-3-030-30806-3_16
15. Tredup, R., Rosenke, C.: Narrowing down the hardness barrier of synthesizing elementary net systems. In: CONCUR. LIPIcs, vol. 118, pp. 16:1–16:15. Schloss Dagstuhl - Leibniz-Zentrum fuer Informatik (2018). https://doi.org/10.4230/LIPIcs.CONCUR.2018.16
16. Tredup, R., Rosenke, C., Wolf, K.: Elementary net synthesis remains NP-complete even for extremely simple inputs. In: Khomenko, V., Roux, O.H. (eds.) PETRI NETS 2018. LNCS, vol. 10877, pp. 40–59. Springer, Cham (2018). https://doi.org/10.1007/978-3-319-91268-4_3

Parameterized Dynamic Variants of Red-Blue Dominating Set

Faisal N. Abu-Khzam[1], Cristina Bazgan[2], and Henning Fernau[3(✉)]

[1] Department of Computer Science and Mathematics, Lebanese American University, Beirut, Lebanon
faisal.abukhzam@lau.edu.lb
[2] Université Paris-Dauphine, PSL University, CNRS, LAMSADE, 75016 Paris, France
cristina.bazgan@lamsade.dauphine.fr
[3] Universität Trier, Fachber. 4 – Abteilung Informatikwissenschaften, 54286 Trier, Germany
fernau@uni-trier.de

Abstract. We introduce a parameterized dynamic version of the Red-Blue Dominating Set problem and its partial version. We prove the fixed-parameter tractability of the dynamic versions with respect to the (so called) edit-parameter while they remain $\mathcal{W}[2]$-hard with respect to the increment-parameter. We provide a complete study of the complexity of the problem with respect to combinations of the various parameters.

Keywords: Dynamic problems · Reoptimization · Parameterized complexity · Set cover · Hitting set

1 Introduction

In the Red-Blue Dominating Set problem (henceforth *RBDS*), we are given a graph $G = (R \cup B, E)$ such that $R \cap B = \emptyset$, together with an integer $s \geq 0$, and we are asked whether R contains a subset S of cardinality at most s such that every element of B has at least one neighbor in S. In this case the elements of R and B are called red and blue vertices respectively, and S is a red-blue dominating set of G. We shall also refer to G as a *red-blue graph.* It is well-known that *RBDS* is equivalent to Set Cover as well as to Hitting Set.[1]

In this paper we are interested in the parameterized complexity of dynamic versions of *RBDS* and some of its variants. In such dynamic settings, originally defined by Downey et al. in [12] in the context of the Dominating Set problem, we assume the edges of the input graph G can appear or disappear with time, so an initially feasible *RBDS* solution S (not necessarily optimal) may no longer

[1] The problem should not be confused with Red-Blue Set Cover as studied in [7].

This work was supported by the bilateral research cooperation CEDRE between France and Lebanon (grant number 40334SG).

A. Chatzigeorgiou et al. (Eds.): SOFSEM 2020, LNCS 12011, pp. 236–247, 2020.
https://doi.org/10.1007/978-3-030-38919-2_20

dominate all of B and we want to construct another solution S' so that the Hamming distance between S and S' is minimized. The problem is formally defined as follows.

Dynamic Red-Blue Dominating Set (*DRBDS*)
Given: Two red-blue graphs $G = (R \cup B, E)$ and $G' = (R \cup B, E')$, a subset S of R that is a red-blue dominating set of G, integers k and r such that $d_H(E, E') \leq k$.
Question: Is there a subset S' of R such that $d_H(S, S') \leq r$ and S' is a red-blue dominating set of G'?
By analogy to operations on characteristic vectors, the *Hamming distance* between two sets (subsets of the same, mostly implicitly given ground set) refers to the cardinality of the symmetric set difference, i.e., $d_H(S, S') = |S \Delta S'|$.

Remark 1. Since r is a bound on the Hamming distance between S and S', we are not necessarily interested in deleting elements from S. Therefore we can always assume $S \subseteq S'$ according to the formulation above. Then, $d_H(S, S') = |S'| - |S|$. Hence, r is called the *increment-parameter*. Similarly, in view of the problem at hand, it suffices to consider the special case when $E' \subseteq E$, i.e., E' resulted from E by removing some edges. Then, $d_H(E, E') = |E| - |E'|$. More precisely, if $G = (R \cup B, E)$, $G' = (R \cup B, E')$, S, k and r form an instance of *DRBDS*, then also $\hat{G} = (R \cup B, E \cup (E' \setminus E))$, $G' = (R \cup B, E')$, S, k and r form an instance of *DRBDS* as well, and $S' \subseteq R$ is (or is not) a red-blue dominating set of G' in both cases. We call k the *edit-parameter*.

Notice that *DRBDS* could be also viewed as a reoptimization variant of Set Cover as defined in [3]. However, the more interesting approximation results obtained in that paper rely on the given solution to be optimum, a condition that is relaxed in this paper. So, we are following here the terminology introduced in [1,12], which is not the same as later used in [2] under a similar name.

Motivation

In social networks, there is a growing interest in domination problems that can model the search for influencers in the best way, in particular aiming at people who can spread positive influence and participate in launching a global campaign such as, for example, a non-smoking campaign. In such a context, the network consists of two types of people: the set R of non-smokers (who are not to be convinced but can help in convincing others) and the set B which consists of those who are known to have the smoking habit. The natural social network of friendship relations can be modeled as a graph.

Ideally one would seek a set of individuals $S \subseteq R$ (to serve as influencers) that are friends of all the elements of B. However, in real settings, the objective would be to affect (or, in a more formal sense, to dominate) as many elements of B as possible. This latter objective is modeled using the partial version of *RBDS*, originally defined in [15] in the context of learning theory. Moreover, it would be more realistic to assume someone is influenced by the non-smoking campaign if there are at least a number, say q, of his or her friends who serve as influencers. This gives rise to the *q-RBDS* problem which differs from *RBDS* only in the domination condition requiring each element of B to have at least q neighbors in $S \subseteq R$.[2]

We shall study the partial *q-RBDS* problem, among other variants, and focus on its parameterized dynamic version to cope with settings where the network is changing with time.

Another interesting scenario where dynamic *RBDS* could be useful is when the red and blue vertices correspond to stores and customers, respectively, and the links are based on credit card transactions: an edge uv means customer u frequently purchases items from store v. In such a setting, it would be interesting to have a smallest possible list of stores that are preferred by a large number (or majority) of customers. Moreover, since data from credit card transactions is dynamic, and the interests of customers change with time, some links in the corresponding red-blue graph can appear or disappear, which might require updating a previously computed (partial) *RBDS* solution. The so-called partial *DRBDS* problem is obviously the right model in this case. In fact, the more general partial (dynamic) *q-RBDS* problem could be of interest, as well. This is another direction studied here.

Finally, let us mention another motivation for dynamic problems as considered in the paper: assume we have found (over time, with experience) a *nice* solution to the problem that we are interested in. Here, *nice* does not necessarily mean smallest or largest, but just satisfying a number of properties, out of which some are formalized and are those we wish to keep up even if the situation changes slightly. As the previous solution was nice, we do not want to change it too much. This justifies the general assumption (important in the context of Parameterized Complexity) that the two *change parameters* can be assumed to be small.

Throughout this paper, we adopt common graph-theoretic terminology and notations. Apart from the problems mentioned above, we will discuss quite a number of auxiliary problems that might be of independent interest. The paper is structured as follows. In the next section, we study the complexity of the dynamic version of the Red-Blue Dominating Set problem. The partial version is studied in Sect. 3, while the last section briefly addresses approximability but focuses on open problems.

2 Complexity of Dynamic Red Blue Dominating Set

The fact that *DRBDS* is $\mathcal{NP}$-hard is obvious. It follows immediately from the $\mathcal{NP}$-hardness of *RBDS* itself. To see this, let $(G = (R \cup B, E), s)$ be an instance of *RBDS*; construct two graphs G_1 and G_2 as follows:[3] G_1 is obtained from G by adding a special vertex w and joining it by $|B|$ edges to each vertex of B.

[2] In this model, possibly also the influence of smokers on their smoking friends should be taken into account. This would lead to notions like alliances or monopolies as discussed in [13]. We are not going into this direction in this paper.

[3] We also refer to the general discussions of hardness for dynamic problems in [5].

G_2 is obtained by deleting the $k = |B|$ edges incident on w. Now set $S = \{w\}$ and $r = s$ to obtain the dynamic *RBDS* instance. Obviously, any solution S' is equivalent to a solution to the given *RBDS* instance and vice versa.

We now define the Need-Based Red-Blue Dominating Set problem and use it to obtain some algorithmic results which might be of independent interest.[4]

Need-Based Red-Blue Dominating Set (*NB-RBDS*)

Given: A red-blue graph $G = (R \cup B, E)$ together with integers $s, q \geq 0$ and a function $\eta : B \longrightarrow \{0, 1, \ldots, q\}$.
Question: Does R contain a subset D of cardinality at most s such that every element b of B has at least $\eta(b)$ neighbors in D?

Notice that in approximation algorithms, this type of problems has been studied as a special case of covering integer programs, see [22] as an example reference. Namely, considering $\boldsymbol{\eta}$ as a $|B|$-dimensional vector and thinking of $\boldsymbol{x}$ as a $|R|$-dimensional binary solution vector, as well as A as the $|B| \times |R|$-biadjacency matrix of the bipartite graph G, then *NB-RBDS* asks to find a solution vector $\boldsymbol{x}$ with at most s one-entries that satisfies $A\boldsymbol{x} \geq \boldsymbol{\eta}$. This translation immediately provides an $O(|R|^{O(|R|)})$-algorithm based on Lenstra's results [18]. We present an alternative approach now, which is better roughly in the case when $(q+1)^{|B|} < |R|^{|R|}$.

Notice that vertices with $\eta(v) = 0$ are trivially satisfied, so they can be removed from the instance. Hence, we can (tacitly) assume $\eta : B \longrightarrow \{1, \ldots, q\}$.

Theorem 1. *NB-RBDS is fixed-parameter tractable with respect to $|B|$ and q as parameters.*

Proof. Let $R = \{r_1, \ldots, r_n\}$, $B = \{b_1, \ldots, b_m\}$ and $q = \max_{1 \leq i \leq m}\{\eta(b_i)\}$. We show how to construct a solution S in time $O((q+1)^{|B|})$ by using dynamic programming.

Consider the set of all functions from B to the set $\{0, 1, \ldots, q\}$. There are at most $(q+1)^{|B|}$ many of them. A more precise upper-bound on their number is $\prod_{b \in B}(\eta(b) + 1)$, as each such function can be represented by a vector $(x_1, \ldots, x_m)$, where $x_i \in \{0, 1, \ldots, \eta(b_i)\}$ and $m = |B|$.

In the dynamic programming algorithm that we describe next, our *target vector* is $\boldsymbol{x} = (x_1, \ldots, x_m)$, where x_i is the number of elements of R that are still needed to (finally) cover b_i with $\eta(b_i)$ elements of R. Let $R_j = \{r_1, r_2, \ldots, r_j\}$ and let $C[\boldsymbol{x}, j]$ be the minimum number of elements of R_j needed to be added to S in order to cover each b_i with x_i elements from R_j. Hence, we initialize $C[\mathbf{0}, 0] = 0$ and $C[\boldsymbol{x}, 0] = \infty$ for all $\boldsymbol{x} \neq \mathbf{0}$. Then we have the following recursion for $j > 0$:

$$C[\boldsymbol{x}, j] = \min\{C[\boldsymbol{x}, j-1], 1 + C[\max\{\boldsymbol{x} - \chi_{N(r_j)}, \mathbf{0}\}, j-1]\},$$

[4] The notion of *capacitated domination* is related. Unfortunately, this notion is not used consistently in the literature. While in [14], both capacities and demands are associated to vertices, so that capacities equal to degrees and demands equal to needs would be exactly a need-based variant of domination, in [6,11], there is no demand function. However, we are not going into this direction here, also because for our purposes, the need-based variation is rather an auxiliary problem.

where $\chi_{N(r_j)}$ is the characteristic vector of $N(r_j) \subseteq B$; the maximum operation is understood component-wisely. So in the case where r_j is in a smallest solution from R_j we subtract 1 from each x_i where $b_i \in N(r_j)$. Obviously, the bottom up dynamic programming approach would compute any target vector in time $O((q+1)^m|R|)$. This could be speed up by not considering target vectors with components $x_i > \eta(b_i)$. □

Observe that one could view our problem as a (very) special case of the Hitting Set of Bundles problem considered in [10]. In fact, for each vertex $b \in B$, we would introduce the set system $S(b)$ of all $\eta(b)$-element subsets of $N(b)$ and the question would be to select a subset $s(b) \in S(b)$ for each $b \in B$ so that the set $\bigcup_{b\in B} s(b) \subseteq R$ has cardinality at most s. Unfortunately, we arrive at a very special case of Hitting Set of Bundles, so that the parameterized complexity results known for that problem are not very helpful in our case. In particular, the situation where Damaschke could prove $\mathcal{W}[1]$-hardness, when the bundles have size at most two, we would face the situation when $q = 2$ is a constant and we parameterize by $|B|$ and s, a situation covered (in a much stronger sense) by the previous theorem, leading to an $\mathcal{FPT}$ result in our scenario.

Note that the solution size, s, was not treated as a parameter in the proof above, so the dynamic programming algorithm finds a solution of minimum size in R. We can (therefore) obtain the following result. Notice that we now fix $q \geq 1$ as part of the problem definition.

Corollary 1. *Dynamic q-RBDS is fixed-parameter tractable with respect to the edit-parameter k.*

Proof. Consider two red-blue graphs $G = (R \cup B, E)$ and $G' = (R \cup B, E')$, a subset S of R that is a red-blue dominating set of G, integers k and r such that $d_H(E, E') \leq k$. By Remark 1, we can assume that $E' \subseteq E$. Let $B' \subseteq B$ be the set of elements of B that have less than q neighbors in S after the deletion of at most k edges when moving from E to E'. Let $B' = \{b_1, b_2, \ldots, b_{k'}\}$, $k' \leq k$ (since at most k edges are deleted). Let $\eta(b_i) = \max\{q - |N_{G'}(b_i) \cap S|, 0\}$. In other words, $\eta(b_i)$ is the number of elements of R that are still needed to dominate b_i with q (red) neighbors. Let $q' = \max_{1\leq i\leq k'}\{\eta(b_i)\}$. Obviously $q' \leq k$.

Now we are left with the instance $(G = (R \cup B', E), r, \eta)$ of the *NB-RBDS* problem, which is solvable in $O((q'+1)^{|B'|})$. This proves our assertion, knowing that both q' and $|B'|$ are bounded above by k. □

The following corollary follows immediately from the above; it corresponds to the case $q = 1$.

Corollary 2. *Dynamic RBDS is fixed-parameter tractable with respect to the edit-parameter k.*

Remark 2. More precisely, the proof of Corollary 1 shows that we can estimate the running time as $O^*(2^k)$, as $\eta : B' \to \{0, 1\}$. Notice that algorithms of the form $O^*(2^{o(k)})$ are not to be expected under the *Set Cover Conjecture*, see [9], because it is possible to formulate Set Cover as Dynamic *RBDS*. In essence, this is also done in the proof of the next theorem, in an even more general setting.

Now we turn our attention to the increment-parameter, r. It was shown in [1,12] that Dynamic Dominating Set is $\mathcal{W}[2]$-hard when parameterized by the increment-parameter r only. We show the same for q-$RBDS$.

Theorem 2. *For any $q \geq 1$, Dynamic q-$RBDS$ is $\mathcal{W}[2]$-hard with respect to the increment-parameter r.*

Proof. By reduction from the $\mathcal{W}[2]$-hard $RBDS$ problem. Let $(G = (R \cup B, E), r)$ be an instance of $RBDS$. We construct an instance (G_1, G_2, S, k, r) of Dynamic q-$RBDS$ as follows.
G_1 is obtained from G by adding q red vertices forming the set S that, together with B, induces a complete bipartite subgraph. Let $S = \{w_1, w_2 \ldots, w_q\}$. Then $G_1 = (R' \cup B, E_1)$ where $R' = R \cup S$ and $E_1 = E \cup \{vw_i : v \in B, 1 \leq i \leq q\}$. Let $G_2 = (R' \cup B, E_2)$ where $E_2 = E_1 \setminus \{vw_1 : v \in B\}$ and $k = |B|$. In other words, every element of B is dominated in G_1 by the q vertices of S. However, in G_2 every element of B is dominated by the $q-1$ vertices of $S \setminus \{w_1\}$. A solution S' of this Dynamic q-$RBDS$ instance must contain at most r vertices from R that dominate B. □

Remark 3. It is not very difficult to design a multi-tape Turing machine that solves Dynamic q-$RBDS$ by first guessing the (at most) r vertices to be added to the existing red-blue dominating set and then verifying this guess by using one tape per vertex (better said neighborhood) in the spirit of [8]. The only difference to the classical approach is that some head positions have to be individually set by the reduction machine that constructs this Turing machine, based on the information how the given set S already dominates other vertices.

Corollary 3. *Dynamic $RBDS$ is $\mathcal{W}[2]$-complete with increment-parameter r.*

3 The Partial Dynamic RBDS Set Problem

In the Partial Red-Blue Dominating Set problem, or Partial $RBDS$ for short, we are given an additional parameter t, the *budget parameter*, and the objective is to find (whether there is) a subset S of R with $|S| \leq s$ that dominates at least t elements of B. The dynamic version is defined as follows.

Dynamic Partial Red-Blue Dominating Set ($PRBDS$)
Given: Two red-blue graphs $G = (R \cup B, E)$ and $G' = (R \cup B, E')$; an integer $t \geq 0$; a subset S of R satisfying $|N_G(S)| \geq t$; integers $k, r \geq 0$ such that $d_H(E, E') \leq k$.
Question: Is there a subset S' of R such that $d_H(S, S') \leq r$ and $|N_{G'}(S')| \geq t$?

Similar comments as collected in Remark 1 apply here as well: we may hence assume that $E' \subseteq E$ and that $S' \supseteq S$.

By enforcing $t = |B|$, it is not hard to see that the previously obtained hardness results for the increment-parameter transfer (see Theorem 2).

Corollary 4. *Dynamic PRBDS is $\mathcal{W}[2]$-hard with increment-parameter r.*

This observation lets us focus on the other two natural parameters of this problem, the edit-parameter k and the *target-parameter* t.

The Partial *RBDS* problem is known to be fixed-parameter tractable with respect to t. This was (equivalently) formulated in [4] in terms of Partial (Set) Cover. The currently fastest algorithm runs in randomized time $O^*(2^t)$, using polynomial space, as shown by Koutis and Williams [17] for the related Partial Dominating Set problem. To keep the paper self-contained, we are going to describe how this type of algorithm would look like for Partial *RBDS* next. The key is a reduction to a problem that is based on the following algebraic setting. Let X denote a set of variables. A monomial of degree d is a product of d variables from X, with multiplication assumed to be commutative. A monomial is called multilinear if no variable appears twice or more in the product. A polynomial $P(X)$ (over the semiring of nonnegative integers $\mathbb{N}$) is a linear combination of monomials with coefficients from $\mathbb{N}$. Such polynomials, along with addition and commutative multiplication, form a commutative semiring, denoted by $\mathbb{N}[X]$. The maximum degree among all monomials of $P(X)$ is called the degree of $P(X)$. An arithmetic circuit over $\mathbb{N}$ and X is a directed acyclic graph. Each node of in-degree zero is an input gate, which is labeled either with an element from $\mathbb{N}$ or with a variable from X. The graph contains a single output node of out-degree zero. Each other node is either an addition or a multiplication gate. Arithmetic circuits are representations for polynomials from $\mathbb{N}[X]$. A polynomial $P(X) \in \mathbb{N}[X]$ contains a certain monomial if the monomial appears with a nonzero coefficient in the linear combination that constitutes $P(X)$.
Multilinear Monomial Detection (*MlD*)

Given: An arithmetic circuit C representing a polynomial $P(X)$ over $\mathbb{N}$, an integer $d \geq 0$.
Question: Does $P(X)$, construed as a sum of monomials, contain a multilinear monomial of degree at most d?

Koutis and Williams [17] showed that *MlD*, parameterized by the degree parameter d, is fixed-parameter tractable, by providing a randomized algorithm running in time $O^*(2^d)$, using polynomial space. They used this result to prove that Partial Dominating Set can be solved in randomized $\mathcal{FPT}$ time $O^*(2^t)$.

To showcase this technique, we are first explaining how to derive an analogous result for Partial *RBDS*. Let $G = (R \cup B, E)$ and $k, t \geq 0$ form an instance of Partial *RBDS*. We are going to construct a circuit C for the following polynomial.

$$P_k(X) := \left(\sum_{r \in R} \prod_{b \in N_G(r)} (1 + z \cdot x_b) \right)^k,$$

where the set of variables X consists of one variable x_b for each vertex $b \in B$, as well as one additional variable z. Now, $P_k(X)$ contains a monomial of the form $z^t x_{b_1} \cdots x_{b_t}$ for $B' := \{b_1, \ldots, b_t\}$ forming a t-element subset of B if and only if

B' is dominated by (at most) k elements from R. The intuition is the following: By raising the sum-of-products to the kth power, any monomial is formed by picking k of the product-terms. As the sum ranges over all red vertices, this corresponds to selecting $\ell \leq k$ red vertices, forming $R' = \{r_1, \ldots, r_\ell\}$. Each of these vertices from R' will dominate the whole neighborhood. However, as we need to only dominate t vertices, we may select t vertices (if possible) from $N_G(R')$, and moreover, each vertex b_i from this chosen set $B' \subseteq N_G(R')$ selects one vertex $d(b_i) \in R'$ as its dominator. Consider the monomial

$$\prod_{r_j \in R'} \prod_{b \in B', r_j = d(b)} z \cdot x_b$$

contained in $P_k(X)$. It is obviously of the required form; in particular, it is multilinear (with respect to $X \setminus \{z\}$) and contains z^t but not z^{t+1}.

Moreover, for any other monomial $z^t \xi_1 \cdots \xi_t$, formed in a different way, we necessarily find $1 \leq i < j \leq t$ such that $\xi_i = \xi_j$, i.e., this monomial is not multilinear. Observe that the size of C is polynomial in the size of G, because the term of the sum-of-products need not be repeated in a circuit. Hence, one could use the randomized *MlD*-algorithm to solve Partial *RBDS* in randomized time $O^*(2^t)$ as claimed, where the parameter t becomes the degree parameter.

We show the same applies to the dynamic version, when parameterized by the edit-parameter.

Theorem 3. *Dynamic Partial RBDS is fixed parameter tractable with respect to the edit-parameter k.*

Proof. Let (G, G', S, k, r, t) be an instance of Dynamic Partial *RBDS*, as in the definition above. Assume $E' \subseteq E$ and $S' \supseteq S$. Observe that $t - |N_{G'}(S)| \leq k$, since at most k elements of $N_{G'}(S)$ are affected by at most k edge deletions. So it would be enough to dominate at most $t - |N_{G'}(S)|$ elements of $B \setminus N_{G'}(S)$. We can hence use the presented $\mathcal{FPT}$-algorithm for Partial *RBDS*, applied to the red-blue subgraph induced by $R \cup (B \setminus N_{G'}(S))$, with $t - |N_{G'}(S)|$ $(\leq k)$ as a parameter. □

As we will see, the following seemingly easy generalization cannot be solved by using this algebraic approach. This proves that a rather natural variation of *MlD* cannot be solved in $\mathcal{FPT}$-time. We are now turning our attention towards Partial q-*RBDS* for arbitrary (fixed) $q \geq 1$:
Given: A red-blue graph $G = (R \cup B, E)$, integers $k, t, s \geq 0$.
Question: Is there a subset $S \subseteq R$, $|S| \leq s$, and a subset $N \subseteq B$ with $|N| \geq t$ such that each element in N has at least q neighbors in S?

If q is part of the input, we speak of Partial General *RBDS*. As a natural generalization, we again consider a need-based variation. We will return to the whole family of problems Dynamic Partial q-*RBDS* below. First recall that in the previous section, we used the (more general) Need-Based *RBDS* problem to address the dynamic variant of q-*RBDS* and we showed it to be fixed-parameter tractable. Unfortunately, this is more delicate here, as we will show now.

In the Need-Based Partial Red-Blue Dominating Set problem, we are given an integer q and a function $\eta : B \longrightarrow \{0, 1, \ldots q\}$, and we say that an element v of B is dominated, or henceforth *satisfied*, by a subset D of R if it has $\eta(v)$ neighbors in D. A formal definition follows.

Need-Based Partial Red-Blue Dominating Set (*NB-PRBDS*)
Given: A red-blue graph $G = (R \cup B, E)$ together with integers $s, t, q \geq 0$ and a function $\eta : B \longrightarrow \{0, 1, \ldots q\}$.
Question: Does R contain a subset D of cardinality at most s that satisfies at least t elements of B?

Here, a subset D of R *satisfies* at least t elements of B if there is a set $B' \subseteq B$ such that $|B'| \geq t$ and for all $v \in B'$, $|N(v) \cap D| \geq \eta(v)$.

As in the previous section, we can tacitly assume that $\eta(v) > 0$ for all $v \in B$, as otherwise we can easily satisfy v and hence remove v and decrement t.

It might be tempting to think that the ideas leading to $\mathcal{FPT}$-algorithms in Theorem 3 transfer to this case. We did try to work in this direction, but there seems to be some difficulty because different individuals have different needs, and this cannot be modeled while checking, at the same time, that at least t blue vertices are dominated.

This difficulty can be backed with the following hardness result.

Theorem 4. *NB-PRBDS is $\mathcal{W}[1]$-hard with respect to t and q (and s) as parameters, even if the need function η is constant.*

Proof. First, observe that if $s = t = q$ and η being the constant function $\eta = q$, *NB-PRBDS* asks about a biclique in the bipartite graph $G = (R \cup B, E)$ with exactly $s = q$ vertices in R (as q is a trivial lower bound on the size of any solution) and (at least) $t = q$ vertices in B. Rather recently, Lin proved that it is $\mathcal{W}[1]$-hard to find such a biclique $K_{t,t}$ in a given bipartite graph, see [20] in combination with [19, Lemma 3.1]. □

Corollary 5. *Partial General RBDS is $\mathcal{W}[1]$-hard, parameterized with t, q, or s.*

Observe that usually, the partial variants of domination-like problems tend to be in the class $\mathcal{FPT}$. To the best of our knowledge, this is the first problem variant where this question turns out to be hard. Yet, there is a catch in this assertion, which can be seen by turning to the family of problems *NB-q-PRBDS* whose definition coincides with that of *NB-PRBDS*, apart from the fact that q is no longer part of the input here. We do not have a hardness result in this case, nor do we know of algorithmic results, even not in the case when $q = 2$, the case $q = 1$ having been dealt with (algorithmically) above. Rather, when we look at Dynamic Partial General *RBDS*, where we have q as part of the input, we can show the following result.

Corollary 6. *Dynamic Partial General RBDS is $\mathcal{W}[1]$-hard with respect to t or q as parameters. $\mathcal{W}[1]$-hardness even holds for the combined parameter (t, q, r, k).*

Proof. As above, consider $t = q$. If $G' = (R \cup B, E')$ (together with $t = q$) is an instance of *NB-PRBDS*, then we obtain G by adding a $K_{q,q}$, with $s = q$. The edit-parameter is q^2, the increment-parameter would be q. □

We are now proposing a generalization of *MlD* that we prove to be hard when parameterized with the degree parameter. This result could be of independent interest. As it is not central to the topic of the paper, we omit its proof.

Multilinear Monomial Detection with Partition (*MlDwP*)

Given: An arithmetic circuit C representing a polynomial $P(X)$ over $\mathbb{N}$, a partition of X, an integer $d \geq 0$.
Question: Does $P(X)$, construed as a sum of monomials, contain a multilinear monomial of degree at most d that contains, for each class of the partition, either all variables in that class or no variable from that class?

Theorem 5. *MlDwP, parameterized by the degree parameter, is* $\mathcal{W}$*[1]-hard.*

Remark 4. One could try to alternatively parameterize Partial *RBDS* by k and $k' := |B| - t$. In fact, the problem is easily seen to be $W[2]$-hard, when parameterized by k', because the problem can then be re-formulated as follows (disregarding isolated vertices): Given some red-blue graph $G = (R \cup B, E)$ and integers $k, k' \geq 0$, find subsets $R' \subseteq R$ and $B' \subseteq B$, with $|R'| \leq k$ and $|B'| \leq k'$, such that $N(R') = B \setminus B'$ and $N(B') = R \setminus R'$. Hence, if $k = 0$, the question boils down to *RBDS* itself, with the roles of red and blue being exchanged.

Clearly, this tweak does not change the dynamic version of the problem at all, it is equivalent to Dynamic Partial *RBDS* as studied above.

Remark 5. One could also think of changing the notion of a *solution* in the dynamic partial setting. This would mean the following problem.

Dynamic Partial Red-Blue Dominating Set with Blue Focus (*RBDSBF*)
Given: Two red-blue graphs $G = (R \cup B, E)$ and $G' = (R \cup B, E')$; an integer $t \geq 0$; a subset T of B satisfying $|T| \geq t$, dominated by $S \subseteq R$; integers $k, r \geq 0$ such that $d_H(E, E') \leq k$.
Question: Is there a subset T' of B, dominated by some S' with $|S'| \leq |S|$, such that $d_H(T, T') \leq r$ and $|T'| \geq t$?

$\mathcal{NP}$-hardness of this variant is easily seen by starting from some *RBDS* instance $G' = (R \cup B, E')$ and a bound s on the size of the red-blue dominating set. Construct G from G' by selecting a subset $S \subseteq R$ with $|S| = s$ and adding $|B|$ edges (to form E) so that S is a red-blue dominating set of G. With $t = k = |B|$ and $r = 0$, we have defined all ingredients of the equivalent Dynamic *RBDSBF* instance. This also proves that the problem is para-$\mathcal{NP}$-hard for the increment-parameter r.

4 Concluding Remarks

In this paper, we undertook a multivariate analysis of Red-Blue Dominating Set. Clearly, one could also consider further parameters, for instance the *loss parameter*, which is the difference between $|N(S)|$ and $|N(S')|$. As most of our results are negative ones, there is surely a need for further parameterizations. Also, it would be very helpful to know if *NB-q-PRBDS* belongs to $\mathcal{FPT}$ for any fixed $q > 1$, as this would also help classify the dynamic variants of *PRBS* for fixed $q > 1$. This is the most interesting open problem in that area in our opinion.

We completely neglected approximability issues so far. The more classical *DRBDS* problem (as Set Cover reoptimization) was previously considered in [3,21]. Let us at least mention one positive result, concerning the natural maximization variant of Red-Blue Dominating Set, which we call BUDGETED RED-BLUE k-DOMINATING SET, following the tradition of the literature of these problems. Here, we search for a subset of k red vertices that dominate a maximum number of blue vertices. Khuller et al. [16] considered the approximability of the BUDGETED CONNECTED k-DOMINATING SET. In this problem there is a budget k on the number of vertices we can select, and the goal is to dominate as many vertices as possible with a connected set.

Theorem 6. BUDGETED RED-BLUE k-DOMINATING SET *is polynomial-time* $\frac{1}{13}(1-\varepsilon)$ *approximable, for any* $\varepsilon > 0$.

Proof. Given any instance $G = (R \cup B, E)$ (and k) of BUDGETED RED-BLUE k-DOMINATING SET, we construct an instance of BUDGETED CONNECTED k-DOMINATING SET by turning R into a clique. This results in a split graph G' on which we can use the algorithm of Khuller et al. to compute an approximate solution. In any solution containing a vertex b from B, b can be easily replaced by any neighbor of b, so that we can assume that the solution S is a subset of R. This is also true for an optimum solution. Hence, maximum solutions to the instance (G, k) correspond to maximum solutions of (G', k) and vice versa. Thus, the approximation factor of $\frac{1}{13}(1-\varepsilon)$, shown in [16] for BUDGETED CONNECTED k-DOMINATING SET on split graphs, also applies to our problem. □

References

1. Abu-Khzam, F.N., Egan, J., Fellows, M.R., Rosamond, F.A., Shaw, P.: On the parameterized complexity of dynamic problems. Theoret. Comput. Sci. **607**, 426–434 (2015)
2. Alman, J., Mnich, M., Williams, V.V.: Dynamic parameterized problems and algorithms. In: Chatzigiannakis, I., Indyk, P., Kuhn, F., Muscholl, A. (eds.) 44th International Colloquium on Automata, Languages, and Programming, ICALP. LIPIcs, vol. 80, pp. 41:1–41:16. Schloss Dagstuhl - Leibniz-Zentrum für Informatik (2017)
3. Bilò, D., Widmayer, P., Zych, A.: Reoptimization of weighted graph and covering problems. In: Bampis, E., Skutella, M. (eds.) WAOA 2008. LNCS, vol. 5426, pp. 201–213. Springer, Heidelberg (2009). https://doi.org/10.1007/978-3-540-93980-1_16
4. Bläser, M.: Computing small partial coverings. Inf. Process. Lett. **85**(6), 327–331 (2003)
5. Böckenhauer, H.-J., Hromkovič, J., Mömke, T., Widmayer, P.: On the hardness of reoptimization. In: Geffert, V., Karhumäki, J., Bertoni, A., Preneel, B., Návrat, P., Bieliková, M. (eds.) SOFSEM 2008. LNCS, vol. 4910, pp. 50–65. Springer, Heidelberg (2008). https://doi.org/10.1007/978-3-540-77566-9_5

6. Bodlaender, H.L., Lokshtanov, D., Penninkx, E.: Planar capacitated dominating set is $W[1]$-hard. In: Chen, J., Fomin, F.V. (eds.) IWPEC 2009. LNCS, vol. 5917, pp. 50–60. Springer, Heidelberg (2009). https://doi.org/10.1007/978-3-642-11269-0_4
7. Cai, Z., Miao, D., Li, Y.: Deletion propagation for multiple key preserving conjunctive queries: approximations and complexity. In: International Conference on Data Engineering, ICDE, pp. 506–517. IEEE (2019)
8. Cesati, M.: The turing way to parameterized complexity. J. Comput. Syst. Sci. **67**(4), 654–685 (2003)
9. Cygan, M., et al.: On problems as hard as CNF-SAT. ACM Trans. Algorithms **12**(3), 41:1–41:24 (2016)
10. Damaschke, P.: Parameterizations of hitting set of bundles and inverse scope. J. Comb. Optim. **29**(4), 847–858 (2015)
11. Dom, M., Lokshtanov, D., Saurabh, S., Villanger, Y.: Capacitated domination and covering: a parameterized perspective. In: Grohe, M., Niedermeier, R. (eds.) IWPEC 2008. LNCS, vol. 5018, pp. 78–90. Springer, Heidelberg (2008). https://doi.org/10.1007/978-3-540-79723-4_9
12. Downey, R.G., Egan, J., Fellows, M.R., Rosamond, F.A., Shaw, P.: Dynamic dominating set and turbo-charging greedy heuristics. Tsinghua Sci. Technol. **19**(4), 329–337 (2014)
13. Fernau, H., Rodríguez-Velázquez, J.A.: A survey on alliances and related parameters in graphs. Electron. J. Graph Theory Appl. **2**(1), 70–86 (2014)
14. Kao, M., Chen, H., Lee, D.: Capacitated domination: problem complexity and approximation algorithms. Algorithmica **72**(1), 1–43 (2015)
15. Kearns, M.J.: Computational Complexity of Machine Learning. ACM Distinguished Dissertations. MIT Press, Cambridge (1990)
16. Khuller, S., Purohit, M., Sarpatwar, K.K.: Analyzing the optimal neighborhood: algorithms for budgeted and partial connected dominating set problems. In: Symposium on Discrete Algorithms (SODA), pp. 1702–1713. SIAM (2014)
17. Koutis, I., Williams, R.: Limits and applications of group algebras for parameterized problems. ACM Trans. Algorithms **12**(3), 31:1–31:18 (2016)
18. Lenstra Jr., H.W.: Integer programming with a fixed number of variables. Math. Oper. Res. **8**(4), 538–548 (1983)
19. Lin, B.: The parameterized complexity of k-biclique. In: Symposium on Discrete Algorithms (SODA), pp. 605–615. SIAM (2015)
20. Lin, B.: The parameterized complexity of the k-biclique problem. J. ACM **65**(5), 34:1–34:23 (2018)
21. Mikhailyuk, V.A.: Reoptimization of set covering problems. Cybern. Syst. Anal. **46**(6), 879–883 (2010)
22. Srinivasan, A.: Improved approximation guarantees for packing and covering integer programs. SIAM J. Comput. **29**(2), 648–670 (1999)

Refined Parameterizations for Computing Colored Cuts in Edge-Colored Graphs

Nils Morawietz(✉), Niels Grüttemeier, Christian Komusiewicz, and Frank Sommer

Fachbereich Mathematik und Informatik, Philipps-Universität Marburg, Marburg, Germany
{morawietz,niegru,komusiewicz,fsommer}@informatik.uni-marburg.de

Abstract. In the COLORED (s,t)-CUT problem, the input is a graph $G = (V, E)$ together with an edge-coloring $\ell : E \to C$, two vertices s and t, and a number k. The question is whether there is a set $S \subseteq C$ of at most k colors, such that deleting every edge with a color from S destroys all paths between s and t in G. We continue the study of the parameterized complexity of COLORED (s,t)-CUT. First, we consider parameters related to the structure of G. For example, we study parameterization by the number ξ_i of edge deletions that are needed to transform G into a graph with maximum degree i. We show that COLORED (s,t)-CUT is W[2]-hard when parameterized by ξ_3, but fixed-parameter tractable when parameterized by ξ_2. Second, we consider parameters related to the coloring ℓ. We show fixed-parameter tractability for three parameters that are potentially smaller than the total number of colors $|C|$ and provide a linear-size problem kernel for a parameter related to the number of edges with a rare edge color.

1 Introduction

The design of networks that are robust against failure of network components is an important step in the quest for secure communication systems [9]. Since current communication networks are in fact multilayer networks, it is important to consider *multiple failure* scenarios where a failure of a single layer may affect direct connections between many different nodes at once–even if these nodes are spread widely throughout the network [1,5]. Thus, it has been proposed to use edge-colored graphs consisting of a graph $G = (V, E)$, a color set C, and an edge-coloring $\ell : E \to C$ to model the layers. If a network layer fails, then all edges with the corresponding color become unavailable for communication. In other words, we may think of these edges as being removed from the graph. One measure for the vulnerability of a network in this model is the number of layers that have to fail in order to disconnect two given important nodes s and t. To compute this vulnerability measure, one needs to solve the following computational problem [1,5].

Some of the results of this work are also contained in the first author's Master thesis.
F. Sommer—was supported by the DFG, project MAGZ (KO 3669/4-1).

A. Chatzigeorgiou et al. (Eds.): SOFSEM 2020, LNCS 12011, pp. 248–259, 2020.
https://doi.org/10.1007/978-3-030-38919-2_21

COLORED (s,t)-CUT
Input: An edge-colored graph $(G=(V,E),C,\ell)$, two vertices s and t, and a positive integer k.
Question: Is there a subset of colors $S \subseteq C$ with $|S| \leq k$ such that s and t are not in the same connected component in $G' := (V, E \setminus E_S)$, where $E_S := \{e \in E \mid \ell(e) \in S\}$?

COLORED (s,t)-CUT is NP-hard [1,5]. Motivated by this hardness, we study the parameterized complexity of the problem.

Known Results and Related Work. To our knowledge, COLORED (s,t)-CUT was first introduced in a directed version in the context of the analysis of directed attack graphs [7,10]. It was shown, by a reduction from HITTING SET, that in this setting computing (s,t)-cuts with few colors is NP-hard [7,10]. While the graph is directed in this case, the reduction can be easily adapted to show NP-hardness of the undirected case by discarding all edge directions in the constructed graph G. Moreover, this reduction also implies that COLORED (s,t)-CUT is W[2]-hard when parameterized by k. In later work, this reduction from HITTING SET and the above-mentioned hardness results were also discovered directly for COLORED (s,t)-CUT [6,8,11,12]. When the same reduction is from VERTEX COVER, the special case of HITTING SET where every hyperedge has size two, then the resulting instance of COLORED (s,t)-CUT has a vertex cover of size two [12] making the problem NP-hard even in this very restricted case. Moreover, COLORED (s,t)-CUT is NP-hard even if G is a complete graph [11].

On the positive side, by considering all possible choices for choosing the k colors that shall be removed, COLORED (s,t)-CUT can be solved in $n^{\mathcal{O}(k)}$ time. This implies an $n^{\mathcal{O}(\Delta)}$-time algorithm, where Δ is the maximum degree of G, since instances with $\Delta \leq k$ are trivial yes-instances. Moreover, COLORED (s,t)-CUT can be solved in $\mathcal{O}(2^c \cdot (n+m))$ time, where $c := |C|$ is the number of colors. COLORED (s,t)-CUT can be solved in polynomial time when each edge color appears in at most two (s,t)-paths [8,11] and if every edge color has span one. Herein, the *span* of a color is the number of connected components in the subgraph of G that contains only the edges of this color and their endpoints [1]. The latter result was later extended to an algorithm with running time $2^{c_{\text{span}}} \cdot n^{\mathcal{O}(1)}$ where c_{span} is the number of edge colors that have span at least two [2,8,11]. COLORED (s,t)-CUT is fixed-parameter tractable (FPT) with respect to the combination of $p_{\max}$ and k where $p_{\max}$ is the number of edges of a longest simple path between s and t [13]. Finally, COLORED (s,t)-CUT is FPT with respect to the number of (s,t)-paths in G [8]. For all known nontrivial parameters that lead to FPT algorithms, that is, for c, $p_{\max}+k$, c_{span}, and for the number of (s,t)-paths, COLORED (s,t)-CUT does presumably not admit a polynomial problem kernel [8,11].

Our Results. We study new parameterizations for COLORED (s,t)-CUT. Recall that it is known that COLORED (s,t)-CUT is W[2]-hard for the budget parameter k and that COLORED (s,t)-CUT is NP-hard even when G has a vertex cover of size two. The latter result excludes tractability for most standard parameterizations

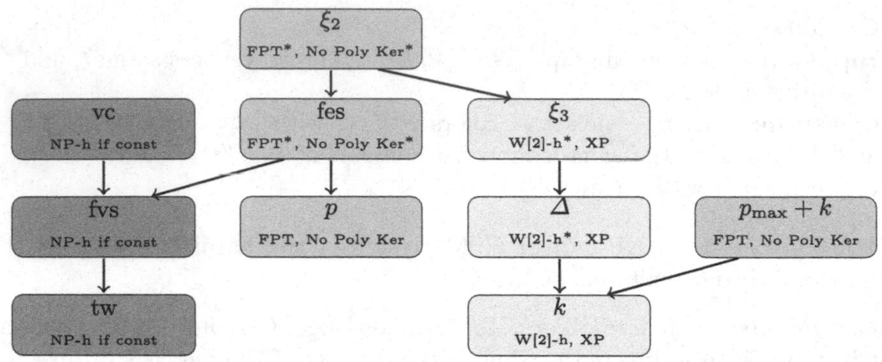

Fig. 1. The parameterized complexity of COLORED (s,t)-CUT for structural graph parameters; vc, fes, fvs, and tw denote the vertex cover number, feedback edge set number, feedback vertex set number, and treewidth, respectively. New results are marked by an asterisk. An arrow $a \to b$ symbolizes that $a \geq g(b)$ for some function g in all graphs. Note that $\xi_2 \to$ fes holds only for connected graphs; for COLORED (s,t)-CUT we assume that G is connected.

that are related to the structure of G, for example for the treewidth of G, the vertex deletion distance to forests (known as feedback vertex set number), or the vertex deletion distance to graphs with maximum degree i: the corresponding parameters are never larger than the size of a smallest vertex cover of G. Thus, we first consider parameters that are related to the *edge deletion distance* to tractable cases of COLORED (s,t)-CUT. Our results are shown in Fig. 1.

Since COLORED (s,t)-CUT can be solved in polynomial time on graphs with constant maximum degree Δ, we consider parameterization by ξ_i, the number of edges that need to be deleted in order to transform G into a graph with maximum degree i. We show that for all $i \geq 3$, COLORED (s,t)-CUT is W[2]-hard for ξ_i. This also implies W[2]-hardness for the parameter Δ: For a vertex of degree $\Delta \geq i$, at least ξ_i incident edges have to be deleted to decrease its degree to i. Hence, $\Delta \leq \xi_i + i$. Consequently, our result strengthens the W[2]-hardness for the parameter k, as $k \leq \Delta$ in all non-trivial instances. Hence, the known $n^{\mathcal{O}(\Delta)}$-time algorithm for graphs with constant maximum degree cannot be improved to an algorithm with running time $f(\Delta) \cdot n^{\mathcal{O}(1)}$. We then show an FPT algorithm for parameterization by ξ_2. This algorithm is obtained via the FPT algorithm for the parameter "number p of simple (s,t)-paths in G". The latter algorithm also gives an FPT algorithm for parameterization by the feedback edge set number of G, the number of edges that need to be removed to transform G into a forest. We also observe that COLORED (s,t)-CUT does not admit a polynomial kernel for ξ_2 and for the feedback edge set number of G.

We then study parameterizations that are related to the edge-coloring ℓ of G; our results are shown in Fig. 2. Assume that $C = \{\alpha_1, \ldots, \alpha_c\}$ and there are at least as many edges with color α_i as with color α_{i+1} for all $i < c$. For any number q, we let the parameter $m_{>q} := |\{e \in E \mid \ell(e) = \alpha_j \text{ for } j > q\}|$ denote

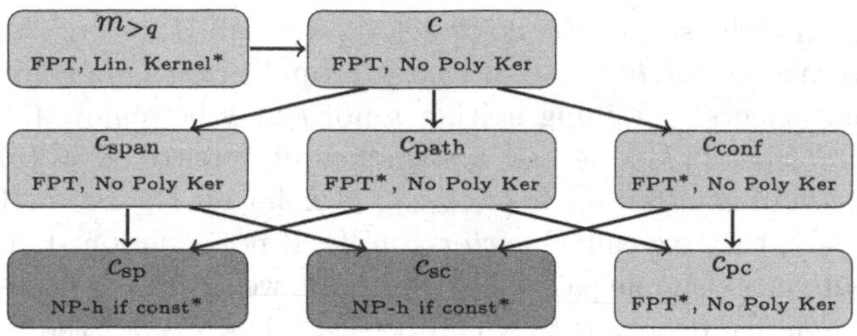

Fig. 2. An overview of the parameterized complexity of COLORED (s,t)-CUT for color-related parameters as analyzed in Sects. 3 and 4. New results are marked by an asterisk. An arrow $a \rightarrow b$ between two parameters a and b, symbolizes that $a \geq g(b)$ for some function g in all instances.

the number of edges with a color that is not among the q most frequent colors. Note that $c \leq m_{>q} + q$ and $m_{>q} \leq m$. Hence, for constant q, the parameter $m_{>q}$ is an intermediate parameter between c and m. We show that for all constant q, COLORED (s,t)-CUT admits a problem kernel of size $\mathcal{O}(m_{>q})$.

We then provide a general framework to obtain FPT algorithms for parameters that are potentially smaller than c, the number of colors. To formulate the framework, we identify certain properties of color sets in the input instances that directly give an FPT algorithm for the parameterization by the size of this color set. We then provide four applications of this framework. The first application is for c_{span}, the number of colors with span at least two. For this parameterization, an FPT algorithm is already known [2,8,11], and an algorithm with the same running time can be obtained by applying our framework. The second application is for parameterization by the number c_{path} of colors that appear in at least three (s,t)-paths. Using our framework, we extend the known polynomial-time algorithm for $c_{\text{path}} = 0$, to an FPT algorithm with running time $2^{c_{\text{path}}} \cdot n^{\mathcal{O}(1)}$. The third application is for the parameterization by c_{conf} which we define as follows. Two colors i and j are in *conflict* if G contains some (s,t)-path containing i and j. Then, c_{conf} is the number of colors i that are in conflicts with at least three other colors. We show by applying our framework, that COLORED (s,t)-CUT can be solved in $2^{c_{\text{conf}}} \cdot n^{\mathcal{O}(1)}$ time. Finally, we strengthen the latter two results by showing an FPT algorithm for the parameter c_{pc} counting the number of colors which are in at least three paths and in at least three conflicts. The parameter c_{pc} can be seen as an "intersection" of c_{path} and c_{conf}. We also show that COLORED (s,t)-CUT is NP-hard even when every color has span one or occurs in at most two paths, and NP-hard even when every color has span one or occurs in at most two conflicts. Thus, an FPT algorithm is unlikely for the intersection of c_{span} with c_{path} or c_{conf}, denoted by c_{sp} and c_{sc}, respectively.

Preliminaries. An *edge-colored graph* is a triple $\mathcal{H} = (G = (V,E), C, \ell : E \rightarrow C)$ where G is an undirected graph, C is a set of colors and $\ell : E \rightarrow C$ is an *edge coloring*. We extend the definition of ℓ to edge sets $E' \subseteq E$ by defining $\ell(E') := \{\ell(e) \mid e \in E'\}$. We let n and m denote the number of vertices and edges in G, respectively, and c the size of the color set C. We call $|I| := m + n$ the *size* of an instance I. We assume $k < m$ and that all input graphs are connected, since connected components containing neither s nor t may be removed.

In a graph $G = (V, E)$, we call a sequence of vertices $P = (v_1, \dots, v_k) \in V^k, k \geq 1$, a *path* of length $k-1$ if $\{v_i, v_{i+1}\} \in E$ for all $1 \leq i < k$. If $v_i \neq v_j$ for all $1 \leq i < j \leq k$, then we call P *vertex-simple*. If not mentioned otherwise, we only talk about vertex-simple paths. Furthermore, we say that a path $(v_1, \dots, v_k)$ is a (v_1, v_k)*-path*. We denote with $V(P) := \{v_i \mid 1 \leq i \leq k\}$ the vertices of P and with $E(P) := \{\{v_i, v_{i+1}\} \mid 1 \leq i < k\}$ the edges of P. Hence, $\ell(E(P))$ denotes the set of colors of a path P in a colored graph $(G = (V, E), C, \ell)$. Given two paths $P_1 = (v_1, \dots, v_k)$ and $P_2 = (w_1, \dots, w_r)$ in G, we define the *concatenation* as $P_1 \cdot P_2 := (v_1, \dots, v_k, w_1, \dots, w_r)$. Note that $P_1 \cdot P_2$ is a path if $\{v_k, w_1\} \in E$. Let $\mathcal{H} = (G, C, \ell)$ be a colored graph and let $s, t \in V$ be two vertices in G. We say that $\tilde{C} \subseteq C$ is a *colored* (s,t)*-cut* in G if for every (s,t)-path P in G, $\ell(E(P)) \cap \tilde{C} \neq \emptyset$. We denote by $\mathcal{C}(\mathcal{H}) := \{\ell(E(P)) \mid P \text{ is an } (s,t)\text{-path in } G\}$ the collection of sets of colors of vertex simple (s,t)-paths in G. Note that $\tilde{C} \subseteq C$ is a colored (s,t)-cut in G if and only if $\tilde{C} \cap C' \neq \emptyset$ for all $C' \in \mathcal{C}(\mathcal{H})$. The following lemma implies that we can efficiently compute an "or"-composition of many COLORED (s,t)-CUT instances.

Lemma 1 ([11]). *Let $I_1, I_2, \dots, I_i$ be a set of* COLORED (s,t)-CUT *instances with the same budget k. Then, we can compute in linear time an instance I' with budget k such that I' is a yes-instance if and only if I_j is a yes-instance for at least one $j \in \{1, \dots, i\}$ and $|I'| \leq \sum_{j=1}^{i} |I_i|$.*

For the standard notions on parameterized complexity, refer to [3,4]. Due to lack of space, several proofs are deferred to the full version of this paper.

2 Structural Graph Parameters

As discussed above, COLORED (s,t)-CUT is unlikely to be FPT for vertex deletion parameters. We thus consider edge deletion parameters.

Definition 1. *Let $G = (V, E)$ be a graph and $i \geq 0$ be an integer. Further, let $\xi_i := \min\{|E'| \mid E' \subseteq E, G - E'$ has a maximum degree of $i\}$ be the* edge deletion distance to a maximum degree of i.

Since COLORED (s,t)-CUT can be solved in polynomial time for constant Δ, the parameter ξ_i measures the distance to a trivial case. Since $\Delta \leq \xi_i + i$, COLORED (s,t)-CUT parameterized by ξ_i is in XP when i is constant. The larger i, the smaller the parameter value ξ_i will be in most instances. We now show that even for small i, namely for $i = 3$, an FPT algorithm for ξ_i is unlikely.

Theorem 1. COLORED (s,t)-CUT *parameterized by ξ_3 is* W[2]*-hard even on planar graphs.*

We now show that this result is tight by showing an FPT algorithm for ξ_2 which is obtained via an FPT algorithm for p, the number of (s,t)-paths in G.

Proposition 1 ([8]). Colored (s,t)-cut *is* FPT *parameterized by* p *and does not admit a polynomial kernel unless* NP $\subseteq$ coNP/poly.

For a graph $G = (V, E)$, we call $F \subseteq E$ a *feedback edge set* if $G - F$ is a forest. We define with $\text{fes} := \min\{|F| \mid F \text{ is a feedback edge set}\}$ the *feedback edge set number.* The following can be obtained by applying Proposition 1.

Proposition 2. Colored (s,t)-cut *is* FPT *parameterized by* fes *or* ξ_2 *does not admit a polynomial kernel for* fes $+\xi_2$, *unless* NP $\subseteq$ coNP/poly.

3 A Kernel for the Number of Edges with Rare Colors

In this section, we give a linear problem kernel for Colored (s,t)-cut parameterized by the number of edges whose color is not among the top-q most frequent colors. More precisely, we define a family of parameters $m_{>q}$ for every $q \in \mathbb{N}$ as follows. For a Colored (s,t)-cut-instance I with color set C, let $(\alpha_1, \alpha_2, \ldots, \alpha_c)$ be an ordering of the colors in C such that the number of edges with color α_i is not smaller than the number of edges with color α_{i+1} for all $i \in \{1, \ldots, c-1\}$. For a given constant q, let $\tilde{C} \subseteq C$ be the set of the q most frequent colors. We then define $m_{>q}$ as the number of edges that are not assigned to a color in $\tilde{C}$. In the following, we show a linear problem kernel for $m_{>q}$ for every q.

Informally, the kernel is based on the following idea: Since q is a constant, we may try all possible partitions of $\{\alpha_1, \ldots, \alpha_q\}$ into a set of colors C_r that we want to remove and a set of colors C_m that we want to keep. Fix one partition (C_r, C_m). Under the assumption posed by this partition, we can simplify the instance as follows. The edges of C_r can be deleted. Moreover, all vertices that are connected by a path P in G, such that $\ell(E(P)) \subseteq C_m$ cannot be separated anymore under this assumption. Thus, all vertices of P can be merged into one vertex. To formalize this merging, we give the following definition. For a colored graph $(G = (V, E), \ell)$ and a set $C_m \subseteq C$, we define $[v]_{C_m} := \{u \in V \mid \exists P = (v, \ldots, u) \text{ in G} : \ell(E(P)) \subseteq C_m\}$ as the set of vertices that are connected to v by a path only colored in C_m. If C_m is clear from the context, we may only write $[v]$. The instance that can be built for specific sets C_r and C_m is defined as follows.

Definition 2. *Let* $I = (G, C, \ell, s, t, k)$ *be a* Colored (s,t)-cut *instance and let* $C_r, C_m \subseteq C$ *with* $C_r \cap C_m = \emptyset$. *The* remove-merge-instance *of* I *with respect to* (C_r, C_m) *is* $\text{rmi}(I, C_r, C_m) := (G' = (V', E'), C', \ell', [s], [t], k - |C_r|)$, *where* $C' := C \setminus (C_r \cup C_m)$, $V' := V_1' \cup V_2'$, *and*

$$
\begin{aligned}
V_1' &:= \{[v] \mid v \in V\}, \\
V_2' &:= \{v^{\alpha}_{\{[u],[w]\}} \mid [u],[w] \in V_1', \alpha \in C', [u] \neq [w], \\
&\qquad \exists u' \in [u], w' \in [w] : \{u', w'\} \in E, \ell(\{u', w'\}) = \alpha\}, \\
E' &:= \{\{[w], v^{\alpha}_{\{[u],[w]\}}\}, \{[u], v^{\alpha}_{\{[u],[w]\}}\} \mid v^{\alpha}_{\{[u],[w]\}} \in V_2'\}, \text{ and} \\
\ell'(\{[u], v^{\alpha}_{\{[u],[w]\}}\}) &:= \alpha.
\end{aligned}
$$

The vertices of V_2' only exist to prevent G' from having parallel edges. We first show that a remove-merge-instance can be computed efficiently.

Proposition 3. *Let* $I = (G = (V, E), C, \ell, s, t, k)$ *be a* Colored (s,t)-cut *instance, and let* $I' = \text{rmi}(I, C_r, C_m)$ *be the remove-merge-instance of* I *for some* $C_r, C_m \subseteq C$ *such that* $C_r \cap C_M = \emptyset$. *Then,* $|I'| \in \mathcal{O}(|I|)$ *and* I' *can be computed in* $\mathcal{O}((|C_r| + |C_m|) \cdot m)$ *time.*

We now show that for any $\tilde{C} \subseteq C$, we can solve the original instance by creating and solving all possible remove-merge-instances for subsets of $\tilde{C}$.

Lemma 2. *Let* $I := (G = (V, E), C, \ell, s, t, k)$ *be a* Colored (s,t)-cut *instance and let* $\tilde{C} \subseteq C$, *then* I *is a yes-instance if and only if there is a subset* $C_r \subseteq \tilde{C}$ *such that the remove-merge-instance* $I' := \text{rmi}(I, C_r, \tilde{C} \setminus C_r)$ *is a yes-instance.*

Theorem 2. *For every constant* $q \in \mathbb{N}$, Colored (s,t)-cut *admits a problem kernel of size* $\mathcal{O}(m_{>q})$ *that can be computed in* $\mathcal{O}(|I|)$ *time.*

Proof. Let $I = (G = (V, E), C, \ell, s, t, k)$ be an instance of Colored (s,t)-cut and let $\tilde{C} = \{\alpha_1, \alpha_2, \ldots, \alpha_q\} \subseteq C$ be the set of the q most-frequent colors. We first describe how to compute an equivalent instance I' from I in linear time and afterwards we show that $|I'| \in \mathcal{O}(m_{>q})$.

Construction of I'. We start by computing the set $\mathcal{I} = \{\text{rmi}(I, C_r, \tilde{C} \setminus C_r) \mid C_r \subseteq \tilde{C}\}$ containing for every $C_r \subseteq \tilde{C}$, the remove-merge instances of I with respect to $(C_r, \tilde{C} \setminus C_r)$. Note that $|\mathcal{I}| = 2^q \in \mathcal{O}(1)$. We write $\mathcal{I} = \{I_1, I_2, \ldots, I_{2^q}\}$ and let $I_i =: (G_i = (V_i, E_i), C_i, \ell_i, [s]_i, [t]_i, k_i)$ denote each instance $I_i \in \mathcal{I}$. By Proposition 3 we can compute every $I_i \in \mathcal{I}$ in $\mathcal{O}(q \cdot |I|) = \mathcal{O}(|I|)$ time. Therefore, we can compute $\mathcal{I}$ in $\mathcal{O}(|I|)$ time. Note that $\max_{i \in \{1,\ldots,2^q\}} k_i = k$ and that $C_i = C \setminus \tilde{C}$ for every $i \in \{1, \ldots, 2^q\}$.

Next, we apply the algorithm of Lemma 1 on all instances of $\mathcal{I}$. Note that the budgets k_i of the instances $I_i \in \mathcal{I}$ might not be equal. Thus, in order to apply Lemma 1 we transform every instance $I_i \in \mathcal{I}$ into an instance I_i^* by adding auxiliary vertices $v_1, \ldots, v_{k-k_i}$ to V_i and auxiliary edges $\{[s]_i, v_j\}$ and $\{[t]_i, v_j\}$ for every $j \in \{1, \ldots, k-k_i\}$ to E_i. Let V_i^* and E_i^* be the resulting sets. Finally, we set $k_i^* = k$ and $\ell_i^*(e) = \ell_i(e)$ if $e \in E_i$ and $\ell_i^*(\{[s]_i, v_j\}) = \ell_i^*(\{[t]_i, v_j\}) = \alpha_j$ for every $j \in \{1, \ldots, k-k_i\}$. Note that we added at most $k-k_i$ vertices and $2(k-k_i$ edges to every instance I_i and that $k-k_i \leq q$. Since q is a constant, $|I_i^*| \in \mathcal{O}(|I_i|)$ and I_i^* can be computed from I_i in $\mathcal{O}(|I_i|)$ time.

Let $\mathcal{I}^* = \{I_1^*, \ldots, I_{2^q}^*\}$ be the resulting set of instances. Note that the budget is k in all instances in $\mathcal{I}^*$. Therefore, we can apply Lemma 1 on the 2^q instances in $\mathcal{I}^*$ and compute an instance I' in $\mathcal{O}(|I|)$ time, such that I' is a yes-instance if and only if there exists some $i \in \{1, \ldots, 2^q\}$ such that I_i^* is a yes-instance. We defer the proof of the equivalence of I and I'.

Size of I'. It remains to give a bound for the size of I'. By Definition 2 of remove-merge-instances, every $I_i \in \mathcal{I}$ contains no edges with a color in $\tilde{C}$, and subdivides every other edge of I. Therefore, every $I_i \in \mathcal{I}$ contains at most $2m_{>q}$ edges. Since $|I_i^*| \in \mathcal{O}(|I_i|)$ we conclude $|I_i^*| \in \mathcal{O}(m_{>q})$ for every $I_i^* \in \mathcal{I}$. Finally, by Lemma 1 it holds that $|I'| \leq \sum_{i=1}^{2^q} |I_i^*| \in \mathcal{O}(m_{>q})$, since $2^q \in \mathcal{O}(1)$. □

4 Parameterization by Color Subsets

In this section we present a general framework for color parameterizations of Colored (s,t)-cut leading to an FPT algorithm. To apply our framework, one has to check two properties of the parameterization.

Definition 3. *A function π that maps every instance $I = (G, C, \ell, s, t, k)$ of* Colored (s,t)-cut *to a subset $\pi(I) \subseteq C$ of the colors of I is called a* color parameterization. *If for every* Colored (s,t)-cut *instance I, $\pi(I)$ can be computed in polynomial time and I can be solved in polynomial time if $\pi(I) = \emptyset$, then π is called a* polynomial color parameterization.

In the following, we will only deal with polynomial color parameterizations. Next, we will use remove-merge-instances to transform an instance I of Colored (s,t)-cut to a set $\mathcal{I}$ of remove-merge-instances of Colored (s,t)-cut such that $\pi(I') = \emptyset$ for each $I' \in \mathcal{I}$ and $\mathcal{I}$ has size $f(\pi(I))$ for some computable function f. Each I' can be solved in polynomial-time since π is polynomial and $\pi(I') = \emptyset$. This leads to an FPT algorithm.

Definition 4. *A color parameterization π has the* strong remove-merge property *if for every* Colored (s,t)-cut *instance I, every $\tilde{C}$ and every $C_r \subseteq \tilde{C}$ it holds that $\pi(I') \subseteq \pi(I)$ where $I' := \text{rmi}(I, C_r, \tilde{C} \setminus C_r)$. Further, π has the* weak remove-merge property *if for every* Colored (s,t)-cut *instance I and every $C_r \subseteq \pi(I)$ it holds that $\pi(I') = \emptyset$ where $I' := \text{rmi}(I, C_r, \pi(I) \setminus C_r)$.*

Lemma 3. *If π has the strong remove-merge property, π also has the weak remove-merge property.*

Lemma 4. *Let π be a polynomial color parameterization with the weak remove-merge property. Then, any instance I of* Colored (s,t)-cut *can be solved in $2^{|\pi(I)|}|I|^{\mathcal{O}(1)}$ time and* Colored (s,t)-cut *does not admit a polynomial kernel for $|\pi(I)|$, unless* NP $\subseteq$ coNP/poly.

Proof. First, we present an FPT algorithm with the claimed running time. Let I be an instance of Colored (s,t)-cut. We compute $\pi(I)$ and the set $\mathcal{I}$ of all remove-merge-instances for G with respect to $\pi(I)$ and answer yes if and only if there is some $I' \in \mathcal{I}$ such that I' is a yes-instance. This algorithm is correct due to Lemma 2. Since π is a polynomial color parameterization, we can compute $\pi(I)$ in polynomial time. Since $|\mathcal{I}| = 2^{|\pi(I)|}$, we can compute $\mathcal{I}$ in $2^{|\pi(I)|}|I|^{\mathcal{O}(1)}$ time. Since π is a polynomial color parameterization that has the weak remove-merge property, we can solve each $I' \in \mathcal{I}$ in $|I|^{\mathcal{O}(1)}$ time. Hence, this algorithm runs in $2^{|\pi(I)|}|I|^{\mathcal{O}(1)}$ time. The kernel lower bound follows from the fact that in every instance I of Colored (s,t)-cut it holds that $|\pi(I)| \leq c$ and Colored (s,t)-cut admits no kernel when parameterized by c, unless NP $\subseteq$ coNP/poly. □

Next, we apply Lemma 4 to three polynomial color parameterizations. The proof for the parameterization by c_{span} is deferred to the full version.

4.1 Number of Path-Frequent Colors

This parameter counts the number of colors occurring on many (s,t)-paths.

Definition 5. *Let $I = (G = (V,E), C, \ell, s, t, k)$ be a* COLORED (s,t)-CUT *instance. A color $\alpha \in C$ is called* path-frequent *if there exist at least three vertex-simple (s,t)-paths such that at least one edge on each path has color α.*

By C_{path} we denote the function that maps each COLORED (s,t)-CUT instance I to the set of path-frequent colors of I. Further, for a fixed instance I, let $c_{\text{path}} := |C_{\text{path}}(I)|$. For a fixed color α one can test in polynomial time whether α is path-frequent [11]. Further, an instance I of COLORED (s,t)-CUT can be solved in polynomial time if $C_{\text{path}}(I) = \emptyset$. [11]. Thus, the following holds.

Lemma 5. *The function C_{path} is a polynomial color parameterization. Moreover, for every α that is contained in at most two (s,t)-paths we can compute all these (s,t)-paths in polynomial time.*

Lemma 6. *The function C_{path} has the strong remove-merge property.*

Proof. Let $I = (G, C, \ell, s, t, k)$ be an instance of COLORED (s,t)-CUT, let $\tilde{C} \subseteq C$, let $C_r \subseteq \tilde{C}$ be the colors which will be removed and let $I' = (G', C', \ell', [s], [t], k - |C_r|) := \text{rmi}(I, C_r, \tilde{C} \setminus C_r)$ be the resulting remove-merge-instance. We show that $C_{\text{path}}(I') \subseteq C_{\text{path}}(I)$. Assume towards a contradiction that there is a color $\alpha \in C_{\text{path}}(I') \setminus C_{\text{path}}(I)$. Thus, there are three vertex-simple $([s],[t])$-paths P_i for $i = \{1,2,3\}$ in G' such that $\ell'(E(P_i)) \subseteq C \setminus C_{\text{path}}(I)$ and each path contains an edge of color α. By construction of G', we can assume without loss of generality that $P_i = ([v_1], v^{\alpha_1}_{[v_1],[v_2]}, [v_2], \ldots, [v_{i_r}])$ for some $i_r \in \mathbb{N}$ where $s \in [v_1]$ and $t \in [v_{i_r}]$. By definition of G', it follows that there exists some $v_i^{j_{\text{in}}} \in [v_i]$ and some $v_i^{j_{\text{out}}} \in [v_{i+1}]$ such that $e_i^j := \{v_i^{j_{\text{in}}}, v_i^{j_{\text{out}}}\} \in E$ with $\ell(e_i^j) = \alpha_i$ for each j, $1 \leq j < i_r$, where $\alpha_i \in C \setminus \tilde{C}$. Further, we set $v_i^{1_{\text{in}}} = s$ and $v_{i_r}^{j_{\text{out}}} = t$, and since $v_i^{j_{\text{in}}}, v_i^{j_{\text{out}}} \in [v_i]$, we can conclude that there is a path P_i^j from $v_i^{j_{\text{in}}}$ to $v_i^{j_{\text{out}}}$ in G such that $\ell(E(P_i^j)) \subseteq \tilde{C} \setminus C_r$. Then $P^i := P_i^1 \cdot P_i^2 \cdot \ldots \cdot P_i^{i_r}$ is a vertex-simple (s,t)-path in G such that $\ell(E(P^i)) \subseteq C \setminus C_r$. Hence, there exist at least three paths from s to t such that at least one edge has color α, a contradiction. □

Lemmas 4, 5, and 6 now give an FPT algorithm.

Theorem 3. COLORED (s,t)-CUT *can be solved in $\mathcal{O}(2^{c_{\text{path}}}|I^{O(1)}|)$ time.*

4.2 Number of Colors in at Least Three Conflicts

The next parameter concerns colors which occur on vertex-simple (s,t)-paths with many different colors.

Definition 6. *Let $I = (G = (V,E), C, \ell, s, t, k)$ be a* COLORED (s,t)-CUT *instance. Two colors $\alpha, \beta \in C$ form a* conflict *if there exists an (s,t)-path such that at least one edge on this path has color α and at least one edge has color β.*

By C_{conf} we denote the function that maps an instance I of Colored (s,t)-cut to the set of colors of I which are in conflict with at least three different colors. Further, for a fixed instance I, let $c_{\text{conf}} := |C_{\text{conf}}(I)|$.

Lemma 7. *Let $D \subseteq C$ be a color set of size at most three, then we can determine in polynomial time if there is an (s,t)-path P on G such that $D \subseteq \ell(E(P))$.*

Lemma 8. *The function C_{conf} is a polynomial color parameterization.*

Lemma 9. *The function C_{conf} has the strong remove-merge property.*

Proof. Let $I = (G, C, \ell, s, t, k)$ be an instance of Colored (s,t)-cut, let $\tilde{C} \subseteq C$, let $C_r \subseteq \tilde{C}$ be the colors which will be removed and let $I' = (G', C', \ell', [s], [t], k - |C_r|) := \text{rmi}(I, C_r, \tilde{C} \setminus C_r)$ be the resulting remove-merge-instance. We show that $C_{\text{conf}}(I') \subseteq C_{\text{conf}}(I)$. Assume towards a contradiction that there exist a color $\alpha \in C_{\text{conf}}(I')$ such that $\alpha \notin C_{\text{conf}}(I)$ and α forms conflicts with colors $\beta_1, \beta_2, \beta_3$. Let $P = ([v_1], v^{\alpha_1}_{[v_1],[v_2]}, [v_2], \dots, [v_x])$ for some $x \in \mathbb{N}$ be a vertex-simple (s,t)-path in G' containing at least one edge of color α and at least one edge of color β_i for some $i \in \{1,2,3\}$, where $s \in [v_1]$ and $t \in [v_x]$. By definition of G' it follows that there exist some $v_j^{\text{in}} \in [v_j]$ and some $v_j^{\text{out}} \in [v_{i+1}]$ such that $e_j := \{v_j^{\text{in}}, v_j^{\text{out}}\} \in E$ with $\ell(e_j) = \alpha_j$ for each $1 \leq j < x$ where $\alpha_i \in C \setminus \tilde{C}$. Further, we set $v_1^{\text{in}} = s$ and $v_x^{\text{out}} = t$. Since $v_j^{\text{in}}, v_j^{\text{out}} \in [v_j]$ we can conclude that there is a path P_j from v_j^{in} to v_j^{out} in G such that $\ell(E(P_j)) \subseteq \tilde{C} \setminus C_r$. Then $P^* := P_1 \cdot P_2 \cdot \ldots \cdot P_x$ is a vertex-simple (s,t)-path in G such that P^* contains at least one edge of color α and at least one edge of color β_i. Hence, color α forms conflicts with each β_i, a contradiction. □

Lemmas 4, 8, and 9 now give an FPT algorithm.

Theorem 4. Colored (s,t)-cut *can be solved in $\mathcal{O}(2^{c_{\text{conf}}}|I|^{O(1)})$ time.*

4.3 Parameter Intersections

In the following we study Colored (s,t)-cut parameterized by the pairwise intersection of all three parameters of the previous sections.

Theorem 5. *Let I be an instance of* Colored (s,t)-cut *and let π, ϕ be color parameterizations with the strong remove-merge property. Then the intersected parameter $\rho(I) := \pi(I) \cap \phi(I)$ also has the strong remove-merge property.*

Proof. Fix a set $\tilde{C} \subseteq C$, fix a set $C_r \subseteq \tilde{C}$ and let $I' = \text{rmi}(I, C_r, \tilde{C} \setminus C_r)$ be the resulting remove-merge-instance. We have to show that $\rho(I') \subseteq \rho(I)$. By definition, $\rho(I') = \pi(I') \cap \phi(I')$. Since π and ϕ are strong, we have $\pi(I') \subseteq \pi(I)$ and $\phi(I') \subseteq \phi(I)$. Hence, $\rho(I') \subseteq \pi(I) \cap \phi(I) = \rho(I)$. □

We now study the pairwise intersection of color parameterizations.

Definition 7. *Let $C_{\text{pc}}(I) := C_{\text{path}}(I) \cap C_{\text{conf}}(I)$ denote the function that maps an instance I of* COLORED (s,t)-CUT *to the set of colors of I which are path-frequent and contained in at least three conflicts. Further, let $c_{\text{pc}} := |C_{\text{pc}}(I)|$.*

Theorem 6. COLORED (s,t)-CUT *can be solved in $\mathcal{O}(2^{c_{\text{pc}}}|I|^{O(1)})$ time.*

Proof. We will prove this theorem by applying Lemma 4. First, we observe that C_{pc} has the weak remove-merge property: Since C_{path} and C_{conf} both have the strong remove-merge property, C_{pc} also has the strong remove-merge property due to Theorem 5.

Second, we show that C_{pc} is polynomial. According to Lemmas 5 and 8 it can be determined in polynomial time whether a color α is in $C_{\text{path}}(I)$ or in $C_{\text{conf}}(I)$. Thus, $C_{\text{pc}}(I)$ can be computed in polynomial time.

It remains to show that an instance $I = (G = (V,E), C, \ell, s, t, k)$ can be solved in polynomial time if $C_{\text{pc}}(I) = \emptyset$. Recall that $\mathcal{C}(I) := \{\ell(E(P)) \mid P \text{ is a vertex-simple } (s,t)\text{-path in } G\}$. Without loss of generality we can assume that each set $D \in \mathcal{C}(I)$ has size at least two. We first show that $\mathcal{C}(I)$ can be computed in polynomial time when $C_{\text{pc}}(I) = \emptyset$. Let $\alpha \in C \setminus C_{\text{path}}(I)$, then there exist at most two paths containing an edge with color α. Both paths can be computed in polynomial time according to Lemma 5. Let $\alpha \in C \setminus C_{\text{conf}}(I)$. In other words, α forms conflicts with at most two other colors β and γ. The colors β and γ can be computed according to Lemma 8. Hence, $\mathcal{C}(I)$ contains at most three sets containing α. Each subset $D \in \mathcal{C}(I)$ can be computed as follows: If color α forms a conflict only with one other color β, then $\{\alpha, \beta\}$ is the unique set in $\mathcal{C}(I)$ containing α. This set can be computed in polynomial time. Now, assume color α forms conflicts with colors β and γ. Next, test if $T := \{\alpha, \beta_1, \beta_2\} \in \mathcal{C}(I)$. This can be done in polynomial time due to Lemma 7. If $T \notin \mathcal{C}(I)$, $\{\alpha, \beta_1\}, \{\alpha, \beta_2\} \in \mathcal{C}(I)$ and there is no other set $D \in \mathcal{C}(I)$ such that $\alpha \in D$. If $T \in \mathcal{C}(I)$, then test for each $i \in \{1,2\}$ whether s and t are connected in $G[\ell^{-1}(\{\alpha, \beta_i\})]$. If yes, the set $\{\alpha, \beta_i\}$ is contained in $\mathcal{C}(I)$.

From $\mathcal{C}(I)$, we now construct an instance $\mathcal{I} := (\mathcal{G} = (\mathcal{V}, \mathcal{E}), C, \ell', s, t, k)$ of COLORED (s,t)-CUT as follows: For each $D \in \mathcal{C}(I)$ create an (s,t)-path P with $\ell'(P) = D$. Note that S is a colored (s,t)-cut for G if and only if S is a colored (s,t)-cut for $\mathcal{G}$.

Now, we show that a colored (s,t)-cut S with $|S| \leq k$ can be computed in polynomial time for $\mathcal{I}$. Let $\alpha \in C_{\text{path}}(\mathcal{I})$. Hence, $\alpha \in C \setminus C_{\text{conf}}(\mathcal{I})$. Hence, $\mathcal{C}(I)$ contains exactly three sets $T_1 = \{\alpha, \beta_1, \beta_2\}, T_2 = \{\alpha, \beta_1\}$ and $T_3 = \{\alpha, \beta_2\}$ containing color α. Note that if there is a fourth set $D \in \mathcal{C}(I)$ such that $\beta_j \in D$ and $D \setminus T_1 \neq \emptyset$ for some $j \in \{1,2\}$, then $\beta_j \in C_{\text{path}}(I) \cap C_{\text{conf}}(I)$, that is, β_j is in at least four paths in G and β_j forms conflicts with at least three different colors. This contradicts the assumption $C_{\text{pc}} = \emptyset$. Hence, such a set $D \in \mathcal{C}(I)$ does not exist. In other words, there is no color γ such that γ forms a conflict with β_j for $j \in \{1,2\}$. The only possible further set containing β_1 or β_2 can be $T_4 := \{\beta_1, \beta_2\}$. First, assume $T_4 \in \mathcal{C}(I)$. Then each colored (s,t)-cut S of G contains at least two of α, β_1, and β_2. Without loss of generality, add α and β_1 to S. Second, if $T_4 \notin \mathcal{C}(I)$, adding α to S covers each T_i for $i \in \{1,2,3\}$.

Afterwards, for each color α we have $\alpha \notin C_{\text{path}}(I')$ and we can apply Lemma 5. Hence, if $C_{\text{pc}}(I) = \emptyset$, I can be solved in polynomial time. □

As in Definition 7, one can define $C_{\text{ps}}(I) := C_{\text{path}}(I) \cap C_{\text{span}}(I)$ and $C_{\text{sc}}(I) := C_{\text{span}}(I) \cap C_{\text{conf}}(I)$. We show that both of them are not polynomial.

Proposition 4. Colored (s, t)-cut *is* NP*-hard even for instances* I *where* $C_{\text{ps}}(I) = \emptyset$ *and* $C_{\text{sc}}(I) = \emptyset$.

References

1. Coudert, D., Datta, P., Perennes, S., Rivano, H., Voge, M.: Shared risk resource group complexity and approximability issues. Parallel Process. Lett. **17**(2), 169–184 (2007)
2. Coudert, D., Pérennes, S., Rivano, H., Voge, M.: Combinatorial optimization in networks with shared risk link groups. Discret. Math. Theor. C. **18**(3) (2016)
3. Cygan, M., et al.: Parameterized Algorithms. Springer, Cham (2015). https://doi.org/10.1007/978-3-319-21275-3
4. Downey, R.G., Fellows, M.R.: Fundamentals of Parameterized Complexity. TCS. Springer, London (2013). https://doi.org/10.1007/978-1-4471-5559-1
5. Faragó, A.: A graph theoretic model for complex network failure scenarios. In: Proceedings of the Eighth INFORMS Telecommunications Conference (2006)
6. Fellows, M.R., Guo, J., Kanj, I.A.: The parameterized complexity of some minimum label problems. J. Comput. Syst. Sci. **76**(8), 727–740 (2010)
7. Jha, S., Sheyner, O., Wing, J.: Two formal analyses of attack graphs. In: Proceedings of 15th IEEE Computer Security Foundations Workshop, pp. 49–63. IEEE (2002)
8. Klein, S., Faria, L., Sau, I., Sucupira, R., Souza, U.: On colored edge cuts in graphs. In: Sociedade Brasileira de Computaçao, Editor, Primeiro Encontro de Teoria da Computaçao–ETC. CSBC (2016)
9. Pióro, M., Medhi, D.: Routing, Flow, and Capacity Design in Communication and Computer Networks. Morgan Kaufmann, Burlington (2004)
10. Sheyner, O., Haines, J.W., Jha, S., Lippmann, R., Wing, J.M.: Automated generation and analysis of attack graphs. In: Proceedings 2002 IEEE Symposium on Security and Privacy, pp. 273–284. IEEE Computer Society (2002)
11. Sucupira, R.A.: Problemas de cortes de arestas maximos e mínimos em grafos. Ph.D. thesis, Universidade Federal do Rio de Janeiro (2017)
12. Wang, Y., Desmedt, Y.: Edge-colored graphs with applications to homogeneous faults. Inf. Process. Lett. **111**(13), 634–641 (2011)
13. Zhang, P., Fu, B.: The label cut problem with respect to path length and label frequency. Theor. Comput. Sci. **648**, 72–83 (2016)

Simple Distributed Spanners in Dense Congest Networks

Leonid Barenboim[1(✉)] and Tzalik Maimon[2]

[1] The Open University of Israel, Raanana, Israel
leonidb@openu.ac.il
[2] Ben-Gurion University of The Negev, Beer-Sheva, Israel
tzali.maimon@bgu.post.ac.il

Abstract. The problem of computing a sparse spanning subgraph is a well-studied problem in the distributed setting, and a lot of research was done in the direction of computing spanners or solving the more relaxed problem of connectivity. Still, efficiently constructing a linear-size spanner deterministically remains a challenging open problem even in specific topologies.

In this paper we provide several simple spanner constructions of linear size, for various graph families. Our first result shows that the connectivity problem can be solved deterministically using a linear size spanner within constant running time on graphs with *bounded neighborhood independence*. This is a very wide family of graphs that includes unit-disk graphs, unit-ball graphs, line graphs, claw-free graphs and many others. Moreover, our algorithm works in the $\mathcal{CONGEST}$ model. It also immediately leads to a constant time deterministic solution for the connectivity problem in the Congested-Clique.

Our second result provides a linear size spanner in the $\mathcal{CONGEST}$ model for graphs with bounded *diversity*. This is a subtype of graphs with bounded neighborhood independence that captures various types of networks, such as wireless networks and social networks. Here too our result has constant running time and is deterministic. Moreover, the latter result has an additional desired property of a small stretch.

Keywords: Spanners · Distributed computing · Diversity

1 Introduction and Related Work

In the distributed setting we have an input graph in which each vertex represents a processor and each edge is a communication line. In this setting there are several models which are of interest. In the $\mathcal{LOCAL}$ model, the running time is counted as the number of rounds one needs to perform in order to achieve some task. The size of messages is not limited in this model and local computations are not counted towards the running time. Another well-studied model is the $\mathcal{CONGEST}$ model, which is much like the $\mathcal{LOCAL}$ model, but with a limit on

Research supported by ISF grant 724/15 and Open University of Israel research fund.

A. Chatzigeorgiou et al. (Eds.): SOFSEM 2020, LNCS 12011, pp. 260–272, 2020.
https://doi.org/10.1007/978-3-030-38919-2_22

the size of each message that can be passed on each edge in the graph. This limit is usually considered to be $O(\log n)$ bits. Recently, single-hop congest networks have been intensively studied. Specifically, the Congested-Clique model, which is sometimes referred to as *all-to-all communication*, is one where, like the $\mathcal{CONGEST}$ model, the size of messages is limited, but each vertex is connected with a communication line to all other vertices in the graph, thus forming a clique. Sometimes the input in the Congested-Clique model is a subgraph of the clique on which one wishes to solve a certain problem.

Given a connected input graph $G = (V, E)$ and a subgraph $G' = (V, E' \subseteq E)$, the connectivity problem is to decide whether all pairs of vertices in V are connected by paths in G'. The problem of spanning a graph is a well-studied problem in the distributed setting, and a lot of research was done in the direction of solving spanners or the more relaxed problem of connectivity [7,12,14–16,19, 21,23,25]. An (α, β)-*spanner* H of a graph G is one where the distance between each two vertices v, u is such that $d_H(v, u) \leq \alpha \cdot d_G(v, u) + \beta$, where d_G is the distance between two vertices in G. One can consider a spanner where $\alpha = 1$, called an *additive spanner* or one where $\beta = 0$, called a *multiplicative spanner*. In this paper we discuss multiplicative spanners, and use the shorthand notation k-*spanner* for spanners with $d_H(v, u) \leq k \cdot d_G(v, u)$, for all $(u, v) \in E$.

Graph spanners were introduced by Peleg and Ullman [23] and Peleg and Schaffer [21]. They proved that for an integer $k \geq 1$, there is an $O(k)$-spanner with $O(n^{1+\frac{1}{k}})$-edges. They showed the existence of a $(4k-1)$-spanner of size $O(n^{1+\frac{1}{k}})$ edges, where $1 < k < \log n$. They also devised a polynomial time algorithm for constructing such a spanner. Elkin and Peleg [12] showed that for any constants $\epsilon, \lambda > 0$, there exists a constant $\beta = \beta(\epsilon, \lambda)$ such that for every n-vertex graph G there is an efficiently constructible $(1 + \epsilon, \beta)$-spanner of size $O(n^{1+\lambda})$. Woodruff [25] showed that a $(2k-1)$-additive spanner has a size of $\Omega(n^{1+\frac{1}{k}})$.

Awerbuch, Berger, Cowen and Peleg work [1] provides a deterministic construction of a $64k$-multiplicative-spanner of size $\tilde{O}(n^{1+\frac{1}{k}})$ within $\tilde{O}(n^{1+\frac{1}{k}})$ time in the $\mathcal{CONGEST}$ model. Grossman and Parter [13] devised another deterministic construction of a $(2k-1)$-spanner of size $O(kn^{1+\frac{1}{k}})$ within $O(2^k n^{\frac{1}{2}-\frac{1}{k}})$ time also in the $\mathcal{CONGEST}$ model. We elaborate on additional results in Sect. 1.3 below. As one can see, constructing a linear size spanner deterministically seems to be a challenging task. Obtaining such spanners in various graph topologies is the subject of the current paper. In particular, we consider *graphs of bounded neighborhood independence*. In such graphs, each neighborhood contains at most c independent vertices, for a parameter c. Notable examples of such graphs are line graphs and claw-free graphs (for which $c \leq 2$), and unit-disk graphs (for which $c \leq 6$). Various graphs that model wireless and social networks have bounded neighborhood independence as well. Consequently, this graph family is of great interest in the distributed computing field and has been extensively studied [2,4,5,17,24].

In this paper we present three results. The first result is spanning the input graph G within k communication rounds using a spanner of size of $O(kn)$ edges, where $k = c$ denotes the neighborhood independence of G (see Subsect. 1.1). For graphs with $k = O(1)$, this is a constant time algorithm that spans a graph using a linear size spanner. To the best of our knowledge, this is the first linear size spanner result with constant time, for any family of graphs. This is helpful in the Congested-Clique model since one can send all the edges of such a spanner to a centralized vertex in the clique using Lenzen's routing scheme [18] and compute a spanning tree locally, which is of independent interest.

The second result is a simple derivation for spanning graphs with bounded diversity in the $\mathcal{LOCAL}$ model. Specifically, we notice that a 2-spanner of linear size which terminates in a single communication round can be constructed for such graphs. (See Sect. 1.1) In general, our result shows a simple spanner of size $O(Dn)$ and works for any graph within a single round of communication. That is, the diversity only affects the size of the spanner, as long as the message size is not bounded. The running time is not affected and nor does the stretch of the spanner.

The third problem we address is where we wish to achieve a spanner of a bounded stretch in congested networks. We show that an $O(D)$-multiplicative-spanner of $O(D^2 n)$ size can be computed deterministically within $O(D^2)$ running time, where D denotes the diversity of the input graph. For graphs with constant diversity D, this offers the first constant-time constant-stretch solution with linear size, for any family of graphs.

1.1 On Diversity and Neighborhood Independence

Diversity is a relatively new graph parameter, introduced in [4] and was there utilized for graph coloring problems. An exact definition of diversity can be found in Sect. 2. Although the well-defined parameter was presented in that paper, the concept was already known much earlier in the study of line-graphs in graph theory. It is long known that if G is a line-graph then there exists a clique cover of the vertices of G, such that each vertex belongs to at most 2 cliques. Recently, it was shown in [5] that it is also useful for solving maximal matching and ruling-sets. Maximal matching is another core symmetry-breaking problem in this setting. In this paper we add spanners to the list of problems which bounded diversity helps resolve. Furthermore, we do so in the bandwidth-restricted $\mathcal{CONGEST}$ model.

The number of independent vertices in each neighborhood of a graph with diversity k is at most k, since each clique contains just one independent vertex. Thus, graphs with bounded diversity have bounded neighborhood independence. However, the family of graphs with bounded neighborhood independence is wider. The study of distributed algorithms for graphs with bounded neighborhood independence was initiated in 2013 by Barenboim and Elkin [2], who devised coloring algorithms for these graphs. Very recently, improved coloring algorithms for graphs with bounded neighborhood independence were obtained by Kuhn [17].

1.2 Quick Review of Our Results

As mentioned above, in this paper we provide three results. We give a quick intuitive description of each of them.

The first algorithm for computing a spanner with an unbounded stretch in the $\mathcal{CONGEST}$ model spans a graph by choosing subsets of independent neighboring vertices for each vertex v in the input graph G. This allows to choose a subset of independent neighbors of v and connect v to them. The independent vertices are chosen in a specific order, which guarantees that paths emanating from neighbors will eventually meet. Consequently, all neighbors in the original graph become connected in the solution. Thus, the solution is a spanner. For graphs with bounded neighborhood independence k, this gives us a linear size spanner within k rounds of communications.

The second algorithm is even simpler, and within one round of communication finds a linear size 2-spanner in graphs with bound diversity in the $\mathcal{LOCAL}$ model. All vertices share their 1-hop neighborhood with all their neighbors. Then a master for each maximal clique Q can be calculated internally in each vertex which belongs to Q. Then each vertex connects to its masters in each of the maximal cliques it belongs to.

The third algorithm computes a small stretch linear size spanner in the $\mathcal{CONGEST}$ model and is provided in two stages. First, we show that in the Congested-Clique model and for graphs with bounded diversity D, one can compute a linear size spanner with a constant stretch in constant time. The algorithm proceeds in phases. The number of phases is bounded by the diversity of the input graph. In each phase each vertex handles one of the cliques it resides on. This is done as follows. In each phase vertices select neighbors that have not been handled yet. Each vertex identifies the subset of unhandled vertices selected by its neighbors that also reside within its neighborhood. These subsets provide certain information to vertices to allow each of them to handle a clique it belongs to that has not been handled yet. Consequently, within D phases all vertices handle all cliques they reside on. To this end, in each phase a certain network structure is maintained. This is a subset of a spanner that extends in each phase, until the desired spanner is achieved and the procedure terminates. In the second stage, we extend our algorithm to work also in the more general $\mathcal{CONGEST}$ model.

1.3 Related Work

The publication of Lenzen's routing scheme [18] brought much focus to the research of many problems in the Congested-Clique, among them the MST problem. Later, Hegeman et al. [14] obtained reductions between the MST problem and the connectivity problem in the congested clique, which result in $O(\log\log\log n)$-time randomized algorithms for these problems. Then, Korhonen [16] showed a deterministic variant to a randomized procedure used in [14]. However, due to certain assumptions required to employ this procedure, the deterministic running time becomes $O(\log\log n)$. This matches the time of the deterministic MST algorithm for congested cliques of Lotker et al. [19]. But still, to the best of our knowledge, no constant time deterministic algorithm exists for solving the connectivity problem in the Congested-Clique model. When considering randomized solutions the situation is much better. In fact, a randomized constant time solution for the MST problem itself was recently published by Jurdzinski and Nowicki [15], which constitutes an optimal randomized solution in this model for both MST and connectivity. We note though that the problem of a small stretch spanner remains open. And yet the best deterministic result for MST (and connectivity) in the Congested-Clique is that of Lotker et al. [19].

Derbel et al. [7] devised a k-round algorithm for constructing $(2k-1)$-spanners with optimal size. We note that all the mentioned deterministic results were achieved using message of size $O(n)$ or unbounded. The situation of deterministic algorithms using small-sized messages is more complicated. Barenboim et al. [3] showed a construction of an $O(\log^{k-1} n)$-spanner of size $O(n^{1+1/k})$ and running time of $O(\log^{k-1} n)$. Derbel, Mosbah and Zemmari [9] showed a $(2k-1)$-spanner of optimal size but in $O(n^{1-1/k})$ time. These results use messages of small size. Grossman and Parter [13] showed that a 3-multiplicative-spanner can be computed in $\tilde{O}(1)$ time in the $\mathcal{CONGEST}$ model with size $O(n^{3/2})$ deterministically. Their more generalized result is a $(2k-1)$-spanner of size $O(n^{1+\frac{1}{k}})$ with running time of $O(\sqrt{n})$ for some constant $k > 2$ for the unweighted case. Much earlier, Erdos [11] conjectured the same lower bound for the multiplicative spanner case which by many is regarded to be true although still remains unproven. If indeed the Erdos conjecture is correct, then this will also close the search for a linear size small stretch multiplicative spanner for the general case making the existence of such spanners in certain families of graphs much more interesting.

2 Preliminaries

An **(α, β)-spanner** is a subgraph H of the input graph G where for every two vertices v and u we have $d_H(v,u) \leq \alpha \cdot d_G(v,u) + \beta$.

The **neighborhood independence** of a vertex v is the size of the largest independent subset of vertices one can choose from $\Gamma(v)$ where $\Gamma(v)$ is the 1-hop neighborhood of v. Naturally, the neighborhood independence of a graph G is the maximum between all neighborhood independences of all vertices of G. The **diversity**[1] of a vertex v is the number of maximal cliques v belongs to in the input graph. The diversity of a graph $G = (V, E)$ is defined as the maximum between diversities of all vertices in V.

3 Spanning Graphs with Bounded Neighborhood Independence in the $\mathcal{CONGEST}$ Model

Let G be a graph in the $\mathcal{CONGEST}$ model which one wants to span with a spanning forest, a tree for each connected component. For graphs with neighborhood independence k we devise a deterministic k rounds algorithm for constructing

[1] Diversity can be also defined with respect to a clique cover of a given graph. Then, the diversity of a vertex is the number of maximal cliques in the cover that the vertex belongs to. In this paper, however, we do not employ clique covers, and so the diversity is defined as the number of maximal cliques in the input graph.

a spanning subgraph of size $O(k \cdot n)$. For graphs with constant neighborhood independence, this gives a simple linear size spanner which can be used to solve the connectivity problem in the Congested Clique.

Each vertex initializes a set of vertices $L_v = \Gamma(v)$ and an empty set of edges $\hat{E}_v = \emptyset$. The sets $\hat{E}_v$ are going to be used for storing the solution. The algorithm proceeds to executing k iterations. In each iteration, each vertex v chooses the vertex with the highest ID, denoted as $c(v)$, out of L_v and adds the edge $(v, c(v))$ to $\hat{E}_v$. Then v sends $ID(c(v))$ to all of its neighbors. Each neighbor reports to v if it is connected to $c(v)$. For each neighbor u of v which is connected to $c(v)$, v removes u from L_v. The vertex v also removes $c(v)$ from L_v. This concludes the description of the algorithm. After k iterations, v returns $\hat{E}_v$ as the result.

Denote G' as the subgraph induced by the subset of edges $\{\cup \hat{E}_v | v \in G\}$. Since there are k iterations, each of which adds a single edge to $\hat{E}_v$ for each $v \in V$, we obtain $|G'| = O(kn)$. The following lemma shows that G' spans G.

Lemma 1. *G' spans G.*

Proof. Let (v, u) be an edge in G. Denote $v = v_0, u = u_0$. If for some iteration $c(v_0) = u_0$ then it is clear that $(v_0, u_0) \in G'$. Otherwise, let v_1 be the vertex v_0 selects in some iteration that had u_0 removed from L_{v_0}. Thus, v_1 is a common neighbor of v_0 and u_0. Therefore, $ID(v_1) > ID(u_0)$ and $(v_0, v_1) \in G'$.

Now, either the edge $(v_1, u_0) \in \hat{E}_{u_0}$ and thus belongs to G', or, during some iteration, the vertex u_0 selects a neighbor u_1 that had v_1 removed from L_{u_0}. Thus we have $ID(u_1) > ID(v_1) > ID(u_0)$, and $(u_0, u_1) \in G'$. Now, again, either the edge $(u_1, v_1) \in \hat{E}_{v_1}$ and thus belongs to G', or, during some iteration, the vertex v_1 selects a neighbor v_2 that had u_1 removed from L_{v_1}. Thus we have $ID(v_2) > ID(u_1) > ID(v_1) > ID(u_0)$ and $(v_1, v_2) \in G'$. This cannot last infinitely since each vertex has a different ID and the paths we build have descending order. Thus, at some point the path $v_0, v_1, \ldots$ and the path $u_0, u_1, \ldots$ either meet, or there is an edge $(v_i, u_i) \in G'$ or an edge $(u_i, v_{i+1}) \in G'$. In either case we have a path from v to u. Since this is true for every edge in G, the subgraph G' spans G. □

In the Congested Clique model, it is now possible for each vertex v to send $\hat{E}_v$ to some central vertex. One can thus obtain an $O(k)$-running-time deterministic algorithm for computing a spanning tree of a subgraph G in the Congested Clique. We note that, to the best of our knowledge, our result is the first to be independent of n, and thus the running time does not increase with the growth of the number of vertices. We summarize the results.

Theorem 1. *There is an $O(k)$ running time deterministic algorithm for computing a spanning subgraph of size $O(kn)$ in the $\mathcal{CONGEST}$ model where k is the neighborhood independence of the graph.*

Corollary 1. *There is an $O(k)$ running time deterministic algorithm for solving the connectivity problem in the congested clique model by finding a spanning tree of the input subgraph where k is the neighborhood independence of the graph.*

4 A Small Size Small Stretch Spanner in Bounded Diversity Graphs

It is also of importance to find a spanning subgraph with bounded stretch. Even though spanners have been well-studied, even in the congested-clique model, to the best of our knowledge, there is no known algorithm for constructing a spanner of linear size and a small stretch for any family of graphs deterministically. The algorithm we devised in the previous section did not guarantee a bounded stretch. We start with a simple result to show that in the $\mathcal{LOCAL}$ model diversity is strongly tied to spanning. We note that this result is trivial but it only comes to show that diversity can be utilized to find an underlying connection in the graph topology. This is evident as spanners are of importance in this manner. We achieve an $O(Dn)$ size 2-stretch spanner within one round of communication. This is done as follows. Each vertex shares its 1-hop neighborhood with all its neighbors. Thus, each vertex can compute locally what are the maximal cliques it belongs to and what is the highest ID vertex in each such clique (which we call the *master* of the clique). Each vertex then connects only to the master in each clique. If (v, u) is an edge in G then v and u are connected with a path at most 2 through the master of the clique containing the edge (v, u) (if there are more than one such clique then v and u are surely connected through the master with the highest ID among all the masters of these cliques). The size of such spanner is $O(Dn)$ as each vertex is connected to at most D masters.

Theorem 2. *In the $\mathcal{LOCAL}$ model one can compute a linear size 2-spanner within one round of communication in graphs with constant diversity.*

We now show the more interesting result which is that for graphs with diversity D one can find a $(D+1)$-stretch spanner of size $O(D^2n)$ within $O(D^2)$ running time in the $\mathcal{CONGEST}$ model. For simplification, we first show our algorithm in the Congested Clique and later we will show that the all-to-all communication is not required.

We begin each iteration with each vertex v choosing a neighbor with the highest ID, denoted $c(v)$, from all vertices it is yet to be connected to in the solution. v sends $c(v)$ to all vertices in G. Let u be a neighbor of v and $c(u)$ the chosen vertex of u. We denote $C = \{c(u) \mid \{u, c(u)\} \in \Gamma(v)\}$ to be all the choices v receives from its neighbors in G such that they, the choices, are also neighbors of v in G. Let P denote all neighbors of v which made the choices in C. That is $P = \{u \mid c(u) \in C\}$. In other words, if a neighbor w of v chose a vertex $c(w)$ which is not a neighbor of v, that is $c(w) \notin \Gamma(v)$, then w is not in P. Next, we would like to connect v to all vertices in P with path of size at most $i+1$ where i denotes the current iteration. We will show that v needs only to choose D vertices out of $C \cup P$ to achieve this. After choosing a subset $\hat{E}$, v broadcasts $\hat{E}$ to all vertices in G. All vertices maintain their own copy of the result so far, denoted R, which represents all the edges accumulated so far. It is this R that each vertex uses to choose a small subset of $C \cup P$. This completes the description of the algorithm. Its pseudocode appears below. We prove its correctness in the following lemmas.

Algorithm 1. CongestSpanner(G, D)

$S = \Gamma(v)$
$R = \emptyset$
for D iterations **do**
  Choose the highest ID vertex $c(v)$ in S
  Add the edge $(v, c(v))$ to R.
  Send $ID(c(v))$ to all neighbors.
  Remove $c(v)$ from S.
  C = all choices $c(u)$ sent to v by its neighbors such that $c(u) \in \Gamma(v)$.
  P = all neighbors u of v which have a choice in C.
  Choose the smallest subset $T \subseteq C \cup P$ with the following condition: Let $E(T)$ be the set of edges between v and the vertices in T. $E(T) \cup R$ connects v to all vertices in P.
  Send the chosen subset of T to all vertices in G. Add T to R. /* This requires $|T|$ rounds. */
  Remove all vertices in P from S.
end for
return R.

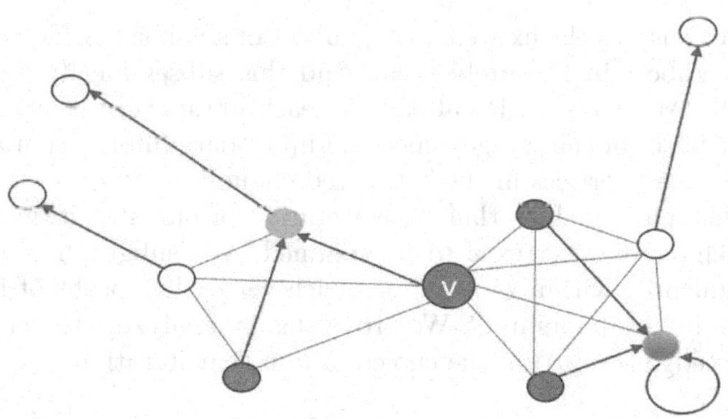

Arrows show choices. Set C in Orange. Set P in Green.

Lemma 2. *Let u, w be neighbors in G that select $c(u), c(w)$, respectively, in iteration i. If $u, w, c(u), c(w)$ belong to the same clique and $c(u) \neq c(w)$, then there is a path in R between u and w of length $i + 1$ in the end of iteration i.*

Proof. The proof is by induction on the iteration i.
Base ($i = 1$): in the first iteration, let two vertices u, w be in the same clique. Let their choices $c(u), c(w)$ be within that clique. Then both u, v select the vertex with the maximum ID in that clique. Consequently, $c(u) = c(w)$ and u is connected to w in R by a path of length 2.
Step: In iteration i, let u, w make choices in the same clique, but u selects a neighbor $c(u)$ with a smaller ID than that of the selection $c(w)$ of w. This means that u already has a path in R to $c(w)$. This path was computed in a previous

iteration, and so its length is at most i. Hence, by the end of iteration i, the vertex u is connected to w through $c(w)$ by a path of length at most $i+1$. □

Lemma 3. *At each iteration i, there is a subset in $C \cup P$ of size at most D which v can choose that connects v to each $u \in P$ with a path of size at most $i+1$.*

Proof. Denote $L = G(C \cup P)$ as the subgraph of G induced by the subset of vertices $C \cup P$ in iteration i. The diversity of L is less than or equals that of G since each distinct clique in L is a sub-clique of a distinct clique in G. Thus, v can be connected to at most D distinct maximal cliques in L. Let Q be a clique containing v. Denote all pairs of neighbors $u, c(u)$ which are in Q as $(u_1, c(u_1)), \ldots, (u_q, c(u_q)))$. Note that P consists of $u_1, u_2, \ldots, u_q$. W.L.O.G, assume that $ID(c(u_1)) \leq \cdots \leq ID(c(u_q))$. Then by Lemma 2, for each $1 \leq j \leq q$ there is a path of length at most i between u_j and $c(u_q)$. Thus by connecting v to $c(u_q)$ we obtain paths from v to all the vertices $u_1, \ldots, u_q$ of length at most $i+1$ each. The above means that for each clique v belongs to in L, we need to add only a single edge, $(v, c(u_q))$, to the solution. Hence at most D edges are added to the solution per iteration for each vertex. □

Since Lemma 3 ensures the existence of a subset of small size which connects v to all of its neighbors in P, surely v can find this subset locally and use it or maybe a smaller size subset. (Recall that in each iteration, v is aware of the entire solution R for G in that stage, since the up-to-date information about R is made available to all vertices in the congested clique).

In the next lemmas we show that each iteration of our algorithm makes a certain distinct clique of a vertex v to be spanned by a subgraph of bounded diameter. Consequently, within D iterations each vertex has paths of bounded length in R to all its neighbors in G. We prove this by analyzing the clique of v that contains the edge $(v, c(v))$ of the choice of v in that iteration.

Lemma 4. *Let Q be a maximal clique containing the edge $(v, c(v))$ in iteration i. Then, within D rounds (performed during the execution of line 11), in the output R there is a path from v to any vertex in Q of length at most $i+1$.*

Proof. Let u be a vertex in Q. Since $(v, c(v)) \in Q$, the vertex u is connected to v in G and also is connected to $c(v)$ in G. Hence, $v \in P$ as computed by u. By Lemma 3, u connects to all vertices in P. Specifically, u makes sure to connect to v with a path of length at most $i+1$. □

Lemma 5. *For $i = 1, 2, \ldots, D$, within i iterations (of the outer loop), there are paths in the output R between v and all vertices of i cliques containing v.*

Proof. Denote $c_i(v)$ as the choice v makes in iteration i. Denote $Q_1, \ldots, Q_{i-1}$ the maximal cliques which contain the edges $(v, c_1(v)), \ldots, (v, c_{i-1}(v))$ respectively. By Lemma 4, at each iteration of the outer loop the output contains a path between v and all of its neighbors in at least one maximal clique containing v. At each of the following iterations, v chooses the vertex with the highest ID

$c(v)$ such that v is not yet connected to in the output. Hence, the edge $(v, c_i(v))$ cannot belong to any of the cliques $Q_1, \ldots, Q_{i-1}$, since v already knows of a connection to all vertices in these cliques. Thus, another distinct maximal clique Q_i must contain the edge $(v, c_i(v))$. Again, by Lemma 4, the entire clique Q_i will be connected to v in iteration i. Hence, within i iterations (of the outer loop), there are paths in the output between v and all vertices of i distinct maximal cliques containing v. Specifically, after D such iterations, v will be left with no maximal cliques to choose a neighbor from, and will be connected to all of its neighbors. □

Since at each inner round each vertex adds at most D edges to the solution and since there are D outer iterations, we conclude with the following result.

Theorem 3. *There is an $O(D^2)$ running time deterministic algorithm for computing a spanner of size $O(D^2 n)$ and stretch at most $D + 1$ where D is the diversity of the graph.*

Now we note that the information used by a vertex v in order to decide on a subset $\hat{E}$ of the graph L is whether or not a neighbor u is already connected to the vertex $c(w)$ in a clique in L that contains the vertices $u, c(u), w, c(w)$ and v. In the Congested Clique model, this knowledge can be aggregated at v in each iteration. This, however, is not necessary as v can query u about the reason why it chose $c(u)$ instead of $c(w)$ in case their IDs are different; is it because u is not connected to $c(w)$ in G or is it because u is already connected to $c(w)$ in the output. To achieve this, each vertex is required to register locally the neighbors it is already connected to in the output. Unlike before, at each iteration i, R will contain the neighbors v is connected to in the output with a path of length at most $i+1$.

Another difference is that Another difference is that v is now required to build up the subset $\hat{E}$ edge by edge for $O(D)$ rounds depending on the responses it gets from querying. (Note that we refer to this inner loop as rounds and to the outer loop as iterations.) It will first choose the vertex with highest ID from C, $c(w)$, and then query all vertices in P whether they are connected to $c(w)$ already. For all those who are, v will remove their choices from C, register them in R as connected to v with a path of length at most $i + 1$ as v adds the edge $(v, c(w))$ to $\hat{E}$, and choose the next available vertex with the highest ID in C. The vertex v needs to repeat this for at most D rounds until we can be sure it is connected to all vertices in P with paths of length at most $i+1$ in accordance to Lemma 3. We provide the pseudocode of this variant for the $\mathcal{CONGEST}$ model in Algorithm 2 below. We summarize this discussion with Theorem 4.

Algorithm 2. CongestSpanner(G, D)

```
S = Γ(v)
R = ∅
for D iterations do
    Choose the highest ID vertex c(v) in S
    Add the edge (v, c(v)) to R.
    Send ID(c(v)) to all neighbors.
    Remove c(v) from S.
    C = all choices c(u) sent to v by its neighbors such that c(u) ∈ Γ(v).
    P = all neighbors u of v which have a choice in C.
    for D rounds do
        Choose the highest ID vertex w from C.
        Remove w from S and from C.
        Add (v, w) to R.
        Query all vertices in P if they already registered w as connected to them in
        their solution.
        For each neighbor u which replied 'yes', remove u from S and from P. Register
        that v is connected to u.
        C = all choices of vertices which are still in P.
        Register that v is connected to w in the solution R.
    end for
end for
return R.
```

Theorem 4. *There is an $O(D^2)$ runtime deterministic algorithm for computing a spanner of size $O(D^2 n)$ and stretch at most $D+1$ in the $\mathcal{CONGEST}$ model where D is the diversity of the graph.*

5 Conclusion

As we showed in this paper, bounded diversity is helpful for computing efficiently a linear size 2-spanner in the $\mathcal{LOCAL}$ model as well as a linear size small stretch spanner in the $\mathcal{CONGEST}$ model. As noted these results are added to several previously-known distributed algorithms for graphs with bounded diversity for well-studied problems [4,5].

Hence we believe that bounded diversity graphs are indeed an interesting family of graphs as this family covers many important network topologies and yet has deterministic polylogarithmic solutions for many of the well-studied problems in the distributed setting. Currently, out of the four major symmetry breaking problems in this setting, namely, vertex-coloring, edge-coloring, Maximal Independent Set and Maximal Matching, only the Maximal Independent Set problem remains open without such a solution for graphs with bounded diversity. We believe it would be of interest to find such a solution.

Acknowledgments. The authors are grateful to Michael Elkin for fruitful discussions and helpful remarks.

References

1. Awerbuch, B., Berger, B., Cowen, L., Peleg, D.: Near-linear cost sequential and distribured constructions of sparse neighborhood covers. In: FOCS 1993, pp. 638–647 (1993)
2. Barenboim, L., Elkin, M.: Distributed deterministic edge coloring using bounded neighborhood independence. Distrib. Comput. **26**(5–6), 273–287 (2013)
3. Barenboim, L., Elkin, M., Gavoille, C.: A fast network-decomposition algorithm and its applications to constant-time distributed computation. In: Scheideler, C. (ed.) Structural Information and Communication Complexity. LNCS, vol. 9439, pp. 209–223. Springer, Cham (2015). https://doi.org/10.1007/978-3-319-25258-2_15
4. Barenboim, L., Elkin, M., Maimon, T.: Deterministic distributed $(\Delta + o(\Delta))$-edge-coloring, and vertex-coloring of graphs with bounded diversity. In: PODC 2017, pp. 175–184 (2017)
5. Barenboim, L., Maimon, T.: Distributed symmetry breaking in graphs with bounded diversity. In: IPDPS 2018, 723–732 (2018)
6. Baswana, S., Sen, S.: A simple linear time algorithm for computing a $(2k - 1)$-spanner of $O(n^{1+1/k})$ size in weighted graphs. In: Baeten, J.C.M., Lenstra, J.K., Parrow, J., Woeginger, G.J. (eds.) ICALP 2003. LNCS, vol. 2719, pp. 384–396. Springer, Heidelberg (2003). https://doi.org/10.1007/3-540-45061-0_32
7. Derbel, B., Gavoille, C., Peleg, D., Viennot, L.: On the locality of distributed sparse spanner construction. In: PODC 2008, pp. 273–282 (2008)
8. Dor, D., Halperin, S., Zwick, U.: All-pairs almost shortest paths. In: FOCS, pp. 452–461 (1996)
9. Derbel, B., Mosbah, M., Zemmari, A.: Sublinear fully distributed partition with applications. Theory Comput. Syst. **47**(2), 368–404 (2010)
10. Elkin, M.: Computing almost shortest paths. ACM Algorithms **1**(2), 283–323 (2005)
11. Erdos, P.: Extremal problems in graph theory. In: Theory of Graphs and its Applications. Proceedings of Symposium Smolenice, pp. 29–36 (1963)
12. Elkin, M., Peleg, D.: $(1 + \epsilon, \beta)$-spanner constructions for general graphs. In: STOC 2001, pp. 173–182 (2001)
13. Grossman, O., Parter, M.: Improved deterministic distributed construction of spanners. In: DISC 2017, 24:1–24:16 (2017)
14. Hegeman, J.W., Pandurangan, G., Pemmaraju, S.V., Sardeshmukh, V.B., Scquizzato, M.: Toward optimal bounds in the congested clique: graph connectivity and MST. In: PODC 2015, pp. 91–100 (2015)
15. Jurdzinski, T., Nowicki, K.: MST in $O(1)$ Rounds of the Congested Clique (2017). https://arxiv.org/abs/1707.08484
16. Korhonen, J.H.: Deterministic MST Sparsification in the Congested Clique (2016). https://arxiv.org/pdf/1605.02022.pdf
17. Kuhn, F.: Faster Deterministic Distributed Coloring Through Recursive List Coloring (2019). arxiv.org/abs/1907.03797
18. Lenzen, C.: Optimal deterministic routing and sorting on the congested clique. In: PODC 2013, pp. 42–50 (2013)
19. Lotker, Z., Pavlov, E., Patt-Shamir, B., Peleg, D.: MST construction in $O(\log \log n)$ communication rounds. In: SPAA 2003, pp. 94–100 (2003)
20. Peleg, D.: Distributed computing: a locality-sensitive approach. In: SIAM 2000 (2000)
21. Peleg, A., Schaffer, A.: Graph spenners. J. Graph Theory **13**(1), 99–116 (1989)

22. Peleg, D., Solomon, S.: Dynamic $(1+\epsilon)$-approximate matchings: a density-sensitive approach. In: SODA 2016, pp. 712–729 (2016)
23. Peleg, D., Ullman, J.: An optimal synchronizer for the hypercube. In: PODC, pp. 77–85 (1987)
24. Schneider, J., Wattenhofer, R.: A log-star distributed Maximal Independent Set algorithm for Growth Bounded Graphs. In: Proceedings of the 27th ACM Symposium on Principles of Distributed Computing, pp. 35–44 (2008)
25. Woodruff, D.P.: Lower bounds for additive spanners, emulators, and more. In: FOCS 2006, pp. 389–398 (2006)

The Order Type of Scattered Context-Free Orderings of Rank One Is Computable

Kitti Gelle and Szabolcs Iván(✉)

Department of Computer Science, University of Szeged, Szeged, Hungary
{kgelle,szabivan}@inf.u-szeged.hu

Abstract. A linear ordering is called context-free if it is the lexicographic ordering of some context-free language and is called scattered if it has no dense subordering. Each scattered ordering has an associated ordinal, called its rank. It is known that the isomorphism problem of context-free orderings is undecidable in general. In this paper we show that it is decidable whether a context-free ordering is scattered with rank at most one, and if so, its order type is effectively computable.

1 Introduction

If an alphabet Σ is equipped by a linear order $<$, this order can be extended to the lexicographic ordering $<_\ell$ on Σ^* as $u <_\ell v$ if and only if either u is a proper prefix of v or $u = xay$ and $v = xbz$ for some $x, y, z \in \Sigma^*$ and letters $a < b$. So any language $L \subseteq \Sigma^*$ can be viewed as a linear ordering $(L, <_\ell)$. Since $\{a, b\}^*$ contains the dense ordering $(aa+bb)^*ab$ and every countable linear ordering can be embedded into any countably infinite dense ordering, every countable linear ordering is isomorphic to one of the form $(L, <_\ell)$ for some language $L \subseteq \{a, b\}^*$. A linear ordering (or an order type) is called *regular* or *context-free* if it is isomorphic to the linear ordering (or, is the order type) of some language of the appropriate type. It is known [1] that an ordinal is regular if and only if it is less than ω^ω and is context-free if and only if it is less than ω^{ω^ω}. Also, the Hausdorff rank [11] of any scattered regular (context-free, resp.) ordering is less than ω (ω^ω, resp) [5,8].

It is known [7] that the order type of a well-ordered language generated by a prefix grammar (i.e. in which each nonterminal generates a prefix-free language) is computable, thus the isomorphism problem of context-free ordinals is decidable if the ordinals in question are given as the lexicograpic ordering of *prefix* grammars. Also, the isomorphism problem of regular orderings is decidable as well [2,13], even in polynomial time [10]. At the other hand, it is undecidable for a context-free grammar whether it generates a dense language, hence the isomorphism problem of context-free orderings in general is undecidable [4].

Algorithms that work for the well-ordered case can in many cases be "tweaked" somehow to make them work for the scattered case as well: e.g. it is decidable whether $(L, <_\ell)$ is well-ordered or scattered [3] and the two algorithms are quite similar. In an earlier paper [6] we showed that it is decidable

A. Chatzigeorgiou et al. (Eds.): SOFSEM 2020, LNCS 12011, pp. 273–284, 2020.
https://doi.org/10.1007/978-3-030-38919-2_23

for a context-free ordering whether it is a well-ordering strictly below ω^2 and if so, the Cantor normal form of its order type is computable.

In the current paper we extend this result by showing that if the rank of a scattered context-free ordering is at most one (we also show that this property is decidable as well), then its order type is effectively computable (as a finite sum of the order types 1, ω and $-\omega$).

2 Notation

Orderings. A *linear ordering* is a pair $(Q, <)$, where Q is some set and the $<$ is a transitive, irreflexive and connex (that is, for each $x, y \in Q$ exactly one of $x < y$, $y < x$ or $x = y$ holds) binary relation on Q. The pair $(Q, <)$ is also written simply Q if the ordering is clear from the context. A (necessarily injective) function $h : Q_1 \to Q_2$, where $(Q_1, <_1)$ and $(Q_2, <_2)$ are some linear orderings, is called an *(order) embedding* if for each $x, y \in Q_1$, $x <_1 y$ implies $h(x) <_2 h(y)$. If Q_1 can be embedded into Q_2, then this is denoted by $Q_1 \preceq Q_2$. If h is also surjective, h is an *isomorphism*, in which case the two orderings are *isomorphic*. An isomorphism class is called an *order type*. The order type of the linear ordering Q is denoted by $o(Q)$.

For example, the class of all linear orderings contain all the finite linear orderings and the orderings of the integers ($\mathbb{Z}$), the positive integers ($\mathbb{N}$) and the negative integers ($\mathbb{N}_-$) whose order type is denoted ζ, ω and $-\omega$ respectively. Order types of the finite sets are denoted by their cardinality, and $[n]$ denotes $\{1, \ldots, n\}$ for each $n \geq 0$, ordered in the standard way.

The ordered sum $\sum_{x \in Q} Q_x$, where Q is some linear ordering and for each $x \in Q$, Q_x is a linear ordering, is defined as the ordering with domain $\{(x, q) : x \in Q, q \in Q_x\}$ and ordering relation $(x, q) < (y, p)$ if and only if either $x < y$, or $x = y$ and $q < p$ in the respective Q_x. If each Q_x has the same order type o_1 and Q has order type o_2, then the above sum has order type $o_1 \times o_2$. If $Q = [2]$, then the sum is usually written as $Q_1 + Q_2$.

If $(Q, <)$ is a linear ordering and $Q' \subseteq Q$, we also write $(Q', <)$ for the subordering of $(Q, <)$, that is, to ease notation we also use $<$ for the restriction of $<$ to Q'.

A linear ordering $(Q, <)$ is called *dense* if it has at least two elements and for each $x, y \in Q$ where $x < y$ there exists a $z \in Q$ such that $x < z < y$. A linear ordering is *scattered* if no dense ordering can be embedded into it. It is well-known that every scattered sum of scattered linear orderings is scattered, and any finite union of scattered linear orderings is scattered. A linear ordering is called a *well-ordering* if it has no subordering of type $-\omega$. Clearly, any well-ordering is scattered. Since isomorphism preserves well-orderedness or scatteredness, we can call an order type well-ordered or scattered as well, or say that an order type embeds into another. We also write $o_1 \preceq o_2$ to denote o_1 embeds into o_2.

For standard notions and facts about linear orderings see e.g. [11] or [12].

Hausdorff classified the countable scattered linear orderings with respect to their rank. We will use the definition of the Hausdorff rank from [5], which slightly differs from the original one (in which H_0 contains only the empty ordering and the singletons, and the classes H_α are not required to be closed under finite sum, see e.g. [11]). For each countable ordinal α, we define the class H_α of countable linear orderings as follows. H_0 consists of all finite linear orderings (including the empty ordering), and when $\alpha > 0$ is a countable ordinal, then H_α is the least class of linear orderings closed under finite ordered sum and isomorphism which contains all linear orderings of the form $\sum_{i\in\mathbb{Z}} Q_i$, where each Q_i is in H_{β_i} for some $\beta_i < \alpha$.

By Hausdorff's theorem, a countable linear order Q is scattered if and only if it belongs to H_α for some countable ordinal α. The *rank* $r(Q)$ of a countable scattered linear ordering is the least ordinal α with $Q \in H_\alpha$. (Observe that the notion of rank from [11] for some scattered ordering Q is either $\mathrm{r}(Q)$ or $\mathrm{r}(Q)+1$.)

As an example, ω, ζ, $-\omega$ and $\omega+\zeta$ or any finite sum of the form $\sum\limits_{i\in[n]} o_i$ with $o_i \in \{\omega, -\omega, 1\}$ for each $i \in [n]$ each have rank 1 while $(\omega+\zeta)\times\omega$ has rank 2.

Languages. Let Σ be an alphabet (a finite nonempty set) and let Σ^* (Σ^+, resp) stand for the set of all (all nonempty, resp) finite words over Σ, ε for the empty word, $|u|$ for the length of the word u, $u \cdot v$ or simply uv for the concatenation of u and v. A *language* is an arbitrary subset L of Σ^*. If $L \subseteq \Sigma^*$ is a language and $u \in \Sigma^*$ is a word, then $u^{-1}L$ denotes the (left) quotient $\{v \in \Sigma^* : uv \in L\}$.

We assume that each alphabet is equipped by some (total) linear order. Two (strict) partial orderings, the strict ordering $<_s$ and the prefix ordering $<_p$ are defined over Σ^* as follows:

- $u <_s v$ if and only if $u = u_1 a u_2$ and $v = u_1 b v_2$ for some words $u_1, u_2, v_2 \in \Sigma^*$ and letters $a < b$,
- $u <_p v$ if and only if $v = uw$ for some nonempty word $w \in \Sigma^*$.

The union of these partial orderings is the lexicographical ordering $<_\ell = <_s \cup <_p$. We call the language L well-ordered or scattered, if $(L, <_\ell)$ has the appropriate property and we define the rank $r(L)$ of a scattered language L as $r(L, <_\ell)$. The order type $o(L)$ of a language L is the order type of $(L, <_\ell)$. For example, if $a < b$, then $o\Big(\{a^k b : k \geq 0\}\Big) = -\omega$ and $o\Big(\{(bb)^k a : k \geq 0\}\Big) = \omega$.

An *ω-word* over Σ is an ω-sequence $a_1 a_2 \ldots$ of letters $a_i \in \Sigma$. The set of all ω-words over Σ is denoted Σ^ω. The orderings $<_p$, $<_s$ and $<_\ell$ are extended to $\Sigma^* \cup \Sigma^\omega$ in the expected way. For an ω-word w over Σ, we let $\mathbf{Pref}(w)$ stand for the set $\{u \in \Sigma^* : u <_p w\}$ of all finite prefixes of w. An ω-word w is called *regular* if $w = uv^\omega = uvvvv\ldots$ for some finite words $u \in \Sigma^*$ and $v \in \Sigma^+$. When w is a (finite or ω-) word over Σ and $L \subseteq \Sigma^*$ is a language, then $L_{<w}$ stands for the language $\{u \in L : u <_\ell w\}$. Notions like $L_{\geq w}$, $L_{<_s w}$ are also used as well, with the analogous semantics.

A *context-free grammar* is a tuple $G = (N, \Sigma, P, S)$, where N is the alphabet of the *nonterminal symbols*, Σ is the alphabet of *terminal symbols* (or *letters*) which is disjoint from N, $S \in N$ is the *start symbol* and P is a finite set of *productions* of the form $A \to \alpha$, where $A \in N$ and α is a *sentential form*, that is, $\alpha = X_1 X_2 \ldots X_k$ for some $k \geq 0$ and $X_1, \ldots, X_k \in N \cup \Sigma$. The derivation relations $\Rightarrow$ and $\Rightarrow^*$ are defined as usual. The *language generated by* a grammar G is defined as $L(G) = \{u \in \Sigma^* \mid S \Rightarrow^* u\}$. Languages generated by some context-free grammar are called *context-free languages.* The class of context-free languages enjoys several effective closure properties, e.g. if L is context-free and w is a regular word, then $L_{<w}$ is effectively context-free as well. For any set Δ of sentential forms, the language generated by Δ is $L(\Delta) = \{u \in \Sigma^* \mid \alpha \Rightarrow^* u \text{ for some } \alpha \in \Delta\}$. As a shorthand, we define $o(\Delta)$ as $o(L(\Delta))$. When $X, Y \in N \cup \Sigma$ are symbols of a grammar G, we write $Y \preceq X$ if $X \Rightarrow^* uYv$ for some words u and v; $X \approx Y$ if $X \preceq Y$ and $Y \preceq X$ both hold; and $Y \prec X$ if $Y \preceq X$ but not $X \preceq Y$. A production of the form $X \to X_1 \dots X_n$ with $X_i \prec X$ for each $i \in [n]$ is called an *escaping production.* A nonterminal $X \in N$ is called *left recursive* if $X \Rightarrow^+ X\alpha$ for some sentential form α and is *recursive* if $X \Rightarrow^+ \alpha X \beta$ for some α and β.

For standard notions on regular and context-free languages the reader is referred to any standard textbook, such as [9].

Linear orderings which are isomorphic to the lexicographic ordering of some context-free (regular, resp.) language are called *context-free (regular, resp.) orderings.*

3 Limits of Languages

In this section we introduce the notion of a limit of a language and establish a connection: the main contribution of this concept is that one can decide whether a context-free language has a finite number of limits and if so, one can effectively compute the limits themselves (Lemma 15), and that a language has a finite number of limits if and only if its order type is scattered of rank at most one (Theorem 1).

Firstly, we recall (and prove for the sake of completeness) that $\Sigma^\infty = \Sigma^\omega \cup \Sigma^*$ forms a complete lattice with the partial ordering $\leq_\ell$. Thus we can take e.g. the supremum for any set $X \subseteq \Sigma^\infty$.

Lemma 1. *$(\Sigma^\infty, \leq_\ell)$ is a complete lattice.*

Proof. Let L be an infinite language. If L has a maximal element, then it is the supremum, otherwise we generate the word $a_1a_2a_3\ldots \in \Sigma^\omega$ in the following way: let $u_0 = \varepsilon$ and $u_i = a_1 \ldots a_i$. We choose the largest possible letter a_{i+1} with $(u_ia_{i+1})^{-1}L$ being nonempty. The word generated by this way is the supremum of L. □

3.1 Limits in General

Though limit (point)s of a language could be defined as limits of Cauchy sequences in a particular metric space, the following definition is more convenient for our purposes.

Definition 1. *The word $w \in \Sigma^\omega$ is a* limit *of a(n infinite) language L, if for each $w_0 <_p w$ there exists a word $u \in L$ such that $w_0 <_p u$.*

If $L \subseteq \Sigma^*$ is a language, then we denote the set of limits of L by $\boldsymbol{Lim}(L)$.

Lemma 2. *If w is a limit of an infinite language L, then for each $w_0 <_p w$ there exist infinitely many words $u \in L$ with $w_0 <_p u$.*

Proof. We construct two sequences $w_0 <_p w_1 <_p \ldots <_p w$ and $u_0,\ u_1, \ldots \in L$ such that $w_i <_p u_i$ for each i with mutual induction. Now w_0 is given. By definition for each i there exists an $u_i \in L$ such that $w_i <_p u_i$ and $w_i \in \mathbf{Pref}(w)$ is constructed such that $|u_{i-1}| + 1 < |w_i|$.

It is clear the words u_i are pairwise different, since each w_i has different length. So we get that w_0 is a prefix of each u_i, so w_0 is a prefix of infinitely many words in L. □

Clearly, if L is finite, then $\boldsymbol{Lim}(L)$ is empty. For the converse we have:

Lemma 3. *For each infinite language L, the set $\boldsymbol{Lim}(L)$ is nonempty.*

Proof. We construct a limit word $w = a_1a_2\ldots \in \Sigma^\omega$. Let $u_0 = \varepsilon$ and $u_i = a_1 \ldots a_i$ and we choose $a_{i+1} \in \Sigma$ such that u_ia_{i+1} is a prefix of infinitely many words in L. Since $u_i^{-1}L$ is infinite by construction, there exists such a letter. Thus we can construct an infinite word which is a limit of L. □

It is not difficult to see that any supremum or infimum of a chain of words of a language is a limit of the language (hence the name "limit").

Lemma 4. *If $w_0, w_1, \ldots$ is a $<_\ell$ (or $>_\ell$ respectively) chain in L, then its supremum (infimum, resp.) is a limit of L.*

Now we recall from [6] that for any context-free language, we can compute a supremum or infimum of some chain within the language.

Lemma 5 ([6], Lemma 1). *For each sentential form α with $L(\alpha)$ being infinite, we can generate a sequence $w_0, w_1, \ldots \in L(\alpha)$ having the form $w_i = u_1u_2^iu_3u_4^iu_5$ and a regular word $w \in \Sigma^\omega$ satisfying one of the following cases:*

$$(i)\ w_1 <_\ell w_2 <_\ell \ldots \text{ and } w = \bigvee_{i\geq 0} w_i \quad \text{or } ii)\ w_1 >_s w_2 >_s \ldots \text{ and } w = \bigwedge_{i \geq 0} w_i$$

Hence, whenever L is an infinite context-free language, one of its limits can be effectively computed, and this particular limit will be a regular word.

Next, we show how to compute limits of unions and products:

Lemma 6. *For any languages K and L, $\boldsymbol{Lim}(K \cup L) = \boldsymbol{Lim}(K) \cup \boldsymbol{Lim}(L)$.*

Proof. Assume w is a limit of L. Then for each $w_0 <_p w$, there exists some $u \in L$ with $w_0 <_p u$. Since then $u \in K \cup L$ as well, w is a limit of $K \cup L$ as well.

For the other direction, assume w is a limit of $K \cup L$. Then for each $w_0 <_p w$, there exists some $u \in K \cup L$ with $w_0 <_p u$. Thus, either there exists infinitely many prefixes w_0 of w for which there exists some $u \in K$ with $w_0 <_p u$ or there exists infinitely many prefixes w_0 of w for which there exists some $u \in L$ with $w_0 <_p u$. In the former case, $w \in \boldsymbol{Lim}(K)$, in the latter, $w \in \boldsymbol{Lim}(L)$. □

Lemma 7. *$\boldsymbol{Lim}(KL) = \boldsymbol{Lim}(K) \cup K\boldsymbol{Lim}(L)$ if $K, L \neq \emptyset$.*

Proof. Let $u \in K$ be a word and w be a limit of L. To prove that uw is a limit of KL we only have to show that for each prefix w' of uw there exist a word $w'' \in KL$ with $w' <_p w''$. Let $w' \in \mathbf{Pref}(uw)$, and since it is enough to see the prefixes which are longer than u, the word w' can be written as uw_0. Since w is a limit of L there exists a word $v \in L$ such that $w_0 <_p v$. Thus there is a word $uv \in KL$ such that $w' <_p uv$.

Now let w be a limit of K. To prove that w is a limit of KL, let u be a word in L and w_0 be a prefix of w. Since w is a limit of K there exists a word $v \in K$ with $w_0 <_p v$ by definition. So $vu \in KL$ and w_0 is a prefix of vu as well.

For the other direction, we have to prove that there are no more limits of KL. Let w be a limit of KL and w_i be the prefix of w with length i. Since w is a limit of KL there exist a word $u_i v_i \in KL$ such that $w_i <_p u_i v_i$ where $u_i \in K$ and $v_i \in L$. Consider for each $i > 0$ the lengths of these words u_i. There are two cases: either there is a finite upper bound on $|u_i|$ or there is not. If $|u_i|$ is not upperbounded, then w is a limit of K, since for each prefix w_i there exists some long enough u_j with $w_i <_p u_j$.

In the case where the lengths of these u_i words is bounded, let $\ell = \max |u_i|$ be the maximal length. Since there are only finitely many words of length at most ℓ, there has to be some $u = u_j$ such that $u = u_i$ for infinitely many indices i. Hence in particular, $w = uw'$ for some $w' \in \Sigma^\omega$. We show that $w' \in \boldsymbol{Lim}(L)$, yielding $w \in K\boldsymbol{Lim}(L)$. Indeed, if $w'' <_p w'$ is a prefix of w', then $uw'' <_p w$ and thus there exists some v_i with $uw'' <_p uv_i$, that is, $w'' <_p v_i$ and so w' is a limit of L. □

Corollary 1. *For any language $L \subseteq \Sigma^*$ and words $u, v \in \Sigma^*$, $\boldsymbol{Lim}(uLv) = u \cdot \boldsymbol{Lim}(L)$.*

Lemma 8. *It is decidable for any context-free language L and regular word $w = uv^\omega$ whether w is a limit of L.*

Proof. Let $L \subseteq \Sigma^*$ be a context-free language and consider the generalized sequential mapping $f : \Sigma^* \to a^*$ defined as

$$f(x) = \begin{cases} a \cdot f(y) & \text{if } x = vy \text{ for some } y \in \Sigma^* \\ \varepsilon & \text{otherwise.} \end{cases}$$

Now for any word x, $f(x) = a^n$ for the unique n such that $x = v^n y$ for some y not having v as prefix. Thus, v^ω is a limit of a language L' if and only if $f(L')$ is infinite; hence, by Corollary 1, $w = uv^\omega$ is a limit of L if and only if $f(u^{-1}L)$ is infinite. Since the class of context-free languages is effectively closed under left quotients and generalized sequential mappings, and their finiteness problem is decidable, the claim is proved.

Lemma 9. *Assume K is a context-free language and v is a nonempty word. Then it is decidable whether Kv^ω is finite and if so, its members (which are regular words) can be effectively enumerated.*

Proof. Consider the generalized sequential mapping $f : \Sigma^* \to \Sigma^*$ defined as

$$f(x) = \begin{cases} f(y) & \text{if } x = v^R y \text{ for some } y \in \Sigma^* \\ x & \text{otherwise.} \end{cases}$$

(Here v^R is the reverse $a_n \dots a_1$ of the word $v = a_1 \dots a_n$.)

Now for any word x, $f(x) = y$ for some y not having v^R as prefix such that $x = (v^R)^k y$ for some $k \geq 0$, that is, f strips away the leading v^Rs of its input. So, we have that $\big(f(K^R)\big)^R$ consists of those words we get from members of K, stripping away their trailing vs. Now $u \in \big(f(K^R)\big)^R$ if and only if u does not end with v and $uv^\omega \in Kv^\omega$. Moreover, Kv^ω is finite if and only if there exist some $u_1, u_2, \dots, u_n \in \Sigma^*$ such that for each $i \in [n]$ the word u_i does not end with v and $Kv^\omega = \{u_i v^\omega \mid i \in [n]\}$. So we get that Kv^ω is finite if and only if so is $\big(f(K^R)\big)^R$ which is decidable since the class of context-free languages is effectively closed under reversal and generalized sequential mappings, and their finiteness problem is also decidable. In this case, members of $\big(f(K^R)\big)^R = \{u_1, \dots, u_n\}$ can also be effectively enumerated and $Kv^\omega = \{u_j v^\omega : j \in [n]\}$. □

3.2 Finitely Many Limits

In this part we establish the decidability of the problem whether a context-free language L has finitely many limits and that this property corresponds exactly to L being scattered of rank at most one.

In the rest of the paper when grammars are involved, we assume the grammar $G = (N, \Sigma, P, S)$ contains no left recursive nonterminals, and for each $X \in N$, $L(X) \subseteq \Sigma^+$ is infinite and $S \Rightarrow^* uXv$ for some $u, v \in \Sigma^*$. Moreover, each nonterminal but possibly S is assumed to be recursive. Any context-free grammar can effectively be transformed into such a form without changing the order type of the generated language, see e.g. [7].

It is also known [3] that if the context-free grammar G generates a scattered language, then for each recursive nonterminal X there exists a unique (and computable) primitive nonempty word u_X such that whenever $X \Rightarrow^+ uX\alpha$ for some $u \in \Sigma^*$ and sentential form α, then $u \in u_X^+$. Moreover, for each pair $X \approx Y$ of recursive nonterminals there exists a (computable) word $u_{X,Y} \in \Sigma^*$ such that whenever $X \Rightarrow^+ uYv$, then $u \in u_{X,Y} u_Y^*$.

Lemma 10. *If L has a unique limit w, then $o(L_{<w}) \preceq \omega$ and $o(L_{>w}) \preceq -\omega$. Moreover, in this case both $o(L_{<w})$ and $o(L_{>w})$ are effectively computable (and hence so is $o(L) = o(L_{<w}) + o(L_{>w})$).*

Proof. In the $o(L_{<w})$ case, if $L_{<w}$ is finite (which is decidable and if so, its size is computable), we are done. Otherwise $L_{<w}$ is infinite, which means it has a limit, and this limit has to be the w (since it is unique for L). Now for any word $u \in L_{<w}$, the language $(L_{<w})_{<u} = L_{<u}$ has to be finite, otherwise by Lemma 3 it would have a limit $w' \leq \bigvee L_{<u} < w$ and hence L would have at least two limits. Thus, each $u \in L_{<w}$ has finitely many predecessors and so $o(L_{<w}) = \omega$.

Analogously, we get that each $u \in L_{>w}$ has finitely many upper bounds in L and so $o(L_{>w})$ is either $-\omega$ or some finite number, which is decidable. □

Before proceeding to the case of concatenation, we recall the notion of prefix chains from [6]. A language $L \subseteq \Sigma^*$ is called a *prefix chain* if $L \subseteq \mathbf{Pref}(w)$ for some ω-word w. Lemma 2 from [6] states that it is decidable for any context-free language L whether L is a prefix chain and if so, a suitable regular $w = uv^\omega \in \Sigma^\omega$ can be effectively computed. For ease of notation, we use $o(X)$ and $\boldsymbol{Lim}(X)$ for $o(L(X))$ and $\boldsymbol{Lim}(L(X))$.

Lemma 11. *If X is a recursive nonterminal, then $u_X^\omega \in \boldsymbol{Lim}(X)$ and if some $w \neq u_X^\omega$ is also a member of $\boldsymbol{Lim}(X)$, then $\boldsymbol{Lim}(X)$ is infinite.*

Proof. Since $X \Rightarrow^* u_X^n X\alpha$ holds for each recursive nonterminal for some $n > 0$ and sentential form α, there exists a word $v \in L(X)$ for each $u \in \mathbf{Pref}(u_X^\omega)$ such that $u <_p v$: we only have to take a word generated from $(u_X^n)^k X\alpha^k$ for a sufficiently large k.

If $w \neq u_X^\omega$ is also a limit of $L(X)$, then it can be written as $w = ubw'$, where $u <_p u_X^\omega$, $ub \not<_p u_X^\omega$ and $w' \in \Sigma^\omega$. So if we consider a derivation of the form $X \Rightarrow^* u_X^n X\alpha$, we get that each $(u_X^n)^k ubw'$, $k > 0$ is a limit, thus $\boldsymbol{Lim}(X)$ is infinite as these words are pairwise different, since in each such word the marked occurrence of b is the first position where $(u_X^n)^k ubw'$ differs from u_X^ω. □

Lemma 12. *Assume K and L are context-free languages such that $\boldsymbol{Lim}(K) = \{u_i v_i^\omega \mid 1 \leq i \leq k\}$ and $\boldsymbol{Lim}(L) = \{u'_j w_j^\omega \mid 1 \leq j \leq \ell\}$ are finite sets of regular words. Then it is decidable whether $\boldsymbol{Lim}(KL)$ is finite and if so, then it is a computable (finite) set of regular words.*

Proof. By Lemma 7, $\boldsymbol{Lim}(KL) = \boldsymbol{Lim}(K) \cup K\boldsymbol{Lim}(L)$. Since $\boldsymbol{Lim}(K)$ is a finite set (and is of course computable since it is given as input), we only have to deal with $K \cdot \boldsymbol{Lim}(L)$. Since

$$K \cdot \boldsymbol{Lim}(L) = K \cdot \{u'_j w_j^\omega : j \in [\ell]\} = \bigcup_{j \in [\ell]} (Ku'_j) w_j^\omega$$

and this union is finite if and only if so is each language $(Ku'_j)w_j^\omega$, which is decidable by Lemma 9 (since the languages Ku'_j are each context-free), we get decidability and even computability if each of them is finite. □

Corollary 2. *Assume $n \geq 0$ and $X_1, \ldots, X_n \in N \cup \Sigma$ are symbols so that for each X_i, $\boldsymbol{Lim}(X_i)$ is a known finite set. Then it is decidable whether $L(X_1 \ldots X_n)$ has a finite number of limits and if so, $\boldsymbol{Lim}(X_1 \ldots X_n)$ is effectively computable.*

Proof. Let us introduce the fresh nonterminals $Y_1, \ldots, Y_{n-1}$ and productions $Y_1 \to X_1 Y_2$, $Y_2 \to X_2 Y_3, \ldots$, $Y_{n-1} \to X_{n-1} X_n$. Applying Lemma 12 or Corollary 1 (depending on whether X_i is a nonterminal or a letter) for the nonterminals Y_{n-1}, Y_{n-2}, $\ldots$, Y_1 in this order we can decide whether each $L(Y_i)$ has a finite number of limits, and if so, we compute $\boldsymbol{Lim}(Y_i)$ as well, proving the statement since $L(X_1 \ldots X_n) = L(Y_1)$. □

Lemma 13. *Assume X is a recursive nonterminal, $L(X)$ is not a prefix chain and for some nonterminal $X' \approx X$ there exists a production $X' \to \alpha X''\beta$ with β containing at least one nonterminal. Then $L(X)$ has infinitely many limits.*

Proof. Let $u <_s v$ be members of $L(X)$. Since β contains a nonterminal, $L(\beta)$ is infinite and has a limit w by Lemma 3. By the conditions on the recursive nonterminal X, we get $X \Rightarrow^* u_1 X u_2 \beta u_3$ for some words $u_1, u_2, u_3 \in \Sigma^*$. By Corollary 1, both $u_1 u u_2 w$ and $u_1 v u_2 w$ are limits of $L(X)$ and they are distinct by $u <_s v$. Applying Lemma 11 we get $\boldsymbol{Lim}(X)$ is infinite. □

Lemma 14. *The word $w \in \Sigma^\omega$ is the unique limit of an infinite language L if and only if for each $w_0 <_p w$ there exists only finitely many words $u \in L$ such that $u <_s w_0$ or $w_0 <_s u$.*

Proof. We will see just the case where $w_0 <_s u$, the other one can be done analogously.

Suppose for the sake of contradiction there exist infinitely many words $u \in L$ with $w_0 <_s u$. Let w_0 be the shortest such word, it can be written as $w_0 = w_0' a$. Since there are just finitely many words $u \in L$ with $w_0' <_s u$, it has to be the case that $w_0' <_p u$ and $w_0' a <_s u$ for infinitely many $u \in L$. Then there exists a letter $b \in \Sigma$ such that $b > a$ and infinitely many words $u \in L$ such that $w_0' b <_p u$. But any limit of these words is in $w_0' b \Sigma^\omega$ (and by Lemma 3 at least one limit exists), which cannot be equal to w, so the language L has two different limits which is a contradiction. □

Lemma 15. *Assume $L(G)$ is scattered. Then it is decidable for each nonterminal X whether $L(X)$ has finitely many limits and if so, $\boldsymbol{Lim}(X)$ is a computable set of regular words.*

Proof. We prove the statement by induction on $\prec$. So let X be a nonterminal and assume we already know for each $Y \prec X$ whether $\boldsymbol{Lim}(Y)$ is finite and if so, we already explicitly computed the set $\boldsymbol{Lim}(Y)$ itself.

If X is nonrecursive, and $X \to \alpha_1 \mid \ldots \mid \alpha_k$ are all the alternatives of X, then applying Corollary 2 and Lemma 6 we get both decidability and computability.

So let X be a recursive nonterminal. If $L(Y)$ has at least two limits for some $Y \prec X$, then by Corollary 1 so does $L(X)$, thus by Lemma 11 $\boldsymbol{Lim}(X)$ is infinite and we are done. So we can assume from now on that each $L(Y)$ with $Y \prec X$ has exactly one limit which is already computed.

If $L(X)$ is a prefix chain (which is decidable by Lemma 2 of [6]), then its supremum is its unique limit, we can compute it by Lemma 5. So we can assume that $L(X)$ is not a prefix chain. Now if there exist some production of the form $X' \to \alpha X''\beta$ with $X' \approx X'' \approx X$ and β containing at least one nonterminal, then by Lemma 13, $\boldsymbol{Lim}(X)$ is infinite and we can stop.

Otherwise, we can assume that each non-escaping production in the component of X has the form $X' \to \alpha X'' u$ for some $u \in \Sigma^*$, $X' \approx X'' \approx X$. Since $L(X)$ is scattered, for each such α it has to be the case that $L(\alpha) \subseteq u_{X',X''} u_{X''}^*$.

Now let $X' \approx X$ be a nonterminal and $\alpha_1, \ldots, \alpha_k$ all the escaping alternatives of X'. Applying Corollary 2 and Lemma 6 we can decide whether

$\boldsymbol{Lim}(L(\{\alpha_1,\ldots,\alpha_k\}))$ is finite, and if so, we can compute this set of regular words. Now if $L(\alpha_1,\ldots,\alpha_k)$ has at least two limits, then so does $L(X')$ and $L(X)$ as well, hence $\boldsymbol{Lim}(X)$ is infinite by Lemma 11 and we are done. Otherwise, if $L(\{\alpha_1,\ldots,\alpha_k\})$ is infinite, then its unique limit is a computable word. On the other hand, for each recursive nonterminal X' the word $u_{X'}^{\omega}$ is a limit of $L(X')$, and by $L(\{\alpha_1,\ldots,\alpha_k\}) \subseteq L(X')$, the two limits has to coincide. If these regular words are not the same (which is decidable), then again, $L(X)$ has infinitely many limits and we can stop.

Hence we can assume that for each $X' \approx X$, the language $L(X')$ has the limit $u_{X'}^{\omega}$ which is the same as the unique limit of $L(\{\alpha_1,\ldots,\alpha_k\})$ if this latter language is infinite.

We claim that in this case, $L(X)$ has the unique limit u_X^{ω}. To see this, we apply Lemma 14 and show that for each prefix w_0 of u_X^{ω}, there are only finitely many words $u \in L(X)$ with either $w_0 <_s u$ or $u <_s w_0$.

Assume to the contrary that $w_0 <_p u_X^{\omega}$ and there are infinitely many words $u \in L(X)$ with either $w_0 <_s u$ or $u <_s w_0$. Each word $u \in L(X)$ can be derived from X using a leftmost derivation sequence resulting in a sentential form $u_X^t u_{X,X'}\alpha v$ for some $t \geq 0$ so that $X' \to \alpha$ is an escaping production from the component of X and $u \in u_X^t u_{X,X'} L(\alpha) v$. Since u and $w_0 <_p u_X^{\omega}$ are not related by $<_p$, we have an upper bound for t, which, as G does not contain left-recursive nonterminals, places an upper bound for $|v|$. Hence, there are only finitely many possibilities for picking $t \geq 0$, X', α and v, thus for some combination of them, there are infinitely many such words u belonging to the same language $u_X^t u_{X,X'} L(\alpha) v$. So we can write each such u as $u = u_X^t u_{X,X'} u' v$ with $u' \in L(\alpha)$, and we can write w_0 as $w_0 = u_X^t u_{X,X'} w_0'$, that is, $w_0' <_p u_{X'}^{\omega}$. This yields that $u'v <_s w_0'$ or $w_0' <_s u'v$ for infinitely many words $u' \in L(\alpha)$. Thus, there are infinitely words $u' \in L(\alpha)$ of length at least $|w_0'|$ with either $u'v <_s w_0'$ or $w_0' <_s u'v$, hence with either $u' <_s w_0'$ or $w_0' <_s u'$, which is a contradiction, since by Lemma 14 this would yield that $L(\alpha)$ has at least two limits, which we already handled in a former case. □

Theorem 1. *Suppose L is a context-free language having a finite number of limits. Then $o(L)$ is effectively computable and is scattered of rank at most one.*

Proof. We prove the statement by induction on the number of limits.

If L has no limits, then it is finite by Lemma 3, and so $o(L) = |L|$ is computable.

If L has a unique limit (which is decidable by Lemma 15), then $o(L)$ can be embedded into $\omega + -\omega$ and is computable by Lemma 10.

Now assume L has at least two limits. Since L is infinite, we can compute a regular limit of the form $w = uv^{\omega}$ for L by Lemma 5. By Lemma 8, it is decidable whether w is a limit of either $L_{<w}$ or $L_{>w}$ or both of them. (By Lemma 6, w is a limit of at least one of them.) If w is not a limit of $L_{<w}$ ($L_{>w}$, resp.), then this language has a smaller number of limits than L and we can proceed by induction. Suppose now w is a limit of $L_{<w}$ – it has to be $w = \bigvee L_{<w}$ then. If L has a limit which is larger than w (that is, $L_{>w}$ is infinite and either w is not a limit of $L_{>w}$ or $L_{>w}$ has at least two limits – this is decidable as well), then $L_{<w}$ has a smaller number of limits than L (since no limit of $L_{<w}$ can be strictly larger than its supremum) and we can proceed again by induction and get that $L_{<w}$ is computable. It is also decidable whether $L_{<w}$ has only one limit and if so, its order type is also computable and we are done.

The last case is when $w = \bigvee L_{<w}$ is the largest limit of L and $L_{<w}$ has at least two limits. Thus, there exists some limit w' of $L_{<w}$ and an integer $n \geq 0$ such that $w' <_s uv^n$, or equivalently, $L_{<uv^n}$ is infinite for some $n \geq 0$. We can compute (say, the least) such n by starting from $n = 0$ and iterating, eventually we will find an integer n with this property. Then, $L_{<uv^n}$ has a smaller number of limits than $L_{<w}$ so we can use induction and compute $o(L_{<uv^n})$; also, $(L_{<w})_{\geq uv^n}$ has a smaller number of limits than $L_{<w}$ (since w' is missing) and we can apply induction to this half as well and compute its order type. Then, $o(L_{<w})$ is the sum of the two already computed order types.

Repeating the same argument (by appropriate modifications: taking infimum instead of supremum, splitting the case when w is the least limit of L) we get that $o(L_{>w})$ is also computable, and $o(L)$, being the sum $o(L_{<w}) + o(L_{>w})$, is hence computable as well.

We also got that the order type of such a language has to be a finite sum of the order types ω, $-\omega$ and 1, that is, has to have rank at most 1. □

Corollary 3. *Suppose L is a scattered context-free language of rank at most one. Then $o(L)$ is effectively computable.*

Proof. If $o(L) \in \{\omega, -\omega\}$, then L has one limit, while if $o(L)$ is finite, then it has no limits. Since scattered order types of rank at most one are finite sums of the order types ω, $-\omega$ and 1, thus scattered languages of rank at most one are finite unions of languages of order type ω, $-\omega$ or 1, by Lemma 6 we get that such languages have a finite number of limits, and thus their order type is effectively computable by Theorem 1. □

Corollary 4. *For any context-free language L, it is decidable whether L is a scattered language of rank at most one, and if so, $o(L)$ can be effectively computed (as a finite sum of the order types 1, ω and $-\omega$).*

4 Conclusion

We showed that it is decidable whether a context-free ordering is scattered of rank at most one, and if so, then its order type is effectively computable as a finite sum of the order types 1, ω and $-\omega$. This extends our earlier result [6] which was applicable only for well-orderings.

An interesting question for further study is whether the rank of a scattered context-free ordering is computable. Another, maybe easier one is to determine which rank-two scattered orderings are context-free (as there are uncountably many such orderings, the vast majority of them cannot be context-free), and whether their isomorphism problem is still decidable.

A related notion is that of tree automatic orderings: these are the order types of regular tree languages equipped with the lexicographic ordering (on trees). Through derivation trees, there is a tight connection between context-free string languages and regular tree languages but as the two orderings differ (lexicographic ordering of trees vs their frontiers), it is unclear whether there is a nontrivial inclusion between these two classes of orderings (or at least for the scattered case). (Observe that the lexicographic order of the *frontier words* of trees is not an automatic relation.)

Acknowledgements. Ministry of Human Capacities, Hungary grant 20391-3/2018/FEKUSTRAT is acknowledged. Szabolcs Iván was supported by the János Bolyai Scholarship of the Hungarian Academy of Sciences. Kitti Gelle was supported by the ÚNKP-19-3-SZTE-86 New National Excellence Program of the Ministry of Human Capacities.

References

1. Bloom, S.L., Ésik, Z.: Algebraic ordinals. Fundam. Inform. **99**(4), 383–407 (2010)
2. Bloom, S.L., Ésik, Z.: The equational theory of regular words. Inf. Comput. **197**(1), 55–89 (2005)
3. Ésik, Z.: Scattered context-free linear orderings. In: Mauri, G., Leporati, A. (eds.) DLT 2011. LNCS, vol. 6795, pp. 216–227. Springer, Heidelberg (2011). https://doi.org/10.1007/978-3-642-22321-1_19
4. Ésik, Z.: An undecidable property of context-free linear orders. Inf. Process. Lett. **111**(3), 107–109 (2011)
5. Ésik, Z., Iván, S.: Hausdorff rank of scattered context-free linear orders. In: Fernández-Baca, D. (ed.) LATIN 2012. LNCS, vol. 7256, pp. 291–302. Springer, Heidelberg (2012). https://doi.org/10.1007/978-3-642-29344-3_25
6. Gelle, K., Iván, S.: On the order type of scattered context-free orderings. In: The Tenth International Symposium on Games, Automata, Logics, and Formal Verification, September 2–3, 2019, pp. 169–182 (2019)
7. Gelle, K., Iván, S.: The ordinal generated by an ordinal grammar is computable. Theoret. Comput. Sci. **793**, 1–13 (2019)
8. Heilbrunner, S.: An algorithm for the solution of fixed-point equations for infinite words. RAIRO - Theoret. Inf. Appl. Informatique Théorique et Applications **14**(2), 131–141 (1980)
9. Hopcroft, J.E., Ullman, J.D.: Introduction to Automata Theory, Languages, and Computation. Addison-Wesley Publishing Company, Reading (1979)
10. Lohrey, M., Mathissen, C.: Isomorphism of regular trees and words. Inf. Comput. **224**, 71–105 (2013)
11. Rosenstein, J.: Linear Orderings. Pure and Applied Mathematics. Elsevier Science, Amsterdam (1982)
12. Stark, J.A.: Ordinal arithmetic (2015). https://jalexstark.com/notes/OrdinalArithmetic.pdf
13. Thomas, W.: On frontiers of regular trees. ITA **20**(4), 371–381 (1986)

Up-to Techniques for Branching Bisimilarity

Rick Erkens[1], Jurriaan Rot[2,3], and Bas Luttik[1](✉)

[1] Eindhoven University of Technology, Eindhoven, The Netherlands
{r.j.a.erkens,s.p.luttik}@tue.nl
[2] University College London, London, UK
[3] Radboud University Nijmegen, Nijmegen, The Netherlands

Abstract. Ever since the introduction of behavioral equivalences on processes one has been searching for efficient proof techniques that accompany those equivalences. Both strong bisimilarity and weak bisimilarity are accompanied by an arsenal of up-to techniques: enhancements of their proof methods. For branching bisimilarity, these results have not been established yet. We show that a powerful proof technique is sound for branching bisimilarity by combining the three techniques of up to union, up to expansion and up to context for Bloom's BB cool format. We then make an initial proposal for casting the correctness proof of the up to context technique in an abstract coalgebraic setting, covering branching but also η, delay and weak bisimilarity.

1 Introduction

Bisimilarity is a fundamental notion of behavioral equivalence between processes [13]. To prove that processes P, Q are bisimilar it suffices to give a bisimulation relation $\mathcal{R}$ containing the pair (P, Q). But bisimulations can become quite large, which makes proofs long. To remedy this issue, up-to techniques were proposed [13,19]. They are used, for example, in the π-calculus, where even simple properties about the replication operator are hard to handle without them [21], but also in automata theory [5] and other applications, see [4,17] for an overview.

For weak bisimilarity the field of up-to techniques is particularly delicate. Milner's weak bisimulations up to weak bisimilarity cannot be used to prove weak bisimilarity [20] and the technique of up-to context is unsound for many process algebras, most notably some that use a form of choice. Up-to techniques for weak bisimilarity have been quite thoroughly studied (e.g., [15,17]). The question remains whether such techniques apply also to other weak equivalences.

In this paper, we study branching, delay and η bisimilarity, and propose general criteria for the validity of two main up-to techniques. We make use of the general framework of enhancements due to Pous and Sangiorgi [17,19], and prove that the relevant techniques are *respectful*: this allows to modularly combine them in proofs of bisimilarity (recalled in Sect. 3).

This work was partially supported by a Marie Curie Fellowship (grant code 795119).

A. Chatzigeorgiou et al. (Eds.): SOFSEM 2020, LNCS 12011, pp. 285–297, 2020.
https://doi.org/10.1007/978-3-030-38919-2_24

We start out by recasting the up-to-expansion technique, which has been proposed to remedy certain issues in up-to techniques for weak bisimilarity [20], to branching bisimilarity. Then, we study up-to-context techniques, which can significantly simplify bisimilarity proofs about processes generated by transition system specifications. Up-to context is not sound in general, even for strong bisimilarity. For the latter, it suffices that the specification is in the GSOS format [4]. For weak bisimilarity one needs stronger assumptions. It was shown in [4] that Bloom's *simply WB cool format* [3,9] gives the validity of up-to context. We adapt this result to branching, η and delay bisimilarity, making use of each of the associated "simply cool" formats introduced by Bloom. These were introduced to prove congruence of weak equivalences; our results extend to respectfulness of the up-to-context technique, which is strictly stronger in general [17].

For the results on up-to-context, we give both a concrete proof for the case of branching bisimilarity, and a general coalgebraic treatment that covers weak, branching, η and delay in a uniform manner (Sect. 5). This is based on, but also simplifies the approach in [4], by focusing on (span-based) simulations, avoiding technical intricacies in the underlying categorical machinery. Our coalgebraic results are essentially about respectfulness of simulation, suitably instantiated to weak simulations and subsequently extended to bisimulations via the general framework of [17]. We conclude with some directions for future work in Sect. 6.

2 Preliminaries

A Labelled Transition System (LTS) is a triple $(\mathbb{P}, \mathbb{A}, \rightarrow)$ where $\mathbb{P}$ is a set of states, $\mathbb{A}$ is a set of actions with $\tau \in \mathbb{A}$ and $\rightarrow \subseteq \mathbb{P} \times \mathbb{A} \times \mathbb{P}$ is a set of transitions. We denote a transition (P, α, P') by $P \xrightarrow{\alpha} P'$. For any α we consider $\xrightarrow{\alpha}$ a binary relation on $\mathbb{P}$. With this in mind let $\Longrightarrow$ denote the transitive reflexive closure of $\xrightarrow{\tau}$. By $P \xrightarrow{(\alpha)} P'$ we mean that $P \xrightarrow{\alpha} P'$ or $\alpha = \tau$ and $P = P'$. The capital letters P, Q, X, Y, Z range over elements of $\mathbb{P}$. The letters α, β denote arbitrary elements from $\mathbb{A}$ and with lowercase letters a we denote arbitrary elements of $\mathbb{A} \backslash \{\tau\}$, so the action a is not a silent action.

The set of relations between sets X and Y is denoted by $\mathsf{Rel}_{X,Y}$; when $X = Y$ we denote it by Rel_X, ranged over by $\mathcal{R}, \mathcal{S}$. Relation composition is denoted by $\mathcal{R} ; \mathcal{S} = \{(P, Q) \mid \exists X.\, P \mathrel{\mathcal{R}} X \text{ and } X \mathrel{\mathcal{S}} Q\}$, or simply by $\mathcal{RS}$. For any set X, the partial order $(\mathsf{Rel}_X, \subseteq)$ forms a complete lattice, where the join and meet are given by union $\bigcup X$ and intersection $\bigcap X$ respectively. A function $f \colon \mathsf{Rel}_X \rightarrow \mathsf{Rel}_X$ is monotone iff $\mathcal{R} \subseteq \mathcal{S}$ implies $f(\mathcal{R}) \subseteq f(\mathcal{S})$. The set $[\mathsf{Rel}_X \rightarrow \mathsf{Rel}_X]$ of such monotone functions is again a complete lattice, ordered by pointwise inclusion, which we denote by $\leq$. Thus, join and meet are pointwise: $\bigvee F = \lambda \mathcal{R}. \bigcup\{f(\mathcal{R}) \mid f \in F\}$ and $\bigwedge F = \lambda \mathcal{R}. \bigcap\{f(\mathcal{R}) \mid f \in F\}$.

Bisimulation. Consider the function $\mathsf{brs}(\mathcal{R}) = \{(P, Q) \mid$ for all P' and for all α, if $P \xrightarrow{\alpha} P'$ then there exist Q', Q'' s.t. $Q \Longrightarrow Q' \xrightarrow{(\alpha)} Q''$ and $P \mathrel{\mathcal{R}} Q'$ and $P' \mathrel{\mathcal{R}} Q''\}$. We say that $\mathcal{R}$ is a *branching simulation* if $\mathcal{R} \subseteq \mathsf{brs}(\mathcal{R})$. Moreover we define $\mathsf{br} = \mathsf{brs} \wedge (\mathsf{rev} \circ \mathsf{brs} \circ \mathsf{rev})$ where $\mathsf{rev}(\mathcal{R}) = \{(Q, P) \mid P \mathrel{\mathcal{R}} Q\}$ and say that

$\mathcal{R}$ is a *branching bisimulation* if $\mathcal{R} \subseteq \mathsf{br}(\mathcal{R})$. Since brs and hence br are monotone and $(\mathcal{P}(\mathbb{P}\times\mathbb{P}), \subseteq)$ is a complete lattice, br has a greatest fixed point. We denote it by $\asymp$ and refer to it as *branching bisimilarity*. To prove $P \asymp Q$, it suffices to provide a relation $\mathcal{R}$ that contains the pair (P, Q) and show that $\mathcal{R} \subseteq \mathsf{br}(\mathcal{R})$; the latter implies $\mathcal{R} \subseteq \asymp$. Up-to techniques strengthen this principle (Sect. 3).

Delay (bi)similarity is defined analogously through the function ds, defined as brs but dropping the condition $P \mathrel{\mathcal{R}} Q'$. Weak simulations are defined using the map $\mathsf{ws}(\mathcal{R}) = \{(P, Q) \mid$ for all P' and for all α, if $P \xrightarrow{\alpha} P'$ then there exist Q', Q'', Q''' such that $Q \Longrightarrow Q' \xrightarrow{(\alpha)} Q'' \Longrightarrow Q'''$ and $P' \mathrel{\mathcal{R}} Q'''\}$. Finally, for η simulation, we have hs, defined as ws but adding the requirement $P \mathrel{\mathcal{R}} Q'$.

Intuitively, the four notions of bisimilarity defined above vary in two dimensions: first, *branching* and *delay* bisimilarity consider internal activity (represented by τ-steps) only before the observable step, whereas η and *weak* bisimilarity also consider internal activity after the observable step; second, *branching* and η bisimilarity require that the internal activity does not incur a change of state, whereas for *delay* and *weak* this is not required.

GSOS and Cool Formats. *GSOS* is a rule format that guarantees strong bisimilarity to be a congruence [3]. Bloom introduced *cool languages* as restrictions of GSOS, forming suitable formats for weak, branching, η and delay bisimilarity [2].

A signature Σ is a set of operators that each have an arity denoted by $\mathsf{ar}(\sigma)$. We assume a set of variables $\mathbb{V}$ and denote the set of terms over a signature Σ by $\mathbb{T}(\Sigma)$. For a term t we denote the set of its variables by $\mathsf{vars}(t)$. A term t is closed if $\mathsf{vars}(t) = \emptyset$. A *substitution* is a partial function $\rho : \mathbb{V} \rightharpoonup \mathbb{T}(\Sigma)$. We denote the application of a substitution to a term t by t^ρ. A substitution is *closed* if ρ is a total function such that $\rho(x)$ is closed for all $x \in \mathbb{V}$.

Definition 2.1. *A positive GSOS language is a tuple (Σ, R) where Σ is a signature and R is a set of transition rules of the form $\frac{H}{\sigma(x_1,\ldots,x_{\mathsf{ar}(\sigma)}) \xrightarrow{\alpha} t}$ where t is a term, $x_1, \ldots, x_{\mathsf{ar}(\sigma)}$ are distinct variables and H is a set of premises such that each premise in H is of the form $x_i \xrightarrow{\beta} y_i$ where the left-hand side x_i occurs in $x_1, \ldots, x_{\mathsf{ar}(\sigma)}$; the right-hand sides y_i of all premises are distinct; the right-hand sides y_i of all premises do not occur in $x_1, \ldots, x_{\mathsf{ar}(\sigma)}$; the target t only contains variables that occur in the premises or in the source.*

The (not necessarily positive) GSOS format also allows negative premises, that is, premises of the form $x_i \stackrel{\beta}{\not\rightarrow}$. In this paper we do not consider those.

An *LTS algebra* for a signature Σ consists of an LTS $(\mathbb{P}, \mathbb{A}, \rightarrow)$ together with a Σ-indexed family of mappings on $\mathbb{P}$ of corresponding arity. We denote the mapping associated with an element of Σ by the same symbol, i.e., for all $\sigma \in \Sigma$ there is a map $\sigma : \mathbb{P}^{\mathsf{ar}(\sigma)} \rightarrow \mathbb{P}$. If $t \in \mathbb{T}(\Sigma)$ and $\rho : \mathbb{V} \rightarrow \mathbb{P}$ is an assignment of states to variables, then we denote by t^ρ the interpretation of t in $\mathbb{P}$.

Now, let $\mathcal{L} = (\Sigma, R)$ be a GSOS language. Then an LTS algebra for Σ is a *model* for $\mathcal{L}$ if it satisfies the rules in R, i.e., if for every rule $\frac{H}{\sigma(x_1,\ldots,x_n) \xrightarrow{\alpha} t} \in R$

and for every assignment ρ we have that whenever $\rho(x_i) \xrightarrow{\beta_i} \rho(y_i)$ for every premise $x_i \xrightarrow{\beta_i} y_i \in H$ then also $\sigma(\rho(x_1), \ldots, \rho(x_n)) \xrightarrow{\alpha} t^\rho$.

The *canonical model* for a GSOS language $\mathcal{L} = (\Sigma, R)$ has as states the set of closed Σ-terms and for all closed terms P and P', a transition $P \xrightarrow{\alpha} P'$ if, and only if, there is a rule $\frac{H}{\sigma(x_1,\ldots,x_n) \xrightarrow{\alpha} t} \in R$ and a substitution ρ such that $P = \sigma(\rho(x_1), \ldots, \rho(x_n))$, $P' = t^\rho$, and $\rho(x_i) \xrightarrow{\beta_i} \rho(y_i)$ for all premises $x_i \xrightarrow{\beta_i} y_i \in H$. The mapping associated with an n-ary element $\sigma \in \Sigma$ maps every sequence $t_1, \ldots, t_n$ for closed terms to the closed term $\sigma(t_1, \ldots, t_n)$.

Bloom's *cool formats* [2] rely on some auxiliary notions. A rule of the form $\frac{x_i \xrightarrow{\tau} y_i}{\sigma(x_1,\ldots,x_n) \xrightarrow{\tau} \sigma(x_1,\ldots,y_i,\ldots,x_n)}$ is called a *patience rule* for the ith argument of σ. A rule is *straight* if the left-hand sides of all premises are distinct. A rule is *smooth* if, moreover, no variable occurs both in the target and the left-hand side of a premise. The ith argument of $\sigma \in \Sigma$ is *active* if there is a rule $\frac{H}{\sigma(x_1,\ldots,x_n) \xrightarrow{\alpha} t}$ in which x_i occurs at the left-hand side of a premise. A variable y is *receiving* in the target t of a rule r in $\mathcal{L}$ if it is the right-hand side of a premise of r. The ith argument of $\sigma \in \Sigma$ is *receiving* if there is a variable y and a target t of a rule in $\mathcal{L}$ s.t. y is receiving in t, t has a subterm $\sigma(v_1, \ldots, v_n)$ and y occurs in v_i.

For instance, CCS [13] has the rule $\frac{x_1 \xrightarrow{\alpha} y_1}{x_1 + x_2 \xrightarrow{\alpha} y_1}$ for the binary choice operator $+$. The first argument is active, but the semantics does not allow a patience rule for it. The issue can be mitigated by guarded sums, replacing choice by infinitely many rules of the form $\Sigma_{i \in I} : \alpha_i.x_i \xrightarrow{\alpha_i} x_i$. These rules have no premises; therefore there are no active arguments and no patience rule is needed.

Definition 2.2. *A language* $\mathcal{L} = (\Sigma, R)$ *is* simply WB cool *if it is positive GSOS and 1. all rules in* $\mathcal{L}$ *are straight; 2. only patience rules have* τ*-premises; 3. for each operator every active argument has a patience rule; 4. every receiving argument of an operator has a patience rule; and 5. all rules in* $\mathcal{L}$ *are smooth. The language* $\mathcal{L}$ *is* simply BB cool *if it satisfies 1, 2, and 3. It is* simply HB cool *if it satisfies 1, 2, 3, and 4. It is* simply DB cool *if it satisfies 1, 2, 3, and 5.*

In [9], van Glabbeek presents four lemmas, labelled BB, HB, DB and WB, respectively, that are instrumental for proving that branching, η, delay and weak bisimilarity are congruences for the associated variants of cool languages. In [9] these lemmas are established for the canonical model, but they have straightforward generalisations to arbitrary models; these generalisations will be instrumental for our results in Sects. 4 and 5. We only present the generalisations of WB and BB here; the generalisations of HB and DB proceed analogously.

Lemma 2.1. *Let* $\mathcal{L}$ *be a simply WB cool language, let* $(\mathbb{P}, \mathbb{A}, \rightarrow)$ *be a model for* $\mathcal{L}$*, let* $\eta : \mathbb{V} \rightarrow \mathbb{P}$ *be an assignment and let* $\frac{H}{\sigma(x_1,\ldots,x_n) \xrightarrow{\alpha} t}$ *be a rule in* $\mathcal{L}$*. If for each* $x \xrightarrow{\beta} y$ *in* H *we have* $\eta(x) \Longrightarrow\xrightarrow{(\beta)}\Longrightarrow \eta(y)$*, then* $\sigma(x_1, \ldots, x_n)^\eta \Longrightarrow\xrightarrow{(\alpha)}\Longrightarrow t^\eta$.

Lemma 2.2. *Let $\mathcal{L}$ be a simply BB cool language, let $(\mathbb{P}, \mathbb{A}, \rightarrow)$ be a model for $\mathcal{L}$, and let $\frac{\{x_i \xrightarrow{\beta_i} y_i | i \in I\}}{\sigma(x_1,\ldots,x_n) \xrightarrow{\alpha} t}$ be some rule in $\mathcal{L}$. If $\eta, \theta : \mathbb{V} \rightarrow \mathbb{P}$ are assignments s.t. for all $i \in I$ it holds that $\eta(x_i) \Longrightarrow \theta(x_i) \xrightarrow{(\beta_i)} \theta(y_i)$ and for every $x \notin \{x_i, y_i \mid i \in I\}$ we have $\eta(x) = \theta(x)$, then $\sigma(x_1, \ldots, x_n)^\eta \Longrightarrow \sigma(x_1, \ldots, x_n)^\theta \xrightarrow{(\alpha)} t^\theta$.*

3 The Abstract Framework for Bisimulations

We recall the lattice-theoretical framework of up-to techniques proposed by Pous and Sangiorgi [16], which allows to obtain enhancements of branching bisimilarity and other coinductively defined relations in a modular fashion. Throughout this section, let $f, b, s \colon \mathsf{Rel}_X \rightarrow \mathsf{Rel}_X$ be monotone maps.

We think of $\mathsf{gfp}(b)$ as the coinductive object of interest (e.g., bisimilarity); then, to prove $(P, Q) \in \mathsf{gfp}(b)$ it suffices to prove $(P, Q) \in \mathcal{R}$ for some $\mathcal{R} \subseteq b(\mathcal{R})$ (e.g., a bisimulation). The aim of using up-to techniques is to alleviate this proof obligation, by considering an additional map f, and proving instead that $\mathcal{R} \subseteq b(f(\mathcal{R}))$; such a relation is called a *b-simulation up to f* (e.g., a bisimulation up to f). Typically, this map f will increase the argument relation. Not every function f is suitable as an up-to technique: it should be sound.

Definition 3.1. *We say that f is b-sound if $\mathit{gfp}(b \circ f) \subseteq \mathit{gfp}(b)$.*

When one proves $\mathcal{R} \subseteq b(f(\mathcal{R}))$ it follows that $\mathcal{R} \subseteq \mathsf{gfp}(b \circ f)$. Soundness is indeed the missing link to conclude $\mathcal{R} \subseteq \mathsf{gfp}(b)$. Unfortunately the composition of two b-sound functions is not b-sound in general [17, Exercise 6.3.7]. To obtain compositionality we use the stronger notion of respectfulness.

Definition 3.2. *A function f is b-respectful if $f \circ (b \wedge \mathit{id}) \leq (b \wedge \mathit{id}) \circ f$.*

This originates from Sangiorgi [19], and was used to prove that up-to context is sound for strong bisimilarity, for faithful contexts. Lemma 3.1 states that respectful functions are sound, and gives methods to combine them. It summarises certain results from [16,17] about *compatible* functions: f is b-compatible if $f \circ b \leq b \circ f$. Thus, respectfulness simply means $b \wedge \mathsf{id}$-compatibility. While compatibility is stronger than respectfulness, this difference disappears if we move to the *greatest* compatible function, given as the join of all b-compatible functions.

Lemma 3.1. *Consider the* companion *of b, defined by* $\mathsf{t} = \bigvee\{f \mid f \circ b \leq b \circ f\}$.

1. *for all respectful functions f it holds that $f \leq \mathsf{t}$;*
2. *for all sets F such that $f \leq \mathsf{t}$ for every $f \in F$, we have $\bigvee F \leq \mathsf{t}$;*
3. *for any two functions $f, g \leq \mathsf{t}$ it holds that $g \circ f \leq \mathsf{t}$;*
4. *if $\mathcal{S} \subseteq b(\mathcal{S})$ then, for $\lambda\mathcal{R}.\mathcal{S}$ the constant-to-$\mathcal{S}$ function, $\lambda\mathcal{R}.\mathcal{S} \leq \mathsf{t}$;*
5. *if $f \leq \mathsf{t}$ then $f(\mathit{gfp}(b)) \subseteq \mathit{gfp}(b)$;*
6. *if $f \leq \mathsf{t}$ then f is b-sound.*

Lemma 3.1 is used to obtain powerful proof techniques for branching bisimilarity and other coinductive relations. If f is below the companion t, it can safely be used as an up-to technique; moreover, such functions combine well, via composition and union. The above lemma gives some basic up-to techniques for free: for instance, the function $f(\mathcal{R}) = \mathcal{R} \cup \mathsf{gfp}(b)$ is below t (for any b). We will focus on up-to-expansion and up-to-context. Especially the latter requires more effort to establish, but can drastically alleviate the effort in proving bisimilarity.

We conclude with two useful lemmas. The first states that for symmetric techniques it suffices to prove respectfulness for similarity, and the second is a proof technique for respectfulness (and, in fact, the original characterisation).

Lemma 3.2. *Let $b = s \wedge (\mathsf{rev} \circ s \circ \mathsf{rev})$. If f is symmetric (i.e. $f = \mathsf{rev} \circ f \circ \mathsf{rev}$) and s-respectful, then f is b-respectful.*

Lemma 3.3. *The function f is b-respectful if and only if for all $\mathcal{R}, \mathcal{S}$ we have that $\mathcal{R} \subseteq \mathcal{S}$ and $\mathcal{R} \subseteq b(\mathcal{S})$ implies $f(\mathcal{R}) \subseteq b(f(\mathcal{S}))$.*

4 Branching Bisimilarity: Expansion and Context

Up-to Expansion. The first up-to technique for strong bisimilarity was reported by Milner [13]. It is based on the enhancement function $\lambda\mathcal{R}.{\sim}\mathcal{R}{\sim}$ where $\sim$ denotes strong bisimilarity. It is well known that a similar enhancement function $\lambda\mathcal{R}.{\approx}\mathcal{R}{\approx}$ is unsound for weak bisimilarity [17,20], and the same counterexample shows that the enhancement function $\lambda\mathcal{R}.{\asymp}\mathcal{R}{\asymp}$ is unsound for branching bisimilarity: the relation $\{(\tau.a, 0)\}$ on CCS processes [13] is a branching bisimulation up to $\lambda\mathcal{R}.{\asymp}\mathcal{R}{\asymp}$, using that $a \asymp \tau.a$, but clearly $\tau.a$ is not branching bisimilar to 0. The function $\lambda\mathcal{R}.{\sim}\mathcal{R}{\sim}$ is br-respectful. But it turns out that one can do slightly better, using an efficiency preorder called expansion [1,17]. We proceed to define such a preorder for branching bisimilarity and show that it results in a more powerful up-to technique than strong bisimilarity.

Definition 4.1. *Consider the function $\mathsf{br}_{\succcurlyeq} : \mathcal{P}(\mathbb{P} \times \mathbb{P}) \to \mathcal{P}(\mathbb{P} \times \mathbb{P})$ defined as $\mathsf{br}_{\succcurlyeq}(\mathcal{R}) = \{(P, Q) \mid$ for all P' and all α, if $P \xrightarrow{\alpha} P'$ then there exists Q' such that $Q \xrightarrow{(\alpha)} Q'$ and $P' \mathcal{R} Q'$; and for all Q' and all α, if $Q \xrightarrow{\alpha} Q'$ then there exist P', P'' such that $P \Longrightarrow P' \xrightarrow{\alpha} P''$ with $P' \mathcal{R} Q$ and $P'' \mathcal{R} Q'\}$. We say $\mathcal{R}$ is a* branching expansion *if $\mathcal{R} \subseteq \mathsf{br}_{\succcurlyeq}(\mathcal{R})$. Denote $\mathsf{gfp}(\mathsf{br}_{\succcurlyeq})$ by $\succcurlyeq$.*

Informally $P \succcurlyeq Q$ means that P and Q are branching bisimilar and P always performs at least as many τ-steps as Q. Similar notions of expansion can be defined for η and delay bisimilarity. Examples are at the end of this section.

Lemma 4.1. *The function $\lambda\mathcal{R}.{\succcurlyeq}\mathcal{R}{\preccurlyeq}$ is br-respectful.*

The proof of Lemma 4.1 is routine if we use Lemmas 3.2 and 3.3. It suffices to show that if $\mathcal{R} \subseteq \mathsf{brs}(\mathcal{S})$ and $\mathcal{R} \subseteq \mathcal{S}$ then ${\succcurlyeq}\mathcal{R}{\preccurlyeq} \subseteq \mathsf{brs}({\succcurlyeq}\mathcal{S}{\preccurlyeq})$. This inclusion can be proved by playing the branching simulation game on the pairs in ${\succcurlyeq}\mathcal{R}{\preccurlyeq}$.

Up-to Context. Next, we consider LTSs generated by GSOS languages. Here, an up-to-context technique enables us to use congruence properties of process algebras in the bisimulation game: it suffices to relate terms by finding a mutual context for both terms. We show that if $\mathcal{L}$ is a language in the simply BB cool format, then the closure w.r.t. $\mathcal{L}$-contexts is br-respectful.

Definition 4.2. *Let $\mathcal{L} = (\Sigma, R)$ be a positive GSOS language and let $\mathcal{R}$ be a relation on closed $\mathcal{L}$-terms. The closure of $\mathcal{R}$ under $\mathcal{L}$-contexts is denoted by $\mathcal{C}_{\mathcal{L}}(\mathcal{R})$ and is defined as the smallest relation that is closed under the following inference rules:* $\frac{P\mathcal{R}Q}{P\mathcal{C}_{\mathcal{L}}(\mathcal{R})Q}$ *and* $\frac{P_1\mathcal{C}_{\mathcal{L}}(\mathcal{R})Q_1 \quad \dots \quad P_{\mathsf{ar}(\sigma)}\mathcal{C}_{\mathcal{L}}(\mathcal{R})Q_{\mathsf{ar}(\sigma)}}{\sigma(P_1,\dots,P_{\mathsf{ar}(\sigma)})\mathcal{C}_{\mathcal{L}}(\mathcal{R})\sigma(Q_1,\dots,Q_{\mathsf{ar}(\sigma)})}$.

Theorem 4.1. *Let $\mathcal{L}$ be a simply BB cool language. Then $\mathcal{C}_{\mathcal{L}}$ is br-respectful.*

For the proof, we use Lemma 3.3 and show that if $\mathcal{R} \subseteq \mathsf{br}(\mathcal{S})$ and $\mathcal{R} \subseteq \mathcal{S}$ then $\mathcal{C}_{\mathcal{L}}(\mathcal{R}) \subseteq \mathsf{br}(\mathcal{C}_{\mathcal{L}}(\mathcal{R}))$. The proof is by induction on elements of $\mathcal{C}_{\mathcal{L}}(\mathcal{R})$, using Lemma 2.2, which essentially states that a suitable saturation of the canonical model of $\mathcal{L}$ (Sect. 2) is still a model of $\mathcal{L}$. This is generalised in Sect. 5.

The following two examples use a variant of CCS [13] with replication (!); we refer to [17] for its syntax and operational semantics.

Example 4.1. We show that $!\tau.(a|\bar{a}) \asymp !(\tau.a + \tau.\bar{a})$. Consider the relation $\mathcal{R}$ containing just the single pair of processes. It suffices to prove that $\mathcal{R}$ is a branching bisimulation up to $\lambda\mathcal{R}.{\succcurlyeq}\mathcal{C}_{\mathcal{L}}(\mathcal{R}){\preccurlyeq}$ since both $\mathcal{C}_{\mathcal{L}}$ and $\lambda\mathcal{R}.{\succcurlyeq}\mathcal{R}{\preccurlyeq}$ are br-respectful. In the proof one can use properties for strong bisimilarity like $!P|P \sim P$ and $P|Q \sim Q|P$. Since $\sim \subseteq \succcurlyeq$ these laws also apply to expansion. Then the expansion law $P|\tau.Q \succcurlyeq P|Q$ ensures that $\mathcal{R}$ suffices.

Example 4.2. We show that $!(a + b) \asymp !\tau.a|!\tau.b$. The relation $\mathcal{R}$ containing just the single pair of processes is a branching bisimulation up to $\lambda\mathcal{R}.{\sim}\mathcal{C}_{\mathcal{L}}(\mathcal{R}){\sim}$. This is sufficient: since $\lambda\mathcal{R}.{\succcurlyeq}\mathcal{R}{\preccurlyeq}$ is br-respectful and $\lambda\mathcal{R}.{\sim}\mathcal{R}{\sim} \leq \lambda\mathcal{R}.{\succcurlyeq}\mathcal{R}{\preccurlyeq}$, the function $\lambda\mathcal{R}.{\sim}\mathcal{R}{\sim}$ is below the companion of br, and therefore it can be combined with $\mathcal{C}_{\mathcal{L}}$ to obtain a br-sound technique.

A similar result as Theorem 4.1 is established for weak bisimilarity in [4]. In fact, one can use the lemmas at the end of Sect. 2 to treat η and delay bisimilarity as well. We develop a uniform approach in the following section.

5 Respectfulness of Up-to Context: Coalgebraic Approach

We develop conditions for respectfulness of contextual closure that instantiate to variants for branching, weak, η and delay bisimilarity. In each case, the relevant condition is implied by the associated simply cool GSOS format.

The main step is that contextual closure is respectful for *similarity*, for a relaxed notion of models of positive GSOS specifications. The case of weak, branching, η and delay are then obtained by considering simulations between

LTSs and appropriate saturations thereof.[1] We use the theory of coalgebras; in particular, the respectfulness result for simulations is phrased at an abstract level. We assume familiarity with basic notions in category theory. Further, due to space constraints, we only report basic definitions; see, e.g., [11,18] for details.

The abstract results in this section are inspired by, and close to, the development in [4]. Technically, however, we simplify in two ways: (1) focusing on simulations rather than on (weak) bisimulations directly through functor lifting in a fibration; and (2) avoiding the technical sophistication that arises from the combination of fibrations and orderings, by using a (simpler) span-based approach in the proofs. Still, we use a number of results from [4], connecting monotone GSOS specifications to distributive laws. The cases of branching, η and delay bisimilarity, which we treat here, were left as future work in [4]. Note that we do not propose a general coalgebraic theory of weak bisimulations, as introduced, e.g., in [6], but focus on LTSs, which are the models of interest here.

Coalgebra. We denote by Set the category of sets and functions. Given a functor $B\colon \mathsf{Set} \to \mathsf{Set}$, a *$B$-coalgebra* is a pair (X, f) where X is a set and $f\colon X \to B(X)$ a function. A *coalgebra homomorphism* from a B-coalgebra (X, f) to a B-coalgebra (Y, g) is a map $h\colon X \to Y$ such that $g \circ h = Bh \circ f$.

Let $\mathbb{A}$ be a fixed of labels with $\tau \in \mathbb{A}$. Labelled transition systems are (equivalent to) coalgebras for the functor B given by $B(X) = (\mathcal{P}X)^{\mathbb{A}}$. Indeed, a B-coalgebra consists of a set of states X and a map $f\colon X \to (\mathcal{P}X)^{\mathbb{A}}$ mapping a state $x \in X$ to its outgoing transitions; we write $x \xrightarrow{\alpha}_f y$ or simply $x \xrightarrow{\alpha} y$ for $y \in f(x)(\alpha)$. In this section we mean coalgebras for this functor, when referring to LTSs. The notations $\Longrightarrow_f$ and $\xrightarrow{(\alpha)}_f$, defined in Sect. 2, are used as well.

To define (strong) bisimilarity of coalgebras we make use of *relation lifting* [11], which maps a relation $R \subseteq X \times Y$ to a relation $\mathsf{Rel}(B)(R) \subseteq BX \times BY$. This is given by $\mathsf{Rel}(B)(R) = \{(u, v) \mid \exists z \in B(R). B(\pi_1)(z) = u \text{ and } B(\pi_2)(z) = v\}$. Now, given B-coalgebras (X, f) and (Y, g), a relation $R \subseteq X \times Y$ is a bisimulation if for all $(x, y) \in R$, we have $f(x)\ \mathsf{Rel}(B)(R)\ g(y)$. In case of labelled transition systems, this amounts to the standard notion of strong bisimilarity.

Algebra. An algebra for a functor $H\colon \mathsf{Set} \to \mathsf{Set}$ is a pair (X, a) where X is a set and $a\colon H(X) \to X$ a function. An algebra morphism from (X, a) to (Y, b) is a map $h\colon X \to Y$ such that $h \circ a = b \circ Hh$. While coalgebras are used here to represent variants of labelled transition systems, we will also make use of algebras, to speak about operations in process calculi. In order to do so, we first show how to represent a signature Σ as a functor $H_\Sigma\colon \mathsf{Set} \to \mathsf{Set}$, such that H_Σ algebras are interpretations of the signature Σ. Given Σ, this functor H_Σ is defined by: $H_\Sigma(X) = \coprod_{\sigma \in \Sigma} \{\sigma\} \times X^{\mathsf{ar}(\sigma)}$. On maps $f\colon X \to Y$, H_Σ is defined pointwise, i.e., $H_\Sigma(f)(\sigma(x_1, \ldots, x_{\mathsf{ar}(\sigma)})) = \sigma(f(x_1), \ldots, f(x_{\mathsf{ar}(\sigma)}))$.

We denote by $T_\Sigma\colon \mathsf{Set} \to \mathsf{Set}$ the *free monad* of H_Σ. Explicitly, $T_\Sigma(X)$ is the set of terms over Σ with variables in X, as generated by the grammar

[1] Note that this is fundamentally different from reducing weak bisimilarity to strong bisimilarity on a saturated transition system; there, a challenging transition is weak as well. Here, instead, strong transitions are answered by weak transitions.

$t ::= x \mid \sigma(t_1, \ldots, t_{\mathsf{ar}(\sigma)})$ where x ranges over X and σ ranges over Σ. In particular, $T_\Sigma(\emptyset)$ is the set of closed terms. The set $T_\Sigma(X)$ is the carrier of a *free algebra* $\kappa_X \colon H_\Sigma T_\Sigma(X) \to T_\Sigma(X)$: there is an arrow $\eta_X \colon X \to T_\Sigma(X)$ (the unit of the monad T_Σ) such that, for every algebra $b\colon H_\Sigma(Y) \to Y$ and map $f \colon X \to Y$, there is a unique algebra homomorphism $f^\sharp \colon T_\Sigma(X) \to Y$ s.t. $f^\sharp \circ \eta_X = f$. In particular, we write $b^* \colon T_\Sigma(Y) \to Y$ for $\mathsf{id}_Y^\sharp$. Intuitively, b^* inductively extends the algebra structure b on Y to terms over Y.

Simulation of Coalgebras. We recall how to represent simulations [12], based on ordered functors. This enables speaking about weak simulations (Sect. 5.3). As before, by Lemma 3.2, relevant respectfulness results extend to bisimulations.

An *ordered functor* is a functor $B \colon \mathsf{Set} \to \mathsf{Set}$ together with, for every set X, a preorder $\sqsubseteq_{BX} \subseteq BX \times BX$ such that, for every map $f \colon X \to Y$, $Bf \colon BX \to BY$ is monotone. Equivalently, it is a functor B that factors through the forgetful functor $U \colon \mathsf{PreOrd} \to \mathsf{Set}$ from the category of preorders and monotone maps. For maps $f, g \colon X \to BY$, we write $f \sqsubseteq_{BY} g$ for pointwise inequality, i.e., $f(x) \sqsubseteq_{BY} g(x)$ for all $x \in X$. Throughout this section we assume B is ordered.

To define simulations, we recall from [12] the *lax relation lifting* $\mathsf{Rel}_\sqsubseteq(B)$, defined on a relation $R \subseteq X \times Y$ as $\mathsf{Rel}_\sqsubseteq(B)(R) = \sqsubseteq_{BX} ; \mathsf{Rel}(B)(R) ; \sqsubseteq_{BY}$.

Definition 5.1. *Let (X, f) and (Y, g) be B-coalgebras. Define the following monotone operator $s \colon \mathsf{Rel}_{X,Y} \to \mathsf{Rel}_{X,Y}$ by $s(R) = (f \times g)^{-1}(\mathsf{Rel}_\sqsubseteq(B)(R))$. A relation $R \subseteq X \times Y$ is called a* simulation *if it is a post-fixed point of s.*

Example 5.1. The functor $B(X) = (\mathcal{P}X)^{\mathbb{A}}$ is ordered, with $u \sqsubseteq_{BX} v$ iff $u(a) \subseteq v(a)$ for all $a \in \mathbb{A}$. The associated lax relation lifting maps $R \subseteq X \times Y$ to $\mathsf{Rel}_\sqsubseteq(B)(R) = \{(u, v) \mid \forall a \in \mathbb{A}. \forall x \in u(a). \exists y \in v(a). (x, y) \in R\}$. A relation $R \subseteq X \times Y$ between (the underlying state spaces of) LTSs is a simulation in the sense of Definition 5.1 iff it is a simulation in the standard sense: for all $(x, y) \in R$: if $x \xrightarrow{\alpha} x'$ then $\exists y'. y \xrightarrow{\alpha} y'$ and $(x', y') \in R$.

5.1 Abstract GSOS Specifications and Their Models

An *abstract GSOS specification* [22] is a natural transformation of the form $\lambda \colon H_\Sigma(B \times \mathsf{Id}) \Rightarrow BT_\Sigma$. Let X be a set, let $a \colon H_\Sigma(X) \to X$ be an algebra, and let $f \colon X \to BX$ be a coalgebra; the triple (X, a, f) is a *λ-model* if the diagram on the right commutes.

$$\begin{array}{ccccc}
H_\Sigma(X) & \xrightarrow{H_\Sigma\langle f, \mathsf{id}\rangle} & H_\Sigma(BX \times X) & \xrightarrow{\lambda_X} & BT_\Sigma(X) \\
{\scriptstyle a}\downarrow & & & & \downarrow{\scriptstyle Ba^*} \\
X & & \xrightarrow{\quad f \quad} & & BX
\end{array}$$

In our approach to proving the validity of up-to techniques for weak similarity, it is crucial to relax the notion of λ-model to a *lax model*, following [4]. A triple (X, a, f) as above is a *lax λ-model* if we have that $f \circ a \sqsubseteq_{BX} Ba^* \circ \lambda_X \circ H_\Sigma\langle f, \mathsf{id}\rangle$, and an *oplax λ-model* if, conversely, $f \circ a \sqsupseteq_{BX} Ba^* \circ \lambda_X \circ H_\Sigma\langle f, \mathsf{id}\rangle$. Since $\sqsubseteq_{BX}$ is a preorder, (X, a, f) is a λ model iff it is both a lax and an oplax model.

Taking the algebra $\kappa_\emptyset\colon H_\Sigma T_\Sigma \emptyset \to T_\Sigma \emptyset$ on closed terms, there is a unique coalgebra structure $f\colon T_\Sigma\emptyset \to BT_\Sigma\emptyset$ turning $(T_\Sigma\emptyset, \kappa_\emptyset, f)$ into a λ-model. We sometimes refer to this coalgebra structure as the *operational model* of λ.

We say λ is *monotone* if for each component λ_X, we have

$$\frac{u_1 \sqsubseteq_{BX} v_1 \qquad \dots \qquad u_n \sqsubseteq_{BX} v_n}{\lambda_X(\sigma((u_1,x_1),\dots,(u_n,x_n))) \sqsubseteq_{BT_\Sigma X} \lambda_X(\sigma((v_1,x_1),\dots,(v_n,x_n)))}$$

for every operator $\sigma \in \Sigma$, elements $u_1, \dots u_n, v_1, \dots, v_n \in BX$ and $x_1, \dots, x_n \in X$, with $n = \mathsf{ar}(\sigma)$. Informally, if premises have 'more behaviour' (e.g., more transitions) then we can derive more behaviour from the GSOS specification.

Example 5.2. If $BX = (\mathcal{P}X)^{\mathbb{A}}$, then a monotone λ corresponds to a positive GSOS specification (Definition 2.1). In that case, an algebra $a\colon H_\Sigma(X) \to X$ together with a B-coalgebra (i.e., LTS) is a λ-model if, for every $P \in X$, we have that $P \xrightarrow{\alpha} P'$ iff there is a rule $\frac{H}{\sigma(x_1,\dots,x_n)\xrightarrow{\alpha} t}$ and a map $\rho\colon V \to X$ (with V the set of variables occurring in the rule) such that $P = a(\sigma(\rho(x_1), \dots, \rho(x_n)))$, $P' = \rho^\sharp(t)$ (recall that $\rho^\sharp$ denotes the unique algebra homomorphism associated with ρ) and for all premises $x_i \xrightarrow{\beta_i} y_i \in H$ we have that $\rho(x_i) \xrightarrow{\beta_i} \rho(y_i)$. This coincides with the interpretation in Sect. 2. A lax model only asserts the implication from right to left (transitions are closed under application of rules) and an oplax model asserts the converse (every transition arises from a rule).

5.2 Respectfulness of Contextual Closure

We prove a general respectfulness result of contextual closure w.r.t. simulation. First we generalise contextual closure as follows [4]. Given algebras $a\colon H_\Sigma(X) \to X$ and $b\colon H_\Sigma(Y) \to Y$, the *contextual closure* $\mathcal{C}_{a,b}\colon \mathsf{Rel}_{X,Y} \to \mathsf{Rel}_{X,Y}$ is defined by $\mathcal{C}_{a,b}(R) = a^* \times b^*(\mathsf{Rel}(T_\Sigma)(R)) = \{(a^*(u), b^*(v)) \mid (u,v) \in \mathsf{Rel}(T_\Sigma)(R)\}$. For $X = Y = T_\Sigma(\emptyset)$ and $a = b = \kappa_\emptyset\colon H_\Sigma T_\Sigma(\emptyset) \to T_\Sigma(\emptyset)$, $\mathcal{C}_{a,b}$ coincides with the contextual closure $\mathcal{C}$ of Definition 4.2. This allows us to formulate the main result of this section, giving sufficient conditions for respectfulness of the contextual closure with respect to s from Definition 5.1. In fact, this result is slightly more general than needed: we will always instantiate (X, a, f) below with a λ-model.

Theorem 5.1. *Suppose that (X, a, f) is an oplax model of a monotone abstract GSOS specification λ, and (Y, b, g) is a lax model. Then $\mathcal{C}_{a,b}$ is s-respectful.*

5.3 Application to Weak Similarity

Let (X, f) be an LTS. Define a new LTS $(X, \overline{f})$ by $x \xrightarrow{\alpha}_{\overline{f}} x'$ iff $x \Longrightarrow_f \xrightarrow{(\alpha)}_f \Longrightarrow_f x'$. We call $(X, \overline{f})$ the *wb-saturation* of (X, f). Let $s_{wb}\colon \mathsf{Rel}_X \to \mathsf{Rel}_X$ be the functional for simulation (Definition 5.1) between (X, f) and $(X, \overline{f})$. Then $R \subseteq s_{wb}(R)$ precisely if R is a weak simulation on (X, f).

Proposition 5.1. *Let (X, a, f) be a model of a positive GSOS specification, and suppose $(X, a, \overline{f})$ is a lax model. Then $\mathcal{C}_{a,a}$ is s_{wb}-respectful.*

The condition of being a lax model is exactly as in Lemma 2.1. Hence, the contextual closure of any simply WB-cool GSOS language is s_{wb}-respectful. To obtain an analogous result for delay similarity, we simply adapt the saturation to *db-saturation*, and the appropriate functional s_{db}.

Branching Similarity. To capture branching simulations of LTSs in the coalgebraic framework, we will work again with saturation. It is not immediately clear how to do so: we encode branching simulations by slightly changing the functor, in order to make relevant intermediate states observable.

Let $B'(X) = (\mathcal{P}(X \times X))^{\mathbb{A}}$. A B'-coalgebra is similar to an LTS, but transitions take the form $x \xrightarrow{\alpha} (x', x'')$, i.e., to a pair of next states. We will use this to encode branching similarity, as follows. Given an LTS (X, f), define the *bb-saturation* as the coalgebra $(X, \overline{f})$ where $x \xrightarrow{\alpha}_{\overline{f}} (x', x'')$ iff $x \Rightarrow x' \xrightarrow{(\alpha)} x''$. Further, note that every LTS (X, f) gives a B' coalgebra (X, f') by setting $x \xrightarrow{a}_{f'} (x', x'')$ iff $x' = x$ and $x \xrightarrow{a}_{f} x''$.

For an LTS (X, f), consider the functional $s_{bb} \colon \mathsf{Rel}_X \to \mathsf{Rel}_X$ for B'-simulation between (X, f') and $(X, \overline{f})$ (Definition 5.1). Then a relation $R \subseteq X \times X$ is a branching simulation precisely if $R \subseteq s_{bb}(R)$.

To obtain the desired respectfulness result from Theorem 5.1, the last step is to obtain a GSOS specification for B' from a given positive GSOS specification (for B). This is possible if all operators are straight. In that case, every rule is of the form $\frac{\{x_i \xrightarrow{\beta_i} x_i'\}_{i \in I}}{\sigma(x_1, \ldots, x_{\mathsf{ar}(\sigma)}) \xrightarrow{\alpha} t}$ for some $I \subseteq \{1, \ldots, \mathsf{ar}(\sigma)\}$. This is translated to

$$\frac{\{x_i \xrightarrow{\beta_i} (x_i'', x_i')\}_{i \in I}}{\sigma(x_1, \ldots, x_{\mathsf{ar}(\sigma)}) \xrightarrow{\alpha} (t^\rho, t)} \qquad \text{where } \rho(x) = \begin{cases} x_i'' & \text{if } x = x_i \text{ for some } i \in I \\ x & \text{otherwise} \end{cases}$$

If the original specification is presented as an abstract GSOS specification λ, then we denote the corresponding abstract GSOS specification (for B') according to the above translation by λ'. (It is currently less clear how to represent this translation directly at the abstract level; we leave this for future work.)

Proposition 5.2. *Let (X, a, f) be a model of a positive GSOS specification λ with only straight rules. Then (X, a, f') is a model of λ', defined as above; and if $(X, a, \overline{f})$ is a lax model, with $(X, \overline{f})$ the bb-saturation, then $\mathcal{C}_{a,a}$ is s_{bb}-respectful.*

We recover Theorem 4.1 from Proposition 5.2 and Lemma 2.2 (and Lemma 3.2 to move from similarity to bisimilarity). Again, to obtain respectfulness for η-similarity, one simply adapts the notion of saturation.

6 Conclusion and Future Work

We have seen two main up-to techniques, that can be combined: expansion and, most notably, contextual closure. In particular, we have shown that for any language defined by a simply cool format, the contextual closure is respectful for the associated equivalence; this applies to weak, branching, η and delay bisimilarity. The latter follows from a general coalgebraic argument on simulation.

There are several avenues left for future work. First, we have treated up-to-expansion on a case-by-case basis; it would be useful to have a uniform treatment of this technique that instantiates to various weak equivalences. Second, it would be interesting to investigate up-to context for *rooted* and *divergence-sensitive* versions of the weak behavioural equivalences. Associated 'cool' rule formats have already been proposed [3,9]. Third, the current treatment of up-to context heavily relies on positive formats; whether our results can be extended to rule formats with negative premises is left open. Perhaps the *modal decomposition* approach to congruence results [7,8] can help—investigating the relation of this approach to up-to techniques is an exciting direction of research. Finally, extension of the formats to languages including a recursion construct would be very interesting, especially since the proofs that weak and branching bisimilarity are compatible with this construct use up-to techniques [10,14].

Acknowledgements. We thank Filippo Bonchi for the idea how to encode branching bisimilarity coalgebraically, and the reviewers for their useful comments.

References

1. Arun-Kumar, S., Hennessy, M.: An efficiency preorder for processes. Acta Inf. **29**(8), 737–760 (1992)
2. Bloom, B.: Structural operational semantics for weak bisimulations. TCS **146**(1&2), 25–68 (1995)
3. Bloom, B., Istrail, S., Meyer, A.R.: Bisimulation can't be traced. In: POPL, pp. 229–239. ACM (1988)
4. Bonchi, F., Petrisan, D., Pous, D., Rot, J.: A general account of coinduction up-to. Acta Inf. **54**(2), 127–190 (2017)
5. Bonchi, F., Pous, D.: Hacking nondeterminism with induction and coinduction. Commun. ACM **58**(2), 87–95 (2015)
6. Brengos, T.: Weak bisimulation for coalgebras over order enriched monads. Log. Methods Comput. Sci. **11**(2), 1–44 (2015)
7. Fokkink, W., van Glabbeek, R.: Divide and congruence II: from decomposition of modal formulas to preservation of delay and weak bisimilarity. Inf. Comput. **257**, 79–113 (2017)
8. Fokkink, W., van Glabbeek, R., Luttik, B.: Divide and congruence III: from decomposition of modal formulas to preservation of stability and divergence. Inf. Comput. **268**, 104435 (2019). https://doi.org/10.1016/j.ic.2019.104435. Article no. 31 pages
9. van Glabbeek, R.: On cool congruence formats for weak bisimulations. TCS **412**(28), 3283–3302 (2011)
10. van Glabbeek, R.J.: A complete axiomatization for branching bisimulation congruence of finite-state behaviours. In: Borzyszkowski, A.M., Sokołowski, S. (eds.) MFCS 1993. LNCS, vol. 711, pp. 473–484. Springer, Heidelberg (1993). https://doi.org/10.1007/3-540-57182-5_39
11. Jacobs, B.: Introduction to Coalgebra: Towards Mathematics of States and Observation. Cambridge Tracts in Theoretical Computer Science, vol. 59. Cambridge University Press, Cambridge (2016)

12. Jacobs, B., Hughes, J.: Simulations in coalgebra. ENTCS **82**(1), 128–149 (2003)
13. Milner, R.: Communication and Concurrency. PHI Series in Computer Science. Prentice Hall, Upper Saddle River (1989)
14. Milner, R.: A complete axiomatisation for observational congruence of finite-state behaviors. Inf. Comput. **81**(2), 227–247 (1989)
15. Pous, D.: New up-to techniques for weak bisimulation. TCS **380**(1–2), 164–180 (2007)
16. Pous, D.: Coinduction all the way up. In: LICS, pp. 307–316. ACM (2016)
17. Pous, D., Sangiorgi, D.: Enhancements of the bisimulation proof method (2012)
18. Rutten, J.: Universal coalgebra: a theory of systems. TCS **249**(1), 3–80 (2000)
19. Sangiorgi, D.: On the proof method for bisimulation. In: Wiedermann, J., Hájek, P. (eds.) MFCS 1995. LNCS, vol. 969, pp. 479–488. Springer, Heidelberg (1995). https://doi.org/10.1007/3-540-60246-1_153
20. Sangiorgi, D., Milner, R.: The problem of "weak bisimulation up to". In: Cleaveland, W.R. (ed.) CONCUR 1992. LNCS, vol. 630, pp. 32–46. Springer, Heidelberg (1992). https://doi.org/10.1007/BFb0084781
21. Sangiorgi, D., Walker, D.: The Pi-Calculus - A Theory of Mobile Processes. Cambridge University Press, Cambridge (2001)
22. Turi, D., Plotkin, G.: Towards a mathematical operational semantics. In: LICS, pp. 280–291. IEEE (1997)

Foundations of Data Science and Engineering – Regular Papers

Crowd Detection for Drone Safe Landing Through Fully-Convolutional Neural Networks

Giovanna Castellano, Ciro Castiello, Corrado Mencar, and Gennaro Vessio(✉)

Department of Computer Science, University of Bari, Bari, Italy
{giovanna.castellano,ciro.castiello,corrado.mencar, gennaro.vessio}@uniba.it

Abstract. In this paper, we propose a novel crowd detection method for drone safe landing, based on an extremely light and fast fully convolutional neural network. Such a computer vision application takes advantage of the technical tools some commercial drones are equipped with. The proposed architecture is based on a two-loss model in which the main classification task, aimed at distinguishing between crowded and non-crowded scenes, is simultaneously assisted by a regression task, aimed at people counting. In addition, the proposed method provides class activation heatmaps, useful to semantically augment the flight maps. To evaluate the effectiveness of the proposed approach, we used the challenging VisDrone dataset, characterized by a very large variety of locations, environments, lighting conditions, and so on. The model developed by the proposed two-loss deep architecture achieves good values of prediction accuracy and average precision, outperforming models developed by a similar one-loss architecture and a more classic scheme based on MobileNet. Moreover, by lowering the confidence threshold, the network achieves very high recall, without sacrificing too much precision. The method also compares favorably with the state-of-the-art, providing an effective and efficient tool for several safe drone applications.

Keywords: Unmanned aerial vehicles · Crowd detection · Public safety · Safe landing · Computer vision · Convolutional neural networks

1 Introduction

Unmanned aerial vehicles (UAVs), commonly known as drones, are increasingly used in a wide range of domains, from fast delivery, to video surveillance and aerial photography [14,15]. Their increasing popularity is mainly due to the commercial availability of a large variety of drones, even at very low prices. However, as the use of drones increases, the need of mechanisms for public safety accordingly grows. In particular, although current regulations generally forbid drones from flying over a crowd, unpredictable problems, such as adverse environmental conditions, could make the drone's emergency landing unavoidable, even in the presence of a crowd. For example, in Italy, a drone must never fly over a crowd;

A. Chatzigeorgiou et al. (Eds.): SOFSEM 2020, LNCS 12011, pp. 301–312, 2020.
https://doi.org/10.1007/978-3-030-38919-2_25

moreover, it can operate only at a safe horizontal distance of at least 50 m from a crowd.[1] In light of this, there is a pressing need for safety mechanisms able to detect human crowds, in order to define *no-fly* zones for dynamically adapting the flight plan. For example, the given flight plan can be adapted by computing safe way-points on geo-referenced data [2]. This makes the approach desirable not only for remotely controlled drones but specifically for autonomous UAVs, which appear to be the next generation drones [9].

Many commercial drones are equipped with on-board cameras and embed cheap and powerful GPUs which allow to address the problem of crowd detection using advanced computer vision approaches such as Convolutional Neural Networks (CNNs). Despite the advancements obtained with CNN-based techniques in a large variety of computer vision tasks, these are not always optimal for dealing with image sequences captured from drones, because of the various challenges posed by aerial images [17]. A computer vision-based strategy for crowd detection has been recently adopted by Tzelepi and Tefas [12,13], which instructed a fully-convolutional network (FCN) to automatically distinguish between crowded and non-crowded scenes captured from drones. FCNs were used to provide a "light-weight" model, as imposed by the computational limits of the drone. In addition, this model handles images of arbitrary dimension—which is important for processing possibly low resolution images—, and provides estimated heatmaps to semantically enrich the flying zones.

To the best of our knowledge, the research in [12,13] is the only one addressing the problem of detecting crowds from drones through deep learning algorithms. Hence, potentialities and limitations of this approach have yet to be investigated in depth. A first issue concerns the need to evaluate the robustness of the FCN-based approach against a larger set of images/frames corresponding to a larger variability of captured scenes. Indeed, due to the difficulties in collecting and manually annotating large datasets, the data used in [12,13] are limited both in size and covered scenarios. Moreover, that work reduces the problem of crowd detection to a binary one, discriminating between crowded (i.e., with an arbitrary high number of people) and non-crowded (i.e., with no people) scenes. Conversely, there is the need to evaluate the effectiveness of the method in detecting the presence of crowd also in "less crisp" situations, where the distinction between crowd and non-crowd is less clear.

To this end, the present paper contributes to advance the state-of-art on drone-based crowd detection by overcoming the above-mentioned limitations. Firstly, we consider a wide and complex dataset, namely the VisDrone benchmark dataset [17]. These benchmark data are composed by a large set of frames and images captured by various drones, covering a wide spectrum of locations, environments, objects and density, in different scenarios and under different weather and lighting conditions. The use of these data is meant to inject additional variance: this makes the risk of overfitting a serious concern requiring a careful tuning of the network's hyper-parameters. Using such a benchmark

---

[1] https://www.enac.gov.it/sites/default/files/allegati/2018-Lug/Regulation_RPAS_Issue_2_Rev_4_eng.pdf.

dataset, we train and evaluate different FCN architectures. Specifically, we propose two light-weight architectures, the first one being a model based on a classic cross-entropy loss; while the second one is a multi-output model characterized by a joint loss which combines the cross-entropy to a regression loss, based on the people count. Experimental results show that the latter approach slightly outperforms the former. Both models provide better results than MobileNet, which is a more complex pre-trained FCN. Moreover, both our models outperform the FCN architecture proposed in [13], when applied to the same data.

The rest of this paper is organized as follows. Section 2 reviews the related works. Section 3 presents the proposed approach. Section 4 describes the data used for the present study and provides experimental results. Section 5 concludes the paper and sketches future developments of the present research.

2 Related Work

There has been little research on the problem of drone safe landing through the use of on-board cameras. Some works make use of the camera to detect a marker on the ground; others to detect a safe region to land on. Concerning the former approach, it is aimed at guiding the drone towards the central position of the marker by using either hand-crafted features [6,10] or features automatically extracted by a CNN [9]. In [9], the authors proposed a light-weight CNN, called lightDenseYOLO, which predicts the marker direction from each input image. As for the latter approach, the goal is to estimate the safety of the landing area in order to avoid obstacles. Also in this case, the use of CNN-based features [7] may be preferred to more traditional ones based on colour and texture features [8]. In [7], the authors used a small, embedded CNN, which is trained on synthetic aerial data to learn the segmentation of images into safe and obstacle regions.

A different way to look at the problem of safe landing concerns the detection of crowds. Besides the already-mentioned works of Tzelepi and Tefas [12,13], crowd detection from drones is still an unexplored research direction. In [12], the authors firstly adapted a pre-trained model by discarding the fully-connected higher layers in favor of an extra convolutional layer, making it an FCN: this is done in order to reduce the parameters to be learnt and the computational cost. Then, a so-called two-loss convolutional model was proposed, including a softmax layer and an extra layer based on the linear discriminant analysis method to improve the between-class separability. Due to the lack of available datasets, experiments were performed on the *Crowd-Drone* dataset, purposely designed by the authors. Specifically, *Crowd-Drone* was created by querying YouTube with keywords describing crowded scenes captured from drones, resulting in 11, 840 crowded and non-crowded images. The two-loss convolutional model outperformed the one-loss model, achieving an accuracy of ~95% in the binary discrimination crowded vs. non-crowded scenes. The proposed approach is suitable for safety applications since it can output heatmaps that semantically enrich the flight maps by defining "fly" and "no-fly" zones. Each heatmap was obtained by feeding the network with the corresponding image labeled as "crowd" and by extracting the output of the last convolutional layer, which is the desired heatmap. In [13], the authors extended their previous work by proposing a novel regularization technique, based on the graph embedding framework, which is applicable not only to crowd detection but also to generic classification problems. This approach, however, provided little improvement in terms of classification performance. The investigation of a similar approach against larger and more variable data is the main goal of the present work.

Some recent works have successfully applied deep CNNs to the problem of crowd counting and crowd density estimation. This topic is receiving growing attention due to its applicability to the context of video surveillance for purposes of metropolis security. Zhang et al. [16] proposed an iterative switching process where the density estimation and the count estimation tasks are alternately optimized, through backpropagation: in this way, the two related tasks assist each other and achieve a lower loss. In addition, since a model trained on a specific scene can be hardly used in other scenes, the authors proposed a data-driven method to select samples from the training set to fine-tune the pre-trained CNN: the model is thus more apt to the unseen target scenes it is asked to estimate. The proposed crowd CNN model outperformed classic approaches based on hand-crafted features on a challenging dataset. In [1], Boominathan et al. proposed CrowdNet: a deep CNN-based framework for estimating crowd density from images of highly dense crowds (more than one thousand people). Highly dense crowds typically suffer from severe occlusion and are characterized by non-uniform scaling: for instance, an individual near the camera is captured in great detail, while an individual away from the camera can be represented as a head blob. To address this issue, CrowdNet uses a combination of a shallow and a deep architecture which simultaneously operate at a high semantic level (face detection) and at the head blob low-level. The model is made robust to scale variations by using a data augmentation technique based on patches cropped from a multi-scale pyramidal representation of each training image. However, these works do not consider aerial images taken from drones.

To the best of our knowledge, the problem of people counting in images captured from drones has been tackled only by Küchhold et al. [5]. In contrast to the recent trend, the authors used features based on the luminance channel and kernel density estimation, showing that this approach can be faster and more accurate than a CNN-based method. Nevertheless, in this work, as well as in other works taking into account people counting with images from traditional cameras, the presence of a crowd is already assumed in the scene. Instead, in our research we confront the problem of first determining the presence or absence of a crowd for the purposes of drone safe landing.

3 Proposed Approach

As previously stated, light architectures are required to meet the computational limitations imposed by the UAVs' hardware. Therefore, we propose a light-weight FCN for crowd detection in video frames acquired from drones. Relying only on

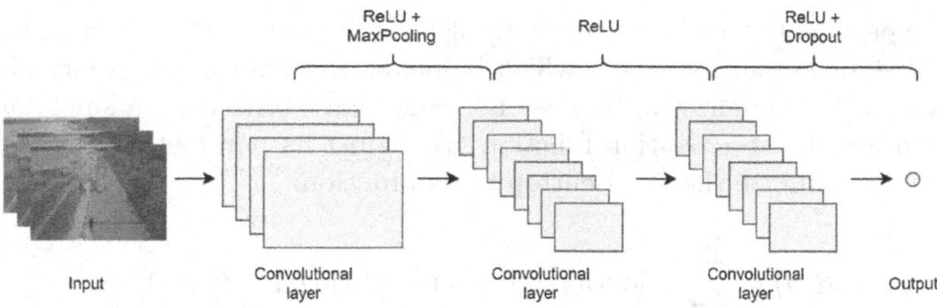

Fig. 1. The proposed FCN architecture. The output layer can be either a single output layer or a multi-output layer.

convolutional layers reduces considerably the amounts of parameters to be learnt, as the fully connected layers typically stacked on top of the convolutional base contribute the most to the overall computational cost. Another advantage is that the network can be fed with images of arbitrary dimensions, as only the fully connected layers expect inputs having a fixed size. Finally, the convolutional layers preserve the spatial information which is instead destroyed by the fully connected layers, because of their connection to all input neurons.

The proposed FCN architecture is depicted in Fig. 1. To speed up calculation, without sacrificing too much capacity, our model takes as input 128×128 three-channels images, normalized in the range $[0, 1]$ before training. The input is then propagated through a convolutional layer having 32 filters, with kernel size 5×5 and stride 1. This layer is followed by a commonly used ReLU non-linearity. This configuration is intended to preserve the initial input information. Then, the output of ReLU is down-sampled by a max pooling layer, which divides each spatial dimension by a factor of 2. Next, there are two consecutive convolutional layers, having 64 filters each with kernel size 3×3. The number of filters in these layers is higher mainly because the number of low level features (i.e., circles, edges, lines, etc.) is typically low, but there are many ways to combine them to obtain higher level features. Moreover, it is quite common to have layers with a doubled number of filters, since the pooling layer allows each feature map to be reduced, thus avoiding a computational explosion. Each of these two convolutional layers is followed by a ReLU activation. Finally, there is one output layer preceded by a dropout layer, with dropout rate of 50%, to mitigate overfitting. We conceived this architecture mainly for two reasons: (*i*) it provides a very light model to meet the strict computational requirements of the UAV; (*ii*) it is complex enough to avoid underfitting the data.

The proposed architecture is smaller than the one proposed in [13]. The latter is characterized by six convolutional layers, each one, except for the last layer, followed by a parametric ReLU as activation function. The output of the last convolutional layer is fed to an output layer with a *softmax* activation. The first and fifth convolutional layers are followed by max pooling layers to reduce their input size. The first pooling layer is followed by a response-normalization layer to improve generalization. Finally, a dropout layer, with dropout rate of 50%, follows the fifth convolutional layer to reduce overfitting.

We experimented with two variants of the proposed architecture. The first variant is a single output model which is meant to perform the binary classification crowd vs. non-crowd. To this end, it is characterized by a single output layer with a *sigmoid* activation function. To adjust its parameters, the network attempts to minimize the cross-entropy loss function:

$$\mathcal{H}(\theta) = \sum_{i=1}^{N} y_i^c \log(h_\theta^c(x_i)) + (1 - y_i^c)\log(1 - h_\theta^c(x_i)),$$

where N is the number of samples, y_i^c is true class label and $h_\theta^c(x_i)$ is the predicted class label (c stands for classification). Since the separation between the two classes may be relatively vague, to strengthen the model we propose a two-loss convolutional network which attempts to simultaneously predict the class of the input image together with its precise people count. In other words, the model is asked to predict also the precise cardinality of the crowd. While multi-output models are typically used to provide different outputs from the same input, we added the regression task so that it can assist the classification task when the model is less confident with the label to be assigned to the input image. To evaluate the cardinality of the crowd, the mean absolute error loss is considered:

$$\mathcal{L}(\theta) = \frac{1}{N}\sum_{i=1}^{N} |y_i^r - h_\theta^r(x_i)|,$$

being y_i^r and $h_\theta^r(x_i)$ the actual and estimated people counts (r stands for regression). Hence, the proposed FCN model minimizes the following joint loss function, combining the cross-entropy loss and the mean absolute error loss:

$$\mathcal{J}(\theta) = \mathcal{H}(\theta) + \mathcal{L}(\theta).$$

Once the proposed FCN is trained, its last convolutional layer is used to obtain heatmaps of class activation over the input images. To do this, we use the implementation of class activation map (CAM) described in [11]. Given an input image, this technique extracts the output feature map of the last convolutional layer and weights every channel by the gradient of the class with respect to that channel. In this way, a class activation map is obtained, indicating how intensely the input image activates the class.

4 Experiment

To test the effectiveness of our method in accurately detecting the presence of a crowd, we partly re-arranged the VisDrone dataset, as described in Sec. 4.1. As a baseline for a fair comparison we employed a MobileNet model [3] pre-trained on ImageNet [4]: MobileNet is a light architecture which is well suited to mobile and embedded computer vision applications. This architecture introduced the so-called depthwise separable convolutions, which perform a single convolution over each colour channel rather than combining all of them. This significantly reduces the numbers of parameters to be learned. To perform transfer learning on the VisDrone dataset, we used the common practice to remove the top level classifier, which is very specific for the original classification problem, and to stack a custom layer to be trained on our task. In addition, we compared the proposed method to our implementation of the architecture proposed in [13]. The main implementation difference concerns the use of a classic ℓ_2 regularization term, applied to every convolutional layer to further mitigate overfitting. In [13], this regularization technique provided slightly lower performance than the one proposed by the authors.

In the following subsections, we firstly describe the dataset preparation, as well as implementation details; then, we report the results obtained. Qualitative results are also reported based on the crowd heatmaps provided by the method.

4.1 Dataset Preparation

Developing a large crowd dataset from a drone perspective is a very time consuming and expensive process. To overcome this issue, we used an adaptation of the VisDrone benchmark dataset,[2] collected by the AISKYEYE team at the Laboratory of Machine Learning and Data Mining, Tianjin University, China. The data have been used for the VisDrone 2018 and 2019 challenge. To date, VisDrone is the largest dataset of aerial images from drones ever published. The original dataset consists of 288 video clips, with 261,908 frames and 10,209 additional static images: they were acquired by various drone platforms, across 14 different cities in China [17]. The captured scenes cover various weather and lighting conditions, environment (urban and country), objects (pedestrians, vehicles, etc.) and density (sparse and crowded scenes). The maximum resolutions of video clips and static images are 3840×2160 and 2000×1500, respectively (sample images are shown in Fig. 2). Frames and images were manually annotated with more than 2.6 million bounding boxes of targets (this kind of ground truth is available only for the training and validation sets). The object categories involve human and vehicles: pedestrians, persons, cars, vans, buses, and so on.

The benchmark data embedded in VisDrone have been originally conceived to tackle different kinds of tasks, ranging from object detection in images/videos to single or multi-object tracking. For our purposes, the compilation of a crowd dataset was necessary, therefore we profited from the VisDrone annotations of pedestrians and persons. Since a precise definition of "crowdedness" is unpractical, we considered the presence of at least 10 pedestrians/persons as a crowd. In this way, we were able to collect a subset of images from VisDrone, all of them labeled as *crowd* or *non-crowd*, depending on the count of the involved pedestrians/persons. The arranged dataset is described in Table 1: it is composed by well-balanced classes and it lacks of a hold-out validation set (mainly because holding out a fraction of the test set would have resulted in a too small test set).

[2] http://aiskyeye.com.

Fig. 2. Sample images from the VisDrone dataset.

Table 1. Characteristics of the arranged crowd dataset.

Class	Training set (size)	Test set (size)
Non-crowd (<10)	15,591	1,634
Crowd (≥10)	15,081	1,760
Total	30,672	3,394

4.2 Experimental Results

Experiments were run on an Intel Core i5 equipped with the NVIDIA GeForce MX110, with dedicated memory of 2GB: this GPU has similar performance, or slightly lower, compared to the NVIDIA Jetson TX2 typically mounted on drones for several applications. Thus, it allowed us to estimate the real-time capacity of the model. As deep learning framework, we used TensorFlow 2.0 and the Keras API. The proposed models were trained from scratch by performing stochastic gradient descent with randomly sampled mini-batches of 64 images and learning rate of 0.01. As previously mentioned, to reduce the computational cost the input images were resized to 128×128 and they were normalized within the range $[0, 1]$.

To assess the effectiveness of our models, we made a comparison with other existing FCN architectures: the MobileNet architecture and the FCN architecture proposed in [13]. Concerning the MobileNet model, we used a lower learning rate of 0.0001 in order to prevent the previously learned weights from being destroyed. The network was initialized with the parameter α equals to 0.50, which proportionally decreases the number of filters in each layer, making the model lighter. Moreover, it is worth remarking that we used larger input images of shape 224×224, so as to address the higher capacity of the network, and each input channel was re-scaled to the range $[-1, 1]$, as this is the input expected by the network. Finally, concerning our replica of the model proposed in [13],

Table 2. Results.

Model	Input	Accuracy	AP	Size (MB)	Speed (fps)
Present replica of [13]	128×128	79.13%	79.76%	~17.3	54.74
Pre-trained MobileNet	224×224	83.11%	76.98%	~3.6	64.03
Proposed FCN (one-loss)	128×128	84.03%	77.54%	~1.1	70.70
Proposed FCN (two-loss)	128×128	86.80%	82.68%	~2.0	48.48

it was set with the same parameters described in the paper. More precisely: images were resized to 128×128 pixels; the learning rate was set to 10^{-5}, with momentum of 0.9; the batch size was set to 64.

We did not perform fine tuning of the MobileNet model, as we noticed that this was detrimental to prediction accuracy. This can be explained considering that the dataset we used is very challenging and noisy, thus the trained model learns soon the irrelevant patterns in the training data, resulting in overfitting. For the same reason, we trained all the models for few epochs (less than 5 epochs, requiring about one hour of training time), as the models began to overfit soon. This was expected, since managing such a complex dataset makes the risk of overfitting a serious concern.

Experimental results are provided in Table 2: they are expressed in terms of prediction accuracy and average precision. We also provide measures of size (HDF5 format) and speed (frames per second) of the experimented models.

As shown, our replica of the architecture proposed in [13] achieved the worst results in terms of prediction accuracy. This may be due to the architecture which has still too capacity for the classification problem at hand: in fact, it has the biggest size among the experimented models. MobileNet achieved better results in terms of accuracy than the previous model, but with a lower average precision. This was partly expected since, although the model was tuned to have half of the filters in each layer to be adjusted, it is still too complex for the problem too. Moreover, we must consider that, even if it is not drastically different, the ImageNet dataset MobileNet was trained on is characterized by a number of photographic scenes which are very different from aerial images captured from drones. In other words, a perspective problem arises. It is worth noting that different values for α were tested, obtaining slightly lower performance.

The proposed one-loss model achieved superior performance with the best size (~1.1 MB) and speed (~70.70 fps). These features make the model extremely fast and useful for a safety application deployed on drones. Finally, the best classification results were obtained by the proposed two-loss model, outperforming all other models. In particular, an accuracy of 86.80% and an average precision of 82.68% were achieved. By looking at the one-loss model's misclassifications, we noticed that the two-loss model was better in correctly classifying images from a perspective orthogonal to the ground; in other words, when the drone was hovering. This suggests that the addition of the regression task can be beneficial to those cases where the detection of people is more difficult, i.e. when individuals

Fig. 3. From left to right: the original image, the corresponding heatmap and the superimposed image for four test crowded and non-crowded scenes. For a better visualization, the images have been re-scaled to the original proportions.

look only as head blobs. On the other hand, this model exhibits a size which is a trade off between the MobileNet and the one-loss method. Nevertheless, the two-loss network is still extremely light and fast.

Comparing our results with the state-of-the-art reported in literature [13], a lower classification accuracy has been here observed, i.e. 86.80% vs. 95.46%. However, as previously stated, the state-of-the-art results were obtained on a smaller dataset, characterized by a lower variance. Indeed, tested on our crowd dataset, the state-of-the-art method was less able to generalize to previously unseen examples. The main goal of our research was to test the applicability and scalability of the FCN-based approach to more challenging scenarios. In this sense, the results obtained are encouraging for the purposes of our research, especially if we consider that by lowering the confidence threshold to 0.35, the proposed two-loss model is able to achieve a very high recall of 97.50%, while maintaining a precision as high as 78.35%. In this way, the method is excellent in detecting all crowded scenes, without sacrificing too much precision, i.e. without suffering too much from false positives. Moreover, it is worth to note that both derived models are extremely light and fast.

The proposed method can be used to output heatmaps to semantically enrich the flying zones. Examples of heatmaps, superimposed to the original images, are provided in Fig. 3. It can be seen that the model is able to distinguish the safer zones, i.e. trees, buildings, streets with no pedestrian, etc., from the risky zones where people are standing or walking. It is worth to note that these heatmaps were obtained by feeding the network with the images re-sized to 128×128: this allows the network to output the heatmap while providing the prediction for the original image. Conversely, in [13] heatmaps were obtained by feeding the network with higher resolution images of 1024×1024. While this approach is able to generate higher quality heatmaps, on the other hand it slows down the processing speed, as higher resolution images require much more computation.

5 Conclusion

In this paper, we have addressed the problem of human crowd detection from drones. This is important for UAV applications, as unpredictable problems can cause the drone to fly over a crowd, thus safety mechanisms are required to automatically adjust the flight plan to prevent the drone from landing on a risky zone. In order to cope with the strict computational requirements of a drone, we have proposed a very light-weight FCN classification model, trained to distinguish between crowded and non-crowded scenes by leveraging the camera and GPU typically mounted on several currently available drones. The classification model has been trained from scratch on the very challenging VisDrone dataset. The proposed method is based on minimization of a joint loss function combining two loss terms, one for classification and one for regression, so that the regression task assists the classification one by trying to estimate the people count. This model outperforms a one-loss similar architecture and a more complex method based on the well-known MobileNet architecture. A deep network pre-trained on ImageNet can be less tailored to distinguish among aerial images, mainly because of their different perspective against traditional photographic scenes. The proposed method not only discriminates between crowded and non-crowded scenes, but also provides heatmaps that can be used to semantically enrich the flight maps. The approach also compares favorably with the state-of-the-art, providing an extremely fast, light and with high recall tool.

An open issue of the present research is the lack of a well-defined concept of crowdedness which forced us to propose a simple concept based only on the people count. This may have affected the classification task, specially when the people count was around 10. Moreover, it may have caused the network to overlook some other patterns which can possibly improve performance. Future work should investigate more refined concepts of crowdedness, based for example on the spatial density of the crowd.

Acknowledgement. The research is supported by Ministero dell'Istruzione, del-l' Università e della Ricerca (MIUR) under grant PON ARS01_00820 "RPASInAir – Integrazione dei Sistemi Aeromobili a Pilotaggio Remoto nello spazio aereo non segregato per servizi".

References

1. Boominathan, L., Kruthiventi, S.S., Babu, R.V.: CrowdNet: a deep convolutional network for dense crowd counting. In: Proceedings of the 24th ACM International Conference on Multimedia, pp. 640–644. ACM (2016)
2. Castelli, T., Sharghi, A., Harper, D., Tremeau, A., Shah, M.: Autonomous navigation for low-altitude UAVs in urban areas. arXiv preprint arXiv:1602.08141 (2016)

3. Howard, A.G., Zhu, M., Chen, B., Kalenichenko, D., Wang, W., Weyand, T., Andreetto, M., Adam, H.: MobileNets: efficient convolutional neural networks for mobile vision applications. arXiv preprint arXiv:1704.04861 (2017)
4. Krizhevsky, A., Sutskever, I., Hinton, G.E.: ImageNet classification with deep convolutional neural networks. In: Advances in Neural Information Processing Systems, pp. 1097–1105 (2012)
5. Küchhold, M., Simon, M., Eiselein, V., Sikora, T.: Scale-adaptive real-time crowd detection and counting for drone images. In: 2018 25th IEEE International Conference on Image Processing (ICIP), pp. 943–947. IEEE (2018)
6. Lin, S., Garratt, M.A., Lambert, A.J.: Monocular vision-based real-time target recognition and tracking for autonomously landing an UAV in a cluttered shipboard environment. Auton. Robots **41**(4), 881–901 (2017)
7. Marcu, A., Costea, D., Licaret, V., Pirvu, M., Slusanschi, E., Leordeanu, M.: SafeUAV: learning to estimate depth and safe landing areas for UAVs from synthetic data. In: Proceedings of the European Conference on Computer Vision (ECCV), pp. 0–0 (2018)
8. Mukadam, K., Sinh, A., Karani, R.: Detection of landing areas for unmanned aerial vehicles. In: 2016 International Conference on Computing Communication Control and Automation (ICCUBEA), pp. 1–5. IEEE (2016)
9. Nguyen, P., Arsalan, M., Koo, J., Naqvi, R., Truong, N., Park, K.: LightDenseYOLO: a fast and accurate marker tracker for autonomous UAV landing by visible light camera sensor on drone. Sensors **18**(6), 1703 (2018)
10. Polvara, R., Sharma, S., Wan, J., Manning, A., Sutton, R.: Towards autonomous landing on a moving vessel through fiducial markers. In: 2017 European Conference on Mobile Robots (ECMR), pp. 1–6. IEEE (2017)
11. Selvaraju, R.R., Cogswell, M., Das, A., Vedantam, R., Parikh, D., Batra, D.: Grad-CAM: visual explanations from deep networks via gradient-based localization. In: Proceedings of the IEEE International Conference on Computer Vision, pp. 618–626 (2017)
12. Tzelepi, M., Tefas, A.: Human crowd detection for drone flight safety using convolutional neural networks. In: 2017 25th European Signal Processing Conference (EUSIPCO), pp. 743–747. IEEE (2017)
13. Tzelepi, M., Tefas, A.: Graph embedded convolutional neural networks in human crowd detection for drone flight safety. IEEE Trans. Emerg. Top. Comput. Intell. (2019). https://doi.org/10.1109/TETCI.2019.2897815
14. Valavanis, K.P., Vachtsevanos, G.J.: Handbook of Unmanned Aerial Vehicles. Springer, Heidelberg (2015)
15. Zeng, Y., Zhang, R., Lim, T.J.: Wireless communications with unmanned aerial vehicles: opportunities and challenges. IEEE Commun. Mag. **54**(5), 36–42 (2016)
16. Zhang, C., Li, H., Wang, X., Yang, X.: Cross-scene crowd counting via deep convolutional neural networks. In: Proceedings of the IEEE Conference on Computer Vision and Pattern Recognition, pp. 833–841 (2015)
17. Zhu, P., Wen, L., Bian, X., Ling, H., Hu, Q.: Vision meets drones: a challenge. arXiv preprint arXiv:1804.07437 (2018)

Explaining Single Predictions: A Faster Method

Gabriel Ferrettini(✉), Julien Aligon, and Chantal Soulé-Dupuy

Université de Toulouse, UT1, IRIT, (CNRS/UMR 5505), Toulouse, France
{gabriel.ferrettini,julien.aligon,chantal.soule-dupuy}@irit.fr

Abstract. Machine learning has proven increasingly essential in many fields. Yet, a lot obstacles still hinder its use by non-experts. The lack of trust in the results obtained is foremost among them, and has inspired several explanatory approaches in the literature. In this paper, we are investigating the domain of single prediction explanation. This is performed by providing the user a detailed explanation of the attribute's influence on each single predicted instance, related to a particular machine learning model. A lot of possible explanation methods have been developed recently. Although, these approaches often require an important computation time in order to be efficient. That is why we are investigating about new proposals of explanation methods, aiming to increase time performances, for a small loss in accuracy.

Keywords: Machine learning · Explanation model · predictive model

1 Introduction

Many explanation methods exist in the literature, to overcome the "black box" problem of model prediction results. These methods are mainly devoted to explain a predictive model in a global way. These methods are not relevant when a domain expert user (for instance a biologist) has to study the behavior of particular dataset instances over a predictive model (for instance in the context of cohort study). In this direction, previous studies offer the possibility of explaining single instance prediction, over a model, as in [12] and [2]. One major problem of these contributions is the complexity of the proposed algorithms ($O(n^2)$). Thus, it is illusory for domain experts wishing to apply this method to study the behavior of a set of instances. This complexity makes them very slow to calculate on datasets with a large number of attributes. Our work fits the general ambition to help a domain expert user to (re)find motivation to get involved in data analysis operations. In particular, our goal is to rely as much as possible on her/his area of expertise, while limiting knowledge in data analysis. In this paper, we aim to facilitate the use of predictive models by explaining their predictions in a way balancing information and computing time.

A. Chatzigeorgiou et al. (Eds.): SOFSEM 2020, LNCS 12011, pp. 313–324, 2020.
https://doi.org/10.1007/978-3-030-38919-2_26

The contributions presented in this paper include:

- A comparison between two selected prediction explanation approaches. This is done to decide which method to use as a basis, among the two closest to our scope.
- Two new methods for classification explanations, based on [12], and adapted to achieve a better calculation time, without losing too much information.

The paper is organized as follows. Section 2 explores some work already done in the domain of prediction explanation. In particular, the literature helps us to identify an explanation method as close as possible to our scope: helping non expert users to understand the inner workings of a predictive model. Then, in Sect. 3, we propose improvements of the selected method to achieve a better calculation time, without losing too much information. Finally, we experiment our proposals to check their interest in terms of computation time and their impacts in terms of loss of accuracy.

2 Related Works

Explaining the influence of each attribute (of a dataset) on the output of a predictive model have been explored largely. An example of the works pertaining to global attribute importance on a model can be seen here: [1]. The most recent methods are based on swapping the values of attributes in the dataset and analysing which swap affect the trained model predictions the most. The more modifying the attributes values affects the predictions, the most this attribute is considered important for the model, as a whole. These methods are often used during feature selection, allowing to opt out attributes not used by the model. Many ways of explaining single predictions have been explored but these methods often struggle between being too simplistic, or too complex to be interpreted by a human, notwithstanding the problem of computation time, which can become problematic for more advanced methods. The possible applications of prediction explanations have been investigated by [8]. According to their paper, the interest for explaining a predictive model is threefold:

- First, it can be seen as a mean to understand how a model works in general, by peering at how it behaves in diverse points of the instance space.
- Second, it can help a non expert user to judge of the quality of a prediction and even pinpoint the cause of flaws in its classification. Correcting them would then lead the user to perform some intuitive feature engineering operations.
- Third, it can allow the user to decide the type of model preferable to another one, even if he has no knowledge of the principles underlying each of them.

A great number of works pertaining to prediction explanation led to [6], which theorized a category of explanation methods, named *additive* methods, and produced an interesting review of the different methods developed in this category. Some of these methods are described in detail in [3] and [10]. They are summarized in [6] as methods attributing for a given prediction, a weight to each attribute of the dataset. This creates a very simple "predictive model", mimicking the original model's behavior locally. Thus, we have a simple interpretable linear model which gives information on the original model's inner working in a small vicinity of the predicted instance. The methods from which these weights are attributed to each attributes varies between the different *additive* methods, but the end result is always this vector of weights. This article has highlighted several interesting properties about these methods, which make it a very useful theoretical object: Local precision: The system describes precisely the model in the close vicinity of the explained instance. "Missingness": If an attribute is missing for the prediction, the method does not give it a weight, or gives it a weight of zero. Consistence: If the explained model changes in a way that makes an attribute more important, or does not change its importance, its attributed weight is not diminished. This property is important, as some of the early prediction explanation methods could have an erratic behavior in some cases, as shown in an example of [6]. Other lines of reasoning have been explored, as in [2], which explored prediction explanation in the point of view of model performance. Meaning that their metric shows which feature improves the performance of the model, rather than which feature the model consider as important for its prediction. If this line of reasoning is really interesting for the model explanation field, it does not correspond to our scope as well as other methods, as we are aiming to help users understand how a model works, and not how to improve it. In this paper, we are aiming to facilitate the understanding of any machine learning models for user without particular knowledge on data analysis or machine learning. Thus, it is more relevant to focus on the works as [12] or [3], cited as *additive* methods, as they generate a simple set of importance weights for each attribute. This set of weights is easy to interpret, even for someone without expertise on machine learning. Yet, these methods have a major deterrent: their computation time makes them difficult to use for the average user. That is why [6] explored methods to generate explanations faster, but at the cost of very restricting hypotheses, as the Independence of each attributes of the dataset, or the linearity of the model, which is not always the case. Thus, we are aiming for a simplification to reduce computational time of methods like [12], but applicable in a more generic way than [6]. With this work, we want to facilitate the generation of prediction explanation, without having to restrict ourselves to a given set of models. The ability to explain the prediction of any model thus appears to be a key point for allowing a broader public (non expert) to access and use machine learning models. This need led us to consider the diverse explanation systems, developed in the literature, as having a major interest for giving more autonomy to domain experts performing data analysis tasks. Yet, the computational load found in the most generic methods can be a hindrance to their use. In this paper, we seek to select a prediction explanation method as generic as possible and try lowering its computing time without loosing too much information.

3 Choosing a Basic Explanation Method

In order to start developing a faster *additive* explanation method, we have to select an algorithm from the literature and reduce its complexity without losing too much information. For this, we compare two methods developed by the authors of [11] and [12], as they are classical and similar in their design, but different in their interpretation.

3.1 Prediction Explanation Methods

Given a dataset D of instances and a set of n attributes $A = \{a_1, .., a_n\}$, each attribute being either continuous or nominal, its possible values are then integers or real number. Each instance $x \in D$ is defined by the values of each of its attributes: $x = \{x_1, ..., x_n\}, \forall i \in 1..n, x_i \in \mathbb{N} \vee x_i \in \mathbb{R}$. We want to explain a predictive model, based on the function $f : D \rightarrow [0, 1]$, whose result is the confidence score in the classification of the instance x for a class C, as predicted by the model.

Information Loss Method. One of the first definition for classification explanation is proposed in [11]. According to their method, the influence of an attribute a_i on the classification of a given instance is defined as the difference between the classifier prediction (with a_i) and its prediction without the knowledge of attribute a_i. Thus, given a dataset of instances described along the attributes of A, the influence of the attribute a_i on the classification of an instance x by the classifier confidence function f on the class C can be represented as:

$$inf^C_{f,a_i}(x) = f(x) - f(x \backslash a_i) \tag{1}$$

Where $f(x \backslash a_i)$ represents the probability distributions for a classification of the instance x by the classifier f without knowledge of the attribute a_i. We name this method as the *information loss method* (shortened as *loss method*).

Information Gain Method. In more recent works, as in [12], another possible formula is based on the information brought by an attribute in the dataset:

$$inf^C_{f,a_i}(x) = f(x_{a_i}) - f(\varnothing) \tag{2}$$

Where $f(x_{a_i})$ represents the probability that the instance x is included in the class C with only the knowledge of the attribute a_i (according to the predictive model). We name this method as the *information gain method* (shortened as *gain method*). In order to simulate the absence of an attribute, the authors of [11] theorize possible approaches, among which we selected to retrain the classifier without the corresponding attribute.

Comparing the Two Methods: Toy Example on a Basic Dataset. As an illustration, and in order to ease interpretations, we apply these two methods in a simple ID3 decision tree [7], trained on the well-known Fisher's *Iris* dataset[1].

[1] Iris, Fisher: https://en.wikipedia.org/wiki/Iris_flower_data_set.

As a decision tree is a naturally interpretable model, and Iris dataset is well studied in the literature, it is easy to compare and interpret the two methods and detect eventual problems. We use Weka [4] and OpenML [13] to perform the data management and model training while ensuring the reproductibility of all experiments. OpenML is a machine learning collective platform. It includes a repository of datasets and workflows, in which each user can upload any new dataset and run any data mining task on them. To estimate the reliability of *loss* and *gain* methods, we apply a 5-fold cross validation on the *Iris* dataset. The explanations of both methods are generated on the validation set, for each iteration of the cross-validation. We generate thus prediction explanations of Weka's *J48* tree classifier, for the whole *Iris* dataset. Then, we compare those explanations to the decision tree, and get a general sense of the explanation accuracy. Each instance of the Iris dataset is composed of four attributes: *petal length*, *petal width*, *sepal length* and *sepal width*. Each instance is included in one of these three classes: *Iris Setosa*, *Versicolor* or *Virginica*.

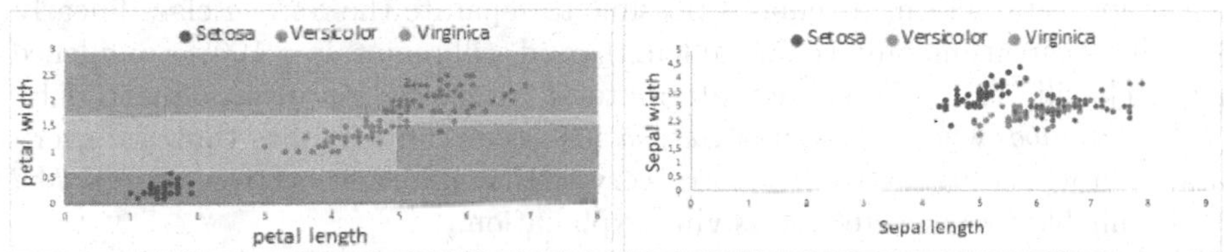

Fig. 1. Repartition of the 3 different classes by petal length and width, with their corresponding generalization according to the decision tree

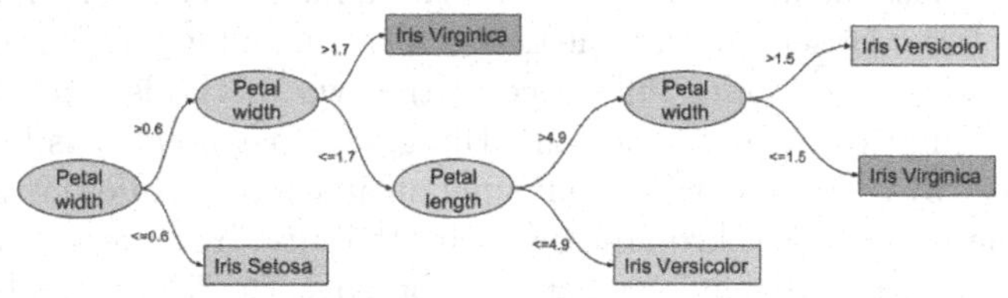

Fig. 2. A decision tree trained on Iris

By a simple look at the trained decision tree Fig. 2, we see the influence of the *loss method* should be zero for the sepal length and width attributes, as they are not used by the tree at all. Moreover, the *Setosa* class instances should only be influenced by petal width, as it is the only attribute used to classify them. For *Iris Virginica* and *Versicolor*, we can expect a high influence from petal width, and a lower from petal length influence, yet still significant as these two attributes are used. We can now compare these expectations to the results

Table 1. Average influence of the different attributes according to each explanation method, for each class of instances

	Loss method				Gain method			
	Sepal length	Sepal width	Petal length	Petal width	Sepal length	Sepal width	Petal length	Petal width
Setosa	**0**	**0**	**0**	**0**	0.211	0.104	0.316	0.316
Versicolor	0	0	0.835	0.135	0.037	0.077	0.308	0.373
Virginica	0	0	0.047	0.691	0.150	0.019	0.355	0.375
Average	0	0	0.305	0.284	0.131	0.066	0.326	0.356

indicated Table 1. We see the *loss method* does not behave as expected for the *Setosa* instances, as all the attributes are being given an influence equals to 0. This can be understood by looking at the representation of the concept learned by the tree Fig. 1. We note that, by removing one of the attributes between petal length and petal width, it remains possible to separate the *Setosa* class linearly from the others using only the petal length, and still maintain a 100% confidence in the classification. Thus, each attribute is considered as inconsequential by the *loss method*, when classifying *Setosa* instances. This implies that, for every dataset in which two attributes carry very similar information, the *loss method* will be unable to generate a satisfying explanation.

The *gain method*, on the other hand, gives an importance to all the four attributes about *Setosa* instances. A minimal importance is given for sepal length and width, unlike the petal length and width. It is easily understandable by observing the graphs of repartition of the different classes (Fig. 1). We note the petal length and width can easily separate the three classes, but the sepal attributes are less defined in their separation. This is especially true for the *Versicolor* class, as it is mixed with its two adjacent classes. Thus, the *gain method* seems to be closer to what the decision tree is doing when being trained. Finally, we conclude the main difference of the two methods relies in the fact they are not trying to calculate the same thing: the *loss method* is based on the information lost by the model when removing an attribute, while the *gain method* is based on the information brought by each attribute. Yet, we remark the *loss method* has an aberrant behavior when confronted with two attributes bringing the same information. Thus, the *gain method* seems the best proposal for a prediction explanation. But this method has its flaws: it only takes into account the information brought by each attribute, independently. In many datasets, the attributes are often interdependent. Our next objective is to consider the influence of group of attributes as described in the next section.

4 Toward a More Efficient Method

In order to answer the problems of interaction between attributes, we propose to take inspiration from the work of [12]. We are here in a framework close to the situation of a game called "coalitions", where each group of attributes can have an influence on the prediction of the model. Therefore, we cannot consider each attribute as independent, but all the possible combinations of attributes. The influence of an attribute is measured according to its importance in each coalition. We can then refer to the coalition games as defined by Shapley in [9]: A coalitional game of N players is defined as a function mapping subsets of players to gains $g : 2^N \mapsto \mathbb{R}$. The parallel can easily be drawn with our situation, where we wish to assess the influence of a given attribute *in every possible coalition of attributes.* We then look at not only the influence of the attribute, but also its use in all subsets of attributes. We thus define the *complete influence* of an attribute $a_i \in A$ on the classification of an instance x (the notations remain the same as in Sect. 3.1):

$$\mathcal{I}^C_{a_i}(x) = \sum_{A' \subseteq A \setminus a_i} p(A', A) * (inf^C_{f,(A' \cup a_i)}(x) - inf^C_{f,A'}(x)) \tag{3}$$

With $p(A', A)$ a penalty function accounting for the size of the subset A'. Indeed, if an attribute changes a lot the result of a classifier, depending of a lot of attributes, it can be considered as very influential compared to the others. On the opposite, an attribute changing the result of a classifier, whereas this classifier is based on a few number of attributes, cannot be considered to have a decisive influence. The Shapley value [9] is a promising candidate, and defines this penalty as:

$$p(A', A) = \frac{|A'|! * (|A| - |A'| - 1)!}{|A|!} \tag{4}$$

This *complete influence* of an attribute now takes into consideration its importance among all the possible attribute configurations, which is closer to the original intuition behind attributes' influence. However, computing the *complete influence* of a single instance is extremely computationally expensive, with a complexity in $\bigcirc(2^n * l(n, x))$, with n the number of attributes, x the number of instances in the dataset and $l(n, x)$ the complexity of training the model to be explained. It is then not practical to use the *complete influence*. Consequently, it becomes necessary to seek a more efficient way to explain predictions. Although the *complete influence* is too computationally heavy, it can be considered as an excellent baseline [12]. Thus, we can evaluate other explanation methods by studying their differences with the *complete influence*.

4.1 Finding New Estimators of the *Complete Influence*

An approximation of the *complete influence* has to remain accurate and practical, as much as possible. For this we cannot fully rely on recent works (e.g. [12] and [6]), as explained in Sect. 2. In particular, looking for a subset of all the subgroups could be more practical in terms of complexity. This solution should produce explanation, a priori, more accurate than the basic consideration of independent attributes (*linear influence*). We consider then the *depth-k complete influence* defined as:

$$\mathcal{I}_{a_i}^{C_k}(x) = \sum_{A' \subseteq A \backslash a_i | A'| \leq k} p_k(A', A) * (inf_{f,(A' \cup a_i)}^{C}(x) - inf_{f,A'}^{C}(x)) \quad (5)$$

$$p_k(A', A) = \frac{|A'|! * (|A| - |A'| - 1)!}{k * (|A| - 1)!} \quad (6)$$

In particular, we can note that the *linear* influence is actually identical to the *depth-1 complete influence*. The intuition behind this approach is to eliminate the larger groups, which have a lesser impact on the shapley value, while being the most costly to calculate. We then hope to achieve a better calculation time without losing too much information.

Another possible approach is to identify the attributes having a correlation between them. We can obtain a grouping such as:
$G = \{\{a_1, a_3\}, \{a_2, a_5, a_8\}, \{a_4\}...\}$. We then only have to calculate the grouped influence of these attributes groups, without having to consider every possible attributes' combination. We then obtain a *coalitionnal influence* of an attribute $a_i \in g, g \in G$:

$$simple\mathcal{I}_{a_i}^{C}(x) = \sum_{g' \subseteq g \backslash a_i} p(g', g) * (inf_{f,(g' \cup a_i)}^{C}(x) - inf_{f,g'}^{C}(x)) \quad (7)$$

Given the fact we can set a maximum cardinal c for our subgroups, the complexity is, in the worst case, $O(2^c * \frac{n}{c} * l(n, x)) \approx O(n * l(n, x))$. This method calculates less groups than the *depth-k complete influence*, but tries to make up for it by only grouping the attributes actually related to each other. In order to determine which attributes seem to be related, we use an automated correlation detection algorithm, as proposed in [5]. In order to determine if it is possible to generate a satisfactory approximation of the influence of an attribute with the new *depth-k complete influence* and the *coalitionnal influence*, it is necessary to assess the number of attribute's combinations we need to take into account before being sufficiently near to the *complete influence* defined in Eq. 3. Moreover, we need to assess if the results of the *depth-k complete influence* produce better explanations than *linear* and *coalitionnal* influences, in view of its higher computation cost. These are the objectives of our next section.

4.2 Evaluating the Two New Heuristics

In this section we aim to evaluate the value of the *coalitional* and *depth-k complete* influences, considering their precision when compared to the *complete* influence, and their computational time.

Experimental Protocol. Our experiments are run on the OSIRIM[2] cluster. This cluster is equipped with 4 AMD Opteron 6262HE processors with 16 ×

[2] http://osirim.irit.fr/site/en.

1,6 GHz cores, for a total of 64 cores, and 10 × 512 GB of RAM. Our tests are realized from the data available in the Openml platform [13]. We selected the biggest collection of datasets[3] on which classification tasks have been run. We also consider six classification tasks: naïve Bayes, nearest neighbors, J34 decision tree, J34 random forest, bagging naïve Bayes and support vector machine. Due to the heavy computational cost of the complete influence (considered as the reference of our experiments), we selected the datasets having at most nine attributes. Thus, a collection of 324 datasets is obtained. Considering the six types of workflows, we have a total of 1944 runs. For each of those runs, we generate each type of influence proposed in this paper, for each instance of the 324 datasets: the *complete* influence for the baseline, along with the *linear*, *coalitional* and *k-complete* influences. The *k-complete* influences are generated for every possible values of k (from 2 up to the number of attributes of the dataset). The *coalitional* influences are generated using subproups of attributes. Here, these subgroups are produced using the algorithm described in [5], which is based on an $\alpha \in [0, 0.5]$ parameter (small values of α resulting in smaller subgroups, and high values in bigger ones). We generate the possible subgroups with 5 different values of α to study the influence of subgroup size. To compare the different explanation methods, we consider the explanation results as a vector of attribute influences noted $\mathcal{I}(x) = [i_1, ..., i_n]$ with n the number of attributes in the dataset. Thus, each of the attributes a_k is given an influence $i_k \in [0, 1]$ by the method $\mathcal{I} : \forall k \in [1..n], i_k = \mathcal{I}_{a_i}(x)$. We then define a difference between two vectors of influences i, j as the normalised euclidian distance:

$$d(i, j) = \frac{1}{2\sqrt{n}} \sum_{k=1}^{n} \sqrt{(i_k - j_k)^2} \tag{8}$$

Considering this formula, we define an error score based on the difference between an explanation method and the *complete* influence method. Given an instance x, an explanation method $\mathcal{I}(x)$, and the *complete influence* method $\mathcal{I}^C(x)$:

$$err(\mathcal{I}, x) = d(\mathcal{I}(x), \mathcal{I}^C(x)) \tag{9}$$

For each instance of each dataset, we generate the error score of every method, allowing us to compare their performances across the different datasets we collected. Each error score is the distance of the method from the *complete* method. Thus, lesser error is indicative of a more precise estimation of the *complete* method.

Results and Interpretations. Figure 3 indicates the computation time of all the explanation methods. This time takes in account every step of each method: the training of the models, the predictions necessary to calculate the influences, and the constitution of the correlated groups for the coalitional method. As expected, the *coalitionnal* influences are much more efficient than the *k-complete*

[3] Available in https://www.openml.org/s/107/tasks.

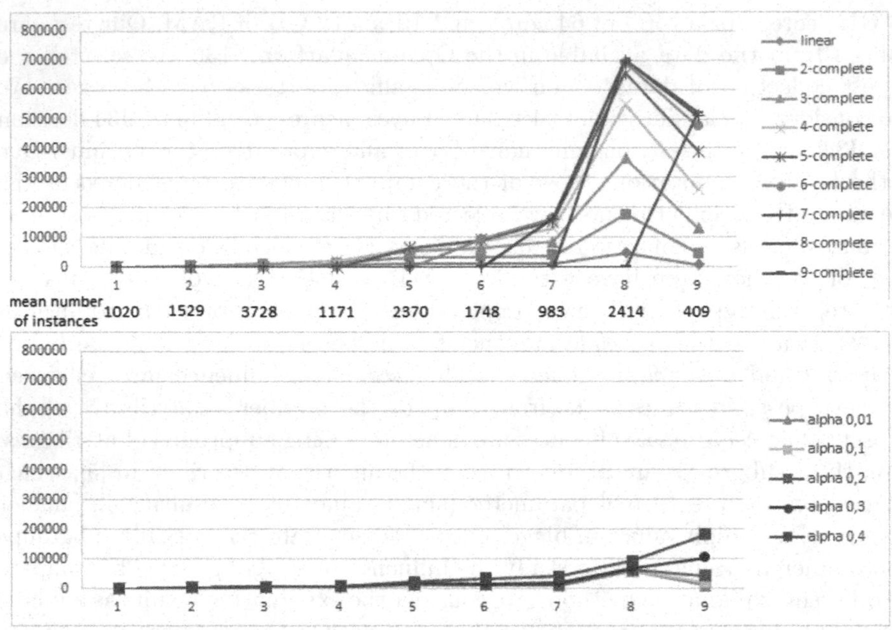

Fig. 3. Execution time, in milliseconds, of each explanation method depending on the number of attributes in the dataset. The mean number of instances is added for comparison.

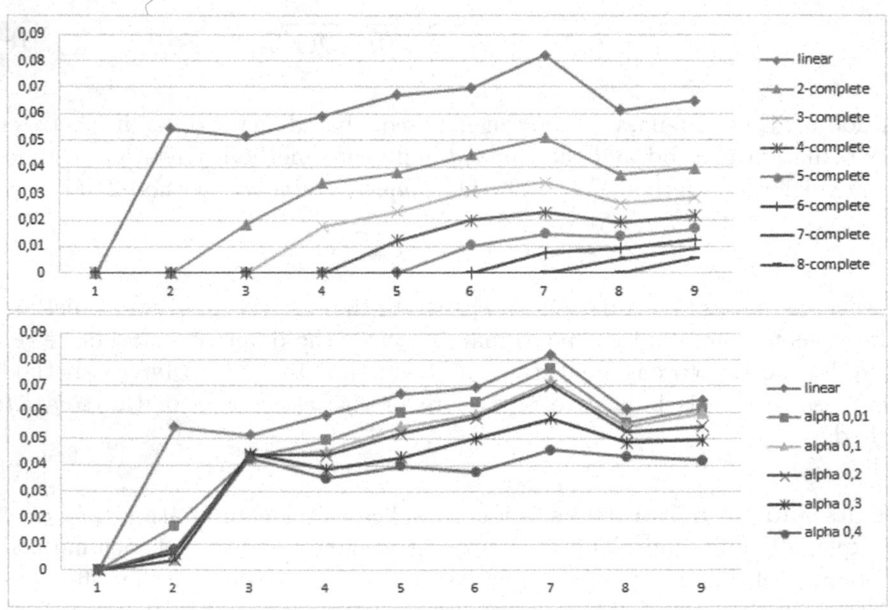

Fig. 4. Error score between each explanation method and the *complete influence* depending on the number of attributes in the dataset.

influences (less than 200 s for the first ones compared to 700 s for the other ones). The decrease in computation time for 9 attributes is explained by the important decrease in the mean number of instances. This makes each retraining faster to do, even if there are twice more subgroups to take into account. Figure 4 depicts the mean error score, aggregating the error score (Eq. 9) of each explanation method for each of our 324 datasets. In this figure, the lowest curve is the closest to the *complete* influence method, and thus is performing the best. As expected, the *linear* influence gives the worst results. This is explained by the fact it only considers single attributes, which is far from all the possible groups of attributes considered by the *complete* influence. The comparison between the *k-complete* influences and the *coalitionnal* influences (represented by their alpha parameters) is more delicate. Certainly, the *k-complete* influences outperform the other influences, in the majority of the cases, but with a high cost in execution time. Also, it does not mean the *coalitionnal* influences are less interesting for our case. They generate a smaller subset of groups of attributes while preserving an acceptable error score. Overall, the *coalitional* methods are not as satisfactory as the *k-complete* in term of effectiveness. But their comparatively very low execution time make them far more desirable when confronted with large datasets with an important number of attributes. Considering these results, it seems the *k-complete* influence is preferable with a relatively small k, and a dataset having few attributes, while the *coalitional* influence seems to become preferable with a higher number of attributes. Obviously, larger subgroups seems to increase the methods precision, but in the case of the *coalitional* method, their impact on computation time seems to be relatively small when compared to the performance gain. Besides, our study about the groups generated by the grouping algorithm shows that the number of groups stays relatively small, even for large alphas. As an example, for the datasets of 9 attributes, the mean size of the biggest generated group is 4, using an alpha of 0.4. This means that the *coalitional* influence is working with far less information than the *k-complete* one. Studying the influence of different ways to generate the coalitions of attributes on the *coalitional* influence could be a good aim for the near future. With new methods, it could be possible to find more relevant groups, bettering the precision of the explanation without an important computational cost. Moreover, it would be interesting to investigate overlapping attributes coalitions, using algorithms allowing for attributes to be in different coalitions, allowing to explore more subgroups if necessary.

5 Conclusion

We proposed in this paper a new way to explain the predictions of a single instance, aiming to reduce their cost of computation without losing too much information. The explaining methods we relate are a step further toward the goal of helping users employing machine learning tools. Our first experiment, in comparing the *loss* and *gain* methods, has led us to pinpoint flaws in their initial design, especially the lack of consideration about attribute combinations. We proposed two adequate methods for prediction explanation, i.e. the *k-complete* and *coalitional*. These methods are faring better in term of calculation time, with a relatively small loss in accuracy. A short term perspective will be to investigate the different subgroups generation methods and the results will help us to focus our efforts on the most promising candidates. A more long term perspective is to implement a complete tool on this basis, with the goal of guiding a domain expert through the building and exploitation of a machine learning model. Another longer-term perspective also focuses on the problem of an objective evaluation of explaining methods existing in the literature. Indeed, to the best of our knowledge, there is no benchmark for objectively evaluating these methods.

References

1. Altmann, A., Tolosi, L., Sander, O., Lengauer, T.: Permutation importance: a corrected feature importance measure. Bioinformatics **26**(10), 1340–1347 (2010)
2. Casalicchio, G., Molnar, C., Bischl, B.: Visualizing the Feature Importance for Black Box Models. arXiv e-prints, April 2018
3. Datta, A., Sen, S., Zick, Y.: Algorithmic transparency via quantitative input influence: theory and experiments with learning systems. In: 2016 IEEE Symposium on Security and Privacy (SP), pp. 598–617, May 2016
4. Hall, M., Frank, E., Holmes, G., Pfahringer, B., Reutemann, P., Witten, I.H.: The weka data mining software: an update. ACM SIGKDD Explor. Newslett. **11**(1), 10–18 (2009)
5. Henelius, A., Puolamaki, K., Boström, H., Asker, L., Papapetrou, P.: A peek into the black box: exploring classifiers by randomization. Data Min. Knowl. Discov. **28**(5–6), 1503–1529 (2014). qC 20180119
6. Lundberg, S., Lee, S.I.: A unified approach to interpreting model predictions. In: NIPS (2017)
7. Quinlan, J.: Induction of decision trees. Mach. Learn. **1**(1), 81–106 (1986)
8. Ribeiro, M.T., Singh, S., Guestrin, C.: "Why should i trust you?": explaining the predictions of any classifier. In: Proceedings of the 22nd ACM SIGKDD International Conference on Knowledge Discovery and Data Mining, KDD 2016, pp. 1135–1144. ACM, New York (2016)
9. Shapley, L.S.: A value for n-person games. In: Contributions to the Theory of Games, vol. 28, pp. 307–317 (1953)
10. Shrikumar, A., Greenside, P., Kundaje, A.: Learning important features through propagating activation differences. In: Proceedings of the 34th International Conference on Machine Learning, ICML 2017, vol. 70, pp. 3145–3153 (2017)
11. Štrumbelj, E., Kononenko, I.: Towards a model independent method for explaining classification for individual instances. In: Song, I.-Y., Eder, J., Nguyen, T.M. (eds.) DaWaK 2008. LNCS, vol. 5182, pp. 273–282. Springer, Heidelberg (2008). https://doi.org/10.1007/978-3-540-85836-2_26
12. Strumbelj, E., Kononenko, I.: An efficient explanation of individual classifications using game theory. J. Mach. Learn. Res. **11**, 1–18 (2010)
13. Vanschoren, J., van Rijn, J.N., Bischl, B., Torgo, L.: OpenML: networked science in machine learning. SIGKDD Explor. **15**(2), 49–60 (2013)

Inferring Deterministic Regular Expression with Unorder

Xiaofan Wang[1,2] and Haiming Chen[1(✉)]

[1] State Key Laboratory of Computer Science, Institute of Software, Chinese Academy of Sciences, Beijing 100190, China
{wangxf,chm}@ios.ac.cn

[2] University of Chinese Academy of Sciences, Beijing, China

Abstract. Schema inference has been an essential task in database management, and can be reduced to learning regular expressions from sets of positive finite-sample. In this paper, we extend the single-occurrence regular expressions (SOREs) to single-occurrence regular expressions with unorder (uSOREs), and give an inference algorithm for uSOREs. First, we present an *unorder-countable finite automaton* (uCFA). Then, we construct an uCFA for recognizing the given finite sample. Next, the uCFA runs on the given finite sample to count the number of occurrences of the subexpressions (connectable via unorder) for every possibly repeated matching. Finally we transform the uCFA to an uSORE according to the above results of counting. Experimental results demonstrate that, for larger samples, our algorithm can efficiently infer an uSORE with better generalization ability.

Keywords: Schema inference · Regular expressions · Automata · Unorder

1 Introduction

The schemata in database, such as JSON (JavaScript Object Notation) Schema [2], DTD (Document Type Definitions) and XSD (XML Schema Definitions) [21], facilitate query processing, automatic data integration, and static analysis of transformations [1,11,16,17], etc. However, in practice, many data exchange formats, such as XML documents, are not accompanied by a schema [4,18], or a valid schema [5,6]. Therefore, it is essential to devise algorithms for schema inference. In this paper, we focus on inferring schema from XML documents.

Schema inference can be reduced to learning regular expressions from sets of positive finite-sample. Single-occurrence regular expressions (SOREs) [7,8] are widely used in DTD files and XSD files. There are many works focusing on learning SOREs or subclasses of SOREs [7,8,13,14]. However, SOREs, which are just defined on standard regular expressions, do not support unorder. Regular expressions with unorder are extended from standard regular expressions with unorder

Work supported by National Natural Science Foundation of China under Grant No. 61872339.

A. Chatzigeorgiou et al. (Eds.): SOFSEM 2020, LNCS 12011, pp. 325–337, 2020.
https://doi.org/10.1007/978-3-030-38919-2_27

operators %. The unorder % is first used in Standard Generalized Markup Language (SGML) [19], and later in a limited form in XML Schema [21]. Unordered XML, which facilitates query optimization and set-oriented parallel processing [3], has also been studied recently with respect to schema language definitions [9,12,20]. In this paper, we propose a class of single-occurrence regular expressions with unorder (uSOREs). For 426,135 regular expressions extracted from XSD files and 43,326 regular expressions extracted from SGML files, our experiments (see Table 2) showed that the proportions of uSOREs are 90.32% and 93.26%, respectively. This indicates the practicality of uSOREs. Therefore, it is necessary to study the inference algorithm for uSORE.

For inference algorithms of SOREs, Bex et al. [8] proposed the algorithms RWR and RWR^2_ℓ [8]. Freydenberger et al. [14] presented the algorithm *Soa2Sore* [14]. For a given finite sample, the result of *Soa2Sore* can be more precise than that of RWR or its variants [14]. For inferring deterministic regular expressions with unorder, Ciucanu et al. [12] proposed two subclasses disjunction multiplicity expression (DME) and disjunction free multiplicity expression (ME) [12], ME does not use the disjunction ('|') operators. Ciucanu et al. [12] presented the algorithm $learner^{+}_{DME}$ [12] for learning DME. The learnt DME, where every symbol occurs at most once, is a minimal schema consistent with the examples. However, the unorder of two words u_1 and u_2 in [12] is defined as the multiset union $u_1 \uplus u_2$. So does for the unorder defined in [9] and [20], which is used in disjunctive interval multiplicity expressions (DIMEs) [9] and single-occurrence regular bag expressions (SORBEs) [20], respectively. For instance, let $u_1 = aa$ and $u_2 = b$, $u_1 \uplus u_2 = \{aab\}$, while $u_1 \% u_2 = \{aab, baa\}$. Additionally, so far there does not exist algorithm for learning an expression, which supports unorder (%). Therefore, we propose a new subclass uSORE and the corresponding learning algorithm. For larger samples, our algorithm can efficiently infer an uSORE with better generalization ability.

The main contributions of this paper are as follows. First, we define unorder-countable finite automaton (uCFA). Then, we present an inference algorithm for uSOREs. The main steps are as follows: (1) Construct an uCFA by converting the SOA, which is built for a given finite sample; (2) The uCFA runs on the given finite sample to count the number of occurrences of the subexpressions (connectable via unorder) for every possibly repeated matching; and (3) Transform the uCFA to an uSORE according to the above counting results.

The paper is structured as follows. Section 2 gives the basic definitions. Section 3 describes the uCFA and provides an example of such an automaton. Section 4 presents the inference algorithm of the uSORE. Section 5 presents experiments. Section 6 concludes the paper.

2 Preliminaries

2.1 Regular Expression with Unorder

Let Σ be a finite alphabet of symbols. A standard regular expression over Σ is inductively defined as follows: ε and $a \in \Sigma$ are regular expressions, for any regular expressions r_1 and r_2, the disjunction $(r_1|r_2)$, the concatenate $(r_1 \cdot r_2)$, and the Kleene-star r_1^* are also regular expressions. Usually, we omit concatenation operators in examples. The regular expressions with unorder are extended from standard regular expressions by adding the unorder $r_1\%r_2$. For regular expressions $r_1, r_2, \cdots, r_k$, $\mathcal{L}(r_1\%r_2\%\cdots\%r_k) = \bigcup_{\{\tau_1,\tau_2,\cdots,\tau_k\}\in Perm(\{1,2,\cdots,k\})} \mathcal{L}(r_{\tau_1})\cdots\mathcal{L}(r_{\tau_k})$, where $1 \leq i, \tau_i \leq k$ $(k \geq 2)$ and $Perm(\{1,2,\cdots,k\})$[1] is the set of permutations of $\{1,2,\cdots,k\}$. Note that r^+ and $r?$ are used as abbreviations of rr^* and $r|\varepsilon$, respectively.

For a regular expression r, $|r|$ denotes the length of r, which is the number of symbols and operators occurring in r. Let $\mathbb{N}=\{1,2,\cdots\}$ and $\mathbb{N}_0=\{0,1,2,\cdots\}$. For regular expressions $r_1, r_2, \cdots, r_k$ $(k\geq 2)$, $\%\{r_1, r_2, \cdots, r_k\} = r_1\%r_2\%\cdots\%r_k$, $[\{r_1, r_2, \cdots, r_k\}] = (r_1|r_2|\cdots|r_k)$. Let $\%\{r_1\} = r_1$ and $[\{r_1\}] = r_1$. For a finite sample S, $|S|$ denotes the number of stings in S. $\varnothing$ denotes empty set. For a matrix $M_{x\times y}$, let $[\mathbb{N}_0]_{x\times y}$ denote $M(i,j)\in\mathbb{N}_0$ $(1\leq i\leq x, 1\leq j\leq y)$. For a constant t, $\{t\}_{|T|}$ denote a set consisting of $|T|$ elements t. For space consideration, all omitted proofs can be found at http://github.com/GraceFun/InfuSORE.

2.2 SORE, uSORE, SOA

SORE is defined as follows.

Definition 1 (SORE [7,8]**).** *Let Σ be a finite alphabet. A single-occurrence regular expression (SORE) is a standard regular expression over Σ in which every terminal symbol occurs at most once.*

Since $\mathcal{L}(r^*) = \mathcal{L}((r^+)?)$, in this paper, a SORE does not use the Kleene-star operation.

Example 1. $(ab)^+$ is a SORE, while $(ab)^+a$ is not.

uSORE extends SORE with unorder and does not use the Kleene-star operation, is defined as follows.

Definition 2 (uSORE). *Let Σ be a finite alphabet. A single-occurrence regular expression with unorder (uSORE) is a regular expression with unorder over Σ in which every terminal symbol occurs at most once.*

According to the definition of deterministic regular expressions [10], uSOREs are deterministic by definition.

Example 2 $a?b(c^+\%d?)(e^+)?$, $c\%d$, and $(a?b)\%(c|d)^+e$ are uSOREs, while $a(b|c)^+a$ is not a SORE, therefore not an uSORE.

Definition 3 (SOA [8,14]**).** *Let Σ be a finite alphabet, and let q_0, q_f be distinct symbols that do not occur in Σ. A single-occurrence automaton (SOA) over Σ is a finite directed graph $\mathscr{A} = (V, E)$ such that (1) $\{q_0, q_f\} \in V$, and $V = \Sigma \cup \{q_0, q_f\}$. (2) q_0 has only outgoing edges, q_f has only incoming edges, and every $v \in V$ lies on a path from q_0 to q_f.*

A string $a_1 \cdots a_n$ $(n \geq 0)$ is accepted by an SOA $\mathscr{A}$, if and only if there is a path $q_0 \rightarrow a_1 \rightarrow \cdots \rightarrow a_n \rightarrow q_f$ in $\mathscr{A}$.

[1] For instance, $Perm(\{1,2,3\}) = \{\{1,2,3\}, \{1,3,2\}, \{2,1,3\}, \{2,3,1\}, \{3,1,2\}, \{3,2,1\}\}$.

3 Unorder-Countable Finite Automaton (*u*CFA)

An *u*CFA is defined to count the number of occurrences of the subexpressions (connectable via unorder) for every possibly repeated matching by accepting the given finite sample. *u*CFA is a variant of the CFA defined in [22], which is defined to count the minimum and maximum number of repetitions of the subexpressions (derivable from CFA) by accepting the given finite sample.

3.1 Counter States and Update Instructions

For any given finite sample as input, counter states are the states that are associated with update instructions to compute the number of occurrences of the subexpressions (connectable via unorder) for every possibly repeated matching. Update instructions are introduced as follows:

Let Q_c denote the set of counter states. Let $\mathcal{C}$ denote the set of counter variables, and let $c_q \in \mathcal{C}$ ($q \in Q_c$) denote a counter variable. The mapping θ: $\mathcal{C} \mapsto \mathbb{N}$ is the function assigning a value to each counter variable in $\mathcal{C}$. θ_1 denotes that $c_q = 1$ for each $q \in Q_c$. Let $I = \{i_q\}_{q \in Q_c}$. Each $i_q \in I$ denotes that the i_qth time for the subexpression associated counter state q that can be repeatedly matched by the substrings in the given finite sample. Let partial mapping λ: $I \mapsto \mathbb{N}$ be a function assigning value to each index $i_q \in I$. Let partial mapping λ': $\wp(I) \mapsto \{\mathbb{N}\}_{|I|}$ be a function assigning value to each index $i_q \in I'$ ($I' \in \wp(I)$). λ_1 denotes that $i_q = 1$ for each $q \in Q_c$. An update instruction is defined by the partial mapping β: $\wp(I) \times \mathcal{C} \mapsto \{\emptyset, \mathbf{max}\} \times \{\mathbf{res}, \mathbf{inc}\}$ ($\emptyset$ for empty instruction, **res** for reset, **inc** for increment, and **max** for solving the maximum of a set of values). β also defines mapping g_β between mappings $\lambda' \times \theta$. $+_j$ ($j \in \mathbb{N}$) is the state that can be directly transited from q for each $i_q \in I'$. If $\beta(I', c_q) = (\mathbf{max}, \mathbf{inc})$, then $g_\beta(\lambda', \theta)(I', c_q) = (\{\mathbf{max}(\bigcup_{i' \in I' \cup \{+_j\}} \lambda'(\{i'\}))\}_{|I'|}, \theta(c_q) + 1)$. If $\beta(I', c_q) = (\emptyset, \mathbf{res})$, then $g_\beta(\lambda', \theta)(I', c_q) = (\lambda'(I'), 1)$. If $\beta(I', c_q) = (\emptyset, \mathbf{inc})$, then $g_\beta(\lambda', \theta)(I', c_q) = (\lambda'(I'), \theta(c_q) + 1)$. If $\beta(I', c_q) = (\mathbf{max}, \mathbf{res})$, then $g_\beta(\lambda', \theta)(I', c_q) = (\{\mathbf{max}(\bigcup_{i' \in I' \cup \{+_j\}} \lambda'(\{i'\}))\}_{|I'|}, 1)$.

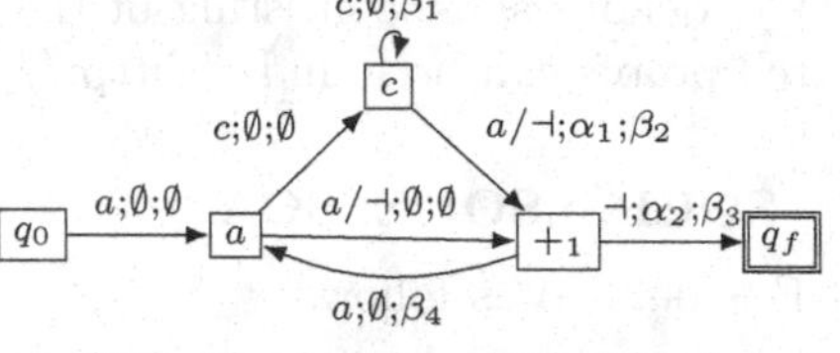

$\alpha_1 : (M(c, i_c), i_c) \mapsto (plus(M(c, i_c), c_c), \mathbf{inc})$. $\alpha_2 : (M(+_1, i_{+_1}), i_{+_1}) \mapsto (plus(M(+_1, i_{+_1}), c_{+_1}), \mathbf{inc})$. $\beta_1 : c_c \mapsto \mathbf{inc}$; $\beta_2 : c_c \mapsto \mathbf{res}$; $\beta_4 : c_{+_1} \mapsto \mathbf{inc}$; $\beta_3 : (\{i_c\}, c_{+_1}) \mapsto (\mathbf{max}, \mathbf{res})$.

Fig. 1. The *u*CFA $\mathcal{A}$ for regular language $\mathcal{L}((a(c^+)?)^+)$. The label of the transition edge is $(y; \alpha_i; \beta_j)$ ($i, j \in \mathbb{N}$), y ($y \in \Sigma \cup \{\dashv\}$) is a current letter, α_i is an update instruction for elements in M and I, respectively. β_j is an update instruction for elements in $\wp(I)$ and a counter variable in $\mathcal{C}$. Note that, for β_j: $(I'/\varnothing, c_q) \mapsto (\emptyset, \mathbf{inc}/\mathbf{res})$, β_j is abbreviated as $c_q \mapsto \mathbf{inc}/\mathbf{res}$.

Let $I_m = \mathbf{max}(\bigcup_{i_q \in I} \lambda(i_q))$. For each counter state $q \in Q_c$, let $M_{|Q_c| \times I_m}$ denote a matrix, $M(q, i_q)$ denotes the number of occurrences of the subexpression associated with counter state q, after it is the i_qth time for that subexpression has been repeatedly matched by the substrings in the given finite sample.

We also define partial mapping γ: $M \mapsto [\mathbb{N}_0]_{|Q_c| \times I_m}$ as a function assigning values to the elements in M. γ_0 denotes that every element in M is initialized to 0; Let partial mapping α: $M(Q_c \times I) \times I \mapsto plus(M(Q_c \times I), \mathcal{C}) \times \mathbf{inc}$. For a counter state $q \in Q_c$, the function $plus(M(Q_c \times I), \mathcal{C})$ specifies that $M(q, i_q) := M(q, i_q) + c_q$. α also defines the partial mapping f_α: $\gamma \times \lambda \times \theta \mapsto \gamma \times \lambda$, such that $\alpha(M(q, i_q), i_q) = (plus(M(q, i_q), c_q), \mathbf{inc})$: $f_\alpha(\gamma, \lambda, \theta)((M(q, i_q), i_q), c_q) = (\gamma(M(q, i_q)) + \theta(c_q), \lambda(i_q) + 1)$. Let $g_\emptyset(\lambda', \theta) = (\lambda', \theta)$ and $f_\emptyset(\gamma, \lambda, \theta) = (\gamma, \lambda)$.

3.2 Unorder-Countable Finite Automaton

Definition 4 (Unorder-Countable Finite Automaton). *An Unorder-Countable Finite Automaton (uCFA) is a tuple $(Q, Q_c, \Sigma, \mathcal{C}, q_0, q_f, \Phi, M, I)$. The members of the tuple are described as follows:*

- *Σ is a finite alphabet (non-empty).*
- *q_0 and q_f : q_0 is the initial state, q_f is the unique final state.*
- *Q is a finite set of states. $Q = \Sigma \cup \{q_0, q_f\} \cup \{+_i\}_{i \in \mathbb{N}}$.*
- *$Q_c \subset Q$ is a finite set of counter states. Counter state is a state $+_i$ or a state q ($q \in \Sigma$) with loop that can be directly transited from a state $+_j$ ($i, j \in \mathbb{N}$).*
- *$\mathcal{C}$ is finite set of counter variables that are used for counting the number of occurrences of the subexpressions connectable via unorder. $\mathcal{C} = \{c_q | q \in Q_c\}$, for each counter state q, we also associate a counter variable c_q.*
  M is a matrix. An element $M(q, i_q)$ in M denotes the number of occurrences of the subexpression associated with counter state q, after it is the i_qth time for that subexpression has been repeatedly matched by the substrings in the given finite sample.
- *$I = \{i_q\}_{q \in Q_c}$. $i_q \in I$ denotes that the i_qth time for the subexpression associated counter state q that can be repeatedly matched by the substrings in the given finite sample.*
- *Φ maps each state $q \in Q$ to a set of pairs consisting of a state $p \in Q$ and two update instructions. Φ: $Q \mapsto \wp(Q \times ((M(Q_c \times I) \times I \mapsto plus(M(Q_c \times I), \mathcal{C}) \times \boldsymbol{inc}) \cup \{\emptyset\}) \times ((\wp(I) \times \mathcal{C} \mapsto \{\emptyset, \boldsymbol{max}\} \times \{\boldsymbol{res}, \boldsymbol{inc}\}) \cup \{\emptyset\}))$.*

For Q, Σ, q_0 and q_f, they are the same with the corresponding definitions in CFA [22]. The configuration of an uCFA is defined as follows.

Definition 5 (Configuration of an uCFA). *A configuration of an uCFA is a triple $(q, \gamma \times \lambda, \lambda' \times \theta)$, where $q \in Q$ is the current state, $\gamma \times \lambda$: $M \times I \mapsto [\mathbb{N}_0]_{|Q_c| \times I_m} \times \mathbb{N}$, $\lambda' \times \theta$: $\wp(I) \times \mathcal{C} \mapsto \{\mathbb{N}\}_{|I|} \times \mathbb{N}$. The initial configuration is $(q_0, \gamma_0 \times \lambda_1, \lambda' \times \theta_1)$, and a configuration is final if and only if $q = q_f$.*

The transition function of an uCFA is defined as follows:

Definition 6 (Transition Function of an uCFA). *The transition function δ of an uCFA $(Q, Q_c, \Sigma, \mathcal{C}, q_0, q_f, \Phi, M, I)$ is defined for any configuration $(q, \gamma \times \lambda, \lambda' \times \theta)$ and the letter $y \in \Sigma \cup \{\dashv\}$, where $\dashv$ denotes the end symbol of a string.*

(1) $y \in \Sigma : \delta((q, \gamma \times \lambda, \lambda' \times \theta), y) = \{(z, f_\alpha(\gamma, \lambda, \theta), g_\beta(\lambda', \theta)) | (z, \alpha, \beta) \in \Phi(q) \wedge (z = y \vee ((y, \alpha, \beta) \notin \Phi(q) \wedge z \in \{+_i\}_{i \in \mathbb{N}}))\}$.

(2) $y = \dashv$: $\delta((q, \gamma \times \lambda, \lambda' \times \theta), \dashv) = \{(z, f_\alpha(\gamma, \lambda, \theta), g_\beta(\lambda', \theta)) | (z, \alpha, \beta) \in \Phi(q) \wedge (z = q_f \vee z \in \{+_i\}_{i \in \mathbb{N}})\}$.

The construction of an uCFA and $\Phi(q)$ will be given in Sect. 4.1. uCFA and CFA have the same way for string recognition [22].

Definition 7 (Deterministic uCFA). *An uCFA $(Q, Q_c, \Sigma, \mathcal{C}, q_0, q_f, \Phi, M, I)$ is deterministic if and only if $|\delta((q, \gamma \times \lambda, \lambda' \times \theta), y)| \leq 1$ for any $q \in Q$, $y \in \Sigma \cup \{\dashv\}$ and $\gamma \times \lambda$: $M \times I \mapsto [\mathbb{N}_0]_{|Q_c| \times I_m} \times \mathbb{N}$, $\lambda' \times \theta$: $\wp(I) \times \mathcal{C} \mapsto \{\mathbb{N}\}_{|I|} \times \mathbb{N}$.*

Example 3. Let $\Sigma = \{a, c\}$, $Q = \{q_0, a, c, +_1, q_f\}$, $Q_c = \{c, +_1\}$, $\mathcal{C} = \{c_c, c_{+_1}\}$ and $I = \{i_c, i_{+_1}\} = \{1, 1\}$, $M = [M(c, 1), M(+_1, 1)]^{\mathrm{T}}$. Figure 1 illustrates a deterministic uCFA $\mathcal{A} = (Q, Q_c, \Sigma, \mathcal{C}, q_0, q_f, \Phi, M, I)$ recognizing the language $\mathcal{L}((a(c^+)?)^+)$.

4 Inference of uSOREs

Our inference algorithm works in the following steps.

(1) We construct an uCFA by converting the SOA, which is built for the given finite sample. (2) For the given finite sample used in step (1) as input and the uCFA obtained from step (1), the uCFA counts the number of occurrences of the subexpressions (connectable via unorder) for every possibly repeated matching. (3) We transform the uCFA to an uSORE according to the above results of counting.

Algorithm 1. *InfuSORE*

Input: a finite sample S;
Output: an uSORE $r_\%$: $\mathcal{L}(r_\%) \supseteq S$;
SOA $\mathscr{A}$ =2T-INF(S);
uCFA $\mathcal{A} = Soa2uCfa(\mathscr{A})$;
if *Running*($\mathcal{A}, S$) **then**
    $r_\% = GenuSORE(\mathcal{A})$;
return $r_\%$;

Algorithm 1 is the framework of our inference algorithm. Algorithm 2T-INF [8] constructs the SOA for the sample S. Algorithm $Soa2uCfa$ is given in Sect. 4.1, algorithm *Running* [22] is used to run the uCFA in Sect. 4.2, algorithm *GenuSORE* is presented in Sect. 4.3.

4.1 Constructing uCFA

In this section, we present the construction of an uCFA. Since uCFA and CFA [22] have the same way for string recognition, the state-transition diagram of an uCFA is also same with the corresponding that of a CFA. Thus, we can first construct the state-transition diagram of an uCFA by using the corresponding algorithm in [22]. We then give the detailed descriptions of the uCFA.

Algorithm 2 constructs the state-transition diagram (a finite directed graph G) of an uCFA by using algorithm $Construct_G$ [22], which is used to construct G by modifying the SOA $\mathscr{A}$, which is built for the given finite sample S [8]. After the state-transition diagram G of an uCFA was constructed, the detailed descriptions of the uCFA $\mathcal{A}$ are as follows.

Algorithm 2. *Soa2uCfa*

Input: SOA $\mathscr{A}(V, E)$;
Output: an uCFA $\mathcal{A}$;
1: $G = Construct_G(\mathscr{A}, \mathscr{A})$;
2: uCFA $\mathcal{A} = (Q, Q_c, \Sigma, \mathcal{C}, q_0, q_f, \Phi, M, I)$;
3: **return** $\mathcal{A}$;

$\mathcal{A} = (Q, Q_c, \Sigma, \mathcal{C}, G.q_0, G.q_f, \Phi, M, I)$ where $\Sigma = G.V \setminus (\{q_0, q_f\} \cup \{+_i\}_{i \in \mathbb{N}})$, $Q = G.V$, $Q_c = \{q | q \in G. \succ (q) \wedge q \in \Sigma\} \cup \{G.+_i\}_{i \in \mathbb{N}}$, $\mathcal{C} = \{c_q | q \in Q_c\}$, $I = \{i_q\}_{q \in Q_c}$, and $M_{|Q_c| \times I_m} = [M_1, M_2, \cdots, M_{I_m}]$, $M_i = [M(q_1, i), M(q_2, i), \cdots, M(q_{|Q_c|}, i)]^{\mathrm{T}}$ $(1 \le i \le I_m$, $q_j \in Q_c$ and $1 \le j \le |Q_c|)$. Here we present $\Phi(q)$:

(1) $q = q_0 : \Phi(q) = \{(p, \emptyset, \emptyset) | p \in G. \succ (q_0)\}$.
(2) $q \in \Sigma : \Phi(q) = \{(p, \{(M(q, i_q), i_q) \mapsto (plus(M(q, i_q), c_q), \mathbf{inc}) | q \neq p \wedge q \in G. \succ (q)\} \cup \{\emptyset\}, \{(\varnothing, c_q) \mapsto (\emptyset, \mathbf{res}) | q \neq p \wedge q \in G. \succ (q)\} \cup \{(\varnothing, c_q) \mapsto (\emptyset, \mathbf{inc}) | q = p\} \cup \{\emptyset\}) | p \in G. \succ (q)\}$.
(3) $q \in \{+_i\}_{i \in \mathbb{N}} : \Phi(q) = \{(p, \{(M(q, i_q), i_q) \mapsto (plus(M(q, i_q), c_q), \mathbf{inc})\}, \{(\{i_l | i_l \in I, l \in G. \prec (q) \wedge l \notin G. \prec (+_j), +_j \neq q, j \in \mathbb{N}\}, c_q) \mapsto (\mathbf{max}, \mathbf{res})\}) | p \in G. \succ (q) \wedge (p \in \{+_i\}_{i \in \mathbb{N}} \cup \{q_f\} \vee p \notin \mathcal{R}_q)\} \bigcup \{(p, \emptyset, \{(\varnothing, c_q) \mapsto (\emptyset, \mathbf{inc})\}) | p \in G. \succ (q) \wedge p \in \Sigma \wedge p \in \mathcal{R}_q\}$.

Note that, the set of each $\mathcal{R}_q$ ($q \in \{+_i\}_{i \in \mathbb{N}}$) is a global variable in algorithm $Construct_G$. G is obtained, then we can obtain the set of each $\mathcal{R}_q$. Each $\mathcal{R}_q$ is established to specify the transition entrances for state q to count the number of occurrence of the corresponding subexpression connectable via unorder.

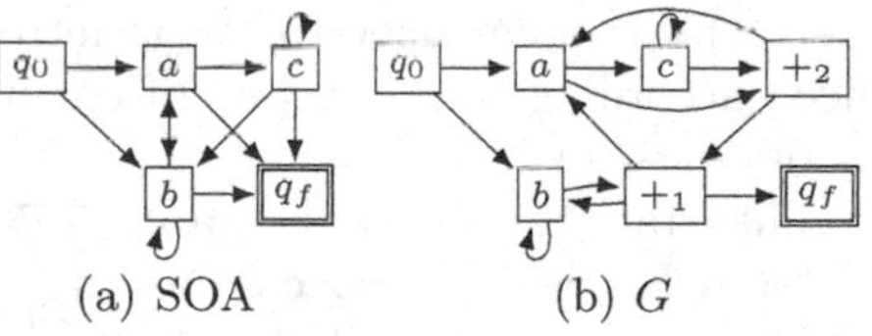

Fig. 2. The SOA (a) for the finite sample $S = \{a, acc, acbb, bab\}$. The state-transition diagram G of the uCFA (b) is constructed by modifying the SOA (a).

Suppose the SOA $\mathscr{A}$ uses n_s alphabet symbols and contains t_s transitions ($t_s > n_s$). Since the time complexity of constructing a CFA in [22] is $\mathcal{O}(n_s t_s)$, the time complexity of constructing an uCFA is also $\mathcal{O}(n_s t_s)$.

Example 4. For the sample $S = \{a, acc, acbb, bab\}$, the SOA $\mathscr{A}$ is showed in Fig. 2(a). *Soa2uCfa* converts the SOA $\mathscr{A}$ into the uCFA $\mathcal{A}$, the corresponding state-transition diagram of the uCFA $\mathcal{A}$ is demonstrated in Fig. 2(b). For space consideration, we illustrate the uCFA recognizing $\mathcal{L}((a(c^+)?)^+)$ in Example 3.

Theorem 1. *For any given finite sample S, if the uCFA $\mathcal{A}$ is constructed from the SOA $\mathscr{A}$ =2T-INF(S), then the uCFA is deterministic and $\mathcal{L}(\mathcal{A}) \supseteq S$.*

4.2 Counting with uCFA

Given a finite set of strings as input, the uCFA counts the number of occurrences of the subexpressions (connectable via unorder) for every possibly repeated matching. The constructed uCFA still runs on the finite sample, which is used to build the SOA that is the input of algorithm $Soa2uCfa$. Counting rules are given by transition functions. We use the algorithm *Running* proposed in [22] to run the uCFA. If *Running* returns *true*, then the running is terminated that the counting results (in matrix $M_{|\mathcal{A}.Q_c| \times I_m}$) are obtained.

For the given finite sample S, the number of strings is N and $\overline{L}$ is the average length of the sample strings. The time complexity of running with CFA is $\mathcal{O}(N\overline{L})$ [22]. Then, the time complexity of counting with uCFA is also $\mathcal{O}(N\overline{L})$.

Example 5. For sample $S = \{a, acc, acbb, bab\}$, the uCFA $\mathcal{A}$ is constructed in Sect. 4.1, *Running* returns *true*. Table 1 lists the results of $M(q, i_q)$ ($q \in \{c, b, +_2, +_1\}$, $i_q \in \{1, 2, 3, 4, 5\}$) after the M is obtained.

Table 1. The results of $M(q, i_q)$ after running the uCFA $\mathcal{A}$ ($q \in \{c, b, +_2, +_1\}$).

$M(q, i_q)$					
q	i_q				
	1	2	3	4	5
c	0	2	1	0	0
b	0	0	2	1	1
$+_2$	1	1	1	1	0
$+_1$	1	1	2	3	0

4.3 Generating uSORE

In this section, we transform the uCFA constructed in Sect. 4.1 to an uSORE, where the unorder operator % is introduced according to the results of counting obtained in Sect. 4.2.

The state-transition diagram of the uCFA can be respected as an SOA if alphabet includes $+_i$ ($i \in \mathbb{N}$). A SORE can be derived from an SOA by using the algorithm *Soa2Sore* [14]. Then the SORE containing symbols $+_i$ can be obtained from uCFA. In order to obtain an uSORE, first, we use the algorithm *Soa2Sore* which inputs the state-transition diagram of the uCFA to generate a SORE. Then, for each subexpression r of the SORE, we rewrite r to the form $r^{(l_q, u_q)}$ if r is associated by a counter state q. Finally, an uSORE is obtained by introducing the unorder operators and providing a normal form.

Algorithm 3. *GenuSORE*

Input: An uCFA $\mathcal{A}$;
Output: An uSORE $r_{\%}$;
Let G be the state-transition diagram of the uCFA $\mathcal{A}$;
SORE $r_s = Soa2Sore(G)$;
Search all subexpressions r_b from r_s:
  if $r_b = a^+$ ($a \in \Sigma$ and $a \in \mathcal{A}.Q_c$) **then**
    Replace r_b by $a^{(l_a, u_a)}$;
  if $r_b = (e+_i)^+$ (for expression e) **then**
    Replace r_b by $(e)^{(l_{+_i}, u_{+_i})}$;
$r_{\%} = add_{\%}(r_s)$; $r_{\%} = NormalForm(r_{\%})$; **return** $r_{\%}$;

Let $l_q = \min\{M(q, i_q) | M(q, i_q) \neq 0, 1 \leq i_q \leq I_m\}$ and $u_q = \max\{M(q, i_q) | M(q, i_q) \neq 0, 1 \leq i_q \leq I_m\}$. According to the tuples $\{(l_q, u_q)\}_{q \in \mathcal{A}.Q_c}$

and the values in $\{M(q,i_q)|q \in \mathcal{A}.Q_c, 1 \le i_q \le I_m\}$, subroutine $add_{\%}$, which is used to introduce the unorder (%) into an expression, is described as follows.

$add_{\%}(r)$. r is the expression possibly containing the subexpressions of form $e = (e_1|e_2|\cdots|e_k)^{(l_q,u_q)}$, where $k \ge 2$ and the subexpression $(e_1|e_2|\cdots|e_k)$ is associated with the counter state $q \in \mathcal{A}.Q_c$.

(1) $k=l_q=u_q$. Let $e=e_1\%e_2\%\cdots\%e_k$.
(2) $k<u_q$. Let $E=\{e_1,e_2,\cdots,e_k\}$, where e_l $(1\le l\le k)$ is associated with the counter state $q_l \in \mathcal{A}.Q_c$. Let $F=\{e_{j_1},e_{j_2},\cdots,e_{j_t}\}$ $(1\le t,j_t\le k)$. The set F with maximum size is extracted from E, such that there exists $i_q \in I$: $M(q_l,i_q)>0$ for each $q_l \in \mathcal{A}.Q_c$. Let $E=E\backslash F$. We repeatedly extract the set F from the set E till $F=\varnothing$. Let F_i denote ith time for extracting F from E. Let E' denote the set of the remainders in E. Then, let $e = (\%\{r_1\}_{r_1\in F_1}|\%\{r_2\}_{r_2\in F_2}|\cdots|\%\{r_i\}_{r_i\in F_i}|\cdots|[\{r'\}_{r'\in E'}])$. If $l_q<u_q$ or $2\le l_q=u_q$, $e=e^+$.
(3) $k\ge u_q$. Let $e=(e_1|e_2|\cdots|e_k)$. If $l_q<u_q$ or $2\le l_q=u_q$, $e=e^+$.

The unorder (%) is introduced into the expression, however, there are many subexpressions with the tuples in $\{(l_q,u_q)\}_{q\in\mathcal{A}.Q_c}$. Then, we provide the subroutine *NormalForm*, which converts an expression with unorder into a defined form of uSORE (a normal form).

NormalForm(r). Let $r_{b1}=e^{(l_q,u_q)}$, where the expression e is associated with the counter state q_e. (1) Let $r_b=(\cdots\%r_{b1}\%\cdots)$. If r_b is a subexpression in r, then let $r_{b1}=e$, if $l_{q_e}<u_{q_e}$ or $2\le l_{q_e}=u_{q_e}$, $r_{b1}=r_{b1}^+$.L Assume that r_b is associated with a counter state p. If there exists a counter state $q'\in G.\prec(p)$ and $i_{q_e},i_{q'}\in I$ such that $M(q_e,i_{q_e})=0$ and $M(q',i_{q'})=1$, then let $r_{b1}=r_{b1}?$. (2) Let $r_b=(\cdots|r_{b1}|\cdots)^+$. If r_b is a subexpression in r, then let $r_{b1}=e$. (3) If r_{b1} is a subexpression in r, then let $r_{b1}=e$. If $l_{q_e}<u_{q_e}$ or $2\le l_{q_e}=u_{q_e}$, $r_{b1}=r_{b1}^+$.

Assume that, the state-transition diagram G of the input uCFA contains n_g nodes and t_g transitions. Then, *Soa2Sore* takes $\mathcal{O}(n_g t_g)$ time to infer a SORE r_s. It takes $\mathcal{O}(|r_s|)$ time to transform the SORE r_s to the expression with the tuples in $\{(l_q,u_q)\}_{q\in\mathcal{A}.Q_c}$. For subroutines $add_{\%}$ and *NormalForm*, each of them takes $\mathcal{O}(|r_s|)$ time to process an expression. Thus, the time complexity of algorithm *GenuSORE* is $\mathcal{O}(n_g t_g+|r_s|)=\mathcal{O}(n_g t_g)$.

Example 6. The state-transition diagram (G) of the uCFA $\mathcal{A}$ is shown in Fig. 2(b). For each counter state $q\in\mathcal{A}.Q_c$, $M(q,i_q)$ $(1\le i_q\le I_m)$ is illustrated in Table 1. Then the set $C=\{(l_q,u_q)\}_{q\in\mathcal{A}.Q_c}$ is computed. $(l_c,u_c)=(1,2)$, $(l_b,u_b)=(1,2)$, $(l_{+_2},u_{+_2})=(1,1)$ and $(l_{+_1},u_{+_1})=(1,3)$. The algorithm *Soa2Sore* infers an expression $r_s=((((a(c^+)?)+_2)^+|b^+)+_1)^+$, which is rewritten to $r_s=((a(c^{(1,2)}?))^{(1,1)}|b^{(1,2)})^{(1,3)}$ by introducing tuples in C. Let $r_{\%}=add_{\%}(r_s)$, $r_{\%}=((a(c^{(1,2)}?))^{(1,1)}\%b^{(1,2)})^+$. Then, $r_{\%}$ is transformed to a normal form $((a(c^+)?)?\%(b^+)?)^+$. Thus, the finally obtained uSORE is $((a(c^+)?)?\%(b^+)?)^+$.

Theorem 2. *For any finite sample S, let $r_{\%}$:=InfuSORE(S), then $\mathcal{L}(r_{\%})\supseteq S$.*

5 Experiments

In this section, we first analyse the practicability of uSOREs, then we evaluate our algorithm on XML data in terms of generalization ability and time performance. Since *Soa2Sore* is the most efficient algorithm to infer a precise SORE [14], our algorithm is mainly compared with the algorithm *Soa2Sore*.

Table 2. Proportions of uSOREs and SOREs.

Subclasses	uSOREs	SOREs
% of SGML	93.26	85.81
% of XSD	90.32	81.45

31,386 XSD files and 52,567 SGML files were grabbed from Maven and GitHub. For 426,135 regular expressions extracted from XSD files and 43,326 regular expressions extracted from SGML files, Table 2 showed that the proportions of uSOREs are 90.32% and 93.26%, respectively. While, the proportions of SOREs are 82.45% and 85.81%, respectively. This indicates the practicability of uSOREs. Then, we evaluate our algorithm on XML data.

5.1 Generalization Abilities

We evaluate our algorithm *InfuSORE* by computing the precision and recall according to the given sample. We specify that, the learnt expression with higher precision and recall has better generalization ability. The average precision and average recall, which are as functions of sample size, respectively, are averaged over 1000 expressions.

We randomly extracted the 1000 expressions from XSDs and SGML, which were grabbed from GitHub. To learn each extracted expression e_0, we randomly generated corresponding XML data by using ToXgene[2]. The samples are extracted from the XML data, each sample size is that listed in Fig. 3. And we define precision (p) and recall (r). Let positive sample (S_+) be the set of the all strings accepted by e_0, and let negative sample (S_-) be the set of the all strings not accepted by e_0. Let e_1 be the expression derived by *InfuSORE*. Let $\mathcal{L}(e_1)^{\leq n}$ denote the set of strings, where a string is accepted by e_1 and has a length not over $n{=}2|e_1|{+}1$. A true positive sample (S_{tp}) is the set of the strings, which are in S_+ and in $\mathcal{L}(e_1)^{\leq n}$. While a false negative sample (S_{fn}) is the set of the strings, which are in S_+ and not in $\mathcal{L}(e_1)^{\leq n}$. Similarly, a false positive sample (S_{fp}) is the set of the strings, which are in S_- and in $\mathcal{L}(e_1)^{\leq n}$. While a true negative sample (S_{tn}) is the set of the strings, which are in S_- and not in $\mathcal{L}(e_1)^{\leq n}$. Then, let $p = \frac{|S_{tp}|}{|S_{tp}|+|S_{fp}|}$ and $r = \frac{|S_{tp}|}{|S_{tp}|+|S_{fn}|}$. Note that, we can construct automata (receptors) [15] for e_0 and e_1, respectively. Then we can obtain

[2] http://www.cs.toronto.edu/tox/toxgene/.

$|S_{tp}|$, $|S_{fp}|$ and $|S_{fn}|$. According to above given computations for precision and recall, we can also evaluate the result of algorithm *Soa2Sore*.

The plots in Fig. 3(a) show that, the precision for the expression derived by *InfuSORE* is consistently higher than that for the expression learnt by *Soa2Sore*. However, the plots in Fig. 3(b) illustrate that, for a larger sample (sample size ≥ 600), the recall for the expression derived by *InfuSORE* is higher than that for the expression learnt by *Soa2Sore*. In general, for larger samples, *InfuSORE* has better generalization ability such that its result has higher precision and recall.

5.2 Time Performance

To illustrate the efficiency of algorithm *InfuSORE*, we provide the statistics about running time in different size of samples and different size of alphabets. Table 3(a) shows the average running times in seconds for *InfuSORE* with different inputs of sample size. We randomly extracted 1000 expressions of alphabet size 10 from the above XSDs and SGML. To learn each expression, we randomly generated corresponding XML data by using ToXgene, the samples are extracted from the XML data, each sample size is that listed in Table 3(a). The running times listed in Table 3(a) are averaged over 1000 expressions of that sample size. Table 3(b) shows the average running times in seconds for *InfuSORE* as a function of alphabet size. For each alphabet size listed in Table 3(b), we also randomly extracted 1000 expressions of that alphabet size from the above XSDs and SGML. To learn each expression, we also randomly generated corresponding XML data by using ToXgene, but for each sample extracted from the XML data, the sample size is 1000. The running times listed in Table 3(b) are averaged over 1000 expressions of that alphabet size. According to above given computations for average running time, we can also evaluate the algorithm *Soa2Sore*.

Table 3(a) and (b) illustrate that, for each given sample size and alphabet size, the running times for *InfuSORE* are closer to that for *Soa2Sore*, respectively. *Soa2Sore* is the fast algorithm to infer a SORE [14]. Thus, this implies that, the algorithm *InfuSORE* is suitable for processing larger samples and generating the uSOREs with more alphabet symbols.

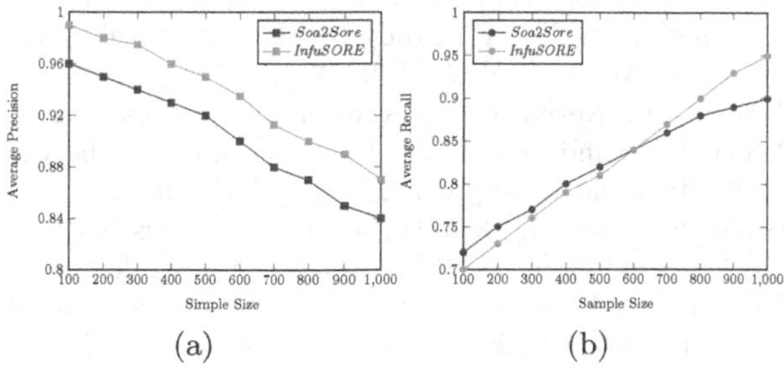

Fig. 3. (a) and (b) are average precision and average recall as functions of the sample size for each algorithm, respectively.

Table 3. (a) and (b) are average running times in seconds for *InfuSORE* and *Soa2Sore* as the functions of sample size and alphabet size, respectively.

(a)			(b)		
Sample size	time(s) ($\|\Sigma\| = 10$)		Alphabet size	time(s) ($\|S\| = 1000$)	
	InfuSORE	*Soa2Sore*		*InfuSORE*	*Soa2Sore*
100	0.034	0.021	5	0.056	0.034
1000	0.054	0.048	10	0.062	0.049
10000	0.208	0.197	20	0.074	0.056
100000	1.809	1.750	50	0.200	0.171
1000000	21.222	19.183	100	1.150	0.873

6 Conclusion

This paper proposed a series of strategies for inferring uSOREs. The main strategies include: use $Soa2uCfa$ to construct an uCFA from the SOA built for the given finite sample; use *Running* to run the uCFA to obtain the number of occurrences of the subexpressions (connectable via unorder) for every possibly repeated matching; and use *GenuSORE* to transform the uCFA to the uSORE according to the above results of counting. For larger samples, our algorithm can efficiently infer an uSORE with better generalization ability. For future works, we can extend the uSORE with counting, and study the inference algorithms. We can also extend the inference algorithm for uSOREs to infer JSON Schema, which has more superiorities than other schemas.

References

1. The JSON query language. http://www.jsoniq.org
2. json-schema.org: The home of JSON Schema. http://json-schema.org/
3. Abiteboul, S., Bourhis, P., Vianu, V.: Highly expressive query languages for unordered data trees. Theory Comput. Syst. **57**(4), 927–966 (2015)
4. Barbosa, D., Mignet, L., Veltri, P.: Studying the XML Web: gathering statistics from an XML sample. World Wide Web **9**(2), 187–212 (2006)
5. Bex, G.J., Martens, W., Neven, F., Schwentick, T.: Expressiveness of XSDs: from practice to theory, there and back again. In: Proceedings of the 14th International Conference on World Wide Web, pp. 712–721. ACM (2005)
6. Bex, G.J., Neven, F., Van den Bussche, J.: DTDs versus XML Schema: a practical study. In: Proceedings of the 7th International Workshop on the Web and Databases: Colocated with ACM SIGMOD/PODS 2004, pp. 79–84. ACM (2004)
7. Bex, G.J., Neven, F., Schwentick, T., Tuyls, K.: Inference of concise DTDs from XML data. In: International Conference on Very Large Data Bases, Seoul, Korea, pp. 115–126, September 2006
8. Bex, G.J., Neven, F., Schwentick, T., Vansummeren, S.: Inference of concise regular expressions and DTDs. ACM Trans. Database Syst. **35**(2), 1–47 (2010)

9. Boneva, I., Ciucanu, R., Staworko, S.: Schemas for unordered XML on a DIME. Theory Comput. Syst. **57**(2), 337–376 (2015)
10. Brüggemann-Klein, A., Wood, D.: One-unambiguous regular languages. Inf. Comput. **142**(2), 182–206 (1998)
11. Che, D., Aberer, K., Özsu, M.T.: Query optimization in XML structured-document databases. VLDB J. **15**(3), 263–289 (2006)
12. Ciucanu, R., Staworko, S.: Learning schemas for unordered XML. arXiv preprint arXiv:1307.6348 (2013)
13. Freydenberger, D.D., Kötzing, T.: Fast learning of restricted regular expressions and DTDs. In: Proceedings of the 16th International Conference on Database Theory, pp. 45–56. ACM (2013)
14. Freydenberger, D.D., Kötzing, T.: Fast learning of restricted regular expressions and DTDs. Theory Comput. Syst. **57**(4), 1114–1158 (2015)
15. Hovland, D.: The membership problem for regular expressions with unordered concatenation and numerical constraints. In: Dediu, A.-H., Martín-Vide, C. (eds.) LATA 2012. LNCS, vol. 7183, pp. 313–324. Springer, Heidelberg (2012). https://doi.org/10.1007/978-3-642-28332-1_27
16. Manolescu, I., Florescu, D., Kossmann, D.: Answering XML queries on heterogeneous data sources. In: VLDB, vol. 1, pp. 241–250 (2001)
17. Martens, W., Neven, F.: Typechecking top-down uniform unranked tree transducers. In: Calvanese, D., Lenzerini, M., Motwani, R. (eds.) ICDT 2003. LNCS, vol. 2572, pp. 64–78. Springer, Heidelberg (2003). https://doi.org/10.1007/3-540-36285-1_5
18. Mignet, L., Barbosa, D., Veltri, P.: The XML web: a first study. In: Proceedings of the 12th International Conference on World Wide Web, pp. 500–510. ACM (2003)
19. International Organization for Standardization: Information Processing: Text and Office Systems: Standard Generalized Markup Language (SGML). ISO (1986)
20. Staworko, S., Boneva, I., Gayo, J.E.L., Hym, S., Prud'Hommeaux, E.G., Solbrig, H.: Complexity and expressiveness of ShEx for RDF. In: 18th International Conference on Database Theory (ICDT 2015) (2015)
21. Thompson, H., Beech, D., Maloney, M., Mendelsohn, N.: XML Schema Part 1: Structures, 2nd Edn. W3C Recommendation (2004)
22. Wang, X., Chen, H.: Inferring deterministic regular expression with counting. In: Trujillo, J., et al. (eds.) ER 2018. LNCS, vol. 11157, pp. 184–199. Springer, Cham (2018). https://doi.org/10.1007/978-3-030-00847-5_15

POI Recommendation Based on Locality-Specific Seasonality and Long-Term Trends

Elena Stefancova(✉) and Ivan Srba

Slovak University of Technology in Bratislava, Ilkovicova 2, 84104 Bratislava, Slovakia
{elena.stefancova,ivan.srba}@stuba.sk

Abstract. This work deals with time-aware recommender systems in a domain of location-based social networks, such as Yelp or Foursquare. We propose a novel method to recommend Points of Interest (POIs) which considers their yearly seasonality and long-term trends. In contrast to the existing methods, we model these temporal aspects specifically for individual geographical localities instead of globally. According to the results achieved by the experimental evaluation on Yelp dataset, locality-specific seasonality can significantly improve the recommendation performance in comparison to its global alternative. We found out that it is helpful mostly within recommendations for highly-active users (it has a smaller influence for the novice users) and as expected, in localities with a strong seasonal weather variation. Another interesting finding is that in contrast to seasonality, we did not observe an improvement in case of locality-specific long-term trends.

Keywords: Recommender systems · Points of interest · Time-aware recommendation · Seasonality · Long-term trends

1 Introduction

Recommender systems are an important part of various web applications and their popularity is on the rise due to their crucial role in keeping customers satisfied and to help customers with overcoming information overload. Taking context (e.g., time, location) into account in some cases was proven to significantly improve the accuracy of provided recommendations.

Temporal context-aware recommendation reflects recency (long/short-term trends related to items and their categories or user interests and preferences, e.g. popularity of comedy movies can be increasing) and periodicity/seasonality (behaviour pattern based on time during the year, day of the week, etc., e.g. a user goes for a coffee in the morning and to a pub in the evening) [2,6].

We focus on the domain of *Location-Based Social Networks (LBSNs)*, which provide users with a possibility to search and rate so called *Points of Interest (POIs)* (e.g. shops, restaurants). This kind of social networks is well known for importance of geographical and temporal context in recommendation.

While taking geographical context into account, we focus specifically on temporal context, namely on *seasonality* and *long-term trends* in a society.

A. Chatzigeorgiou et al. (Eds.): SOFSEM 2020, LNCS 12011, pp. 338–349, 2020.
https://doi.org/10.1007/978-3-030-38919-2_28

Some previous research works have already focused on these temporal aspects within POI recommendation, but usually only on a *global granularity* and they have not *explicitly* considered *locality specifics* (i.e., seasonality and long-term trends specific for a particular geographic area, influenced by e.g. local climate). We decided to address this open problem in our work.

By means of data pre-filtering and context-aware modelling (based on matrix factorization), we incorporated locality-specific temporal aspects (seasonality and long-term trends) into POI recommendation. Locality-specific temporal aspects are modeled individually for each category (each POI is typically assigned into one or several categories, such as bars, restaurants, or doctors) in each of the localities. We compare such locality-specific temporal aspects with global temporal aspects, which are built without taking locality into consideration. Since geographical context is crucial in LBSNs, in order to improve the final results, we used a geographical post-filtering as well – recommendations with temporal aspects are reordered and POIs situated geographically close to the previously visited items are preferred.

Our main contribution is a research of the dependence of the temporal aspects on the geographical locality. In particular:

- we explicitly model locality-specific temporal aspects in the process of context-aware recommendation,
- we do not restrict the method to one particular temporal aspect, but we investigate seasonality as well as long-term trends,
- we address a more challenging and more valuable task to recommend POI reviews (usually corresponding to visits of previously unvisited POIs) instead of POI check-ins (corresponding to repetitive visits of already known POIs),
- we evaluate the proposed method by using an extensive real-world dataset from the Yelp system[1].

2 Background and Related Work

Within the classical recommendation techniques, context is not taken into account. However, in many cases its inclusion is significantly beneficial [1]. The most common types of the context are time, locality or company of other people. In the case of context-aware recommendation, the recommended items are the result of function of users, items, as well as context (Eq. 1, U stands for users, I for items, C for context) [1].

$$R : U \times I \times C \rightarrow Rating \tag{1}$$

The incorporation of the context can be done by [2]:

- *Pre-filtering* – a segment of user ratings on items is filtered according to relevance of the context, then selected part is used for recommendation process. This approach can lead to an item splitting (splitting an item to more items

[1] https://www.yelp.com/.

according to the context) [4]. Despite the increased data sparsity, item splitting is beneficial, when some contextual feature separates the item ratings into more homogeneous rating groups. However, if the contextual feature is not influential the splitting technique may produce a minor decrease of the precision and recall.
- *Post-filtering* – first the ratings are generated, then reordered based on the context.
- *Context-aware modelling* – information of context is used within the recommendation process itself.

Temporal and geographical context is present in many online services. One type of them are Location-Based Social Networks[2] (LBSNs) [5], e.g. Foursquare[3] (popular due to real-time location sharing and checking in) or Yelp[4] (focused on crowd-sourced reviews about businesses). The basic element of LBSNs (item) is a Point of Interest (POI). Check-ins (corresponding to POI visits of a particular user at the given time) [9,10,12] or reviews of POIs [11] serve as transactions. While users usually create check-ins for the same POI repetitively, reviews are created only once – typically after the first POI visit.

Since LBSNs try to motivate users to visit a lot of new localities and share the experience, it is crucial to make the recommendation as relevant as possible. Context aspects, such as locality (e.g. a user is willing to visit POIs only in some areas or in a certain distance from his/her usual places of occurrence) and time (e.g. preferences of users are strongly cyclic during the day/week/year), are necessary [10,11]. Previous results showed that the best model granularity (with respect to the interaction of context and items) is to group items by their categories [3]. Similarly also in LBSNs, POIs are organized into thematic categories, which are commonly utilized during the recommendation process.

Zhang et al. [11] introduced a POI recommendation approach called GeoSoCa through exploiting geographical, social and categorical correlations among users and POIs. It takes into account the popularity of a POI in the corresponding category and models the weighed popularity as a power-law distribution to leverage the categorical correlations between POIs.

For employing the locality, so called *Home Locality* (home address or the most visited area) [9] or *Personal Functional Regions* (several visited areas, sometimes connected to particular type of activities) [10] are used. Temporal aspects are often used in combination with collaborative filtering, since people tend to have similar cyclic patterns [10]. Another popular approach is to employ a decay (recency) aspect of the item or rating.

A typical problem in POI recommendation is data sparsity – low frequency of check-ins [11] results from the nature of this kind of explicit feedback. In case of huge sparsity of the data, methods revealing latent properties are popular, such as a state-of-the-art matrix factorization [8]. Matrix factorization decomposes the $U \times I$ interaction matrix into two lower dimensional matrices.

[2] https://en.wikipedia.org/wiki/Geosocial_networking.
[3] https://foursquare.com/.
[4] https://www.yelp.com.

For the purpose of evaluation, datasets from LBSNs (Yelp, Foursquare, Gowalla) are commonly used. Especially popular is a dataset from Yelp system, as it is published officially and updated regularly. The most used evaluation methods are precision@k and recall@k [10,12].

While many research papers have already focused on temporal aspects in POI recommendation, to the best of our knowledge, none of them did explicitly take into consideration locality specifics (a locality of an item can be determined at a different level of granularity, e.g. city, region, country). Nevertheless, temporal aspects could be highly influenced by such locality influences. A cyclic pattern of a year could be affected by the seasonal weather, holidays etc. POIs or whole POI categories could have a different increasing or decreasing popularity trend in different localities (e.g. a McDonald's has a pretty stable popularity in the USA, but its popularity could be growing in a new market).

3 Method Proposal

We propose a novel method of time-aware recommendation to explore how locality-specific seasonality and long-term trends can influence the performance of context-aware recommender systems. Our hypotheses are as follows:

- H1: Incorporating locality-specific seasonal patterns of POI categories will increase the recommendation performance.
- H2: Incorporating locality-specific long-term trends of POI categories will increase the recommendation performance.

Unlike the majority of the existing works, we consider as a rating (transaction) a POI review instead of a check-in. A reason for this decision is a potentially high number of missing check-ins since users many times skip check-in and thus check-in data may not optimally reflect actual POI repetitive visits. In addition, since a review often corresponds to the first visit of such POI, our method is applicable to recommend new POIs to a user (POIs which have not been visited before), what is a more difficult but also a more valuable task.

The overall scheme of the proposed method and its evaluation is depicted in Fig. 1. It is a hybrid approach based on a combination of collaborative filtering and content-based recommendation. The collaborative filtering part employs a matrix factorization. The content-based recommendation part is supplementing the user-item matrix with an additional item features matrix (matrix of items and their features, e.g. a category).

Context is incorporated in the proposed method by a hybrid approach as well. We propose two versions of the method corresponding to two temporal aspects we are interested in – seasonality version and long-term trend version. In both method versions, the core of the recommendation process is a context-aware modelling approach. In addition, a temporal pre-filter (for seasonality only version of the method) and a geographical post-filter are utilized.

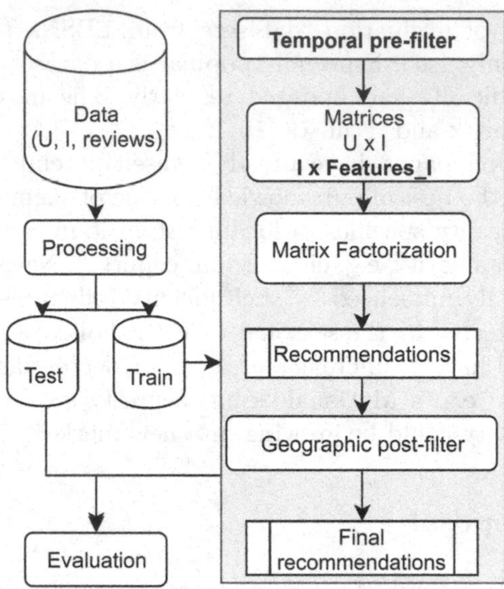

Fig. 1. Scheme of the proposed method and its evaluation.

3.1 Hybrid Matrix Factorization Model

The core part of the proposed method consists of adapted hybrid LightFM model previously proposed in [7]. This model represents users and items as linear combinations of their context features' latent factors (embeddings).

Let U be the set of users, I be the set of items (POIs), and F^I the set of item features. In contrast to the original LightFM model, we do not use user features in our method (LightFM model uses internally an identity matrix as user-feature matrix). As a result, each $u \subset U$ is represented only by one latent vector similarly to the standard matrix factorization model.

Each POI i is described by a set of features $f_i \subset F^I$. The features are known in advance and represent POI contextual metadata (particular features capturing contextual data for seasonality and long-term trend versions of the method are described in more details in the following subsections).

The model is parameterised in terms of d-dimensional item feature embeddings $\boldsymbol{e}_f^I$ for each item feature f_i. Each item feature is also described by a scalar bias term b_f^I.

The latent representation of item i is given by the sum of its features' latent vectors:

$$\boldsymbol{p}_i = \sum_{j \in f_i} \boldsymbol{e}_j^I \tag{2}$$

The bias term for item i is given by the sum of the features' biases:

$$b_i = \sum_{j \in f_i} b_j^I \tag{3}$$

Since, we do not use user-feature matrix in the proposed method, the linear combination of features' latent vectors is not necessary to obtain the latent representation $\boldsymbol{q}_u$ and the bias term b_u for user u.

The model's prediction for user u and item i is then given by the dot product of user and item representations, adjusted by user and item feature biases:

$$\widehat{r}_{ui} = f\left(\boldsymbol{q}_u \cdot \boldsymbol{p}_i + b_u + b_i\right) \tag{4}$$

The optimization function of LightFM model requires a binary feedback. Each user interacts with a number of items, either in a favourable way (a positive interaction), or in an unfavourable way (a negative interaction). For this reason, the given POI review ratings are normalized for every user individually as a negative rating (if rating is in the lower quartile of his/her ratings) and positive rating (otherwise).

In the proposed method, we decided to use the WARP loss function – it is suitable for optimization of the highest items on the recommendation list. As an adaptive learning rate method, AdaDelta is used – it gives better results in combination with low-number-of-epochs WARP, so for the less time cost version.

3.2 Seasonality Pre-filtering and Modelling

In the first version of the method, we explore the locality-specific seasonality. We consider a year as a cycle length and every month as an individual seasonality unit. This version of the method combines temporal pre-filtering and modelling.

The first step is an item splitting by applying the temporal pre-filter. Transactions of every POI are divided according to a month and because of that, one original item (POI) splits into up to 12 new items.

For the purpose of context-aware modelling, item feature vectors are used as a storage of contextual item information. Namely, the following item features are used: an original POI ID (month instances of the original POI are connected through this common item feature), a month (encoded either as an ordinal number and as a cyclic ordinal value in the form of sine and cosine value), a POI locality (one-hot encoded country), POI categories (one-hot encoded) and POI seasonality scores.

The POI seasonality scores are calculated as a weighted arithmetic mean of category seasonality scores across all POI categories. The category weight is calculated as an inverse number of its assigned POIs (the less frequent category, the higher is its influence).

The category seasonality scores are calculated from the category popularity at the given locality and in the given period. It is calculated from all previous years as a ratio of an average number of reviews per month in the given period to an average number of reviews per all months during the year. As the given period, we experimented with two options: current month, when the recommendation is created, and current month plus two neighbour-months (a wider season).

3.3 Long-Term Trends Modelling

The second version of the method, which takes long-term trends into account, is based on context-aware modelling only.

The following item features are used: a POI locality (one-hot encoded country), POI categories (one-hot encoded) and POI trend scores.

Similarly to POI seasonality scores, the POI trend scores are calculated as a weighted arithmetic mean of category trend scores across all POI categories.

The category trend scores reflect the changes in category popularity during current and three previous seasons (one year is considered to be a season). The more recent season, the higher priority is given to its popularity changes (the most recent season: 1/2 of the score, the second one: 1/3, the third one: 1/6). Besides the "raw" difference in the number of reviews, we calculate also a difference in a percentage (positive or negative) and a difference as a gradient.

3.4 Geographical Post-filtering

After generating the time-aware recommendations (in both versions of the method), we reorder the recommended POIs based on their locality – the POIs, which are located within the certain distance d from any of the previously visited POIs, are prioritized. These prioritized items are moved to the top of the list (in the same order as they were in the original list). If the number of these items is not enough to cover necessary number of recommendations, the rest of the (not geographically preferred) items would complete the new list. The radius from the previously visited items d, which is considered during this process, is supposed to be selected experimentally.

4 Experimental Evaluation

4.1 Dataset

We chose official Yelp dataset[5] for the method experimental evaluation. We created two dataset samples for the purpose of evaluation:

- a dataset of cities with strong (e.g., Mississauga, Champaign) as well as weak (e.g., Goodyear, Glendale) seasonal weather variation,
- a dataset of cities (e.g., Mississauga, Champaign) with strong seasonal weather variation only.

The dataset with strong seasonal weather variation contains approx. 3 200 POIs, 39 000 reviews and 3 000 users. The dataset with mixed seasonal weather variation is approx. two times larger. Both datasets contain data collected during 13 years (from 2005 until 2018).

Exploratory analysis of the dataset revealed an important precondition supporting the Hypothesis 1 – we can observe a stronger locality-specific bias in

[5] https://www.yelp.com/dataset.

the number of reviews for summer and winter months in cities with a strong seasonal weather variation (an example is provided on Fig. 2).

The exploratory analysis supports also a theory about long-term trends of categories in a society. We calculated a ratio (per each year) between the number of reviews in the corresponding category and the number of all reviews in the system for several selected categories. E.g. ratio of sushi restaurants tended to grow and after few years the growth stopped and the numbers stabilized for several years. Interesting trends can be observed also for tobacco shops (an increase after the introduction of vapes was followed by a later decrease and a recent stabilization).

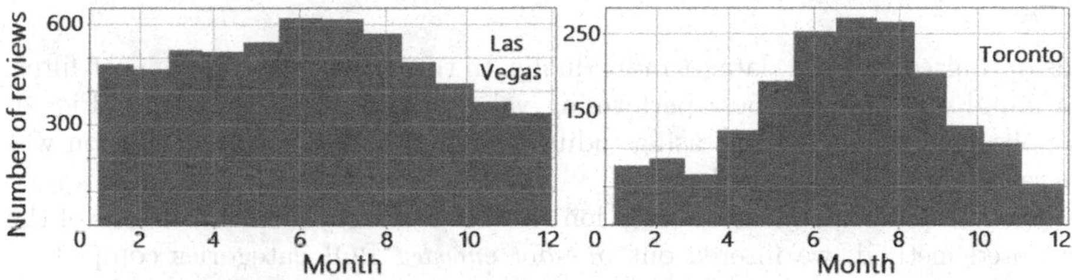

Fig. 2. Number of reviews for "ice cream" category in a city with weak (Las Vegas)/strong (Toronto) winters divided according to months.

4.2 Experiment Setup

During the preprocessing step, we removed from both datasets low-activity users, POIs and categories. Only users with 5 and more reviews are included. Similarly POIs must have at least 3 reviews. Finally, categories containing less than 10 reviews per year in average are ignored.

The original reviews (a 1–5 stars rating) are converted to three values: 0 (an unvisited POI), −1 (a visited POI but the rating is in the lowest quantile of the user's ratings), 1 (the rest of more positive ratings).

The context-aware recommendation was implemented with a use of LightFM[6] library. As the evaluation metrics, we selected commonly used precision@k (Eq. 5) and recall@k (Eq. 6), where I_U corresponds to the set of recommended POIs and I_R to the set of relevant POIs.

$$precision@k = \frac{|I_U \cap I_R|}{min(k, |I_R|)} * 100 \tag{5}$$

$$recall@k = \frac{|I_U \cap I_R|}{|I_U|} * 100 \tag{6}$$

The proposed method contains several parameters which were set by means of hyperparameter tuning. In case of context-aware modelling, the number of components (dimensionality) of latent vectors was set to 30, the number of epochs

[6] https://github.com/lyst/lightfm.

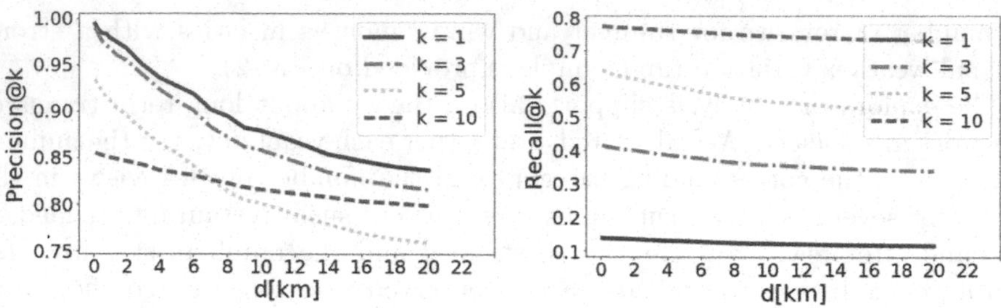

Fig. 3. Dependency of precision@k and recall@k on a radius d (in kilometers) used in the geographical post-filtering (without any other contextual influence).

was optimized for each dataset individually. In the case of geographic post-filter, we found out that the best performing value of the radius is 0.1 km (Fig. 3). We observed that with increasing radius, the precision of recommendation was decreasing for all values of k.

We also performed feature selection on item features. In both versions of the proposed method, we filtered out *one-hot encoded* POI categories completely, that did not improve results. In case of seasonality version, we used month encoded as a cyclic value and as the given period we used the current month with neighbour-months. Furthermore, during the experimentation, we discovered that it is more efficient to consider only categories with some considerable seasonality present and represent these categories in a form of seasonality/trend score.

We compared our method (denoted as *local seasonality/local trends*) with two baselines:

- *without seasonality/without trends* – basic matrix factorization without item features (i.e., without temporal context),
- *global seasonality/global trends* – the method with item features calculated globally (i.e., locality was ignored), what corresponds to the approach, which is currently employed in the existing solutions.

In the case of seasonality, the last year was divided to twelve test sets (corresponding to each month, the presented results are afterwards calculated as an average from all twelve test sets), while the train set was created from all previous reviews. In the case of long-term trends, the last year of reviews was used as the test set, while all previous reviews were used as the train set.

4.3 Results

H1: Seasonality. We found out that the proposed pre-filtering as well as context-aware modelling individually improved the recommendation. The best performance was finally achieved by their combination.

Table 1. The comparison of results achieved by the proposed method (local seasonality) and by the baselines. Note: Bold font highlights the best performance, italic font highlights results that overcome matrix factorization without temporal features.

Method	Without seasonality		Global seasonality		Local seasonality	
k	precision@k	recall@k	precision@k	recall@k	precision@k	recall@k
	Cities with mixed seasonal weather variation					
1	1.01	0.52	**1.32**	**1.10**	*1.11*	*0.83*
3	1.50	1.41	1.15	1.15	1.22	1.22
5	2.11	2.07	1.15	1.15	1.22	1.22
10	2.36	2.36	1.15	1.15	1.31	1.31
	Cities with strong seasonal weather variation					
1	1.78	0.41	*2.42*	*1.77*	**3.29**	**2.41**
3	1.12	0.84	*2.41*	*2.41*	**2.82**	**2.82**
5	1.30	1.23	*2.50*	*2.50*	**2.82**	**2.82**
10	1.79	1.78	*2.68*	*2.68*	**2.82**	**2.82**

Table 2. The comparison of results achieved by the proposed method (local trends) and by the baselines.

Method	Without trends		Global trends		Local trends	
k	precision@k	recall@k	precision@k	recall@k	precision@k	recall@k
	Cities with mixed seasonal weather variation					
1	1.33	0.46	0.60	0.19	0.60	0.22
3	1.45	1.11	0.44	0.26	0.75	0.53
5	2.28	2.08	0.64	0.56	1.04	0.93
10	3.22	3.16	1.47	1.44	2.77	2.71

The results in Table 1 show that locality-specific seasonality performed significantly better than global one for the cities, where the weather is more varied during the year. For all cities (with mixed seasonal weather variation), taking seasonality into account improved performance for $k = 1$. The decrease in performance for higher values of k can be explained by the cities without seasonal weather variation pushing the numbers down. We found out that the seasonality works better for users with higher number of previous reviews, while the performance for novice users is similar with the popularity-based recommendation.

From these results, we can conclude that locality-specific seasonality can indeed achieve better performance than the global seasonality and thus we can confirm our first hypothesis H1. Nevertheless, this finding is not applicable in general and in all cases. As it can be expected, the significant improvement was observed for localities and categories with significant seasonal weather variation. Therefore, we recommend to use locality-specific seasonality exclusively in case of localities and categories with considerable seasonality patterns.

We compared our results with the work of Zhang et al. [11], whose GeoSoCa employs the geographical context, the social correlations between users and the categorical correlation modeling. Their dataset contained Yelp data of Pheonix, Arizona, USA. They obtained precision@k of 1

H2: Long-term trends. As the results in Table 2 show, employing features representing locality-specific as well as global long-term trends did not improve overall results of recommendation. The features for locality-specific long-term trends performed slightly better than the global ones. Nevertheless, both methods (with features for locality-specific as well as global long-term trends) performed worse than recommendation without any contextual features and thus we cannot confirm our second hypothesis H2. The best performing features of long-term trends were the "raw" differences between seasons.

This result can be explained by an undesired overfitting on provided temporal features and by a low diversity of long-term trends in the selected cities. We found out that the more geographically diverse cities are associated with the better results achieved by the method using locality-specific long-term trends (than the method using the global trends). Although the selected cities in the experimental dataset are located in different countries, they are all located in North America. It seems that in order to achieve better performance with locality-specific long-term trends, it would be necessary to use a more varied mix of localities.

5 Conclusion and Future Work

While temporal context is widely analyzed in the domain of POI recommender systems, we did not identify any work that would attempt to explicitly model locality-specific temporal aspects. Therefore, we proposed and experimentally verified the recommendation system that explicitly model locality-specific seasonality and long-term trends and compare them with their global versions.

At first, we can confirm a positive influence of locality-specific yearly seasonality. In case of cities with strong seasonal weather variation, we can tap the full potential of locality-specific seasonality as it significantly improved the performance in all cases (particularly for users with a higher level of activity). In case of all cities, we observed improvement on the top of the recommendation list, what is the successful result as well. Secondly, the consideration of long-term trends turned out to be less successful. While locality-specific features overcome global ones, these features in general caused overfitting and cannot outperform the standard matrix factorization without temporal features.

The findings provide opportunities for a future research. It would be interesting to experiment with different locality units (regions, countries) and seasonality lengths (month, week, day). Employing clustering of localities based on seasonal similarities could help as well. Moreover, since the proposed method is not dependent on a particular dataset or a domain, it can be applied at datasets from other LBSNs (e.g. Foursquare) or even in other domains.

Acknowledgments. This work was partially supported by the Slovak Research and Development Agency under the contracts No. APVV-15-0508 and APVV SK-IL-RD-18-0004, by the Scientific Grant Agency of the Slovak Republic under the contracts No. VG 1/0667/18 and VG 1/0725/19, and by the student grant provided by Softec Pro Society.

References

1. Adomavicius, G., Sankaranarayanan, R., Sen, S., Tuzhilin, A.: Incorporating contextual information in recommender systems using a multidimensional approach. ACM Trans. Inf. Syst. **23**(1), 103–145 (2005). https://doi.org/10.1145/1055709.1055714
2. Aggarwal, C.C.: Recommender Systems. Springer, Cham (2016). https://doi.org/10.1007/978-3-319-29659-3
3. Baltrunas, L., Ludwig, B., Ricci, F.: Matrix factorization techniques for context aware recommendation. In: Proceedings of the Fifth ACM Conference on Recommender Systems, pp. 301–304. RecSys 2011. ACM, New York (2011). https://doi.org/10.1145/2043932.2043988
4. Baltrunas, L., Ricci, F.: Context-dependent recommendations with items splitting, vol. 560, pp. 71–75 (2010)
5. Bao, J., Zheng, Y., Mokbel, M.F.: Location-based and preference-aware recommendation using sparse geo-social networking data. In: Proceedings of the 20th International Conference on Advances in Geographic Information Systems, pp. 199–208. SIGSPATIAL 2012, ACM, New York (2012). https://doi.org/10.1145/2424321.2424348
6. Basilico, J., Raimond, Y.: DéJà Vu.: The Importance of Time and Causality in Recommender Systems. In: Proceedings of the Eleventh ACM Conference on Recommender Systems, p. 342 (2017). https://doi.org/10.1145/3109859.3109922
7. Kula, M.: Metadata embeddings for user and item cold-start recommendations. CoRR abs/1507.08439 (2015)
8. Peña, F.J.: Unsupervised context-driven recommendations based on user reviews. In: Proceedings of the Eleventh ACM Conference on Recommender Systems - RecSys 2017, pp. 426–430 (2017). https://doi.org/10.1145/3109859.3109865
9. Wang, W., Yin, H., Chen, L., Sun, Y., Sadiq, S., Zhou, X.: St-sage: a spatial-temporal sparse additive generative model for spatial item recommendation. ACM Trans. Intell. Syst. Technol. 8(3), 1–25 (2017). https://doi.org/10.1145/3011019
10. Yang, D., Zhang, D., Zheng, V.W., Yu, Z.: Modeling user activity preference by leveraging user spatial temporal characteristics in LBSNs. IEEE Trans. Syst. Man Cybern. Part A Syst. Hum. **45**(1), 129–142 (2015). https://doi.org/10.1109/TSMC.2014.2327053
11. Zhang, J.D., Chow, C.Y.: GeoSoCa: exploiting geographical, social and categorical correlations for point-of-interest recommendations. In: Proceedings of the 38th International ACM SIGIR Conference on Research and Development in Information Retrieval. SIGIR 2015, pp. 443–452. ACM, New York (2015). https://doi.org/10.1145/2766462.2767711
12. Zhao, S., Zhao, T., King, I., Lyu, M.R.: Geo-teaser: geo-temporal sequential embedding rank for point-of-interest recommendation. In: WWW (2017)

Selection of a Green Logical Data Warehouse Schema by Anti-monotonicity Constraint

Issam Ghabri[1,2(✉)], Ladjel Bellatreche[2], and Sadok Ben Yahia[1,3]

[1] Faculty of Sciences of Tunis, LIPAH-LR11ES14, University of Tunis El Manar, El Manar, 2092 Tunis, Tunisia
ghabry.issam@gmail.com
[2] LIAS/ISAE-ENSMA - Poitiers University, Poitiers, France
bellatreche@ensma.fr
[3] Department of Software Science, Tallinn University of Technology, Akadeemia tee 15a, 12618 Tallinn, Estonia
sadok.ben@taltech.ee

Abstract. In the era of social media and big data, many organizations and countries are devoting considerable effort and money to reduce energy consumption. Despite that, current research mainly focuses on improving performance without taking into account energy consumption. Recently, great importance has been attached to finding a good compromise between energy efficiency and performance in data warehouse (DW) applications. For a given DW, multiple logical schemes may exist due to the presence of dependencies and hierarchies among the attributes. In this respect, it has been shown that varying the logical schema has an impact on energy saving. In this paper, we introduce a new approach for efficient exploration of the different logical schemes of a DW. To do so, we prune the search space by relying on anti-monotonicity based constraint to swiftly find the most energy-efficient logical schema. The carried out experiments show the sharp impact of the logical design on energy saving.

Keywords: Data warehouse · Logical schema · Variability · Anti-monotonicity · Energy consumption · Green computing

1 Introduction

The initial intention of computer systems design was to improve performance. With the significant increase in computer power consumption, its reduction in the context of data deluge has become an active research topic over the last decade [1]. Recently, the research community recognized the urgency of the integration of the energy dimension when designing software, hardware, systems, and applications. Several initiatives have been proposed to improve energy efficiency (EE). The latter is the ratio of computing work done per energy unit. To do so, we can: (i) either improving performance with the same power or (ii) reducing power consumption without sacrificing too much performance, to wit, with reasonable performance degeneration [7].

A. Chatzigeorgiou et al. (Eds.): SOFSEM 2020, LNCS 12011, pp. 350–361, 2020.
https://doi.org/10.1007/978-3-030-38919-2_29

Nowadays, any IT company is questing for data to increase its added-value. As mentioned in The Economist *"the world's most valuable resource is no longer oil, but data"*[1]. Like any oil, it pollutes as mentioned in the latest Blog entry of the Martin Tisné published on July 24, 2019 (*"Data isn't the new oil, it's the new CO2"*[2]). This pollution is caused by storing and processing this data.

As database researchers, we are then obliged to propose actions related to energy savings of the whole database environment by considering *small* and *big* initiatives.

Historically, the energy dimension has been considered in the context of data-centers as one of the major energy-consuming components of IT applications [12]. This consumption is associated to their servers, storage devices, networks and infrastructure facilities such as cooling systems and power conditioning [9,16]. In 2000s, several studies have been proposed to integrate energy in DBMS components such as query optimizers [13,17–19]. This scenario is feasible and has to be defended since actually several small and medium-sized enterprises intensively own DBMSs. According to Gartner's predictions, DBMSs showed the most dramatic growth in the entire infrastructure software market[3]. A quick visit of the DB-Engine Website[4] dedicated to rank DBMSs according to their popularity, classical DBMSs (Oracle, MySQL, SQL Server, PostgreSQL, DB2) are the top 5 of the most popular systems. In the middle of 2010, researchers moved to database applications, where the energy dimension has been integrated within the physical design phase, seen as one of the most important phases of the database life cycle [12]. The physical design is the crucial phase of the database life cycle since it can be seen as a funnel of the remaining phases [6], since it integrates entries and parameters from first phases (conceptual and logical steps). Based on these entries, it selects optimization structures such as indexes and materialized views to optimize the workloads.

Most recent studies have highlighted the necessity to integrate the energy dimension on other phases such as logical [6] and code generation [2]. In Bouarar et al. [6], the authors attempted to integrate energy dimension in the logical phase of the *DW* life cycle thanks to the variability. It can be defined as the description of possible variations of a system by points of variation. The variation point identifies and locates where variability occurs. In the context of *DW* applications, the presence of hierarchies between attributes and different relations between them may contribute to varying the logical schema. For instance, the snowflake schema is a variation of the star schema, by denormalizing its dimension tables. It should be noticed that optimization techniques such as join implementations (e.g., nested loop, sort-merge join, hash join, etc.) are applied on the logical *DW* schema [10].

[1] https://www.economist.com/leaders/2017/05/06/the-worlds-most-valuable-resource-is-no-longer-oil-but-data.

[2] https://luminategroup.com/posts/blog/data-isnt-the-new-oil-its-the-new-co2.

[3] http://www.lgcnsblog.com/features/dbms-in-the-center-of-the-it-market-for-big-data-management/.

[4] https://db-engines.com/en/ranking.

In this respect, finding an Eco-friendly logical schema, for a given *DW* application, is the furthest from being an easy task. Indeed, it requires the exploration of the whole research space of logical schemes. In [6], the identification of such schema has been done in a brute-force manner. Roughly speaking, their approach takes a *DW* schema obtained by a designer. Based on different relationships that may exist among attributes of the dimension tables, it straightforwardly considers all possible schemes. Based on a mathematical cost model aiming in evaluating a workload, it estimates the consumed energy for each schema and finally, it retains the schema having the minimum cost. The straightforward enumeration becomes impractical to completely explore the search space. Indeed, the size of the search space is exponential in terms of the number of facts. To speed up this selection, we propose to prune the research space. To do so, we propose an anti-monotonicity based constraint. Thus, we partition dimension tables through its hierarchical attributes starting by the big sized one and only smart parts of it are explored. If the generated schema does not improve the current optimal solution we abstain from exploring the rest of the hierarchy levels and we move on to the next dimension table. So on we prune the research space.

The remainder of this paper is organized as follows. Some definitions about *DW* and the interest of the variability are provided in Sect. 2. In Sect. 3, we thoroughly describe our approach. We detail the underlying algorithm in Sect. 4 and we also provide an illustrative example. We highlight the results of the experiments that we carried out to assess the quality of our approach in Sect. 5. We finally conclude this paper and provide research perspectives in Sect. 6.

2 Background

A *DW* is a central repository of integrated data from various, heterogeneous, distributed and disparate sources. It is a structure (similar to a database) that aims to support decision's making. Roughly speaking, it is a huge pile of information historized, organized, purified, integrated and coming from several sources of data, used for analysis and decision support [8]. The data in *DW* is organized in an easy and simple way to facilitate the analysis by the decision-maker. The data manipulated in this context is represented in a multidimensional model, which is better suited for supporting analysis processes. *DW*s are usually modelled by relational schemes like star and snowflake schemes. The former incorporates a single large fact table, which is related to multiple dimension tables. These latter are relatively small compared to the fact table and they are rarely normalized [3].

In a dimension table, many correlations may exist among attributes, such as hierarchies and functional dependencies. A dimension table may contain several hierarchies. A hierarchy contains several related levels (e.g. day, month, year). The levels in a hierarchy allow analyzing data at various granularities. These levels impact query processing and optimization.

The variability is the ability of a system to be efficiently extended, changed, customized or configured for use in a particular context [15]. The variability is present throughout the life cycle of the *DW*. In fact, this latter owns different variation points. Since, for the conceptual design various formalisms exists

(UML, MERISE, etc.). For the logical design, the attribute correlation allows the logical schema to have many variants through the hierarchies. For a *DW* composed of a fact table and n dimension tables, $D = (F, D_1, D_2, ..., D_n)$, $\prod_{d=1}^{n} 2^{h_d - 1}$ different logical schema can be generated, where n is the number of dimension tables, h_d is the number of hierarchies in the d dimension. For the physical design, many variation points may exist, such as the platform (centralized/distributed) or the optimization structures (index/materialized views).

3 The Proposed Approach: LS-Energy

In this paper, we try to enhance the energy consumption of a data warehouse by choosing the most eco-friendly logical schema. To achieve our goal, we use an anti-monotone constraint fulfilled by a logical scheme energy consumption to prune the research space. We start by sketching the basic concepts related to energy consumption in DBMS.

Definition 1. *The energy: is the total amount of work performed by a system over a period of time.*

Definition 2. *The power: is the rate at which the system performs the work.*

Energy is usually measured in *Joules*, while power is measured in *Watts*. Formally, energy and power can be defined as:

$$E = P \times T \tag{1}$$

$$P = \frac{W}{T} \tag{2}$$

Where P, T, W and E represent respectively, a power, a period, the total work performed in that period of time, and the energy consumed.

The power consumption of a given system can be split into two parts *(i)* baseline power; and *(ii)* active power. The former is the power dissipation when the machine is idle, and the active power is the power dissipation due to the execution of the workload. There exist two concepts of power that have to be considered during the evaluation of power utilization in DBMS. Average power representing the average power consumed during the query execution and peak power representing the maximum power [4].

According to Roukh et al. [11] and Bouarrar et al. [6], for some queries, reading large data files requires more I/O operations than processing by the CPU. Then, running these queries on small data sizes sometimes is more energy-consuming than those running on big size databases. Thus, when intermediate results can not be stored in memory, they will be written to disk and read later. Doing so leads to more CPU greediness since the query spends more time on reading/writing its data than processing its records. Nevertheless, queries dominated by I/O operations have less power consumption. On the other hand, when the data size is small, the reading of data ends quickly and in the remaining time, the execution of the query is dominated by the processing of the processor, which is energy greedy.

In their work, Bouarar et al. [6] shed light on two worth mentioning facts:

- Varying the logical schema has an impact on energy consumption, and even better, it shows that the star schema is far from being the most ecological.
- The normalization of the smallest size tables in the presence of CPU-intensive operations represents a definite disadvantage in terms of energy consumption.

Based on the explanations previously described, we propose a new approach called LS-Energy for generating logical schemes that will ultimately allow us to choose the most eco-friendly logical schema. Nevertheless, exploring the whole search space composed of all the possible logical schemes is simply unfeasible. That's why we need to efficiently prune the search space. To do so, we rely on the anti-monotonicity of energy consumption constraint. Let us recall what is meant by anti-monotone constraint.

Definition 3. *A constraint C is anti-monotone, if for any high-level cell* c_h *and a low-level cell* c_l *covered by* c_h*, the following must hold:* c_h *violates* $C \Rightarrow c_l$ *violates C [20].*

Our hypothesis is based on the constraint of anti-monotonicity, which will allow us to prune the research space. Hence, if a level of the hierarchy does not improve the current optimal solution, then, it will not be interesting to continue to explore this hierarchy, then, we move on to another dimension. Thus, our heuristic is, to start with partitioning big-sized tables.

The steps of our approach and their interactions are glanced in Fig. 1. Our approach begins by evaluating the energy consumption of the star schema and considering it as the optimal solution. Next, we will sort in descending order all dimension tables in respect to their sizes. After that, for each dimension table, we break down its hierarchy and generate its logical schema, rewrite the queries and execute them, then we take its energy consumption and we compare it to the current optimal solution.

- If the new value is better than the optimal solution, then, it becomes the optimal solution. Then, we jump to the next level of the hierarchy and we do the same work again until we explore the whole hierarchy.
- If the new value is greater than the current optimal solution, we abstain from the decomposition and we move on to the next dimension and at this level, the anti-monotonicity constraint appears.

To evaluate the energy consumption of each schema, we adjust the mathematical cost model introduced by Roukh et al. [11]. The adjustment was made to make the cost model more generic and to consider all variants of the logical schemes. Since the cost model has been constructed by assuming a *DW* with a star schema. The adjustment mainly concerns the training phase [6]. Owe to this cost model, we can assess theoretically the energy consumed of each logical schema.

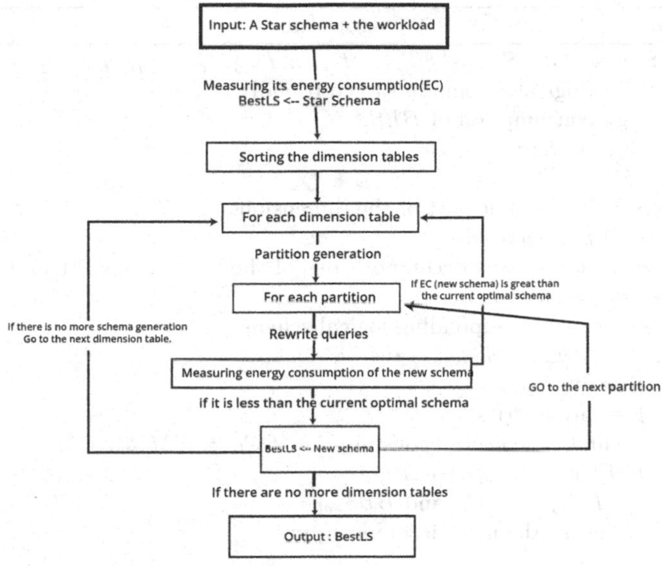

Fig. 1. The LS-Energy approach at a glance

It is worth mentioning that for every schema the generation process is based on attributes correlations and the queries of the workload were rewritten to be suitable for the new schema as explained in [5].

4 Algorithm

Our algorithm, whose pseudo-code is given in Algorithm 1, inputs a star schema of a DW. The first step is to evaluate its energy consumption $CM_{power}(SS)$ and assign this value to EC_{opt} which is the optimal energy consumption of a logical schema (line 2). Next, we sort in descending order the dimension tables with respect to their respective sizes (line 4). After that, for each dimension, we generate a set of decompositions (lines 5–6). Then, for each decomposition, we generate the corresponding logical schema (lines 7–8). In the next step, for each logical schema, we rewrite the queries to be suitable for the schema, execute them, and capture the energy consumption (lines 9–12). Next, we compare the energy consumption of this schema (EC_{ls}) with the optimal energy consumption (line 13). If (EC_{ls}) is less than EC_{opt}, then (EC_{ls}) becomes the optimal consumption EC_{opt}, the schema ls becomes the optimal logical schema and we move to the next level of hierarchy (lines 14–15). If (EC_{ls}) is greater than EC_{opt}, then, we refrain from testing the remaining decompositions of this dimension, and we move on to the next dimension (line 17). At the closing, the algorithm returns the optimal logical schema as well as its energy consumption.

Algorithm 1: THE LS-ENERGY algorithm

Input: A Star schema $SS = \{FT, D_1, D_2, ..., D_n\}$; $Q = \{q_1, q_2, ..., q_m\}$;
Output: BLS: Logical schema most eco-friendly,
EC_{opt} : Energy consumption of BLS.

$EC_{opt} = CM_{power}(SS)$;
$BLS_{opt} = SE$;
$Dimensions$ = Descending sort of the dimensions of (SS);
for *each* $d \in Dimensions$ **do**
    generate the tree of the decompositions of the hierarchical attributes
    for *each decomposition dec* **do**
        generate the corresponding logical schema.
        for *each logical schema ls* **do**
            Rewrite queries;
            Execute queries;
            evaluate the energy consumption $EC_{ls} = CM_{power}(ls)$
            if $EC_{ls} < EC_{opt}$ **then**
                $EC_{opt} = EC_{ls}$ and $BLS_{opt} = ls$
                go to the next decomposition.
            else
                go to next dimension.
return BLS_{opt} *: The optimal logical schema*
EC_{opt} *: Optimal energy consumption*

Example 1. Let us consider the star schema depicted in Fig. 2. The latter contains two dimensions tables *Date*, *Customer* and a fact table *Sales*. The *Customer* table has three hierarchical attributes (*city, nation* and *region*). The *Date* table also has three hierarchical attributes (*month, quarter* and *year*). The size of the *Customer* table is bigger than that of the *Date* table. We start by assessing the energy consumption of the star schema (Fig. 2).

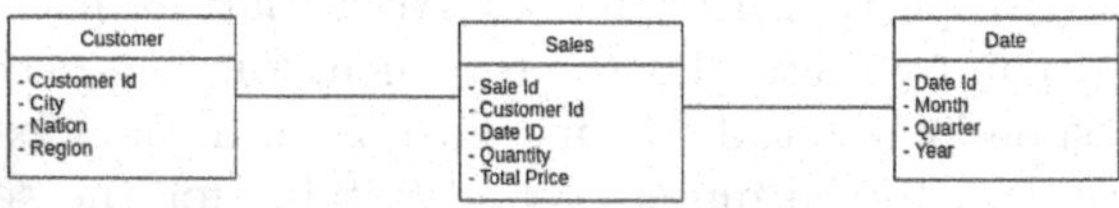

Fig. 2. The star schema

Then, we partition the biggest table by putting its hierarchical attributes in another table. We partition table *Customer* as shown in Fig. 3 to get a new schema denoted $LS1$ and we assess its energy consumption. As its energy consumption is less than that of the star schema, we keep it as the current optimal solution and we continue the partitioning to have a new schema denoted $LS2$ (Fig. 4). We assess its energy consumption. Since the latter has not been

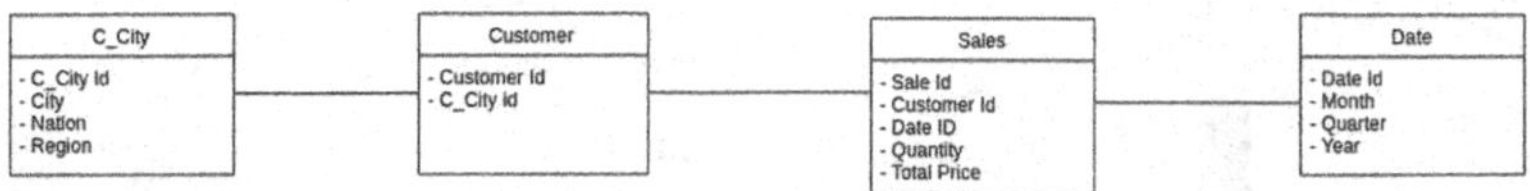

Fig. 3. The schema LS1

improved, we refrain from continuing the decomposition of the table and we jump to the other dimension table i.e, the *Date* Table. Thus, we keep the schema $LS1$ as the optimal solution and partition the *Date* table to get a new schema denoted $LS3$ (Fig. 5). We measure its energy consumption and we compare it to the current optimal solution. Since the energy consumption of $LS3$ is not better than $LS1$, and since *Date* table is the last one, then, we come to an end of the partitioning process and we consider $LS1$ as the final solution, which is the most eco-friendly logical schema.

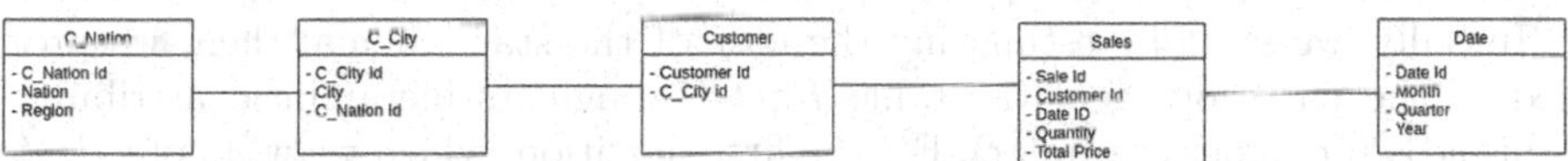

Fig. 4. The schema LS2

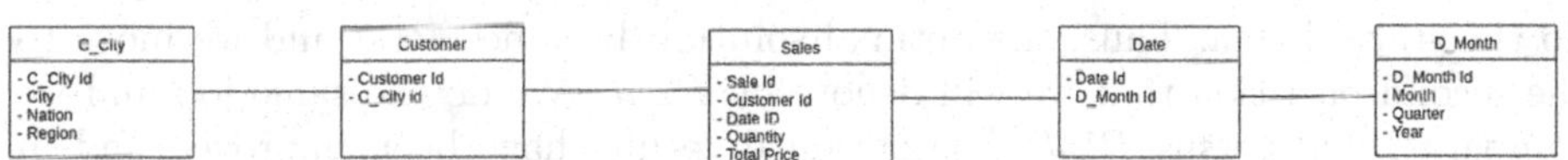

Fig. 5. The schema LS3

5 Experimental Study

In order to show the relevance and efficiency of our proposal, we have conducted several experiments. Thus, we have carried out tests on a *DW* implemented within PostgreSQL on a machine of 8 GB main memory and a 1 TB hard drive under Ubuntu Server 14.04 LTS. Our experimental study was carried out on the *DW* resulting from the benchmark *SSB*[5] with a scale factor of 10 GB. *SSB* is composed of 1 fact table i.e., *Lineorder* and 4 dimension tables (*Part, Customer, Supplier* and *Date*). According to the attribute correlation, 256 logical schemes could be generated.

We also disabled with unnecessary background tasks. Furthermore, the system and the cache are flushed after each query execution. Our experimental study is carried out on a machine which is equipped with a "Watts UP? Pro

[5] https://www.cs.umb.edu/poneil/StarSchemaB.PDF.

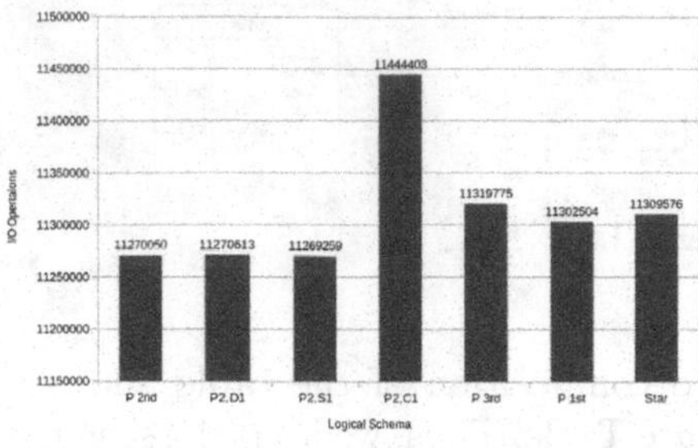

Fig. 6. I/O operations

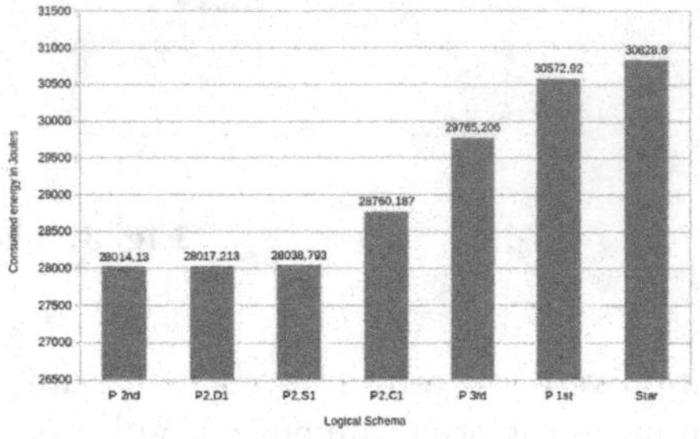

Fig. 7. Total energy consumption

ES" power-meter. It is worth mentioning that the queries are executed in an isolated way. Owe to the cost model, we were able to estimate the I/O cost and the CPU cost necessary to execute the workload. We were also able to measure the energy consumed and the average power of each query since the power is not stable during query processing [11].

Initially, we start by estimating the cost of the star schema, then, we proceed to the partitioning of the table *Part* through its hierarchical attributes (p_brand1, p_catgory & p_mfgr). For the first partition, which we will note *P1st*, we decompose the table *Part* by putting its hierarchical attributes in a separate table and we measure the necessary cost to execute the workload. Then we compare its value with that of the star schema.

The results show that *P*1*st* consumes less power to run the workload than do the star schema. Thus, the optimal solution becomes *P*1*st* and we move to the second partition that we will denote by *P2nd*. We do the same job and we compare *P2nd* versus *P*1*st*. The obtained results have been improved and in the meanwhile, the power decreases. Thus, *P2nd* becomes the optimal solution and we proceed to the last partitioning of the table *Part*. The latter becomes completely normalized. Note that the queries are rewritten to be conform to the target schema.

This time, the values have not improved, hence, *P2nd* is maintained as the optimal solution. As table *Part* was fully normalized, we move to the next table which is *Customer*. The latter has 3 hierarchical attributes (c_city, c_nation & c_region). We have partitioned the *Customer* table by putting its hierarchical attributes into another one while keeping the second decomposition of the *Part* table. We ran the workload of the new schema which we denote *P*2*C*1 and we compare it with the actual optimal schema, to wit, *P2nd*. The results show that the values have not improved compared to the optimal solution but nevertheless still better than those of the star schema ones. Hence, we abstain from generating the other decompositions of the *Customer* table and jump to the next table. The *Supplier* table has 3 hierarchical attributes (s_city, s_nation & s_region). Similar to what we did with the table *Customer* we break it down and we compare its energy consumption to that of *P2nd*. The generated schema, denoted *P*2*S*1, was not better than the optimal solution. As a result, we move to the last table,

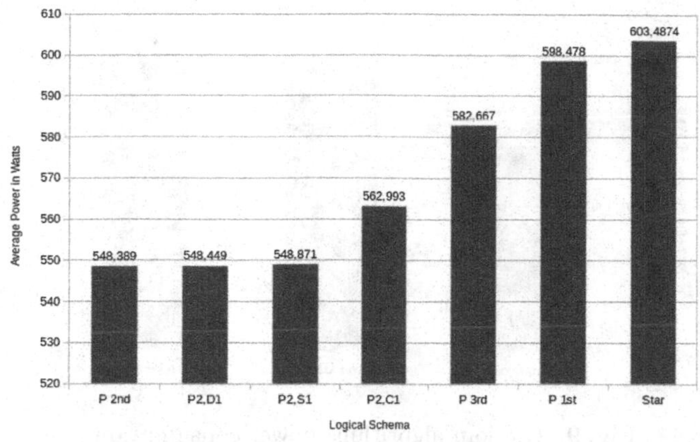

Fig. 8. Average power consumption

Date, which has 3 hierarchical attributes (d_dayofweek, d_month & d_year). By executing the queries of the decomposition of the table *Date* denoted *P2D1*, the energy consumption decreases compared to other decompositions but it is not better than that of *P2nd*. Thus, *P2nd* is maintained as the final solution. Owe to the anti-monotonicity constraint, the search space is pruned by sharply decreasing the number of "*tested*" schemes from 256 to only 7 schemes.

The values of the I/O operations, the average power and the total energy consumed of the different schemes are given respectively in Figs. 6, 7 and 8. It is worth mentioning that the schema size changes from a schema to another. As expected, and as already mentioned in [6], normalizing large dimension tables reduces storage space and the star schema is no more the smallest one.

As we can see in Figs. 6, 7 and 8, the star schema is far from being the best in terms of energy consumption since it needs more power as plotted in Fig. 8 and it needs more CPU cost to run the workload. We did not put the CPU cost chart due to lack of space. Nevertheless, it may be better than other logical schemes for the I/O cost as we can see in Fig. 6.

To get further insights into these results, we observed the execution plan of the different queries of the different schemes. During this observation, we focused on the join order [14] and the used join algorithms. Thus, we unveiled that the join order of the tables changes while processing queries after each partitioning, and the query optimizer starts by joining the small tables containing the selection predicates. Doing so, it generates a gain in both I/O operations and CPU cost and by therefore leads to lowering the energy consumption. On the other hand, changing the join algorithm also influences energy consumption. We noticed that the *query optimizer* alternates between the NESTED LOOP and the HASH JOIN. Interestingly enough, this can be explained by the fact that the Nested loop is mainly indicated for joins with a small number of rows. The HASH JOIN is usually used when the number of table rows becomes important.

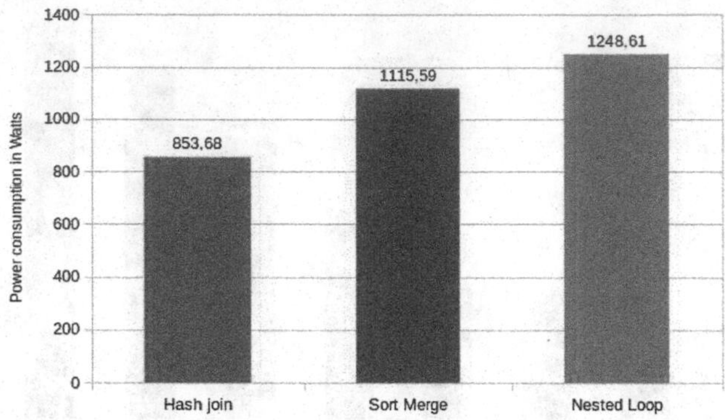

Fig. 9. The join algorithms power consumption

We conducted an additional experimental study in which we ran the *SSB* benchmark workload by forcing the query optimizer to use only one join algorithm from the 3 algorithms (Hash join, Sort-Merge and Nested Loop). The experiments highlight that the Hash join was the best in terms of execution time and also in terms of energy consumption compared to Sort-Merge and Nested Loop respectively. The results of the power consumption of the join algorithms are plotted in Fig. 9. It is worth mentioning that some queries consume less power while processing the Nested loop than the Hash join or the Sort-Merge.

According to Xu et al. the join in NESTED LOOP uses 15 to 17 % of additional processor cycles. The NESTED LOOP join is selected only when the table is small enough to be placed in the database buffer [18]. Whenever the table is large enough, the cost of energy for accessing the hash table is relatively low. Thus, it is more likely that the choice of HASH JOIN is made in large table joins than in smaller table joins [18]. Therefore, we can say that the join algorithms and the join order have a paramount influence on energy consumption.

6 Conclusion

Through this work, the aim was to better stress on the importance of the logical design phase through the assessment of the variability impact on energy saving. In this paper, we introduced an approach called LS-ENERGY that heavily relies on the anti-monotonicity constraint. The results show that trying to normalize the big sized tables can be an interesting solution to decrease energy consumption. Likewise, the join algorithms and join ordering can influence energy consumption. In the near future, we will put the focus on studying the energy in the database field from a higher level which is the conceptual design. Moreover, we plan to deal with energy consumption as a constraint during the physical design phase.

References

1. Abadi, D., et al.: The beckman report on database research. Commun. ACM **59**(2), 92–99 (2016)
2. Acar, H., Alptekin, G.I., Gelas, J., Ghodous, P.: The impact of source code in software on power consumption. Int. J. Electron. Bus. Manag. **14** (2016). http://ijebm-ojs.ie.nthu.edu.tw/IJEBM_OJS/index.php/IJEBM/article/view/693
3. Bellatreche, L., Missaoui, R., Necir, H., Drias, H.: A data mining approach for selecting bitmap join indices. J. Comput. Sci. Eng. **1**, 177–194 (2007)
4. Bellatreche, L., Roukh, A., Bouarar, S.: Step by step towards energy-aware data warehouse design. In: Marcel, P., Zimányi, E. (eds.) eBISS 2016. LNBIP, vol. 280, pp. 105–138. Springer, Cham (2017). https://doi.org/10.1007/978-3-319-61164-8_5
5. Bouarar, S., Bellatreche, L., Jean, S., Baron, M.: Do rule-based approaches still make sense in logical data warehouse design? In: Manolopoulos, Y., Trajcevski, G., Kon-Popovska, M. (eds.) ADBIS 2014. LNCS, vol. 8716, pp. 83–96. Springer, Cham (2014). https://doi.org/10.1007/978-3-319-10933-6_7
6. Bouarar, S., Bellatreche, L., Roukh, A.: Eco-data warehouse design through logical variability. In: Steffen, B., Baier, C., van den Brand, M., Eder, J., Hinchey, M., Margaria, T. (eds.) SOFSEM 2017. LNCS, vol. 10139, pp. 436–449. Springer, Cham (2017). https://doi.org/10.1007/978-3-319-51963-0_34
7. Guo, B., Yu, J., Liao, B., Yang, D., Lu, L.: A green framework for DBMS based on energy-aware query optimization and energy-efficient query processing. J. Netw. Comput. Appl. **84**, 118–130 (2017)
8. Inmon, W.H.: Building the Data Warehouse. Wiley, New York (1992)
9. Liebert, E.: Five strategies for cutting data center energy costs through enhanced cooling efficiency. White paper (2007)
10. Pitoura, E.: Query optimization. In: Liu, L., Özsu, M.T. (eds.) Encyclopedia of Database Systems. Springer, New York (2018). https://doi.org/10.1007/978-1-4614-8265-9_861
11. Roukh, A., Bellatreche, L.: Eco-processing of OLAP complex queries. In: Madria, S., Hara, T. (eds.) DaWaK 2015. LNCS, vol. 9263, pp. 229–242. Springer, Cham (2015). https://doi.org/10.1007/978-3-319-22729-0_18
12. Roukh, A., Bellatreche, L., Boukorca, A., Bouarar, S.: Eco-physic: eco-physical design initiative for very large databases. Inf. Syst. **68**, 44–62 (2017)
13. Roukh, A., Bellatreche, L., Ordonez, C.: Enerquery: energy-aware query processing. In: Proceedings of the 25th ACM International on Conference on Information and Knowledge Management, pp. 2465–2468. ACM (2016)
14. Steinbrunn, M., Moerkotte, G., Kemper, A.: Heuristic and randomized optimization for the join ordering problem. VLDB J. **6**(3), 191–208 (1997)
15. Svahnberg, M., van Gurp, J., Bosch, J.: A taxonomy of variability realization techniques: research articles. Softw. Pract. Exper. **35**(8), 705–754 (2005)
16. Tsirogiannis, D., Harizopoulos, S., Shah, M.A.: Analyzing the energy efficiency of a database server. In: SIGMOD, pp. 231–242 (2010)
17. Tu, Y.C., Wang, X., Zeng, B., Xu, Z.: A system for energy-efficient data management. ACM SIGMOD Record **43**(1), 21–26 (2014)
18. Xu, Z., Tu, Y., Wang, X.: Online energy estimation of relational operations in database systems. IEEE Trans. Comput. **64**(11), 3223–3236 (2015)
19. Xu, Z., Tu, Y.C., Wang, X.: PET: reducing database energy cost via query optimization. Proc. VLDB Endow. **5**(12), 1954–1957 (2012)
20. Yu, P.S., Han, J., Faloutsos, C.: Link Mining: Models, Algorithms, and Applications, 1st edn. Springer, Heidelberg (2010)

The HyperBagGraph DataEdron: An Enriched Browsing Experience of Datasets
Track: Foundation of Data Science and Engineering

Xavier Ouvrard[1,2(✉)], Jean-Marie Le Goff[1], and Stéphane Marchand-Maillet[2]

[1] CERN, 1 Esplanade des Particules, Meyrin, Switzerland
xavier.ouvrard@cern.ch
[2] University of Geneva, Carouge, Switzerland

Abstract. Traditional verbatim browsers give back information linearly according to a ranking performed by a search engine that may not be optimal for the surfer. The latter may need to assess the pertinence of the information retrieved, particularly when s·he wants to explore other facets of a multi-facetted information space. Simultaneous facet visualisation can help to gain insights into the information retrieved and call for further refined searches. Facets are potentially heterogeneous co-occurrence networks, built choosing at least one reference type, and modeled by HyperBag-Graphs—families of multisets on a given universe. References allow to navigate inside the dataset and perform visual queries. The approach is illustrated on Arxiv scientific pre-prints searches.

Keywords: Hyper-Bag-Graphs · Knowledge discovery · Visual queries · Information retrieval

1 Introduction

When browsing a textual database, traditional verbatim browsers give back linear information in the form of ranked list of short reference description. To increase the pertinence of this information, the surfer has often to perform additional searches either by refining the original search terms s·he used or by using other pertinent queries that can help her·him to refine the retrieved information.

In an information space, meaningful information can be regrouped by hierarchical classification or—non exclusive—by semantically cohesive categories that

This work is part of the PhD of X. Ouvrard, done at UniGe and funded by a doctoral position at CERN, co-supervised by Pr. S. Marchand-Maillet and Dr J.M. Le Goff. The authors are really thankful to Tullio Basaglia (CERN Library).

Electronic supplementary material The online version of this chapter (https://doi.org/10.1007/978-3-030-38919-2_30) contains supplementary material, which is available to authorized users.

A. Chatzigeorgiou et al. (Eds.): SOFSEM 2020, LNCS 12011, pp. 362–374, 2020.
https://doi.org/10.1007/978-3-030-38919-2_30

are combined to express concepts, called facets [1]. Those facets are linked by the physical entities contained in the search output. Choosing a type of reference linked to these entities enables the construction of a co-occurrence network per facet and enhance navigation in the information space. For instance, in scientific publications different information are linked in an article: the article reference, the authors, the main keywords... All this metadata can potentially give insights into the information space and can be chosen as reference to build co-occurrences. Choosing as reference for instance the article id, facets depict co-occurrence networks, either of homogeneous type, such as co-authors or co-keywords, or of heterogeneous types, i.e. combining multiple types together. Co-occurrences can potentially contain repetitions or require an individual weighting: modeling it requires multisets instead of sets.

We propose in this article a new way to explore an information space by using hyper-bag-graphs (hb-graphs for short)—families of multisets on a universe called the vertex set—a mathematical structure we introduced in [2]. Hb-graphs are a separate mathematical category from the one of hypergraphs. This is an important difference as hb-graphs store extra-information that can not be kept with hypergraphs and have different algebra operations. Moreover, we have shown in [3] that hb-graphs enhance exchange-based diffusion over co-occurrence networks, providing a fine vertex and hb-edge ranking.

We propose four extensions of the hypergraph framework of [4]. First, the visualisation part is extended to support hb-graphs: it is an important mathematical generalization that supports redundancy and hb-edge based weighting of vertices that requires multiset families (hb-graphs) instead of subset families (hypergraphs). Second, the new framework supports navigation of heterogeneous co-occurrence networks; in the former framework only homogeneous co-occurrences where allowed. Third, multi-references for building co-occurrences is tackled. Fourth, an application is given with Arxiv search, by the implementation of a 2.5D interface to perform visual queries and visualize the Arxiv information space.

Section 2 lists the related work and the mathematical background. Section 3 presents the hb-graph framework. Section 4 gives results and Sect. 5 concludes.

2 Related Work and Mathematical Background

2.1 Information Space Discovery

Discovering knowledge in an information space requires to gather meaningful information, either hierarchically or semantically. Semantics provide support to the definition of facets within an information space [1].

Navigation and visualisation of the information space facets have been achieved previously in many different ways. [5] uses a pivot to stroll between three facets; the approach, based on a tripartite graph, is limited to the visualisation of a small amount of pivots at the same time. In [6], an interactive exploration of implicit and explicit relations in faceted datasets is proposed. The space of visualisation is shared between different metadata with cross findings between metadata, partitioning the space in categories. [7] proposes a visual analytics graph-based framework for exploring an information space. The labeled graph representing the dataset is explored by retrieving paths with same type vertices going through reference vertices. Visualisation facets are navigable graphs of pairwise collaborations.

2.2 Co-occurrence Networks

Data mining is only one step in the knowledge discovery processing chain. If numerical data allows rich statistics on the instances, non numerical data mining consists often in summarizing data as occurrences. Alternately, techniques using data instance similarities such as k-nearest neighbors can be used to link different occurrences: however in high dimensionality, they are limited by the curse of dimensionality, even if some techniques limit its effect [8]. Retrieving links through the dataset itself is another way of detecting co-occurrences.

If the dataset reflects existing links—as group of friends in social networks—the job is easier since an inherent co-occurrence/collaboration network can be built through the data instances. Nonetheless, links are often neither direct nor tangible: thus co-occurrences need to be built or processed from the dataset.

A dataset can be a set of physical references, stored as rows in traditional relational databases. Each physical reference has metadata instances attached to it. Metadata instance types can be either interesting for visualisation or processing additional information. The set of physical references and metadata instances used for visualisation provide the types of the network, each type being seen either as a reference or a facet of the information space. This allows—as it will be explained in the next section—the retrieval of co-occurrences in one facet, based on one reference type—which can differ from the physical reference.

2.3 Multisets and Hb-Graphs

Co-occurrences seen as collaborations are m-adic relationships of occurrences, often modeled as hypergraphs, i.e. families of subsets of a given vertex set. But hypergraphs, as they are subsets, do not support neither hyperedge-based repetition nor hyperedge-based weighting of vertices. Hb-graphs—introduced newly in [2]—as multiset families naturally allow them.

Multisets—also known as bags or msets—have been used for a long time in many domains such as text representation and image. Multisets support the individual weighting of their elements by using a **multiplicity function** on a set called the **universe**. The elements that have non-zero multiplicity value belong to the **support** of the multiset. A **natural multiset** occurs when the multiplicity function has its range in the non-negative integers[1].

More information on hypergraphs and multisets, with additional references, can be found in [3].

[1] We denote $\mathfrak{A}_m = \{x_i^{m_i} : i \in \llbracket n \rrbracket\}$ where $m_i = m(x_i)$ a mset $\mathfrak{A}_m = (A, m)$ of universe $A = \{x_i : i \in \llbracket n \rrbracket\}$, of multiplicity function m and of support $\mathfrak{A}_m^\star = \{x_i : m_i \neq 0\}$.

Table 1. Synthesis of the framework

METADATA	Schema hypergraph ↓	Related to database structure	$\mathcal{H}_{\text{Sch}} = (V_{\text{Sch}}, E_{\text{Sch}})$
	Extended schema hypergraph ↓	Store possible additional processings	$\overline{\mathcal{H}_{\text{Sch}}} = (\overline{V_{\text{Sch}}}, \overline{E_{\text{Sch}}})$
	Extracted extended schema hypergraph ↓	U: set of metadata of interest (visualisation and reference)	$\mathcal{H}_X = (V_X, E_X)$ where $V_X = U$, $E_X = \{e \cap U : e \in \overline{E_{\text{Sch}}}\}$
	Reachability hypergraph ↙↓↘	Hyperedges are connected components E_{cc} ($\subset V_X$) of $\mathcal{H}_X$	$\mathcal{H}_R = (V_R, E_R)$ $V_R = V_X$ $E_R = \{E_{\text{cc}} : E_{\text{cc}} \text{ c.c. of } H_X\}$
	Navigation hypergraph ↓	Choose: $e_r \in E_R$ references $R_{\text{ref}} \subset e_r$	$\mathcal{H}_N = (V_N, E_N)$ $V_N = V_R \backslash R_{\text{ref}}$ $E_N = \{e_r \backslash R : R \subseteq R_{\text{ref}} \wedge R \neq \emptyset\}$
DATA	Facet visualisation hb-graphs	Co-occurrence networks as hb-graphs	

Following [2], a **hb-graph** $\mathfrak{H} = (V, \mathfrak{E})^2$ is a family of multisets called hb-edges $\mathfrak{E} = (\mathfrak{e}_i)_{i \in [\![p]\!]}$ having the same universe $V = \{v_1, ..., v_n\}$ called the vertex set. Each hb-edge $\mathfrak{e}_i \in \mathfrak{E}$ has its own multiplicity function: $m_{\mathfrak{e}_i} : V \to \mathbb{W}$ where $\mathbb{W} \subset \mathbb{R}^+$. A hb-edge can be seen as a dependent weighted system of vertices. A hb-graph with only natural multisets as hb-edges is said **natural**. A **hypergraph** appears as a particular case of natural hb-graph with a binary value—0 or 1—for each hb-edge multiplicity.

The **support hypergraph** $\overline{\mathfrak{H}} = (V, E)$ of a hb-graph $\mathfrak{H} = (V, \mathfrak{E})$ is the hypergraph of same vertex set V and of hyperedges $E = (\mathfrak{e}_i^\star)_{i \in [\![p]\!]}$. The support hypergraph is unique for a given hb-graph. But reconstructing the hb-graph from a support hypergraph generates an infinite number of hb-graphs, showing that the information contained in a hb-graph is denser than in a hypergraph.

Hb-graph unnormalized extra-node representation is obtained by adding an extra-node per hb-edge linked to each hb-edge support vertex with a link thickness proportional to the vertex multiplicity. Figure 2.b(ii) shows an example.

3 Hb-Graph Framework

3.1 Enhancing Navigation

For the sake of clarity, we briefly summarize in Table 1 the enhancement of navigation of [4], achieved by defining different hypergraphs at the metadata level. We take as thumbnail an example based on a publication dataset. Possible metadata types are: *publication id*, title, abstract, *authors*, affiliations, addresses, *author keywords*, *publication categories*, *countries*, *organizations*, and eventually some

[2] We use fraktur font for multisets and hb-graphs: $\mathfrak{A} : A, \mathfrak{e} : e, \mathfrak{E} : E, \mathfrak{H} : H$.

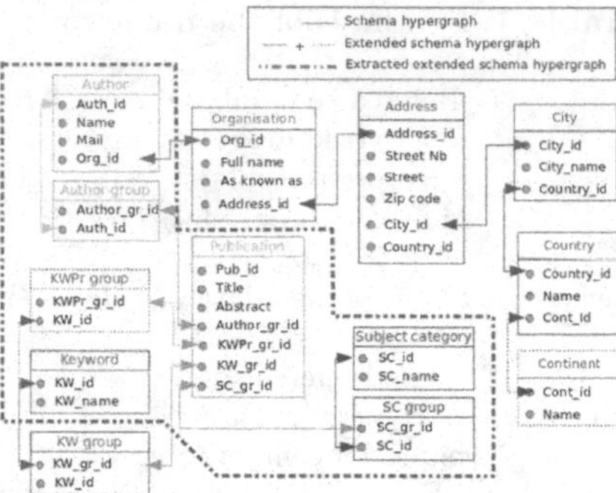

Fig. 1. Schema hypergraph, extended schema hypergraph, Extracted extended schema hypergraph: exploded view shown on an example of publication dataset

processed metadata types such as *processed keywords*, *continent*, ...[3] Enhancing navigation supposes first to define the **schema hypergraph** reflecting the relationships between the database metadata instances. We give the possibility to extend it into **an extended schema hypergraph** to store potential additional processings. Out of the latter an **extracted extended schema hypergraph** $\mathcal{H}_X = (V_X, E_X)$ is enhanced that keeps metadata instances of interest to build the co-occurrences and to be visualized; it might require some intermediate hyperedge bundling. Figure 1 shows the different hypergraphs.

The **reachability hypergraph** $\mathcal{H}_R = (V_R, E_R)$ reflects the connected components of $\mathcal{H}_X$, with $V_R = V_X$: its hb-edges do not intersect. Hence, if $\mathcal{H}_R$ has only one hyperedge, the whole dataset is navigable. We assume that in each hyperedge of the reachability hypergraph, there is at least one metadata type or a combination of metadata types that can be chosen as the **physical reference**. The data instance related to this reference is supposed to be unique. For instance, in a publication dataset the physical reference is the publication id of the publication itself. In the example, the extracted hypergraph has only one component {publication id, authors, processed keywords, subject categories}.

Each hyperedge $e_r \in E_R$ of $\mathcal{H}_R$ leads to one new **navigation hypergraph** $\mathcal{H}_N = (V_N, E_N)$ by choosing a non-empty subset R_{ref} of e_r of possible reference types of interest. The choice of a subset R of R_{ref} allows to consider the remaining vertices of $e_r \backslash R$ as visualisation vertex types, that will be used to generate the facet visualisation hb-graphs and are called the visualisation types. Hence: $E_N = \{e_r \backslash R : R \subseteq R_{\text{ref}} \wedge R \neq \emptyset\}$. When there is only one reference of interest selected at a time in R_{ref} we denote $E_{N/1}$ for E_N. In the publication database example, many navigation hyperedges are possible; the navigation hyperedge choosing as reference publication ids is {authors, publication categories, processed keywords} while using processed keywords as reference is: {authors, publication category, publication ids}.

[3] Metadata of interest for visualisation or referencing are in italic.

3.2 Facet Visualisation Hb-Graphs

In [4], we use sets to store co-occurrences. Nonetheless in many cases, it is worth storing additional information by joining a multiplicity—with nonnegative integer or real values—to the elements of co-occurrences. A small example emphasizes the interest of moving towards multisets: we consider the publication network of Fig. 2. In this example, building co-occurrences accounting the occurrence multiplicity induces not only a refined visualisation, with distinguishable hb-edges in between some of the vertices (augmented reality and 3D) but also yields to refined rankings of both vertices and hb-edges, as mentioned in [3]. As some parts relies on a mathematic description they have been put in Appendice. The reader can always refer to Fig. 2 for an illustration of the concepts where we choose the keywords as reference.

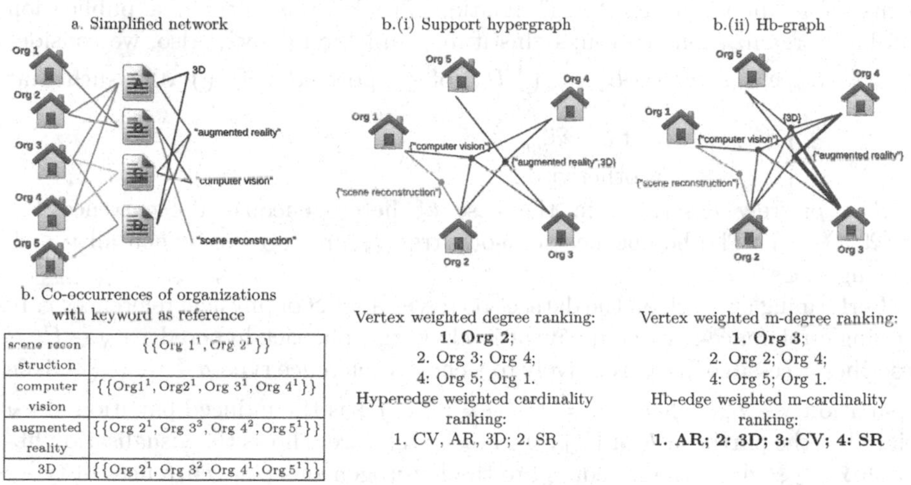

scene reconstruction	$\{\{\text{Org }1^1, \text{Org }2^1\}\}$
computer vision	$\{\{\text{Org1}^1, \text{Org2}^1, \text{Org }3^1, \text{Org }4^1\}\}$
augmented reality	$\{\{\text{Org }2^1, \text{Org }3^3, \text{Org }4^2, \text{Org }5^1\}\}$
3D	$\{\{\text{Org }2^1, \text{Org }3^2, \text{Org }4^1, \text{Org }5^1\}\}$

Fig. 2. A simplified publication network with publication id, organizations, keywords.

Each physical entity d of a dataset $\mathcal{D}$ corresponds to a unique physical reference r. d is described by a set of data instances of different types that are in $\alpha \in \overline{V_{\text{Sch}}}$. We write $\mathcal{I}$ the set of data instances in $\mathcal{D}$, and t the type application that gives the type of an instance.

Hb-graphs requires a common universe taken as vertex set. We consider for each type α, its instance set $U_\alpha = \{i : i \in \mathcal{I} \wedge t(i) = \alpha\}$ of instances of $\mathcal{D}$ of type α. The common universe for the visualisation hb-graph depends on the search.

We write $\mathfrak{A}_{\alpha,r} = (U_\alpha, m_{\alpha,r})$ the **multiset** of universe U_α, of the values of type α—possibly none—that are attached to d, the physical entity of reference r. The support of $\mathfrak{A}_{\alpha,r}$ is $\mathfrak{A}^\star_{\alpha,r} = \{a_{i_1}, ..., a_{i_{k_r}}\}$. Hence, we abusively write: $\mathfrak{A}_{\alpha,r} = \left\{a_{i_1}^{m_{\alpha,r}(a_{i_1})}, ..., a_{i_{k_r}}^{m_{\alpha,r}(a_{i_{k_r}})}\right\}$ omitting the elements of U_α that have a zero multiplicity in $\mathfrak{A}_{\alpha,r}$.

d is entirely described by its reference r and the family of multisets, corresponding to homogeneous co-occurrences of the different types α in $\overline{V_{\text{Sch}}}$ linked to the physical reference, i.e. $\left(r, (\mathfrak{A}_{\alpha,r})_{\alpha \in \overline{V_{\text{Sch}}}}\right)$.

In Fig. 2, the publication id is the physical reference. Taking as reference the publication id, the co-occurences for the Publication A of organisations are: $\left\{\text{Org } 2^1, \text{Org } 3^1, \text{Org } 4^1\right\}$ and of keywords are: $\left\{3\text{D}^1, \text{augmented reality}^1\right\}$. The example in Fig. 2.b shows a reference that is not the physical reference.

Type heterogeneity in co-occurrences can enable simultaneous view of different types in a single facet. To allow type heterogeneity in co-occurrences, we consider a partition Γ of the different types in $\overline{V_{\text{Sch}}}$. Each type belonging to an element ν of the partition Γ will be visualized simultaneously in a co-occurrence: it enriches the navigation process, allowing heterogeneous co-occurrences. An interesting case is when ν has a semantic meaning and elements of ν appear as an "is a" relationship. For instance in a publication database organizations regroups "institute" and "company". Also, we consider $\mathfrak{A}_{\nu,r} \triangleq (U_\nu, m_{\nu,r})$, where $U_\nu \triangleq \bigcup_{\alpha \in \nu} U_\alpha$, of support $\mathfrak{A}^\star_{\nu,r} \triangleq \bigcup_{\alpha \in \nu} \mathfrak{A}^\star_{\alpha,r}$ such that $m_{\nu,r}(a) \triangleq \begin{cases} m_{t(a),r}(a) & \text{if } a \in \mathfrak{A}^\star_{\nu,r} \\ 0 & \text{otherwise} \end{cases}$.

d is entirely described in the case of heterogeneous co-occurrences by $\left(r, (\mathfrak{A}_{\nu,r})_{\nu \in \Gamma}\right)$. The homogeneous co-occurrences are retrieved when all $\nu \in \Gamma$ are singletons.

Performing a search on the dataset retrieves a set $\mathcal{S}$ of physical references r. In the single-reference-restricted navigation hypergraph, each hyperedge $e_N \in E_{N/1}$ describes accessible facets relatively to a chosen reference type $\rho \in V_N \backslash e_N$. Given a partition $\gamma \in \Gamma_N$, where $\Gamma_N \triangleq \{\nu \cap e_N : \nu \in \Gamma\}$ is the induced partition of e_N related to the partition Γ of $\overline{V_{\text{Sch}}}$, the associated facet shows the visualisation hb-graph $\mathfrak{H}_{\gamma/\rho,\mathcal{S}}$ where the hb-edges are the heterogeneous co-occurrences of types in γ relatively to reference instances of type ρ (γ/ρ as short) retrieved from the different references in $\mathcal{S}$.

We then build the co-occurrences γ/ρ by considering the set of all values of type ρ attached to all the references $r \in \mathcal{S}$: $\Sigma_\rho \triangleq \bigcup_{r \in \mathcal{S}} \mathfrak{A}^\star_{\rho,r}$. Each element s of Σ_ρ is mapped to a set of physical references $R_s \triangleq \left\{r : s \in \mathfrak{A}^\star_{\rho,r}\right\} \in \mathcal{P}(\mathcal{S})$ in which they appear: we write r_ρ the mapping. The multiset of values $\mathfrak{e}_{\gamma,s}$ of types $\alpha \in \gamma$ relatively to the reference instance s is $\mathfrak{e}_{\gamma,s} \triangleq \biguplus_{r \in R_s} \mathfrak{A}_{\gamma,r}$.

The **raw visualisation hb-graph** for the facet of heterogeneous co-occurrences γ/ρ attached to the search $\mathcal{S}$ is defined as: $\mathfrak{H}_{\gamma/\rho,\mathcal{S}} \triangleq \left(\bigcup_{r \in \mathcal{S}} \mathfrak{A}^\star_{\gamma,r}, (\mathfrak{e}_{\gamma,s})_{s \in \Sigma_\rho}\right)$. Fig. 2.b(ii) gives an example of such a raw visualisation hb-graph.

Since some hb-edges can possibly point to the same sub-mset of vertices, we build a reduced visualisation weighted hb-graph from the raw visualisation hb-graph. To achieve it we define: $g_\gamma : s \mapsto \mathfrak{e}_{\gamma,s}$ and $\mathcal{R}$ the equivalence relation such that: $\forall s_1 \in \Sigma_\rho, \forall s_2 \in \Sigma_\rho : s_1 \mathcal{R} s_2 \Leftrightarrow g_\gamma(s_1) = g_\gamma(s_2)$.

Considering a quotient class $\overline{s} \in \Sigma_\rho / \mathcal{R}$[4], we write $\overline{\mathfrak{e}_{\gamma,\overline{s}}} \stackrel{\Delta}{=} g_\alpha(s_0)$ where $s_0 \in \overline{s}$.

$\overline{E_\gamma} \stackrel{\Delta}{=} \left\{ \overline{\mathfrak{e}_{\gamma,\overline{s}}} : \overline{s} \in \Sigma_\rho / \mathcal{R} \right\}$ is the support set of the multiset $\{\{\mathfrak{e}_{\gamma,s} : s \in \Sigma_\rho\}\}$: $\overline{\mathfrak{e}_{\gamma,\overline{s}}} \in \overline{E_\gamma}$ is of multiplicity $w_\gamma\left(\overline{\mathfrak{e}_{\gamma,\overline{s}}}\right) = |\overline{s}|$ in this multiset.

It yields: $\{\{\mathfrak{e}_{\gamma,s} : s \in \Sigma_\rho\}\} = \left\{ \overline{\mathfrak{e}_{\gamma,\overline{s}}}^{w_\gamma\left(\overline{\mathfrak{e}_{\gamma,\overline{s}}}\right)} : \overline{s} \in \mathcal{S}_\rho / \mathcal{R} \right\}$.

Let $\tilde{g_\gamma} : \overline{s} \in \Sigma_\rho / \mathcal{R} \mapsto \mathfrak{e} \in \overline{E_\gamma}$, then $\tilde{g_\gamma}$ is bijective. $\tilde{g_\gamma}^{-1}$ allows to retrieve the class associated to a given hb-edge; hence the associated values of Σ_ρ to this class—which will be important for navigation. The references associated to $\mathfrak{e} \in \overline{E_\gamma}$ are $\bigcup_{s \in \tilde{g_\gamma}^{-1}(\mathfrak{e})} r_\rho(s)$. The **reduced visualisation weighted hb-graph** for the search $\mathcal{S}$ is defined as $\mathfrak{H}_{\gamma/\rho, w_\gamma, \mathcal{S}} \stackrel{\Delta}{=} \left(\bigcup_{r \in \mathcal{S}} \mathfrak{A}^\star_{\gamma,r}, \overline{E_\gamma}, w_\gamma \right)$.

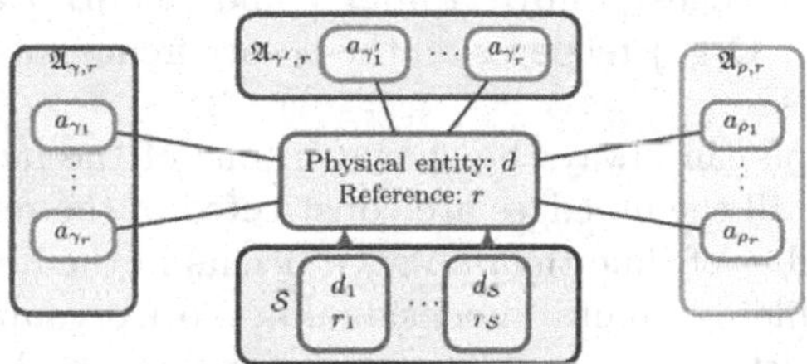

Fig. 3. Navigating between facets of the information space

Using the support hypergraph of the visualisation hb-graphs retrieves the results given in the case of homogeneous co-occurrences in [4]: hence [4] appears as a particular case of the new hb-graph framework.

3.3 Navigability Through Facets

As for a given search $\mathcal{S}$ and a given reference ρ, the sets Σ_ρ and $R_s, s \in \Sigma_\rho$ are fixed, the navigability can be ensured between the different facets. We consider a group of types γ, its visualisation hb-graph $\mathfrak{H}_{\gamma/\rho, w_\gamma}$ and a subset A of the vertex set of $\mathfrak{H}_{\gamma/\rho, w_\gamma}$. We target another group of types γ' of heterogeneous co-occurrences referring to ρ for visualisation. Figure 3 illustrates the navigation.

We suppose that the user selects elements of A as vertices of interest from which s·he wants to switch facet. Hb-edges of $\overline{E_\gamma}$ which contain at least one element of A are gathered in $\overline{E_\gamma}\big|_A \stackrel{\Delta}{=} \left\{ \mathfrak{e} : \mathfrak{e} \in \overline{E_\gamma} \wedge (\exists x \in \mathfrak{e} : x \in A) \right\}$. Using the application $\tilde{g_\gamma}^{-1}$ we retrieve the corresponding class of references of type ρ associated to the elements of $\overline{E_\gamma}\big|_A$, to build the set of references $\overline{V}\big|_A$ of type ρ involved in the building of co-occurrences of type γ'. Each of the classes in $\overline{V}\big|_A$

[4] $\Sigma_\rho / \mathcal{R}$ is the quotient set of Σ_ρ by $\mathcal{R}$.

contains instances of type ρ that are gathered in a set $\mathcal{V}_{\rho,A}$. Each element of $\mathcal{V}_{\rho,A}$ is linked to a set of physical references by r_ρ. Hence we obtain the physical reference set involving elements of A: $\mathcal{S}_A \triangleq \bigcup_{s \in \mathcal{V}_{\rho,A}} R_s$.

The raw visualisation hb-graph $\mathfrak{H}_{\gamma'/\rho}\big|_A \triangleq \left(\bigcup_{r \in \mathcal{S}_A} \mathfrak{A}^\star_{\gamma',r}, (\mathfrak{e}_{\gamma',s})_{s \in \mathcal{V}_{\rho,A}} \right)$ in the targeted facet is now enhanced using $\mathcal{S}_A$ as search set instead of set $\mathcal{S}$. To obtain the reduced weighted version we use the same approach as above. The multiset of co-occurrences retrieved includes all occurrences that have co-occurred with the references attached to one of the elements of A selected in the first facet. Of course if $A = A_{\gamma,\mathcal{S}}$ the reduced visualisation hb-graph contains all the instances of type γ' attached to physical entities of the search $\mathcal{S}$.

In Fig. 2.b(ii), with $A = \{\text{Org1}\}$, allows to retrieve two hb-edges: computer vision—attached to PubB and PubC—and scene reconstruction—PubB. Hence: $S_A = \{\text{PubB}, \text{PubC}\}$. Switching to the Publication facet and keeping as reference keywords, two hb-edges $\{\text{PubB}^1, \text{PubC}^1\}$ and $\{\text{PubB}^1\}$ are retrieved. The same with $A = \{\text{Org1}, \text{Org2}\}$ retrieves all the co-occurences of Publications with reference to keywords.

The reference type can always be shown in one of the faces as a visualisation hb-graph where all the hb-edges are constituted of the reference itself with multiplicity the number of time the reference occurs in the hb-graph.

Ultimately, by building a multi-dimensional network organized around groups of types, one can retrieve very valuable information from combined data sources. This process can be extended to any number of data sources as long as they share one or more types. Otherwise the reachability hypergraph is not connected and only separated navigations are possible.

3.4 The Case of Multiple References

Extending co-occurrences to multiple references chosen in $e_R \in E_R$ is not straightforward. There are two ways of doing so: a disjunctive and a conjunctive way. We consider the set $R \subset e_R$ of references and $e_N = e_R \backslash R$ the visualisation types.

In the disjunctive way, each co-occurrence is built using the same approach than before considering successively each type $\rho \in R$. This is particularly adapted for types that are partitioning the physical references. It is the case for instance in the aggregation of two databases on two different kind of physical data, such as publication and patent, and the co-occurrence of the chosen navigation type is built referring in this case either to a publication or (non-exclusive) to a patent. The hb-graphs obtained are built by extending the family of hb-edges.

In the conjunctive approach, we start by building the cross product of instances of the references and retrieve co-occurrences of elements for which the data d is attached to the corresponding values of cross-reference instances. Hence co-occurrences are restricted to the simultaneous presence of reference instances attached to the physical entity.

3.5 The DataHbEdron[5]

The DataHbEdron provides soft navigation between the different facets of the information space. Each facet of the information space corresponding to a visualisation type includes a visualisation hb-graph viewed in its 2D extra-node representation with a normalised thickness on hb-edges [2]. The different facets are embedded in a 2.5D representation called the DataHbEdron. The DataHbEdron can be toggled between a cube with six faces—Figure 4—and a carousel shape with n faces—not shown here due to the lack of space—to ease navigation between facets. The reference face shows a traditional verbatim list of references corresponding to the search output.

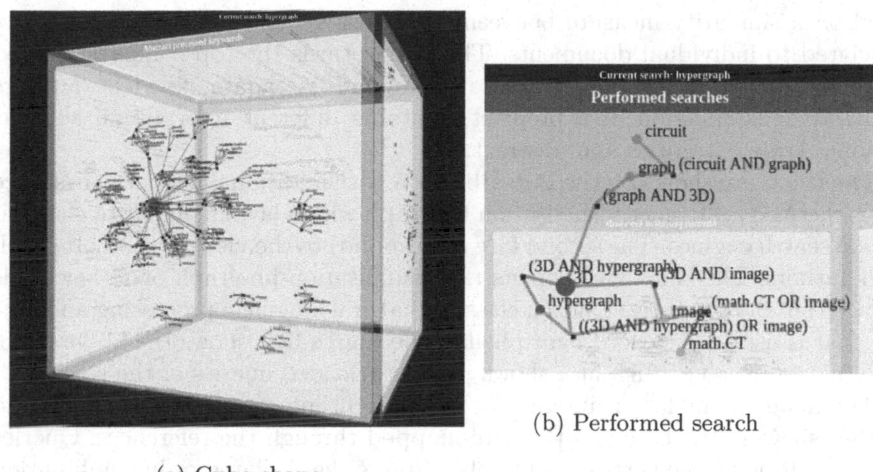

(a) Cube shape

(b) Performed search

Fig. 4. DataHbEdron: cube shape.

Individual faces of the DataHbEdron show different facets of the information space: the underlying visualisation hb-graphs support the navigability through facets. Hb-edges can be selected interactively between the different facets; since each hb-edge is linked to a subset of the references, the corresponding references can be used to highlight information in the different facets as well as in the face containing the reference visualisation hb-graph.

4 Results, Evaluation and Conclusion

4.1 Use Case

We applied this framework to perform searches and visual queries on the Arxiv database allowing simultaneous visualisation of the different facets of the information space constituted by authors, extracted keywords and subject categories.

[5] A video demo is available on: https://www.infos-informatique.net.

The tool developed is now part of the Collaboration Spotting family[6]. When performing a search, the standard Arxiv API[7] is used to query the Arxiv database. The queries can be formulated either by a text entry or done interactively directly using the visualisation: queries include single words or multiple words, with possible Boolean query operators—AND, OR and NOT—and parenthesis groupings. The querying history is stored and presented as an interactive hb-graph to visualize the construction of complex queries including refinement of the queries already performed. Each time a new query is formulated, the corresponding metadata is retrieved by the Arxiv API.

When performing a search on Arxiv, the query is transformed into a vector of words. Arxiv relies on Lucene's built-in Vector Space Model of information retrieval and the Boolean model. The most relevant documents are retrieved based on a similarity measure between the query vector and the word vectors associated to individual documents. The API returns the top n highest scored document metadata associated to the document. Metadata, filled by authors during their submission of a preprint, contains different information such as authors, Arxiv categories and abstract.

The facets are shown on the DataHbEdron with additional faces: the first face shows the Arxiv reference visualisation hb-graph with a layout similar to classical textual search engines. The second face corresponds to the visualisation hb-graph of co-authors. The third face depicts the visualisation hb-graph of co-keywords extracted from the abstracts using classical natural language processing and TF-IDF that is used as keyword multiplicity. The fourth face shows the hb-graph of Arxiv categories. The fifth face shows past or reloaded queries of the session.

Any node on any face is interactive to highlight information from one face to another showing the hb-edges that are mapped through the references. Queries can be built using the vertices of the hb-graph, either isolated or in combination with the current search using AND, OR and NOT. The first query is the only one required to be typed in. Merging queries of different users is immediate as they correspond to hb-edges of a hb-graph. Queries are evolving, gathered, stored and re-executable months later. The surfer has the possibility to display additional contextual information related to authors using DBLP, to keywords using DuckDuckGo for disambiguation and Wikipedia.

4.2 Evaluation

The validity of our framework is asserted by the mathematical construction completeness and robustness: we have achieved the possibility to navigate inside the dataset by showing co-occurrences in a sufficient refined way to support all the information extracted. As this model has been instantiated through a user interface in the use case of Arxiv, but, also, as mentioned previously, on some other sample data using csv files, its versatility is ensured. We have gathered in Table 2 some of the non-exhaustive features that allows to compare our solution

[6] http://collspotting.web.cern.ch/.

[7] https://arxiv.org/help/api/index.

Table 2. Elements of comparison (see text for details)

	Verbatim browser	PivotPath [5]	PivotSlice [6]	CS core [7]	DataEdron cube [4]	DataHbEdron
output	linear	tripartite graph	graph	graph	linear & hypergraph	linear & hb-graph
#facets	1	3	many	many	4	many
view per facet	no	no	no	yes	yes	yes
simultaneous facet views	no	yes	yes	no	yes	yes
heterogeneous co-occurrences	x	no	no	yes	no	yes
multiple references	x	no	no	disjunctive	no	conjunctive, disjunctive
zoom in data	new query	no	yes	yes	no	yes
filter data	new query	no	yes	yes	no	by visual queries
visual query	no	no	yes, restricted to current search	yes, restricted to current search	no	yes, even with new search
redundancy in co-occurrences	x	no		no	no	yes
information extraction	limited	pivot change	elaborated questions	elaborated questions	elaborated questions	elaborated questions
combination of facets	no	no	yes	yes	yes	yes
type of ranking	binary cosine similarity	no		number of references per vertex	hyperedges and vertices	hb-edges and vertices

with others. The user interface uses a 2.5D approach, but it is out of the scope of this article to make any claim on the quality of the interactions a user can have with such an interface.

5 Future Work and Conclusion

The framework supports dataset visual queries, possibly contextual, that either result from searches on related subjects or refine the current search: it enables full navigability of the information space. It provides powerful insights into datasets using simultaneous facet visualisation of the information space constructed from the query results. This framework is versatile enough to enhance user insight into many other datasets, particularly textual and multimedia ones.

References

1. Ranganathan, S.R.: Elements of Library Classification. Asia Publishing House, Mumbai (1962)
2. Ouvrard, X., Le Goff, J.-M., Marchand-Maillet, S.: Adjacency and tensor representation in general hypergraphs. part 2: multisets, hb-graphs and related e-adjacency tensors. arXiv preprint arXiv:1805.11952 (2018)

3. Ouvrard, X., Le Goff, J.-M., Marchand-Maillet, S.: Diffusion by exchanges in hb-graphs: highlighting complex relationships extended version. arXiv:1809.00190v2 (2019)
4. Ouvrard, X., Le Goff, J., Marchand-Maillet, S.: Hypergraph modeling and visualisation of complex co-occurence networks. Electron. Notes Discrete Math. **70**, 65–70 (2018)
5. Dörk, M., Riche, N.H., Ramos, G., Dumais, S.: PivotPaths: strolling through faceted information spaces. IEEE Trans. Vis. Comput. Graph. **18**(12), 2709–2718 (2012)
6. Zhao, J., Collins, C., Chevalier, F., Balakrishnan, R.: Interactive exploration of implicit and explicit relations in faceted datasets. IEEE Trans. Vis. Comput. Graph. **19**(12), 2080–2089 (2013)
7. Agocs, A., Dardanis, D., Le Goff, J.-M., Proios, D.: Interactive graph query language for multidimensional data in collaboration spotting visual analytics framework. ArXiv e-prints, December 2017
8. Indyk, P., Motwani, R.: Approximate nearest neighbors: towards removing the curse of dimensionality. In: Proceedings of the Thirtieth Annual ACM Symposium on Theory of Computing, pp. 604–613. ACM (1998)

Towards the Named Entity Recognition Methods in Biomedical Field

Anna Śniegula[1], Aneta Poniszewska-Marańda[2(✉)], and Łukasz Chomątek[2]

[1] Department of Informatics in Economy, University of Lodz, Lodz, Poland
anna.sniegula@uni.lodz.pl

[2] Institute of Information Technology, Lodz University of Technology, Lodz, Poland
{aneta.poniszewska-maranda,lukasz.chomatek}@p.lodz.pl

Abstract. Natural Language Processing (NLP) is very important in modern data processing taking into consideration different sources, forms and purpose of data as well as information in different areas our industry, administration, public and private life. Our studies concern Natural Language Processing techniques in biomedical field. The increasing volume of information stored in medical health record databases both in natural language and in structured forms is creating increasing challenges for information retrieval (IR) technologies. The paper presents the comparison study of chosen Named Entity Recognition techniques for biomedical field.

Keywords: Machine learning · Natural Language Processing · Recurrent neural networks · Named Entity Recognition · Conditional Random Fields · Long-Short Term Memory · Genia corpus

1 Introduction

Natural Language Processing (NLP) is a very important branch of Artificial Intelligence (AI) enhancing the performance of data processing and information extraction from unstructured datasets. AI and Machine Learning play an increasingly important role in medicine – successful integration of AI solutions can improve its efficiency and decrease the costs. AI is widely used to support developing drugs, gene editing, for personalised treatment, disease diagnosis and clinical decision support. More ambition systems involve the combination of multiple sources. The increasing volume of information stored in medical health record databases both in natural language and in structured forms is creating increasing challenges for information retrieval (IR) technologies.

Clinical data are often stored in natural language form. Narrative language is convenient for doctors to express events and medical concepts, unfortunately it makes the data difficult for searching, summarisation, decision support or statistical analysis. In order to perform above tasks the information has to be extracted with various natural language processing (NLP) techniques [10].

A. Chatzigeorgiou et al. (Eds.): SOFSEM 2020, LNCS 12011, pp. 375–387, 2020.
https://doi.org/10.1007/978-3-030-38919-2_31

Our studies concern NLP techniques in biomedical field. Named Entity Recognition (NER) is a fundamental NLP task to extract entities of interest (e.g., disease names, medication names and lab tests) from clinical narratives. The study aims to compare most commonly used Named Entity Recognition techniques and investigate how well they can perform the task of identifying a large number of classes that are strongly related with each other.

The paper is structured as follows. Section 2 presents the current state of art in the field of Named Entity Recognition in biomedicine. Section 3 gives a theoretical description of natural language processing techniques (focusing on NER task). Section 4 describes the research methodology, conducted comparative tests and applied evaluation techniques while Sect. 5 deals with obtained results and conclusions.

2 Related Works of Named Entity Recognition in Biomedicine

Named Entity Recognition (NER) is a subfield of information extraction [10]. NER is a task of recognizing words or phrases that should be categorised as expressions denoting entities. Example entity names in medical field are diseases, drugs, treatment, genes, cancer, protein and RNA [1,5,6,14,16].

Recent research in biomedical informatics is focused on named entity recognition. NER can be solved with the use of many techniques that can be divided into several groups [2]: *dictionary based approach*, *rule based approach*, *machine learning (ML) approach*, *deep learning approach* and *hybrid approach.* In the next section they are described in greater detail.

Recent papers concentrate on *deep learning approach* applying *recurrent neural networks (RNNs)* such as *Long-Short Term Memory (LSTM)* [12] or *Gated Recurrent Units (GRU)*. Common trend is to combine *deep learning* with other *ML* method on top of the recurrent layers. *CRF* is most commonly used method in this hybrid approach. It ensures that the optimal sequence of tags over the entire sentence is obtained. The authors of [13] combined *Residual Dilated Convolutional Neural Network* with *CRF*. *Dictionary-based techniques* are also still widely used in biomedical field, mostly as a support of some more complex machine learning methods. Most of the works use UMLS Metathesaurus or Genia corpus as knowledge database.

Clinical NER attempts receive lower performance measures values (best F_1 score equals 91.32, it was obtained by [13]) in comparison to similar trials with corpuses in non-technical fields, where recently [3] obtained 93.5 F_1 score on the CoNLL 2003 corpus. The researchers obtained lower results in Biomedical NER because biomedical texts involve multiple challenges that are listed below. Firstly, the data available for researchers in the biomedical field are limited, mostly due to the patient privacy and confidentiality requirements. The available annotated databases are usually insufficient for NER task [19]. Secondly, the medical texts are written in a specific manner different from ordinary language, often written shorthand, contain incomplete sentences, informal grammar and littered with misspellings and non-standard abbreviations and acronyms. Moreover medical language is characterised by long phrases containing special characters and dashes. Furthermore, the medicine is a rapidly expanding field, large number of researches conducted contribute to the constantly growing number of medical concepts. That makes extremely difficult to keep the medical dictionaries up to date. What is more, concepts in medicine often carry ambiguous meaning, it implies the NER models to keep the word context information along the training process [11,17].

3 Natural Language Processing

The term *Natural language* is used to describe any language used by human beings, to distinguish it from "artificial" languages used by computer, for instance programming languages and data representation languages [10]. Natural language processing (NLP) term describes computational techniques that process spoken and written human language [9].

Natural Language Processing (NLP) include data preprocessing techniques like data cleaning, tokenization, normalization (stemming, lemmazation or other form standardization). Preparing text requires choosing the optimal tools, however it helps to improve accuracy of proceeding NLP tasks.

Other tasks of NLP concentrate on extracting statistical features like term frequency (TF), inverse document frequency (IDF) or syntactical features including Part Of Speech (POS) tagging. NLP techniques are tools to achieve superior task. Among the most applied tasks is *Information Extraction (IE)* involving searching for relevant information in documents.

Named Entity Recognition. *Named entity recognition (NER)* is a stage of IE. It is one of the key NLP tasks that helps to convert unstructured text into computer readable structured data [17]. NER refers to the task of recognizing expressions denoting entities (i.e. Named Entities) such as diseases, drugs or people's names in documents containing natural language texts. NER can be solved with the use of many techniques that can be divided into several groups [2]: dictionary based approach, rule based approach, statistical approach, deep learning approach, hybrid approach.

Dictionary based methods store the dictionaries of NER phrases on lists called *gazetteers*. This simple approach has some restrictions connected with rapidly expanding medical knowledge base – it is hard to keep the gazetteers up to date. *Rule based approach* is based on a set of manually defined rules and patterns that can describe the whole sentences. Therefore the context of the phrase is available during NER process. This approach is very time consuming, it requires defining all the rules separately by the user.

Machine learning approach involves mostly supervised or semi-supervised techniques. Standard classification methods are applicable, however sequence-based methods that use the whole sentences as sequences of words instead of sets of single words, are more widely used. *Deep learning* is a sub-domain of machine learning that uses neural network architectures, in NER especially valuable are architectures that capture long-term dependencies and operate on data sequences.

Hybrid methods combine multiple approaches, e.g. dictionary based technique with statistical method. The advantage of this approach is that it is versatile, as the user can easily improve its performance with no retraining [15]. Other popular approach combines statistical and deep learning methods, – e.g. LSTM, that is capable of capturing long distance dependency relations among words and entities, and CRF in the output layer, that guarantees optimal sequence labelling [12]. The presented comparison studies sequential ML technique (CRF) and recurrent neural network (LSTM) as most popular machine learning NER techniques that analyse sequences of words.

Conditional Random Fields. Conditional Random Fields (CRF) is designed for predicting an output sequence of tags corresponding to a sequence input. It is widely applied in natural language processing, computer vision and bioinformatics [18]. It is a discriminative statistical machine learning model that relies on decision boundary between classes. CRF is based on conditional probability, that calculates the probability of Y when random variable X has specific value. Probability is calculated based on the output values of the feature functions. Probability is normalized with a constant value Z so that its distributions summed to 1. The general formula for CRF can be stated as follows 1:

$$P(X \mid Y) = \frac{1}{Z(X)} exp \sum_{i=0}^{N-1} \sum_{k=1}^{K} \lambda_k f_k(X, i, y_i, y_{i-1}) \tag{1}$$

where N is a number of words in a Vector; i is current word position in a vector; K is the number of feature functions; λ are weights calculated for each function during the training process.

The output label is the one that achieves the maximum probability.

LSTM – Recurrent Neural Network. *Long short term memory network (LSTM)* is a type of recurrent neural network (RNN) algorithm. RNN can store historical information in a hidden layer. LSTM was designed specifically to keep information from longer period than traditional RNN. They advantage of gate units that protect the stored memory contents from perturbation by irrelevant inputs [7]. Hidden layer consists of memory cell that allows to keep the cell state and transfers information along the sequence chain. Gates (input gate, forget gate and output gate) control which information is saved and which should be forgotten [12]. The variation Bi-LSTM consists of forward-LSTM and backward-LSTM that are stacked on top of each other [12]. The extension makes it possible to process each sequence in two directions in separate recurrent nets that are connected to the same output layer. This combination ensures that at any point of the sequence the algorithm keeps the information both about the previous and the future elements. That is why it can understand the context better than the standard unidirectional LSTM.

4 Research Methodology and Implementation of Selected Methods

Dataset Characteristics. We performed NER task on GENIA corpus. Genia is commonly used corpus by researchers both as dictionary and as base corpus to perform NER task, it is available in different versions and different formats. We used version 3.0.2 that consists of 1999 abstract records from MEDLINE database and is a taxonomy of 34 biologically relevant categories. We reduced the number of classes to 32. The entity *RNA_Substructure* was omitted during the tests, because it occurred in only one abstract. Also the entities labelled *other_name* was omitted as the label scope is too wide.

Table 1 presents all entities in the corpus, the number of abstracts in which they are present, number of their occurrences and the number of unique phrases. The most common category is protein molecule that appears over 20000 times in the dataset. The entities occurrences are unevenly distributed among classes. Almost 90% of any entity occurrences belong to one third of the classes. 20 classes appear less than a 1000 times and 4 classes appear less than 100 times in the corpus.

Original files are in XML based mark-up format – for NER task purpose it was transformed to BIO format with words tagged with "*O*" do not belong to any entity or *B_entity_name* (B stands for beginning) and *I_entity_name* (I stands for inside) to indicate the first and subsequent words belonging to the entities.

The dataset was split into train and test subsets and created three train-test pairs, the first train – test pair was created with an alphabetical split, the second split was random, both datasets divided the abstracts in the proportion of 70/30. Random training subset contains 46969 term occurrences and alphabetically split training subset contains 47251 term occurrences. The last dataset was created in a random split (because with the randomly split subsets we achieved better results), however, this time the training subset was bigger (it contained 85% of all abstracts). Table 2 presents complete entity terms distribution among all the subsets. The last column presents how many times more each entity occurred in the 85/15 split train subset in comparison to the 70/30 split train subset.

Implementation of Selected Techniques. The purpose of the research was to compare most commonly used NER techniques (CRF and LSTM) – to check how well these methods can detect large number (34) of Named Entities with the uneven frequency distribution. Moreover, to check if integrating CRF with the information from UMLS MetaThesaurus can increase the general performance of NER. The total number of performed tests was 8.

During the comparison we wanted to check if the task can be done with the use of existing tool with minimal user effort. There are many open source libraries available in different programming languages. Most of them are for general NER extraction purpose for extracting traditional entities like "Person", "Location", "Organisation". One of the most appreciated library is *Stanford Named Entity*

Table 1. Genia Corpus 3.0.2 entities and their distribution among abstracts

Class name	No of genia abstracts	No of entity occurrence	No of unique phrases
protein_molecule	1774	20855	4005
other_name	1979	13132	6658
protein_family_or_group	1754	7665	2691
cell_type	1637	6844	1867
DNA_domain_or_region	1145	6368	3245
other_organic_compound	738	3938	1179
cell_line	1091	3472	1812
lipid	398	2345	378
protein_complex	674	2167	546
virus	398	2065	366
multi_cell	496	1660	452
DNA_family_or_group	709	1341	745
protein_domain_or_region	342	889	583
protein_subunit	251	817	295
amino_acid_monomer	227	765	179
tissue	290	656	366
cell_component	331	622	207
peptide	146	492	249
body_part	195	432	191
DNA_molecule	278	413	277
atom	115	331	65
inorganic	71	250	64
polynucleotide	150	242	175
RNA_molecule	280	241	134
nucleotide	86	236	59
RNA_family_or_group	185	236	88
mono_cell	76	221	89
other_artificial_source	85	167	94
protein_substructure	73	122	85
DNA_substructure	73	99	79
carbohydrate	21	92	44
protein_N/A	77	86	61
DNA_N/A	36	47	34
RNA_substructure	1	2	2

Recognizer based on CRF algorithm – it is written in JAVA but there are plug-ins available in multiple languages (Python, .NET/F#/C#, PHP, Ruby and more). Other well known NER tools is spaCy – open-source library implemented in Python. Research [8] shows that spaCy's NER tool performs second best among

Table 2. Dataset split description

Classname	AlphabetTrain (70/30 split)	AlphabetTest (70/30 split)	RandomTrain (70/30 split)	RandomTest (70/30 split)	RandomTrain (85/15 split)	RandomTest (85/15 split)	Train dataset increment in 85/15 split
amino_acid_monomer	604	161	573	192	648	117	75
atom	229	102	256	75	298	33	42
body_part	327	105	296	136	359	73	63
carbohydrate	84	8	73	19	85	7	12
cell_component	398	224	425	197	515	107	90
cell_line	2332	1140	2487	985	3041	431	554
cell_type	5113	1731	4753	2091	5852	992	1099
DNA_domain_or_region	4024	2344	4528	1840	5427	941	899
DNA_family_or_group	936	405	859	482	1130	211	271
DNA_molecule	243	170	277	136	361	52	84
DNA_N/A	23	24	31	16	34	13	3
DNA_substructure	38	61	71	28	85	14	14
inorganic	204	46	171	79	224	26	53
lipid	1353	992	1618	727	2017	328	399
mono_cell	182	39	143	78	184	37	41
multi_cell	1052	608	1149	511	1416	244	267
nucleotide	146	90	163	73	195	41	32
other_artificial_source	144	23	95	72	146	21	51
other_organic_compound	2720	1218	2721	1217	3342	596	621
peptide	405	87	346	146	411	81	65
polynucleotide	133	109	190	52	222	20	32
protein_complex	1929	238	1611	556	1894	273	283
protein_domain_or_region	645	244	591	298	763	126	172
protein_family_or_group	5528	2137	5414	2251	6517	1148	1103
protein_molecule	15822	5033	14903	5952	17046	2909	3043
protein_N/A	33	53	54	32	67	19	13
protein_substructure	71	51	92	30	109	13	17
protein_subunit	498	319	578	239	697	120	119
RNA_family_or_group	144	92	157	79	204	32	47
RNA_molecule	127	114	160	81	219	22	59
tissue	511	145	450	206	557	99	107
virus	1253	812	1461	604	1700	365	239
total	47273	18948	46720	19501	56710	9511	9990

four well-established open-source NER tools regarding accuracy (Stanford NER performance was slightly better) and that it is the fastest in processing speed, however there is no detailed information provided in its documentation which models are implemented in the background.

The purpose was to find a tool already allocated in medical domain. That is why Python library "CliNER" was chosen – an open-source natural language processing system for named entity recognition in clinical text of electronic health records [4]. The authors of the library report that they achieved 0.83% F_1 with the NER task with the data from i2b2/VA 2010 challenge. The library supports all the methods needed for our study, it provides two NER extraction techniques: CRF and LSTM. Moreover it has implemented basic UMLS integration. The library is oriented to find medical terms like TEST, PROBLEM and TREATMENT. The idea was to adapt the library slightly to recognize entities from Genia corpus.

First test was performed with the use of CRF-based classifier. CRF is calculated with the CRFSuite library. CliNER CRF classifier implements linguistic features such as word unigram, part of speech tag (generated with *nltk pos_tagger*), last two characters, word shape, previous and next features, previous and next 3 unigrams, regex or units.

In the second test the CRF was extended with the knowledge-based features based on the semantic types of the phrases obtained from the UMLS Metathesaurus database. Phrases belonging to Genia entities found their representatives in 119 UMLS different semantic types. Table 4 presents mapping statistics including how many UMLS semantic types were assigned to each Genia class representatives and the most frequent UMLS semantic type for each Genia class.

The third classification test was performed with the bidirectional LSTM. The CliNER library implements both character level and word level Bi-LSTM. Character sequence embeddings feed into word level LSTM. This approach is sensitive to misspellings [4]. LSTM algorithm is realised by Keras python deep learning library.

The purpose was to investigate how much the splitting order influence the quality of training the model, therefore, we repeated all the three tests described above twice, once with the alphabetically split dataset, once with the randomly split dataset. Finally, to estimate the potential of the models to improve when train subset increases, the tests on 85/15 randomly split dataset were performed. This time only one CRF model (UMLS-based) and LSTM model were chosen.

Performance Evaluation. To evaluate the performance of the methods confusion matrix data of each test case was collected. A *confusion matrix* is a standard way of displaying classification results to designate the efficiency of a classification model. Generally, columns correspond to the true classificatory state, while the rows correspond to the algorithm results. Confusion matrix values can be used to calculate many measures of classification performance – it was used accuracy (2), precision (3), recall (4) and F_1 score (5). These metrics were calculated according to the following formulas:

$$Accuracy = \frac{TP + TN}{TP + FP + TN + FN} \tag{2}$$

$$Precision = \frac{TP}{TP + FP} \tag{3}$$

$$Recall = \frac{TP}{TP + FN} \tag{4}$$

$$F_1 Score = 2 \cdot \frac{Precision \cdot Recall}{Precision + Recall} \tag{5}$$

where: *FP* is the number of false positive; *TN* is the number of true negative; *TP* are true positive; *FN* are false negatives.

The precision is used to evaluate the correct degree of prediction power of the model defining the proportion of the correct positive identifications. Recall represents the proportion of positive cases that were correctly identified F-measure calculates the harmonic mean of precision and recall resulting in achieving the balance between precision and recall.

The above matrices can be used to evaluate individual classes in multi-class classification. Quality of the overall classification is usually assessed in two ways: macro-averaging and micro-averaging. Macro-averaging is a mean value of each measure – as the result it treats all the classes equally. Micro-averaging performance measures are calculated with the above formulas, obtaining the cumulative sums of FP, TN, TP, FN. It favours classes with the bigger number of representatives.

5 Results and Discussion

We achieved better results with the randomly split dataset. CRF tests with random dataset achieved micro F_1 score over 4% better than with the alphabetically split dataset. For LSTM the improvement was even better, almost 7%.

The best approach (among 70/30 split subsets) appeared to be the combination of UMLS with CRF with which we achieved the micro F_1 score equal to 57.53%. The Table 3 presents micro values of accuracy, accuracy Non-O, precision, recall and F_1) calculated for each test we performed.

In the discussion we present mainly F_1 score values to compare the tests results, as it is the most balanced measure. As far as the other metrics are concerned, all the tests achieved better precision than recall. The difference between two measures ranged between 6–18%.

Macro measures are lower, because of the uneven entity occurrences and the very small recognition of the classes with small number of representatives. The entities with the lowest number of occurrences achieved 0 correctly classified phrases and it significantly lowers macro values. For example, for the test no. 4 (random CRF-UMLS) we achieved 57.53% micro F_1 score, while the macro F_1 score values 40.08%. 3 classes were not classified correctly at all.

Table 3. Evaluation of the tests – micro measures

Test no.	Split method	Train test split	Research method	Correctly classified	f1 general	Accuracy	Non 0	Precision	Recall
1	Alphabetical	70% - 30%	CRF	9117	52.29%	85.04%	52.47%	57.25%	48.12%
2	Random	70% - 30%	CRF	10281	56.72%	86.06%	55.62%	61.29%	52.79%
3	Alphabetical	70% - 30%	CRF UMLS	9370	53.21%	85.16%	53.06%	57.58%	49.45%
4	Random	70% - 30%	CRF UMLS	10479	57.53%	86.29%	56.26%	61.80%	53.80%
5	Alphabetical	70% - 30%	LSTM	5851	35.07%	79.67%	22.43%	40.56%	30.88%
6	Random	70% - 30%	LSTM	7272	42.02%	81.45%	26.86%	48.05%	37.34%
7	Random	85% - 15%	CRF UMLS	5309	58.91%	86.86%	58.21%	62.36%	55.82%
8	Random	85% - 15%	LSTM	3491	43.80%	81.50%	24.65%	54.28%	36.70%

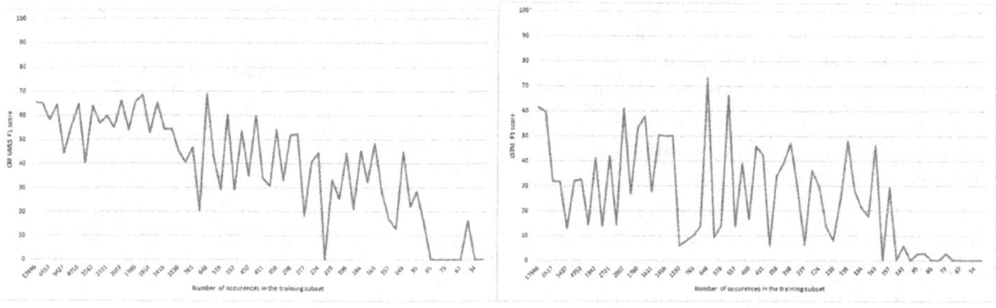

Fig. 1. The influence of the entity occurrence number on F1 score value; CRF_UMLS test and LSTM test

Figure 1 presents the relation between number of the entities occurrences in the training subset and the achieved F_1 score value. Left chart presents CRF_UMLS test results, right chart presents LSTM results. LSTM results are less stable – for neural-network method we probably have not enough data to observe regular and stable increase. On the other hand, the minimum number of the entity occurrence that the model needs for training to be able to recognise any entity correctly is similar for both tests (slightly in a favour of CRF). CRF needed minimum 85–95 occurrences, LSTM needed 145–150 occurrences. For entities with about 600 occurrences CRF model was able to achieve F_1 score value round 70%. We did not achieve any further progress with CRF algorithm. For LSTM it is difficult to draw conclusions, it also reached F_1 score value round 70% with about 600 entity occurrences. However, after that the F_1 value suddenly collapsed, we need further research with the larger dataset to check if we will be able to improve the F_1 score value.

Table 4. Genia to UMLS semantic type mapping

Genia class	Umls most frequent semantic type	No. of different UMLS types	UMLS influence on F1
RNA_molecule	Intellectual Product	16	10.15
body_part	Body Part, Organ, or Organ Component	32	9.05
mono_cell	Disease or Syndrome	14	6.86
other_organic_compound	Organic Chemical	60	6.64
atom	Element, Ion, or Isotope	14	4.71
protein_substructure	Spatial Concept	20	4.57
amino_acid_monomer	Biologically Active Substance	14	4.33
nucleotide	Nucleic Acid, Nucleoside, or Nucleotide	9	3.5
peptide	Biologically Active Substance	31	2.95
lipid	Organic Chemical	27	2.06
multi_cell	Finding	43	1.72
protein_domain_or_region	Spatial Concept	38	1.63
protein_N/A	Functional Concept	18	1.59
RNA_family_or_group	Nucleic Acid, Nucleoside, or Nucleotide	21	1.14
protein_subunit	Intellectual Product	24	1.02
cell_type	Cell	54	1
DNA_family_or_group	Qualitative Concept	46	0.64
protein_molecule	Enzyme	73	0.39
cell_component	Cell Component	29	0,17
protein_family_or_group	Enzyme	71	0,13
carbohydrate	Organic Chemical	10	0
DNA_N/A	Intellectual Product	15	0
DNA_substructure	Qualitative Concept	15	0
cell_line	Cell	54	−0,17
inorganic	Element, Ion, or Isotope	19	−0,33
virus	Virus	30	−0.52
protein_complex	Intellectual product	30	−0.58
polynucleotide	Nucleic Acid, Nucleoside, or Nucleotide	12	−1.01
DNA_domain_or_region	Intellectual Product	73	−1.17
tissue	Body Part, Organ or Organ Component	39	−1.65
other_artificial_source	Functional Concept	20	−2.68
DNA_molecule	Qualitative Concept	32	−3.55

During the UMLS Metathesaurus knowledge acquisition process we have classified words into 108 different UMLS semantic types. Knowledge base from UMLS Metathesaurus increased macro F_1 score by 0.75% for alphabetically split dataset and 1.64% for randomly split dataset. It is generally less then we have expected. Table 4 presents which UMLS semantic type was most commonly mapped for each Genia class and how many different UMLS types were associated with words annotated as Genia classes. The last column presents the F_1 difference between plain CRF test and CRF with the UMLS knowledge. Presented statistics are related to tests number 2 and 4 (with the randomly split dataset as it achieved better performance).

For the randomly split dataset the UMLS integration had positive effect for classification of 20 Genia classes, negative effect for 9 classes and it did not affect 3 classes. Three classes (with the lowest number of entities in the train set) achieved F_1 score equal to zero. The UMLS integration had negative or no effect on DNA related classes (four out of five). The most common UMLS types identified for them are qualitative concept and intellectual product.

We observed the most significant falloff (−3.55%) for class *DNA_molecule*. At the same time F_1 score for *RNA_molecule* was increased by 11.15%.

One of the probable reasons for dictionary classification difficulties is ambiguity of words, one word can be classified as different semantic types depending on a context that in dictionaries is unavailable. For instance a word "family" can be classified as *multi_cell* ("asymptomatic family members"), but it can also be *protein_family_or_group* (for instance phrase "chemokine family"), *other_organic_compound* ("NSAID family members"), or even *dna_or_region* ("Egr family binding element").

6 Conclusions

LSTM classification results were lower than CRF, although in the state of the art it usually outperforms CRF. Neural network methods generally require very large training datasets, significantly larger in comparison to other supervised machine learning techniques. This study showed that Genia corpus does not contain enough data yet to benefit from neural networks, at least when we want to train the model to find all the Genia semantic types. We were able to achieve best value of 42.02% micro F_1 score with LSTM (for the 70-30 split, for the test number 6 Table 3). LSTM method occurred to be the most vulnerable for the different data spiting, changing the split from the alphabetical to random increased the F_1 score by almost 7%, in comparison with CRF and CRF_UMLS the increase was 3.15% and 4.32% respectively.

Last two tests were performed with the bigger training subset (containing 85% of abstracts). To allow better comparison of the results, the training subset contains all the abstracts from the smaller training random subset. Rest 15% of the abstracts were moved randomly from the test random subset. We performed LSTM test and the CRF with UMLS knowledge (as it achieved a little better results than the plain CRF during the 70% – 30% test). Enlarging the training subset improved the F_1 score by almost 2% with the LSTM method and by 1.48% with the CRF_UMLS. The CRF_UMLS still outperformed LSTM. We achieved the best F_1 micro result 58.91% (with the test no. 7, Table 3).

The results were generally lower as expected. However, we still believe that it is possible to classify multiple number of entities and achieve satisfactory results with the use of the current state of art methods. To verify this thesis we would need a dataset with a much larger number of tagged entities. The Genia training subset that contained seven entities that had less than 100 representatives, 17 entities between 100–1000 representatives, and only 10 entities with over 1000 occurrences turned out to contain not enough tagged phrases for this challenging task. Other factor that makes the tagging difficult is that in Genia corpus some entities are related with each other (for instance DNA-related, RNA-related, cell-related). For these entities we were unable to tag phrases correctly. We believe that the general performance would be better if all the entities were disjunctive. In the future we plan to extend the experiment including other data corpuses.

References

1. Abacha, A.B., Zweigenbaum, P.: Medical entity recognition: a comparison of semantic and statistical methods. In: Proceedings of BioNLP 2011 Workshop, BioNLP 2011, pp. 56–64 (2011)
2. Allahyari, M., et al.: A Brief Survey of Text Mining: Classifiation, Clustering and Extraction Techniques (2017)
3. Baevski, A., Edunov, S., Liu, Y., Zettlemoyer, L., Auli, M.: Cloze-driven Pretraining of Self-attention Networks. http://arxiv.org/abs/1903.07785
4. Boag, W., Sergeeva, E., Kulshreshtha, S., Szolovits, P., Rumshisky, A., Naumann, T.: CliNER 2.0: Accessible and Accurate Clinical Concept Extraction. http://arxiv.org/abs/1803.02245
5. Finkel, J.R., Grenager, T., Manning, C.: Incorporating non-local information into information extraction systems by Gibbs sampling. In: Proceedings of the 43rd Annual Meeting on Association for Computational Linguistics, ACL 2005, pp. 363–370 (2005)
6. Hatzivassiloglou, V., Dubou, P.A., Rzhetsky, A.: Disambiguating proteins, genes, and RNA in text: a machine learning approach. Bioinformatics **17**(Suppl. 1), S97–S106 (2001). ISSN 1367-4803
7. Hochreiter, S., Schmidhuber, J.: Long short-term memory. Neural Comput. **9**(8), 1735–1780 (1997). ISSN 0899-7667
8. Jiang, R., Banchs, R.E., Li, H.: Evaluating and Combining Name Entity Recognition System, pp. 21–27. https://aclweb.org/anthology/papers/W/W16/W16-2703/
9. Jurafsky, D., Martin, J.H.: Speech and Language Processing, 2nd edin. Prentice Hall, Upper Saddle River (2009). ISBN 978-0-13-187321-6
10. Meystre, S.M., Savova, G.K., Kipper-Schuler, K.C., Hurdle, J.F.: Extracting information from textual documents in the electronic health record: a review of recent research. Yearb. Med. Inf. **17**, 128–144 (2008). ISSN 0943-4747
11. Pradhan, S., et al.: Evaluating the state of the art in disorder recognition and normalization of the clinical narrative. J. Am. Med. Inf. Assoc. **22**(1), 143–154 (2014). ISSN 1527-974X

12. Qin, Y., Zeng, Y.: Research of clinical named entity recognition based on Bi-LSTM-CRF. J. Shanghai Jiaotong Univ. (Sci.) **23**(3), 392–397 (2018)
13. Qiu, J., Wang, Q., Zhou, Y., Ruan, T., Gao, J.: Fast and accurate recognition of chinese clinical named entities with residual dilated convolutions. In: Proceedings of IEEE International Conference on Bioinformatics and Biomedicine (BIBM), pp. 935–942 (2018)
14. Quimbaya, A.P., et al.: Named entity recognition over electronic health records through a combined dictionary-based approach. Procedia Comput. Sci. **100**, 55–61 (2016)
15. Sasaki, Y., Tsuruoka, Y., McNaught, J., Ananiadou, S.: How to make the most of NE dictionaries in statistical NER. BMC Bioinform. **9**(11), S5 (2008). ISSN 1471-2105
16. Song, Y.-J., Jo, B.-C., Park, C.-Y., Kim, J.-D., Kim, Y.-S.: Comparison of named entity recognition methodologies in biomedical documents. BioMed. Eng. OnLine **17**(2), 158 (2018)
17. Sun, W., Cai, Z., Li, Y., Liu, F., Fang, S., Wang, G.: Data processing and text mining technologies on electronic medical records: a review. J. Healthc. Eng. **2018**, 4302425 (2018)
18. Sutton, C., McCallum, A.: An Introduction to Conditional Random Fields. arXiv:1011.4088 [stat], November 2010
19. Zhang, J., et al.: Category multi-representation: a unified solution for named entity recognition in clinical texts. In: Phung, D., Tseng, V.S., Webb, G.I., Ho, B., Ganji, M., Rashidi, L. (eds.) PAKDD 2018. LNCS (LNAI), vol. 10938, pp. 275–287. Springer, Cham (2018). https://doi.org/10.1007/978-3-319-93037-4_22

Vietnamese Punctuation Prediction Using Deep Neural Networks

Thuy Pham[1], Nhu Nguyen[1], Quang Pham[2], Han Cao[3], and Binh Nguyen[1,3,4(✉)]

[1] University of Science, Vietnam National University in Ho Chi Minh City, Ho Chi Minh City, Vietnam
ngtbinh@hcmus.edu.vn

[2] Singapore Management University, Singapore, Singapore

[3] Inspectorio Research Lab, Ho Chi Minh City, Vietnam

[4] AISIA Research Lab, Ho Chi Minh City, Vietnam

Abstract. Adding appropriate punctuation marks into text is an essential step in speech-to-text where such information is usually not available. While this has been extensively studied for English, there is no large-scale dataset and comprehensive study in the punctuation prediction problem for the Vietnamese language. In this paper, we collect two massive datasets and conduct a benchmark with both traditional methods and deep neural networks. We aim to publish both our data and all implementation codes to facilitate further research, not only in Vietnamese punctuation prediction but also in other related fields. Our project, including datasets and implementation details, is publicly available at https://github.com/BinhMisfit/vietnamese-punctuation-prediction.

Keywords: Punctuation prediction · BiLSTM · Conditional random field · Attention model

1 Introduction

Punctuation is a system of symbols indicating the structure of a sentence where one needs to slow down, notice, or express emotion. Punctuation marks are vital to understand and disambiguate the meaning of sentences. Most automatic speech recognition systems usually do not provide punctuation in their outputs. Therefore, it is essential to assign appropriate punctuation marks to transcribed text so that it can be understood correctly.

In literature, punctuation prediction has been extensively studied during the last two decades, especially in the English language. Beerferman et al. [3] propose a lightweight approach for constructing a punctuation annotation system by relying on a trigram language model and Viterbi algorithm. Huang and Zweig [6] model the punctuation annotation problem as a sequence tagging problem where each word is tagged with appropriate punctuation. Lu et al. [12] present a new punctuation prediction approach for transcribed conversational speech texts using the dynamic conditional random field model on both Chinese and English.

A. Chatzigeorgiou et al. (Eds.): SOFSEM 2020, LNCS 12011, pp. 388–400, 2020.
https://doi.org/10.1007/978-3-030-38919-2_32

Cuong et al. [14] propose efficient inference algorithms to capture long-range dependencies among punctuations using high-order semi-Markov conditional random fields. Peitz [16] formulate the punctuation prediction as machine translation instead of using a language model based punctuation prediction method. Zhang et al. [21] study a new technique in punctuation prediction for the stream of words in transcribed speech texts with excellent accuracy in both test datasets of IWSLT [15] and TDT4 [19]. Regarding neural network methods, Tilk et al. [20] introduce a two-stage recurrent neural network using LSTM units to predict suitable punctuation for automatic speech recognition systems. Ballesteros and Wanner [2] investigate a novel LSTM-based model for predicting punctuation marks into raw text material. Recently, Li et al. [9] introduce an efficient generative model for punctuation prediction without observing the underlying punctuation marks and reconstructing the tree's underlying punctuation. Regarding the Vietnamese language, there have been various works in different fields such as word segmentation [4,13] and Part-of-Speech (POS) tagging [17].

In this work, we aim at building a large-scale dataset and providing an extensive benchmark for predicting punctuation in the Vietnamese language. Notably, we collect over 40,000 articles from the Vietnamese news and novels to build two datasets with a total of over 900,000 sentences. Different from previous works of [14,18], which assume inputs are already segmented into sentences; we make a general assumption that inputs can contain several sentences without punctuation information. Therefore, we train our model on paragraphs, which is more realistic and challenging. To provide a comprehensive benchmark for this task, we consider both traditional methods using CRF [18] and deep neural networks. Generally, the punctuation distribution in the text is highly imbalanced: most words are followed by a space that makes training punctuation prediction systems even more difficult. To address this challenge, we propose to train deep neural networks with the *focal loss* [10], which can give more weights to rare classes. While the focal loss shows promising results with our experiments on the Vietnamese Novels dataset, the class imbalance nature of this task is still a challenging problem and becomes an important research direction. We strongly believe that different languages have divergent challenges to build an efficient punctuation prediction system. As a consequence, our work can be considered as an additional contribution to the problem for the Vietnamese language, where there is little publication using a deep learning approach.

Since training with paragraph requires a strong text representation and the model's ability to remember long-range dependencies, we argue that the traditional CRF based methods are not suitable for this setting. Mainly, each CRF model treats each word as a one-hot vector, thus does not exploit its rich semantic meaning. Moreover, CRFs, primarily linear CRFs, only consider the relationship among words in a small window, thus ignoring information from distant words, which can be potentially informative. To address the above limitations, we propose a deep LSTM with an attention model to predict punctuations from the text. Our model learns to represent words by embedding vectors to exploit their semantic relationship. LSTM [5] can model long term relationships in sequences, which is used as the base of our model to accumulate knowledge in the paragraph. However, LSTM may remember information from a too far distant, which may be noisy and hinder the overall performance. Therefore, we equip LSTM with an attention layer so that it can selectively choose which information in the past is useful for the current prediction.

2 Punctuation Prediction as Sequence Tagging

2.1 Problem Formulation

Similar to the previous works for English and Chinese [11,22], we model the punctuation prediction task as a sequence labeling problem. Remarkably, we label each word by its immediately following punctuation, where label O denotes a space. In this study, we aim at considering seven main types of punctuation marks in the Vietnamese language including the period (.), the comma (,), the colon (:), the semicolon (;), the question mark (?), the exclamation mark (!), and the space. By modeling punctuation prediction as a sequence tagging problem, conventional methods such as conditional random fields (CRF) and neural networks can be applied directly without any significant modification. In the simple case, we use the label O to indicate that a word is not followed by any punctuation. For example, one can consider the following sentence in the Vietnamese language[1].

> Biển tạo ra 1/2 lượng oxy con người hít thở, giúp lưu chuyển nhiệt quanh Trái Đất và hấp thụ một lượng lớn CO2.

> (The ocean produces a half of the amount of oxygen that humans can breathe, and help to circulate heat around the Earth and absorb large amounts of CO2.)

This paragraph can be labeled as follows.

biển/O tạo/O ra/O 1/2/O lượng/O oxy/O con/O người/O hít/O thở/Comma giúp/O lưu/O chuyển/O nhiệt/O quanh/O trái đất/O và/O hấp/O thụ/O một/O lượng/O lớn/O co2/Period

It is worth noting that all the words are in lower case since the word case information is usually not available for the punctuation prediction task. For instance, when the texts are transcribed from speeches, we do not have the case information for the words.

2.2 Punctuation Prediction with Conditional Random Field

By formulating the punctuation prediction as a sequence labeling problem, a simplified approach is employing Conditional Random Field (CRF) [8], which has been applied successfully in the literature [14,18]. As our work is closely related to [18], we consider CRF as a baseline and implement CRF with three feature templates, as suggested in [18].

[1] https://vnexpress.net/khoa-hoc/dai-duong-can-thiet-voi-su-song-tren-trai-dat-the-nao-3976195.html.

3 Neural Networks for Punctuation Prediction

3.1 Network Architectures

Semantic Representation of Syllables. In this section, we describe our proposed approach to obtain the semantic vector of each syllable in a sequence. First, we initialize two embedding matrices for syllable and character as $E_s \in \mathbb{R}^{d \times S}$ and $E_c \in \mathbb{R}^{d \times C}$, where S and C are respectively the numbers of syllables and characters in the vocabulary, and d is the embedding dimension. For simplicity, here we use the same embedding dimension d for both syllables and characters. Given a sequence of L syllables $x = \{x_1, \ldots, x_L\}$, each of which is represented as a one-hot vector, we calculate the sequence of syllable embedding as:

$$\begin{aligned} e^s_x =& \{e^s_{x_1}, \ldots, e^s_{x_L}\} \text{ satisfying that} \\ e^s_{x_i} =& E_s \cdot x_i, \end{aligned} \tag{1}$$

where $(\cdot)$ is the matrix-vector dot product and $e^s_{x_i} \in \mathbb{R}^d$. Each element of $e^s_{x_i}$ is a semantic representation of the syllable x_i. However, a common practice is that we usually map rare words into the same vector corresponding to an "out of vocabulary (OOV)" token, which may lose useful information and hinder the performance. Therefore, we propose to enhance the semantic vectors $e^s_{x_i}$ with the semantic information from the character constructing x_i. Without loss of generality, we assume that each syllable x_i is itself a sequence of N characters $x_i = \{c_1, \ldots, c_N\}$. Similarly, we can obtain the sequential character representation of x_i as $se^c_{x_i} \in \mathbb{R}^{d \times N}$

$$se^c_{x_i} = \{e^c_{c_1}, \ldots, e^c_{c_N}\} \text{ satisfying that} \tag{2}$$

$$e^c_{c_i} = E_c \cdot c_i \tag{3}$$

Since characters in a syllable have short-range dependencies, we can learn such dependencies in $e^c_{x_i}$ by applying a convolution layer defined as

$$c_j = f(W \otimes e^c_{c_j:c_{j+h-1}}), \tag{4}$$

where $W \in \mathbb{R}^{d \times h}$ is the convolution parameter with length h and $\otimes$ denotes the convolution operation. By applying the operations defined in Eq. (3) on $e^c_{x_i}$, we get the character dependences in $e^c_{x_i}$ as $c = [c_1, \ldots, c_{N-h+1}]$, where each c_i represents the relationship among h consecutive characters in x_i. To obtain a fixed representation of the semantic vector built from characters, we apply the max pooling over c to compute $e^c_{x_i} \in \mathbb{R}^d$, and then, we combine it with $e^s_{x_i}$ for achieving a syllable representation as follows:

$$e_{x_i} = e^s_{x_i} \oplus e^c_{x_i}, \tag{5}$$

where $\oplus$ is the vector element-wise summation.

Predicting Punctuation with Deep Neural Networks. To this end, we have the semantic representation $e = \{e_{x_1}, \ldots, e_{x_L}\}$ of the original sequence x. In the next step, we use a Bidirectional LSTM to read the sequence e from both ends and obtain a sequence of hidden states, each of which is a concatenation of each individual LSTM's hidden state: $h_i = [\overrightarrow{h_i}, \overleftarrow{h_i}], i = 1, \ldots, L$ and $h_i \in \mathbb{R}^{2h}$, where h is the hidden size of one component LSTM. Subsequently, the model predicts the distribution of punctuations over the syllable x_i as

$$\hat{y}_i = \text{softmax}(R \cdot h_i), \tag{6}$$

where $R \in \mathbb{R}^{|Y| \times 2h}$ is the parameter of the softmax layer and $|Y|$ denotes the total number of punctuations in the vocabulary. Given the true punctuation prediction y, we can compute a loss (e.g. cross-entropy) between y and $\hat{y}$ and backprop to update all the parameters: $E_s, E_c, W, LSTM$, and R end-to-end.

Also, we consider two improved models that can potentially capture more complex structures of the data. First, we enhance the fully connected layer with the attention mechanism [1]. It means the model can focus on particular syllables in the past while predicting the current punctuation mark, and we refer to this as the **BiLSTM + Attention** model.

Finally, we replace the softmax classification layer by a CRF layer, which is a traditional method for this task [8]; this model is denoted as **BiLSTM + CRF**. It is important to remark that BiLSTM + Attention can be regarded as an additional improvement over BiLSTM due to the attention mechanism. Similarly, we also consider BiLSTM + CRF as an improvement over CRF for the reason that CRF models can use learned features from BiLSTM instead of manually designed features, as mentioned in [18].

3.2 Training with Focal Loss

A standard training procedure is to randomly sample a mini-batch from the training data, train the model, and then repeat until the convergence happens. As long as we model the punctuation prediction as a tagging problem, a nature choice of the loss function is the cross entropy loss between the predicted punctuation and the true punctuation. However, one main drawback of the classical cross entropy loss is that it has the same penalty for both easy and difficult classes, which is problematic as a result of the distribution of punctuation marks in natural languages is highly imbalanced. To address this problem, we propose to use the *focal loss* [10] that can give more weights to rare classes in the data:

$$FL\left(p_t\right) = -\alpha_t (1 - p_t)^{\gamma} \log\left(p_t\right). \tag{7}$$

Equation (7) shows the formula of the focal loss, where α_t is the balance factor of class t and γ is the *focusing factor*. Focal loss has been successfully used in the object detection problem where the training dataset is highly imbalanced with the background class. Nonetheless, focal loss has not been applied in natural language processing to the best of our knowledge (Fig. 1).

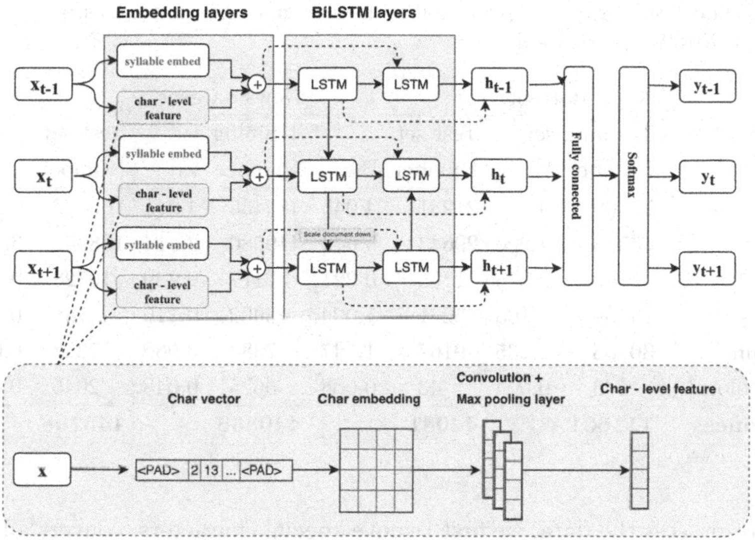

Fig. 1. A neural network architecture for the punctuation prediction problem.

4 Datasets for Vietnamese Punctuation Prediction

To investigate punctuation prediction for the Vietnamese language, we build two large-scale datasets from Vietnamese novels[2] and newspapers[3] with a total of over 900,000 sentences. Table 1 shows the different distribution of punctuation marks in these two datasets.

There are 734244 sentences in the Vietnamese Newspapers dataset, while the Vietnamese Novels dataset only has 183734. Although the top two punctuation marks having the most significant percentage of occurrence are comma and period in both datasets, the distribution of remaining ones is quite different. For instance, the appearance rate of the colon mark is 0.26% in Vietnamese newspapers, nearly three times bigger than the corresponding rate (0.092%) in Vietnamese novels. From Table 1, there exist much more (about 32 times) exclamative sentences in novels (1.894%) than newspapers (0.059%).

Similarly, we observe that authors prefer using interrogative sentences in Vietnamese novels (0.994%) rather than Vietnamese newspapers (0.113%). However, the occurrence rates of both colon and semicolon marks in newspapers are much larger than novels. These rates for both colon and semicolon marks in newspapers are 0.260% and 0.047%, respectively. Meanwhile, the corresponding values are 0.092% and 0.004% in novels. It turns out that Vietnamese novelists rarely use semicolon mark in their work. As a result, we decide not to merge two datasets owing to their inherently different sources, thus having different punctuation distributions. Therefore, it is worth seeing how proposed models perform on entirely different datasets.

[2] https://gacsach.com/tac-gia/nguyen-nhat-anh.html.
[3] https://baomoi.com.

Table 1. The distribution of punctuation marks in the training, testing sets from Vietnamese Novels and News dataset.

	Novel dataset				News dataset			
Punctuation	Training set		Test set		Training set		Test set	
	Number	%	Number	%	Number	%	Number	%
Comma (,)	50909	3.77	21231	4.045	482435	4.041	160472	4.054
Period (·)	66519	4.926	29643	5.648	419580	3.514	138967	3.51
Colon (:)	742	0.055	1153	0.221	32177	0.269	10728	0.271
Qmark (?)	14899	1.103	5271	1.004	13902	0.116	4468	0.113
Exclam (!)	30183	2.235	9167	1.747	7384	0.062	2333	0.059
Semicolon (;)	48	0.004	43	0.008	5675	0.048	2045	0.052
Sentences	**111601**		**44081**		**440866**		**145768**	

To pre-process the data, we first remove special characters, convert all words into lower cases, and standardize URLs, emails, and hashtags. Then, we remove sentences that do not contain any punctuation mark, do not end with a punctuation mark, or the ending punctuation is not a period, a question mark, or an exclamation mark. Different from previous works [14,18] assuming data are already segmented into sentences, here we do not make such assumptions and allow each model to work on arbitrary paragraphs of the text. Therefore, as most of the lengths of sentences on our datasets are smaller than or equal to 100, we decide to split the data into segments of length 100 and label them using the format as described in Sect. 2.1 and [14,18].

Finally, we divide the data into training, validation and testing sets with the ratio 60%–20%–20%. The distribution of punctuation marks among the training and testing sets for two datasets (Vietnamese Novels and Vietnamese Newspapers) can be found in Table 1.

5 Experiments

In this section, we present our experiments on two datasets described in Sect. 4. We consider both traditional CRF models as described in Sect. 2, and deep learning models (**BiLSTM, BiLSTM+Attention**, and **BiLSTM+CRF**) trained with both focal loss and normal cross-entropy loss. For deep learning models, we initialize the character embedding randomly and use Fasttext[4] as an initialization for the syllable embedding and syllables that are not in Fasttext have their embeddings initialized randomly. Both character and syllable embedding matrices are updated during the training process. All hyper-parameters such as the learning rate and focal loss hyperparameters are cross-validated from the validation set. We use the CRF++ toolkit[5] and implement other models with

[4] https://fasttext.cc/.
[5] https://taku910.github.io/crfpp/.

Tensorflow[6]. For deep learning models, we set the LSTM's hidden dimension to be 300 and train using Adam optimizer [7] for 30 epochs.

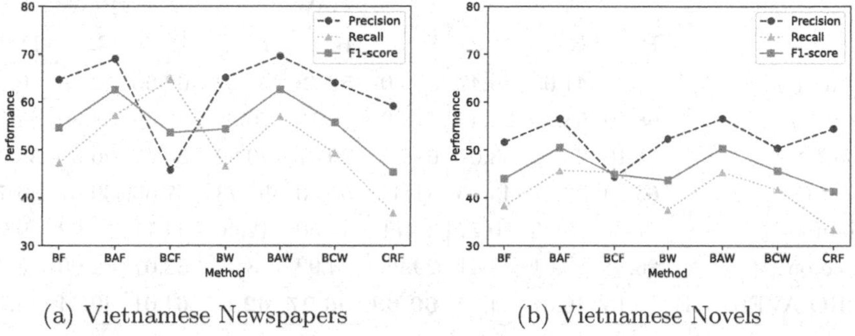

(a) Vietnamese Newspapers (b) Vietnamese Novels

Fig. 2. The performance comparison by micro precision, recall and F_1 score on the testing set (B: BiLSTM, A: Attention, C: CRF, F: trained with focal loss, W: trained without focal loss).

Table 2. Experimental results on the Vietnamese Newspapers dataset with focal loss (B: BiLSTM, A: Attention, C: CRF, F: trained with focal loss, W: trained without focal loss).

Punctuation	BF			BAF			BCF		
	P	**R**	**F**	**P**	**R**	**F**	**P**	**R**	**F**
Comma (,)	62.99	41.33	49.91	66.96	53.46	59.45	42.57	62.65	50.69
Period (·)	66.90	60.32	63.44	72.51	67.20	69.76	50.12	74.54	59.94
Colon (:)	59.49	21.71	31.81	58.59	32.00	41.39	54.25	24.29	33.56
Qmark (?)	58.86	33.68	42.85	61.12	49.40	54.64	47.75	42.82	45.15
Exclam (!)	34.51	4.20	7.49	43.03	5.96	10.47	34.88	4.20	7.50
Semicolon (;)	25.58	2.52	4.58	32.48	4.35	7.67	24.85	1.60	3.01
MICRO AVERAGE	64.67	47.29	54.63	**69.01**	**57.23**	**62.57**	45.79	64.73	53.64

Tables 2, 3, 6, and 7 show the performance of different deep learning based methods in terms of Precision (P), Recall (R), and F1-score (F) using cross-entropy loss or focal loss in different methods for datasets. Here, **B** stands for BiLSTM, **W** stands for the case not using focal loss, **A** stands for the Attention model, and **C** stands for the CRF model. Finally, the performance of CRF models are reported in Tables 4 to 5. Due to space constraints, we refer to [18] for details for the three templates.

As the data are highly imbalanced among punctuation marks, we opt to use micro averaged precision (P), recall (R), and F_1 score [18] to evaluate these

[6] https://www.tensorflow.org/.

Table 3. Experimental results on the Vietnamese Newspapers dataset without using focal loss (B: BiLSTM, A: Attention, C: CRF, F: trained with focal loss, W: trained without focal loss).

Punctuation	BW			BAW			BCW		
	P	**R**	**F**	**P**	**R**	**F**	**P**	**R**	**F**
Comma (,)	62.30	41.03	49.47	68.30	52.42	59.32	62.90	42.10	50.44
Period (·)	68.80	58.84	63.43	72.09	68.13	70.06	65.69	63.89	64.77
Colon (:)	58.10	23.12	33.07	61.54	29.87	40.22	56.62	26.35	35.96
Qmark (?)	63.10	33.48	43.75	61.01	51.30	55.73	57.66	39.32	46.76
Exclam (!)	37.27	5.96	10.27	35.71	7.50	12.40	44.71	5.62	9.98
Semicolon (;)	26.51	3.01	5.41	29.25	4.92	8.43	32.07	2.90	5.32
MICRO AVERAGE	65.13	46.61	54.34	**69.63**	**56.97**	**62.67**	64.01	49.34	55.72

Table 4. Experimental results on the Vietnamese Newspapers dataset using CRF models.

Punctuation	Template 1			Template 2			Template 3		
	P	**R**	**F**	**P**	**R**	**F**	**P**	**R**	**F**
Comma (,)	50.22	14.03	21.93	58.07	34.77	43.50	58.50	33.13	42.31
Period (·)	60.46	24.86	35.23	60.95	43.54	50.80	62.22	42.40	50.43
Colon (:)	47.02	8.68	14.65	53.01	17.00	25.75	52.86	16.31	24.93
Qmark (?)	46.91	11.37	18.30	55.43	19.32	28.65	54.92	19.47	28.75
Exclam (!)	29.84	3.90	6.90	32.58	4.93	8.56	38.49	5.23	9.21
Semicolon (;)	20.00	0.76	1.47	26.80	1.56	2.96	27.97	1.26	2.41
MICRO AVERAGE	55.05	17.80	26.90	**59.16**	**36.86**	**45.42**	59.96	35.49	44.59

models. Models' hyper-parameters are cross-validated on the validation set and we report the best setting on the test set. Figure 2 shows the results of various models we considered. First, we observe that deep learning methods outperform the traditional CRF model significantly on both datasets. Moreover, BiLSTM+Attention achieves the highest performance overall. Second, on the Vietnamese Novels dataset, we observe that, except BiLSTM+CRF, models trained with focal loss have a modest improvement over the traditional cross-entropy loss. However, on the Vietnamese Newspapers dataset, training with focal loss results in nearly identical performance. One possible reason is that it is much more difficult to perform hyper-parameter selection on the Vietnamese Newspapers dataset, which results in the non-optimal setting for focal loss. Overall, experimental results show that class imbalance is a challenging problem in punctuation prediction, and focal loss can become a promising strategy to alleviate this difficulty.

Table 5. Experimental results on the Vietnamese Novels dataset using CRFs.

Punctuation	Template 1			Template 2			Template 3		
	P	**R**	**F**	**P**	**R**	**F**	**P**	**R**	**F**
Comma (,)	42.47	15.92	23.16	51.69	25.92	34.53	52.66	35.26	42.23
Period (·)	44.94	21.25	28.86	52.77	34.72	41.88	51.78	26.22	34.81
Colon (:)	21.43	0.24	0.47	30.77	0.32	0.62	27.27	0.24	0.47
Qmark (?)	58.34	34.19	43.11	71.20	49.48	58.38	72.00	49.55	58.70
Exclam (!)	44.90	27.23	33.90	54.90	41.99	47.58	54.79	41.79	47.42
Semicolon (;)	0.00	0.00	0.00	0.00	0.00	0.00	0.00	0.00	0.00
MICRO AVERAGE	45.52	20.77	28.52	54.40	33.09	41.15	**54.39**	**33.40**	**41.38**

Table 6. Experimental results on the Vietnamese Novels dataset with focal loss (B: BiLSTM, A: Attention, C: CRF, F: trained with focal loss, W: trained without focal loss).

Punctuation	BF			BAF			BCF		
	P	**R**	**F**	**P**	**R**	**F**	**P**	**R**	**F**
Comma (,)	49.00	29.74	37.01	56.10	38.45	45.63	36.71	47.26	41.32
Period (·)	50.20	41.74	45.58	55.86	47.33	51.24	46.56	45.73	46.14
Colon (:)	50.00	0.24	0.47	21.43	0.95	1.81	0.00	0.00	0.00
Qmark (?)	69.56	55.45	61.71	70.34	65.60	67.89	60.90	67.41	63.99
Exclam (!)	51.48	45.79	48.47	52.09	54.30	53.18	59.47	33.52	42.88
Semicolon (;)	0.00	0.00	0.00	0.00	0.00	0.00	0.00	0.00	0.00
MICRO AVERAGE	51.64	38.35	44.02	**56.52**	**45.67**	**50.52**	44.37	45.43	44.89

Table 7. Experimental results on the Vietnamese Novels dataset without using focal loss (B: BiLSTM, A: Attention, C: CRF, F: trained with focal loss, W: trained without focal loss).

Punctuation	BW			BAW			BCW		
	P	**R**	**F**	**P**	**R**	**F**	**P**	**R**	**F**
Comma (,)	52.04	27.05	35.60	56.13	38.35	45.57	48.53	32.46	38.90
Period (·)	49.69	41.72	45.36	55.51	47.01	50.91	49.22	44.45	46.71
Colon (:)	26.67	0.32	0.62	66.67	0.63	1.25	14.71	0.39	0.77
Qmark (?)	69.05	56.69	62.26	71.67	64.69	68.00	68.39	61.09	64.54
Exclam (!)	51.81	45.63	48.53	52.20	53.66	52.92	47.86	52.03	49.86
Semicolon (;)	0.00	0.00	0.00	0.00	0.00	0.00	0.00	0.00	0.00
MICRO AVERAGE	52.30	37.49	43.67	**56.52**	**45.34**	**50.31**	50.34	41.71	45.62

Detailedly, for the best model using BiLSTM + Attention, from the Vietnamese Newspapers dataset, using the focal loss could achieve a slightly lower (about 0.1%) F1-score than without using it (62.57% vs. 62.67%). Meanwhile, for the Vietnamese Novels dataset, using the focal loss could obtain a slightly higher F1-score than without using the focal loss (50.52% vs. 50.31%). The experimental results in both datasets are a little bit different due to the difference between the distribution of punctuation marks in these two datasets, and especially, some punctuation marks rarely occur in Vietnamese novels rather than Vietnamese newspapers. In addition, we perform grid search on the pairs (α, γ) with α in $\{0.1, 0.25, 0.5, 0.75, 0.99\}$ and γ in $\{0.1, 0.5, 1.0, 2.0, 5.0\}$ for focal loss hyper-parameter selection. In future work, we plan to increase the grid size and tune these parameters carefully to achieve better performance. For BiLSTM-CRF, the focal loss is originally developed for *softmax* classifiers on the top of deep networks. It may be one reason explaining the performance drop observed in BiLSTM-CRF models.

Finally, regardless of the training loss, our results (Fig. 2) show that BiLSTM with Attention is the best among all the models considered. Furthermore, training with the focal loss can provide modest improvement to BiLSTM and BiLSTM with Attention.

6 Conclusion and Future Work

We have studied the punctuation prediction problem for the Vietnamese language. We collect two large-scale datasets and conduct extensive experiments with both traditional method (using CRF models) and a deep learning approach. We address the class imbalance problem in this task and show promising results using the focal loss on the Vietnamese Newspapers data.

In future work, we plan to use word embeddings and other techniques (ELMO, BERT, or word segmentation) for data pre-processing. Also, we do different experiments with more challenging datasets using Vietnamese speech/spoken- conversation transcripts. For instance, datasets from the IWSLT evaluation campaigns can be used to construct an efficient method for Vietnamese punctuation prediction. Another research direction is combining the punctuation prediction problem with other classical NLP tasks such as word segmentation or named entity recognition. For example, if one could correctly tokenize a paragraph into words and label each token with a named entity, the disambiguation level of this paragraph would reduce. It turns out that the punctuation prediction system would be easier to train. However, existing tokenizer and NER systems trained with punctuation information available is not the case in our problem. Therefore, directly applying a tokenizer might be a suboptimal solution. We strongly believe that learning these two tasks together will offer a better solution. Eventually, both the data and the implementation are publicly available at https://github.com/BinhMisfit/vietnamese-punctuation-prediction for further research.

Acknowledgement. We would like to thank The National Foundation for Science and Technology Development (NAFOSTED), University of Science, Inspectorio Research Lab, and AISIA Research Lab for supporting us throughout this paper.

References

1. Bahdanau, D., Cho, K., Bengio, Y.: Neural machine translation by jointly learning to align and translate. CoRR abs/1409.0473 (2015)
2. Ballesteros, M., Wanner, L.: A neural network architecture for multilingual punctuation generation. In: Proceedings of the 2016 Conference on Empirical Methods in Natural Language Processing, pp. 1048–1053. Association for Computational Linguistics, Austin, November 2016. https://doi.org/10.18653/v1/D16-1111. https://www.aclweb.org/anthology/D16-1111
3. Beeferman, D., Berger, A., Lafferty, J.: Cyberpunc: a lightweight punctuation annotation system for speech. In: Proceedings of the 1998 IEEE International Conference on Acoustics, Speech and Signal Processing, vol. 2, pp. 689–692, May 1998. https://doi.org/10.1109/ICASSP.1998.675358
4. Dien, D., Hoang, K., Toan, N.V.: Vietnamese word segmentation. In: NLPRS (2001)
5. Hochreiter, S., Schmidhuber, J.: Long short-term memory. Neural Comput. **9**(8), 1735–1780 (1997)
6. Huang, J., Zweig, G.: Maximum entropy model for punctuation annotation from speech, January 2002
7. Kingma, D.P., Ba, J.: Adam: a method for stochastic optimization. arXiv preprint arXiv:1412.6980 (2014)
8. Lafferty, J., McCallum, A., Pereira, F.C.: Conditional random fields: probabilistic models for segmenting and labeling sequence data. In: Proceedings of the 18th International Conference on Machine Learning 2001 (ICML 2001), pp. 282–289 (2001)
9. Li, X.L., Wang, D., Eisner, J.: A generative model for punctuation in dependency trees, pp. 357–373, July 2019
10. Lin, T.Y., Goyal, P., Girshick, R.B., He, K., Dollár, P.: Focal loss for dense object detection. In: 2017 IEEE International Conference on Computer Vision (ICCV), pp. 2999–3007 (2017)
11. Lu, W., Ng, H.T.: Better punctuation prediction with dynamic conditional random fields. In: Conference on Empirical Methods in Natural Language Processing (2010)
12. Lu, W., Tou Ng, H.: Better punctuation prediction with dynamic conditional random fields, pp. 177–186, January 2010
13. Nguyen, C.T., Nguyen, T.K., Phan, X.H., Nguyen, L.M., Ha, Q.T.: Vietnamese word segmentation with CRFs and SVMs: an investigation. In: PACLIC (2006)
14. Nguyen, V.C., Ye, N., Lee, W.S., Chieu, H.L.: Conditional random field with high-order dependencies for sequence labeling and segmentation. J. Mach. Learn. Res. **15**, 981–1009 (2014)
15. Paul, M.: Overview of the IWSLT 2009 evaluation campaign. In: International Workshop on Spoken Language Translation (IWSLT) 2009, pp. 1–18 (2009)
16. Peitz, S., Freitag, M., Mauser, A., Ney, H.: Modeling punctuation prediction as machine translation. In: IWSLT (2011)
17. Pham, D.D., Tran, G.B., Pham, S.B.: A hybrid approach to Vietnamese word segmentation using part of speech tags. In: 2009 International Conference on Knowledge and Systems Engineering, pp. 154–161 (2009)

18. Pham, Q.H., Nguyen, B.T., Cuong, N.V.: Punctuation prediction for Vietnamese texts using conditional random fields. In: ACML Workshop: Machine Learning and Its Applications in Vietnam, pp. 1–9 (2014)
19. Stephanie, S., Kong, J., Graff, D.: TDT4 multilingual text and annotations LDC2005T16 (2005)
20. Tilk, O., Alumae, T.: LSTM for punctuation restoration in speech transcripts. In: INTERSPEECH 2015, pp. 683–687 (2015)
21. Zhang, D., Wu, S., Yang, N., Li, M.: Punctuation prediction with transition-based parsing. In: ACL (2013)
22. Zhao, Y., Wang, C., Fu, G.: A CRF sequence labeling approach to Chinese punctuation prediction. In: Pacific Asia Conference on Language, Information and Computation (2012)

Foundations of Software Engineering – Regular Papers

A Light-Weight Tool for the Self-assessment of Security Compliance in Software Development – An Industry Case

Fabiola Moyón[1,2(✉)], Christoph Bayr[2], Daniel Mendez[3,4], Sebastian Dännart[5], and Kristian Beckers[1]

[1] Siemens CT Munich, Munich, Germany
{fabiola.moyon,kristian.beckers}@siemens.com
[2] Technical University of Munich, Munich, Germany
christoph.bayr@tum.de
[3] Blekinge Institute of Technology, Karlskrona, Sweden
daniel.mendez@bth.se
[4] fortiss GmbH, Munich, Germany
[5] INFODAS GmbH, Cologne, Germany
s.daennart@infodas.de

Abstract. Companies are often challenged to modify and improve their software development processes in order to make them compliant with security standards. The complexity of these processes renders it difficult for practitioners to validate and foresee the effort required for compliance assessments. Further, performing gap analyses when processes are not yet mature enough is costly and involving auditors in early stages is, in our experience, often inefficient. An easier and more productive approach is conducting a self-assessment. However, practitioners, in particular developers, quality engineers, and product owners face difficulties to identify security-relevant process artifacts as required by standards. They would benefit from a proper and light-weight tool to perform early compliance assessments of their processes w.r.t. security standards before entering an in-depth audit. In this paper, we report on our current effort at Siemens Corporate Technology to develop such a light-weight assessment tool to assess the security compliance of software development processes with the IEC 62443-4-1 standard, and we discuss first results from an interview-based evaluation.

Keywords: Security standards · Secure software engineering · Security assessment · Secure development process · Tool-support

1 Introduction

In context of software process improvement journeys, security is becoming more and more important. For some industries, achieving secure development is

A. Chatzigeorgiou et al. (Eds.): SOFSEM 2020, LNCS 12011, pp. 403–416, 2020.
https://doi.org/10.1007/978-3-030-38919-2_33

imposed by regulations, e.g. the IEC 62443-4-1 standard (the 4-1) [9] when developing for the industrial systems domain, health care or mobility. Such standards describe secure software development practices and artifacts and dictate a continuous software process improvement as part of the development life cycle [12].

The desired outcome of security compliance to such standards is two-fold: a more mature development organization which manages security risks and more trustable software products which target state-of-the-art security measures. To demonstrate compliance, companies evaluate gaps between the current development process and the chosen standard. Such evaluation is known as security compliance assessment (SCA).

Although security compliance is compulsory, we argue that in practice it is perceived as costly and with uncertain return of investment [21,23,24,26]. Software projects stakeholders need to plan and execute SCA while controlling development resource allocations.

In our experience – covering among all authors a total 14 years of practice conducting industrial audits – SCA is commonly performed by experts outside the development area. Either auditors conduct official assessments, as part of financial or quality audits, or security experts help project participants in preparing for a proper (external) compliance audit. SCA can be enforced by central departments or security officers as part of their policies, e.g. at Siemens [15], at Airbus [1], or at Microsoft [21]. In worst-case scenarios, when a software product was attacked, an SCA is executed as a post-mortem activity with the pressure of clients, authorities, and/or social media. Again, externals are involved to point out security deficiencies of the software development process, some of these issues even reaching the public (see, e.g., [17,27]).

Attempts for self-assessments by development practitioners are limited due to the complexity and ambiguity of security standards. Standards describe secure development with sets of process requirements, hence not only applicable to technology, but also to organizational structures.

In previous research, we proposed the Security-standard Compliant Assessment Model (S^2C-AM), an approach to perform SCA with the 4-1 standard as baseline [8]. In this paper, we extend this work by providing a light-weight tool that allows development practitioners to perform SCA themselves and to foresee the readiness of their processes for an external assessment. A self-assessment using our tool suits in a three-day workshop only with local resources. Moreover, in the context of formal audits, the results can be explicitly tracked to the 4-1 requirements delivering a common ground for auditors and project participants. Through a self-assessment, development stakeholders can not only obtain insights into the effort required for compliance but they can also get an overview in early stages based on previously defined measurements and threshold values. In principle, our hope is that practitioners benefit from such a systematic approach that reports in a light-weight manner and visually via a process modeling language where improvement opportunities can be found.

In the following, we describe the 4-1 standard and relevant previous work on SCA in Sect. 2. Section 3 describes a self-assessment and Sect. 4 presents the light-weight tool. In Sect. 5, we discuss a preliminary evaluation of the tool.

2 Fundamentals and Related Work

2.1 The IEC 62443-4-1 Security Standard

Standards provide established best practices collected from experts from different industries [11]. Security standards have been described as ambiguous and often complicated to read and understand [4]. Commonly accepted standards for secure development are the ISO 27034 [12] and the IEC 62443-4-1 [9].

The 4-1 standard concerns secure product development for industrial automation and control systems (IACS). It describes a secure development life-cycle with eight *practices*: Security Management, Specification of Security Requirements, Secure by Design, Secure Implementation, Management of security-related issues, Security update management, Security guidelines. Each practice consists of several *process requirements*. In previous work, we made the 4-1 more readable through process models. Such process models are basis for results presentation in the presented lightweight tool [22].

2.2 The Security Standard Compliance Assessment Model - S^2C-AM

In previous work, we proposed an approach to measure compliance with the 4-1 standard, the S^2C-AM[8]. It evaluates compliance with 4-1 standard process requirements using two dimensions: process maturity and artifact maturity. *Process maturity* is defined as the 4-1 states: the maturity levels of the Capability Maturity Model Integration for Development [7]. *Artifact maturity* is then defined based on artifact's *completeness* (content of recommended elements) *timeliness* (definition of being up-to-date through the time).

The global reporting of S^2C-AM is a two-dimension matrix where every 4-1 process requirement is located. Location in the matrix is based on the results of the so-called *Requirement Card*. A *Requirement Card* contains metrics to grasp the compliance status of the development process area, subject to assessment, with the respective 4-1 secure development practice. Our lightweight tool consists on instances of S^2C-AM's Requirement Cards for each of the 4-1 practices's requirements and 4-1 process models. To facilitate the self-assessment, we pursue that our metrics fulfill ideal attributes [2,3,5,13] e.g.: simplicity, so everyone involved in the topic can apply the measurement; cost-effectiveness, so cost of data collection does not exceed the resulting benefit; consistency and automation, so the results are the same regardless of who perform the assessment; threshold adjustment, so the development stakeholders can set their compliance appetite based on available resources.

2.3 Other Security Assessment Approaches

The Building Security In Maturity Model (BSIMM) [16] and the Software Assurance Maturity Model (SAMM) [6] support secure product development assessments. In particular, BSIMM provides a baseline of commonly applied security activities. An assessment yields a scorecard which shows the results of each activity compared to baseline. SAMM, on the other hand, is supported by the Open Web Application Security Project (OWASP) and provides a list of security practices with instructions on how to measure security-activities performance. Both BSIMM and SAMM are, however, not based on regulatory standards. Our lightweight tool closes this gap by supporting compliance assessments according to the 4-1 standard. Our process model further makes explicit potential gaps between the 4-1 and chosen development process areas. Hence, users conducting self-assessments do not need to have knowledge of the 4-1 standard before hand.

3 Security Compliance Self-assessment

Compliance with security-standards and actual security integration in development processes are hard to measure. Two major organizational groups deal with this task and, therefore, executing security compliance assessments (SCA). Auditors, for example, are used to deal with blurry standard requirements and manage long checklists of compliance evidences. Security officers want to know if they meet regulations, and also where are compliance breaches. Moreover, they have to ensure accurate investment on security initiatives.

Both groups have notion of a well-established secure development life-cycle. They are aware of common implementation pitfalls at the product and process level and tend to develop own assessment matrices and questionnaires when tools are not available. This requires proficiency in reporting to different organizational levels using the same assessment tool as source.

In contrast, development practitioners, those involved in the development process, are not used to deal with security standards. They deal with requirements which reflect the functionality of the product, but the design of the development process is not in scope of their daily duty. They execute security-related tasks like testing, configuration, selection of external components with limited awareness of the relation to security standards.

To perform security compliance self-assessments, development practitioners need preset tools describing the standard secure development process with simple language and providing easy reporting – the light-weight assessment concept. As assessors, development practitioners conduct assessments by following tool's instructions. They provide information to the tool based on the current process without expert knowledge on the standard secure development process. However, during assessments, awareness on the ideal secure development lifecycle should be increased. Finally, reporting should show deficiencies in the software product and the development process levels. An assessor should be able to establish a road map to solve deficiencies. The tool may provide a visualization of criticality levels to guide prioritization.

The lightweight tool proposed in the following section is intended to enable the assessor to perform the self-assessment, generate the road map overview, and facilitate a compliance track.

4 Tool Implementation for a Self-assessment of Security Compliance

The tool consists of eight assessment sheets each corresponding to one practice of the 4-1 standard. Assessors can evaluate compliance with the whole 4-1 standard (based on experience, a three-day workshop is recommended) or with a specific practice (2 h for data collection plus 30 min for results sum up are recommended). This is a pragmatic approach for self-assessments, as development practitioners can make trial runs with processes they have more expertise. An assessment sheet is both the evaluation guide and the reporting tool –the light-weight assessment concept. At the end of SCA, the assessment sheet reflects the compliance status of the current process in relation to the ideal process demanded by the 4-1 practice.

Assessment sheets contain the following: the *practice process model*, the corresponding *requirement cards* and *evaluation results*[1]. Figure 1 presents an overview of an assessment sheet for the 4-1 practice Secure Implementation (SI). An extended description of the elements is the following:

Practice Process Model: This model represents the 4-1 practice processes with explicit activities flows and artifacts using the Business Process and Modeling Notation (BPMN). Models' accuracy and usability by non-security practitioners were evaluated in previous work [22]. Development practitioners found models precise and easy to read in comparison to the standard text. Models served to design requirement cards and also as a visual report of the compliance status. Figure 3 shows an example for the SI practice of the 4-1 where the artifacts *SI-a1 Secure coding standards* and *SI-a5 Security coding error* as well as where the task *SI-t8 Examine threats and ability to exploit interfaces, trust boundaries and assets* are both highlighted as defective.

Requirement Cards: These implement the S^{2}C-AM's cards for each of the 47 requirements of the 4-1 (c.f. Sect. 2.2, [8]). To arise awareness of the standard, the original *4-1 requirement text* is stated. However, SCA aims at providing values to the *metrics*. Process metrics are separated from software product metrics in order to improve precision. We preferred numerical metrics since they enable assessors to set a compliance goal using *thresholds*. In addition, metrics are related to process models' *tasks* and *input/output artifacts*. Assessors register the status by selecting the items. Figure 2 shows an instance of the requirement *SI-1 - Secure*

[1] The tool is implemented as a Microsoft Excel file with embedded Visio models and vba macros.

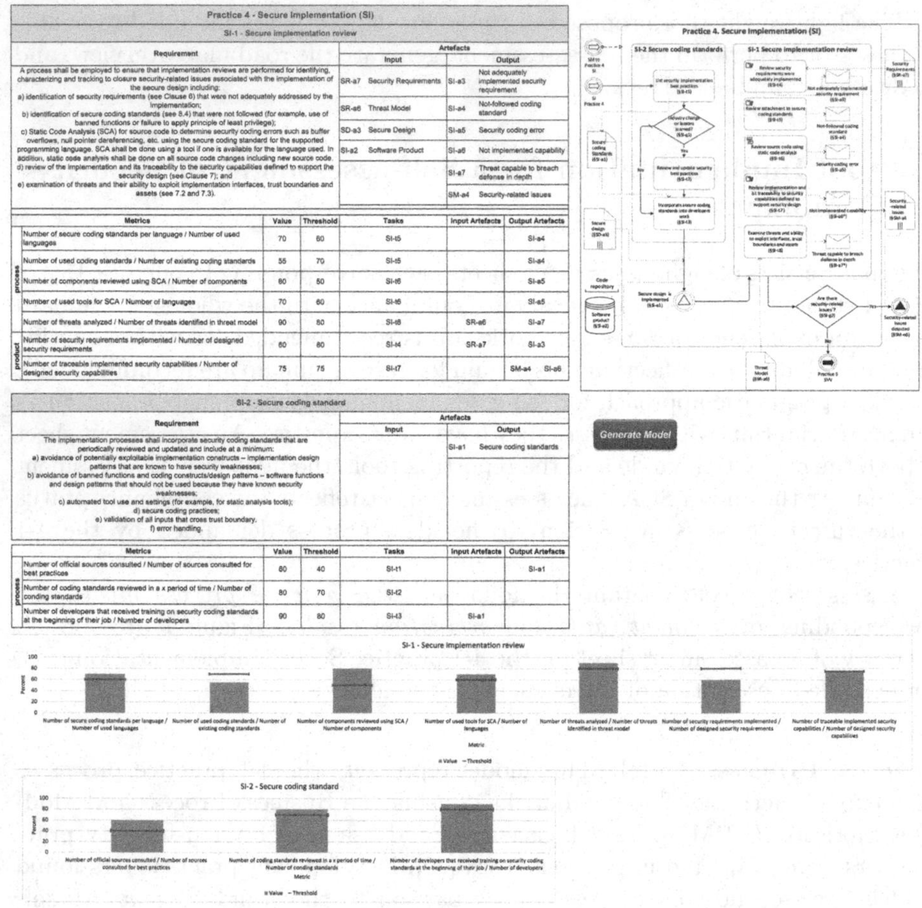

Practice 4 - Secure Implementation (SI)

SI-1 - Secure implementation review

Requirement	Artefacts: Input		Artefacts: Output	
A process shall be employed to ensure that implementation reviews are performed for identifying, characterizing and tracking to closure security-related issues associated with the implementation of the secure design including: a) identification of security requirements (see Clause 6) that were not adequately addressed by the implementation; b) identification of secure coding standards (see 8.4) that were not followed (for example, use of banned functions or failure to apply principle of least privilege); c) Static Code Analysis (SCA) for source code to determine security coding errors such as buffer overflows, null pointer dereferencing, etc. using the secure coding standard for the supported programming language. SCA shall be done using a tool if one is available for the language used. In addition, static code analysis shall be done on all source code changes including new source code. d) review of the implementation and its traceability to the security capabilities defined to support the security design (see Clause 7); and e) examination of threats and their ability to exploit implementation interfaces, trust boundaries and assets (see 7.2 and 7.3).	SR-a7	Security Requirements	SI-a3	Not adequately implemented security requirement
	SR-a6	Threat Model	SI-a4	Not-followed coding standard
	SD-a3	Secure Design	SI-a5	Security coding error
	SI-a2	Software Product	SI-a6	Not implemented capability
			SI-a7	Threat capable to breach defense in depth
			SM-a4	Security-related issues

	Metrics	Value	Threshold	Tasks	Input Artefacts	Output Artefacts
process	Number of secure coding standards per language / Number of used languages	70	60	SI-t5		SI-a4
process	Number of used coding standards / Number of existing coding standards	55	70	SI-t5		SI-a4
process	Number of components reviewed using SCA / Number of components	80	50	SI-t6		SI-a5
process	Number of used tools for SCA / Number of languages	70	60	SI-t6		SI-a5
process	Number of threats analyzed / Number of threats identified in threat model	90	80	SI-t8	SR-a6	SI-a7
product	Number of security requirements implemented / Number of designed security requirements	60	80	SI-t4	SR-a7	SI-a3
product	Number of traceable implemented security capabilities / Number of designed security capabilities	75	75	SI-t7		SM-a4 SI-a6

SI-2 - Secure coding standard

Requirement	Artefacts: Input	Artefacts: Output	
The implementation processes shall incorporate security coding standards that are periodically reviewed and updated and include at a minimum: a) avoidance of potentially exploitable implementation constructs – implementation design patterns that are known to have security weaknesses; b) avoidance of banned functions and coding constructs/design patterns – software functions and design patterns that should not be used because they have known security weaknesses; c) automated tool use and settings (for example, for static analysis tools); d) secure coding practices; e) validation of all inputs that cross trust boundary. f) error handling.		SI-a1	Secure coding standards

	Metrics	Value	Threshold	Tasks	Input Artefacts	Output Artefacts
process	Number of official sources consulted / Number of sources consulted for best practices	60	40	SI-t1		SI-a1
process	Number of coding standards reviewed in a x period of time / Number of conding standards	80	70	SI-t2		
process	Number of developers that received training on security coding standards at the beginning of their job / Number of developers	90	80	SI-t3	SI-a1	

Fig. 1. Overview of the assessment sheet for the (*Practice 4 - Secure Implementation (SI)*). It shows *requirements cards* for the two requirements in the practice (upper left), the *practice process model* visualization of the complete practice (upper right), and the *evaluation results* (bottom). See details in Figs. 2 and 3

implementation review of the Secure Implementation practice. The artifact *SI-a5 Security coding error* and the task *SI-t8* were marked as defective. At the end of the assessment, the process model is updated showing in red the missing tasks or artifacts. This representation enables the assessors to easily identify where exactly a deviation from the standard takes place and where improvement measures should be taken.

Evaluation Results: As global summary, the tool provides a bar chart showing the compliance status versus the desired *Threshold*. This visualization is useful for security or IT risk management areas.

Practice 4 - Secure Implementation (SI)				
SI-1 - Secure implementation review				
Requirement	Artefacts			
	Input		Output	
A process shall be employed to ensure that implementation reviews are performed for identifying, characterizing and tracking to closure security-related issues associated with the implementation of the secure design including: a) identification of security requirements (see Clause 6) that were not adequately addressed by the implementation; b) identification of secure coding standards (see 8.4) that were not followed (for example, use of banned functions or failure to apply principle of least privilege); c) Static Code Analysis (SCA) for source code to determine security coding errors such as buffer overflows, null pointer dereferencing, etc. using the secure coding standard for the supported programming language. SCA shall be done using a tool if one is available for the language used. In addition, static code analysis shall be done on all source code changes including new source code. d) review of the implementation and its traceability to the security capabilities defined to support the security design (see Clause 7); and e) examination of threats and their ability to exploit implementation interfaces, trust boundaries and assets (see 7.2 and 7.3).	SR-a7	Security Requirements	SI-a3	Not adequately implemented security requirement
	SR-a6	Threat Model	SI-a4	Not-followed coding standard
	SD-a3	Secure Design	SI-a5	Security coding error
	SI-a2	Software Product	SI-a6	Not implemented capability
			SI-a7	Threat capable to breach defense in depth
			SM-a4	Security-related issues

	Metrics	Value	Threshold	Tasks	Input Artefacts	Output Artefacts
process	Number of secure coding standards per language / Number of used languages	70	60	SI-t5		SI-a4
	Number of used coding standards / Number of existing coding standards	55	70	SI-t5		SI-a4
	Number of components reviewed using SCA / Number of components	80	50	SI-t6		SI-a5
	Number of used tools for SCA / Number of languages	70	60	SI-t6		SI-a5
	Number of threats analyzed / Number of threats identified in threat model	90	80	SI-t8	SR-a6	SI-a7
product	Number of security requirements implemented / Number of designed security requirements	60	80	SI-t4	SR-a7	SI-a3
	Number of traceable implemented security capabilities / Number of designed security capabilities	75	75	SI-t7		SM-a4 SI-a6

Fig. 2. Requirement Card of *SI-1 - Secure implementation review* in *Practice 4 - Secure Implementation*. (Color figure online)

5 Preliminary Evaluation at Siemens

With this preliminary evaluation, we aim to understand if our tool contains the elements that assessment practitioners would include in a SCA of the 4-1 security standard. In particular, we test the usability of the *light-weight mode* of the tool by the target users – non-experts whose background is the actual software development field. Our results will allow us to the analyze feasibility of the light-weight self-assessment concept. Further, we can distill some hints on future work before officially completing and disseminating the tool in industrial environments both in projects or training concepts. Our evaluation is guided by good practices on empirical studies [19] and constraint to the following research questions:

RQ 1. How can we measure compliance of organizations with the IEC 62443-4-1 standard?
RQ 2. How can the assessment results be presented effectively?
RQ 3. What are the challenges in assessing compliance using this tool?

5.1 Design

We chose semi-structured interviews to analyze subjects' experiences and opinions as openly as possible [25]. Our evaluation is preliminary in a sense that we

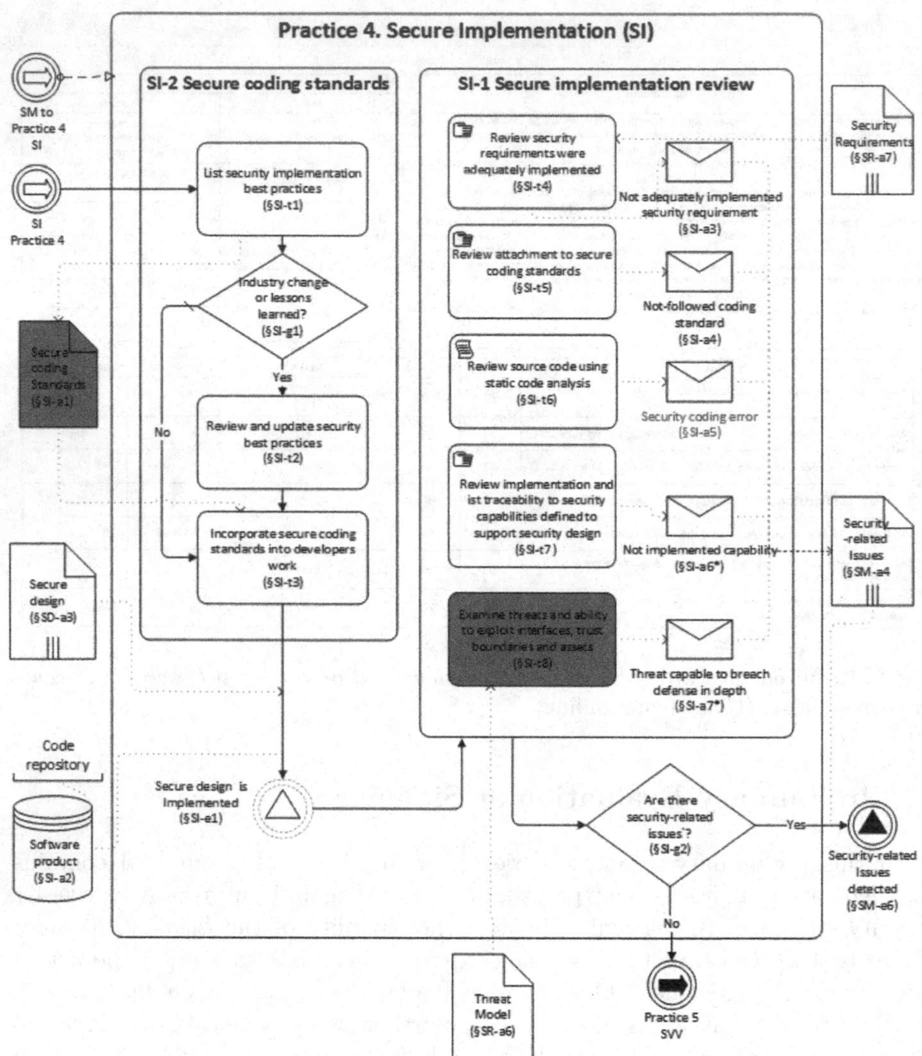

Fig. 3. Process model of *Practice 4 - Secure Implementation* with marked defective components.

evaluate our tool in an early stage without actual pilot in a case study. Yet, we still opt for methodological rigor in keys aspects such as the selection of subjects, the design and review of our interview instrument, and the analysis of results considering both the answers to our interview questions as well as the comments from the interviewees.

Subject Selection. Part of the subjects have expertise in security compliance assessments and knowledge of the 4-1 standard. To stress the ability of the tool to support self-assessments in a light-weight mode, we chose subjects with development background and basic knowledge of the 4-1 standard. Table 1 characterises the participants.

Table 1. Overview of the evaluation participants.

Id	Knowledge of the IEC 62443-4-1 standard	Expertise in security compliance assessment	Background
P1	Advanced	Expert	- Security consultant in projects involving the 4-1 standard and software development frameworks - Previously auditor and security officer
P2	Expert	Expert	- Lead security expert and researcher for industrial environments - Lead project manager in compliance assessments of the IEC 62443 standards family
P3	Advanced	Expert	- Security consultant in projects of ISO 27k compliance norms - Knowledge of assessment frameworks: Cobit, CMMI, BSMMI
P4	Beginner	None	- Software development - Works with 4-1 standard process models
P5	Beginner	None	- Software development - Works with standard process models

Survey Instrument. During the interviews, we relied on a questionnaire shown below. Our questions are grouped into three blocks. The first block deals with the measurement of compliance and covers RQ1. The second block asks for the graphical representation of the results in the tool and targets RQ2. RQ3 is covered in the last block with general questions on security compliance assessments.

Interview. We set a time-frame of 1 h for each interview. Initially, the participants were asked about their background and experience with the IEC 62443-4-1 standard in general as well as with assessments in the context of the standard or other security-related standards.

Table 2. Survey instrument

What do you think about the metrics? (RQ1)
1. Do the individual metrics make sense?
2. Do the metrics correctly represent the requirements?
3. Can the individual metrics be answered well?
Usability and graphical representation of the reporting results (RQ2)
1. Is the representation by the process models clear?
2. Are the metric-diagrams helpful?
3. Is the graphical representation useful for the evaluation of the results?
4. What is your overall opinion of the tool: usability and design?
Can the presented tool be used in an assessment? (RQ3)
1. What is most important to you in an assessment?
2. How would you handle the results from this tool?
3. How would you let yourself be influenced by the result?
4. Would you use such a tool in your daily work?
5. What needs to be improved to meet your requirements?

Subsequently, we gave a short overview of the topic and the intention of this work. We selected as basis the practice 4 Secure Implementation from the 4-1 Standard [9]. We presented the tool with all its capabilities based on this practice. Afterwards, the interviewees used the tool and were allowed to ask questions. After clarification of doubts, we started with the actual questionnaire. During the interviews, the participants were able to consult the tool at any time.

5.2 Results

We group the answers to our research questions in the following paragraphs. Answers to closed questions in the questionnaire were mapped to a five-point Likert scale, see Fig. 4. In this case, interviewees were additionally asked for a rationale in order to grasp free opinions and extra comments (Table 2).

RQ1. How can we measure compliance of organizations with the IEC 4-1 standard? The interviewees evaluated the quality of the created metrics. Most of the metrics were easy to understand. However, some required explanation. After an explanation, each participant agreed that the metrics make sense and can serve as a good basis for an assessment. P2 suggests to have a "*Clarification of what exactly is meant*" while P1 describes "*They are specific*". The participants with development knowledge but no assessment experience (P4 and P5) find the metrics easy to answer with numbers. Among the assessment experts, the opinions vary, P1 stated that, e.g., "There are some things you can't count e.g. number of secure coding standards" while P2 and P3 find it possible. Except for P4, the participants agreed that the requirements of 4-1 standard are

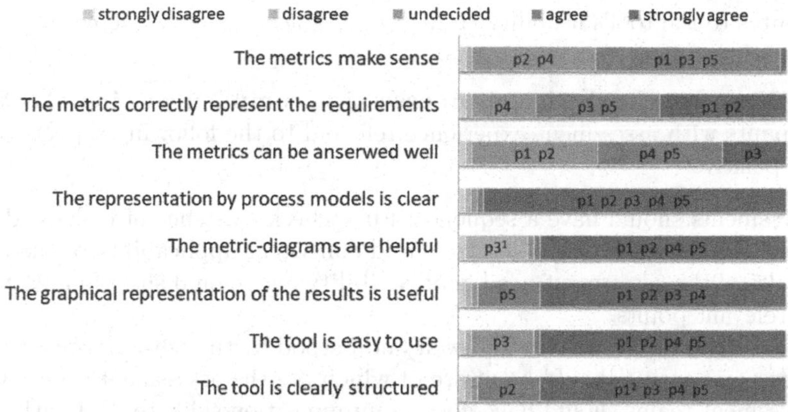

Fig. 4. Distribution of responses to closed questions in the questionnaire.

well represented by the metrics. For P4, quality aspects of the 4-1 standard may be overseen when metrics focus only on quantitative aspects. For quantitative aspects, the metrics mostly refer to artifacts, P2 remarked "It is hard to break down exactly. You can go on as long as you want".

RQ2. How can the assessment results be presented effectively? Overall, participants especially appreciate the compliance report based on process models. In the example presented, everyone understood the compliance gap showed by the models (see Fig. 3). The process models can be used to understand the standard and the metrics. Some quotes from participants are: "*You have to look at it in detail, then it's obvious*" - P2; "*If you don't know the model, it's very good*" - P3.

Regarding the metric diagrams at the end of each practice, participants stated that they are a useful in addition to the metrics and models. P3 pointed out management as an audience "*Yeah, it's helpful, but not right there*". In terms of content, the participants seemed very satisfied by the diagrams, since they can easily determine how close they are to the respective threshold.

All participants agreed on a good usability of the tool. P1 noted, however, that it would be difficult to handle without the introduction of the author. Therefore, instructions for using the tool should be included. In particular, the participants agreed that the tool is structured in a meaningful and clear way.

The fact that the tool had been implemented in an Excel file was well received by all participants although poses extra work to provide signed macros in environments where macros are automatically blocked: "*Excel is something that everybody has.*" - P1. In this context, P3 (security expert) noted that Excel is a common tool and someone may misuse it. For example, the metrics could be changed by the user. Versioning and proper access control may reduce the risk in a real scenario. To sum up, the simple procedure during the assessment contributes to the tool's usability (P2: *"I only have to enter value and threshold and then I can mark the components."*).

RQ3. What are the challenges in assessing compliance using this tool? Participants with assessment experience referred to the following aspects of the tool as positive:

- Assessments should have a sequence: First check existence of tasks and artifacts, then check quality of artifacts and consistent applicability of the tasks.
- Results of the assessment need good visibility and should show in a easy way the relevant points.
- Metrics should be plausible and well understood by the self-assessment team.
- Assessment teams should be aware of which are the assessment objectives.
- Assessment teams should have good communication skills to deal with internal and external stakeholders. Internal during the assessment, external to elevate reports and claim resources for implementation.

Participants reported that they would use the results of the tool to take action for improvements. P1 might use the tool to discuss the status quo and missing components with customers in assessment projects. P3 would process and summarize results to communicate them both at management and development level. P2 may use process models to support the results of metrics.

Overall, the results of our preliminary evaluation thereby strengthen our confidence in that this (Excel-based) tool set is a proper means for a lightweight assessment by non-experts while still making explicit some points for improvement.

These points for improvement are: instructions to use the tool (P1), global overview with am structured summary of results (P1, P2, P3). P2 recommends to apply the tool in several projects and track issues to update the metrics. Further, metrics require unique id numbers. This will support communication and tracking.

Threats to Validity. As any qualitative study with a preliminary character, we face various threats to validity such as a possible bias from the expert interviewees emerging from their local context. We compensated for the bias by considering a larger sample size relative to the context including non-expert interviewees as well to gain a sense or usability from a non-security practitioners perspective. In any case, we were particularly interested in harvesting opinionated (experience-based) views and perceptions as they provide a suitable ground to steer future evaluations in differing contexts (and following different empirical approaches).

6 Conclusion

In this paper, we have reported on a technique including tooling for a lightweight self-assessment of security compliance in the development process. One particular focus was set on covering both the artifact view and the process view alike as we consider both imperative in context of software process improvement [18,20]. The transformation from a regular development process to a secure development process can be further tracked with metrics based on artifacts and tasks. The gap is reported directly by the metric report and can be also globalized in the S[2]C-AMframework which reports in a compliance matrix.

In a world that has already significantly adopted lean and agile development practices, we believe security compliance assessment techniques need to be aligned for self-learning organizations. We consider this light-weight approach to be essential. Our tool supports cross-functional teams in fulfilling security requirements in the development process. Our intention is to help these teams to become mature w.r.t. their secure product development compliance to IEC 62443-4-1.

In particular, our contribution helps practitioners to:

- Understand security compliance goals fast by using process models
- Use practical metrics to assess the level of as-is compliance
- Make gaps transparent and easy to communicate by annotating process
- Output results also in the format of other process maturity frameworks such as CMMI

For researchers, we contribute the following:

- A blueprint of how a complex secure development process such as IEC 62443-4-1 can be operationalized for practitioners. This work can serve as blueprint for operationalizing further process quality standards as well
- Metrics aligned with process models for transparent visualization of problems. This can serve research as a inspiration of how to communicate process quality problems to management

Our work has the following limitations. Interview research poses threats to validity, the most important for this preliminary evaluation relate to possible selection and confirmation bias. We tried to mitigate this through interview preparation and rigor, allowing participants to answer freely and asking them to use the tool themselves. In addition, choosing participants that might not be directly influenced by the results of this research.

As future steps, we will align further with frameworks for IT governance such as COBIT [10]. Moreover we will provide a holistic approach of how governance can be combined with large scale agile software engineering methods such as SAFe [14]. We will also conduct further studies with practitioners and certification bodies to assess the possible adaption of our approach in practice.

References

1. Airbus cybersecurity. https://airbus-cyber-security.com/products-and-services/
2. Basili, V., Caldiera, G., Rombach, H.: The goal question metric approach. Encycl. Softw. Eng. 528–532 (1994)
3. Basili, V., Weiss, D.: A methodology for collecting valid software engineering data. IEEE Trans. Softw. Eng. **SE–10**(6), 728–738 (1984)

4. Beckers, K.: Pattern and Security Requirements: Engineering-Based Establishment of Security Standards. Springer, Cham (2015). https://doi.org/10.1007/978-3-319-16664-3
5. Böhme, R., Freiling, F.C.: On metrics and measurements. In: Eusgeld, I., Freiling, F.C., Reussner, R. (eds.) Dependability Metrics. LNCS, vol. 4909, pp. 7–13. Springer, Heidelberg (2008). https://doi.org/10.1007/978-3-540-68947-8_2
6. Chandra, P.: Software assurance maturity model v1.5 (2017)
7. CMMI Product Team: CMMI for development, version 1.3. Technical report, CMU/SEI-2010-TR-033, Software Engineering Institute, Carnegie Mellon University (2010)
8. Dännart, S., Constante, F.M., Beckers, K.: An assessment model for continuous security compliance in large scale agile environments. In: Giorgini, P., Weber, B. (eds.) CAiSE 2019. LNCS, vol. 11483, pp. 529–544. Springer, Cham (2019). https://doi.org/10.1007/978-3-030-21290-2_33
9. IEC: 62443-4-1. Security for industrial automation and control systems Part 4–1. Product security development life-cycle requirements (2018)
10. ISACA: Cobit 5 (2012)
11. ISO: The main benefits of ISO standards. www.iso.org/benefits-of-standards
12. ISO/IEC: 27034. Information technology - security techniques - application security (2011)
13. Jaquith, A.: Security Metrics: Replacing Fear, Uncertainty, and Doubt. Pearson Education, London (2007)
14. Leffingwell, D., Yakyma, A., Knaster, R., Jemilo, D., Oren, I.: SAFe reference guide (2017)
15. Maidl, M., Kröselberg, D., Christ, J., Beckers, K.: A comprehensive framework for security in engineering projects based on IEC 62443. In: ISSRE Workshops, USA, 15–18 October 2018 (2018)
16. McGraw, G., Migues, S., Chess, B.: Building security in maturity model. www.bsimm.com
17. Mello, J.: Cybercrime diary, Q2 2019 who's hacked (2019). cybersecurity-ventures.com
18. Fernández, D.M., et al.: Artefacts in software engineering: a fundamental positioning. J. Syst. Softw. **18**, 2777–2786 (2019)
19. Fernández, D.M., Passoth, J.: Empirical software engineering: from discipline to interdiscipline. CoRR abs/1805.08302 (2018). http://arxiv.org/abs/1805.08302
20. Méndez Fernández, D., Wagner, S.: A case study on artefact-based RE improvement in practice. In: Abrahamsson, P., Corral, L., Oivo, M., Russo, B. (eds.) PROFES 2015. LNCS, vol. 9459, pp. 114–130. Springer, Cham (2015). https://doi.org/10.1007/978-3-319-26844-6_9
21. Microsoft Corporation iSEC Partners: Microsoft SDL: return-on-investment (2009)
22. Moyon, F., Beckers, K., Klepper, S., Lachberger, P., Bruegge, B.: Towards continuous security compliance in agile software development at scale. In: RCoSE. ACM (2018)
23. Ponemon Institute LLC: The true cost of compliance study (2017)
24. PWC: Compliance on the forefront: setting the pace for innovation (2019)
25. Shull, F., Singer, J., Sjøberg, D.I.: Guide to Advanced Empirical Software Engineering. Springer, New York (2007). https://doi.org/10.1007/978-1-84800-044-5
26. Thomson Reuters: Costs of compliance report 2018 (2018)
27. U.S. House of Representatives: The equifax data breach, majority staff report (2018)

A Novel Hybrid Genetic Algorithm for the Two-Stage Transportation Problem with Fixed Charges Associated to the Routes

Ovidiu Cosma, Petrica C. Pop(✉), and Cosmin Sabo

Department of Mathematics and Computer Science, Technical University of Cluj-Napoca, North University Center at Baia Mare, Baia Mare, Romania
{ovidiu.cosma,petrica.pop}@cunbm.utcluj.ro, sabo.cosmin@gmail.com

Abstract. This paper concerns the two-stage transportation problem with fixed charges associated to the routes and proposes an efficient hybrid metaheuristic for distribution optimization. Our proposed hybrid algorithm incorporates a linear programming optimization problem into a genetic algorithm. Computational experiments were performed on a recent set of benchmark instances available from literature. The achieved computational results prove that our proposed solution approach is highly competitive in comparison with the existing approaches from the literature.

1 Introduction

This work deals with a variant of the transportation problem, namely the fixed-charges transportation problem in a two-stage supply chain network consisting of a set of manufacturers, a set of distribution centers (DC's) and a set of customers, whose scope is to identify and select the manufacturers and the distribution centers fulfilling the demands of the customers under minimal costs. The main characteristic of the two-stage fixed-charges transportation problem (TSFCTP) is that a fixed charge is associated with each route that may be opened in addition to the variable transportation cost which is proportional to the amount of goods shipped.

The fixed-charges transportation problem (FCTP) generalizes the classical transportation problem and it was introduced by Balinski [1]. Guisewite and Pardalos [6] showed that the fixed-charges transportation problem is NP-hard. For more information on the FCTP, including a review of exact and heuristic algorithms developed for solving the problem, we refer to Buson et al. [2]. For a review on the variants of the fixed-charges transportation problem and related problems to the investigated TSFCTP, we refer to Cosma et al. [4] and Pop et al. [9].

The existing literature regarding the two-stage transportation problem with fixed-charges associated to the routes is rather scarce. The form investigated in

A. Chatzigeorgiou et al. (Eds.): SOFSEM 2020, LNCS 12011, pp. 417–428, 2020.
https://doi.org/10.1007/978-3-030-38919-2_34

our paper was introduced by Jawahar and Balaji [7]. They described a mathematical model of the transportation problem as a mixed integer linear programming, a heuristic approach based on genetic algorithm (GA) with a specific coding scheme suitable for two-stage transportation problems, as well as a set of 20 benchmark instances of various sizes and capacities. Their obtained computational results have been compared to lower bounds and approximate solutions obtained by relaxing the integrality constraints. Raj and Rajendran [10] proposed two scenarios of the two-stage transportation problem: the first one, called Scenario 1, considers fixed charges associated to the routes in addition to unit transportation costs and unlimited capacities of the DCs, while the second one, called Scenario 2, considers the opening costs of the DCs in addition to unit transportation costs. In the case of Scenario 1, which coincides with the form considered in our paper, they described a two-stage genetic algorithm in order to solve the problem. They also proposed a solution representation that allows a single-stage genetic algorithm to solve it. The major feature of these GA's is a compact representation of the chromosomes based on permutations. Pop et al. [8] proposed a hybrid algorithm that combines a steady-state genetic algorithm with a local search procedure for solving the problem. Recently, Calvete et al. [3] described a matheuristic approach for the problem that incorporates an optimization problem within an evolutionary algorithm and proposed a set of 20 larger randomly generated instances and Cosma et al. [5] developed an efficient multi-start Iterated Local Search procedure for the total distribution costs minimization of the TSFCTP, which constructs an initial solution, uses a local search procedure to increase the exploration, a perturbation mechanism and a neighborhood operator in order to diversify the search.

Our novel solution approach has some important and original features that differentiate it from the existing ones from the literature. We used an integer chromosome representation in which the genes have integer values that represent estimates of the number of units to be transported on each transportation link of the model. For an efficient exploration of the solutions space and in order to avoid evolution stalling due to local minima, several chromosome populations have been created, evolving separately to different offspring, which are finally merged into the populations.

We organized the remainder of the paper as follows: in Sect. 2, we give some notations and definitions related to the two-stage transportation problem with fixed-charges associated to the routes that will be used throughout the paper and we also present a mixed integer formulation of the problem. The novel solution approach for solving the considered transportation problem is described in Sect. 3 and some preliminary computational experiments and the achieved results are presented and analyzed in Sect. 4. Finally, we conclude our work and discuss our plans for future work in Sect. 5.

2 Definition of the Two-Stage Fixed-Charges Transportation Problem

In order to define the considered two-stage fixed-charges transportation problem, we start by defining the related sets, decision variables and parameters:

p	the number of manufacturers and i is the manufacturer identifier;
q	the number of distribution centers (DCs) and j is the DC identifier;
r	the number of customers and k is the customer identifier;
S_i	the capacity of manufacturer i;
D_k	the demand of customer k;
f_{ij}	the fixed charge for the link from manufacturer i to DC j
g_{jk}	the fixed charge for the link from DC j to customer k;
b_{ij}	the unit cost of transportation from manufacturer i to DC j;
c_{jk}	the unit cost of transportation from DC j to customer k.
x_{ij}	the number of units transported from manufacturer i to DC j,
y_{jk}	the number of units transported from DC j to customer k,
z_{ij}	is 1 if the route from manufacturer i to DC j is used and 0 otherwise,
w_{jk}	is 1 if the route from DC j to customer k is used and 0 otherwise

Given a set of p manufacturers, a set of q distribution centers (DC's) and a set of r customers with the following properties:

1. Each manufacturer may ship to any of the q DCs at a transportation cost b_{ij} per unit from manufacturer i, where $i \in \{1, ..., p\}$, to DC j, where $j \in \{1, ..., q\}$, plus a fixed charge f_{ij} for operating corresponding the route.
2. Each DC may ship to any of the r customers at a transportation cost c_{jk} per unit from DC j, where $j \in \{1, ..., q\}$, to customer k, where $k \in \{1, ..., r\}$, plus a fixed charge g_{jk} for operating the corresponding route.
3. Each manufacturer $i \in \{1, ..., p\}$ has S_i units of supply and each customer $k \in \{1, ..., r\}$ has a given demand D_k.

The aim of the two-stage fixed-charges transportation problem is to determine the routes to be opened and corresponding shipment quantities on these routes, such that the customer demands are fulfilled, all shipment constraints are satisfied, and the total distribution costs are minimized.

An illustration of the investigated TSFCTP is presented in the next figure (Fig. 1).

The TSFCTP can be modeled as the following mixed integer problem described by Raj and Rajendran [10]:

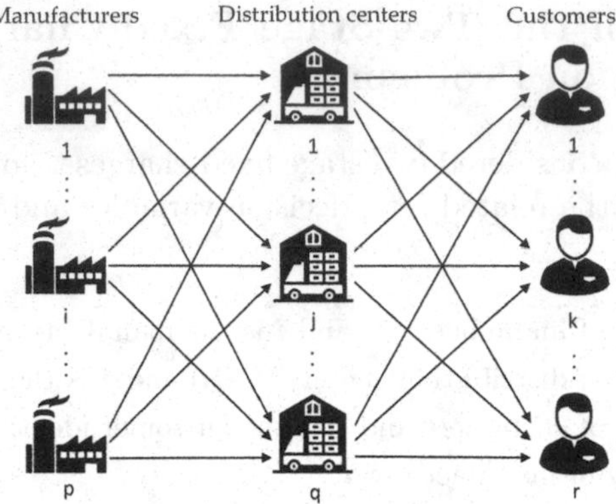

Fig. 1. Illustration of the two-stage fixed-charges transportation problem

$$\min \ Z = \sum_{i=1}^{p}\sum_{j=1}^{q}(b_{ij}x_{ij} + f_{ij}z_{ij}) + \sum_{j=1}^{q}\sum_{k=1}^{r}(c_{jk}y_{jk} + g_{jk}w_{jk}) \tag{1}$$

$$s.t. \quad \sum_{j=1}^{q} x_{ij} \leq S_i, \qquad \forall\, i \in \{1, ..., p\} \tag{2}$$

$$\sum_{j=1}^{q} y_{jk} = D_k, \qquad \forall\, k \in \{1, ..., r\} \tag{3}$$

$$\sum_{i=1}^{p} x_{ij} = \sum_{k=1}^{r} y_{jk}, \qquad \forall\, j \in \{1, ..., q\} \tag{4}$$

$$x_{ij} \geq 0, \qquad \forall\, i \in \{1, ..., p\},\ \forall\, j \in \{1, ..., q\} \tag{5}$$

$$y_{jk} \geq 0, \qquad \forall\, j \in \{1, ..., q\}, \forall\, k \in \{1, ..., r\} \tag{6}$$

$$z_{ij} = \begin{cases} 1, & x_{ij} > 0 \\ 0, & x_{ij} = 0 \end{cases} \qquad \forall\, i \in \{1, ..., p\},\ \forall\, j \in \{1, ..., q\} \tag{7}$$

$$w_{jk} = \begin{cases} 1, & y_{jk} > 0 \\ 0, & y_{jk} = 0 \end{cases} \qquad \forall\, j \in \{1, ..., q\},\ \forall\, k \in \{1, ..., r\} \tag{8}$$

The objective function minimizes the total distribution cost: the fixed charges and the transportation per-unit costs. Constraints (2) guarantee that the quantity shipped out from each manufacturer does not exceed the available capacity, constraints (3) guarantee that the total shipment received from DCs by each customer is equal to its demand and constraints (4) are the flow conservation conditions and they guarantee that the units received by a DC from manufacturers are equal to the units shipped from the DCs to the customers. The last four constraints ensure the integrality and non-negativity of the decision variables.

The considered two-stage transportation problem with fixed charges associated to the routes is a *NP*-hard optimization problem because it extends the fixed-charges transportation problem, which have been shown to be NP-hard by Guisewite and Pardalos [6]. That is why in order to tackle the two-stage transportation problem with fixed charges associated to the routes, we proposed an efficient hybrid genetic algorithm.

3 Description of the Hybrid Metaheuristic Algorithm

Our proposed hybrid metaheuristic approach consists of a genetic algorithm (GA) whose operation is based on solving a set of linear optimization problems.

GAs are search heuristic methods inspired from the theory of natural evolution. They can deliver good solutions efficiently, making them attractive for solving difficult optimization problems.

One of the most important elements of a GA is the chromosome representation scheme. In our genetic algorithm, each chromosome contains $p \times q + q \times r$ genes, that represent the transportation links in the distribution system. There are $p \times q$ links from the p manufacturers to the q DCs, and $q \times r$ links from the q DCs to the r customers. The value of each gene represents an estimate of the number of units transported along the corresponding transportation link, in the optimal solution of the TSFCTP. The gene corresponding to the link from manufacturer i to DC j is denoted $\tilde{x}_{ij}$, and the gene corresponding to the link from DC j to customer k is denoted $\tilde{y}_{jk}$. The genes are used to estimate the total cost for transporting one unit on the corresponding link, so that it includes the correct fraction of the fixed charge required for opening that link. These estimated costs denoted $\tilde{b}_{ij}$ and $\tilde{c}_{jk}$ are computed according to relations (9) and (10), where $\tilde{b}_{ij}$ correspond to the links from manufacturers to DCs and $\tilde{c}_{jk}$ correspond to the links from DCs to customers.

$$\tilde{b}_{ij} = \begin{cases} b_{ij} + \dfrac{f_{ij}}{\tilde{x}_{ij}}, & \text{if } \tilde{x}_{ij} > 0 \\ b_{ij} + f_{ij}, & \text{if } \tilde{x}_{ij} = 0 \end{cases} \tag{9}$$

$$\tilde{c}_{jk} = \begin{cases} c_{jk} + \dfrac{g_{jk}}{\tilde{y}_{jk}}, & \text{if } \tilde{y}_{jk} > 0 \\ c_{jk} + g_{jk}, & \text{if } \tilde{y}_{jk} = 0 \end{cases} \tag{10}$$

The value of the genes that form the initial population chromosomes are chosen randomly. However, it is improbable that the random estimates will be close to reality. For improving the quality of the chromosomes, we developed a simple algorithm called Estimates Correction, that will be used also for improving each new chromosome created by the crossover operator.

For defining the Estimates Correction algorithm, we consider the following linear optimization problem:

$$\min \sum_{i=1}^{p} \sum_{j=1}^{q} \tilde{b}_{ij} x_{ij} + \sum_{j=1}^{q} \sum_{k=1}^{r} \tilde{c}_{jk} y_{jk} \tag{11}$$

$$s.t. \ (2), (3), (4), (5), (6)$$

Algorithm 1. Algorithm Estimates Correction

```
input: chromosome(x̃_ij, ỹ_jk)
output: corrected chromosome
 1: Z̃ ← ∞
 2: Solve the linear optimization problem (11)
 3: Calculate Z based on relation (1)
 4: if Z̃ < Z or the solution is a duplicate then
 5:     Result ← saved chromosome
 6:     STOP
 7: else
 8:     Z̃ ← Z
 9:     Save the chromosome
10:     Update the estimates: x̃_ij ← x_ij,  ỹ_jk ← y_jk
11:     goto 2.
12: end if
```

where $\tilde{b}_{ij}$ and $\tilde{c}_{jk}$ are the total unit transportation cost estimates, calculated according to (9) and (10).

This is a well-known optimization problem, namely the Minimum Cost Flow Problem, for which there are several algorithms that solve it efficiently. For the results presented in this paper, we used the Network Simplex algorithm.

The Estimates Optimization algorithm works as follows:

Step 1 initializes the total cost of distribution $\tilde{Z}$. The linear optimization problem (11) is optimally solved in step 2. The amounts x_{ij} and y_{jk} determined in step 2 are used in step 3 to calculate the total cost Z of the TSFCTP. Steps 2 and 3 may repeat because of the loop created by the jump in step 11, but there is no guarantee that the solutions will always be improved. The decision to stop or continue the algorithm is taken in step 4. The algorithm is stopped if the last iteration worsened the solution, or the resulting chromosome is a duplicate. Two chromosomes are considered identical, if the corresponding linear optimization problems (11) have identical solutions, even if the $\tilde{x}_{ij}$ and $\tilde{y}_{jk}$ estimates are different. The last saved chromosome represents the result of the algorithm. If step 7 is reached, it means that the last iteration improved the TSFCTP solution.

Algorithm 2. Algorithm Genetic Evolution

```
input: population
repeat
    repeat
        p1 ← Tournament Selection ( population )
        p2 ← Tournament Selection ( population )
        new offspring ← Mutation ( Crossover ( p1, p2 ))
    until new generation is completed
    population ← Admission ( population, new generation )
until evolution stalls
```

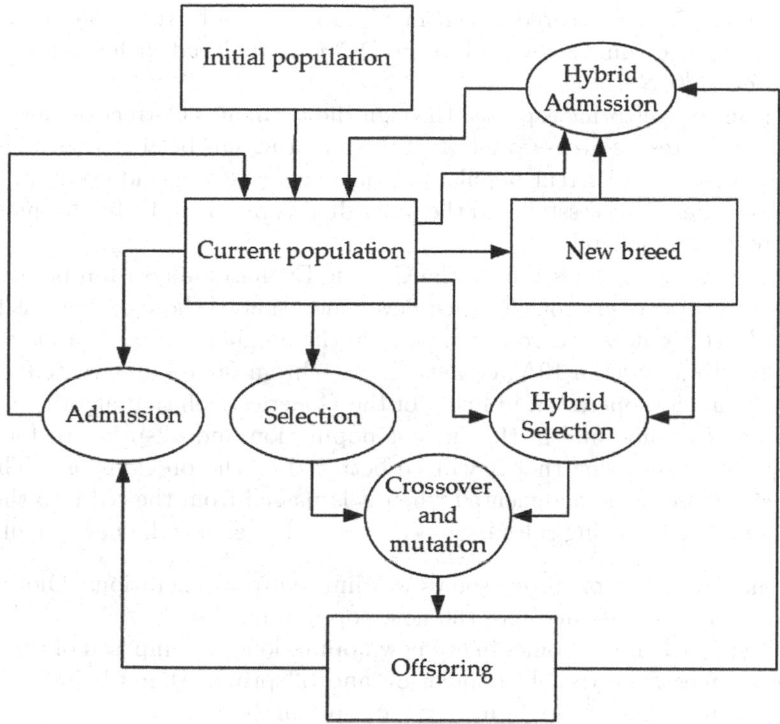

Fig. 2. The operating principle of our genetic optimization algorithm

In this case, the chromosome is saved, the $\tilde{x}_{ij}$ and $\tilde{y}_{jk}$ estimates are updated in step 10, and the algorithm continues with a jump to step 2.

The operating principle of our genetic optimization algorithm is shown in Fig. 2.

The process of evolution for a chromosome population is presented in Algorithm 2.

The selection operator chooses two parent chromosomes $p1$ and $p2$ from the current population, for mating. The two parents are chosen using the tournament selection strategy (lines 3, 4). The number of participants in each tournament is randomly selected between 2 and 10.

The crossover operator combines the genes of the two parents, to form the chromosome of the offspring (line 5). The genes of the offspring are taken either from $p1$ or $p2$ with equal probabilities. This results in an offspring carrying equal genetic information from both parents.

Each new chromosome can suffer a mutation, with 0.01 probability. The mutation operator chooses randomly a client k, and clears all the estimates for its links $\tilde{y}_{jk}$. Then maximum 5 DCs are randomly chosen and the estimates of their links to client k are replaced with random values in the interval $[0, D_k]$. Next a DC j is randomly chosen and all the estimates for the links from j to the manufacturers $\tilde{x}_{ij}$ are cleared. Then maximum 5 manufacturers are randomly chosen and their estimates for the links to DC j are replaced with random values in the interval $[0, S_i]$.

The resulting offspring is passed through the Estimates Correction algorithm, that also evaluates its fitness value Z. If the offspring has better fitness than the last individual in the current population, then the crossover and eventual mutation are considered successful, and the individual is retained. Unfit chromosomes are destroyed immediately.

The internal loop (lines 2–6) of the Genetic Evolution algorithm performs at least $3N$ crossover operations for each new generation. If those operations fail to create at least $2N$ new fit chromosomes, then the number of crossover operations is increased to maximum $10N$, or until $2N$ fit chromosomes are created.

The Admission operation in line 7 of the Genetic Evolution algorithm takes some of the chromosomes in the current population and offspring to form the new population, hopping that it will be better than the previous one. The age of each chromosome is incremented when it is passed from the old into the new population. The following rules have been applied to create the new population:

- The maximum age of chromosomes was limited to 3 generations. Older chromosomes are not allowed into the new population.
- The first 2/3N chromosomes in the new population are composed of the fittest chromosomes from the old population and offspring. At most, half of these chromosomes may originate from the old population.
- The remaining 1/3N chromosomes are randomly chosen from the rest of the old population and offspring.

The main loop of the Genetic Evolution algorithm ends when the best chromosome is no longer improved in the last three consecutive generations. This chromosome could represent the optimal solution to the TSFCTP, but usually for complex problems it is only a local minimum.

Our main optimization procedure is presented in Algorithm 3.

Algorithm 3. Main

```
randomly generate population
Genetic Evolution(population)
repeat
    new breed ← population
    randomly generate population
    Genetic Evolution(population)
    repeat
        p1 ← Tournament Selection ( population )
        p2 ← Tournament Selection ( new breed )
        new offspring ← Mutation ( Crossover ( p1, p2 ))
    until new generation is completed
    population ← Hybrid Admission ( population, new breed, new generation )
    Genetic Evolution(population)
until the running time limit is exceeded
```

The initial population is composed of $N = min(\frac{pq+qr}{5}, 500)$ randomly generated chromosomes. The $\tilde{x}_{ij}$ estimates are chosen randomly from the interval $[0, S_i]$, and the $\tilde{y}_{jk}$ estimates are chosen randomly from the interval $[0, D_k]$. Each random chromosome in the initial population is passed through the Estimates Correction algorithm.

The chromosome population is improved by calling the Genetic Evolution procedure. This procedure ends when the optimal solution of TSFCTP is found or is reached in a local minimum. The population thus obtained represents a new breed of chromosomes, the evolution of which is stopped, because any subsequent gains would appear far too slow.

The main loop of the algorithm (lines 3 14) creates new breeds of chromosomes that are merged together using hybrid selection and the crossover operator, for the best possible coverage of the solutions space. The loop ends when the running time limit of the algorithm is exceeded. Each breed evolution starts with a population of random chromosomes.

The Loop on lines 7 11 creates a new chromosome population by merging two different breeds. The selection process organizes one tournament in each of the two breeds that are merged together Thus each offspring will contain genetic information from both breeds. In the merging operation, a minimum of $5N$ crossover operations are performed, that result in at least $4N$ new fit chromosomes. If this is not possible, then the number of crossover operations is extended up to a maximum of $15N$.

The Hybrid Admission operation performed at the end of the loop, creates a new population, taking chromosomes from the two merged populations and their offspring, based on the rules described above. The newly created population follows the normal process of evolution.

4 Computational Results

In order to analyze the performance of our proposed algorithm, we tested it on a set of benchmark instances that was proposed by Calvete et al. [3]. We performed 5 independent runs for each instance, as there were performed by Calvete et al. [3].

The computational results obtained by our proposed solution approach in comparison to the matheuristic approach proposed by Calvete et al. [3] are presented in Table 1. The first column in Table 1 gives the number of the instance and the second one provides its size. The next two columns contain the optimal solution obtained by CPLEX when it is available and the corresponding time. The running time of CPLEX was limited to 3600 s. The instances for which CPLEX could not find the optimal solution within the running time, are marked with an asterisk. The last columns provide the results reported by Calvete et al. [3] and our achieved results. The following information is provided: the minimum and maximum objective function values obtained in the five runs of each instance (Z_{min}, and Z_{max}), the average gap and the average time spent by the algorithms for finding the best solution. The gap was calculated as proposed by

Calvete et al. [3]: $gap = 100 \times (Z - Z_{min})/Z_{min}$, where Z is the average of the solutions objective value.

The computational times are reported in seconds. The results written in bold represent cases for which the best results have been achieved either by CPLEX, or using the mat heuristic proposed by Calvete et al. [3], or by our novel solution approach.

Table 1. Computational results achieved by our proposed soft computing approach compared to existing methods

No.	Size			CPLEX		Calvete et al. [3]				Our approach			
	p	q	r	Z_{opt}	T_{cplex}	Z_{min}	Z_{max}	gap	Time	Z_{min}	Z_{max}	gap	Time
1.	2	4	6	71484	0.2	opt	opt	0.00	0.0	opt	opt	0.00	0.006
2.	2	4	8	102674	0.4	opt	opt	0.00	0.0	opt	opt	0.00	0.004
3.	4	8	12	124253	0.3	opt	opt	0.00	0.2	opt	opt	0.00	0.036
4.	4	8	16	136779	0.3	opt	opt	0.00	0.2	opt	opt	0.00	0.020
5.	6	12	18	150932	0.3	opt	opt	0.00	0.2	opt	opt	0.00	0.101
6.	6	12	24	200998	0.9	opt	opt	0.00	0.2	opt	opt	0.00	0.128
7.	8	16	24	147741	0.5	opt	opt	0.00	0.4	opt	opt	0.00	0.224
8.	8	16	32	196187	2.4	opt	opt	0.00	1.8	opt	opt	0.00	0.993
9.	10	20	30	162660	1.2	opt	opt	0.00	2.2	opt	opt	0.00	0.993
10.	10	20	40	216758	23.1	opt	opt	0.00	2.6	opt	opt	0.00	1.483
11.	20	40	60	235366	11.2	opt	235783	0.07	8.4	opt	opt	0.00	15.3
12.	20	40	80	*424386	>3600	424732	426762	0.12	126.4	**423827**	**423827**	0.00	554.3
13.	30	60	90	296441	2047	296451	297937	0.25	276.6	opt	**296443**	0.0003	299.2
14.	30	60	120	*405231	>3600	405099	405294	0.01	278.6	**404625**	**404625**	0.00	250.0
15.	40	80	120	***346934**	>3600	347381	348277	0.15	592.8	**346934**	**346934**	0.00	1564.8
16.	40	80	160	*554160	>3600	553841	556802	0.19	503.4	**553392**	**554341**	0.047	2664.0
17.	50	100	150	***371799**	>3600	**371799**	372480	0.01	1088.6	371831	371976	0.006	1575.7
18.	50	100	200	*678451	>3600	678150	680430	0.19	814.6	**677708**	**679078**	0.103	2357.2
19.	60	120	180	***370693**	>3600	370905	372023	0.16	1600.8	370809	370961	0.02	2129.9
20.	60	120	240	*577985	>3600	575566	576213	0.05	1125.4	**575241**	576304	0.998	2538.6

Regarding the computational times, it is difficult to make a fair comparison between algorithms, because they did not run on the same computer and they were implemented in different programming languages. In order to be able to make an objective comparison, we will analyze the processing power of the computers that ran the two algorithms, and the efficiencies of the programming languages used for their implementation.

The matheuristic algorithm proposed by Calvete et al. [3] has been run on an Intel Pentium D CPU at 3.0 GHz having 3.2 GB of RAM, while our algorithm on an Intel Core i5-4590 processor at 3.3 GHz with 4GB of RAM. The single thread ratings of the two processors can be found in [12] and we observed that our processor runs 3.03 times faster. As regards the programming languages, we used Java, while the algorithm proposed by Calvete et al. [3] was programmed in C++. A comparison between the two programming languages in terms of efficiency can be found in [11]. The time factor for C++: 1, the time factor for Java 64 bit: 5.8. Therefore, our programming language is 5.8 times slower. In consequence, we considered that the greater speed of the Core i5 processor roughly compensates the slowness of the Java programming language. Because the ratings are always approximate, we did not use any scaling factor. The running times reported in Table 1 are the times measured during the experiments.

Analyzing the computational results reported in Table 1, we can observe that the algorithm developed by Calvete et al. [3] provided the optimal solutions in 10 out of 20 instances and our proposed solution approach obtained the optimal solutions for all the instances for which CPLEX delivered the optimal solution.

Our algorithm provided the optimal solution in all the five runs in less than 1 s for the first ten instances and within 15.3 s in the case of instance 11 and within 299.2 s in the case of instance 13. We should point out that in the case of instance 12, the solution reported by Calvete et al. [3] as to be obtained by CPLEX is wrong. We obtained a different solution using CPLEX which is displayed in Table 1.

In the case of instances 14 and 16, our algorithm provided in all the five runs better solutions compared to the ones delivered by CPLEX within 3600 s and Calvete et al. [3]. For 6 out of 20 instances our proposed approach does not provide the same solution in all the five runs, but we can remark that the gap ranges between 0.0003 and 0.103, fact that proves the stability of our proposed solution approach. Instance 17 is the only one for which the algorithm developed by Calvete et al. [3] delivered a better value of the minimum objective function (Z_{min}) than our proposed algorithm, but our maximum objective value function (Z_{max}) is better.

Overall, the comparison between our proposed solution approach and the algorithm of Calvete et al. [3] can be summarized as follows:

Our algorithm provided the best maximum objective value function (Z_{max}) and the best gap for each of the 20 instances and our algorithm provided the best minimum objective function (Z_{min}) for each of the instances, with only one exception: instance 17, for which our solution is 0.009% weaker. The computation times of our algorithm are better for 10 out of 20 instances. Our algorithm needed longer computation times for some instances, but that is explicable because our algorithm found better solutions for those instances.

Compared with CPLEX, our algorithm found better solutions or the same solutions for 18 out of 20 instances. The exceptions are instances 17 and 19, for which our solutions are 0.009% respectively 0.016% weaker. Our algorithm performs faster than CPLEX for all the instances.

5 Conclusions

In this paper, we described a novel hybrid genetic algorithm for solving the two-stage transportation problem with fixed charges associated to the routes. Our method incorporates a linear programming optimization procedure within the framework of a genetic algorithm. Some important features of our proposed algorithm are: the use of an efficient representation in which the chromosomes are generated in two stages, the use of several chromosome populations that are created and that evolve separately to different offspring, which are finally merged into the populations, giving us the possibility to explore other parts of the solutions space and escaping from local optima.

We evaluated the performance of the proposed solution approach on a set of benchmark instances recently proposed by Calvete et al. [3]. The computational results that we achieved, prove the efficiency of our proposed solution approach in yielding high quality solutions within reasonable running times, besides its superiority against other existing competing methods from the literature.

In future, we plan to improve the developed hybrid genetic algorithm by combining with local search methods and to evaluate the generality and scalability of the proposed solution approach by testing it on larger instances.

References

1. Balinski, M.I.: Fixedcost transportation problems. Nav. Res. Logist. **8**(1), 41–54 (1961)
2. Buson, E., Roberti, R., Toth, P.: A reduced-cost iterated local search heuristic for the fixed-charge transportation problem. Oper. Res. **62**(5), 1095–1106 (2014)
3. Calvete, H., Gale, C., Iranzo, J., Toth, P.: A matheuristic for the two-stage fixed-charge transportation problem. Comput. Oper. Res. **95**, 113–122 (2018)
4. Cosma, O., Danciulescu, D., Pop, P.C.: On the two-stage transportation problem with fixed charge for opening the distribution centers. IEEE Access **7**(1), 113684–113698 (2019)
5. Cosma, O., Pop, P.C., Pop Sitar, C.: An efficient iterated local search heuristic algorithm for the two-stage fixed-charge transportation problem. Carpathian J. Math. **35**(2), 153–164 (2019)
6. Guisewite, G., Pardalos, P.: Minimum concave-cost network flow problems: applications, complexity, and algorithms. Ann. Oper. Res. **25**(1), 75–99 (1990)
7. Jawahar, N., Balaji, A.N.: A genetic algorithm for the two-stage supply chain distribution problem associated with a fixed charge. Eur. J. Oper. Res. **194**, 496–537 (2009)
8. Pop, P.C., Sabo, C., Biesinger, B., Hu, B., Raidl, G.: Solving the two-stage fixed-charge transportation problem with a hybrid genetic algorithm. Carpathian J. Math. **33**(3), 365–371 (2017)
9. Pop, P.C., Matei, O., Pop Sitar, C., Zelina, I.: A hybrid based genetic algorithm for solving a capacitated fixed-charge transportation problem. Carpathian J. Math. **32**(2), 225–232 (2016)
10. Raj, K.A.A.D., Rajendran, C.: A genetic algorithm for solving the fixed-charge transportation model: two-stage problem. Comput. Oper. Res. **39**(9), 2016–2032 (2012)
11. Hundt, R.: Loop recognition in C++/Java/Go/Scala. In: Proceedings of Scala Days (2011). https://days2011.scala-lang.org/sites/days2011/files/ws3-1-Hundt.pdf
12. https://www.cpubenchmark.net/compare/Intel-Pentium-D-830-vs-Intel-i5-4590/1127vs2234

Do People Use Naming Conventions in SQL Programming?

Aggelos Papamichail, Apostolos V. Zarras[(✉)], and Panos Vassiliadis

Department of Computer Science and Engineering, University of Ioannina, Ioannina, Greece
{apapamichail,zarras,pvassil}@cs.uoi.gr

Abstract. In this paper, we investigate the usage of naming conventions in SQL programming. To this end, we define a reference style, consisting of naming conventions that have been proposed in the literature. Then, we perform an empirical study that involves the database schemas of 21 open source projects. In our study, we evaluate the adherence of the names that are used in the schemas to the reference style. Moreover, we study how the adherence of the names to the reference style evolves, during the lifetime of the schemas. Our study reveals that many conventions are followed in all schemas. The adherence to these conventions is typically stable, during the lifetime of the schemas. However, there are also conventions that are partially followed, or even not followed. Over time, the adherence of the schemas to these conventions may improve, decay or remain stable.

Keywords: Naming conventions · Coding styles · SQL programming

1 Introduction

Take a look at the code snippet that is given in Listing 1. It is a typical SQL table definition from the database schema of Joomla (Table 2). There are several naming issues that clutter the definition of the table. For instance, the name of the table, `'#__menu'`, begins with a sequence of special characters. Moreover, the table name and the column names are quoted. In general, the use of special characters and quotes in names is not considered a good practice, for compatibility and portability reasons [9,11]. Several column names consist of multiple terms. Concerning readability, this practice is perfectly fine. However, the way of separating the terms is not consistent. For some multi-term names the terms are separated with underscores (e.g., `'checked_out'`, `'checked_out_time'`), other multi-term names are in camelCase (e.g., `'browserNav'`), while there are also multi-term names without any separation between the constituent terms (e.g., `'menutype'`, `'utaccess'`). Another possible readability problem is the use of acronyms in some column names (e.g., `'lft'`, `'rgt'`) [9,11]. From a lexicographical point of view, table names are typically in plural or in some collective form, while column names are in singular form [9,11]. However, in Listing 1 the name of the table is in singular form.

A. Chatzigeorgiou et al. (Eds.): SOFSEM 2020, LNCS 12011, pp. 429–440, 2020.
https://doi.org/10.1007/978-3-030-38919-2_35

```
--
-- Table structure for table ‘#__menu‘
--

CREATE TABLE ‘#__menu‘ (
  ‘id‘ int(11) NOT NULL auto_increment,
  ‘menutype‘ varchar(75) default NULL,
  ‘name‘ varchar(255) default NULL,
  ‘alias‘ varchar(255) NOT NULL default ’’,
  ‘link‘ text,
  ‘type‘ varchar(50) NOT NULL default ’’,
  ‘published‘ tinyint(1) NOT NULL default 0,
  ‘parent‘ int(11) unsigned NOT NULL default 0,
  ‘componentid‘ int(11) unsigned NOT NULL default 0,
  ‘sublevel‘ int(11) default 0,
  ‘ordering‘ int(11) default 0,
  ‘checked_out‘ int(11) unsigned NOT NULL default 0,
  ‘checked_out_time‘ datetime NOT NULL default ’0000-00-00␣00:00:00’,
  ‘pollid‘ int(11) NOT NULL default 0,
  ‘browserNav‘ tinyint(4) default 0,
  ‘access‘ tinyint(3) unsigned NOT NULL default 0,
  ‘utaccess‘ tinyint(3) unsigned NOT NULL default 0,
  ‘params‘ text NOT NULL,
  ‘lft‘ int(11) unsigned NOT NULL default 0,
  ‘rgt‘ int(11) unsigned NOT NULL default 0,
  ‘home‘ INTEGER(1) UNSIGNED NOT NULL DEFAULT 0,
  PRIMARY KEY  (‘id‘),
  KEY ‘componentid‘ (‘componentid‘,‘menutype‘,‘published‘,‘access‘),
  KEY ‘menutype‘ (‘menutype‘)
) TYPE=MyISAM CHARACTER SET ‘utf8‘;
```

Listing 1. A typical SQL table definition in Joomla.

Using appropriate naming conventions in source code is important for portability, readability and maintainability reasons [14]. In this paper, we investigate the use of naming conventions in SQL programming. Specifically, we perform an empirical study that involves 21 database schemas found in respective free and open source (FOSS) projects. To begin, we introduce *a reference style* that consists of a set of naming conventions, which have been proposed in the literature [9,11]. Then, we focus on two issues: (1) *we assess the adherence of the names that are used in the schemas to the naming conventions of the reference style*; (2) *we investigate the evolution of the schemas, to see if the adherence of the names to the conventions improves, decays or remains stable.* To assess the adherence of a schema to the reference style *we developed a tool*, called *DBSea*, which is available as an open source project[1].

The rest of this paper is structured as follows. In Sect. 2, we discuss related work. In Sect. 3, we detail the reference naming style and the setup of our study. In Sect. 4, we present our findings. Finally, in Sect. 5 we conclude with a summary of our contribution and the future perspectives of this work.

2 Related Work

Several interesting empirical studies have been performed regarding the usage of names in source code. According to these studies, the usage of full word identifiers

[1] github.com/apapamichail/DBsea.

improves source code readability [4,13]. Typically, short identifiers take longer to understand [10]. Nevertheless, in some cases single letter identifiers may convey meaningful information [5]. The styles used for separating multi-term identifiers like CamelCase and underscores are also important for software comprehension [6]. The impact of each style varies depending on the development task and the developers' experience. Further research efforts study naming patterns and anti-patterns for classes, attributes, methods, variables and so on [3,7,8]. Moreover, there are studies that report naming patterns used in visual programming [16]. Another line of research, concerns techniques for the recommendation of class, variable and method names [1,2,12] that can be used to improve the readability of the code.

Differently from the aforementioned efforts, *in this paper we perform an empirical study that concerns the usage of naming conventions in SQL programming.*

3 Setup

In this section, we discuss in detail the naming conventions and the database schemas that we consider in our study.

3.1 Reference SQL Naming Style

The naming conventions that we consider come from Joe Celko's SQL programming style [9], Simon Holywell's SQL style guide [11], and the ISO-11179 naming standard. We do not claim that this list of conventions is complete, neither that it covers the in-house style of every possible organization. However, we believe it is a good starting point for our study as they come from 3 well-known sources that are not specific to any particular DBMS. We do not consider the assumed naming conventions as ground truth. Instead, we assess the extent to which they are actually used in practice.

Table 1, summarizes the naming conventions that we consider. In the table, each convention is introduced with a brief description and an acronym that we use to facilitate the visualization of the results. We categorize the conventions with respect to their scope, which can be tables and/or columns. Moreover, we categorize the conventions with respect to their purpose, which can be to facilitate portability, readability, and maintainability.

For portability reasons between different commercial and open source DBMSs it is better to start SQL elements names with letters (SWL) and end them with letters or numbers (EWL). In addition, it is better to avoid using special characters (ASC), spaces (AUS) and delimiters (AUD). Moreover, it is recommended to use names of a proper length (UPL), not exceeding respective standard upper bound limits. Celko provides a table with various identifier length limits that have been assumed in different DBMSs. Based on this table, the limit that we assume in our study is 30 characters.

Table 1. Reference style.

Purpose	Scope	Acronym	Mnemonic
Portability	Tables & Columns	SWL	Start With Letter
		EWL	End With Letter or number
		ASC	Avoid Special Characters
		AUS	Avoid Using Spaces
		AUD	Avoid Using Delimiters
		UPL	Use Proper Length
Readability, Maintainability		UTS	Uniform Term Separation
		UMW	Use More Words
		ACC	Avoid Camel Case
		ACU	Avoid Consecutive Underscores
		ARW	Avoid Reserved Words

Purpose	Scope	Acrony m	Mnemonic
Readability, Maintainability	Tables only	SWC	Start With Capital
		TIP	Tables In Plural
		ACN	Avoid Concatenating Names
	Columns only	USP	Use Standardized Postfixes
		CIS	Columns In Singular
		DCN	Different Column Names
		ANP	Avoid Names by Place
		AII	Avoid "Id" as Identifier

For readability reasons, it is better to use a uniform term separation style (UTS) for multi-term names. To facilitate understanding, names that consist of multiple terms should contain more words than acronyms (UMW). According to [9], names in CamelCase should be avoided (ACC) because empirical evidence indicates that they disrupt the flow of reading, by making the eye concentrate on case changes. Similarly, the use of consecutive underscores (ACU) and reserved words (ARW) in names is not a good practice.

In the context of an SQL schema, tables are unique concepts that represent collections of related data. Therefore, table names should be treated like proper nouns, starting with a capital letter (SWC). As tables represent collections of related data, it is expected that table names should be in plural (TIP). It would also be good to avoid concatenating table names (ACN) to name relations between them. This is a common practice for naming relations, but the concatenated names do not reveal the purpose of the relation.

Concerning column names, when needed, it would be good to use standardized postfixes (USP); a list of such postfixes is provided in [9]. Columns, represent specific properties of the related data. Hence, it is expected that column names should be in singular form (CIS). Column names should be different from table names (DCN). Defining column names by place should also be avoided (NBP), in the sense that column names should not include table names as prefixes or suffixes. Moreover, using "id" to name primary keys is not a good practice (AII), because it does not reveal the purpose of the keys.

3.2 Database Schemas

In our study we consider a well-established large collection of database schemas, the only available that comprises multiple schema versions. We have used this collection in previous studies to investigate the evolution of database schemas [15, 17, 18]. The collection consists of five scientific projects from CERN, two medical projects and eleven CMS projects. Table 2, gives detailed statistics regarding the

Table 2. Schemas statistics.

Case Study		# Revisions	# Tables at First Known Version	# Tables at Last Known Version	# Columns at First Known Version	# Columns at Last Known Version	Domain
ATLAS	A particle physics experiment at CERN	84	73	56	709	857	**Scientific**
CASTOR	A hierarchical storage system for physics data	192	62	74	632	838	
SRM2	A client system for CASTOR.	58	11	11	54	84	
DQ2	A data management system for ATLAS.	54	10	26	116	184	
EGEE	A project that provides access to high-throughput	16	6	9	34	63	
Ensembl	A project that concerns genome databases.	528	19	75	82	486	**Medical**
BioSQL	A shared database for storing sequence data.	47	21	28	227	731	
Typo3	A CMS for managing any kind of digital content.	98	10	23	122	421	**CMSs**
PhpBB	An Internet forum package.	133	61	65	613	565	
PhpWiki	A wiki that supports multiple storage back-ends.	21	10	10	33	49	
SlashCode	The web site for All Things Slash.	398	42	87	259	610	
Zabbix	An enterprise network monitoring project.	27	47	48	312	313	
e107	A project for the creation of dynamic sites.	17	33	34	261	274	
Coppermine	A photo gallery project.	117	8	22	85	169	
DekiWiki	A platform for content and mashups.	16	28	40	204	315	
Nucleus	A simple CMS for web blogs.	4	20	20	110	112	
OpenCart	An e-commerce platform for online merchants.	165	48	114	74	230	
TikiWiki	A system for wikis, forums and blogs.	153	207	215	1528	1628	
XOOPS	A platform for community websites.	7	31	32	297	129	
MediaWiki	The platform of Wikimedia projects.	322	17	50	100	318	
Joomla	A project for publishing web content.	45	35	36	307	321	

database schemas of the projects. Specifically, for each schema the table provides the number of versions it went through, the total number of tables and the total number of columns in the first and the last known versions of the schema. All the data sets are available at the web site of the DAINTINESS group[2].

4 Research Questions and Answers

In this section we discuss the findings of our study, organized with respect to the research questions that we investigate. To address our questions we define respective metrics that measure the adherence of the table/column names used in the examined schemas to the conventions of the reference style, and the way that the adherence of the names to the naming conventions evolves, during the lifetime of the schemas. Table 3, gives details about the basic notions that we assume for the definition of the metrics.

4.1 Is the Reference Style Followed by the Schemas?

To address our first research question, we define a simple metric, called *Adherence Indicator (AI)*.

[2] github.com/DAINTINESS-Group/EvolutionDatasets.

Table 3. Basic notions and notation.

- $\Omega = \{S^1, S^2, \ldots, S^K\}$ is the overall set of schemas that we consider in our study.
- $R_{NC} = \{nc_1, nc_2, \ldots, nc_N\}$ is the set of naming conventions that constitute the reference naming style.
- $H^{S^j} = \{S^j_f, S^j_{f+1}, \ldots, S^j_\ell\}$ denotes a history of subsequent versions of a database schema $S^j \in \Omega$.
- $\Omega^\ell = \{S^1_\ell, S^2_\ell, \ldots, S^K_\ell\}$ is the set of the last known versions of the examined schemas Ω.
- $S^j_i.N_T$ is the set of table names, used in a schema version $S^j_i \in H^{S^j}$.
- $S^j_i.N_C$ is the set of column names, used in a schema version $S^j_i \in H^{S^j}$.
- $\Omega^\ell.N_T = \{S^1_\ell.N_T, S^2_\ell.N_T, \ldots, S^K_\ell.N_T\}$ denotes the sets of table names, used in the last known versions Ω^ℓ of the examined schemas.
- $\Omega^\ell.N_C = \{S^1_\ell.N_C, S^2_\ell.N_C, \ldots, S^K_\ell.N_C\}$ denotes the sets of column names, used in the last known versions Ω^ℓ of the examined schemas.
- $H^{S^j}_T = \{S^j_f.N_T, S^j_{f+1}.N_T, \ldots, S^j_\ell.N_T\}$ denotes the history of tables names, used throughout the history H^{S^j} of $S^j \in \Omega$.
- $H^{S^j}_C = \{S^j_f.N_C, S^j_{f+1}.N_C, \ldots, S^j_\ell.N_C\}$ denotes the history of column names, used throughout the history H^{S^j} of $S^j \in \Omega$.
- $\Omega^{H_T} = \{H^{S^1}_T, H^{S^2}_T, \ldots, H^{S^K}_T\}$, denotes the table names histories of the examined schemas Ω.
- $\Omega^{H_C} = \{H^{S^1}_C, H^{S^2}_C, \ldots, H^{S^K}_C\}$, stands for the column names histories of the examined schemas Ω.

The subscript **T|C** in the formulas (Definitions 1 to 5) indicates that a formula is applicable to tables (**T**) or columns (**C**). Equivalently, in the text we use the term **table/column** to express this property.

Definition 1. *[Adherence Indicator] The Adherence Indicator is a function $AI(S^j_i.N_{T|C}, nc)$ that takes as input a set of table/column names $S^j_i.N_{T|C}$, used in a schema version $S^j_i \in H^{S^j}$ of a schema $S^j \in \Omega$, and a naming convention $nc \in R_{NC}$. The value of the function gives the percentage of the table/column names that adhere to nc. More formally, $AI(S^j_i.N_{T|C}, nc) = \frac{|S^j_i.N_A|}{|S^j_i.N_{T|C}|} * 100\%$, where $S^j_i.N_A$ is the subset of $S^j_i.N_{T|C}$ that adhere to nc.*

We focus our analysis on the last known versions Ω^ℓ of the examined schemas. Later, (Sect. 4.2) we show that these versions are representative of the schemas' histories Ω^H.

To begin our analysis, we consider the reference style as a whole. Specifically, our goal is to determine the *adherence of table/column names to the overall style.* To achieve this goal, we calculate the values of $AI(S^j_\ell.N_{T|C}, nc)$ for the naming conventions of the reference style R_{NC} and the sets of table/column names $\Omega^\ell.N_{T|C}$. For each set of table/column names $S^j_\ell.N_{T|C}$, we partition the naming conventions of R_{NC} in three subsets, $P_{S^j_\ell.N_{T|C}} = \{CF, PF, NF\}$, containing the naming conventions that are completely followed ($AI(S^j_\ell.N_{T|C}, nc) = 100\%$),

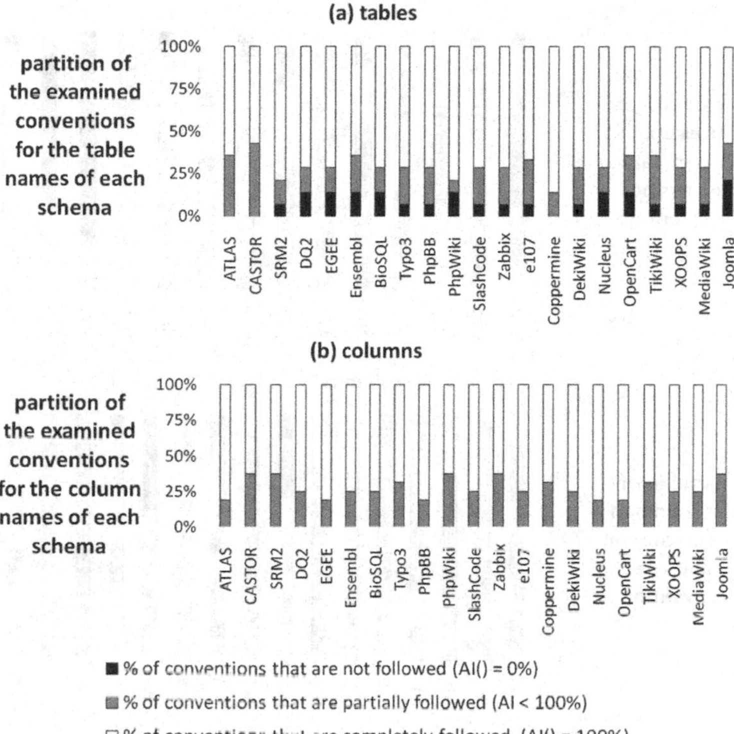

Fig. 1. Adherence of the names used in the schemas to the reference naming style.

partially followed ($AI(S_\ell^j.N_{T|C}, nc) < 100\%$), not followed ($AI(S_\ell^j.N_{T|C}, nc) = 0\%$) by $S_\ell^j.N_{T|C}$, respectively.

Figure 1, shows the results that we obtain. Specifically, for the sets of table/-column names $\Omega^\ell.N_{T|C}$ the figure provides respective stacked bars, describing the partitions of R_{NC}^{CI}. A stacked bar is divided in three parts, each giving the percentage[3] of conventions that belong to a partition subset.

In the results, we observe that *the names used in the schemas do not follow the reference style faithfully.* On the positive side, many conventions are completely followed. Regarding tables, the percentage of naming conventions that are completely followed is higher than 62%, in all schemas. As for columns, the respective percentage of naming conventions is higher than 57%, in all schemas. On the negative side, *several conventions are partially followed* and *few others are not followed at all.* Concerning tables, the percentage of naming conventions that are partially followed ranges from 7.14% to 42.8%, while the percentage of conventions that are not followed varies from 0% to 21.43%. Regarding columns,

[3] Due to the lack of space we use percentages to give an overview of the results. The complete raw results that we obtained in our study can be found in www.cs.uoi.gr/~zarras/SQLNamingConventions/SQLStatisticsSLA.rar.

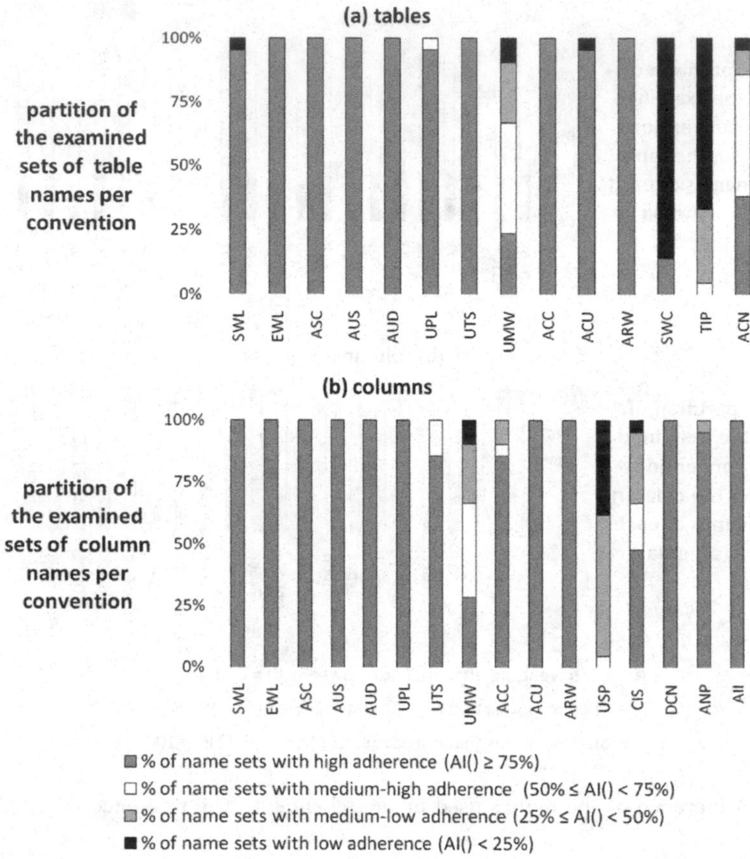

Fig. 2. Adherence of the names used in the schemas to each naming convention.

the percentage of naming conventions that are partially followed ranges from 18.75% to 37.50%.

Next, we focus our analysis on the individual naming conventions. Our objective is to assess the *adherence of table/column names to each naming convention.* To address this issue, for each naming convention $nc \in R_{NC}$ we partition the examined sets of table/column names $\Omega^\ell.N_{T|C}$ in four subsets, $P_{nc} = \{A_H, A_{MH}, A_{ML}, A_L\}$, containing the sets of tables/column names that have high ($AI(S^j_\ell.N_{T|C}, nc) \geq 75\%$), medium-high ($50\% \leq AI(S^j_\ell.N_{T|C}, nc) < 75\%$), medium-low ($25\% \leq AI(S^j_\ell.N_{T|C}, nc) < 50\%$), low ($AI(S^j_\ell.N_{T|C}, nc) < 25\%$) adherence to nc.

Figure 2, shows the results that we obtain. In particular, for the naming conventions of the reference style, the figure provides corresponding stacked bars, describing the partitions of the examined sets of table/column names $\Omega^\ell.N_{T|C}$. A stacked bar is divided in four parts, each giving the percentage of the sets of table/column names that belong to a partition subset.

In general, we observe *high adherence of the sets of table/column names to most naming conventions*. Regarding tables, the percentage of name sets that belong to A_H is higher than 95.24%, in 10 out of 14 conventions. As for columns, the percentage of name sets that belong to A_H is higher than 85.71%, in 13 out of 16 conventions. Concerning tables, the exceptions are UMW, SWC, TIP and ACN. Regarding columns, the exceptions are UMW, USP and CIS. All of these conventions concern the readability of the schemas. In particular, the table/column names that are used in the schemas may comprise more acronyms than words (UMW). Moreover, the schemas may contain concatenated table names (ACN), table names are not in plural form (TIP) and/or table names that do not begin with capital letters (SWL). Similarly, the schemas may contain column names that are not in singular form (CIS) and/or column names with non standardized postfixes (USP).

4.2 Does the Adherence of the Schemas to the Reference Style Evolve?

Having some clear evidence of adherence to the reference style, we move to the next issue that we consider in our study. We investigate the adherence of the table/column names to the reference style, with respect to the history of the examined schemas.

Specifically, we check if the adherence of the table/column names improves, decays or stays the same, between the first and the last known versions of the examined schemas. For this purpose, we employ the *Adherence Progress Indicator (API)* metric, defined below.

Definition 2. [Adherence Progress Indicator] *We define the Adherence Progress Indicator as a function $API(H^{S^j}_{T|C}, nc)$ that takes as input the history of table/column names $H^{S^j}_{T|C}$, used in $S^j \in \Omega$, and a naming convention $nc \in R_{NC}$. The value of the function is the difference between the value of the Adherence Indicator function for the names $S^j_\ell.N_{T|C}$, used in the last known version $S^j_\ell \in H^{S^j}$ of S^j, and the names $S^j_f.N_{T|C}$, used in the first known version $S^j_f \in H^{S^j}$ of S. Formally, $API(H^{S^j}_{T|C}, nc) = AI(S^j_\ell.N_{T|C}, nc) - AI(S^j_f.N_{T|C}, nc)$.*

We calculate the values of $API(H^{S^j}_{T|C}, nc)$ for the naming conventions R_{NC} and the table/column names histories $\Omega^{H_{T|C}}$ that we consider in our study. For each naming convention $nc \in R_{NC}$ we partition the histories in three subsets $P^{H_{T|C}}_{nc} = \{A_I, A_S, A_D\}$, containing histories of table/column names with improved ($API(H^{S^j}_{T|C}, nc) > 0\%$), stable ($API(H^{S^j}_{T|C}, nc) = 0\%$) and decayed ($API(H^{S^j}_{T|C}, nc) < 0\%$) adherence to nc, respectively.

Figure 3 gives the results that we obtain. For the naming conventions of the reference style, the figure provides corresponding stacked bars, describing the partitions of the examined table/column names histories $\Omega^{H_{T|C}}$. A stacked bar is divided in three parts, each giving the percentage of histories that belong to a partition subset.

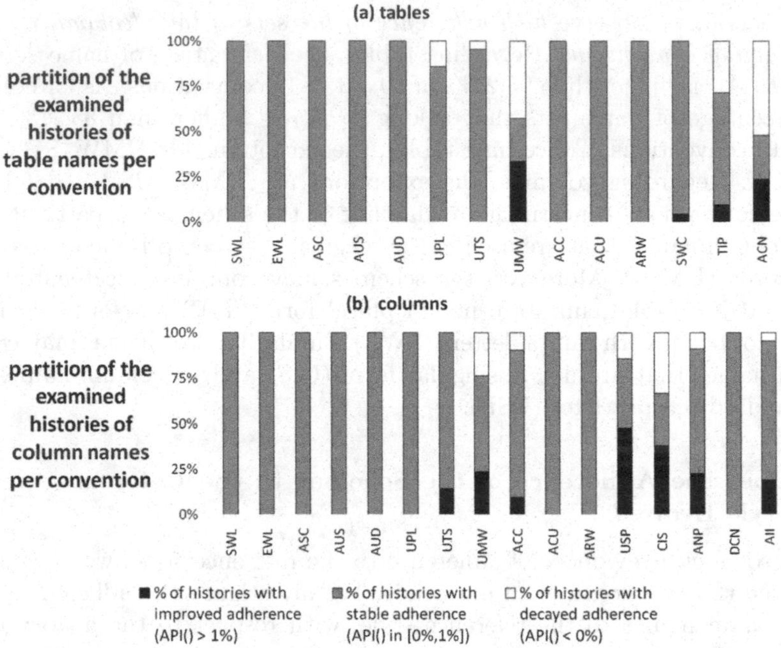

Fig. 3. Progress of adherence to each naming convention.

In the results we see that *the adherence of the examined table/column names to most of the naming conventions is stable.* Regarding tables, the percentage of histories that belong to A_S is higher than 85.71%, in 11 out of 14 conventions. As for columns, the percentage of histories that belong to A_S is higher than 71.43%, in 14 out of 16 conventions.

The adherence of table/column names to the rest of the naming conventions may *improve*, *decay*, or *remain stable*. Specifically, in the largest percentage of histories, the adherence of table names to UMW and ACN decays (Fig. 3(a)). Nevertheless, we also observe considerable percentages of histories with improved and stable adherence. In the case of TIP, the adherence of the table names is stable in a large percentage of histories. However, there is also a notable percentage of histories with decayed adherence, and a small percentage of histories with improved adherence. In the largest percentage of histories, the adherence of column names to USP and CIS improves, while there are also notable percentages of histories with stable and decayed adherence (Fig. 3(b)).

4.3 Threats to Validity

A possible threat to the *construct validity* of our study is deficiencies of the tool that we used for the assessment of the examined schemas. To cope with this threat, we developed DBSea based on well-known open source libraries and tools

(WordNet[4], Apache Commons Math[5], ANTLR[6]). For the validation of the tool, we developed an extensive set of unit tests that covers the naming conventions of the reference style. Moreover, we manually checked the correctness of DBSea by inspecting random samples of the collected data. *Internal validity*, is not an issue in our study, as we do not attempt to establish any particular cause-effect relationships.

Regarding *external validity*, our study has been conducted in a well-defined context, database schemas used in FOSS. We studied a reasonable number of schemas with variance in the respective fields of use. The schemas also vary in size and number of versions. Thus, we believe that the examined schemas are representative for the case of open source projects. Nevertheless, studying more schemas from open source and industrial projects, may reveal further interesting observations.

5 Conclusion

In this paper, we defined a reference style consisting of naming conventions that have been proposed in the literature. Then, we assessed whether these conventions are used in practice in a study that involved 21 schemas used in respective FOSS projects. We observed that many conventions are followed in all schemas, but there are also conventions that are partially followed, or not followed at all. During the lifetime of the schemas, the adherence to the conventions that are generally followed is stable, while the adherence to the rest of the conventions may improve, decay, or remain stable.

Our study is a starting point towards the investigation of further issues concerning the usage of naming conventions in SQL programming. For instance, it would be interesting to examine why some projects follow certain conventions more often than others. Another possible issue is to find reasons that make developers deviate from naming conventions. Using good naming practices and conventions is a basic prerequisite for the development of clean SQL code. Nevertheless, it is not the only one; the structure of the code is also important. Looking for best practices, patterns, and quality metrics in this context is an interesting issue for future research. Another interesting research direction concerns tools and techniques for the refactoring of SQL code.

Acknowledgements. We would like to thank the anonymous reviewers for their useful suggestions and comments.

References

1. Allamanis, M., Barr, E.T., Bird, C., Sutton, C.A.: Learning natural coding conventions. In: Proceedings of the 22nd ACM SIGSOFT International Symposium on Foundations of Software Engineering (FSE), pp. 281–293 (2014)

[4] wordnet.princeton.edu.

[5] commons.apache.org/proper/commons-math/.

[6] www.antlr.org/.

2. Allamanis, M., Barr, E.T., Bird, C., Sutton, C.A.: Suggesting accurate method and class names. In: Proceedings of the Joint 23rd ACM SIGSOFT Symposium on the Foundations of Software Engineering and 15th European Software Engineering Conference (FSE/ESEC), pp. 38–49 (2015)
3. Arnaoudova, V., Penta, M.D., Antoniol, G.: Linguistic antipatterns: what they are and how developers perceive them. Empirical Softw. Eng. **21**(1), 104–158 (2016)
4. Avidan, E., Feitelson, D.G.: Effects of variable names on comprehension an empirical study. In: Proceedings of the 25th International Conference on Program Comprehension (ICPC), pp. 55–65 (2017)
5. Beniamini, G., Gingichashvili, S., Klein-Orbach, A., Feitelson, D.G.: Meaningful identifier names: the case of single-letter variables. In: Proceedings of the 25th International Conference on Program Comprehension (ICPC), pp. 45–54 (2017)
6. Binkley, D., Davis, M., Lawrie, D., Maletic, J.I., Morrell, C., Sharif, B.: The impact of identifier style on effort and comprehension. Empirical Softw. Eng. **18**(2), 219–276 (2013)
7. Butler, S.: Mining Java class identifier naming conventions. In: Proceedings of the 34th IEEE-ACM-SIGSOFT International Conference on Software Engineering (ICSE), pp. 1641–1643 (2012)
8. Butler, S., Wermelinger, M., Yu, Y.: A survey of the forms of Java reference names. In: Proceedings of the 23rd IEEE International Conference on Program Comprehension, (ICPC), pp. 196–206 (2015)
9. Celko, J.: SQL Programming Style. Morgan-Kaufmann, Burlington (2005)
10. Hofmeister, J.C., Siegmund, J., Holt, D.V.: Shorter identifier names take longer to comprehend. Empirical Softw. Eng. **24**(1), 417–443 (2019)
11. Holywell, S.: SQL Style Guide. www.sqlstyle.guide
12. Kashiwabara, Y., Onizuka, Y., Ishio, T., Hayase, Y., Yamamoto, T., Inoue, K.: Recommending verbs for rename method using association rule mining. In: Proceedings of the 21st IEEE International Conference on Software Analysis, Evolution, and Reengineering (SANER), pp. 323–327 (2014)
13. Lawrie, D., Morrell, C., Feild, H., Binkley, D.: What's in a Name? A study of identifiers. In: Proceedings of the 14th IEEE International Conference on Program Comprehension (ICPC), pp. 3–12 (2006)
14. Martin, R.C.: Clean Code - A Handbook of Agile Software Craftsmanship. Prentice Hall, Upper Saddle River (2009)
15. Skoulis, I., Vassiliadis, P., Zarras, A.V.: Growing up with stability: how open-source relational databases evolve. Inf. Syst. **53**, 363–385 (2015)
16. Swidan, A., Serebrenik, A., Hermans, F.: How do scratch programmers name variables and procedures? In: 17th IEEE International Working Conference on Source Code Analysis and Manipulation (SCAM), pp. 51–60 (2017)
17. Vassiliadis, P., Kolozoff, M., Zerva, M., Zarras, A.V.: Schema evolution and foreign keys: a study on usage, heartbeat of change and relationship of foreign keys to table activity. Computing **101**(10), 1431–1456 (2019)
18. Vassiliadis, P., Zarras, A.V., Skoulis, I.: Gravitating to rigidity: patterns of schema evolution - and its absence - in the lives of tables. Inf. Syst. **63**, 24–46 (2017)

Employing Costs in Multiagent Systems with Timed Migration and Timed Communication

Bogdan Aman(✉) and Gabriel Ciobanu

Faculty of Computer Science, Alexandru Ioan Cuza University, Iaşi, Romania
{bogdan.aman,gabriel}@info.uaic.ro

Abstract. We use a process calculus to describe easily multiagent systems with timeouts for mobility and communication, and with assigned costs for agents actions and for the locations of a distributed network. After presenting an operational semantics and some results regarding this calculus, we provide a translation of the multiagent systems to weighted timed automata having a bisimilar behaviour. Such a translation allows the use of an existing software tool for verification of various properties of the multiagent systems, and for optimizing the costs involved in the distributed networks of mobile agents.

1 Introduction

Multiagent systems with mobility and communication between agents present several challenges for formal methods. Each agent that moves in a distributed system and interacts with other agents should confine to some timing constraints, keeping also certain flexibility in performing its migration. These challenges include a specification language able to describe easily the distributed network of mobile agents, a formal semantics able to describe the execution steps, as well as automated verification involving behavioural quantitative aspects (timing, costs). Automated verification is more and more required because the fact that agents move and perform local communication makes these systems not only complicated, but also increasingly used in distributed networks.

In this paper we integrate a rather simple specification language with a rather sophisticated software tool able to simulate and verify various properties involving timeouts and costs for migration and communication. We introduce an extension with costs of the process calculus TiMo (Timed Mobility); TiMo is used for describing distributed systems composed of a finite number of explicit locations [7,8]). The mobile agents (modelled as processes) can migrate between the locations of the distributed network, and also interact with other agents by using local communication. Timeouts attached to migration and communication actions in TiMo offer flexibility to agents behaviour. In the version of TiMo used in this paper, the timeout for migration marks the fact that the movement to another location must be performed after exactly a number of time units, and the new location could be a variable instantiated in previous steps.

It is worth noting that TiMo is able to describe the dynamic topology of the multiagent systems in a compositional way. This aspect is very useful in

A. Chatzigeorgiou et al. (Eds.): SOFSEM 2020, LNCS 12011, pp. 441–453, 2020.
https://doi.org/10.1007/978-3-030-38919-2_36

specifying such a system because the system is typically composed of several mobile agents combining their behaviours for a common goal. We propose a version of TiMo extended with some costs assigned to actions and locations in order to illustrate the effort of performing an action or of staying in a location for a period of time, respectively. Thus, this extension is called cTiMO (with 'c' from 'cost'). To determine a cost by performing a specific arithmetic operation on clock values, we consider that the computations are performed only when the clocks display integer values.

The Paper is Structured as Follows: Section 2 presents the syntax and operational semantics of cTiMO, together with some results regarding the costs. Section 3 briefly presents weighted timed automata, an extension of timed automata with costs added to both edges and locations. These weighted timed automata are then related to cTiMO networks. Based on the relationship, we can use the existing tool UPPAAL [5] to simulate and verify multiagent systems with timeouts and costs. Various quantitative properties of cTiMO networks can be verified in UPPAAL, and so optimize execution time and certain costs.

2 Syntax and Operational Semantics of cTiMO

Syntax of cTiMO is presented in Table 1, where the following are assumed:

- *Loc*, *Chan* and *Id* are sets of locations, communication channels, and process identifiers, respectively;
- for each process identifier $id \in Id$, there is an arity m_{id} and distinct variables u_i $(1 \leq i \leq m_{id})$ such that a unique definition $id(u_1, \dots, u_{m_{id}}) \stackrel{def}{=} P_{id}$ exists;
- $a \in Chan$ denotes a communication channel, and $l \in Loc$ is a location or a location variable; $c \in \mathbb{N}$ denotes a cost, and $t \in \mathbb{N}$ is a *timeout* of an action; u and v denote a tuple of variables, and a tuple of expressions over values, variables and allowed operations, respectively.

Table 1. cTiMO syntax.

Processes	$P, Q ::= \mathsf{go}_c^t\, l \text{ then } P \;\mid$	(move)
	$a_c^{\Delta t}!\langle v\rangle \text{ then } P \text{ else } Q \;\mid$	(output)
	$a_c^{\Delta t}?(u) \text{ then } P \text{ else } Q \;\mid$	(input)
	$\mathbf{0} \;\mid$	(termination)
	$id(v)$	(recursion)
Located processes	$L ::= l[[_c\, P]]$	
Networks	$N ::= L \;\mid\; L \mid N$	

The timeout associated to a move process denotes the amount of time that has to pass before the migration to a new location is performed, while the cost constraint applied to a move process specifies the cost needed to perform the movements between locations. Migration is achieved by using a process $\mathsf{go}_c^t\, l$ then P

describing the movement of process P from the location in which it currently resides to the location l in exactly t time units and with a cost c. As location l can be a variable, its value can dynamically change by performing local communication with other processes. This provides flexibility, as processes can change their behaviour based on certain changes in the multiagent system. A timer is denoted either by t (for migration actions) or Δt (for output and input actions). A timer t associated to a migration process $\mathsf{go}^t_c\, l$ then P indicates that process P should change its location by moving to location l in exactly t time units, while the cost of this movement is c. A timer Δt from an output process $a^{\Delta t}_c!\langle z\rangle$ then P else Q restricts the availability of channel a for sending the value z to at most t time units, and the cost of performing this operation is c. In a similar manner, the timer Δt from an input process $a^{\Delta t}_c?(x)$ then P else Q restricts the availability of channel a for receiving a value to at most t time units. In case of a successful communication by using output or input actions, the previous processes continue by executing P, while an unsuccessful communication continues by executing the alternative process Q; in order the simplify the presentation the cost to perform this switch is the same as performing the communication. Note that performing the switch or the communication are mutual exclusive actions.

The process $\mathbf{0}$ denotes inaction, while a located process $l[[_c\, P]]$ specifies a process P executing at location l where the cost of computing is c. A network N is built from parallel located processes L, and it is well-formed if $fv(N) = \emptyset$.

Note that there is only one binding process, namely $a^{\Delta t}_c?(u)$ then P else Q, in which the variable u is bound within process P (but not within the alternative process Q). All the other variables (including location variables) are free. The sets of free variables for a process P and a network N are denoted by $fv(P)$ and $fv(N)$, respectively, where $fv(P_{id}) \subseteq \{u_1, \ldots, u_{m_{id}}\}$ holds. Processes are defined up to an α-conversion, and the process $P\{v/u, \ldots\}$ means that inside process P all the free occurrences of the variable u are replaced by the value v (possibly after α-converting some names in P in order to avoid clashes). The process $id(v)$ describes recursion, with $id(x) = P_{id}$ being the process definition. The call invokes the process P_{id}, replacing its formal parameters x with values v, namely by launching the process $P_{id}\{v/x\}$.

Structural equivalence $\equiv$ over networks is the smallest congruence given by:

$$N \mid l[[_c\mathbf{0}]] \equiv N \quad N \mid N' \equiv N' \mid N \quad (N \mid N') \mid N'' \equiv N \mid (N' \mid N'').$$

The role of structural relation $\equiv$ is to rearrange a network such that the rules of the *operational semantics* can be applied. Each located process $l_i[[_{c_i}\, P_i]]$ of a network N is called a component.

Operational Semantics of cTiMO is presented in Table 2. A complete step describes individual actions with costs (by $\xrightarrow{\Lambda,C}$), followed by a time step with costs (by $\overset{t,C}{\rightsquigarrow}$). A multiset-labelled transition $N \xrightarrow{\Lambda,C} N'$ indicates that the actions from the multiset Λ are executed in parallel in one step, and the costs of their execution is from the multiset C. The order of the costs in C depends on the order of actions in Λ, as each action has a unique execution cost. If $\Lambda = \{\lambda\}$ and action λ has a cost c, then the notation $N \xrightarrow{\lambda,c} N'$ is used. Given a network N, a time step of length t and costs C is modelled by a transition $N \overset{t,C}{\rightsquigarrow} N'$.

In rule (MOVE0), the migration process $\mathsf{go}^0_{c''}\, l'$ then P moves to location l' in order to behave there as P. The migration cost is c'', and the cost of its computation at the new location l' is computed by using the cost c'.

In rule (COM), an output process $a^{\Delta t}_{c_1}!\langle v\rangle$ then P else Q executing at location l is able to successfully send along channel a a tuple of values v; the input process $a^{\Delta t}_{c_2}?(u)$ then P' else Q' executing at the same location l successfully receives the values on the same channel a. The output and input processes remain at location l and continue their executions as P and $P'\{v/u\}$, respectively. The label $!v@l|?u@l, c_1|c_2$ indicates the fact that the first process performs the output action $!v@l$ with cost c_1, and the second process performs the complementary input action $?u@l$ with cost c_2. The label $!v@l|?u@l$ is an equivalent notation for the label $\{v/u\}@l$ from [3], but is used here because we are interested in the actions taken by each process and not only on the overall evolution as in [3].

Table 2. Operational Semantics of cTiMO.

(DSTOP) $$l[[_c\, \mathbf{0}]] \xrightarrow{t,0} l[[_c\, \mathbf{0}]]$$

(DMOVE) $$l[[_c\, \mathsf{go}^t_{c'}\, l' \text{ then } P]] \xrightarrow{t',c*t'} l[[_c\, \mathsf{go}^{t-t'}_{c'}\, l' \text{ then } P]]\ (t \geq t' \geq 0)$$

(MOVE0) $$l[[_c\, \mathsf{go}^0_{c''}\, l' \text{ then } P]] \mid l'[[_{c'}\mathbf{0}]] \xrightarrow{l \rhd l', c''} l[[_c \mathbf{0}]] \mid l'[[_{c'}\, P]]$$

(COM) $$l[[_c\, a^{\Delta t}_{c_1}!\langle v\rangle \text{ then } P \text{ else } Q]] \mid l[[_c\, a^{\Delta t'}_{c_2}?(u) \text{ then } P' \text{ else } Q']] \xrightarrow{!v@l|?u@l,c_1|c_2} l[[_c\, P]] \mid l[[_c\, P'\{v/u\}]]$$

(DPUT) $$l[[_c\, a^{\Delta t}_{c'}!\langle v\rangle \text{ then } P \text{ else } Q]] \xrightarrow{t',c*t'} l[[_c\, a^{\Delta t-t'}_{c'}!\langle v\rangle \text{ then } P \text{ else } Q]]\ (t \geq t' \geq 0)$$

(PUT0) $$l[[_c\, a^{\Delta 0}_{c'}!\langle v\rangle \text{ then } P \text{ else } Q]] \xrightarrow{a!^{\Delta 0}@l,c'} l[[_c\, Q]]$$

(DGET) $$l[[_c\, a^{\Delta t}_{c'}?(u) \text{ then } P \text{ else } Q]] \xrightarrow{t',c*t'} l[[_c\, a^{\Delta t-t'}_{c'}?(u) \text{ then } P \text{ else } Q]]\ (t \geq t' \geq 0)$$

(GET0) $$l[[_c\, a^{\Delta 0}_{c'}?(u) \text{ then } P \text{ else } Q]] \xrightarrow{a?^{\Delta 0}@l,c'} l[[_c\, Q]]$$

(DCALL) $$\frac{l[[_c\, P_{id}\{v/x\}]] \xrightarrow{t,c*t} l[[_c\, P'_{id}]]}{l[[_c\, id(v)]] \xrightarrow{t,c*t} l[[_c\, P'_{id}]]} \text{ where } id(x) \stackrel{def}{=} P_{id}$$

(CALL) $$\frac{l[[_c\, P_{id}\{v/x\}]] \xrightarrow{id@l,c'} l[[_c\, P'_{id}]]}{l[[_c\, id(v)]] \xrightarrow{id@l,c'} l[[_c\, P'_{id}]]} \text{ where } id(x) \stackrel{def}{=} P_{id}$$

(DPAR), (PAR) $$\frac{N_1 \xrightarrow{t,C_1} N'_1 \quad N_2 \xrightarrow{t,C_2} N'_2}{N_1 \mid N_2 \xrightarrow{t,C_1|C_2} N'_1 \mid N'_2} \qquad \frac{N_1 \xrightarrow{\Lambda_1,C_1} N'_1}{N_1 \mid N_2 \xrightarrow{\Lambda_1,C_1} N'_1 \mid N_2}$$

(DEQUIV) $$\frac{N \equiv N' \quad N' \xrightarrow{t,C} N'' \quad N'' \equiv N'''}{N \xrightarrow{t,C} N'''}$$

(EQUIV) $$\frac{N \equiv N' \quad N' \xrightarrow{\Lambda,C} N'' \quad N'' \equiv N'''}{N \xrightarrow{\Lambda,C} N'''}$$

If an output or input process $a_c^{\Delta 0} *$ then P else Q for $* \in \{!\langle v\rangle, ?(u)\}$ is unable to communicate and its timer reaches the value 0, then the process continues by executing the alternative process Q at the current location (by using the rule (PUT0) or the rule (GET0), respectively). The cost of changing the execution towards the alternative process Q is c, the same as for performing the communication successfully. Rule (CALL) simulates the unfolding of a recursion process. Rules (EQUIV) and (DEQUIV) model the use of the equivalence relation $\equiv$ in order to rearrange a network. Rule (PAR) puts in parallel smaller networks in order to obtain larger networks such that processes have multiple possible evolutions.

The rules starting with the capital letter 'D' are used to model the passing of time. Notice that in the rules (DMOVE), (DPUT) and (DGET) the maximum time that can elapse is limited by the timers of the actions that can be executed next. This implies that in rule (DPAR), the time that can pass is the minimum time that can pass in the two systems. Following the work done on timed π-calculus [16], we only consider non-Zeno behaviours, namely just a finite number of transitions can be executed within a finite amount of time.

A network N' is directly reachable from N if $N \xrightarrow{\Lambda,C} N_1 \overset{t,C'}{\rightsquigarrow} N'$. Since the cost does not influence the behaviour of the system, a network N is guaranteed to proceed even without costs. The functions *erasec*(N) and *erasec*(P) applied to a network N and to a process P, respectively, maps each network and process to the same network and process in which costs are ignored (removed).

Results: We have the following results.

Proposition 1. *1. If $N \xrightarrow{\Lambda,C} N'$, then erasec$(N) \xrightarrow{\Lambda}$ erasec(N').*

2. If $N \overset{t,C'}{\rightsquigarrow} N'$, then erasec$(N) \overset{t}{\rightsquigarrow}$ erasec(N').

Proposition 2. *1. If erasec$(N) \xrightarrow{\Lambda} N'$, then exists N'' and C such that $N \xrightarrow{\Lambda,C} N''$ and erasec$(N'') = N'$.*

2. If erasec$(N) \overset{t}{\rightsquigarrow} N'$, then is N'' and C such that $N \overset{t,C}{\rightsquigarrow} N''$ and erasec$(N'')=N'$.

In what follows we consider the relation $\equiv_c$ that simply rearranges the order of costs depending on the order of components in the network. Precisely, $\equiv_c$ is defined as the smallest relation given by the following equalities:

$$C \equiv_c C \qquad C \mid C' \equiv_c C' \mid C \qquad (C \mid C') \mid C'' \equiv_c C \mid (C' \mid C'').$$

Theorem 1. *For any networks N, N' and N'', the following sentences hold:*

1. If $N \overset{t,C}{\rightsquigarrow} N'$ and $N \overset{t,C'}{\rightsquigarrow} N''$, then $N' \equiv N''$ and $C \equiv_c C'$;

2. $N \overset{(t+t'),C''}{\rightsquigarrow} N'$ if and only if there is a N'' such that $N \overset{t,C}{\rightsquigarrow} N''$ and $N'' \overset{t',C'}{\rightsquigarrow} N'$, where $C'' = C' + C''$ (the sum is component-wise).

The first part of Theorem 1 claims that for any network N, performing only time reductions does not lead to nondeterministic behaviour. The second part claims that if a network N is able to perform a certain time step, then its evolution can be split in smaller time steps.

Since located processes contain single processes, namely not processes running in parallel, then determining the cost of execution for a network reduces to determining the cost of execution for all located processes (and then just adding them up). For determining the cost of the execution of a located process, it should be noticed that each location has associated its own cost (of staying there, per time unit), while each migration or communication action has associated an execution cost. In what follows, if $L_1|L_2 \xrightarrow{!v@l|?u@l,c_1|c_2} L_1'|L_2'$, then we have L_1 and L_2 evolving in parallel such that $L_1 \xrightarrow{!v@l,c_1} L_1'$ and $L_2 \xrightarrow{?u@l,c_2} L_2'$.

Definition 1 (Execution Cost). *Let* $e_L = L \overset{d_0,c_0}{\rightsquigarrow} L_1 \ldots \overset{d_n,c_n}{\rightsquigarrow} L_{n+1}$ *be a finite execution of a located process* $L = l[[_c\, P]]$*. Then* $cost(e_L) = \sum_{0\leq i\leq n} c_i$ *represents the cost of the execution of* e_L*, while* $time(e_L) = \sum_{0\leq i\leq n} d_i$ *represents the execution time of* e_L *.*

For a given process $L' = l'[[_{c'}\, P']]$*, the minimal cost* $mincost_L(L')$ *of reaching* L' *starting from* $L = l[[_c\, P]]$ *is the minimum of the costs of executions starting in* L *and ending in* L'*. Formally,*

$$mincost_L(L') = \begin{cases} min(\{cost(L \overset{d_0,c_0}{\rightsquigarrow} \ldots \overset{d_n,c_n}{\rightsquigarrow} L')\}) & \text{if } L' \text{ reachable from } L, \\ 0 & \text{otherwise.} \end{cases}$$

The next result shows how the execution time is computed by using appropriate values for the costs appearing in the network. This illustrates also the fact that the TiMo formalism is just an instance of the cTiMO formalism.

Proposition 3. *Let* e_L *be a finite execution of a located process* L*. If all location costs are set to* 1 *and all the migration and communication costs are set to* 0*, then* $cost(e_L) = time(e_L)$*.*

Proposition 4. *If a located process* $L = l[[_c\, P]]$ *has its process* P *without any communication action, then for any finite execution* $e_L = L \ldots L'$ *with* $L' = l'[[_{c'}\, 0]]$ *we have* $cost(e_L) = mincost_L(L')$*.*

3 Translating cTiMO into Weighted Timed Automata

Weighted timed automata were defined in [2] by adding cost information to the edges and locations of the classical timed automata. The cost added to a location represents the price per time unit of residing in that location, while the cost added to an edge represents the price of executing the corresponding transition. In this way, for each run of an automaton a global cost can be computed by adding the costs along the run of all delay and discrete transitions.

We consider the time domain $\mathbb{R}_+$ and a finite set X of variables called *clocks*. A *clock valuation* over X is a mapping $v : X \to \mathbb{R}_+$ assigning to each clock x a value from $\mathbb{R}_+$. The set of all clock valuations over the set X is denoted by $\mathbb{R}_+^X$. For $t \in \mathbb{R}_+$, the valuation $v + t$ is defined by $(v + t)(x) = v(x) + t$ for all clocks $x \in X$. For a set $Y \subseteq X$ of clocks, $v[Y \leftarrow 0]$ denotes the valuation assigning the value 0 to any $x \in Y$, and the value $v(x)$ to any $x \in X \setminus Y$. The valuation assigning 0 to every clock $x \in X$ is denoted by $\mathbf{0}$, while $\mathcal{C}(X)$ denotes the set of clock constraints defined as conjunctions of atomic constraints of the form $x \bowtie c$, where $x \in X$, $c \in \mathbb{N}$ and $\bowtie \in \{<, \leq, =, \geq, >\}$. AP denotes the finite set of atomic propositions. For $g \in \mathcal{C}(X)$ and $v \in \mathbb{R}_+^X$, we use $v \vDash g$ to denote that v satisfies g. Let Σ be a set of input and output actions together with τ actions.

Definition 2. *A weighted timed automaton A over X and AP is a tuple $(L, l_0, T, \lambda, cost)$, where L is a finite set of locations, $l_0 \in L$ is the initial location, $T \subseteq L \times \mathcal{C}(X) \times \Sigma \times 2^X \times L$ is a finite set of transitions, $\lambda : L \to 2^{\mathsf{AP}}$ is a labelling function, and $cost : L \cup T \to \mathbb{N}$ assigns costs to locations and transitions.*

The semantics of a weighted timed automaton without costs is similar to that of a timed automaton. It is a timed transition system $(S, s_0, \to)$, where $S = L \times \mathbb{R}_+^X$, $s_0 = (l_0, \mathbf{0})$, and $\to$ contains two types of transitions:

- *delay transitions*: $(l, v) \xrightarrow{\delta(d)} (l, v + d)$ if $d \in \mathbb{R}_+$;
- *discrete transitions*: $(l, v) \xrightarrow{tr} (l', v')$ if there exists a transition $tr = (l, g, \sigma, Y, l') \in T$ such that $v \vDash g$, $v' = v[Y \leftarrow 0]$ and $\sigma \in \Sigma$.

For each step there is associated a cost defined by:

- $cost((l, v) \xrightarrow{\delta(d)} (l, v + d)) = cost(l) \cdot d$;
- $cost((l, v) \xrightarrow{tr} (l', v')) = cost(tr)$.

A run ρ of the weighted timed automaton is a finite or infinite sequence of steps in this transition system. The cost of ρ is denoted by $cost(\rho)$, and it is the accumulated cost of steps along such a run.

In what follows we denote the set of networks by $\mathcal{N}$, and the set of weighted timed automata by $\mathcal{A}$. Their transition systems differ not only in transitions, but also in states. Therefore, we define a specific notion of bisimilarity.

Definition 3. *A relation $\sim$ over* cTiMO *networks and weighted timed automata is a bisimulation if, whenever it holds that $(N, (A, \langle l, v \rangle)) \in \sim$, we have*

- *if $N \xrightarrow{\lambda, c} N'$, then exists tr_λ such that $\langle l, v \rangle \xrightarrow{tr_\lambda} \langle l', v' \rangle$ and $(N', (A, \langle l', v' \rangle)) \in \sim$, where $cost(\langle l, v \rangle \xrightarrow{tr_\lambda} \langle l', v' \rangle) = c$;*
- *if $N \overset{d,c}{\rightsquigarrow} N'$, then $\langle l, v \rangle \xrightarrow{\delta(d)} \langle l', v' \rangle$ and $(N', (A, \langle l', v' \rangle)) \in \sim$, where $v' = v + d$ and $cost(\langle l, v \rangle \xrightarrow{\delta(d)} \langle l', v' \rangle) = c$.*

Using this notion of bisimulation, we get the following result that claims that building a weighted timed automaton for each component of a network N leads to the equivalence between the cTiMO network N and its corresponding weighted timed automaton A in the initial state $\langle l_N, v_N \rangle$ (namely $(A, \langle l_N, v_N \rangle)$).

Theorem 2. *Given a* cTiMO *network N, there exists a weighted timed automaton A_N having a bisimilar behaviour. Formally, $N \sim (A_N, \langle l_N, v_N \rangle)$.*

4 Simulating cTiMO Multiagent Systems by using UPPAAL

Based on the previous translation of the cTiMO networks into weighted timed automata having a bisimilar behaviour, we use UPPAAL to simulate and verify the multiagent systems described in cTiMO. In order to compute the total cost of an execution we should be able to perform a specific arithmetic operation on clock values by reading the value and multiplying with the cost of staying in a location for a time unit. These operations are not directly supported by the theory of timed automata, and so they are not implemented in UPPAAL. To overcome this obstacle, we consider the integer clocks defined in [13], and so a transition can take place only when the clock has integer values by using the select choice $t \in int[0, n]$ to choose an integer t from the interval $[0, n]$.

We illustrate a simulation in UPPAAL by using a simple network composed of the located processes L_1 and L_2 defined as:

$L_1 = l[[{}_2\,\mathsf{go}_1^3\, l' \text{ then } (a_2^{\Delta 2}!\langle v\rangle \text{ then } \mathbf{0} \text{ else } \mathbf{0})]]$, and
$L_2 = l'[[{}_3\, a_1^{\Delta 5}?(x) \text{ then } \mathbf{0} \text{ else } \mathbf{0}]]$.

The simulation is done using the following two templates:

- LocProc1( ) is used to create the located process $L1$ by using the command L1 = LocProc1(). It is also possible to instantiate any number of located processes having the same signature; if we consider parameters, we can define different instantiations. The template is depicted in Fig. 1.
- LocProc2( ) is the template used to create the located process $L2$ by using the command L2 = LocProc2(). The template is depicted in Fig. 2.

The initial cTiMO network is given by the located processes $L1$ and $L2$. We provide here the details regarding the *LocProc*1() template of Fig. 1 (the other construction is similar). This template has four locations: l0, l1, l2 and l3 (the last three locations should be primed, but for readability we avoid using additional

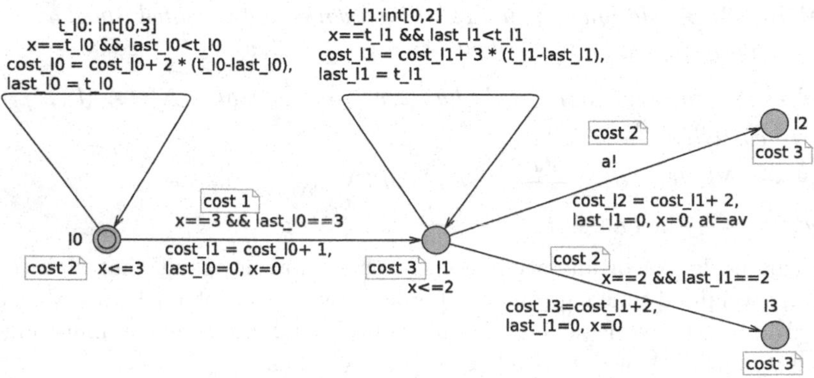

Fig. 1. LocProc1 Template.

symbols). For readability also, we added comments with the cost of each location and transition just to see how we compute the costs using integers.

The initial location is l0 which corresponds to the fact that the initial location of L_1 is l. The location has attached the atomic proposition $x <= 3$ (taken from the cTiMO move action go_1^3) which has the effect that this location must be left within 3 units of time. The loop transition is used to model the passage of time and the increment of the cost to reach location l0, cost denoted by *cost_l0*. We could use only a global *cost* variable, but we prefer to use specific notations to compute the cost to reach each location because it will be easier later in the verification phase. The outgoing transition towards location l1 is guarded by the constraints $x == 3$ and $last\_l0 == 3$ which correspond to the above mentioned move action, and to the fact that at location l0 was computed the cost for three time units spent there. Once moved at location l1, the agent can either synchronize on channel a and then move to location l2 or, migrate to location l3 if channel a expires. For location l1 we have a loop that is used to compute the cost of the process being located here. Assuming that the communication is performed, the current location is l2.

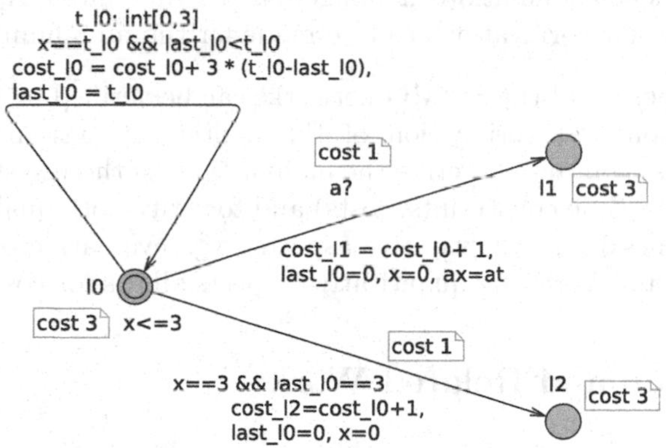

Fig. 2. LocProc2 Template.

According to the results presented in the previous sections, such a description in UPPAAL allows us to verify various properties of the multiagent systems with timeout for migration/communication and with costs, systems specified in a natural way in cTiMO. For verification we use the sophisticated software tool UPPAAL. UPPAAL is commonly used to verify complex properties of networks of timed automata, properties expressed in Computation Tree Logic (CTL). We performed several verifications for multiagent systems described in cTiMO, verification involving quantitative aspects expressed by both time and costs.

- L1.l1 − − > L1.l2

  This formula checks that, once the located process L1 is at location l1, then it will always reach the location l2. This implies that after leaving location l1, the communication on channel a always takes place (due to the presence of the located process $L2$ that has the dual communication transition).

- $\mathsf{E}\langle\,\rangle$ L1.cost_l2 == 9 and L2.l2
This formula checks whether once the located process L1 has the cost *cost_l2* equal to 9, the located process $L2$ reached location $l2$. This means $L1$ communicated on channel a, while $L2$ did not. As expected, this property is not satisfied; it becomes satisfied if we replace $L2.l2$ by $L2.l1$.
- $\mathsf{E}\langle\,\rangle$ L1.cost_l2 + L2.cost_l2 == 19
This formula checks whether the total cost of the evolution to reach locations $L1.l2$ and $L2.l2$ is equal to 19 (the total cost of reaching the two locations by independent transitions). Since we have a communication synchronization, it is impossible to reach the two locations as final ones (as explained above). The property becomes satisfied if we replace *L2.cost_l2* by *L2.cost_l1*.
- $\mathsf{A}[\,]$ L1.cost_l2 <= 9
This formula verifies that, whatever are the interactions between the involved located processes, the cost of reaching location $l2$ by $L1$ is always less than 9.
- $\mathsf{E}[<= 6; 100000]$(max : L1.cost_l2)
This formula is used to estimate the maximum value of *cost_l2* of $L1$ by performing 100000 simulations no longer than 6 time units. As expected, the value is 9. Similar verifications can be made for the minimum value.

Many other properties of the cTiMO networks can be verified by using UPPAAL.

The simulation and verification of the multiagent systems presented as cTiMO networks require to describe the main actions of these systems (mobility, message exchange, time constraints, costs) and to verify both qualitative aspects (e.g., safety, liveness) and quantitative aspects (e.g., evaluation of performances using various costs). Verifying quantitative aspects allows for cost optimization.

5 Conclusion and Related Work

We described multiagent systems with timeouts for migration and communication and with various costs by using a calculus called cTiMO. An operational semantics allows to get interesting observations and results regarding the evolution of these multiagent systems. This approach allows to specify easily complex multiagent systems, and to get deep insights on their dynamics. Moreover, the specification by using a process calculus allows the description of larger and larger systems in a modular way (due to compositionality). We also present a translation of the cTiMO networks to weighted timed automata having a bisimilar behaviour (bisimulation is a notion introduced in process calculi). Based on such a translation, we use UPPAAL to simulate and verify cTiMO networks. Since the multiagent systems with timeouts and costs could have very complex behaviours, the simulation and automated verification become more and more required and necessary. By using the steps described before, we were able to verify various properties of the systems described in cTiMO involving timed migration and costs for actions and locations. This type of (formal) verification could be useful for checking properties involving concurrency and communication protocols of real multiagent systems.

Different notions of cost have been tackled previously in the framework of process calculi. In [4] the cost of computation is expressed in terms of space consumption, and the mobile agents are allowed to migrate only if the target location has sufficient capacity to accommodate it. In [14] a cost is associated with each action of CCS (calculus of communicating systems), but the costs are not compared as only the overall cost counts. The theory of typed bisimulation equivalence for the π-calculus [15] was adapted in [12] to provide an adequate theory of costed process behaviour by defining a very simple variation of the asynchronous π-calculus called π_{cost} in which channels are viewed as resources usable only if sufficient funds are available. The $-calculus (pronounced cost calculus) is another extension with costs of the π-calculus that is used for problem solving by providing support to handle intractability and undecidability, and thus being able to go beyond the Turing Machine model [10]. In the spice calculus [17], an extension based on the π-calculus and spi-calculus [1], annotations attached to transitions are used to keep track of the costs of process executions. Our approach is different, considering costs for actions and locations determined by time consumption. More important, TiMo is much easier to use than the π-calculus (TiMo is a sugared version of the timed distributed π-calculus [9]).

There exist previous approaches for the verification of multiagent systems. In some model-based approaches, a system is represented as a Kripke structure by using an appropriate logic, and the specification is represented by a formula expressed in this logic. The verification consists of checking whether the model satisfies the specification; actually, this is an algorithmic-based technique done automatically. Model checking has been used previously to verify agent-based systems, for instance in [6] and [11]. In [6], the agents specification language AgentSpeak is translated to the language used by a model checker. In [11], the specification and verification of agents is addressed by using a temporal action logic. In both [6] and [11] the agents were equipped with reasoning and communicative abilities expressed by using specific logics, and the systems were modelled by a set of rules allowing communication in various situations. The correctness is checked by verifying that the system satisfies the required properties such as deadlock-free, safety and reachability, as well as by checking that unexpected executions expressed as logical formulas never happen in the system.

Our approach is different from these approaches. We use a specification language for describing multiagent systems with timed migration, timed communication and with costs. The language offers migration and communication primitives allowing to model the explicit localization of the agents, and systems in which the agents can move between locations and then interact locally. The fact that the costs appear on both locations and actions is also a new aspect we did not find in previous approaches. Using cost information provides the opportunity of interesting optimization problems such as: Is it possible to minimize/maximize the cost for reaching a given goal state? We used the powerful software tool Uppaal to answer various questions (including this one), and to obtain a number of critical verification results for cTiMO networks. It is worth to mention that there exists a version of UPPAAL called UPPAAL CORA working on priced timed automata involving costs. However, UPPAAL CORA is used only for finding optimal paths matching goal conditions, and it cannot use cost variables in expressions or verifications.

Acknowledgement. This work was partially supported by the project funded by the Ministry of Research and Innovation within Program 1 - Development of the national RD system, Subprogram 1.2 - Institutional Performance - RDI excellence funding projects, Contract no.34PFE/19.10.2018.

References

1. Abadi, M., Gordon, A.D.: A calculus for cryptographic protocols: the spi-calculus. Inf. Comput. **148**, 1–70 (1999). https://doi.org/10.1006/inco.1998.2740
2. Alur, R., La Torre, S., Pappas, G.J.: Optimal paths in weighted timed automata. Theor. Comput. Sci. **318**, 297–322 (2004). https://doi.org/10.1016/j.tcs.2003.10.038
3. Aman, B., Ciobanu, G.: Verification of critical systems described in real-time TiMo. STTT **19**, 395–408 (2017). https://doi.org/10.1007/s10009-016-0439-9
4. Barbanera, F., Bugliesi, M., Dezani-Ciancaglini, M., Sassone, V.: A calculus of bounded capacities. In: Saraswat, V.A. (ed.) ASIAN 2003. LNCS, vol. 2896, pp. 205–223. Springer, Heidelberg (2003). https://doi.org/10.1007/978-3-540-40965-6_14
5. Behrmann, G., David, A., Larsen, K.G.: A tutorial on UPPAAL. In: Bernardo, M., Corradini, F. (eds.) SFM-RT 2004. LNCS, vol. 3185, pp. 200–236. Springer, Heidelberg (2004). https://doi.org/10.1007/978-3-540-30080-9_7
6. Bordini, R.H., Fisher, M., Pardavila, C., Wooldridge, M.J.: Model checking AgentSpeak. In: The Second International Joint Conference on Autonomous Agents & Multiagent Systems, AAMAS 2003, pp. 409–416 (2003). https://doi.org/10.1145/860575.860641
7. Ciobanu, G., Koutny, M.: Modelling and verification of timed interaction and migration. In: Fiadeiro, J.L., Inverardi, P. (eds.) FASE 2008. LNCS, vol. 4961, pp. 215–229. Springer, Heidelberg (2008). https://doi.org/10.1007/978-3-540-78743-3_16
8. Ciobanu, G., Koutny, M.: Timed mobility in process algebra and Petri nets. J. Log. Algebr. Program. **80**, 377–391 (2011). https://doi.org/10.1016/j.jlap.2011.05.002
9. Ciobanu, G., Prisacariu, C.: Timers for distributed systems. Electr. Notes Theor. Comput. Sci. **164**, 81–99 (2006). https://doi.org/10.1016/j.entcs.2006.07.013
10. Eberbach, E.: The $-calculus process algebra for problem solving: a paradigmatic shift in handling hard computational problems. Theor. Comput. Sci. **383**, 200–243 (2007). https://doi.org/10.1016/j.tcs.2007.04.012
11. Giordano, L., Martelli, A., Schwind, C.: Verifying communicating agents by model checking in a temporal action logic. In: Alferes, J.J., Leite, J. (eds.) JELIA 2004. LNCS (LNAI), vol. 3229, pp. 57–69. Springer, Heidelberg (2004). https://doi.org/10.1007/978-3-540-30227-8_8
12. Hennessy, M., Gaur, M.: Counting the cost in the π-calculus (extended abstract). Electr. Notes Theor. Comput. Sci. **229**, 117–129 (2009). https://doi.org/10.1016/j.entcs.2009.06.042

13. Huang, X., Singh, A., Smolka, S.A.: Using integer clocks to verify clock-synchronization protocols. ISSE **7**, 119–130 (2011). https://doi.org/10.1007/s11334-011-0152-5
14. Kiehn, A., Arun-Kumar, S.: Amortised bisimulations. In: Wang, F. (ed.) FORTE 2005. LNCS, vol. 3731, pp. 320–334. Springer, Heidelberg (2005). https://doi.org/10.1007/11562436_24
15. Milner, R.: Communicating and Mobile Systems - The π-calculus. Cambridge University Press, Cambridge (1999)
16. Saeedloei, N., Gupta, G.: Timed π-calculus. In: Abadi, M., Lluch Lafuente, A. (eds.) TGC 2013. LNCS, vol. 8358, pp. 119–135. Springer, Cham (2014). https://doi.org/10.1007/978-3-319-05119-2_8
17. Tomioka, D., Nishizaki, S., Ikeda, R.: A cost estimation calculus for analyzing the resistance to denial-of-service attack. In: Futatsugi, K., Mizoguchi, F., Yonezaki, N. (eds.) ISSS 2003. LNCS, vol. 3233, pp. 25–44. Springer, Heidelberg (2004). https://doi.org/10.1007/978-3-540-37621-7_2

Maintainability of Automatic Acceptance Tests for Web Applications—A Case Study Comparing Two Approaches to Organizing Code of Test Cases

Aleksander Sadaj, Mirosław Ochodek(✉), Sylwia Kopczyńska, and Jerzy Nawrocki

Poznan University of Technology, Poznań, Poland
sadaj.aleksander@gmail.com, miroslaw.ochodek@cs.put.poznan.pl

Abstract. [Context] Agile software development calls for test automation since it is critical for continuous development and delivery. However, automation is a challenging task especially for tests of user interface, which can be very expensive. [Problem] There are two extreme approaches of structuring the code of test duties for web-applicating, i.e., linear scripting and keyword-driven scripting technique employing the page object pattern. The goal of this research is to compare them focusing on the maintainability aspect. [Method] We develop and maintain two automatic test suites implementing the same test cases for a mature open-source system using these two approaches. For each approach, we measure the size of the testing codebase and the number of lines of code that need to be modified to keep the test suites passing and valid through five releases of the system. [Results] We observed that the total number of physical lines was higher for the keyword-driven approach than for the linear scripting one. However, the number of programmatical lines of code was smaller for the former. The number of lines of code that had to be modified to maintain the tests was lower for the keyword-driven scripting test suite than for the linear-scripting one. We found the linear-scripting technique was more difficult to maintain because the scripts consist only of low-level code directly interacting with a web browser making it hard to understand the purpose and broader context of the interaction they implement. [Conclusions] We conclude that test suites created using the keyword-driven approach are easier to maintain and more suitable for most of the projects. However, the results show that the linear scripting approach could be considered as a less expensive alternative for small projects that are not likely to be frequently modified in the future.

Keywords: Acceptance testing · Keyword-driven testing · Linear scripting · Web applications · Selenium · Cucumber

A. Chatzigeorgiou et al. (Eds.): SOFSEM 2020, LNCS 12011, pp. 454–466, 2020.
https://doi.org/10.1007/978-3-030-38919-2_37

1 Introduction

Quality assurance (QA) and software testing play an important role in the software development process. According to the recent industrial survey of 1700 executives from 32 countries published in 2019, they account for 26% of IT budgets [12]. At the same time, the study shows that roughly 40% of the increase in QA and testing budget was caused by the higher number of iteration cycles resulted from the growth in the adoption of agile and DevOps. These new methods shift attention to test automation [13] since it is critical for continuous integration, development, and delivery [18].

However, test automation is a challenging task. According to Cohn's pyramid [2], automation should be brought to all levels of testing. At the bottom of the pyramid are automated unit tests, which shall be implemented in large quantities, while at the top are end-to-end tests exercising user interface. Unfortunately, the latter tests can be expensive to write, brittle, and slow [17].

When making a decision about automating user-interface tests, the decision-maker has to consider multiple technical, economic, and human and organizational factors [5]. On the one hand, there are undoubted benefits, but on the other, there are costs of implementing and maintaining test cases (i.e., creating new and updating existing ones to work for a new release of the software). There are two extreme options that could be used to develop automatic tests: (1) "quick-and-dirty" way i.e., using the simplest possible approach that does not require intensive training of testers, and the opposite (2) to develop a well-structured testing framework, which could help reducing maintenance costs by increasing the reusability of the test code. The first option could be tempting for smaller organizations that do not have large budgets to spend on test automation. However, the question is: *how* much more they would need to invest in test maintenance if they decided to take the "quick-and-dirty" approach while introducing user-interface test automation to an organization?

Thus, we decided to analyze two extreme approaches to structuring the code of automatic test cases. As representatives of the approaches we chose ***linear scripting*** (LS) and ***keyword-driven scripting*** (KS). The purpose of the analysis is *comparison* of the approaches concerning *maintenance costs* from the viewpoint of *a software tester* in the context of a *small software-development organization developing web applications wanting to introduce end-user test automation to the development of one of its software products.* To achieve the goal, we formulated three research questions to be answered in case study research:

- **RQ1:** *Which approach to create automatic test cases (LS, KS) require more lines of code to be modified while maintaining a test suite?*
- **RQ2:** *What are the main problems one can encounter while maintaining a test suite using both approaches (LS, KS) to create automatic test cases?*
- **RQ3:** *How difficult is it to maintain a test suite using the LS, KS approaches?*

Justification: We limit our focus to web applications taking into account their popularity and high-availability of open-source automation test frameworks.

Since there are numerous frameworks and approaches to develop automatic test cases for web applications and the arbitrary choice of the approaches might have been biased by the author's knowledge in the area, we chose the two approaches based on the results of the survey among practitioners [16]. The respondents perceived as the most reasonable approaches to structuring the code of the test suites in the context of this study the following approaches: *linear scripting* (using Selenium [10] and jUnit [19]) for the "quick-and-dirty" approach and *keyword-driven scripting* (using Cucumber Framework [3], Page Objects [4,14], Selenium [10], and jUnit [19]) for the well-structured approach. We are going to analyze both all physical lines of code needed by a given approach, and those lines that require manual work of a developer to implement test cases, which we call programmatical lines of code further on.

2 Case Study

To answer the research questions stated in Sect. 1, we conducted a case study using the guidelines by Runeson et al. [15]. This kind of study allows to thoroughly examine a selected case while taking into account any related conditions.

2.1 Case Selection

Since we wanted to observe how the evolution of a software product (a web application) affects the test suite of automatic test cases created using two different approaches. Therefore, a mature web application for which we could access the history of releases and the source code of each release was needed. We have chosen *Moodle* [11]—an open-source learning management system (LMS) that had its first version released on 20th August 2002, and is well known worldwide among students, tutors, and other people involved in academia. In order to simulate the process of the constant development of the application, we used the first and the last release of every major release of Moodle, beginning from the *1.9.19* version, i.e., we analyzed versions: 1.9.19, 2.0.0, 2.9.9, 3.0.0, and 3.6.2.

We selected three features to investigate that represent a range of possible cases, i.e., typical and extreme cases [20] from the perspective of maintenance effort. We have decided to pick one feature that would probably not change much during the development of the platform and two that would change a lot. The former feature is the *login feature*, as it is, most of the time, represented by a straightforward form consisting of two fields that a user has to fill in with their user name and password. The latter two features were selected based on the analysis of the history of changes of the *Git* repository. To find the features, we queried Git for the files with the *.feature* extension that contain test-cases written by the developers of Moodle. We assumed that changes made in these files are good indicators (proxies) of the changes made to the system's features (see [16] for details). From the 10 most frequently modified features that were present in all considered versions of Moodle we selected *calendar* and *course management listing*.

2.2 Test Suite Implementation and Maintenance

Each test suite contained the same test cases as the tests implemented by the Moodle team that can be found in the repository together with the source code of the system. However, some of the tests from the original Moodle's test repository were intentionally omitted, because they were testing some technical aspects of the features, for example, tests checking whether some AJAX calls have been made or not, which is outside the scope of the study.

To implement the test cases, we chose the Java language as it is one of the most popular programming languages in the world. We intended to study the code of the test suite explicitly, not the code that is responsible for controlling the web browser. For that reason, to implement the tests using the linear scripting approach, we decided to use one of the most known frameworks that serve this purpose, being Selenium WebDriver [10]. For the keyword-driven scripting approach, we used the Cucumber framework [3], which has its own language, called Gherkin, that is used to express the steps of the test cases in a natural-language-like way. To make it possible to execute the tests automatically, some Jave code needed to be written. In addition to the Cucumber framework, we also decided to use the *Page Object* pattern [4]. It is one of the most recognizable patterns that are used along *Selenium*. This pattern works around the idea of creating objects that are a representation of the User Interface that a developer is going to be testing.

The common part for both types of the tests is that on the low level of understanding, they both use Java and Selenium WebDriver to implement the logic and manipulate the web browser.

The procedure of the study was as follows. Firstly, the test cases were implemented for the first version of the Moodle for each feature in the following order: (1) Login, (2) Calendar, and (3) Category management listing. We implemented a test suite for each feature, and only after finishing one, we switched to the next feature. For instance, we started from implementing the test suite for the login feature with the first considered version of Moodle *1.9.19*, and implemented the tests with the linear scripting approach. Then, we implemented the same tests, but with keyword-driven scripting approach.

Secondly, we switched to the next version of Moodle. We run all of the tests for the currently maintained feature to see how many of the test cases have started to fail. The failures could be triggered by a number of reasons. After identifying the issues with the tests and defining the reasons, we fixed all of the failures and made the suite pass again.

After we completed the whole cycle for one feature, meaning having implemented the test suite for a certain feature in the first version being *1.9.19*, gone through all of the next versions, and finally made sure that all of the feature's tests are passing for the last version *3.6.2*, the next step of the study was to repeat the process for the next feature in order.

2.3 Data Collection and Analysis

Quantitative Metrics. We decided use *lines of code* (LOC) as a proxy of the overall effort needed to maintain a test suite. With the assumption that *effort* = *size* × *productivity* we decided to omit the *productivity* component, because it introduces too much noise and would be impossible to measure in this research, as the small company's developer team which is the case being studied we take into account is purely hypothetical. We did not want to guess the costs of the developers' conceptual thinking about the test suite. We decided to measure the quantity of the physical lines of code present in the code-base separately for the linear scripting tests and keyword-driven ones.

We did not implement the setup of the test cases (e.g., adding test data to the database) because they would be the same for both compared approaches. Therefore, all of the needed accounts, courses, course categories, or user groups had been created manually before the implementation of the tests. The created data is identical for each version of Moodle, so data consistency has been ensured. One of the criteria taken into account while picking the features to write the tests for was that the data setup required for such feature to work should be relatively easy to conduct and it should be a one-time-only operation, meaning that the data would not change along with the tests. Thanks to picking the tested features with that information in mind, the setup could be done manually in an easy way and would have to be done only once. This has allowed lots of unnecessary code not to be written.

The quantitative metrics were gathered after the implementation of all test suites for all versions, and all features have been finished. Since all tests were manged in the Git repository, we used the system measurement mechanism.

Qualitative Analysis. We decided to collect the following qualitative information about the process of creating and maintaining test cases:

- What are the biggest difficulties while maintaining a test suite?
- How difficult is it to write the test suite for a new feature (with no pre-implemented architecture)?
- How difficult is it to debug a failed test case?
- How difficult is it to understand a test case by reading its code?
- Which of the approaches is faster to develop?
- Which of the approaches facilitates test case repairing more?

3 Results and Observations

Based on the results of the study, we made the following observations that answer the research questions.

▷ *RQ1: Which approach to create automatic test cases require more lines of code to be modified while maintaining a test suite?*

Observation 1: The *keyword-driven* approach resulted in producing more lines in total than the *linear scripting* technique.

Justification: Code for the linear scripting approach constitutes of Java code while in the keyword-driven one the test suite is implemented with Java, Gherkin, and Page Objects. The linear scripting approach resulted in total *1111* physical lines of code, while the keyword-driven one in *1514*. It follows from Table 1 that for each version of Moodle the keyword-driven approach required more lines of code than the other one. This difference is because, frequently, it is needed to write more code that will provide enough abstractions when one creates reusable code using the keyword-driven approach.

Note: We decided to split version v2.9.9 into three parts—fixing the existing tests, extending the existing tests to match their full new capabilities, and implementing tests for functionalities that have appeared in this version.

Table 1. Physical lines of code per approach

Approach/Moodle's version	v1.9.19	v2.0.0	v2.9.9 - fixing	v2.9.9 - extension	v2.9.9 - adding missing	v3.0.0	v3.6.2
Linear scripting	658	662	655	761	1112	1112	1111
Keyword-driven scripting	937	929	929	1074	1473	1473	1514
Keyword-driven scripting – parts:							
Cucumber (Gherkin)	170	163	162	217	425	425	425
Cucumber (Java)	335	337	344	381	491	491	492
Page Object (Java)	432	429	433	476	557	557	597

Observation 2: The *keyword-driven* approach resulted in a smaller Java testing codebase than the *linear scripting* technique.

Justification: In the keyword-driven scripting approach the Java codebase consists of the code implementing the Page Objects and Java definitions of test steps (keywords). It follows from Table 1 that the total number of the physical lines of code for the keyword-driven scripting test suite was *1089* physical lines of code. This is a slightly lower number than in the case of the linear scripting test suite, which had *1111* lines of Java code.

Observation 3: The *keyword-driven* approach resulted in a lower number of changed lines of code required to maintain test cases than the *linear scripting* technique.

Justification: Let us define *changed* lines of code metric as a sum of *added*, *modified* and *deleted* lines of code that had to be altered in order to keep a test suite passing, or in other words—maintain it. Figure 1 depicts the number of

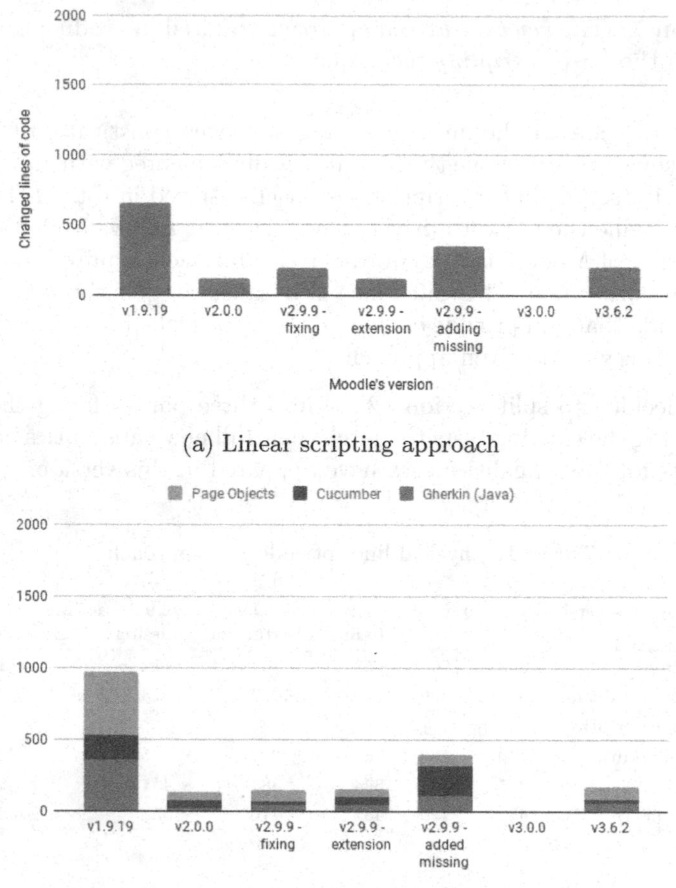

(a) Linear scripting approach

(b) Keyword-driven scripting approach

Fig. 1. Total modified lines of code with linear scripting approach and with keyword-driven scripting one.

changed lines of code for each approach, per each version of the platform. The total amount of changed lines of code for the keyword-driven scripting approach was *964* lines of code, while for the linear scripting one it was *984*. As predicted, investing time and effort in implementing a structured and abstract test suite, such as one written with the help of keyword-driven scripting approach, pays off in the long term of test suite maintenance.

Observation 4: The *keyword-driven* approach resulted in a lower number of modified lines of Java code required to maintain test cases than the *linear scripting* technique.

Justification: A total of *326* modified lines of code come from the *Cucumber (Gherkin)* framework's code. As stated before, it does not contain programmatic code, so we decided to exclude it from the sum of total modified lines of code to see the difference between the Java code. This resulted in *638* modified lines of code in the keyword-driven scripting test suite and *984* in the linear scripting one. This means that *35%* fewer changes were needed to maintain the Java part of the keyword-driven scripting test suite than the linear one.

▷ *RQ2: What are the main problems one can encounter while maintaining a test suite using both approaches (LS, KS) to create automatic test cases?*

Observation 5: *Linear scripts* can be made unreadable quickly.

Justification: In the linear scripting approach, with low effort invested in making the code readable, the code of tests cases quickly became difficult to read. The readability issue had its origin in the fact that test cases implemented in this approach contain a significant amount of low-level code. The code used Selenium WebDriver to execute steps of test scenarios. The purpose of the code was not visible at first glance as the methods had technical names. During the development, to make our code readable, we split the code into blocks, divided by the responsibility or a task that such block of code would execute. We put one line of comment for each one, in which we explained what does the block do. There were many situations, especially in the more complex test cases, in which besides the explanation, such block of code would still be difficult to read and understand. However, overcompensating and putting even more effort into commenting, in our opinion, would hurt the readability, as there would be too many comments and they would clutter the code.

Observation 6: Commenting blocks of code helps in making test cases easier to read and understand in the *linear scripting* approach.

Justification: As we mentioned in the justification of Observation 5, due to the fact that the linear-scripting-approach test cases are dense with the code, they quickly become difficult to read and understand. To mitigate this risk, we decided to split the code into blocks, based on their responsibility, and provide a comment line of code. The comments explained what each block do. The frequency and level of details of the comments is an arbitrary decision. Too little explanation could not bring any value, and too many details could make the test cases challenging to read. It would clutter the code.

Observation 7: Introducing changes to the codebase, after a widely used selector has changed, is tedious work in the *linear scripting* approach.

Justification: In the linear scripting approach methods locating GUI elements (e.g., via link text, XPath, IDs or CSS classes), so-called selectors, are frequently used. When a widely used selector gets changed, the test suite requires to be modified in many places. For example, the login button's selector, which is used at least at the beginning of every test to log in the test user, would require changes in all these places. Moreover, the blocks of code in the linear scripting approach are very dense. It makes the parts of code requiring modification hard to notice, which makes the maintenance a difficult task.

Observation 8: To fully benefit from the *keyword-driven scripting* approach, one needs to implement the *keywords* in a way that will make them reusable across different test cases.

Justification: A keyword is a natural language statement that is used to express some part of a test scenario (e.g., an action of login). It is mapped by a developer to code that during the execution of the test will make the machine execute this part automatically. If there are similar test scenarios across different test cases, the defined keywords can be reused. However, to effectively use them and not to clutter the code, a developer has to have good knowledge on what steps and keywords have already been defined and what responsibilities they have. This might be a difficult aspect of using this approach, as it requires one to know the whole test codebase and understand it well. It might also introduce some problems for new members of the team, especially junior ones, as they would have more effort to be made in order to start contributing to the project.

▷ *RQ3: How difficult is it to maintain a test suite using the LS, KS approaches?*

Observation 9: The *keyword-driven* approach requires more lines of test code to write than the *linear scripting* one, but it is easier to implement.

Justification: As it follows from Table 1, the keyword-driven scripting approach required us to write more lines in Java and Gherkin compared to the linear scripting one. However, in our opinion, it was easier to implement the test suites using the former one. The reason behind it is that the code developed using the keyword-driven scripting approach is more structured. While implementing the code, using semantically correct components, it would give the context of what is the code going to accomplish. Also, this approach required us to write less low-level code using the Selenium WebDriver.

Observation 10: The *keyword-driven* approach requires knowledge of more design patterns, e.g. *Page Object*

Justification: Using complex and abstract approaches to writing the code, such as the keyword-driven scripting approach can easily lead to the need for learning more design patterns, such as the Page Object one [4,14]. The code to implement the Page Object pattern made up for 1/3 of the test suite's code, so it was important to us to know how to use it. Although it would be possible to implement the keyword-driven scripting test suite without the pattern, to fully benefit from the approach, we had to use this and some other design patterns.

4 Threats to Validity

We identified the following threats to validity that could have affected both our study and the results gathered from it.

The first threat concerns the choice of the test techniques investigated in the study. Not to base the decision on the opinion of the researchers, the choice of approaches follows from the results of the survey conducted among practitioners world-wide. Still, some subjectivity might be included.

The second threat we identified is the possibility that our own experience and opinions affecting the study itself. We tried to be as objective as we could while formulating observations and discussed them together. However, there is always a chance that our opinions might slip into either the conclusions drawn from it.

The third threat to validity is the scale of the research. We tried to pick a mature, open-source web-based application, and we selected Moodle. We stand by this decision, as this platform is widely and internationally used, and it has been available on the market for over 10 years. The threat might be caused not only by analyzing only one, however complex system but also by the number and the complexity of features we decided to test. We need to accept the threat. To mitigate this threat the features represent various types. One is a non-complex and stable feature, and two of them are more complex and changing features.

Another threat to validity we identified is the very fact that Moodle is an open-source platform. This has an influence on several aspects of the source code that could directly affect our study. Rotation of contributors, free-of-charge involvement might affect the quality of the source code, and as a result, affect how we created test suites.

5 Related Work

Multiple studies have been conducted in the area of maintainability of test suites for web applications. However, to the best of our knowledge, they are mostly focused on the topic of increasing the robustness and immunity to changes of test suites or proposing some new approaches, e.g., [1,8,9]. Many of them compare the maintainability of test suites, but focus on using different approaches to find web elements from the web applications (e.g., [6,8]).

According to our up-to-date knowledge, there is one empirical case study on the similar topic. Leotta et al. [7] conducted a case study comparing the *Capture-Replay* approach against a *programmable* one. The difference between this and our study is that we decided to use *linear scripting* technique to implement the non-complex test suite to compare the maintainability. The *Capture-Replay* approach the researchers have used is taken from *Selenium IDE* in which a programmer starts recording, clicks through a test case and then saves the test. Selenium IDE then recreates the test, but saving it as lines of code written in the linear scripting manner. The difference is that automatically created code is prone to creating unnecessary lines and lacks intelligence. We think that the fact of the *Capture-Replay*'s code being automatically created by a software tool might be a threat to validity. Based on that, we decided to implement the test cases in our study ourselves, so we can eliminate those threats. Also, the researchers gave some quantitative values considering the cost of implementation, but they did not focus on the qualitative part of the research.

6 Conclusions

The goal of the study we conducted was to compare approaches to structure code of test cases in the context of maintainability of the automatic acceptance tests for web-based applications. The approaches we selected to compare were *linear scripting* and *keyword-driven scripting*. We conducted a case study on the Moodle learning management system within which we created and maintained automatic test suites for three features.
From the results of the study the following observations follows:

- **Observation 1:** The keyword-driven approach resulted in producing more lines in Java and Gherkin than the linear scripting technique
- **Observation 2:** The keyword-driven approach resulted in a smaller Java testing codebase than the linear scripting technique
- **Observation 3:** The keyword-driven approach resulted in a smaller number of modified lines of Java and Gherkin required to maintain test cases than the linear scripting technique
- **Observation 4:** The keyword-driven approach resulted in a smaller number of modified lines of Java code required to maintain test cases than the linear scripting technique
- **Observation 5:** Linear scripts can be made unreadable quickly
- **Observation 6:** Commenting blocks of code helps in making test cases easier to read and understand
- **Observation 7:** Introducing changes to the codebase after a widely used selector has changed is a tedious work
- **Observation 8:** To fully benefit from the keyword-driven scripting approach, one needs to implement the keywords in a way that will make them reusable across different test cases
- **Observation 9:** The keyword-driven approach requires more lines to write (Java and Gherkin), but it is easier to write
- **Observation 10:** The keyword-driven approach requires knowledge of more design patterns, e.g., Page Object

Based on the observations, the following recommendations regarding the choice between the two considered approaches depending on the nature of the software project can be formulated.

Can the size of the application-to-be-tested be predicted? If the size of the application is small and, at best, the complexity is not high using structured and complex scripting techniques might be excessive and we would rather recommend picking approaches like *linear scripting*. It can save time on both implementation and learning. The complex techniques would not have the chance to pay back the time invested in learning them. The test suite codebase would be too small and possibly too simple to profit from it.

Does the project have an end-of-life date set? Given the development team has time and resources for learning a complex technique of structuring the codebase, the recommendation would be to spend the time learning it. In the long run it will be easier for the developers to implement new tests or maintain the existing ones. The longer the project will be developed, the more gains will be drawn from implementing such an approach.

Is it important for the team to broaden knowledge? Implementing test suites using the keyword-driven approach requires a higher level of expertise than for the simple linear scripts. We believe that acquiring this additional knowledge will increase the development's team value as programmers. Their experience will be richer with some new approaches, techniques and might unlock a whole new perspective of looking at software development for them. The benefits of learning new approaches in the testing field might transfer to the regular source code development world.

Is understandability of the tests important? The complex scripting approaches such as the keyword-driven scripting require a larger amount of work and effort to be done at the beginning to grasp their concepts and to understand them. Using such concept forces the codebase to be structured. Thus, it facilitates the process of joining the development team and orienting oneself in the codebase. In our opinion, it will be easier to explain the way how the test suite has been implemented if it has been done in a structured, thought-through, logical way.

It would be worth to continue work in the area. Especially, scaling up the case study to improve the generalizability of the conclusions would be valuable.

References

1. Carvalho, R.: A comparative study of GUI testing approaches (2016)
2. Cohn, M.: Succeeding with Agile: Software Development Using Scrum. Pearson Education, London (2010)
3. Dees, I., Wynne, M., Hellesoy, A.: Cucumber Recipes: Automate Anything with BDD Tools and Techniques. Pragmatic Bookshelf, Raleigh (2013)
4. Fowler, M.: https://martinfowler.com/bliki/PageObject.html. Accessed 25 Oct 2019
5. Garousi, V., Mika, M.: When and what to automate in software testing? A multi-vocal literature review. Inf. Softw. Technol. **76**, 92–117 (2016)
6. Leotta, M., Clerissi, D., Ricca, F., Spadaro, C.: Comparing the maintainability of selenium WebDriver test suites employing different locators: a case study. In: Joining AcadeMiA and Industry Contributions to Testing Automation (JAMAICA) (2013)
7. Leotta, M., Clerissi, D., Ricca, F., Tonella, P.: Capture-replay vs. programmable web testing: an empirical assessment during test case evolution. In: WCRE 2013, Koblenz, Germany, pp. 272–281 (2013)

8. Leotta, M., Ricca, F., Stocco, A., Tonella, P.: Reducing web test cases aging by means of robust XPath locators. IEEE (2013)
9. Leotta, M., Stocco, A., Ricca, F., Tonella, P.: Using multi-locators to increase the robustness of web test cases. IEEE (2015)
10. Mg, R.P.: Learning Selenium Testing Tools. Packt Publishing Ltd., Birmingham (2015)
11. Moodle.org: Moodle – Open-source learning platform (2019). https://moodle.org. Accessed 17 June 2019
12. Natarajan, S., Balasubramaniam, K., Kanitkar, M.: Efficiency and cost containment in quality assurance, 10th edn. Capgemini, Micro Focus, Sogeti, World Quality Report 2018-19 (2019)
13. Ochodek, M., Kopczyńska, S.: Perceived importance of agile requirements engineering practices-a survey. J. Syst. Softw. **143**, 29–43 (2018)
14. pluralsight.com: Getting Started with Page Object Pattern for Your Selenium Tests (2019). https://www.pluralsight.com/guides/getting-started-with-page-object-pattern-for-your-selenium-tests. Accessed 13 June 2019
15. Runeson, P., Host, M., Rainer, A., Regnell, B.: Case Study Research in Software Engineering: Guidelines and Examples. Wiley, Hoboken (2012)
16. Sadaj, A.: Maintainability of automatic acceptance tests for web applications–a case study comparing two approaches to organizing code of test cases. Master's thesis, Poznan University of Technology (2019)
17. Spinellis, D.: State-of-the-art software testing. IEEE Softw. **34**(5), 4–6 (2017)
18. Ståhl, D., Bosch, J.: Modeling continuous integration practice differences in industry software development. J. Syst. Softw. **87**, 48–59 (2014)
19. Tahchiev, P., Leme, F., Massol, V., Gregory, G.: JUnit in Action. Manning Publications Co., Greenwich (2010)
20. Yin, R.: Case Study Research: Design and Methods. SAGE Publications, Thousand Oaks (2003)

Recommending Trips in the Archipelago of Refactorings

Theofanis Vartziotis, Apostolos V. Zarras(✉), Anastasios Tsimakis, and Panos Vassiliadis

Department of Computer Science and Engineering, University of Ioannina, Ioannina, Greece
{tvartzio,zarras,pvassil}@cs.uoi.gr, atsimakis@gmail.com

Abstract. The essence of refactoring is to improve source code quality, in a principled, behavior preserving, one step at the time, process. To this end, the developer has to figure out the refactoring steps, while working on a specific source code fragment. To facilitate this task, the documentation that explains each primitive refactoring typically provides guidelines and tips on how to combine it with further refactorings. However, the developer has to cope with many refactorings and lots of guidelines.

To deal with this problem, we propose *a graph-based model that formally specifies refactoring guidelines and tips* in terms of nodes that correspond to refactorings and edges that denote *part-of*, *instead-of* and *succession relations*. We refer to this model as *the Map of the Archipelago of Refactorings* and we use it as the premise of *the Refactoring Trip Advisor*, a refactoring recommendation tool that facilitates the combination of refactorings. A first assessment of the tool in a practical scenario that involves 16 developers and a limited set of refactorings for composing and moving methods brought out positive results that motivate further studies of a larger scale and scope.

Keywords: Refactoring recommendation · Refactoring graph · Refactoring combination

1 Introduction

Refactoring is a basic prerequisite for keeping our source code clean. The basic idea is to improve source code quality, via a series of small behavior-preserving transformations [4,11].

> *"The biggest problem with Extract Method is dealing with local variables, and temps are one of the main sources of this issue. When I'm working on a method, I like Replace Temp with Query to get rid of any temporary variables that I can remove. If the temp is used for many things, I use Split Temporary Variable first to make the temp easier to replace."*

The previous quote is from Martin Fowler's catalog of refactorings [4]. What is interesting in this quote is that it provides certain guidelines on how to perform

A. Chatzigeorgiou et al. (Eds.): SOFSEM 2020, LNCS 12011, pp. 467–478, 2020.
https://doi.org/10.1007/978-3-030-38919-2_38

the Extract Method refactoring. To make things easier, it suggests to remove temporary variables *before* the method extraction, using Replace Temp with Query. Taking a step further, the quote suggests to use Split Temporary Variable *before* Replace Temp with Query, so as to facilitate the removal of multi-purpose temporary variables.

Observe the next quote, which suggests using Replace Method with Method Object *instead of* Extract Method, in the case of very complex methods.

> *"Sometimes, however, the temporary variables are just too tangled to replace. I need Replace Method with Method Object. This allows me to break up even the most tangled method, at the cost of introducing a new class for the job."*

Moreover, in the following quote we see *part-of* relations, which dictate how to realize Extract Superclass based on more primitive refactorings like Extract Method and Pull Up Method.

> *"Examine the methods left on the subclasses. See if there are common parts, if there are you can use Extract Method followed by Pull Up Method on the common parts."*

Hence, refactorings come along with several informal guidelines that tell us how to combine them into more complex evolution tasks. *What is the problem with that?* On the one hand, *there are way too many refactorings and guidelines* in Fowler's catalog. Specifically, the catalog consists of 68 different refactorings, while the documentation of these refactorings includes more than 100 guidelines and tips [16]. On the other hand, *the state of the art on refactoring* (two detailed surveys can be found in [6] and [2]) *does not provide means that facilitate the effective exploitation of this knowledge.*

To deal with the aforementioned issues, we propose an approach that allows the developers to combine refactorings into more complex evolution tasks via the following key concepts:

- *The Map of the Archipelago of Refactorings, a graph-based model that specifies informal guidelines and tips, found in Fowler's catalog*, in terms of nodes that correspond to refactorings, and edges that signify *part-of*, *instead-of* and *succession relations* between them. In Zarras et al. [16] we introduced a coarse sketch of the map, while in this paper we provide its detailed formal definition.
- *The Refactoring Trip Advisor*, a refactoring recommendation facility that provides *an interactive perspective of the archipelago map*, which makes suggestions regarding which refactoring(s) to use before, after, or instead of a particular refactoring. The Refactoring Trip Advisor further provides *guidelines on how to apply individual refactorings*, and enables the *identification of refactoring opportunities.*

We assess our approach in two steps: (1) we show that the Refactoring Trip Advisor adheres to the basic refactoring tool principles, recommended by

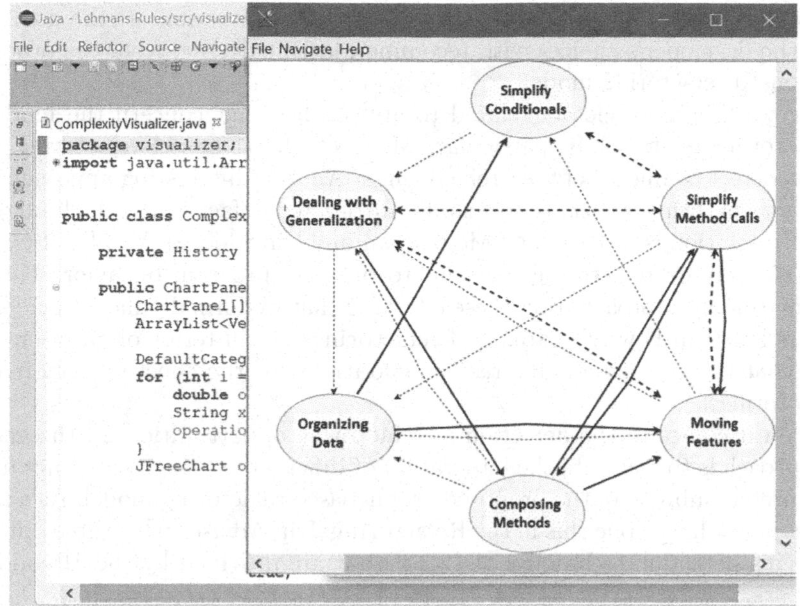

Fig. 1. The archipelago hyper-map.

Murphy-Hill and Black in [10]; (2) we validate that the Refactoring Trip Advisor can be successfully used in a realistic re-engineering scenario, in a study that involves 16 developers with varying profiles.

The rest of this paper is structured as follows. In Sect. 2, we discuss related work. In Sect. 3, we detail the *modus operandi* of the proposed approach. In Sect. 4, we concentrate on the validation of the proposed approach. Finally, in Sect. 5 we summarize our contribution and point out the future directions of this work.

2 Related Work

Opdyke introduced refactoring as a behavior preserving process that changes a software, so as to enable other changes to be made more easily [11]. Mens and Tourwé [6] provide an excellent survey that addresses several different aspects of the refactoring process (e.g., refactoring activities, techniques, supporting tools). A more recent extensive survey that focuses on techniques and tools for the detection of refactoring opportunities is provided by Al Dallal [2].

An important result that is brought out by the empirical study of Kim et al. [5] is the need to combine refactorings in more complex evolution tasks. To deal with this issue, the state of the art comprises a number of interesting search-based refactoring approaches (e.g., [9]) that apply refactorings automatically towards maximizing the software quality improvement, with respect to a set of target quality indicators. Our approach follows a different direction, as the goal is to widen the developer's choices with recommendations derived from the proposed graph-based refactoring model.

Our work is more closely related to approaches that concern the modeling of refactoring relations. In particular, Mens et al. [7] model refactoring relations to detect conflicts between refactorings. Another interesting approach that employs refactoring relations to enable automated refactoring scheduling and conflict resolution is proposed by Moghadam and Cinnéide [8]. Van Der Straeten et al. [15], rely on refactoring relations to preserve program behavior. The key difference of our approach from these efforts is that we formally model guidelines and tips found in Fowler's catalog of refactorings [4], in terms of an interactive model that provides actionable recommendations for the effective combination of refactorings.

When it comes to the detection of refactoring opportunities [2], the goal of our approach is to facilitate the integration of different existing techniques under the common umbrella of the proposed graph-based refactoring model. As a proof of concept, we have done this in the Refactoring Trip Advisor with three different refactoring detection techniques [3,12,13] that are provided by the JDeodorant framework.

3 Refactoring Trip Advisor

In this section, we formally model the map of the archipelago of refactorings. Then, we focus on the recommendation of refactoring trips. Finally, we illustrate the role of our approach in a realistic re-engineering scenario.

3.1 Modelling Refactoring Relations

At a glance, *the Map of the Archipelago of Refactorings* models informal guidelines and tips in terms of different relations between refactorings. Our baseline is Martin Fowler's catalog of refactorings [4]. Nevertheless, the extension of the map with further refactorings and relations is straightforward. The core concept of the map is a graph, with nodes representing refactorings and edges representing the relations between them. As relations are of different kinds, we introduce corresponding types of edges. More formally, we define the overall model as follows.

Definition 1 *[**Archipelago Map**]. The map of the archipelago of refactorings is a directed graph $\mathbb{M}_{\mathbb{R}}(\mathbb{V}_{\mathbb{R}}, \mathbb{E}_{\mathbb{R}})$, s.t. the nodes $\mathbb{V}_{\mathbb{R}}$ represent refactorings, while the edges $\mathbb{E}_{\mathbb{R}}$ denote relations that correspond to guidelines and tips, concerning the combination of refactorings:*

- *[**Node Properties**] $\mathbb{V}_{\mathbb{R}} = \bigcup_{i=1}^{6} V_R^i$ is divided into disjoint subsets, called regions. The regions correspond to the different categories of refactorings (Fig. 1), defined in Fowler's catalog [4].*

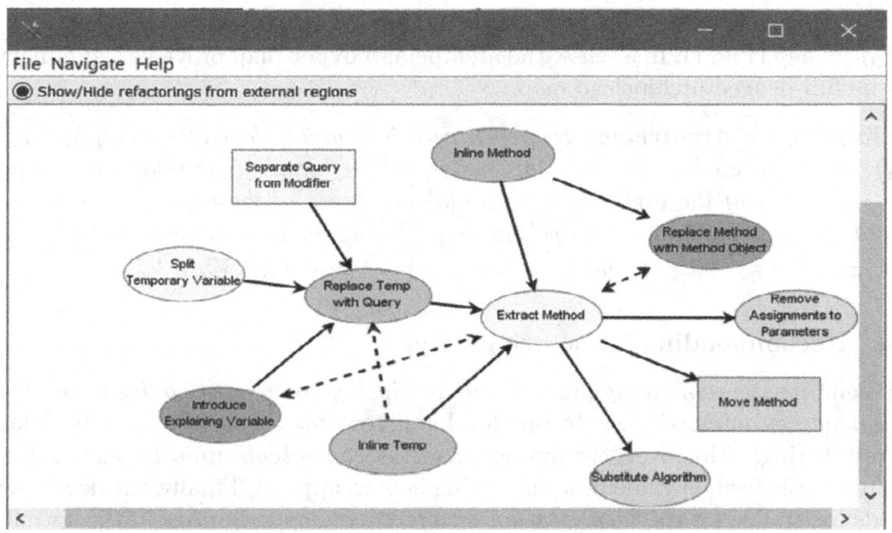

Fig. 2. The *Composing Methods* region map. (Color figure online)

- ***[Edge Properties]*** *The edges* $e(v_i, v_j) \in \mathbb{E}_{\mathbb{R}}$ *are typed, with* $type(e(v_i, v_j)) \in \mathbb{T} = \{succession, part\ of, instead\ of\}$*:*
  - *A succession relation, is represented as a solid unidirectional edge; it denotes that it would be useful to perform the source (resp. target) refactoring* v_i *(resp.* v_j*), before (resp. after) the target (resp. source) refactoring* v_j *(resp.* v_i*).*
  - *A part of relation is represented as a dotted unidirectional edge between two refactorings; it signifies that the source refactoring* v_i*, can be used for the realization of the target refactoring* v_j*.*
  - *A instead of relation, is denoted as a dashed bidirectional edge; it means that either one of the related refactorings can be used, instead of the other.*

The archipelago map is complex, consisting of 68 nodes and 101 edges between them[1]. The complexity of the map further points out the amount of information and the effort that is required from the developer, to exploit refactorings in an effective way. To deal with the complexity of the map, we decompose it into a set of *region maps* (e.g., Composing methods region given in Fig. 2). A region map shows the refactorings of a respective region, along with important refactorings from other regions that are related to them.

Definition 2 ***[Region Map].*** *For every region* V^i_R *of* $\mathbb{V}_{\mathbb{R}}$*, the region map* $M^i_{RG}(V^i_{RG}, E^i_{RG})$ *is an induced sub-graph of* $\mathbb{M}_{\mathbb{R}}$*, s.t.* V^i_{RG} *consists of the refactorings* V^i_R *and related refactorings from other regions, i.e.,* $(V^i_{RG} \supset V^i_R) \wedge (\forall v_j \in V^i_{RG} - V^i_R,\ \exists v_i \in V^i_R$ *s.t.* $(e(v_i, v_j) \in \mathbb{E}_{\mathbb{R}} \vee e(v_j, v_i) \in \mathbb{E}_{\mathbb{R}}))$*.*

[1] The map can be found at: www.cs.uoi.gr/~zarras/RefactoringsArchipelagoWEB/ArchipelagoOfRefactorings.html.

At a higher-level of abstraction the region maps are organized with respect to *a hyper-map* (Fig. 1). In a sense, the archipelago hyper-map provides a summary of the full-fledged archipelago map.

Definition 3 ***[Archipelago Hyper-map].*** *The archipelago hyper-map* $\mathbb{H}_{\mathbb{R}}(\mathbb{V}_{\mathbb{H}}, \mathbb{E}_{\mathbb{H}})$*, is a directed graph, s.t. the nodes of the graph represent the regions* $V_R^1, V_R^1, \ldots V_R^6$ *of the archipelago map and the edges of the hyper-map are produced with respect to the archipelago map* $\mathbb{M}_{\mathbb{R}}(\mathbb{V}_{\mathbb{R}}, \mathbb{E}_{\mathbb{R}})$*, as follows:* $\forall e(v_i, v_j) \in \mathbb{E}_{\mathbb{R}}$, $\exists e(V_R^i, V_R^j) \in \mathbb{E}_{\mathbb{H}}$ *if and only if* $(v_i \in V_R^i \wedge v_j \in V_R^j \wedge V_R^i \neq V_R^j)$.

3.2 Recommending Refactoring Trips

To facilitate the refactoring process, *the archipelago map must go live to provide actionable recommendations* to the developer towards the effective combination of refactorings. Moreover, the developer needs *contextualization* for each refactoring, in the form of guidelines concerning how to apply it. Finally, the developer needs assistance for the *identification of refactoring opportunities* in the specific module (package, class, method, etc.) that he/she is working with.

To deal with the aforementioned issues we developed *the Refactoring Trip Advisor* as an Eclipse plugin[2]. At a glance, *a refactoring trip* begins with the developer selecting a particular refactoring region from the archipelago hyper-map (e.g., the Composing method region given in Fig. 1). As a result, the Refactoring Trip Advisor provides to the developer the map of the selected region. The developer selects the particular refactoring (e.g., Extract Method in Fig. 2) that he/she wants to apply. Then, the Refactoring Trip Advisor highlights in the map the selected refactoring and other related refactorings, which can be used before, after, as part of, or instead of the selected refactoring, with the respective nodes colored in yellow, pink, cyan, purple and tan.

Each refactoring is related with *slideware* that provides guidelines on how to apply it. The slideware consists of three parts: the first part explains the problem solved by the refactoring; the second part, gives a simple example on how to apply the refactoring; the last part, allows the developer to execute available *refactoring detectors* that identify refactoring opportunities in the code. To perform the recommended refactorings the developer can exploit the available IDE refactoring facilities (see the related discussion in Sect. 4.1).

Regarding the refactoring detectors, one of the primary concerns of our approach is *extensibility*. Specifically, our goal is to ease the integration of the tool with (a) *in-house refactoring detectors*, developed specifically for our approach and (b) *external refactoring detectors*, provided by third-party developers. To achieve this goal, we rely on the Three-Steps Refactoring Detector pattern that we introduced in Tsimakis et al. [14]. The pattern facilitates the development of refactoring detectors via a polymorphic hierarchy of template classes that realize a general three-step refactoring detection process. In the Refactoring Trip Advisor we used the pattern for the development of eight in-house refactoring detectors. Moreover, we used the pattern to facilitate the integration of

[2] The plugin source code can be downloaded from github.com/AnastasiosHJW/RefactoringTrip-Advisor.

the Refactoring Trip Advisor with three external refactoring detectors that are provided by the JDeodorant refactoring framework [3,12,13]. More details concerning the usage of the pattern in the Refactoring Trip Advisor can be found in Tsimakis et al. [14].

To discuss the involvement of the Refactoring Trip Advisor in the refactoring process we employ a typical re-engineering scenario that concerns data containers and misplaced responsibilities. Specifically, we focus on an application that analyzes the evolution of Amazon Web Services (AWS) [17]. The heart of the application is the `History` class, a data container that keeps the evolution data of subsequent Web service versions. Around the `History` class there are several client classes that manipulate the evolution data. The code of the client methods is typically long and complicated. Certain parts of the client methods are misplaced, as the `History` class should have been responsible for the manipulation of the evolution data.

```
public ChartPanel[] visualizeGrowth() {
  ChartPanel[] panels = new ChartPanel[2];
  ArrayList<VersionInfo> versionsList = history.getVersions();

  DefaultCategoryDataset operationsGrowth =new DefaultCategoryDataset();
  for (int i = 0; i < versionsList.size(); i++) {
    double opers = versionsList.get(i).getOperationGrowth();
    String xAxis = versionsList.get(i).getId();
    operationsGrowth.setValue(opers, "Operations", xAxis);
  }

  JFreeChart opersGrowthChart = ChartFactory.createLineChart(
    "Growth␣Rate␣Line␣Chart", "Version␣ID",
    "Growth␣Rate", operationsGrowth,
    true, true, false
  );
  panels[0] = new ChartPanel(opersGrowthChart);

  //...............
  return panels;
}
```

Listing 1. Code snippet from `visualizeGrowth()`, before refactoring.

```
public ChartPanel[] visualizeGrowth() {
  ChartPanel[] panels = new ChartPanel[2];

  DefaultCategoryDataset operationsGrowth = history.getOpersGrowth();

  JFreeChart opersGrowthChart = ChartFactory.createLineChart(
    "Growth␣Rate␣Line␣Chart",
    "Version␣ID", "Growth␣Rate",
    operationsGrowth, true, true, false
  );
  panels[0] = new ChartPanel(opersGrowthChart);

  //...............
  return panels;
}
```

Listing 2. Code snippet from `visualizeGrowth()`, after refactoring.

For instance, `GrowthVisualizer` is a client class that visualizes the Web services' growth, in terms of the number of provided operations and data structures. Listing 1 gives a code snippet from the `visualizeGrowth()` method of

`GrowthVisualizer`. Our goal is to make this snippet smaller and simpler. To this end, we begin from the archipelago hyper-map (Fig. 1), which gives general recommendations on how to combine refactorings from different regions. To reorganize the code of the method, we can use refactorings from the Composing Methods region. These refactorings typically result in the extraction of new methods. Therefore, the hyper-map suggests that afterwards it may be useful to consider refactorings from the Moving Features region, to potentially move extracted methods and related fields.

Table 1. Developers profiles (colours highlight boundary values).

	Developer profiles			
Developer ID	Development experience	Refactoring knowledge	Refactoring frequency	Refactoring tool usage
1	low (<2 years)	good	rarely	never
2	low (<2 years)	good	very often	never
3	low (<2 years)	moderate	never	never
4	low (<2 years)	moderate	rarely	very often
5	low (<2 years)	none	never	often
6	low (<2 years)	none	never	rarely
7	medium (3-5 years)	none	never	never
8	medium (3-5 years)	good	often	never
9	medium (3-5 years)	none	never	never
10	medium (3-5 years)	good	very often	never
11	medium (3-5 years)	moderate	rarely	never
12	medium (3-5 years)	good	often	never
13	medium (3-5 years)	good	often	often
14	high (>6 years)	good	often	often
15	high (>6 years)	good	often	never
16	high (>6 years)	good	often	very often

Suppose that we select Extract Method (yellow node in Fig. 2) from the Composing Methods region. The respective refactoring detector [13] suggests to extract a new method for the loop (Listing 1, lines 6–10) that prepares the evolution data-sets for the visualization of the Web service growth. Before the method extraction, the Composing Methods map recommends the use of refactorings like Inline Temp and Replace Temp with Query (pink nodes in Fig. 2) for the removal of local variables. These refactorings shall make the method extraction easier (see first quote in Sect. 1). The Inline Temp detector [14] identifies several such variables (e.g., `versionList`, `opers`, `xAxis`) After the method extraction, the map suggests to consider Move Method (cyan node in Fig. 2) to place the extracted methods (`getOpesGrowth()`) close to the data that it manipulates, while the Move Method detector [12] identifies `History` as the appropriate target class. Overall, the refactored code snippet is given in Listing 2.

4 Validation

To validate the Refactoring Trip Advisor we consider two main issues. First, we assess whether it adheres to the basic refactoring tool principles, introduced by Murphy-Hill and Black [10]. Then, we examine what do the developers actually think about the tool.

4.1 Fitness for Purpose

Murphy-Hill and Black introduced five principles for refactoring tools [10]. Specifically, a refactoring tool should allow the programmer to: (**R1**) Choose the desired refactoring quickly; (**R2**) switch seamlessly between program editing and refactoring; (**R3**) view and navigate the program code while using the tool; (**R4**) avoid providing explicit configuration information; (**R5**) access all the other tools normally while using the tool.

The Refactoring Trip Advisor facilitates the selection of refactorings (**R1**) based on two concepts: (1) it provides the region maps that hide from the developer the complexity of the entire archipelago map; (2) it provides the archipelago hyper-map that summarizes the contents of the archipelago map. The basic functionalities of the tool are provided via a separate frame (Figs. 1 and 2). The developer can activate this frame and switch back to the main IDE frame, at any point, to edit/view/navigate the program code (**R2**, **R3**), and use any other tool that is available through the IDE (**R5**). The recommendation of refactorings based on the region maps does not require any configuration information. On the other hand, certain detectors of refactoring opportunities do. Typically, the required information concerns thresholds that customize the detectors' modus operandi. Nevertheless, the developer can avoid setting these thresholds (**R4**), as the proposed tool assumes default values for them.

4.2 The Developers' Opinions

The *goal* of this study is to let the developers *use* the Refactoring Trip Advisor and get their *feedback* concerning the overall approach. To familiarize the developers with the Refactoring Trip Advisor we employed the re-engineering scenario introduced in Sect. 3. Specifically, we asked the developers to simplify the `visualizeGrowth()` method (a code snippet of the method is given in Listing 1) with the help of the tool. As a starting point, we prompted the developers to focus on refactorings from the Composing Methods region. After this experience, we asked them to perform an overall review of the the Refactoring Trip Advisor, by rating the usefulness of the refactoring relations, slideware, and detection facilities, based on a typical 5-level likert scale. The developers could provide further comments, remarks, and suggestions, concerning our approach.

Our study involves a sample of 16 developers from industry and academia (students and staff members). The selection was based on purposive sampling of heterogeneous instances [1]; the developers were chosen deliberately to reflect diversity on (a) development experience, (b) knowledge about refactoring, (c) frequency of refactoring, and (d) usage of refactoring tools. The detailed profiles of the developers are provided in Table 1.

Key Findings. Table 2 summarizes the results that we obtained. The results are provided in three parts: the first part (Table 2 - left) gives the *statistical breakdown of the refactorings* that have been performed by the developers; the second part (Table 2 - middle) analyzes the *exploitation of the recommendations* that have been provided by the Refactoring Trip Advisor, in terms of the percentage

Table 2. Refactorings & tool feature ratings (colours highlight boundary values).

Developer ID	Refactorings performed				Exploitation of recommendations		Tool features rating		
	Inline Temp	Extract Method	Move Method	Total	% of suggested refactorings that have been perfomed	% of performed refactorings that have been suggested	Relations	Slides	Detectors
1	7	2	2	11	100,00%	100,00%	5	5	5
2	6	2	0	8	72,73%	100,00%	4	5	4
3	2	0	0	2	18,18%	100,00%	5	5	5
4	7	2	2	11	100,00%	100,00%	5	5	4
5	3	2	0	5	45,45%	100,00%	4	5	3
6	3	2	2	7	63,64%	100,00%	4	4	2
7	4	1	0	5	45,45%	100,00%	4	5	3
8	7	2	0	9	81,82%	100,00%	5	5	5
9	7	0	0	7	63,64%	100,00%	5	4	4
10	7	3	2	12	100,00%	91,67%	4	4	4
11	7	2	2	11	100,00%	100,00%	5	5	5
12	7	2	2	11	100,00%	100,00%	5	5	5
13	0	2	2	4	36,36%	100,00%	4	4	4
14	7	2	0	9	81,82%	100,00%	4	5	3
15	7	2	2	11	100,00%	100,00%	4	4	3
16	5	2	2	9	81,82%	100,00%	3	3	3
Average	5,38	1,75	1,13	8,25	74,43%	99,48%	4,38	4,56	3,88
Stdev	2,28	0,77	1,02	3,00	26,60%	2,08%	0,62	0,63	0,96

of recommended refactorings that have been performed by the developers, and the percentage of performed refactorings that have been recommended; finally, the third part (Table 2 - right) details the assessment of the tool's features.

Overall, *the developers managed to use the Refactoring Trip Advisor in the context of a realistic task.* Nevertheless, *the exploitation of the provided recommendations by the developers varies, along with the quality of the results that they produced.* In particular, two developers (3 and 9) just removed local variables from `visualizeGrowth()`, via Inline Temp. Five developers (2, 5, 7, 8, 9) simplified `visualizeGrowth()` even more, using Extract Method. Finally, *nine developers* (1, 4, 6, 10, 11, 12, 13, 15, 16) *solved the problem of misplaced responsibilities*, by moving the extracted methods to the `History` class, via Move Method.

The percentage of recommended refactorings that have been performed by the developers varies from 18.18% to 100%. However, *most of the developers applied a large percentage of the recommended refactorings*; twelve developers performed more than 60% of the recommended refactorings, while the average percentage of refactorings that have been performed is 74.43% with a standard deviation of 26.60%. On the other hand, *the percentage of performed refactorings that have been recommended is 100% for all the developers, except one,* who extracted a method that was not included in the suggestions.

Concerning the assessment of the overall approach, *the developers provided quite high ratings, with the refactoring relations and slides being the most appreciated features*, followed by the refactoring detection facilities. Regarding further comments and remarks, several developers pointed out that the proposed approach *helped them to learn more about refactoring*. Moreover, they mentioned that the proposed approach could be considered both for *development* and *education* purposes. The developers' *suggestions* ranged from *concrete improvements* (e.g., to reduce the number of pop-up windows and make the slideware resizable), to *broader ideas* like making the representation of the map more interactive, allowing the developer to customize it by adding/removing/changing refactorings and relations, adding more slides on the relations between refactorings, complementing the slideware with audio/video, and so on.

Threats to Validity. Two factors that threaten the internal validity of one group experiments are *history* and *maturation*; the longer the time of the experiment, the more likely are these threats [1]. In our study, the overall duration of the tests was reasonably short, ranging from 15 to 65 min. Another threat to internal validity is the social desirability bias, in the sense that the developers simply agreed with the tool recommendations. To deal with this threat, we used anonymous questionnaires. Our assessment relies on a single scenario that focuses on a subset of refactorings for composing and moving methods. These are threats to external validity. We further asked from the developers an overall review, concerning the proposed approach as a whole. Nevertheless, to be able to generalize the results further studies should be performed, involving more developers, subject systems and refactorings.

5 Conclusion

In this paper, we proposed an approach that provides actionable recommendations for the effective combination of refactorings. The recommendations are based on an interactive model that formally specifies respective informal refactoring guidelines. The proposed approach is inline with the fundamental refactoring tool principles. To further assess the approach, we conducted a study that involved 16 developers and a limited set of refactorings for composing and moving methods. The developers found the overall approach useful. The positive results that we obtained encourage follow up studies of a broader scale and scope.

Our approach is currently based on knowledge that is "hidden" in Fowler's catalog. Other sources of information, views and notations can be considered towards its extension. Making the refactoring detection techniques more developer-intuitive and easy to use is also an issue that should be further investigated. In the future, it would also be interesting to investigate the relation between the proposed approach and the issue of technical debt prioritization.

Acknowledgements. We would like to thank the anonymous reviewers for their feedback on the paper.

References

1. Cook, T.D., Campbell, D.T.: Quasi-Experimentation: Design and Analysis Issues for Field Settings. Houghton Mifflin Company, Boston (1979)
2. Al Dallal, J.: Identifying refactoring opportunities in object-oriented code: a systematic literature review. Inf. Softw. Technol. **58**, 231–249 (2015)
3. Fokaefs, M., Tsantalis, N., Stroulia, E., Chatzigeorgiou, A.: Identification and application of extract class refactorings in object-oriented systems. J. Syst. Softw. **85**(10), 2241–2260 (2012)
4. Fowler, M.: Refactoring: Improving the Design of Existing Code. Addison-Wesley, Boston (2000)
5. Kim, M., Zimmermann, T., Nagappan, N.: An empirical study of refactoring challenges and benefits at Microsoft. IEEE Trans. Softw. Eng. **40**(7), 633–649 (2014)
6. Mens, T., Tourwé, T.: A survey of software refactoring. IEEE Trans. Softw. Eng. **30**(2), 126–139 (2004)
7. Mens, T., Taentzer, G., Runge, O.: Analysing refactoring dependencies using graph transformation. Softw. Syst. Model. **6**(3), 269–285 (2007)
8. Moghadam, I.H., Cinnéide, M.Ó.: Resolving conflict and dependency in refactoring to a desired design. e-Informatica **9**(1), 37–56 (2015)
9. Morales, R., Chicano, F., Khomh, F., Antoniol, G.: Efficient refactoring scheduling based on partial order reduction. J. Syst. Softw. **145**, 25–51 (2018)
10. Murphy-Hill, E.R., Black, A.P.: Refactoring tools: fitness for purpose. IEEE Softw. **25**(5), 38–44 (2008)
11. Opdyke, W.F.: Refactoring object-oriented frameworks. Ph.D. thesis, University of Illinois - Urbana Champaign (1992)
12. Tsantalis, N., Chatzigeorgiou, A.: Identification of move method refactoring opportunities. IEEE Trans. Softw. Eng. **99**(3), 347–367 (2009)
13. Tsantalis, N., Chatzigeorgiou, A.: Identification of extract method refactoring opportunities for the decomposition of methods. J. Syst. Softw. **84**(10), 1757–1782 (2011)
14. Tsimakis, A., Zarras, A.V., Vassiliadis, P.: The three-step refactoring detector pattern. In: Proceedings of the 24th European Conference on Pattern Languages of Programs (EuroPLoP) (2019, to appear). www.cs.uoi.gr/~zarras/papers/C36.pdf
15. Van Der Straeten, R., Jonckers, V., Mens, T.: A formal approach to model refactoring and model refinement. Softw. Syst. Model. **6**(2), 139–162 (2007)
16. Zarras, A.V., Vartziotis, T., Vassiliadis, P.: Navigating through the archipelago of refactorings. In: Proceedings of the the Joint 23rd ACM SIGSOFT Symposium on the Foundations of Software Engineering and 15th European Software Engineering Conference (FSE/ESEC), pp. 922–925 (2015)
17. Zarras, A.V., Vassiliadis, P., Dinos, I.: Keep calm and wait for the spike! insights on the evolution of Amazon services. In: Proceedings of the 28th International Conference on Advanced Information Systems Engineering (CAiSE), pp. 444–458 (2016)

String Representations of Java Objects: An Empirical Study

Matúš Sulír(✉)

Technical University of Košice, Letná 9, 042 00 Košice, Slovakia
matus.sulir@tuke.sk

Abstract. String representations of objects are used for many purposes during software development, including debugging and logging. In Java, each class can define its own string representation by overriding the toString method. Despite their usefulness, these methods have been neglected by researchers so far. In this paper, we describe an empirical study of toString methods performed on a corpus of Java files. We are asking what portion of classes defines toString, how are these methods called, and what do they look like. We found that the majority of classes do not override the default (not very useful) implementation. A large portion of the toString method calls is implicit (using a concatenation operator). The calls to toString are used for nested string representation building, exception handling, in introspection libraries, for type conversion, and in test code. A typical toString implementation consists of literals, field reads, and string concatenation. Around one third of the string representation definitions is schematic. Half of such schematic implementations do not include all member variables in the printout. This fact motivates the future research direction – fully automated generation of succinct toString methods.

Keywords: ToString methods · Java · Quantitative study · Qualitative study

1 Introduction

Almost every object-oriented programming language supports a way to represent the state of an object as a text string. Usually, the textual representation is obtained by calling a method such as `String()`, `printOn:`, or `to_s`.

In Java, this method is called `toString`. Its basic implementation is defined in the root of the class hierarchy – the class `Object`. It is therefore callable on an object of any type. However, the default implementation displays only a class name and a hash code of the object, e.g., `java.io.FileWriter@5c29bfd`. Each class can override this method to implement a potentially useful textual representation. For example, a `HashMap` instance can be represented as "{a = 1, b = 2}".

String representations are used during debugging, logging, testing and for many other purposes. For instance, when a developer displays a list of variables

A. Chatzigeorgiou et al. (Eds.): SOFSEM 2020, LNCS 12011, pp. 479–490, 2020.
https://doi.org/10.1007/978-3-030-38919-2_39

in a contemporary debugger, each non-primitive variable is initially shown as a string obtained by calling the `toString` method on it. Only after expanding it, individual member variables are displayed.

Despite their importance, `toString` and related methods have received limited attention by researchers so far. Schwarz [10] performed a very brief analysis of `printOn:` methods in Smalltalk. Although not directly focused on string representation methods, Qiu et al. [8] and Lemay [5] performed large-scale studies of Java language usage, where they found the `toString` method is among the most used standard library calls.

In this paper, we would like to present an empirical study about the prevalence and properties of `toString` methods in Java. Using automated analysis of a source code corpus and manual inspection of selected examples, we will answer the following research questions:

- **RQ1:** What portion of classes defines their own `toString` method?
  - **RQ1.1:** How deep in the inheritance hierarchy are `toString` methods defined?
- **RQ2:** How are `toString` methods called in the code?
  - **RQ2.1:** Are they explicit (method calls) or implicit (concatenations of a string and a non-string)?
  - **RQ2.2:** Are they often called from other `toString` implementations?
  - **RQ2.3:** What are other common usage scenarios of string representations?
- **RQ3:** What do typical `toString` methods look like?
  - **RQ3.1:** What language constructs do they consist of?
  - **RQ3.2:** Do they often call `toString` of the superclass?
  - **RQ3.3:** Are they rather schematic, or do they contain advanced logic?
  - **RQ3.4:** Do they print the values of all member variables or only a portion of them?

The contribution of this paper is twofold. First, we expand the available knowledge in this area by describing the current state of string representation implementation and usage. Second, since these implementations may be repetitive, this paper can act as a basis for their future automated generation. This could, for example, make debugging easier, particularly in situations when the available display space is constrained [12].

In Sect. 2, we briefly describe the method. In Sects. 3, 4, and 5, the method is elaborated and the results of each research question are presented. Finally, we describe the threats to validity (Sect. 6), related work (Sect. 7), conclusion and future work (Sect. 8).

2 Method Outline

To perform the analysis, we needed a corpus of parsable source code files. For some questions or their parts, type resolution was necessary. This means that except for the source files themselves, we needed also their compiled versions along with all their dependencies, such as third-party libraries. Furthermore, although projects in online repositories (e.g., GitHub) are easily accessible in large quantities, they can include homework assignments or personal backups, potentially skewing the analysis results [4]. Therefore, we sought for a curated collection of software projects.

The only well-known corpora fulfilling all of the above criteria are the Qualitas Corpus [13] and corpora based on it, such as Qualitas.class [14]. Since they are outdated, we decided to build a new corpus based on them. From the Qualitas Corpus [13], we selected a random subset of 15 open source projects which are still active (updated in the last year) and successfully buildable. We manually downloaded their current versions and whenever possible, we also automatically downloaded both the source code and binary forms of their dependencies, including the transitive ones. This way, we obtained an up-to-date corpus consisting of 759 artifacts (projects and modules) with 106,473 unique Java files (based on the package and type names). The number of classes in these files, including named inner and nested classes, is 149,057. For more than 97% of them, we were able to successfully resolve all their superclasses up to the root of the type hierarchy.

After constructing the corpus, we executed an automated source code analysis process on it. This analysis was performed by a custom program we have written using the Spoon library [7]. To enable the reproducibility of research, both the corpus-building script and the analysis program are available online[1].

All quantitative results were produced fully automatically by the program. The output of the analysis included also source code examples, some of which we inspected manually, thus producing qualitative results (RQ2.3 and a part of RQ3.1). Details of the method are different for each of the research questions, so they are described in the following sections.

3 ToString Definitions

First of all, we would like to know what proportion of classes defines their own `toString` method and how many of them rely on the default representation. A naive approach answering this might be to count the classes defining `toString` and divide them by the total number of classes. However, we must consider also type hierarchies: For example, a `HashMap` does not define its own `toString`, but its superclass, `AbstractMap`, defines one. A string representation defined in a superclass is usually still useful, but in some cases it may be less specific than the method defined directly in the class.

Therefore, we first counted all concrete (not abstract) classes, excluding enumeration types, anonymous and test classes, considering only classes with resolved supertypes. The result was a list of 94,548 classes. Then for each of them, we constructed a type hierarchy from the class itself up to `Object`. Finally, we determined which `toString` method would be executed when calling this method on the object of the analyzed class. It is always the most specific defined method

[1] https://github.com/sulir/tostring-study.

– e.g., in the mentioned example, for the hierarchy `HashMap` → `AbstractMap` → `Object`, it is the method in `AbstractMap`.

We divided the results into four categories: the most specific `toString` method is defined directly in the given class, in its direct superclass, indirect superclass, or in the root of the class hierarchy (`Object`). In Fig. 1, we can see the results in a graphical form.

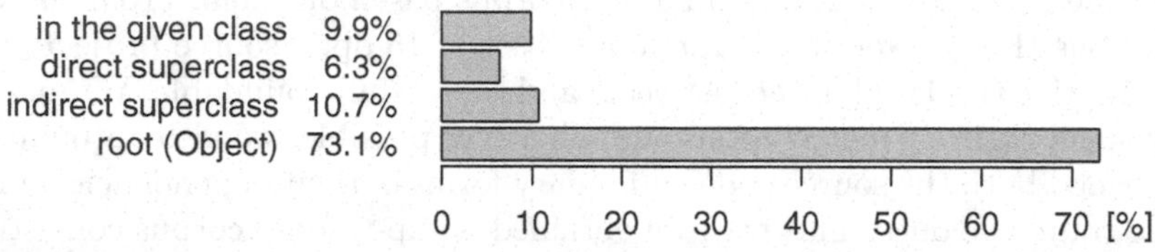

Fig. 1. The most specific toString method definition for classes

Answering **RQ1** and **RQ1.1**, only 9.9% of classes define their own `toString` implementation directly, while 17.0% derive a custom implementation from one of their superclasses. The majority of the analyzed classes (73.1%) does not define a custom string representation at all – it relies on the default one, which is useless in many cases. This highlights a need to offer fully automated generation of string representations. Although modern IDEs (integrated development environments) offer semi-automated (partially manual) generation of `toString` methods, developers evidently do not utilize this feature to a high extent.

4 ToString Invocations

In Java, `toString` methods can be called either explicitly or implicitly. An explicit call is a standard method call visible in the source code, e.g.:

```
String value = someObject.toString();
```

An implicit `toString` call is performed automatically on a non-string object during concatenation of a string and non-string expression:

```
Object nonStringObject;
String value = "string" + nonStringObject;
```

When answering **RQ2** and its sub-questions, we considered all parsable Java files (106,473). We automatically searched for all `toString()` method invocations and categorized them as either implicit or explicit. The implicit calls were recognized by searching for the "+" operator with one string and one non-string operand. The locations of the calls were also noted and categorized as either directly inside another `toString` method definition or outside it. Then, we selected a random subset of the calls, which we inspected manually to gain insights about string representation usage.

For the summary of quantitative results, see Table 1.

Table 1. Explicit vs. implicit toString calls, calls from other toString methods.

Call type	Explicit (`obj.toString()`)	36.7%
	Implicit (concatenation)	63.3%
Call location	From within other `toString` methods	13.0%
	Outside a `toString` method	87.0%

4.1 Explicit and Implicit Calls

First, we will answer **RQ2.1**. A majority of `toString` calls (63.3%) is implicit – i.e., not visible in the source code at a first sight. On average, we found 0.34 explicit and 0.59 implicit `toString` invocations per Java file. This totals to 0.93 `toString` invocations per Java file, which means string representations are fairly commonly utilized in the code.

4.2 Calls from Other ToStrings

Answering **RQ2.2**, we determined that 13.0% of all `toString()` calls were located directly in another `toString` definition. This means string representations are fairly often built recursively from other textual representations. In addition to this, the `toString` method is sometimes called also indirectly through a chain of other auxiliary methods.

4.3 Other Usage Scenarios

Except for building string representations of objects from other string representations, there are other common usage scenarios. Now we will look at the examples of them, thus answering **RQ2.3**.

String representations are commonly used to work with exceptions. There are two recurring patterns which we encountered during the manual inspection of the results. In the first case, an object is converted to a string and then included in the message of the exception being constructed and thrown. The second case is the conversion of a caught exception to a string and passing it to a logger.

Textual representations were used for logging beyond exception handling: Any object can be converted to a string and written to a logger (e.g., a file or console) for debugging purposes.

We also encountered multiple `toString` invocations in introspection libraries, where they were used to convert runtime metadata of the program into strings, so they could be later used for debugging and visualization purposes.

A very common usage of `toString` is for a conversion of a string-like object (e.g., an `XMLString`) to a standard `String`. The most frequent conversion we encountered was from a `StringBuilder` (or related classes), which is a mutable implementation of a text string in Java.

Finally, `toString` calls were present also in unit tests. In assertions, a string representation was often compared with another string representation or a literal.

Particularly the last two applications (type conversion and unit testing) are typical examples of cases when the `toString` method is not used only for debugging and development purposes, but where it has an influence on the correct program functionality. This has an important implication for us. If we wanted to implement an automatically applied string representation generator, we would have to take great care not to break the existing representations. For example, if an object had the `toString` method implemented in an indirect superclass and we wanted to generate a more specific one directly in the given class, we could break the functionality if the program logic relied on the `toString` from the superclass.

5 ToString Contents

To answer **RQ3**, we analyzed all `toString` methods defined inside classes in our corpus (considering only the ones for which we had the source code available and superclasses resolved). In total, 11,302 method definitions were analyzed.

5.1 Language Constructs

Our main goal was to find out whether there exist certain very common forms of `toString` definitions, which are possibly repetitive. To answer **RQ3.1**, for each analyzed method, we obtained a set of all AST (abstract syntax tree) node types present in it. For example, the definition

```
return "a" + "b";
```

contains the AST node set {Return, Literal, BinaryOperator}. We excluded too generic AST types, such as blocks or type references. Then, we counted how often each such node set is present in the collection of the analyzed `toString` methods.

In Table 2, we can see a list of the most frequently occurring node sets, along with the percentages of the `toString` methods where they occur and examples of their source code. This provides us an overview of how typical `toString` methods look, which of them were probably semi-automatically generated with an IDE, and to what extent their generation could be fully automatized.

Almost 14% of `toString` methods consist of a return statement and a mix of literals, field reads on the current object (`this`), and binary operators (most notably, "`+`"). We suppose these definitions were frequently generated using an IDE.

The second most frequent set contains method invocations in addition to these node types. Here we observe less schematic code, since on some variables, various methods were called to more precisely specify the string representation.

The next three node type sets usually represent the same essence – only one member variable is included in the string representation, without any additional

Table 2. The most frequent node type sets in the collection of analyzed toString method definitions.

%	Node type set	Source code example
13.82	{Return, Literal, ThisAccess, FieldRead, BinaryOperator}	`return "PreparedStatementCreator: sql=[" + sql + "]; parameters =" + this.parameters;`
8.84	{Return, Literal, ThisAccess, FieldRead, BinaryOperator, Invocation}	`return "registry[" + this.sessions.size() + "sessions]";`
8.03	{Return, ThisAccess, FieldRead, Invocation}	`return table.toString();`
5.81	{Return, ThisAccess, Invocation}	`return getName();`
5.63	{Return, ThisAccess, FieldRead}	`return flag;`
5.42	{Return, Literal}	`return "Immediate key";`
5.15	{Return, ThisAccess, Invocation, BinaryOperator, Literal}	`return getArtifact() + " < " + getRepositories();`
2.90	{Return, ThisAccess, FieldRead, Invocation, Literal}	`return String.format("%s[value=%s]", getClass().getSimpleName(), value);`

2.62	{Return, LocalVariable, ConstructorCall, VariableRead, Invocation, Literal, ThisAccess, FieldRead}	`final StringBuilder buf = new StringBuilder();`
		`buf.append("[local: ").append(this.local);`
		`buf.append("defaults: ")`
		`.append(this.defaults);`
		`buf.append("]");`
		`return buf.toString();`

2.44	{Return, LocalVariable, ConstructorCall, VariableRead, Invocation, Literal, ThisAccess, FieldRead, BinaryOperator, If}	`StringBuilder sb = new StringBuilder("FactoryCreateRule[");`
		`if (creationFactory != null) {`
		`sb.append("creationFactory=");`
		`sb.append(creationFactory);`
		`}`
		`sb.append("]");`
		`return (sb.toString());`

information. In cases when the given field was not a sole member variable of the class, a question arises how this field was selected and why its name is not printed.

When a `toString` method returns solely a literal, it is often related to the given class name; this is not a rule though. Methods consisting of invocations, binary operators, and literals frequently contained getters – but again, this is not a rule.

The last three lines in Table 2 are mainly just variations of the first one, but using either a string-formatting function `String.format` or mutable strings (`StringBuilder`, with or without null checking) instead of string concatenation.

5.2 Reusing Superclass Implementations

Next, we were interested in whether `toString` implementations reuse the string representation of a superclass in some way – i.e., whether they include a call to `super.toString()`. We analyzed all `toString` methods inside classes derived from a class other than `Object`.

We found out only 10.3% of such methods call `super.toString()`. This is probably caused by the fact that only a small portion of classed directly defines `toString`, which makes reuse difficult.

5.3 Schematic Implementations

Since our future vision is to fully automate the string representation generation, in **RQ3.3** we focused on the schematicity of the `toString` implementations. We define a schematic implementation as a method definition which consists only of one or more of these language constructs: `return`, `this`, string literals, `super.toString()`, direct reading of the fields of the current object, a null-checking `if` statement or ternary operator, and standard string-building operations (a `toString` call, string concatenation, a `String.format` call, basic `StringBuilder` and `StringBuffer` operations).

We found that 33.4% of `toStrings` correspond to our definition of schematic implementation, while 66.6% of them are more complicated. The former group looks promising with respect to the possibility of fully automated generation.

5.4 Member Variables Read

Finally, in **RQ3.4**, we further inspect the schematic implementations found in the previous step. Schematic `toString` definitions often contain a class name and a list of name–value pairs of the object's member variables. Many of them were probably generated using a wizard in an IDE. However, note that the generation process is not fully automated: the programmer must still manually select which member variables will be included in the textual representation and which ones will be omitted. Otherwise the representations might get impractically long, especially if the member variables themselves are non-primitive objects with their own – similarly structured – string representation. Some fields might be also considered irrelevant to the application domain (e.g., logging support). Therefore, we hypothesize only a portion of member variables are included in string representations.

To empirically confirm our intuition, we consider all schematic `toString` implementations inside classes with at least one non-constant (i.e., not static final) member variable. For them, we determine what proportion of non-constant member variables of the given class are read in its `toString` method. For simplicity, we consider only member variables (fields) defined directly in a given class, not in its superclasses.

The results are depicted in Fig. 2. About half of the analyzed classes (51.5%) read all fields in its string representation, while the other half (48.5%) include only some of them or none.

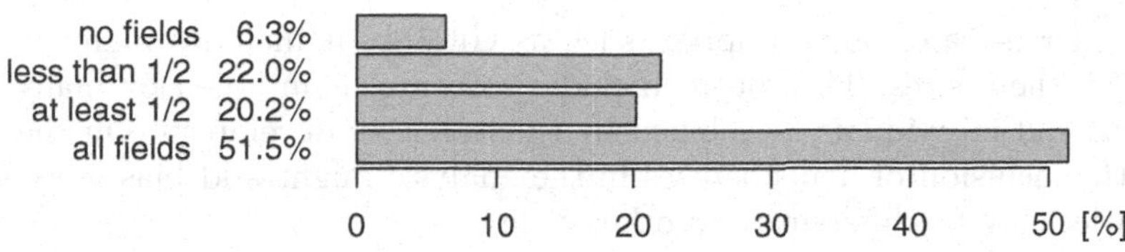

Fig. 2. The proportion of member variables read in schematic toString methods

We consider the latter case to be a suitable candidate for full automation: Using heuristics or machine learning, we could select a subset of fields which should be included in the string representation.

6 Threats to Validity

Now we will describe threats to the validity of our study according to Wohlin et al. [15].

6.1 Construct Validity

During the analysis, we excluded duplicate files based on package names and top-level type names (fully qualified class names). Nevertheless, there still may be duplicate classes present under different names.

In the first research question, we excluded test classes based on standardized directory names, which might not be sufficient. However, it is questionable whether it is useful to have string representations of test classes or not, and what exactly is considered a test class.

For the third research question, we represented methods by the sets of node types they consist of, which may be an oversimplification. The displayed examples might not fully represent the whole node type sets. Nevertheless, RQ3.1 was partially qualitative and the listed exact percentages are mainly supplementary.

Our definition of schematic implementation is rather ad-hoc. However, it was inspired by common `toString` generation templates of the most used Java IDEs.

During the analysis of member variable reads, we did not include member variables defined in the superclasses. However, this improves the clarity of the study since the fields in superclasses are not always accessible (there may be private, package-private and located in a different package, etc.), which would complicate the interpretation of the results.

6.2 External Validity

Findings about the corpus used in this study might not be generalizable, since the corpus may not be representative of all software used in practice. We tried to mitigate this threat by basing it on an existing curated corpus and including dependencies to increase its size. Although it was not constructed directly by crawling large-scale software forges such as GitHub, it includes many projects hosted on these sites. The corpus includes software maintained by many organizations and individuals, developed by thousands of developers. On the other hand, the inclusion of dependencies in the analysis might add bias as the code of libraries may be different from other code.

Although the original corpus was dated, we selected only projects updated in the last year. The list of the 15 base projects therefore includes rather mature projects – but their dependencies can include also newer and smaller ones.

7 Related Work

The most similar study to ours was performed by Schwarz [10]. He found that 28% of Smalltalk repositories in Squeaksource included a definition of a `printOn:` method (a Smalltalk analogy of `toString`), and the average length of a `printOn:` method was 7.1 lines. A few other findings regarding the properties of `printOn:` methods were very briefly described. In contrast to him, our study was performed on Java, uses a larger corpus, defines more precise research questions, and includes a more in-depth analysis.

There exist multiple works studying language and API usage in general. However, they are not specifically focused on `toString` methods in any way. For example, Dyer et al. [3] performed a study of a huge collection of AST nodes to study the usage of Java language features. Three separate studies – by Ma et al. [6], Qiu et al. [8], and Lemay [5] – studied general API call usage on large Java corpora. These three studies consistently found out that the `toString` method is one of the most called standard API methods in source code.

Xu et al. [16] parse `toString` methods to determine the mapping between lines in log files and the source code fragments that produced them. They did not perform any empirical study of the `toString` methods though.

In a study of the Hackage Haskell corpus, `deriving Show`, responsible for the string representation generation, was the most common `deriving` statement [1].

Complementary to string representations, researchers designed also approaches to specify graphical representations of objects: namely DoodleDebug [10], Vebugger [9], and the Moldable Inspector [2]. In our previous work [11], we investigated what properties developers expect from such visual representations.

8 Conclusion and Future Work

In this paper, we described an empirical study of textual object representations in Java. We found that the majority of classes (73%) relies on the default (and often not very useful) `toString` implementation defined in the `Object` class.

A majority of `toString` method calls (63%) is implicit – using a string concatenation operator with one non-string operand. The `toString` methods are called from other representation-building methods (13%), in exception handling and logging code, in introspection utilities, when performing type conversion, and from unit tests. They are therefore consumed not only by developers as a debugging aid, but are sometimes necessary for the software to function properly.

A very common `toString` definition consists of literals, field reads, and string concatenation operators. String representations of superclasses are rarely reused (10% of `toStrings` contain them). A significant portion of `toString` definitions (over 33%) is rather schematic. Half of such schematic implementations read all member variables, the other half excludes some of them.

In the future, we would like to extend our study. First, we could use a larger and more representative source code corpus. Second, we would like to answer our research question in a more in-depth manner, particularly RQ3. Finally, our main future goal is fully automated generation of useful and still succinct `toString` methods without any manual interaction by a developer, particularly by selecting the most important fields to display using machine learning.

Acknowledgments. This work was supported by Project VEGA No. 1/0762/19 Interactive pattern-driven language development. This work was also supported by FEI TUKE Grant no. FEI-2018-57 "Representation of object states in a program facilitating its comprehension".

References

1. Bezirgiannis, N., Jeuring, J., Leather, S.: Usage of generic programming on hackage: experience report. In: Proceedings of the 9th ACM SIGPLAN Workshop on Generic Programming, WGP 2013, pp. 47–52. ACM, New York (2013). https://doi.org/10.1145/2502488.2502494
2. Chiş, A., Nierstrasz, O., Syrel, A., Gîrba, T.: The moldable inspector. In: 2015 ACM International Symposium on New Ideas, New Paradigms, and Reflections on Programming and Software, Onward! 2015, pp. 44–60. ACM, New York (2015). https://doi.org/10.1145/2814228.2814234
3. Dyer, R., Rajan, H., Nguyen, H.A., Nguyen, T.N.: Mining billions of AST nodes to study actual and potential usage of Java language features. In: Proceedings of the 36th International Conference on Software Engineering, ICSE 2014, pp. 779–790. ACM, New York (2014). https://doi.org/10.1145/2568225.2568295
4. Kalliamvakou, E., Gousios, G., Blincoe, K., Singer, L., German, D.M., Damian, D.: The promises and perils of mining GitHub. In: Proceedings of the 11th Working Conference on Mining Software Repositories, MSR 2014, pp. 92–101. ACM, New York (2014). https://doi.org/10.1145/2597073.2597074
5. Lemay, M.J.: Understanding Java usability by mining GitHub repositories. In: 9th Workshop on Evaluation and Usability of Programming Languages and Tools (PLATEAU 2018). OpenAccess Series in Informatics (OASIcs), vol. 67, pp. 2:1–2:9. Schloss Dagstuhl-Leibniz-Zentrum fuer Informatik, Dagstuhl, Germany (2019). https://doi.org/10.4230/OASIcs.PLATEAU.2018.2
6. Ma, H., Amor, R., Tempero, E.: Usage patterns of the Java standard API. In: Proceedings of the XIII Asia Pacific Software Engineering Conference, APSEC 2006, pp. 342–352. IEEE Computer Society, Washington (2006). https://doi.org/10.1109/APSEC.2006.60
7. Pawlak, R., Monperrus, M., Petitprez, N., Noguera, C., Seinturier, L.: Spoon: a library for implementing analyses and transformations of Java source code. Softw. Pract. Exp. **46**(9), 1155–1179 (2016). https://doi.org/10.1002/spe.2346

8. Qiu, D., Li, B., Leung, H.: Understanding the API usage in Java. Inf. Softw. Technol. **73**(C), 81–100 (2016). https://doi.org/10.1016/j.infsof.2016.01.011
9. Rozenberg, D., Beschastnikh, I.: Templated visualization of object state with Vebugger. In: Proceedings of the 2014 Second IEEE Working Conference on Software Visualization, VISSOFT 2014, pp. 107–111. IEEE Computer Society, Washington (2014). https://doi.org/10.1109/VISSOFT.2014.26
10. Schwarz, N.: DoodleDebug, objects should sketch themselves for code understanding. In: 5th Workshop on Dynamic Languages and Applications, DYLA 2011 (2011)
11. Sulír, M., Juhár, J.: Draw this object: a study of debugging representations. In: Proceedings of the Conference Companion of the 3rd International Conference on Art, Science, and Engineering of Programming, pp. 20:1–20:11. ACM (April 2019). https://doi.org/10.1145/3328433.3328454
12. Sulír, M., Porubän, J.: Augmenting source code lines with sample variable values. In: Proceedings of the 2018 26th IEEE/ACM International Conference on Program Comprehension (ICPC), pp. 344–347 (May 2018). https://doi.org/10.1145/3196321.3196364
13. Tempero, E., et al.: The Qualitas Corpus: a curated collection of Java code for empirical studies. In: Proceedings of the 2010 Asia Pacific Software Engineering Conference, APSEC 2010, pp. 336–345. IEEE Computer Society, Washington (2010). https://doi.org/10.1109/APSEC.2010.46
14. Terra, R., Miranda, L.F., Valente, M.T., Bigonha, R.S.: Qualitas.class corpus: a compiled version of the Qualitas Corpus. SIGSOFT Softw. Eng. Notes **38**(5), 1–4 (2013). https://doi.org/10.1145/2507288.2507314
15. Wohlin, C., Runeson, P., Höst, M., Ohlsson, M.C., Regnell, B., Wesslén, A.: Experimentation in Software Engineering. Springer, Berlin (2012). https://doi.org/10.1007/978-3-642-29044-2
16. Xu, W., Huang, L., Fox, A., Patterson, D., Jordan, M.I.: Detecting large-scale system problems by mining console logs. In: Proceedings of the ACM SIGOPS 22Nd Symposium on Operating Systems Principles, SOSP 2009, pp. 117–132. ACM, New York (2009). https://doi.org/10.1145/1629575.1629587

Foundations of Algorithmic Computational Biology – Regular Papers

Fast Indexes for Gapped Pattern Matching

Manuel Cáceres[1], Simon J. Puglisi[2], and Bella Zhukova[2](✉)

[1] Department of Computer Science, University of Chile, Santiago, Chile
mcaceres@dcc.uchile.cl

[2] Department of Computer Science, University of Helsinki, Helsinki Institute for Information Technology (HIIT), Helsinki, Finland
{puglisi,bzhukova}@cs.helsinki.fi

Abstract. We describe indexes for searching large data sets for variable-length-gapped (VLG) patterns. VLG patterns are composed of two or more subpatterns, between each adjacent pair of which is a gap-constraint specifying upper and lower bounds on the distance allowed between subpatterns. VLG patterns have numerous applications in computational biology (motif search), information retrieval (e.g., for language models, snippet generation, machine translation) and capture a useful subclass of the regular expressions commonly used in practice for searching source code. Our best approach provides search speeds several times faster than prior art across a broad range of patterns and texts.

1 Introduction

In the classic pattern matching problem, we are given a string P (the pattern or query) and asked to report all the positions where it occures in another (longer) string T (the text). This problem has been very heavily studied and has applications throughout computer science.

In this paper we consider a variant on the classic pattern matching problem, called variable length gap (VLG) pattern matching. In VLG matching, the query P is not a single string but is composed of $k \geq 2$ strings (subpatterns) that must occur in order in the text. Between each subpattern, a number of characters may be allowed to occur, an upper and lower bound on which is specified as part of the query. Formally, our problem is as follows.

Definition 1 (Variable Length Gap (VLG) Pattern Matching [4]**).** *Let* T *be a string of* n *symbols drawn from alphabet* Σ *and* P *be a pattern consisting of* $k \geq 2$ *subpatterns (i.e. strings)* $p_0, \ldots, p_{k-1}$*, each consisting of symbols also drawn from* Σ*, and having lengths* $m_0, \ldots, m_{k-1}$*, and* $k-1$ *gap constraints* $C_0, \ldots, C_{k-2}$*, such that* $C_i = \langle \delta_i, \Delta_i \rangle$ *with* $0 \leq \delta_i \leq \Delta_i < n$ *specifies the smallest* (δ_i) *and largest* (Δ_i) *allowable distance between a match of* p_i *and* p_{i+1} *in* T*. Find all matches—reported as* k*-tuples* $i_0, \ldots, i_{k-1}$ *where* i_j *is the starting position for subpattern* p_j *in* T*—such that all gap constraints are satisfied.*

This research is supported by Academy of Finland through grant 319454.

A. Chatzigeorgiou et al. (Eds.): SOFSEM 2020, LNCS 12011, pp. 493–504, 2020.
https://doi.org/10.1007/978-3-030-38919-2_40

In computational biology, VLG matching is used in the discovery of and search for *motifs*— i.e. conserved features—in sets of DNA and protein sequences (see, e.g., [18,20]). For example, the following is a protein motif from the rice genome (see [18]) expressed as a VLG pattern:

MT[115, 136]MTNTAYGG[121, 151]GTNGAYGAY.

A similar motif concept in music information retrieval means VLG matching also finds applications in mining and searching for characteristic melodies [7] and other musical structures [8,9] in sequences of musical notes expressed in chromatic or diatonic notation. Bader et al. [1] point out several more applications of VLG matching in information retrieval and related fields such as natural language processing (NLP) and machine translation. For example, Metzler and Croft [17] define a language model in which query terms occuring within as certain window of each other must be found (in NLP, such terms are said to be *colocated*). Locating tight windows of a document containing the set of words contained in a search engine query is the problem of query-biased snippet generation [22]. In machine translation, VLG matching is used to derive rule sets from text collections to boost effectiveness of automated translation systems [15].

VLG matching has a big parameter space and it is easy to think of pathological combinations of pattern and text that lead to an exponential number of matches. Fortunately, however, in practice the problem gets naturally constrained in important ways. The gap constraints are always bound the length of documents under consideration, which in the case of source code or web pages means that usually δ_i and Δ_i (and so their difference) are in (at most) the tens-of-kilobytes range. In genomics and proteomics maximum gaps tend to be around 100 characters or so (see, e.g., [18,20]).

Because of the interesting and useful applications outlined above, VLG matching has received a great deal of attention in the past 20 years. The vast majority of previous work deals with the *online* version of the problem in which both the pattern P and the text T are previously unseen and cannot be preprocessed [2,4,5,9,12,18,21]. Our concern in this paper is the *offline* version of the problem, where T is known in advance and can be preprocessed and an index structure built and stored to later support fast search for previously unseen VLG patterns (the stream of which is assumed to be large, practically infinite). Almost all work on the offline problem is of theoretical interest [3,14]. The exception is the recent work of Bader, Gog, and Petri [1], who develop methods for the *offline* setting that use a combination of suffix arrays [16] and wavelet trees [11,19]. Bader et al. show that their index is an order of magnitude faster at VLG matching than are online methods, and several times faster than q-gram-based indexes, the likes of which were behind Google Code Search [6].

Contribution. Our main contribution in the paper is to show that in practice, on a broad range of inputs typical in real applications of VLG matching, simple algorithms based on intersecting ranges of the suffix array corresponding to subpattern occurrences can be made very fast in practice, and comfortably outperform state-of-the-art methods based on wavelet trees.

We emphasise that none of our new approaches are particularly exotic. They are, however, very fast, and so represent non-trivial baselines by which future (possibly more exotic) indexes for VLG pattern matching and related problems (such as regex matching) can be meaningfully measured.

Roadmap. The remainder of this paper is as follows. Section 2 then looks at a simple method for solving VLG matching that works by sorting and intersecting ranges of the suffix array that contain the occurrences of subpatterns of the VLG pattern. Sections 3 and 4 evolve this basic idea, presenting the results of small illustrative experiments along the way. In Sect. 5 we compare our best performing method to the recent wavelet-tree-based approach of Bader et al., which represents the current state-of-the-art for indexed pattern matching (details of our test machine and data sets can also be found in Sect. 5). Reflections and directions for future work are then offered in Sect. 6.

2 VLG Matching via Sorting and Scanning Suffix Array Intervals

Essential to the methods for VLG matching we will consider in this and later sections is the *suffix array* [16] data structure. The suffix array of T, $|\mathsf{T}| = n$, denoted SA, is an array $\mathrm{SA}[0..n-1]$, which contains a permutation of the integers $0..n$ such that $\mathsf{T}[\mathsf{SA}[0]..n-1] < \mathsf{T}[\mathsf{SA}[1]..n-1] < \cdots < \mathsf{T}[\mathsf{SA}[n]..n-1]$. In other words, $\mathsf{SA}[j] = i$ iff $\mathsf{T}[i..n]$ is the jth suffix of T in ascending lexicographical order. Because of the lexicographic ordering, all the suffixes starting with a given substring p of T form an interval $\mathsf{SA}[s..e]$, which can be determined by binary search in $O(|p| \log n)$ time. Clearly the integers in $\mathsf{SA}[s..e]$ correspond precisely to the distinct positions of occurrence of p in T and once s and e are located it is straightforward to enumerate them in time $O(e-s)$.

The starting point for our approaches is a baseline algorithm from the study by Bader et al. called SA-SCAN, which makes use of the suffix array of T. A pseudo-C++ fragment adapted from Bader et al.'s codebase capturing the main thrust of SA-SCAN is shown in Fig. 1. For ease of reading the code here assumes two subpatterns, but is easy to generalize for $k > 2$.

The operation of SA-SCAN can be summarized as follows. First, search for each of the k subpatterns using SA to arrive at k ranges of the SA containing the subpattern occurrences (in the code listing this is acheived by the two `search` method calls). Next, for each range, allocate a memory buffer equal to the range's size and copy the contents of the range from SA to the newly allocated memory and sort the contents of the buffer (positions of subpattern occurrence) into ascending order. Finally, intersect the positions for subpatterns p_0 and p_1 with respect to the gap constraints. Experimenting with SA-SCAN we observed the time taken to find the ranges of subpattern occurrences in SA constituted less than 1% of the overall runtime, with the vast majority of time spent sorting.

```
SA-scan(string_type p1, string_type p2, int min_gap, int max_gap){
    //1:find intervals of SA containing subpattern occurrences
    std::pair<int,int> interval1 = search(p1,T,SA);
    std::pair<int,int> interval2 = search(p2,T,SA);
    //2: copy positions of subpattern occurrence from SA and sort
    int m1 = interval1.second-interval1.first+1;
    int m2 = interval2.second-interval2.first+1;
    int *A = new int[m1];
    int *B = new int[m2];
    std::memcpy(A,SA+interval1.first,m1);
    std::memcpy(B,SA+interval2.first,m2);
    std::sort(A,A+m1);
    std::sort(B,B+m2);
    //3: intersect according to gap constraints
    for(int i=0,j=0; i<m1 && j<m2; i++){
        while(B[j] < (A[i] + min_gap) && j < m2) j++;
        while(j < m2 && B[j] <= (A[i] + max_gap)){
            result.push_back(B[j]);
            j++;
        }
    }
}
```

Fig. 1. A basic C++ implementation of the SA-SCAN VLG matching algorithm suitable for $k = 2$ subpatterns.

Bader et al. use SA-SCAN as a baseline from which to measure the success of their wavelet-tree-based method. SA-SCAN is natural enough, to be sure, but it does look suspiciously like a straw man. To start with, is `std::sort` really the best we can do for sorting those arrays of integers? We replaced the `std::sort` call with a call to an LSD radix sort of our own implementation (using a radix of 256) and replicated an experiment from Bader et al.'s paper, searching several text collections (including web data, source code, DNA, and proteins—see Sect. 5 for more details) for 20 VLG patterns ($k = 2$, $\delta_i, \Delta_i = \langle 100, 110 \rangle$), composed of very frequent subpatterns drawn from the 200 most common substrings of length 3 in each data set.

Figure 2 shows the results obtained on our test machine (see Sect. 5 for specifications). Using radix sort instead of `std::sort`, SA-SCAN becomes at least two times faster on the Kernel and Proteins datasets, almost twice as fast on CC, and more than 30% faster on Para. A large if algorithmically-somewhat-unexciting leap forward—but further improvements are possible[1].

[1] It is possible that further improvements from sorting alone are possible, using a more heavily engineered sort function that our hand-rolled LSD radix sort. Our point here is that sorting is an important dimension along which SA-SCAN can be optimized.

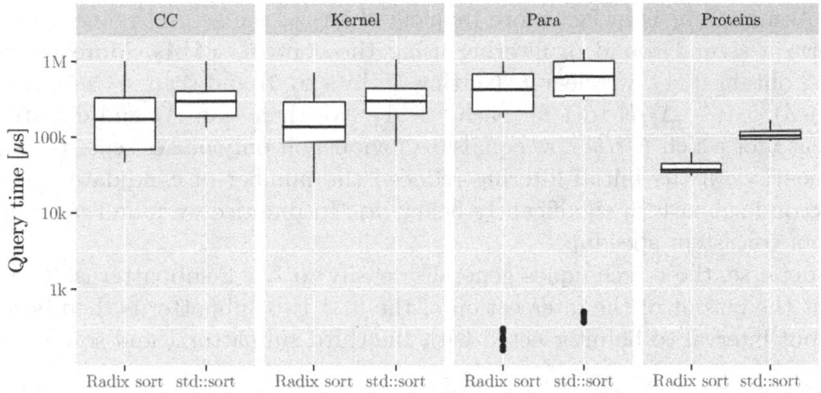

Fig. 2. Time to search a 2GiB subset of the Common Crawl web collection (commoncrawl.org). for 20 VLG patterns ($k = 2$, $\delta_i, \Delta_i = \langle 100, 110\rangle$), composed of very frequent subpatterns drawn from the 200 most common substrings of length 3 in the collection.

3 Filter, Filter, Sort, Scan

Our first serious embellishment to SA-SCAN aims to avoid sorting the full set of subpattern occurrences by filtering out some of the candidate positions that cannot possibly lead to matches. Specifically, we allocate a bitvector F of n/b bits initially all set to 0. We refer to b as the *block size* of the filter. Logically, each bit represents a block of b contiguous positions in the input text, with the ith bit corresponding to the positions $ib..i(b+1)-1$. In describing the use of the filter we assume two subpatterns p_1 and p_2 (with occurrences in $\mathsf{SA}[s_1..e_1]$ and $\mathsf{SA}[s_2..e_2]$, respectively), but the technique is easy to generalize for $k > 2$.

Having allocated F, we scan the interval $\mathsf{SA}[s_1..e_1]$ containing the occurrences of subpattern p_1 and for each element $i = SA[j]$ encountered, we set bits $F[(i+\delta)/b..(i+\Delta)/b]$ to 1 to indicate that an occurrence of p_2 in any of the corresponding blocks of the input is a potential match. During the scan we also copy elements of the interval to an array A_1 of size $m_1 = e_1 - s_1 + 1$. We then scan the interval $\mathsf{SA}[s_2..e_2]$ containing the occurrences of the second subpattern and for each position i encountered we check $F[i/b]$. If $F[i/b] = 0$ then i cannot possibly be part of a match and can be discarded. Otherwise ($F[i/b] = 1$) we add i to a vector A_2 of candidates. We then sort A_1 and A_2 and intersect them with respect to the gap constraints, the same as in the original SA-SCAN algorithm. The hope is that $|A_2|$ is much less than $e_2 - s_2 + 1$, and so the time spent sorting prior to intersection will be reduced.

There are two straightforward refinements to this approach. The first is to make the initial scan not necessarily over $\mathsf{SA}[s_1..e_1]$, but instead over the smaller of intervals $\mathsf{SA}[s_1..e_1]$ and $\mathsf{SA}[s_2..e_2]$. The only difference is that if the interval for p_2 (the second subpattern) happens to be smaller (i.e. p_2 has less occurrences in T than p_1) then we set bits $F[(i-\delta)/b..(i-\Delta)/b]$ (rather than $F[(i+\delta)/b..(i+\Delta)/b]$)

to 1. Assuming p_1 is in fact more frequent than p_2, the second refinement is to perform a second round of filtering using the contents of A_2. More precisely, having obtained A_2, we clear F (setting all bits to 0) and scan A_2 settings bits $F[(i-\delta)/b..(i-\Delta)/b]$ to 1 for each $i \in A_2$. We then scan A_1 and discard any element i for which $F[i/b]$ now equals 0. Obviously it only makes sense to employ this heuristic if the initial filtering reduced the number of candidates, $|A_2|$, of the second subpattern significantly below m_1. In practice we found $m_2 < m_1/2$ led to a consistent speedup.

Of course, these techniques generalize easily to $k > 2$ subpatterns. The idea is that the output of the intersection of the first two subpatterns then becomes an input interval to be intersected with the third subpattern, and so on.

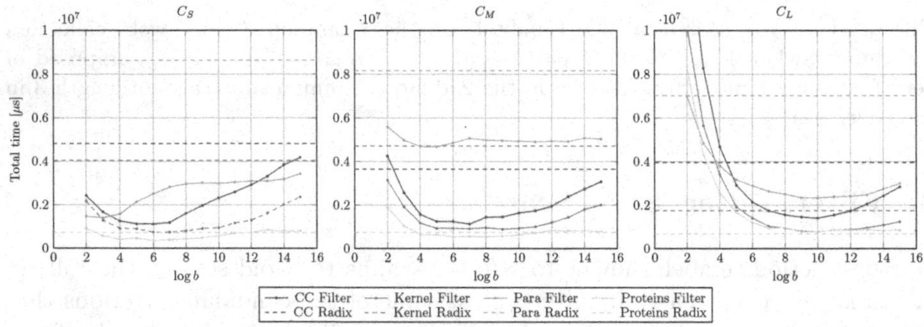

Fig. 3. Effect of filter granularity on search time. Ordinate is runtime in microseconds and mantissa is the logarithm (base 2) of the filter block size. Dashed lines show the time required by SA-SCAN (using radix sort) without any filter. Each plot corresponds to a different set of 20 synthetically generated VLG patterns. All patterns contain 2 of the 200 most frequent subpatterns in each data set. We fixed the gap constraints $C_i = \langle \delta_i, \Delta_i \rangle$ between subpatterns to small ($C_S = \langle 100, 110 \rangle$), medium ($C_M = \langle 1000, 1100 \rangle$), or large ($C_L = \langle 10000, 11000 \rangle$). Section 5 gives more details of data sets and pattern sets.

As Fig. 3 shows, employing F can reduce runtime immensely, but the improvement varies greatly with b. A good choice for b depends on a number of factors. For each occurrence of p_1, we set $\lceil \frac{\Delta-\delta}{b} \rceil$ bits in F. Accesses to F while setting these bits are essentially random (determined by the order of the positions of p_1, which are the lexicographic order of the corresponding suffixes of T), and so it helps greatly if b is chosen so that F, which has size n/b bits, fits in cache. This can be seen in Fig. 3, particularly clearly for the CC, Kernel, and Protein data sets, where performance improves sharply with increasing b until F fits in cache (30 MiB on our test machine) where it quickly stablizes (at $\log b = 3$ for CC and Kernel, and $\log b = 2$ for Protein). Runtimes then remain relatively fast and stable until b becomes so large that the filter lacks specificity, from which point performance gradually degrades. Para has the same trend, though it is not immediately obvious—because the data set is smaller (409MB)

F already fits in L3 cache when $b = 2$. Section 5 gives more details of data sets and pattern sets.

For the large-gap pattern set (C_L), where $\Delta - \delta = 1000$ the optimal choice of b for all data sets is much higher—$b = 1024$ in all cases ($b = 512$ has very similar performance). Here we are seeing the effect of the time needed to set bits in the filter. For example, for Kernel, F already fits in L3 cache when $b = 8$, but at that setting $\lceil \frac{\Delta-\delta}{b} \rceil = 1000/8 = 125$ bits must be set in F per occurrence of p_1. With $b = 1024$ or 512, the number of bits set in F per occurrence of p_1 is just 1 or 2, the same as it is at the optimal setting for the small-gap (C_S) and middle-gap (C_M) pattern sets. This effect can probably be largely alleviated by employing two levels of filters or, alternatively, by implementing a method for setting a word of 1s at a time (effectively reducing the time to set bits from $\lceil \frac{\Delta-\delta}{b} \rceil$ to $\lceil \frac{\Delta-\delta}{b \cdot w} \rceil$, where w is the word size).

4 Direct Text Checking

The filtering ideas described in the previous section can drastically reduce the amount of time spent per subpattern occurrence, but the overall runtime is still $\Omega(occ_1 + occ_2)$, because both subpattern intervals are scanned in full. When the number of occurrences of the less frequent subpattern, say p_1, are significantly less than those of p_2, it is possible to get below that bound by scanning over only the occurrences of p_1, and for each occurrence, i, checking directly in the substring of text $\mathsf{T}[i + \delta..i + \Delta]$ for any occurrences of p_2, each of which corresponds to a match (or valid candidate match in the case $k > 2$). If we use a linear time pattern matching algorithm such as that by Knuth, Morris, and Pratt [13] to search for the occurrences of p_2, runtime (for two subpatterns) becomes $\Theta(occ_1 \cdot (\Delta - \delta))$.

Employed by itself, this kind of text checking can lead to terrible performance when both occ_1 and occ_2 are large. However, when employed in concert with a filter, it can lead to significant performance gains, particularly in later rounds of intersection when $k > 2$. Figure 4 illustrates this for $k = 2$, along with the performance of the other versions of SA-SCAN (Filter and Radix) we have decribed in previous sections. In sum, SA-SCAN has been sped up by more than an order of magnitude on some data sets. In Fig. 5 we see that the text checking heuristic makes an even bigger improvement when the number of subpatterns increases (from $k = 2$ to $k = 4$) because it is employed more often.

5 Experimental Evaluation

In this section we compare the practical performance of our version of SA-SCAN to the wavelet-tree-based method of Bader et al., which is called WT. We use a variety of texts and patterns, which are detailed below (most of these have appeared in experiments described in previous sections). Our methodology in this section closely follows that of [1].

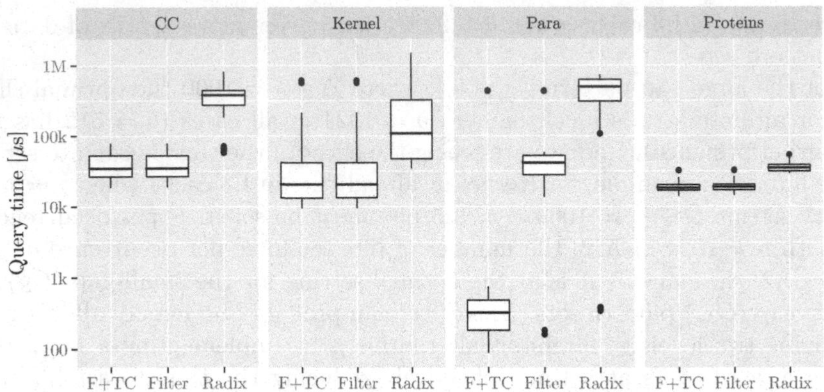

Fig. 4. Direct text checking improves search times further ($k = 2$).

Test Machine and Environment. We used a 2.10 GHz Intel Xeon E7-4830 v3 CPU equipped with 30 MiB L3 cache and 1.5 TiB of main memory. The machine had no other significant CPU tasks running and only a single thread of execution was used. The OS was Linux (Ubuntu 16.04, 64bit) running kernel 4.10.0-38-generic. Programs were compiled using **g++** version 5.4.0.

Texts. We use five datasets from different application domains:

- CC is a 2 GiB prefix of a recent 145TiB web crawl from commoncrawl.org.
- Kernel is a 2 GiB file consisting of source code of all (332) Linux kernel versions 2.2.X, 2.4.X.Y and 2.6.X.Y downloaded from kernel.org. The data set is very repetitive as only minor changes exist between subsequent versions.
- Para is a 410 MiB, which contains 36 sequences of Saccharomyces Paradoxus, is provided by the Saccharomyces Genome Resequencing Project. There are four bases $\{A, C, G, T\}$, but some characters denote an unknown choice among the four bases in which case N is used.
- Proteins is a 1.2 GiB sequence of newline-separated protein sequences (without descriptions, just the bare proteins) obtained from the Swissprot database. Each of the 20 amino acids is coded as one letter.

Patterns. As in [4], patterns were generated synthetically for each data set. We fixed the gap constraints $C_i = \delta_i, \Delta_i$ between subpatterns to small ($C_S = \langle 100, 110 \rangle$), medium ($C_M = \langle 1000, 1100 \rangle$), or large ($C_L = \langle 10000, 11000 \rangle$). VLG patterns were generated by extracting the 200 most common substrings of lengths 3, 5, and 7, which are then used as subpatterns. We then form 20 VLG patterns for each dataset, k (i.e. number of subpatterns), and gap constraint by selecting from the set of 200 subpatterns. We emphasise that the generated patterns, while not specifically designed to be pathological, do represent relatively hard instances for SA-SCAN because of the high frequency of each subpattern.

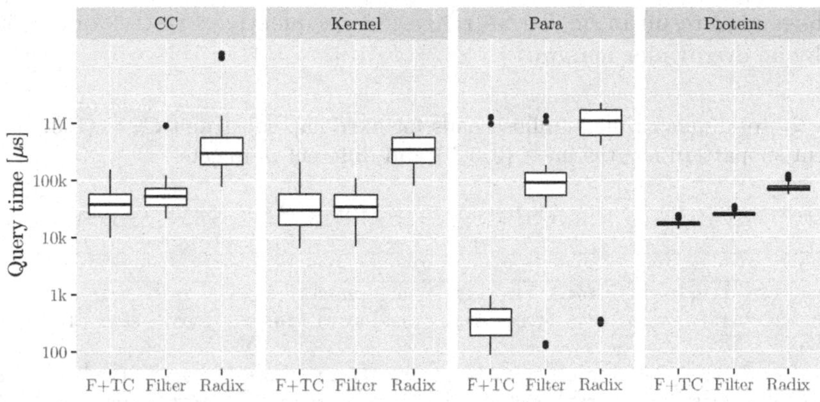

Fig. 5. Direct text checking improves search times further ($k = 4$).

Matching Performance for Different Gap Constraint Bands. Our first experiment aims to elucidate the impact of gap constraint size on query time. We fix the subpattern length $|p_i| = m_i = 3$. Table 1 shows the results from VLG patterns consisting of $k = 2^1, \ldots, 2^5$ subpatterns. Our method, marked FILTER+TC, is always faster than WT, with the exception of the large-gap C_L pattern sets, where on some data sets it yields to WT (most likely due to the text-checking heuristic being less effective on C_L).

Table 1. Total query time in milliseconds on all data sets for fixed $m_i = 3$ and gap constraints $C_S = \langle 100, 110\rangle$, $C_M = \langle 1000, 1100\rangle$, and $C_L = \langle 10000, 11000\rangle$.

Method	CC			Kernel			Para			Proteins		
	C_S	C_M	C_L	C_S	C_M	C_L	C_S	C_M	C_L	C_S	C_M	C_L
k = 2												
Filter+TC	1110	1261	2106	739	823	1023	2372	4696	2335	393	523	1000
WT	14748	18066	41101	7763	8685	26982	14760	57026	99730	8812	11435	24922
k = 4												
Filter+TC	1420	1627	5941	420	1105	4341	5022	12742	16012	418	598	1589
WT	6458	6758	10582	1290	3821	5026	6578	48254	160223	8525	9463	15816
k = 8												
Filter+TC	1109	2857	5705	978	1107	1640	8708	16237	18845	400	597	2070
WT	4641	4358	5439	1255	520	1937	234	357	86996	12708	12866	14054
k = 16												
Filter+TC	1344	1989	4666	1581	1080	1646	3497	4802	13503	547	607	2313
WT	4410	5083	6224	527	513	326	262	260	253	20970	21731	23894
k = 32												
Filter+TC	1344	2176	5835	706	762	1749	6218	6171	18233	393	604	2335
WT	4532	6727	5722	491	668	568	500	540	527	45984	47297	50376

Matching Performance for Different Subpattern Lengths. In our second experiment, we examine the impact of subpattern lengths on query time, fixing the gap constraint to $C_S = 100, 110$. Table 2 shows the results. Larger subpattern lengths tend to result in smaller SA ranges. Consequently, SA-SCAN outperforms WT by an even wider margin.

Table 2. Total query time in milliseconds for fixed gap constraint $C_S = \langle 100, 110 \rangle$ for different subpattern lengths $m_i \in \{3, 5, 7\}$ and different data sets.

Method	CC			Kernel			Para			Proteins		
	3	5	7	3	5	7	3	5	7	3	5	7
k = 2												
Filter+TC	1110	756	654	740	178	46	2372	641	78	393	32	22
WT	14748	8576	6158	7763	1731	93	14760	10176	2502	8812	441	182
k = 4												
Filter+TC	1420	362	310	420	159	53	5022	778	71	417	30	26
WT	6458	1477	637	1290	2182	30	6578	10882	2457	8525	97	67
k = 8												
Filter+TC	1109	683	230	978	206	196	8708	767	153	400	30	23
WT	4641	1380	464	1255	156	47	234	16558	3602	12708	51	33
k = 16												
Filter+TC	1344	541	679	1581	276	164	3497	836	77	547	31	32
WT	4410	922	412	527	155	81	262	29226	6317	20970	83	62
k = 32												
Filter+TC	1344	457	257	706	225	90	6218	730	234	393	33	61
WT	4532	1492	540	491	324	171	500	64070	13813	45984	177	128

Overall Runtime Performance. In a final experiment we explored the whole parameter space (i.e. $k \in \{2^1, \dots, 2^5\}$, $m_i \in \{3, 5, 7\}$, $C \in \{C_S, C_M, C_L\}$). The results are summarized in Fig. 6. Overall out SA-SCAN-based method is faster on average than the wavelet-tree-based one, usually by a wide margin.

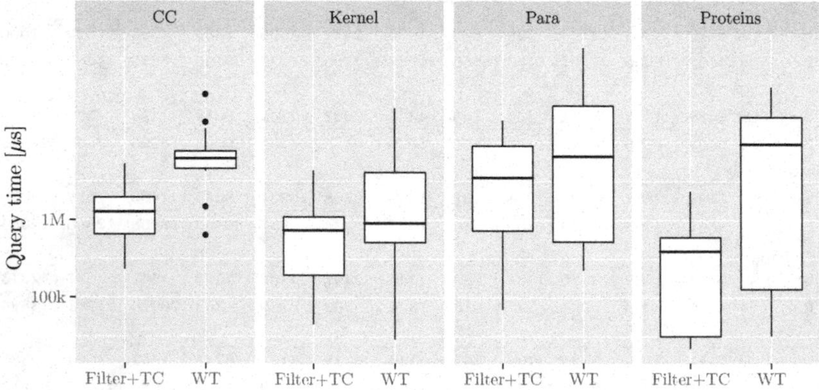

Fig. 6. Overall runtime performance of both methods, accumulating the performance for all $m_i \in \{3, 5, 7\}$ and C_S, C_M, and C_L.

6 Concluding Remarks

We have described a number of simple but highly effective improvements to the SA-SCAN VLG matching algorithm that, according to our experiments, elevate it to be the state-of-the-art approach for the indexed version of problem. We believe better indexing methods for VLG matching can be found, but that our version of SA-SCAN, which makes judicious use of filters, text checking, and subpattern processing order, represents a strong baseline against which the performance of more exotic methods should be measured.

Numerous avenues for continued work on VLG matching exist, perhaps the most interesting of which is to reduce index size. Currently, SA-SCAN uses $n \log n + n \log \sigma$ bits of space for a text of length n on alphabet σ for the suffix array and text, respectively (the WT approach of Bader et al., uses slightly more). Because our methods consist (mostly) of simple scans of SA ranges or scans of the underlying text, they are easily translated to make use of recent results on Burrows-Wheeler-based compressed indexes [10] that allow fast access to elements of the suffix array from a compressed representation of it. Via this observation we derive the first compressed indexes for VLG matching. These indexes use $O(r \log n)$ bits of space, where r is the number of runs in the Burrows-Wheeler transform, a quantity that decreases with text compressibility. On our 2 GiB Kernel data set, for example, the compressed index takes around 20 MiB in practice, and can still support VLG matching in times competitive with the indexes of Bader et al. We plan to explore this in more depth in future work.

Acknowledgments. Our thanks go to Tania Starikovskaya for suggesting the problem of indexing for regular-expression matching to us. We also thank Matthias Petri and Simon Gog for prompt answers to questions about their article and code and the anonymous reviewers for helpful comments. This work was funded by the Academy of Finland via grant 319454 and by EU's Horizon 2020 research and innovation programme under Marie Skłodowska-Curie grant agreement No. 690941 (BIRDS).

References

1. Bader, J., Gog, S., Petri, M.: Practical variable length gap pattern matching. In: Goldberg, A.V., Kulikov, A.S. (eds.) SEA 2016. LNCS, vol. 9685, pp. 1–16. Springer, Cham (2016). https://doi.org/10.1007/978-3-319-38851-9_1
2. Bille, P., Farach-Colton, M.: Fast and compact regular expression matching. Theor. Comput. Sci. **409**(3), 486–496 (2008)
3. Bille, P., Gørtz, I.L.: Substring range reporting. Algorithmica **69**(2), 384–396 (2014)
4. Bille, P., Gørtz, I.L., Vildhøj, H.W., Wind, D.K.: String matching with variable length gaps. Theor. Comput. Sci. **443**, 25–34 (2012)
5. Bille, P., Thorup, M.: Regular expression matching with multi-strings and intervals. In: Proceedings of SODA, pp. 1297–1308. ACM-SIAM (2010)
6. Cox, R.: Regular expression matching with a trigram index or how Google code search worked (2012). https://swtch.com/~rsc/regexp/regexp4.html

7. Crawford, T., Iliopoulos, C.S., Raman, R.: String matching techniques for musical similarity and melodic recognition. Comput. Musicol. **11**, 73–100 (1998)
8. Crochemore, M., Iliopoulos, C.S., Makris, C., Rytter, W., Tsakalidis, A.K., Tsichlas, T.: Approximate string matching with gaps. N. J. Comput. **9**(1), 54–65 (2002)
9. Fredriksson, K., Grabowski, S.: Efficient algorithms for pattern matching with general gaps, character classes, and transposition invariance. Inf. Retr. **11**(4), 335–357 (2008)
10. Gagie, T., Navarro, G., Prezza, N.: Optimal-time text indexing in BWT-runs bounded space. In: Proceedings of SODA, pp. 1459–1477. ACM-SIAM (2018)
11. Grossi, R., Gupta, A., Vitter, J.: High-order entropy-compressed text indexes. In: Proceedings of the SODA, pp. 841–850. ACM-SIAM (2003)
12. Haapasalo, T., Silvasti, P., Sippu, S., Soisalon-Soininen, E.: Online dictionary matching with variable-length gaps. In: Pardalos, P.M., Rebennack, S. (eds.) SEA 2011. LNCS, vol. 6630, pp. 76–87. Springer, Heidelberg (2011). https://doi.org/10.1007/978-3-642-20662-7_7
13. Knuth, D., Morris, J.H., Pratt, V.: Fast pattern matching in strings. SIAM J. Comput. **6**(2), 323–350 (1977)
14. Lewenstein, M.: Indexing with gaps. In: Grossi, R., Sebastiani, F., Silvestri, F. (eds.) SPIRE 2011. LNCS, vol. 7024, pp. 135–143. Springer, Heidelberg (2011). https://doi.org/10.1007/978-3-642-24583-1_14
15. Lopez, A.: Hierarchical phrase-based translation with suffix arrays. In: Proceedings of the EMNLP-CoNLL 2007, pp. 976–985. ACL (2007)
16. Manber, U., Myers, G.: Suffix arrays: a new method for on-line string searches. SIAM J. Comput. **22**(5), 935–948 (1993)
17. Metzler, D., Croft, W.B.: A markov random field model for term dependencies. In: Proceedings of the SIGIR, pp. 472–479. ACM (2005)
18. Morgante, M., Policriti, A., Vitacolonna, N., Zuccolo, A.: Structured motifs search. J. Comput. Biol. **12**(8), 1065–1082 (2005)
19. Navarro, G.: Wavelet trees for all. J. Discrete Algorithms **25**, 2–20 (2014)
20. Pissis, S.P.: MoTeX-II: structured MoTif eXtraction from large-scale datasets. BMC Bioinform. **15**(235), 1–12 (2014)
21. Saikkonen, R., Sippu, S., Soisalon-Soininen, E.: Experimental analysis of an online dictionary matching algorithm for regular expressions with gaps. In: Bampis, E. (ed.) SEA 2015. LNCS, vol. 9125, pp. 327–338. Springer, Cham (2015). https://doi.org/10.1007/978-3-319-20086-6_25
22. Turpin, A., Tsegay, Y., Hawking, D., Williams, H.E.: Fast generation of result snippets in web search. In: Proceedings of the SIGIR 2007, pp. 127–134. ACM (2007)

Linearizing Genomes: Exact Methods and Local Search

Tom Davot[1(✉)], Annie Chateau[1], Rodolphe Giroudeau[1], and Mathias Weller[2]

[1] LIRMM - CNRS UMR 5506, Montpellier, France
{davot,chateau,rgirou}@lirmm.fr
[2] CNRS, LIGM (UMR 8049), Champs-s/-Marne, France
mathias.weller@u-pem.fr

Abstract. In this article, we address the problem of genome linearization from the perspective of Polynomial Local Search, a complexity class related to finding local optima. We prove that the linearization problem, with a neighborhood structure, the neighbor slide, is PLS-complete. On the positive side, we develop two exact methods, one using tree decompositions with an efficient dynamic programming, the other using an integer linear programming. Finally, we compare them on real instances.

1 Introduction

Motivation. When inferring genome sequences from high-throughput sequencing (HTS) data, we obtain (after assembly) fragments of the target sequence called *contigs*[1] without any information on how these contigs are located in the genome. To address this shortcoming, contigs can be linked using external information (usually a read-pairing included in the HTS data), yielding a graph (called *scaffold graph*) whose vertices are contig extremities and edges are either contigs or links between them. The *scaffolding* operation then aims at selecting the best paths in this graph in order to produce longer genomic sequences called scaffolds. Previous work focuses on the production of sequences by solving the so-called SCAFFOLDING problem in this graph [4,14,16]. Scaffolding is a widely studied problem in bioinformatics and can be modeled by numerous, mostly heuristic, methods [8].

Unfortunately, real-world genomes escape the relative simplicity of previous models (that still lead to NP-complete problems). A particular problem is modeling contigs occurring multiple times in the target genome. Such "repeats" and their "multiplicity" (or "copy numbers") vary depending on the species and individual [2]. Due to the conservatism of some assembly methods, a repeat may cover an entire contig which is separated from the other genomic side fragments [11]. Recent methods address this problem, avoiding chimeric reconstruction by using long reads as additional data [3,13]. Unfortunately, most projects on genomic databases are still constituted of short-reads only and are

[1] Contigs are words on a genomic alphabet, usually $\{A, C, G, T\}$.

A. Chatzigeorgiou et al. (Eds.): SOFSEM 2020, LNCS 12011, pp. 505–518, 2020.
https://doi.org/10.1007/978-3-030-38919-2_41

not intended to be resequenced with long-reads technologies in the near future. One motivation of our work it to take care of these kind of projects, and improve assemblies using only the original short-read (though paired-end) data. In this context, a solution to the SCAFFOLDING problem may not be a collection of distinct paths, but rather a graph, called *solution graph* (which is a particular scaffold graph with multiplicities). Transforming such a graph into genomic sequences turns out to be a challenging task. The aim of the present work is to study the problem consisting of removing ambiguities in the solution graph in order to provide longer and error-free genomic sequences with minimal loss of information.

In the following, most of the proofs has been omitted due to space constraints. A full version including the proofs is available in https://hal-lirmm.ccsd.cnrs.fr/lirmm-02332049.

2 Notation and Problem Description

With G denoting a graph, we let $V(G)$ and $E(G)$ be the sets of vertices and edges of G, respectively. A scaffold graph (G, M^*, ω, m') is an edge-weighted, simple, undirected graph G equipped with 1. a perfect matching M^* that corresponds to the contigs, 2. non-contig edges uv whose weights $\omega(uv)$ indicate the likelihood that the contig-extremity u is adjacent to the contig-extremity v in the target genome and 3. a multiplicity m' on contig edges which indicates the desired number of their occurrences (see Fig. 1). An alternating walk $(u_0, \ldots, u_{2\ell-1})$ is a sequence of vertices such that for each $i < \ell$, $u_{2i}u_{2i+1} \in M^*$ and $u_{2i+1}u_{2i+2} \in E(G) \setminus M^*$. If $u_0 = u_{2\ell-1}$, the walk is *closed*. The SCAFFOLDING WITH MULTIPLICITIES problem is stated as follows:

> SCAFFOLDING WITH MULTIPLICITIES (SCAM)
> **Input:** a scaffold graph (G, M^*, ω, m') and $\sigma_p, \sigma_c, k \in \mathbb{N}$
> **Question:** Is there a multiset S of at most σ_c closed and at most σ_p non-closed alternating walks in G such that each $e \in M^*$ occurs $m'(e)$ times across all walks of S and $\sum_{e \in E(S) \setminus M^*} \omega(e) \geq k$?

In this work, we will not focus on the SCAM problem itself (instead, the reader is referred to Weller et al. [14,16,17]). Instead, we assume that we are given a solution S, whose walks then induce a subgraph of the input scaffold graph which we call *solution graph* (G^*, M^*, ω, m). More precisely, given a scaffold graph (G, M^*, ω, m) and a solution S of SCAM, the solution graph (G^*, M^*, ω, m) is obtained by removing the edges that do not belong to S. The multiplicity function m defined on all the edges of the solution graph is the number of times that an edge occurs in S. Note that for each matching edge e, we have $m(e) = m'(e)$. It turns out that, in presence of repeated contigs, a solution graph implies a unique set of sequences if and only if it does not contain so called *ambiguous paths* [15] (see Fig. 1 for an example).

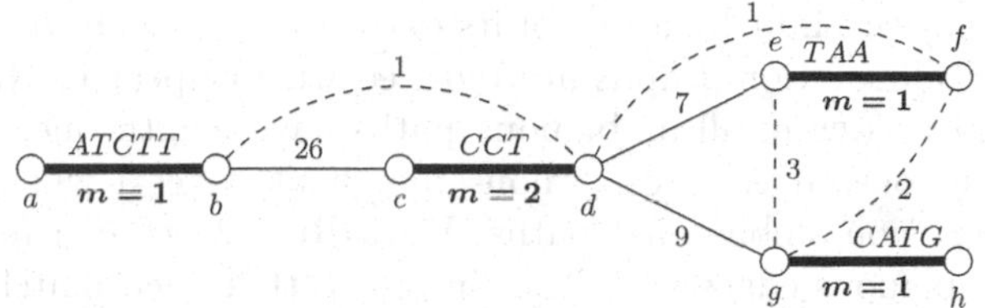

Fig. 1. A scaffold graph and a solution graph obtained with an optimal solution of SCAM. Matching edges are bold and plain edges are part of the solution graph. Edge *cd* has multiplicity two. Other contigs have multiplicity one. Edges of the solution graph also have multiplicity one. Links between contigs are labeled by their weight. Because of the presence of the ambiguous path *cd*, two optimal solutions are possible for SCAM with $\sigma_c = 0$ and $\sigma_p = 2$: $\{(a,b,c,d,e,f),(c,d,g,h)\}$ and $\{(a,b,c,d,g,h),(c,d,e,f)\}$.

Definition 1 (Ambiguous path). *Let p be path with extremities u and v in a solution graph. If, for all vertices x of p, p also contains the matching edge containing x, we call p* alternating. *If all edges of p have the same multiplicity* μ *(that is,* $m(e) = \mu$ *for all* $e \in p$*), then p is called* μ-uniform *(or simply* uniform *is* μ *is unknown). If p is alternating, uniform, and both of u and v are incident with a non-matching edge of multiplicity strictly less than* μ*, then p is called* ambiguous.

To break ambiguous paths, we remove non-contig edges from the solution graph, thereby losing information, and our goal is to minimize this loss. Definition 1 implies that minimal solutions remove all incident non-contig edges from a selected set X of vertices. The "cost" of such a set X can be defined by the following scorings:

Cut score. Pay one per vertex in X: $\text{score}(X) := |X|$.
Path score. Pay one for each multiplicity that is removed:
$\text{score}(X) := \sum\{m(uv) \mid uv \in E \setminus M^* \wedge uv \cap X \neq \varnothing\}$.
Weight score. Pay the total cost of edges that are removed:
$\text{score}(X) := \sum\{m(uv) \cdot \omega(uv) \mid uv \in E \setminus M^* \wedge uv \cap X \neq \varnothing\}$.

Since the Path score and the Weight score are very similar, we study in this paper only the Cut score and the Weight score. The following reduction rules simplify a given instance (solution graph) without changing the solution set X.

Rule 1 ([15]). *Let p be a* μ*-uniform alternating path with extremities u and v. Remove p and add a new contig edge uv with multiplicity* μ.

Rule 2 ([15]). *Let* $uv \in M^*$ *be a contig edge not appearing in ambiguous paths and let u and v have degree at least two. Then, remove uv, add new vertices* u' *and* v' *and add the contig edges* uv' *and* vu' *with multiplicity* $m(uv)$.

Let (G^*, M^*, ω, m) be a solution graph and let $u \in V(G^*)$. We let $N_{G^*}(u) = \{v \mid uv \in E(G^*) \setminus M^*\}$ denote the set of neighbors of u linked to u with a non-matching edge. We say that a vertex u is *clean* if $N_{G^*}(u) = \varnothing$ and a matching edge $uv \in M^*$ is *clean* if at least one of its extremities is clean. In the following, we assume that all solution graphs are reduced with respect to Rule 1, and we observe that, in this case, all ambiguous paths have length one. Thus, we use the term "ambiguous edges" (resp. "non-ambiguous edges") when we speak of ambiguous (resp. non-ambiguous) paths. With Rule 2, we can further assume that all non-ambiguous edges are clean, implying that each matching edge e is ambiguous if and only if e is not clean. Hence, disambiguating a solution means to render all matching edges clean. We can now formulate our problem Semi-Brutal Cut as follows.

Semi-Brutal Cut (SBC)
Input: A solution graph (G^*, M^*, ω, m) and some $k \in \mathbb{N}$
Question: Is there a set X of extremities of ambiguous edges in G^* such that removing all non-matching edges incident to vertices of X renders all matching edges clean, and $\text{score}(X) \leq k$?

For a vertex u of G^*, we let $\omega(u)$ denote the sum of the weights of all non-matching edges incident to u. For a solution X of Semi-Brutal Cut, we let $\omega(X) := \sum_{u \in X} \omega(u)$. We say that u is *cut* if $u \in X$. Since we are not limited in number of cuts for the weight score, we suppose that in a solution X for Semi-Brutal Cut under the weight score, each ambiguous edge of (G^*, M^*, ω, m) contains exactly one vertex in X.

3 Related Work

Problems similar to the linearization of scaffolds are studied in the context of guided, multiple-source assembly problems [12]. However, the model does not integrate multiplicities as a constraint on the structure of the desired paths. In previous work, we show that the variants of Semi-Brutal Cut according to all presented scoring functions are NP-complete [15]. In [10], we explore special classes of graphs, namely bipartite, planar with bounded degree, analyzing complexity and approximability, showing that even in very restricted cases, the problem is hard to solve. We also proposed a 2-approximation algorithm under the weight score and a 4-approximation under the cut score. In the present work, we consider general instances, showing that even finding a locally optimal solution is hard, but propose effective exact methods to linearize genomes.

4 Hardness Using PLS-Reduction

This section is devoted to determine the local-search complexity using the PLS (Polynomial Local Search) class, which models the difficulty of finding a locally optimal solution to an optimization problem [5]. Schäffer and Yannakakis [9] proved several classic combinatorial optimization problems PLS-complete. In the following, we propose a new neighborhood structure called the *neighbor slide* adapted to Semi-Brutal Cut. We recall first some definitions related to PLS.

A *neighborhood structure* N is a function that associates to each solution S a set of solutions $N(S)$. A *local search problem* is a combinatorial optimization problem P for which, given a neighborhood structure N, we want to find a solution S (called *local optimum*), such that no solution in $N(S)$ has a better score. In the following, we let P/N denote a local search problem where P is a combinatorial optimization problem and N a neighborhood structure.

Definition 2 (PLS). *A local search problem P/N is in* PLS *if there are polynomial-time algorithms A_L, B_L, and C_L such that*

(a) for each instance x, A_L gives an initial solution S_{init},
(b) for each solution S, B_L determines the score of S, and
(c) for each solution S, C_L determines if S is a local optimum and, if not, gives a solution with the best score in $N(S)$.

To show that finding a local optimum for a problem P_1/N_1 is at least as difficult as finding a local optimum for a problem P_2/N_2, we use *PLS-reductions.*

Definition 3 (PLS-reduction). *A local search problem P_1/N_1 is PLS-reducible to a local search problem P_2/N_2 if there are polynomial-time computable functions f and g such that:*

(a) If x_1 is an instance of P_1, then $f(x_1)$ is an instance of P_2.
(b) If S_2 is a solution for $f(x_1)$, then $g(x_1, S_2)$ is a solution for x_1.
(c) If S_2 is a local optimum for $f(x_1)$, then $g(x_1, S_2)$ is a local optimum for x_1.

Then, a local search problem P/N is *PLS-complete* if P/N is in *PLS* and every problem in *PLS* can be *PLS*-reduced to P/N. We now introduce the MAX W2SAT problem and the *Flip* neighborhood structure.

MAX W2SAT
Input: A boolean formula φ in conjunctive normal form where each clause C_i has a weight $\omega(C_i)$ and contains exactly two variables.
Task: Find an assignment maximizing the total weight of satisfied clauses.

Definition 4 (Flip). *Let S be a solution for φ. A solution S' is in $N(S)$ if there exists a unique variable x_i such that the assignment of x_i is different in S and S'. We say that S' is obtained by flipping the value of x_i in S.*

Note that MAX W2SAT/Flip is PLS-complete [7]. Let φ be an instance of MAX W2SAT and let S be a solution for φ. We let $\omega(\varphi)$ denote the sum of the weights of all clauses of φ and we let $\omega(S)$ denote the total weight of all clauses that are *not* satisfied by S. From an instance of MAX W2SAT, we build an instance of SBC using the following construction.

Construction 1 (See Fig. 2(left)). *Let φ be an instance of* MAX W2SAT *with n' variables x_i and m' clauses C_j and let $occ(x_i)$ denote the number of occurrences of x_i in φ. We construct the following solution graph (G^*, M^*, ω, m).*

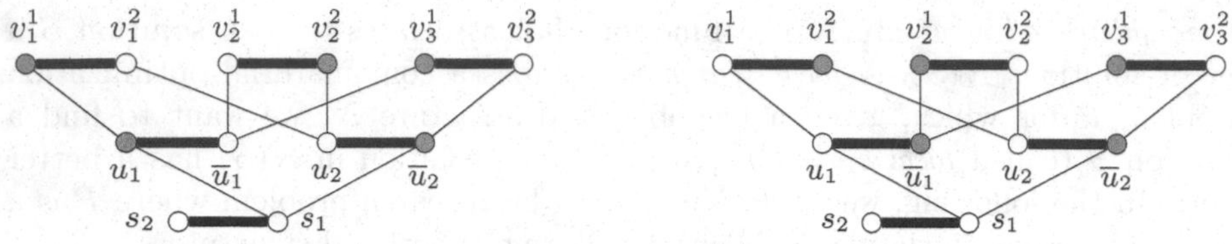

Fig. 2. Left: The graph produced by Construction 1 on input $\varphi = (x_1 \vee x_2) \wedge (\neg x_1 \vee x_2) \wedge (\neg x_1 \vee \neg x_2)$ (each clause has weight one). Matching edges are bold and all non-matching edges have weight one. A solution S with $\omega(S) = 6$ is highlighted in gray. In S, $v_1^1 v_1^2$ is satisfied, $v_2^1 v_2^2$ is unsatisfied and $v_3^1 v_3^2$ is neither satisfied nor unsatisfied. **Right:** A solution of weight 5 produced by a neighbor slide of $u_1\overline{u}_1$.

1. *Construct a matching edge $s_1 s_2$ with $m(s_1 s_2) = 2m'$.*
2. *For each x_i, construct a matching edge $u_i\overline{u}_i$ such that $m(u_i\overline{u}_i) = occ(x_i) + 1$ (*variable edge*).*
3. *For each clause C_j, construct a matching edge $v_j^1 v_j^2$ such that $m(v_j^1 v_j^2) = 2$ (*clause edge*).*
4. *For each clause C_j, let x_k be the tth variable of the clause. If x_k occurs positively in the clause, then add the edge $v_j^t u_k$ with $m(v_j^t u_k) = 1$ and $\omega(v_j^t u_k) = \omega(C_j)$. Otherwise, add the edge $v_j^t \overline{u}_k$ with $m(v_j^t \overline{u}_k) = 1$ and $\omega(v_j^t \overline{u}_k) = \omega(C_j)$.*
5. *Finally, for each matching edge $u_i\overline{u}_i$, if $\omega(u_i) < \omega(\overline{u}_i)$, add an edge $s_1 u_1$ with $m(s_1 u_1) = 1$ and $\omega(s_1 u_1) = \omega(\overline{u}_i) - \omega(u_i)$. If $\omega(u_i) > \omega(\overline{u}_i)$, add an edge $s_1 \overline{u}_1$ with $m(s_1 \overline{u}_1) = 1$ and $\omega(s_1 \overline{u}_1) = \omega(u_i) - \omega(\overline{u}_i)$.*

Note that for each variable edge $u_i\overline{u}_i$, we have $\omega(u_i) = \omega(\overline{u}_i)$. All matching edges except $s_1 s_2$ are ambiguous. If a cut in a clause edge $v_j^1 v_j^2$ is adjacent to a cut in a variable edge, then we say that the clause edge $v_j^1 v_j^2$ is *satisfied.* If no extremity of a clause edge $v_j^1 v_j^2$ is adjacent to a cut vertex in a variable edge, we say that the clause edge $v_j^1 v_j^2$ is *unsatisfied.* Note that a clause edge could be neither satisfied nor unsatisfied. In a graph produced by Construction 1, we simulate the flipping operation with the neighbor slide operation defined as follows:

Definition 5 (Neighbor Slide, see Fig. 2). *Let $S \subseteq V(G^*)$ be a solution for (G^*, M^*, ω, m) and let uv be an unclean matching edge of G^* with $u \in S$.*

The neighbor slide *operation applied to uv produces a new solution S' as follows:*

1. *$S' \leftarrow (S \cup \{v\}) \setminus \{u\}$,*
2. *for each neighbor $n_u \neq s_1$ of u: $S' \leftarrow (S' \cup \{M^*(n_u)\}) \setminus \{n_u\}$, and*
3. *for each neighbor $n_v \neq s_1$ of v: $S' \leftarrow (S' \cup \{n_v\}) \setminus \{M^*(n_v)\}$.*

Thus, a solution S' belongs to $N(S)$ if S' can be produced by applying a neighbor slide operation on S.

Definition 6. *Let φ be an instance of* Max W2SAT*, let (G^*, M^*, ω, m) be the graph produced by Construction 1, and let X be a solution for it. A solution S for φ* corresponds *to X if, for all matching edges $u_i\overline{u}_i$, we have $u_i \in X \Rightarrow S(x_i) = 1$ and $\overline{u}_i \in X \Rightarrow S(x_i) = 0$.*

Note that, after a neighbor slide of a variable edge, all adjacent clause edges are either satisfied or unsatisfied, and that if a clause edge $v_j^1 v_j^2$ is satisfied (resp. unsatisfied), then the corresponding clause C_j is satisfied (resp. unsatisfied).

Lemma 1. *Let X be a solution for (G^*, M^*, ω, m), produced by Construction 1 and let S be the corresponding solution for φ.*

1. *If X is a local minimum, then all clause edges are satisfied or unsatisfied.*
2. *If all clause edges are satisfied or unsatisfied by X, then $\omega(X) = \omega(\varphi) + \omega(S)$.*

Theorem 1. *SBC/Neighbor slide is PLS-complete for the weight score.*

Proof. It is easy to see that SBC/Neighbor slide is in PLS. We propose a PLS-reduction of Max W2SAT/Flip to SBC/Neighbor slide. Let φ be an instance of Max W2SAT. The function defined by Construction 1 produces an instance (G^*, M^*, ω, m) of Semi-Brutal Cut and the function defined in Definition 6 computes a solution for φ from a solution for (G^*, M^*, ω, m). It remains to show that, if a solution X is a local minimum of Semi-Brutal Cut in (G^*, M^*, ω, m), then its corresponding solution S is also a local minimum. By Lemma 1(1) and Lemma 1(2), we have $\omega(X) = \omega(\varphi) + \omega(S)$. Suppose that S is not a local minimum. Then, there is a variable x_i in S such that flipping its value produces a solution S_1' with a smaller weight. Let S_2' be the solution produced by the neighbor-slide operation on the variable edge $u_i\overline{u}_1$ in X. Note that the corresponding solution of S_2' is S_1'. By Lemma 1(1), all clause edges in X are either satisfied or unsatisfied and since the clause edges modified by a neighbor-slide are either satisfied or unsatisfied, all clause edges in S_2' are either satisfied or unsatisfied. Thus, by Lemma 1(2), $\omega(S_2') = \omega(\varphi) + \omega(S_1') < \omega(X)$, contradicting the fact that X is a local minimum.

5 Exact Methods

5.1 Integer Linear Programming

In this section, we propose an integer linear program modeling Semi-Brutal Cut for all scores.

Variables. For each non-matching edge e_k, we define a binary variable x_k which equals 1 if and only if one of its extremities is in the solution, that is, e_k is removed from the graph. For each extremity u_i of an ambiguous edge p, we define two binary variables c_i and n_i. $c_i = 1$ iff u_i is in the solution and $n_i = 1$ if and only if all neighbors $v \neq M^*(u_i)$ of u_i are in the solution.

Constraints.

(1) For any ambiguous matching edge u_iu_j, we force one of the extremities to have degree one by adding the constraint $n_i + n_j + c_i + c_j \geq 1$.
(2) If any extremity u_i is adjacent to a non-ambiguous matching edge, then not all neighbors of u_i can be cut. In this case, we add the constraint $n_\ell = 0$.
(3) For all extremities u_i, we force all neighbors of u_i (except $M^*(u_i)$) to be cut if $n_i = 1$ by adding the constraint $\sum\limits_{u_\ell \in N(u_i)} c_\ell \geq n_i \cdot |N(u_i)|$.
(4) For each extremity u_i of a non-matching edge e_k, we force that e_k is removed from the graph if $u_i = 1$ by adding the constraint $x_k \geq c_i$.

Objective Function. For the cut score, we want to minimize the number of vertices in the solution, that is, the number of variable c_i with value one. Thus, the objective function for the cut score is $\min \sum_i c_i$. For the weight score, we want to minimize the total weight of the edges removed from the graph. Thus, the objective function for the weight score is $\min \sum_{e_k \in E(G) \setminus M^*} x_k \cdot \omega(e_k)$.

5.2 Dynamic Programming on Tree Decompositions

We show that SEMI-BRUTAL CUT can be solved in linear time on classes of graphs that exhibit a constant bound on the treewidth, such as series-parallel or outerplanar graphs. To this end, we present a dynamic programming algorithm, working on nice tree decompositions, that finds an optimal solution in $O(2^{tw} \cdot |E(G)|)$ under the weight score and in $O(5^{tw} \cdot |E(G)|)$ under the cut score, where tw is the treewidth of the input graph.

Definition 7 ([6]). *Given a graph G, a* tree decomposition *for G is a pair $(T, \mathcal{X})$ where T is a tree and $\mathcal{X} = \{B_i \mid i \in V(T)\}$ is a multiset of subsets of $V(G)$ (called "bags") such that*

(a) for each $uv \in E(G)$, there is some i with $uv \subseteq B_i$ and
(b) for each $v \in V(G)$, the bags B_i containing v form a connected subset of T.

The width of $(T, \mathcal{X})$ is $\max_i |B_i| - 1$. Further, $(T, \mathcal{X})$ is called nice *if*

(c) T is rooted at bag B_r, with $B_r = \varnothing$ and each bag has at most two children.
(d) Each bag B_i of T has one of the four types:
  - ***Leaf bag:*** *i has no children and $B_i = \varnothing$.*
  - ***Join bag:*** *i has two children j and k and $B_i = B_j = B_k$.*
  - ***Introduce** u **bag:*** *i has only one child j and $B_j = B_i \setminus \{u\}$.*
  - ***Forget** u **bag:*** *i has only one child j and $B_j = B_i \cup \{u\}$.*

For any bag B_i of T, we let G_i denote the subgraph of G induced by the vertices of G that are introduced "below" B_i (that is, in a bag of the subtree of T that is rooted at i).

Note that for each vertex u of G, $(T, \mathcal{X})$ contains exactly one forget u bag. Further, the root r of a nice tree decomposition is a forget bag and we let r' denote the vertex forgotten by r.

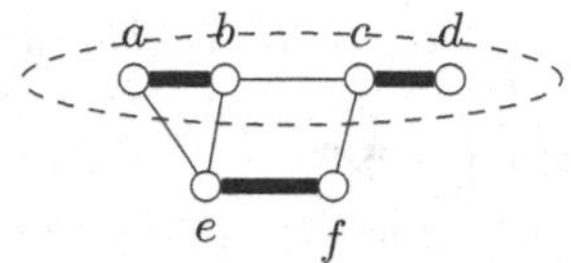

Fig. 3. A subgraph G_i^* with $B_i = \{a, b, c, d\}$ (matching edges in bold). Let $Y_1 = \{(a, \text{“}\varnothing\text{”}), (b, \text{“}\varnothing\text{”}), (c, \text{“}\times\text{”}), (d, \text{“}\times\text{”})\}$, let $Y_2 = \{(a, \text{“}\varnothing\text{”}), (b, \text{“}N\text{”}), (c, \text{“}\varnothing\text{”}), (d, \text{“}\varnothing\text{”})\}$, and let $Y_3 = \{(a, \text{“}N\text{”}), (b, \text{“}\times\text{”}), (c, \text{“}\varnothing\text{”}), (d, \text{“}\varnothing\text{”})\}$. No set vertex set X is eligible for (Y_1, B_i) and (Y_2, B_i) but $\{b, e\}$ is eligible for (Y_3, B_i). The trace of Y_3 is $\mathcal{T}(Y_3) = \{b\}$.

Tree Decompositions Introducing Matching Edges. Let (G^*, M^*, ω, m) be a solution graph and let $B_i \in \mathcal{X}$. In our algorithm, we need $M^*(u) \in B_i$ for each $u \in B_i$. For this reason, we contract all matching edges in (G^*, M^*, ω, m), yielding a graph G' with $V(G') = M^*$. We compute a nice tree decomposition $(T, \mathcal{X})$ of G', then the vertices of G' are expanded, that is, we replace the vertices of G' in the tree decomposition by their corresponding matching edges. Each introduce u bag now introduces the matching edge $uM^*(u)$. We call such a tree decomposition for G^* *M^*-preserving.* In the following, G' refers to the graph with contracted matching edges and G^* refers to the original solution graph.

Signatures. To every (X, V') where X is a solution of a subgraph H and V' is a subset of vertices of H, we associate a signature describing how the vertices of V' are cut in X. The signature of a vertex u can be “$\times$”, “N”, or “$\varnothing$”, depending on whether, respectively, u is cut, all neighbors of u are cut, or u is not cut.

Definition 8 (see Fig. 3). *Let H be a subgraph of G^* such that, for each $u \in V(H)$, we have $M^*(u) \in H$. Let $X \subseteq V(H)$ be a solution for (G^*, M^*, ω, m) in H and let $V' \subseteq V(H)$. A mapping $Y : V' \to \{\text{“}N\text{”}, \text{“}\times\text{”}, \text{“}\varnothing\text{”}\}$ with*

(i) $Y(u) = \text{“}\times\text{”} \Leftrightarrow u \in X$ *and,*
(ii) $Y(u) = \text{“}N\text{”} \Rightarrow N_H(u) \subseteq X$

is called signature *of X in V' and $\mathcal{T}(Y) = \{u \mid Y(u) = \text{“}\times\text{”}\}$ is called* trace *of Y.*

Note that a solution X can be associated to many signatures. Likewise, two different solutions X and X' of H such that $X \cap V' = X' \cap V'$ are associated to the same signatures. In order to minimize the number of signatures, we add some restrictions on the mappings. The main idea is that sub-solutions with the same signature are equivalently suited to construct a complete solution. Thus, for a vertex set V', we define a set of signatures $\mathcal{Y}(V')$ as follows.

Definition 9. *Let V' be a vertex set. We define $\mathcal{Y}(V')$ as the set of all $Y : V' \to \{\text{“}\varnothing\text{”}, \text{“}N\text{”}, \text{“}\times\text{”}\}$ such that, for all $u \in V'$, the three following conditions hold:*

1. *$uM^*(u)$ is clean* $\Leftrightarrow Y(u) = Y(M^*(u)) = \text{“}\varnothing\text{”}$
   (no cut occurs in an already clean matching edge).

2. *if the considered scoring function is the weight score, then:*
   - $Y(u) \neq$ "N" *and,*
   - $Y(u) =$ "$\varnothing$" $\Leftrightarrow Y(M^*(u)) =$ "$\times$"
   (each ambiguous edge contains exactly one cut).
3. *if the considered scoring function is the cut score, then:*
   - $Y(u) =$ "$\varnothing$" $\Rightarrow Y(M^*(u)) \neq$ "$\varnothing$"
   (an ambiguous edge must be clean) and,
   - $Y(u) =$ "N" $\Rightarrow Y(M^*(u)) =$ "$\varnothing$"
   (no need to store a neighbor cut if $M^*(u)$ is cut or has a neighbor cut).

Note that if V' contains a single ambiguous edge, then $|\mathcal{Y}(V')| = 2$ under the weight score and $|\mathcal{Y}(V')| = 5$ under the cut score.

Definition 10. *Let $Y_i : V_i \to \{$"$\varnothing$", "N", "$\times$"$\}$ for $i \in \{1,2\}$ be two signatures such that $V_1 \cap V_2 = \varnothing$. The* union *of Y_1 and Y_2 is the mapping $Y_1 \cup Y_2 : V_1 \cup V_2 \to \{$"$\varnothing$", "$N$", "$\times$"$\}$ with*

$$(Y_1 \cup Y_2)(v) = \begin{cases} Y_1(v) & \text{if } v \in V_1 \\ Y_2(v) & \text{otherwise.} \end{cases}$$

For each bag B_i of a given, M^*-preserving tree decomposition of G^*, we will compute solutions for G_i^*. To this end, we introduce the following definition.

Definition 11 (see Fig. 3). *Let $(\mathcal{X}, T)$ be a nice tree decomposition of G^*, let $X \subseteq V(G_i^*)$, let $B_i \in \mathcal{X}$, let $Y \in \mathcal{Y}(B_i)$, and let $u \in B_i$. Further, let*

(i) Y be the signature of X in B_i and,
(ii) X be a solution for (G_i^*, M^*, ω, m).

Then, we call X eligible *with respect to (Y, B_i).*

If there is no set eligible for a pair (Y, B_i), we say that the signature Y is *incompatible* with G_i^*.

Lemma 2. *Let $B_i \in \mathcal{X}$ and let $Y \in \mathcal{Y}(B_i)$. Y is incompatible with G_i^* if and only if there are $u, v \in B_i$ with $uv \in E(G_i^*) \setminus M^*$ and $Y(u) =$ "N" and $Y(v) \neq$ "$\times$".*

Semantics: *Let $Y : V(G_i^*) \to \{$"$\times$", "N", "$\varnothing$"$\}$. A table entry $[Y]_i$ is some minimum-score solution X that is eligible with respect to (Y, B_i) (and $[Y]_i = \bot$ if no such X exists).*

We set score$(\bot) = \infty$ and $\bot \cup X = \bot$, for any set X.

The Algorithm. Let (G^*, M^*, ω, m) be a solution graph. We compute a M^*-preserving tree decomposition $(\mathcal{X}, T)$ of G^* as described previously. We then traverse $(\mathcal{X}, T)$ from the leaf bags to the root X_r. We compute the table entry for each signature $Y \in \mathcal{Y}(B_i)$ of each bag B_i. Then, we obtain the minimum solution for (G^*, M^*, ω, m) from $[Y]_r$. Let B_j and B_ℓ the children of B_i (if they exist). We compute $[Y]_i$ depending on the type of the bag B_i:

leaf bag: Since $B_i = \varnothing$, the only table entry is $[\varnothing]_i$ and we set $[\varnothing]_i = \varnothing$.
introduce $uv \in M^*$ **bag:** We apply the following routine:

1. First consider that uv is isolated. We copy the table entries of the child B_j and complete them such that all signatures in $\mathcal{Y}(B_i)$ are instantiated:
$$[Y]_i = \operatorname*{argmin}_{Y \in \mathcal{Y}(B_j)} \operatorname*{argmin}_{Y' \in \mathcal{Y}(\{u,v\})} \{\text{score}([Y]_j \cup \mathcal{T}(Y'))\}.$$
2. Then, we introduce successively the non-matching edges incident to uv. If a signature is incompatible with G_i^*, then we set its table entry to $\perp$. Let E' be the set of incident edges to the matching edge uv. For each $xx' \in E'$ and all $Y \in \mathcal{Y}(B_i)$, we set $[Y]_i = \perp$ if
   - if $Y(x) =$ "N" and $Y(x') \neq$ "x" (Lemma 2),
   - if $Y(x') =$ "N" and $Y(x) \neq$ "x" (Lemma 2).

join bag: For all $Y \in \mathcal{Y}(B_i)$, we set $[Y]_i = [Y]_j \cup [Y]_\ell$.
forget $uv \in M^*$ **bag:** For all $Y \in \mathcal{Y}(B_i)$, we set
$[Y]_i = \text{argmin}(\{\text{score}([Y']_j) \mid Y' \in \mathcal{Y}(B_j) \wedge \exists_{Y'' \in \mathcal{Y}(\{u,v\})} Y' = Y \cup Y''\})$.

Lemma 3. *The described algorithm is correct, that is, the computed value of* $[Y]_i$ *corresponds to the semantics.*

In each bag B_i, we have to iterate over all signatures in $\mathcal{Y}(B_i)$. The number of possible values for an ambiguous edge is equal to two under the weight score and to five under the cut score. Thus, the number of signatures in a bag containing tw matching edges is equal to 2^{tw} under the weight score and to 5^{tw} under the cut score. Since the number of bags depends on the number of non-matching edges, we obtain a complexity of $O(2^{tw} \cdot |E(G')|)$ under the weight score and a complexity of $O(5^{tw} \cdot |E(G')|)$ under the cut score. To obtain an optimal solution, we just have to take the value of $[\varnothing]_r$ computed by the algorithm.

Corollary 1. *Given a* M^**-preserving nice tree decomposition with width* tw*,* SEMI-BRUTAL CUT *can be solved in* $O(2^{tw} \cdot |E(G^*)|)$ *time under the weight score and in* $O(5^{tw} \cdot |E(G^*)|)$ *time under the cut score.*

Optimization. As no non-ambiguous matching edge will contain a cut, we can remove these matching edges from the graph before computing the tree decomposition, yielding a reduction of the treewidth. However, we must ensure that each vertex stores its adjacency to a removed matching edge.

Table 1. Results statistics. "ILP" and "Tree Dec." columns indicate the execution times, in seconds.

Data	Treewidth	Cut score			Weight score		
		Score	ILP	Tree Dec.	Score	ILP	Tree Dec.
anopheles	3	1093	4.63	5.10	1387	4.76	4.22
anthrax	2	12	0.42	0.32	17	0.41	0.31
gloeobacter	2	39	0.44	0.36	67	0.46	0.36
lactobacillus	2	13	0.19	0.15	18	0.19	0.14
pandora	1	5	0.25	0.19	6	0.25	0.18
pseudomonas	2	36	0.54	0.42	51	0.53	0.42
rice	2	3	0.01	0.00	3	0.01	0.00
sacchr3	2	3	0.03	0.02	5	0.03	0.02
sacchr12	4	12	0.10	0.07	18	0.09	0.07

6 Experiments

The contribution of the paper being mainly theoretical, we propose implementation and tests on real instances. In order to compare the performance of both algorithms, we tested them on datasets already used in [14]. We can observe that selected instances have a small treewidth. A real instance of SBC is generated from a collection of alternating paths and alternating cycles, thus we may think that such instance has a small treewidth. Our implementation of the tree decomposition based algorithm relies on the HTD library [1] for tree decomposition construction. We use ILOG CPLEX to provide a solution to our integer linear programming formulation. We compare results for both scores, statistics on produced solutions are presented in Table 1. We can see that the tree decomposition algorithm is faster under the weight score, which can be explained by the difference of the theoretical complexity. For the cut score, the dynamic programming is slightly faster than the ILP with one exception for the *anopheles* genome. Since the real instances seem to have a small treewidth and the tree decomposition algorithm uses more the internal structure of the problem, we may think that it remains faster than the ILP.

7 Conclusion

In this paper, we present a novel point of view on a problem dedicated to the production of genomic sequences. The previous exploration of the frontier between tractable and hard cases did not provide a satisfactory polynomial-time algorithm and, thus, we explore here two possible solutions: The first is to position the problem relative to the PLS class, aiming to decide whether local search is easier than global search. The second is to consider natural exact methods. In this context, we studied and implemented a simple and efficient ILP and a tree-decomposition based method, yielding an FPT algorithm with respect to the treewidth of the input graph. Interesting open questions include the existence of polynomial-time approximation algorithms, and whether alternative tools, such as color coding or kernel techniques, allow designing more efficient FPT algorithms. As a more practical perspective, we intend to perform further tests on these algorithms and previous ones, to explore the ability of each method to perform well on various kinds of genomes.

References

1. Abseher, M., Musliu, N., Woltran, S.: htd – a free, open-source framework for (customized) tree decompositions and beyond. In: Salvagnin, D., Lombardi, M. (eds.) CPAIOR 2017. LNCS, vol. 10335, pp. 376–386. Springer, Cham (2017). https://doi.org/10.1007/978-3-319-59776-8_30
2. Biscotti, M.A., Olmo, E., Heslop-Harrison, J.S.: Repetitive DNA in eukaryotic genomes. Chromosome Res. **23**(3), 415–420 (2015)
3. Bongartz, P.: Resolving repeat families with long reads. BMC Bioinform. **20**(232) (2019). https://doi.org/10.1186/s12859-019-2807-4. ISSN 1471-2105
4. Chateau, A., Giroudeau, R.: A complexity and approximation framework for the maximization scaffolding problem. Theor. Comput. Sci. **595**, 92–106 (2015)
5. Johnson, D.S., Papadimitriou, C.H., Yannakakis, M.: How easy is local search? J. Comput. Syst. Sci. **37**(1), 79–100 (1988)
6. Kloks, T.: Treewidth, Computations and Approximations. Lecture Notes in Computer Science, vol. 842. Springer, Heidelberg (1994). https://doi.org/10.1007/BFb0045375
7. Krentel, M.: On finding and verifying locally optimal solutions. SIAM J. Comput. **19**(4), 742–749 (1990)
8. Mandric, I., Lindsay, J., Măndoiu, I.I., Zelikovsky, A.: Scaffolding algorithms (chap. 5). In: Măndoiu, I., Zelikovsky, A. (eds.) Computational Methods for Next Generation Sequencing Data Analysis, pp. 107–132. Wiley, Hoboken (2016)
9. Schäffer, A.A., Yannakakis, M.: Simple local search problems that are hard to solve. SIAM J. Comput. **20**(1), 56–87 (1991)
10. Tabary, D., Davot, T., Weller, M., Chateau, A., Giroudeau, R.: New results about the linearization of scaffolds sharing repeated contigs. In: Kim, D., Uma, R.N., Zelikovsky, A. (eds.) COCOA 2018. LNCS, vol. 11346, pp. 94–107. Springer, Cham (2018). https://doi.org/10.1007/978-3-030-04651-4_7
11. Tang, H.: Genome assembly, rearrangement, and repeats. Chem. Rev. **107**(8), 3391–3406 (2007)
12. Tomescu, A.I., Gagie, T., Popa, A., Rizzi, R., Kuosmanen, A., Mäkinen, V.: Explaining a weighted DAG with few paths for solving genome-guided multi-assembly. IEEE/ACM Trans. Comput. Biol. Bioinform. **12**(6), 1345–1354 (2015)
13. Ummat, A., Bashir, A.: Resolving complex tandem repeats with long reads. Bioinformatics **30**(24), 3491–3498 (2014). https://doi.org/10.1093/bioinformatics/btu437. ISSN 1367-4803
14. Weller, M., Chateau, A., Giroudeau, R.: Exact approaches for scaffolding. BMC Bioinform. **16**(Suppl 14), S2 (2015)

15. Weller, M., Chateau, A., Giroudeau, R.: On the linearization of scaffolds sharing repeated contigs. In: Gao, X., Du, H., Han, M. (eds.) COCOA 2017. LNCS, vol. 10628, pp. 509–517. Springer, Cham (2017). https://doi.org/10.1007/978-3-319-71147-8_38
16. Weller, M., Chateau, A., Dallard, C., Giroudeau, R.: Scaffolding problems revisited: complexity, approximation and fixed parameter tractable algorithms, and some special cases. Algorithmica **80**(6), 1771–1803 (2018)
17. Weller, M., Chateau, A., Giroudeau, R., Poss, M.: Scaffolding with repeated contigs using flow formulations (2018)

Scanning Phylogenetic Networks Is NP-hard

Vincent Berry[1], Celine Scornavacca[2], and Mathias Weller[3](✉)

[1] LIRMM, Université de Montpellier, Montpellier, France
vberry@lirmm.fr
[2] CNRS, Université de Montpellier, Montpellier, France
celine.scornavacca@umontpellier.fr
[3] CNRS, LIGM, Université Paris Est, Marne-la-Vallée, France
mathias.weller@u-pem.fr

Abstract. Phylogenetic networks are rooted directed acyclic graphs used to depict the evolution of a set of species in the presence of reticulate events. Reconstructing these networks from molecular data is challenging and current algorithms fail to scale up to genome-wide data. In this paper, we introduce a new width measure intended to help design faster parameterized algorithms for this task. We study its relation with other width measures and problems in graph theory and finally prove that deciding it is NP-complete, even for very restricted classes of networks.

1 Introduction

Phylogenetic networks are rooted directed acyclic graphs used to depict the evolution of a set of species in the presence of reticulate events such as hybridizations, where two species combine their genetic material to create a new species (see nodes H_1 and H_2 in Fig. 1(left)) [9]. Herein, leaves represent the studied species and the root their most recent common ancestor, from which time flows away (as indicated by the direction of the arcs). Internal vertices represent either speciation events (a single parent) or reticulation events (several parents). Each arc represents the evolution of a species in time, during which each gene in the species genome can change due to mutations, allowing different forms of a gene (*alleles*) to appear among species, and even among individuals within the same species. Though the species history is modeled by a network, the evolution of a single non-recombinant gene can always be depicted by a tree, see Fig. 1(center), embedded in the species network, see Fig. 1(right).

Usually, a species network is inferred from a DNA dataset $S = \{S_1, \ldots, S_L\}$ composed of L genes sequenced from the genome of one or several individuals for each studied species [16]. To find the best phylogenetic network explaining S, a possibility is to sample many different networks N and compute the probability $P(S|N)$ of each N given S. Without giving all details here (they can be found for instance in [16]), $P(S|N)$ can be computed from the individual probabilities $P(G_i|N)$ of gene trees $G_1, \ldots, G_L$ for the L loci given N. In turn, each $P(G_i|N)$ can be computed from the probabilities of all possible embeddings of G_i in N, weighted by their respective probability depending on S_i, i.e. $P(G_i|N, S_i)$.

A. Chatzigeorgiou et al. (Eds.): SOFSEM 2020, LNCS 12011, pp. 519–530, 2020.
https://doi.org/10.1007/978-3-030-38919-2_42

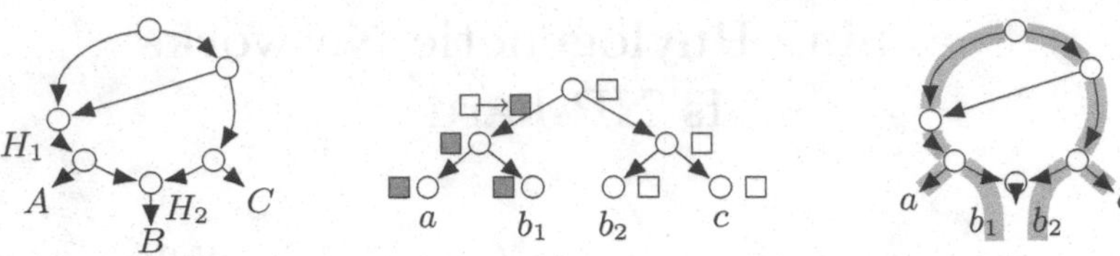

Fig. 1. Left: A phylogenetic network N depicting the evolutionary history of species A, B, and C. **Center:** An evolution scenario for a gene, given the sequences of one individual from species A, one from C and two from B, where different alleles (boxes) are observed: gray for A and for one individual from B, and white for the other individuals. The arc containing the mutation from the white to the gray allele is marked. **Right:** An embedding (gray arcs) in N of this gene evolution scenario.

See Fig. 1(center) for a gene tree and Fig. 1(right) for one of its possible embeddings within the network. Thus, heavy computations are needed to obtain $P(S|N)$ and current algorithms fail to scale to genome-wide data. To design faster algorithms, it is possible to integrate out the possible gene trees and embeddings, as done in [4]. To apply this technique to network inference we designed new partial likelihood formulae to compute $P(S|N)$ and stumbled on a new width parameter for DAGs that clearly puts into evidence why our approach is faster than existing ones, allowing us to handle several real-world datasets within minutes instead of weeks [13]. In this paper, we introduce this new parameter, which we call *scanwidth*, we study its relation with other parameters and problems in graph theory and finally prove that deciding it is NP-hard. A common and intuitive idea when working with phylogenetic networks is to exploit the observation that reticulation should be rare in practice to design algorithms that are fast for only mildly reticulate networks. This tree-likeness is often measured by the tree-width of the input. However, tree decompositions are in no way obligated to follow the leaf-to-root structure that phylogenies naturally impose and this makes dynamic programming on decomposition trees unnecessarily complicated. The scanwidth remedies this problem by forcing the leaves of the network to correspond to the leaves of the decomposition tree, yielding a form of tree-like cutwidth. Thus, our work broadens the arsenal of width measures that can be – and recently have been – used to attack hard problems in phylogenetics [5,8,12]. To get an intuition, imagine a (possibly red) scanner line traversing a network from the leaves to the root; at any moment, its *width* is the number of arcs it cuts. As the line moves up, it traverses nodes, changing the set of arcs it cuts and, hence its width. The cutwidth of the network is the largest width achieved by such a traversing line. Now, consider multiple independent scanner lines, each one scanning an arc incoming to a different leaf of the network. Whenever a node could be passed by two different lines, they are merged to form a single one. This naturally generalizes the cutwidth to a stronger (that is, smaller) width measure that we call *scanwidth.* As with the cutwidth, different orders in which the nodes are passed imply different values of the final width and the goal is to minimize it. In many optimization approaches for phylogenetic networks, a network is traversed from the leaves up to its root, while computing some quantities. For some applications, computations on tree-parts can be done independently for each arc but, when meeting a reticulation node, computations on both arcs entering the node have to be considered jointly. This inter-dependence makes computing the required quantities more time consuming. In such cases, one really wants to process the network while minimizing the numbers of arcs considered jointly. This is captured by the scanwidth parameter.

In this work, we show that deciding the scanwidth of a network relates to an old problem in program optimization called REGISTER SUFFICIENCY (PO1 of Garey and Johnson [7]). Our proof comprises a non-trivial adaptation of an NP-hardness proof [14] for the latter problem to a very restricted class of rooted DAGs, on which REGISTER SUFFICIENCY coincides with deciding the cutwidth and the scanwidth (offset by 1). This hardness proof, as well as the scanwidth parameter itself, may be of independent interest to the design of algorithms for other problems on DAGs.

Note that computing the scanwidth and using it as a parameter for other algorithms are two different pairs of shoes and, though a parameterized algorithm may require a tree extension (see Sect. 2) to be given, there is still hope that the scanwidth can be approximated efficiently. Thus, in analogy with other highly successful (width) parameters such as the treewidth, the hybridization number or the hybridization level [2,3,11,15], we point out that being NP-complete to compute does not hurt the practical usefulness of the scanwidth.

We defer some proofs to a long version of this paper.

2 Preliminaries

Phylogenetic Networks. Let G be a leaf-labelled, directed, acyclic graph with a single source (which is called "root"). The in-degree of a vertex v in G is $\deg_G^-(v)$ and its out-degree is $\deg_G^+(v)$, the sum of those being the degree of v. If all vertices of G have either in-degree one and out-degree zero (*leaves*), in-degree at most one and out-degree at least two (*tree-vertices*), and in-degree at least two and out-degree one (*reticulation*), then G is called *rooted phylogenetic network* (henceforth *network*). Note that the root is a special tree-vertex. We denote the set of leaves of G by $\mathcal{L}(G)$, the set of vertices by $V(G)$ and the tree-vertices by $V_T(G)$. If the root has degree two, the internal vertices have degree three, and the leaves have degree one, then G is called *binary*. If G contains a u-v-path for vertices u and v, we say that u is an ancestor of v (and v is a descendant of u) and we write $v <_G u$.

Vertex Orderings. A linear ordering σ of a subset V' of the vertices of a network G is called *G-respecting* if $u <_G v \Rightarrow u <_\sigma v$ for all $u, v \in V'$. A G-respecting ordering σ over $V(G)$ is called an *extension* (or "reverse topological order") of G, see Fig. 2. We call a tree Γ on $V(G)$ a *tree extension* for G if $x <_G y \Rightarrow x <_\Gamma y$ for all $x, y \in V(G)$. We denote the vertex at position i in σ by $\sigma(i)$ and $\sigma^{-1}(u)$ returns the position of the vertex u in sigma. Since positions and vertices are in bijection, we sometimes use vertices to represent their positions. A position i of

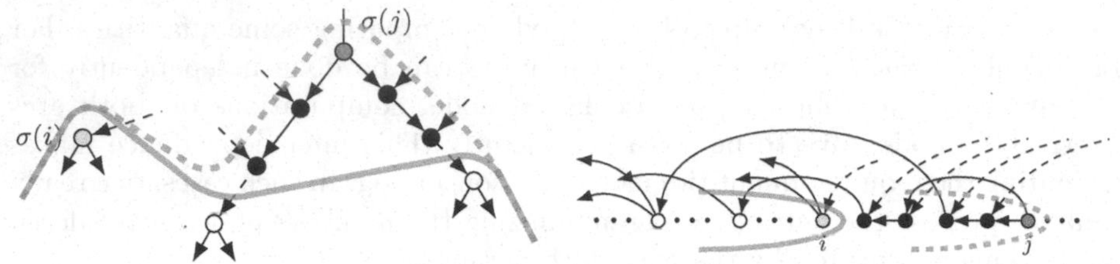

Fig. 2. **Left:** For an extension σ each position i induces a cut through G separating the vertices in $\sigma[1..i]$ (below gray line) from $\sigma[i+1..]$ (above gray line). **Right:** G with vertices linearly arranged according to σ.

σ is called a *milestone* if $\sigma(i)$ is a tree-vertex and σ is called *stable* if all maxima (wrt. $\leq_G$) in $\sigma[1..i]$ for any milestone i are tree-vertices (that is, each reticulation in $\sigma[1..i]$ has a parent in $\sigma[1..i]$). We denote the sub-order of σ restricted to the elements of a set U by $\sigma[U]$ and we abbreviate $\sigma[\{\sigma(i), \sigma(i+1), \ldots, \sigma(j)\}] =: \sigma[i..j]$. For disjoint orders σ and π let $\sigma \circ \pi$ denote the concatenation of σ with π (that is, σ followed by π). For a set X, we let (X) denote any order on the elements of X. Further, for distinct vertices or disjoint vertex sets $X_1, X_2, \ldots$, we abbreviate $(X_1) \circ (X_2) \circ \ldots =: (X_1, X_2, \ldots)$.

(Directed) Cutwidth. For an extension σ of a DAG G and a position i, we will use $\mathrm{C}_i(\sigma)$ to denote the set of arcs from a vertex in $\sigma[i+1..]$ to a vertex in $\sigma[1..i]$ and $\mathrm{cw}_i(\sigma) := |\mathrm{C}_i(\sigma)|$ is called the *cutwidth* of σ at position i. The cutwidth of σ is $\mathrm{cw}(\sigma) := \max_i \mathrm{cw}_i(\sigma)$ and the cutwidth of G, denoted $\mathrm{cw}(G)$, is the minimum of $\mathrm{cw}(\sigma)$ over all extensions σ of G. We allow i to be a vertex instead of a position, as σ is a bijection between the two.

(Directed) Register width. For an extension σ of G and a position i, we will use $\mathrm{RW}_i(\sigma)$ to denote the set of vertices in $\sigma[1..i]$ that have a parent in $\sigma[i+1..]$ and $\mathrm{rw}_i(\sigma) := |\mathrm{RW}_i(\sigma)|$ is called the *register width* (also known as "vertex cut" or "separation" [6]) of σ at position i (again, we allow i to be a vertex instead of a position, as σ is a bijection between the two). The register width of σ is $\mathrm{rw}(\sigma) := \max_i \mathrm{rw}_i(\sigma)$ and the register width of G, denoted $\mathrm{rw}(G)$, is the minimum over all extensions σ for G of $\mathrm{rw}(\sigma)$.

Theorem 1. *For all binary networks G, we have* $\mathrm{cw}(G) = \mathrm{rw}(G) + 1$.

Scanwidth. Let σ be an extension for G and let $i \in \mathbb{N}$. We define $\mathrm{SW}_i(\sigma)$ as the set of all arcs $uv \in \mathrm{C}_i(\sigma)$ for which v and $\sigma(i)$ are weakly connected in $G[\sigma[1..i]]$ (see Fig. 3(left)). $\mathrm{sw}_i(\sigma)$ is defined as $|\mathrm{SW}_i(\sigma)|$, while the scanwidth of σ is $\mathrm{sw}(\sigma) := \max_i \mathrm{sw}_i(\sigma)$ and the scanwidth of G, denoted by $\mathrm{sw}(G)$, is the minimum of $\mathrm{sw}(\sigma)$ over all extensions σ for G. Again, in our notations we allow i to be a vertex instead of a position, as σ is a bijection between the two.

Alternatively, $\mathrm{sw}(G)$ can be defined as follows. For a tree extension Γ for G, we define $\mathrm{GW}_v(\Gamma)$ as the set of arcs $(x, y) \in E(G)$ with $x >_\Gamma v \geq_\Gamma y$. Further,

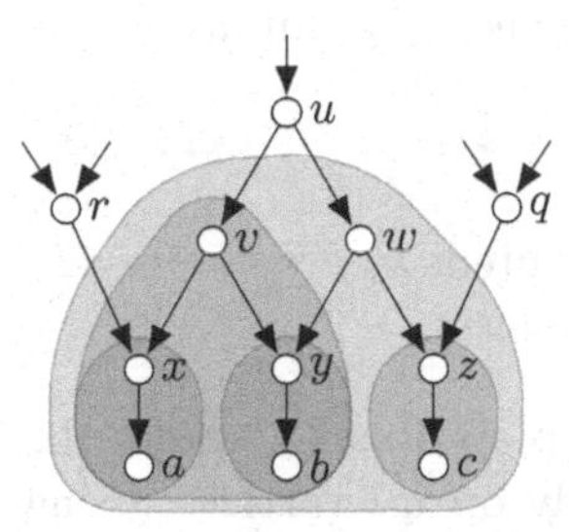

$\sigma(i)$	$\mathrm{SW}_i(\sigma) = \mathrm{GW}_{\sigma(i)}(\Gamma)$
a	$\{xa\}$
b	$\{yb\}$
c	$\{zc\}$
x	$\{rx, vx\}$
y	$\{vy, wy\}$
z	$\{wz, qz\}$
v	$\{rx, uv, wy\}$
w	$\{rx, uv, uw, qz\}$
...	...

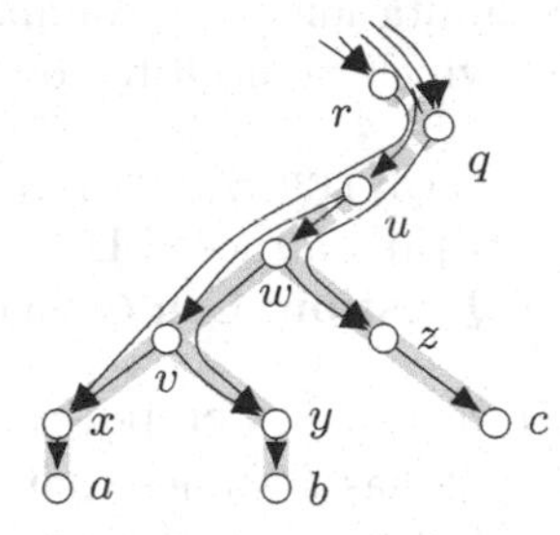

Fig. 3. Illustration of the two definitions of scanwidth. **Left:** Lower part of a graph G where gray zones represent the weakly connected components induced by $\sigma[1..i]$ for $\sigma = (a, b, c, x, y, z, v, w, \ldots)$ and $i \leq 8$. Here, $\mathrm{SW}_v(\sigma) = \{rx, uv, wy\}$ since x, y, and v are weakly connected in $G[a, b, c, x, y, z, v]$. **Middle:** table indicating $\mathrm{SW}_i(\sigma)$ for $i \leq 8$ corresponding to σ. **Right:** part of a tree Γ with $\mathrm{GW}_{\sigma(i)}(\Gamma) = \mathrm{SW}_i(\sigma)$ for all $i \leq 8$. For the extension $\pi = (c, a, z, b, y, x, v, w, \ldots)$, we also have $\mathrm{GW}_{\pi(i)}(\Gamma) = \mathrm{SW}_i(\pi)$ for all $i \leq 8$.

we let $\gamma\mathrm{w}(\Gamma) := \max_v |\mathrm{GW}_v(\Gamma)|$ and $\gamma\mathrm{w}(G) := \min_\Gamma \gamma\mathrm{w}(\Gamma)$. Although a tree extension is defined independently of a (full) extension for G, there is a link between the two notions. Indeed, the sets $\mathrm{GW}_v(\Gamma)$ in an optimal tree extension correspond to the sets $\mathrm{SW}_v(\sigma)$ in one or several optimal extensions σ (see Fig. 3).

Proposition 1. *For any network G, (a) $\gamma\mathrm{w}(G) = \mathrm{sw}(G)$. Further, (b) if G has only one leaf, then $\mathrm{sw}(G) = \mathrm{cw}(G)$. (c) If G is also binary, then $\mathrm{sw}(G) = \mathrm{rw}(G) + 1$.*

Observe that the scanwidth differs largely from the directed path-width [1], which is always zero for DAGs. To relate the scanwidth to established parameters, let us mention that the scanwidth of any level-k network cannot exceed $k + 1$ but it might even be constant. Regarding width-measures, the scanwidth is bounded by the cutwidth from below and the treewidth (of the underlying undirected graph) from above.

3 NP-completeness

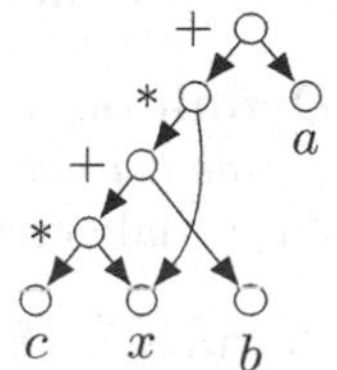

To compute the value of a given algebraic expression such as $(cx+b)x+a$ using a computer, we need to store the values of a, b, c, and x in registers which can then be processed by the CPU. As registers can be overwritten, expressions involving more variables than the number of available registers can be evaluated. The problem of deciding whether a given expression can be evaluated on a CPU with k registers (without recomputing sub-expressions or relying on the costly *spilling* technique) is called REGISTER SUFFICIENCY. We suppose that the input expression is given as a **rooted** DAG of necessary computations. For example, to compute $(cx+b)x+a$, we need to compute $cx+b$, for which we need to compute cx (see figure on the right).

> Register Sufficiency [PO1 in [7]] (RS)
> **Input:** a rooted DAG G of an expression to be computed, $k \in \mathbb{N}$
> **Question:** Can G be computed using at most k registers?

Register Sufficiency can be interpreted as a game played on G, where the player has k stones that have to be placed progressively on *all* vertices, using the following operations [14]:

1. remove a stone from any vertex
2. for a vertex p whose every child contains a stone,
   2a. place an available stone on p or
   2b. move a stone from a child of p to p,

so that each vertex receives a stone exactly *once* during these operations.

Stones represent registers and putting a stone on a vertex of the graph corresponds to computing the vertex and storing the result in that register (this is why we need stones on all children of a vertex when computing it). Removing a stone from a vertex corresponds to forgetting the value of the vertex, which should then be done only if we do not need it in other computations (as vertices cannot be recomputed), i.e. when all its parent vertices have already received a stone.

Winning the game means successfully computing the algebraic expression encoded in the graph while using at most k registers. In this context, an extension σ for a graph G indicates in which order the vertices receive stones. Note that the first stone enters G via applying Rule 2a to a leaf of G. Then, solving the optimization problem associated to Register Sufficiency can be seen as finding an extension of G that minimizes the number k of stones (registers) needed to win the game (compute the expression). As suggested by our formulation, this number equals the previously introduced "register width", $\text{rw}(G)$.

Proposition 2. *A DAG G can be computed using $\leq k$ registers if and only if $\text{rw}(G) \leq k$.*

With Proposition 2, the Register Sufficiency problem can be formulated as: given a rooted DAG G and some integer k, decide if $\text{rw}(G) \leq k$. Following Sethi [14], we will use a special, "initial" vertex in our reduction.

Definition 1. *Let (G, k) be an instance of* Register Sufficiency *such that G has k leaves and all leaves have a common parent ψ. Then, we say that ψ is an* initial vertex *and that (G, k)* has the initial vertex property.

Lemma 1 (See [14]). *Let (G, k) be a yes-instance of* Register Sufficiency *with an initial vertex ψ. Let σ be an extension of G with $\text{rw}(\sigma) \leq k$. Then, $\sigma(k+1) = \psi$ and $\sigma[1..k]$ contains the k leaves of G in any order. Moreover, there is a leaf whose only parent is ψ.*

A corollary of Lemma 1 is that, in a yes-instance (G, k) with the initial vertex property, all children of the initial vertex are leaves. Thus, $\mathrm{rw}_k(\sigma) = k$ for all extensions σ with $\mathrm{rw}(\sigma) \leq k$ and, thus, $\mathrm{rw}(G) = k$ for such yes-instances. Note that 3-SAT reduces to instances (G, k) of REGISTER SUFFICIENCY that have the initial vertex property [14].

Theorem 2 ([14]). *It is* NP*-hard to decide* REGISTER SUFFICIENCY *for instances* (G, k) *that have the initial vertex property.*

Below, we reduce REGISTER SUFFICIENCY on instances with the initial vertex property to REGISTER SUFFICIENCY on rooted, binary, single-leaf DAGs. To this end, we reduce from WEIGHTED 2-SATISFIABILITY instead of 3-SATISFIABILITY and modify parts of the reduction in order to obtain a network that is already bifurcating in some crucial spots. Then, we present a number of polynomial-time executable transformation rules that take one such instance (G, k) of REGISTER SUFFICIENCY having the initial vertex property and replace all remaining high-degree vertices with binary ones without changing the answer for the instance. Finally, a reduction rule is given to ensure that the resulting DAG has a single leaf.

3.1 An Adaptation of a Known NP-hardness Proof

We strengthen the construction presented by Sethi [14] to construct a *binary* DAG with a *single leaf* and *without degree-two vertices.* Our modifications to Sethi's construction come in two stages. First, instead of 3-SAT, we will reduce a 2-SAT variant called MONOTONE WEIGHTED 2-SATISFIABILITY (also known as VERTEX COVER), which is also NP-hard [10]. In this variant, all variables occur non-negated in the instance formula φ, each variable is used at least once, and we ask for an assignment that satisfies φ while setting at most k variables to true. Second, we show how to "binarize" all remaining polytomies and establish a single leaf.

Construction 1 (See Fig. 4). *Given a formula* φ *in monotone 2-CNF on variables* $x_1, \ldots, x_n$ *and clauses* $C_1, \ldots, C_m$*, let* $y_{i,j}$ *denote the* j^{th} *literal in* C_i*. Construct the instance* (G, k')*, where* $k' = 8n + 3m + k + 2$ *and* G *is a rooted DAG on the vertex set* $A' \uplus B' \uplus C \uplus F' \uplus H \uplus P \uplus P' \uplus R' \uplus S' \uplus T' \uplus U \uplus W \uplus X \uplus X' \uplus X^* \uplus Z' \uplus \{\alpha, \psi, d, \rho\}\}$ *where* $R' := \bigcup_{i\in[n]} R'^i$*,* $S' := \bigcup_{i\in[n]} S'^i$*,* $T' := \bigcup_{i\in[n]} T'^i$ *and*

$$A' = \{a_i \mid i \in [2n+1+k]\} \qquad C = \{c_i \mid i \in [m]\} \qquad F' = \{f_{i,1}, f_{i,2} \mid i \in [m]\}$$
$$B' = \{b_i \mid i \in [3n-m]\} \qquad W = \{w_i \mid i \in [n]\} \qquad R'^i = \{r_{i,j} \mid \ j \in [2n-2i+2+k]\}$$
$$U = \{u_{i,1}, u_{i,2}, u_{i,3} \mid i \in [n]\} \qquad X' = \{x'_i \mid i \in [n]\} \qquad S'^i = \{s_{i,j} \mid \ j \in [2n-2i+1+k]\}$$
$$X = \{x_i, \overline{x}_i \mid i \in [n]\} \qquad X^* = \{x^*_i \mid i \in [n]\} \qquad T'^i = \{t_{i,j} \mid \ j \in [2n-2i+1+k]\}$$
$$H = \{h_{i,1}, h_{i,2} \mid i \in [m]\} \qquad P = \{p_i \mid i \in [m]\} \qquad P' = \{p'_i, p''_i, p'''_i \mid i \in [m]\}$$
$$Z' = \{z_i \mid i \in [n+1]\}$$

and the arc set is the union of the following sets:

$$E'_1 = \{\psi v \mid v \in A' \uplus B' \uplus F' \uplus U \uplus H \uplus \{\alpha\}\} \qquad E_5 = \{w_i u_{i,1}, w_i u_{i,2} \mid i \in [n]\}$$
$$E'_2 = \{v\psi \mid v \in R'^i \uplus S'^i \uplus T'^i\} \qquad E'_{7,1} = \{z_{i+1}w_i, z_{i+1}z_i \mid i \in [n]\}$$
$$E_4 = \{x_i z_i, \overline{x}_i z_i, x_i u_{i,1}, \overline{x}_i u_{i,2}, x'_i u_{i,3} \mid i \in [n]\} \qquad E'_{7,2} = \{c_i z_{n+1} \mid i \in [m]\}$$
$$E_{12} = \{r_{i,j} r_{i,j+1} \mid j \in [2n-2i+k+1], i \in [n]\} \qquad E'_9 = \{dv \mid v \in B' \uplus C\}$$
$$E_{13} = \{s_{i,j} s_{i,j+1}, t_{i,j} t_{i,j+1} \mid j \in [2n-2i+k], i \in [n]\} \qquad E_{11} = \{x^*_i x'_i, x'_i x_i \mid i \in [n]\}$$

$$E'_3 = \{\rho v \mid v \in W \uplus X \uplus Z' \uplus \{\psi, d\} \uplus X^* \uplus \{p'_i, p'''_i \mid i \in [m]\} \uplus \{u_{i,3} \mid i \in [n]\} \uplus \{z_{n+1}\}\}$$
$$E'_6 = \{z_i r_{i,j}, x_i s_{i,j}, \overline{x}_i t_{i,j} \mid \ r_{i,j} \in R'^i,\ s_{i,j} \in S'^i,\ t_{i,j} \in T'^i, i \in [n]\}$$
$$E'_8 = \{c_i p_i, p_i f_{i,1}, p_i f_{i,2}, p'''_i f_{i,2}, p'''_i p''_i, p''_i p'_i, p''_i h_{i,2}, p'_i h_{i,1} \mid i \in [m]\}$$
$$E'_{10} = \{x^*_{i,1} f_{i,1}, x^*_{i,2} h_{i,2}, \overline{x}_{i,1} h_{i,1} \mid i \in [m]\}$$

where $x_{i,j}$ *denotes the* j^{th} *variable in* C_i *and* $\overline{x}_{i,j}$ *the negation of* $x_{i,j}$ *(and their corresponding vertices with the same names) and* $x^*_{i,j}$ *the vertex in* X^* *such that* $x^*_{i,j} x_{i,j} \in E_{11}$.

The idea behind Construction 1 is that the "variable-assignment phase" of Sethi [14] still works as before (with k more stones in each step to account for the k additional vertices we have in R'^i, S'^i, and T'^i). In more detail, this process is as follows: in the beginning, all k' stones have to go to all the leaves, at which point ψ is computed using one stone of a vertex in A', while the other $k+2n$ stones of A' are now free (unlike stones on other leaves still having other parents). These $k+2n$ stones need to go to R'^1 (otherwise we will not have enough stones later for these vertices), allowing to compute z_1, who will keep one stone. The $k+2n-1$ other stones from R'^1 are free to go to either S'^1 allowing to compute x_1 or to T'^1 allowing to compute $\overline{x}_1$. The chosen literal allows exactly one stone from U to move to w_1 (e.g. $u_{1,1}$ if x_1 is chosen and $u_{1,2}$, otherwise), who will keep this stone. Thus, $k+2n-2$ stones (from either S'^1 or T'^1) are now free to compute R'^2, followed by z_2. This process continues until w_n receives a stone, at which point we ended Sethi's variable assignment phase. Now, the stone on α moves to z_{n+1} and we are left with the k free stones, coming from either S'^n or T'^n, that we can spend on vertices $x'_j \in X'$ (which then move to $x^*_j \in X^*$) whose corresponding $x_j \in X$ has received a stone before.

Consider what happens if the described "variable assignment" phase chooses k vertices in X satisfying the formula and the k corresponding vertices of X^* receive a stone right after this phase. Consider the gadget corresponding to clause $C_i = (x_j \vee x_\ell)$ and recall that $f_{i,1}$, $f_{i,2}$, $h_{i,1}$, $h_{i,2}$ already hold stones. In analogy with Sethi [14], each c_i receives a stone as follows:

- If C_i is satisfied by x_j, then x^*_j holds a stone, so the stone on $f_{i,1}$ can move to p_i (this is allowed since p_i's children all hold stones).
- Otherwise, both $\overline{x}_j$ and x^*_ℓ hold stones. The first one allows the stone on $h_{i,1}$ to move to p'_i. The second one allows the stone on $h_{i,2}$ to move to p''_i and then to p'''_i, allowing the stone on $f_{i,2}$ to move to p_i.

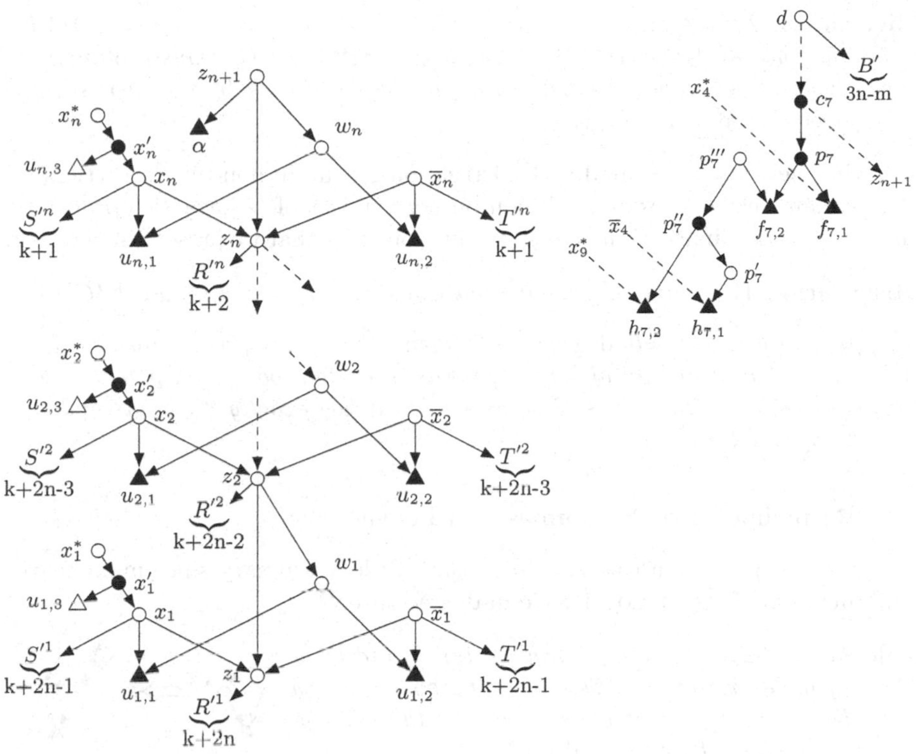

Fig. 4. Illustration of Construction 1; white vertices are children of the root ρ, triangles are leaves and children of ψ. This allows us to omit drawing ψ and ρ. We also omit the leaves in A'. **Left:** "variable-assignment" gadget (arcs of E_2' omitted). **Right:** clause gadget for the clause $C_7 = (x_4 \vee x_9)$. Note that all w_i, p_i, p_i'', p_i''' and c_i are bifurcating.

Thus, in both cases p_i gets a stone which it then passes to c_i. Finally, when all c_i have received a stone, d receives a stone from one of them, freeing up $|B'| = 3n - m$ stones on the vertices in B' and $|C| - 1 = m - 1$ stones on the vertices in C. Since $k \leq n$, these $3n - 1$ stones can then be placed on T'^1 (if x_1 already holds a stone) or S'^1 (if $\overline{x}_1$ already holds a stone) and one of them can then move to $\overline{x}_1$ or x_1, respectively. In this way, all $2n$ vertices x_i and $\overline{x}_i$ progressively receive a stone. Since n of them already got stones in the variable assignment phase, this leaves us with $(3n-1)-n$ stones, $n-k$ of which are then put on the $n-k$ remaining stoneless vertices of X' which immediately move to the remaining vertices of X^*. At this point, all stones on all $h_{i,1}$ and $h_{i,2}$ move to p_i' and p_i'' followed by p_i''' if they did not already do so before. Finally, ρ receives a stone from any of its children.

Theorem 3. *Let $k \in \mathbb{N}$, let φ be a formula in monotone 2-CNF, and let (G, k') be an instance of* Register Sufficiency *constructed by Construction 1 on input (φ, k). Then, φ has a satisfying assignment with $\leq k$ true variables if and only if $\mathrm{rw}(G) \leq k'$.*

Note that networks G created by Construction 1 contain non-binary vertices, as well as many leaves. However, all non-binary vertices of G have nice properties that allow us to "binarize" them using reduction rules that we present in Sect. 3.2.

Observation 1. *Let (G', k') result from Construction 1 and let $u \in V(G')$.*

(a) If u is a non-leaf with $\deg^+(u) \geq 3$, then u has a child with in-degree 1.
(b) If u is a non-leaf with at least 3 parents, then the root ρ is a parent of u.
(c) If u is a leaf with at least 3 parents, then u has exactly 3 parents and ψ is one of them.

3.2 Reducing Nice Polytomies and Leaves

The following reduction rule is used to turn all leaves binary since many leaves constructed in Construction 1 have in-degree three.

Rule 1. *Let (G, k) have an initial vertex ψ and let u be a leaf in G with at least three parents, one of which is ψ. Then, add a new parent v to u, add the arc $v\psi$, and replace all xu by xv except ψu.*

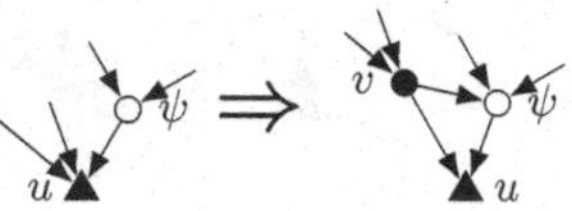

The next rule splits vertices of in- and out-degree at least two into a reticulation and a tree-vertex.

Rule 2. *Let u be a vertex of G, let P and C be its parents and children, respectively, and let $|P| > 1$ and $|C| > 1$. Then, "split" u, i.e. add a new vertex v, add the arc uv and, for all $c \in C$, replace the arc uc by the arc vc.*

Rule 3 (See Fig. 5(left)). *Let u be a vertex with at least three children, let x and y be children of u such that y is a tree-vertex and x is either a tree-vertex or x has a parent $q \neq u$ that is comparable to u in G. Then, "split" u into ru (that is, create a new parent r for u and make all parents of u parents of r instead), subdivide ru with a new vertex r', for all parents q of x with $q >_G u$ replace qx with qr', add the arc rx, subdivide uy with a new vertex w, remove the arc ux and, unless x has a parent $q <_G u$ in G, add the arc wx.*

Correctness proofs of Rules 1–3 are deferred to the full version of this paper. Note that Rule 3 only increases $\deg_G^-(x)$ if x has no parents $q <_G u$ in G. But then, either x is a tree-vertex in G, in which case no new polytomies are created, or x has a parent $q >_G u$, in which case this parent becomes a parent of r' instead. Further, although Rule 3 may introduce degree-two vertices, all of them are parents of tree-vertices and can thus be removed using the following:

The following rule turns polytomous reticulations into binary ones. We make use of the fact that G has a root and an initial vertex (see Definition 1).

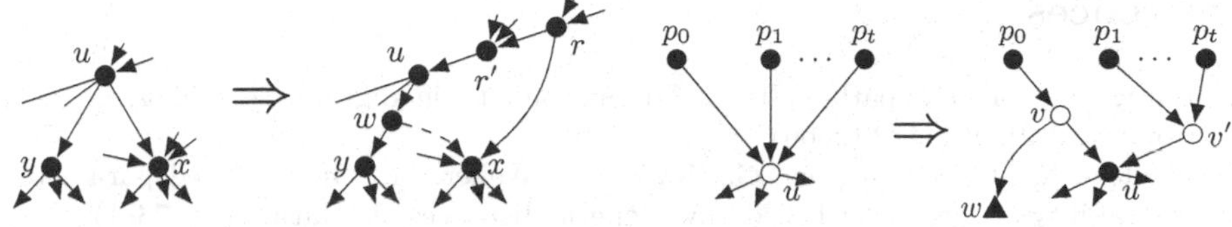

Fig. 5. Illustration of Rule 3 (left) and Rule 4 (right). Triangles are leaves and children of the initial vertex ψ, white vertices are children of the root ρ. Note that, on the left, u has a tree-vertex child before and after the modification.

Rule 4 (see Fig. 5, right). *Let (G, k) have an initial vertex ψ, let ρ be the root of G, let u be a non-leaf with parents $p_0, p_1, \ldots, p_t, p_{t+1} = \rho$ $(t \geq 1)$. Then, add a new leaf w, increase k by one, subdivide p_0u with a vertex v, replace arc ρu by ρv, add a new parent v' of u, replace arc p_iu by p_iv' for all $i \in [t+1]$, and add the arcs $\rho v'$, vw, ψw.*

Note that Rule 4 effectively turns a vertex of in-degree $t + 2$ (for $t \geq 1$) into a vertex of in-degree $t + 1$. Further, note that the instance (G', k') constructed by Construction 1 has an initial vertex ψ.

Rule 5. *Let (G, k) have an initial vertex ψ, let X be the set of leaves of G and let $Y \subseteq X$ contain the leaves that have more than one incoming arc. Let $Y \neq \emptyset$ and let $x_1, x_2, \ldots, x_k$ be an arbitrary total order of X with $x_k \in Y$. Then, turn X into a path by adding the arc $x_{i+1}x_i$ for all i. Further, for all $y \in Y - x_k$, subdivide ψy with a new vertex z, and replace all arcs uy occurring in G by uz.*

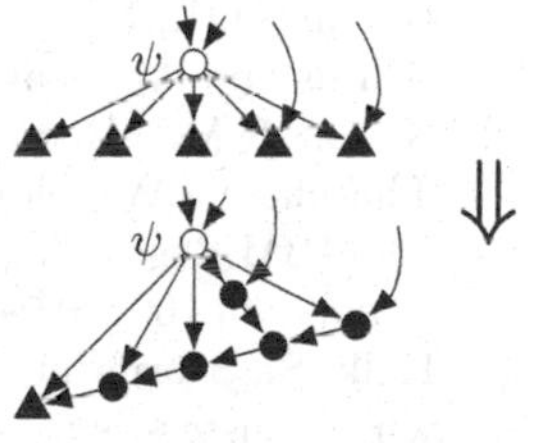

Note that the graphs produced by Construction 1 satisfy $\emptyset \subsetneq Y \subsetneq X$.
Note that Rule 5 destroys the initial vertex property, preventing any further use of Rule 4 and Rule 5. However, Rule 5 does not create new polytomies and, for turning ψ binary by applying Rule 3, it is sufficient that ψ is "primal" (a weaker condition than being initial).

Theorem 4. Register Sufficiency *is* NP*-complete on rooted, single-leaf binary DAGs.*

Acknowledgments. We thank Fabio Pardi to have brought the problem to our attention and the Genome Harvest project, ref. ID 1504-006 ("Investissements d'avenir", ANR-10-LABX-0001-01).

References

1. Barát, J.: Directed path-width and monotonicity in digraph searching. Graphs Comb. **22**(2), 161–172 (2006)
2. Bordewich, M., Scornavacca, C., Tokac, N., Weller, M.: On the fixed parameter tractability of agreement-based phylogenetic distances. J. Math. Biol. **74**(1), 239–257 (2017)
3. Bordewich, M., Semple, C.: Computing the hybridization number of two phylogenetic trees is fixed-parameter tractable. IEEE/ACM Trans. Comput. Biol. Bioinform. **4**(3), 458–466 (2007)
4. Bryant, D., Bouckaert, R., Felsenstein, J., Rosenberg, N.A., RoyChoudhury, A.: Inferring species trees directly from biallelic genetic markers: bypassing gene trees in a full coalescent analysis. Mol. Biol. Evol. **29**(8), 1917–1932 (2012)
5. Bryant, D., Lagergren, J.: Compatibility of unrooted phylogenetic trees is FPT. Theor. Comput. Sci. **351**(3), 296–302 (2006)
6. Díaz, J., Petit, J., Serna, M.: A survey of graph layout problems. ACM Comput. Surv. **34**(3), 313–356 (2002)
7. Garey, M.R., Johnson, D.S.: Computers and Intractability: A Guide to the Theory of NP-Completeness. W.H. Freeman & Co., Ltd., New York City (1979)
8. Grigoriev, A., Kelk, S., Lekić, N.: On low treewidth graphs and supertrees. In: Dediu, A.-H., Martín-Vide, C., Truthe, B. (eds.) AlCoB 2014. LNCS, vol. 8542, pp. 71–82. Springer, Cham (2014). https://doi.org/10.1007/978-3-319-07953-0_6
9. Huson, D.H., Rupp, R., Scornavacca, C.: Phylogenetic Networks: Concepts: Algorithms and Applications. Cambridge University Press, Cambridge (2010)
10. Karp, R.M.: Reducibility among combinatorial problems. In: Miller, R.E., Thatcher, J.W., Bohlinger, J.D. (eds.) Complexity of Computer Computations. The IBM Research Symposia Series, pp. 85–103. Springer, Boston (1972). https://doi.org/10.1007/978-1-4684-2001-2_9
11. Kelk, S., Scornavacca, C.: Constructing minimal phylogenetic networks from softwired clusters is fixed parameter tractable. Algorithmica **68**(4), 886–915 (2014)
12. Kelk, S., Stamoulis, G., Wu, T.: Treewidth distance on phylogenetic trees. Theor. Comput. Sci. **731**, 99–117 (2018)
13. Rabier, C.E., Berry, V., Pardi, F., Scornavacca, C.: On the inference of complicated phylogenetic networks by Markov chain Monte-Carlo (submitted)
14. Sethi, R.: Complete register allocation problems. SIAM J. Comput. **4**(3), 226–248 (1975)
15. Whidden, C., Beiko, R.G., Zeh, N.: Fixed-parameter algorithms for maximum agreement forests. SIAM J. Comput. **42**(4), 1431–1466 (2013)
16. Zhang, C., Ogilvie, H.A., Drummond, A.J., Stadler, T.: Bayesian inference of species networks from multilocus sequence data. Mol. Biol. Evol. **35**(2), 504–517 (2018)

The Maximum Equality-Free String Factorization Problem: Gaps vs. No Gaps

Radu Stefan Mincu[1(✉)] and Alexandru Popa[1,2]

[1] Department of Computer Science, University of Bucharest, Bucharest, Romania
{mincu.radu,alexandru.popa}@fmi.unibuc.ro

[2] National Institute for Research and Development in Informatics, Bucharest, Romania

Abstract. A factorization of a string w is a partition of w into substrings $u_1, \ldots, u_k$ such that $w = u_1 u_2 \cdots u_k$. Such a partition is called equality-free if no two factors are equal: $u_i \neq u_j, \forall i, j$ with $i \neq j$. The *maximum equality-free factorization problem* is to decide, for a given string w and integer k, whether w admits an equality-free factorization with k factors.

Equality-free factorizations have lately received attention because of their application in DNA self-assembly. Condon et al. (CPM 2012) study a version of the problem and show that it is $\mathcal{NP}$-complete to decide if there exists an equality-free factorization with an upper bound on the length of the factors. At STACS 2015, Fernau et al. show that the maximum equality-free factorization problem with a lower bound on the number of factors is $\mathcal{NP}$-complete. Shortly after, Schmid (CiE 2015) presents results concerning the Fixed Parameter Tractability of the problems.

In this paper we approach equality free factorizations from a practical point of view i.e. we wish to obtain good solutions on given instances. To this end, we provide approximation algorithms, heuristics, Integer Programming models, an improved FPT algorithm and we also conduct experiments to analyze the performance of our proposed algorithms.

Additionally, we study a relaxed version of the problem where gaps are allowed between factors and we design a constant factor approximation algorithm for this case. Surprisingly, after extensive experiments we conjecture that the relaxed problem has the same optimum as the original.

Keywords: String factorization · Equality-free · String algorithms · Heuristics

1 Introduction

To factorize a string (or word) means to obtain a partitioning of non-overlapping substrings that reconstitute the original string when concatenated in order. More

This work was supported by project PN19370401 "New solutions for complex problems in current ICT research fields based on modelling and optimization", funded by the Romanian Core Program of the Ministry of Research and Innovation (MCI), 2019–2022.

A. Chatzigeorgiou et al. (Eds.): SOFSEM 2020, LNCS 12011, pp. 531–543, 2020.
https://doi.org/10.1007/978-3-030-38919-2_43

exactly, a *factorization* of a string w is a tuple of strings $(u_1, u_2, \ldots, u_k)$ such that $w = u_1 u_2 \cdots u_k$.

Despite its simple definition, word factorization has a wide number of applications. For instance, finding an occurrence of a string v in a text t can be formulated as t admitting a factorization $t = uvw$. A string v is a prefix of another string t if $t = vw$ and it is a suffix of t if $t = uv$. Moreover, many string problems can be seen as word factorization problems [5] such as: Shortest Common Superstring, Longest Common Subsequence and Shortest Common Supersequence, to name a few. Another example of word factorization problem is the Minimum Common String Partition [1], a problem concerned with identifying factorizations for two strings such that the sequence of factors for one word is the permutation of the other's.

In this paper we focus on the equality-free factorization, a special case of word factorization in which all factors are distinct. The problem is motivated by an application in DNA synthesis [3]. More specifically, it is possible to produce short DNA fragments that will self-assemble into the wanted DNA structure. However, to obtain the desired structure, it is required that no two fragments are identical. Since the fragments must be short, one approach is to split the target DNA sequence into as many distinct pieces as possible.

Previous work. The equality-free factorization problem was first introduced by Condon, Maňuch and Thachuk [3] where it was presented as the *string partitioning problem.* The string partitioning problem asks for a factorization into distinct factors such that each factor is at most of a certain length. The problem was studied in a more general setting where the measure of collision between two factors is either equality or one is a prefix/suffix of the other. Condon et al. showed that these variants are $\mathcal{NP}$-complete. More recently, Fernau, Manea, Mercaş and Schmid [4] presented a similar problem that imposes a lower bound on the number of factors instead of an upper bound on factor length. Fernau et al. showed that this variant is also $\mathcal{NP}$-complete. Afterwards, Schmid [5] studied the Fixed-Parameter Tractability of the two problems. Henceforth, we use the notation of Schmid and refer to the problem variant with a lower bound on the number of factors as Maximum Equality-Free Factorization Size (MaxEFF-s).

Problem definitions. In this paper we consider the optimization version of the MaxEFF-s problem. Additionally, we consider a relaxed variant of the problem in which the factors do not necessarily cover the entire word, a so-called *gapped factorization*, that can be said to emerge from *generalized patterns* [2].

A *gapped factorization* of a string w over some alphabet Σ is a tuple of strings $(u_1, u_2, \ldots, u_k)$ such that $w = \alpha_0 u_1 \alpha_1 u_2 \alpha_2 \cdots \alpha_{k-1} u_k \alpha_k$ with $u_i \in \Sigma^+$ (non-empty substrings) and $\alpha_i \in \Sigma^*$ (possibly empty substrings).

First, let us go over the base definitions:

1. *factorization* of w is a tuple of strings $(u_1, u_2, \ldots, u_k)$ s.t. $w = u_1 u_2 \cdots u_k$.
2. *equality-free factorization* is a factorization with all distinct factors.
3. *gapped factorization* of string w over alphabet Σ is a tuple $(u_1, u_2, \ldots, u_k)$ such that $w = \alpha_0 u_1 \alpha_1 u_2 \alpha_2 \cdots \alpha_{k-1} u_k \alpha_k$, with $u_i \in \Sigma^+$ and $\alpha_i \in \Sigma^*$.

4. *size* of a factorization represents the number of factors.
5. *width* of a factorization represents the length of the longest factor.

Let us define the decision problem MaxEFF-s and its optimization version:

Problem 1 (Maximum Equality-Free Factorization Size - Decision). Does a given string admit an equality-free factorization of size at least k?

Problem 2 (Maximum Equality-Free Factorization Size - Optimization). For a given string w find the largest integer k, such that w admits an equality-free factorization of size k.

In the rest of the paper we refer to Problem 2 as OptEFF-s.

In the relaxed variant we allow gaps between the factors of the string.

Problem 3 (Maximum Gapped Equality-Free Factorization Size - Decision). Does a given string admit a gapped equality-free factorization of size at least k?

Problem 4 (Maximum Gapped Equality-Free Factorization Size - Optimization). For a given string w find the largest integer k, such that w admits a gapped equality-free factorization of size k.

To the best of our knowledge, the gapped version of the equality-free factorization problem has not been studied previously.

In the rest of the paper we refer to Problem 4 as OptGEFF-s.

Our results. We provide heuristic algorithms for computing equality-free factorizations and we also give an approximation ratio guarantee. Additionally, in order to understand how well the algorithms perform it is necessary to compare the solutions of our algorithms with optimum solutions. For this purpose, we choose to build ILP models for OptEFF-s and OptGEFF-s, which we use with the state-of-the-art Gurobi solver to obtain optimum solutions on moderate sized instances.

The paper is organized as follows. In Sect. 2 we introduce our notations and present some observations. One such observation is used to improve the previously known best FPT algorithm for MaxEFF-s (see Sect. 3). Following that, we design a $\frac{1}{2}$-approximation algorithm for OptGEFF-s in Sect. 4. We would have liked to extend this result for OptEFF-s, since we conjecture that the optimum of the two problems is the same. In Sect. 5 we provide an ILP model for both OptEFF-s and OptGEFF-s that was successfully used to give optimum solutions using the Gurobi solver. It is with this model that we have discovered the same optimum for the two problems on each of nearly 300000 instances, leading to the conjecture that their optimum is the same. The design of our proposed heuristic algorithms (dubbed Greedy$k$) is presented in Sect. 6. We prove a $\sqrt{OPT}$ approximation factor for Greedy1 and we give an example where Greedy1 has an approximation factor greater than $\log n$. We study the behavior of our algorithms on genomic data in Subsect. 6.3.

2 Preliminaries

We commonly use the notation $S = s_1 s_2 \dots s_n$ for a string of length n over some alphabet Σ. A substring of S is identified by $S[i..j] = s_i \dots s_j$ and has length $j - i + 1$.

Let there be a string w for which we are given an equality-free factorization of size k. Then, we can construct an equality-free factorization of size $k - 1$ by concatenating one of the longest factors with one of its neighbors. This leads us to the following observation:

Observation 1. *If a string w admits an equality-free factorization of size k, then w admits equality-free factorizations of size i, $\forall i \in \{1, \dots, k\}$.*

One of the implications of the previous observation is that, obtaining the solution of the optimization problem OptEFF-s for a given instance, provides the solutions for the decision problem MaxEFF-s on that respective instance and for all sizes.

Observation 2. *OptEFF-s(w) $\leq$ OptGEFF-s(w): For any input string w, an equality-free factorization of size k gives is also a solution for a gapped factorization of size k.*

Indeed, for $w = \alpha_0 u_1 \alpha_1 u_2 \alpha_2 \cdots \alpha_{k-1} u_k \alpha_k$, if we consider all the gaps α_i to be the empty string ϵ then all equality-free factorizations of size k are also gapped factorizations of size k.

Observation 3. *In a string of length n there exists an equality-free factorization of maximum size with width (i.e. length of the longest factor) at most $\lceil \sqrt{2n} \rceil$.*

The previous observation follows from the fact that, in the worst case, all the factors have different length: $1, 2, \dots, \ell$. This brings us to the well-known finite sum $n = \ell(\ell+1)/2$. Solving for ℓ shows that $width \leq \lceil (\sqrt{1+8n}-1)/2 \rceil \leq \lceil \sqrt{2n} \rceil$. We bring to the attention of the reader that this result is important for two reasons. First, we can use it to reduce the number of variables in our proposed ILP model in Sect. 5 from $O(n^2)$ to $O(n\sqrt{n})$, drastically improving solver computing speed. The second reason is that this result improves the previously known best FPT algorithm for MaxEFF-s.

3 A Better FPT Algorithm for MaxEFF-S

In [5] Schmid shows an FPT algorithm for deciding if a string of length n has an equality-free factorization with k factors with complexity $O((\frac{k^2+k}{2} - 1)^k)$. In this section we design another algorithm with running time $O((k^2 + k)^{\frac{k}{2}})$. The algorithm is similar to the one of Schmid, but uses Observation 3.

By Observation 3, there exists an optimum solution with width at most $\lceil \sqrt{2n} \rceil$. Thus, instead of an $O(n^k)$ algorithm to verify if a string of length n has a factorization with k factors, we obtain an algorithm with complexity $O((2n)^{\frac{k-1}{2}})$, by trying all the possible starting points of the $k-1$ factors (notice that the first factor always starts at position 1).

Finally, the running time of the FPT algorithm follows from the following observation:

Observation 4. *When $n \geq k(k+1)/2$, there always exists a factorization, which means that the problem has a trivial polynomial kernel.*

4 A $\frac{1}{2}$-Approximation Algorithm for OptGEFF-s

In this section we show that there exists a natural reduction from OptGEFF-s to the problem JISPk (the so-called Job Interval Selection Problem with k intervals per job). Moreover, this problem admits a $\frac{1}{2}$-approximation [6].

An instance of JISPk is a set of n k-tuples (also called jobs), containing time intervals. The intervals are of the form $[a, b)$ with a, b integers. Two time intervals $[a, b)$ and $[c, d)$ are said to intersect if $[a, b) \cap [c, d) \neq \emptyset$.

Problem 5 (JISPk). Given n jobs containing k time intervals each, find the maximum number of intervals that can be selected such that (i) no two intervals intersect and (ii) at most one time interval is selected per job.

Theorem 1. *An instance of OptGEFF-s can be transformed into an instance of JISPn, with the same optimum solution.*

Proof. We proceed to construct an instance of JISPn containing $O(n^2)$ jobs from a string w of length n.

Consider the factors in a gapped equality-free factorization of a string w of length n. They are a set of non-overlapping and distinct substrings of w. For each distinct substring of w (which is a possible factor) we create a job in the corresponding JISPn instance. For each job created from a substring s, we add as time intervals $[a, b)$ the start and end indices of all the occurrences $s = w[a \ldots b-1]$ of s in w. At this moment, we have created a set of jobs that are not n-tuples and therefore cannot be said to be a JISPn instance.

To obtain a JISPn instance, we simply pad each tuple by adding an appropriate number of duplicate intervals. This operation constructs an equivalent instance due to the observation that a JISP$(n+t)$ instance with the same optimum as a JISPn instance can be created by adding t duplicates of an arbitrarily selected interval within every job.

Since there are at most n occurrences of a substring in a string and there exist $O(n^2)$ distinct substrings in any given string, we have shown that we can construct a JISPn instance with $O(n^2)$ jobs, from any string of length n.

Moreover, a solution for the JISPn instance that is constructed in the manner described above immediately gives us a solution for OptGEFF-s. Each interval selected from a job corresponds to the occurrence of a factor in the initial string. The intervals are not allowed to intersect and thus the factors are not allowed to overlap. Only one interval may be selected per job and therefore only distinct factors may be selected because the jobs correspond to distinct substrings. As such, we conclude that we can reduce OptGEFF-s to JISPn. □

It is known that JISPn has the following greedy $\frac{1}{2}$-approximation algorithm [6]: at each step, select the time interval with the lowest end time that does not intersect already selected intervals. Using Theorem 1 we have shown that:

Theorem 2. *OptGEFF-s has a* $\frac{1}{2}$*-approximation algorithm.*

With the above results, we may now present a more tidy version of the greedy approximation algorithm for the OptGEFF-s: for each position $j = 1, 2, \ldots, n$ in a string w of length n, select as a factor (if possible) any substring s of w that ends on position j such that (i) s does not overlap the previously selected factor and (ii) s is not equal to any previously selected factor.

5 ILP Formulations for OptEFF-S and OptGEFF-S

We define an ILP model for the problems and then explain the notations:

$$\max \sum_{i=1}^{n} \sum_{j=1}^{n-i+1} x_{ij} \text{ subject to:} \tag{1a}$$

$$\sum_{j=1}^{n-i+1} x_{ij} \leq 1, \forall i = 1, 2, \ldots, n \tag{1b}$$

$$x_{ij} + x_{k\ell} \leq 1, \forall i = 1, 2, \ldots, n-1, \forall j = 1, 2, \ldots, n-i+1 \tag{1c}$$
$$\forall k = i+1, i+2, \ldots, i+j-1 \text{ and } k \leq n, \forall \ell = 1, 2, \ldots, n-k+1$$

$$x_{i\ell} + x_{k\ell} \leq 1, \forall i, k, \ell, \text{ where } S[i..i+\ell] = S[k..k+\ell] \tag{1d}$$

$$x_{ij} - \sum_{\ell=1}^{n-i-j+1} x_{(i+j)\ell} \leq 0, \forall i = 1, 2, \ldots, n-1, \forall j = 1, 2, \ldots, n-i+1 \tag{1e}$$

$$\sum_{\ell=1}^{n} x_{1\ell} \geq 1 \text{ and } \sum_{i=1}^{n} x_{i(n-i+1)} \geq 1 \tag{1f}$$

$$x_{ij} \in \{0, 1\}, \forall i = 1, 2, \ldots, n-1, \forall j = 1, 2, \ldots, n-i+1 \tag{1g}$$

1. The binary variables x_{ij} (see 1g) represent the choice for a factor starting on position i of length j.
2. We need to maximize the number of factors i.e. sum of x_{ij} (see 1a).
3. Only one factor (regardless of length) may begin on any position (see 1b).
4. Factors cannot overlap i.e. begin inside each other (see 1c).
5. Distinct factors: only one of the occurrences of a factor may be selected (1d).

If we want an equality-free gapped factorization, conditions 1–5 are enough. To enforce factorizations without gaps, we add:

Algorithm Greedy1: reads the input string w left-to-right and builds an equality-free factorization F by greedily adding the next shortest factor not yet present in F.
input: string $w[1..n]$; *output:* equality-free factorization F;

```
last ← 1, F ← ∅;
for i ← 1, n step 1 do
    if w[last..i] ∉ F then
        F ← F ∪ {w[last..i]}, last ← i + 1;
if last ≠ n + 1 then
    F ← (F \ {F[−1]}) ∪ {F[−1] · w[last..n]};
return F;
```

1. If a factor u_t is selected, then a factor u_{t+1} of any length that begins immediately after u_t ends must be selected (see 1e).
2. A factor starting on position 1 must be selected; a factor ending on the last position must be selected (see 1f).

Recall that by using Observation 3 we may reduce the number of variables in the formulation to $O(n\sqrt{n})$ by discarding x_{ij} with $j > \lceil\sqrt{2n}\rceil$.

6 Heuristic and Approximation Algorithms for OptEFF-S

In this section we present a family of heuristic greedy-based algorithms for the OptEFF-s problem. We begin with presenting the outline for the algorithms and then we evaluate the performance of the algorithms on datasets composed of randomly generated strings.

The proposed algorithms are based on building a factorization by reading the input left-to-right and greedily adding words to the incumbent solution.

6.1 Description of Greedy1

To illustrate the most basic strategy, consider starting "at the left" and adding the next shortest substring (distinct from the already selected factors) to the incumbent factorization at each step of the algorithm (see Algorithm Greedy1). The only issue is to define the behavior of the algorithm at the end of the string, where we may have a remainder that is already present as a factor in the working solution. We choose to simply concatenate this remainder to the last factor. We mention here that this algorithm is essentially identical to the well-known LZ'78 factorization procedure [7], excepting the handling of the last factor. To prove the correctness of our algorithm, the following property of Greedy1 is of interest:

Property 1 (Prefix property). Let there be two factors u_i and u_j in a factorization $w = u_1 u_2 \cdots u_n$ computed by Algorithm Greedy1. If u_i is a prefix of u_j then $i < j$. In other words, for any factor constructed by Greedy1, its prefixes precede it in the factorization.

Theorem 3. *Greedy1 yields an equality-free factorization in $O(n)$ time.*

Proof. By adding the next distinct substring at each step, the equality-free condition of the factorization is satisfied. The only question is if the behavior of the algorithm at the end of the string is correct or may yield a duplicate factor. If there is a remainder r at the end of the sequence of operations, then it has a duplicate in the factorization. Concatenating this remainder to the last factor v always produces the equality-free factor vr. To prove this we use Property 1: the resulting factor vr must have a duplicate preceding v in the factorization for the procedure to be incorrect. However, v is a prefix of vr and must appear before vr, a contradiction. Therefore the factorization is equality-free.

In the implementation of Greedy1, if the incumbent factorization is a list of starting/ending positions of factors and we use a hash set structure to check for collisions, the average running time is $O(nt)$, with t being the average factor length $t = \frac{n}{OPT}$ and OPT being the size of the optimum. This is because we compute a linear time string hash function n times (i.e. for each $\notin$ operation; see Algorithm Greedy1, line 3). We can optimize by changing the way we compute the hash function (by not discarding previous partial results) or by using a modified insertion in a trie structure instead of using hash sets in order to bring the time down to $O(n)$. □

Theorem 4. *The Greedy1 algorithm is a $\sqrt{OPT}$ approximation for OptEFF-s.*

Proof. A string with n characters can be factorized in at most n factors. The Greedy1 algorithm produces at least $\sqrt{n}$ factors—the case when all the factors have different length. □

In the following paragraphs we focus on the tightness of Greedy1 as an approximation algorithm. In practice Greedy1 can offer very good solutions as can be seen in Fig. 1 from Subsect. 6.3. However, we show that there exists an instance for which the ratio between Greedy1 and OPT is $\Omega(\log n)$. Therefore, in order to obtain a constant factor approximation for OptEFF-s we need to design a different algorithm.

Theorem 5. *Let there be an alphabet $\Sigma = \{x_1, x_2, \ldots, x_n\}$. We build a string $s = X_1 \cdot X_2 \cdots X_n$ by concatenating in order strings $X_i = x_1 x_2 \cdots x_i$. There exists a factorization of s with $\Omega(n \log n)$ factors.*

Proof. A factorization of s is as follows. We begin with X_n and factorize it into n factors: $x_1|x_2|\cdots|x_n$. At each iteration $1 \leq i \leq \lfloor n/2 \rfloor$ we factorize X_{n-i+1} into $\lfloor (n-i+1)/i \rfloor$ factors of length i in left-to-right order: $x_1 \cdots x_i | x_{i+1} \cdots x_{2i} | \cdots$. If X_{n-i+1} is not a multiple of i, then we concatenate the remainder of length $< i$ to the last factor. All of the factors added at iteration i are distinct, but a question remains about the correctness of the algorithm regarding the concatenated factor. Correctness is ensured because the new factor constructed in this manner contains x_{n-i+1}, symbol that cannot appear in subsequent iterations. In the end, the prefix $X_1 X_2 \cdots X_{\lceil n/2 \rceil}$ is transformed into one factor.

The number of factors in this solution is $n + \lfloor (n-1)/2 \rfloor + \lfloor (n-2)/3 \rfloor + \cdots + 1$ which is $\Omega(n + (n/2-1) + (n/3-1) + \cdots + 1) = \Omega(n(1 + 1/2 + 1/3 + \cdots + 1/n)) = \Omega(n \log n)$. □

On string s, Greedy1 produces n factors, i.e. the output factorization is $X_1|X_2|\cdots|X_n$. Using Theorem 5 we conclude that:

Corollary 1. *The approximation ratio of Greedy1 is* $\Omega(\log n)$.

6.2 Description of Greedy*k*

We generalize Algorithm Greedy1 into the family Greedy$k$ in the following way: instead of adding the next distinct substring to the incumbent factorization we consider adding k factors at the same time. In other words, we select the shortest substring that follows such that this substring admits a partition into k distinct factors that have not yet been selected (see Algorithms Greedy$k$ and Factorization). Again, the behavior of the algorithm needs to be defined when it is no longer possible to split a remaining substring into k factors and now it is a little more complicated because Greedy$k$ does not benefit from Property 1:

1. Greedy$k$ will first attempt to partition the remainder r into t distinct and not yet selected factors with $t = k-1, \ldots, 1$.
2. If this is not possible, then we try to discard the last k factors in the factorization $u_i, u_{i-1}, \ldots, u_{i-k+1}$ and re-partition the entire substring $u_{i-k+1} \cdots u_i r$ into t valid factors, $t = k, k-1, \ldots, 1$. This step is added to the algorithm to ensure that Greedy$k$ will always produce an optimum solution when the optimum is known to be k.
3. Finally, if the above steps fail, we append the remainder r to the last factor and proceed to concatenate the last two factors in the solution until duplicates no longer exist.

Function Factorization(*T,F,start,end,w,k*): returns a factorization F with exactly k factors of the substring $w[start..end]$, or *null* if no such factorization is possible. Moreover, the factors in F must not already appear in the incumbent factorization T.
input: factorizations T, F(pre-initialized), string $w[1..n]$, int *start*, int *end*, int k;
output: factorization F (modified from the input F);

```
if end − start + 1 < k then
    return null;
if k = 1 and w[start..end] ∉ T and w[start..end] ∉ F then
    F ← F ∪ {w[start..end]};
    return F;
for i ← start, end − 1 step 1 do
    if w[start..i] ∉ T and w[start..i] ∉ F then
        solution ← Factorization(T, F ∪ {w[start..i]}, i + 1, end, w, k − 1);
        if solution ≠ null then
            return solution
return null;
```

Theorem 6. *The Factorization(T,F,start,end,w,k) recursive function obtains a factorization of size* k*, if it exists in the specified substring of length* ℓ*, in time* $O(\ell^k)$.

Proof. If we denote the length of the substring as $\ell = end - start + 1$, we can observe that we test all of the $\binom{\ell-1}{k-1} = O(\ell^{k-1})$ partitions. The time necessary for the $\notin$ operation depends on the structure used to check for substring collisions. Whether we employ a hash set and compute a hash function or use a trie, the time required is $O(\ell)$ i.e. linear in substring size. Observe that when $k = 1$ the function takes $O(\ell)$ time and that by increasing k by 1, the time is multiplied by $O(\ell)$. By simple induction, the time complexity of the function is $O(\ell^k)$. □

Theorem 7. *Greedyk is correct and runs in $O((n+k)\ell^k)$ average time, with $\ell = k\frac{n}{OPT}$ and OPT being the value of the optimum solution.*

Proof. The equality-free condition of the factorization is satisfied when k distinct factors are added at each step (lines 2–5 in the pseudocode). When it is no longer possible, a substring r may remain:

1. First we try to split r into fewer than k factors (see lines 7–10). If we are successful, then the resulting factorization is equality-free and covers the entire string and is therefore correct.
2. Secondly, we try to discard the last k factors and refactorize the entire remaining substring into k factors or less (see lines 11–14). If we are successful, the resulting factorization is correct.

Algorithm Greedy*k*: reads the input string w from left to right and builds a Factorization F by greedily adding k distinct, not yet selected factors at each step. We use the notations *newEmptyFactorization* to denote creation of a new factorization structure and $F[-1]$ to refer to the last factor inside the factorization structure F.
input: string $w[1..n]$, int k; *output:* Factorization F;

```
1: prev ← 1, last ← 1, F ← ∅, solution ← ∅, lastsol ← ∅, sol2 ← ∅;
2: for i ← 1, n step 1 do
3:     solution ← Factorization(F, newEmptyFactorization, last, i, w, k);
4:     if solution ≠ null then
5:         F ← F ∪ solution, prev ← last, last ← i + 1, lastsol ← solution;
6: if last ≠ n + 1 then
7:     for j ← k − 1, 1 step −1 do
8:         solution ← Factorization(F, newEmptyFactorization, last, n, w, j);
9:         if solution ≠ null then
10:            return F ∪ solution;
11:    for j ← k, 1 step −1 do
12:        sol2 ← Factorization(F \ lastsol, newEmptyFactorization, prev, n, w, j);
13:        if sol2 ≠ null then
14:            return (F \ lastsol) ∪ sol2;
15:    lastFactor ← F[−1], F ← F \ {F[−1]}, lastFactor ← lastFactor · w[last..n];
16:    while lastFactor ∈ F do
17:        lastFactor ← F[−1] · lastFactor;
18:        F ← F \ {F[−1]};
19:    F ← F ∪ {lastFactor};
20: return F;
```

3. Thirdly we employ a fallback where we append the remainder r to the last factor and keep concatenating the last two factors until the factorization is equality-free (see lines 15–19).

In the implementation of Greedy$k$, we traverse the string and attempt to partition a substring of length ℓ into k valid factors. This operation takes $O(\ell^k)$ time (see Theorem 6). There are $O(n+k)$ calls to the partitioning function, therefore the average running time for the algorithm is bounded by $O((n+k)\ell^k)$, with $\ell = k\frac{n}{OPT}$.

6.3 Experimental Results

First, we want to determine the solution quality given by the Greedy$k$ algorithms in practice. The dataset we have selected for this experiment is the RNA string of Saccharomyces cerevisiae narnavirus 23S (obtained from *yeastgenome.org*).

The methodology for the experiment is as follows. For each integer $\ell \in \{4,\ldots,512\}$ we randomly select 10 substrings of length ℓ from the input RNA string (whose length is 2891). We compute the optimum solution for each substring using the ILP formulations in Sect. 5 and the Gurobi solver. We proceed to compute the value of the $\frac{1}{2}$-approximation algorithm for OptGEFF-s from Sect. 4, as well as Greedy$\{1,\ldots,8\}$. Following that, we average the results among the 10 substrings for each length ℓ and plot the fractions $\frac{\frac{1}{2}\text{-approximation}}{\text{OPT}}$, $\frac{\text{Greedy1}}{\text{OPT}}$ and $\frac{\max(\{\text{Greedy1},\ldots,\text{Greedy8}\})}{\text{OPT}}$ in Fig. 1.

We are pleased to report Greedy1 situating within 91% of the optimum (alongside the $\frac{1}{2}$-approximation) and max({Greedy1, ..., Greedy8}) placing within 93% of the optimum. All solutions are within 94% of the optimum on lengths 350–512. In Fig. 2 we also display the running time of Greedy$k$ on longer strings (using whole genomes from *yeastgenome.org* including Saccharomyces bayanus and Saccharomyces cerevisiae). The experiments demonstrate that the Greedy$k$ algorithms are fit for practical usage.

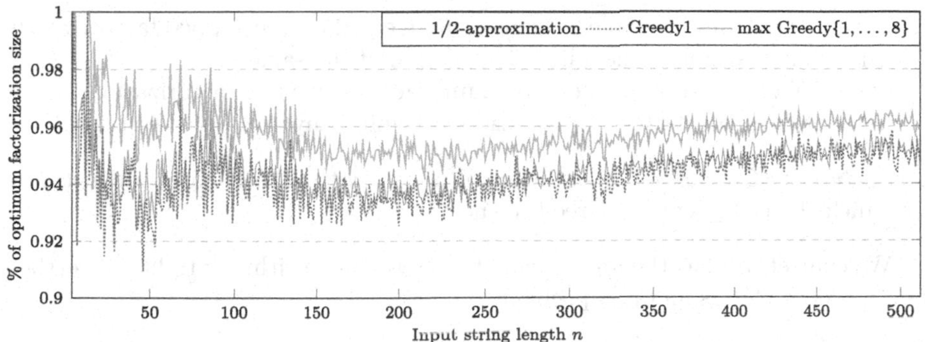

Fig. 1. Plot describing the values of the $\frac{1}{2}$-approximation algorithm for OptGEFF-s, Greedy1, and max({Greedy1, ..., Greedy8}) for all integer string lengths $\in \{4,\ldots,512\}$. All y-axis values are averages of 10 instances per x-axis point, as well as being divided by the average of the optima of the 10 instances.

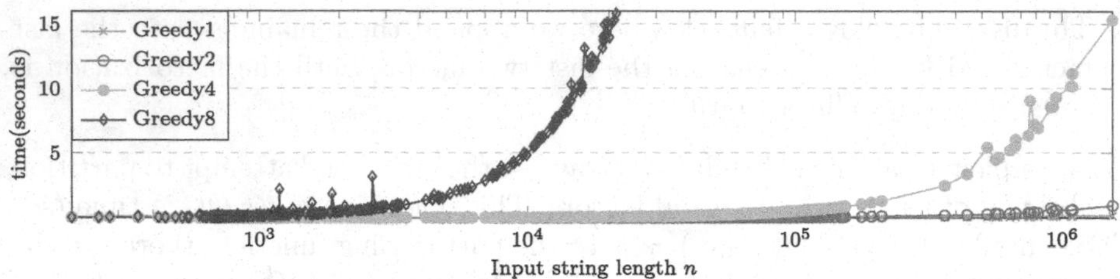

Fig. 2. The running time of Greedy$\{1, 2, 4, 8\}$. The x-axis is logarithmic scaled.

During our testing, we have computed exact solutions using our ILP model for both OptEFF-s and OptGEFF-s on each test instance. We hypothesized that OptGEFF-s would have a higher value solution than OptEFF-s. However, after having evaluated some 300.000 random instances (using various lengths and alphabet size), we observed no difference between the two problems regarding the size of the exact solutions on the same instance. Thus, we believe that the two problems share the same optimum.

7 Conclusions and Open Problems

We have presented heuristic and approximation algorithms for the OptEFF-s and OptGEFF-s problems. Moreover, our experiments show insights into the nature of the problem and provide high quality solutions.

We leave as open problems the following conjectures:

Conjecture 1 (Gaps = No Gaps). OptGEFF-s(w) = OptEFF-s(w), $\forall w$.

We strongly believe that the optimum for the two problems OptGEFF-s and OptEFF-s is one and the same. This result can be achieved if one of the following statements is proven:

1. Given a string, it is possible to transform an equality-free gapped factorization of maximum size into one without gaps and of the same size.
2. The number of factors in the maximum size equality-free factorization for an instance does not decrease if we insert a symbol anywhere in the string.

Conjecture 2 (Greedy1 approximation ratio is tight). There exists an instance for which the ratio between Greedy1 and OPT is $\Theta(\sqrt{n})$.

We conjecture that the analysis of the Greedy1 algorithm is tight. Nevertheless, we leave this as an open problem.

References

1. Bulteau, L., Hüffner, F., Komusiewicz, C., Niedermeier, R.: Multivariate algorithmics for NP-hard string problems. Bull. EATCS **114**, 295–301 (2014)
2. Clifford, R., Harrow, A.W., Popa, A., Sach, B.: Generalised matching. In: Karlgren, J., Tarhio, J., Hyyrö, H. (eds.) SPIRE 2009. LNCS, vol. 5721, pp. 295–301. Springer, Heidelberg (2009). https://doi.org/10.1007/978-3-642-03784-9_29
3. Condon, A., Maňuch, J., Thachuk, C.: The complexity of string partitioning. J. Discrete Algorithms **32**, 24–43 (2015)
4. Fernau, H., Manea, F., Mercas, R., Schmid, M.L.: Pattern matching with variables: fast algorithms and new hardness results. In: 32nd International Symposium on Theoretical Aspects of Computer Science, 4–7 March 2015, Garching, Germany, pp. 302–315 (2015)
5. Schmid, M.L.: Computing equality-free and repetitive string factorisations. Theor. Comput. Sci. **618**, 42–51 (2016)
6. Spieksma, F.: On the approximability of an interval scheduling problem. J. Sched. **2**(5), 215–227 (1999)
7. Ziv, J., Lempel, A.: Compression of individual sequences via variable-rate coding. IEEE Trans. Inf. Theory **24**, 530–536 (1978)

References

Foundations of Computer Science – Short Papers

A Calculus for Language Transformations

Benjamin Mourad(✉) and Matteo Cimini

University of Massachusetts Lowell, Lowell, MA 01854, USA
benjamin_mourad@student.uml.edu, matteo_cimini@uml.edu

Abstract. In this paper we propose a calculus for expressing algorithms for programming languages transformations. We present the type system and operational semantics of the calculus, and we prove that it is type sound. We have implemented our calculus, and we demonstrate its applicability with common examples in programming languages. As our calculus manipulates inference systems, our work can, in principle, be applied to logical systems.

1 Introduction

Operational semantics is a standard de facto to defining the semantics of programming languages [10]. However, producing a programming language definition is still a hard task. It is not surprising that theoretical and software tools for supporting the modeling of languages based on operational semantics have received attention in research [4,5,11]. In this paper, we address an important aspect of language reuse which has not received enough attention: Producing language definitions from existing ones by the application of transformation algorithms. Such algorithms may automatically add features to the language, or switch to different semantics styles. In this paper, we aim at providing theoretical foundations and a software tool for this aspect.

Consider the typing rule of function application below on the left and its version with algorithmic subtyping on the right.

$$\text{(T-APP)}\ \frac{\Gamma \vdash e_1 : T_1 \to T_2 \qquad \Gamma \vdash e_2 : T_1}{\Gamma \vdash e_1\, e_2 : T_2} \quad \overset{f(\text{T-APP})}{\Longrightarrow} \quad \text{(T-APP')}\ \frac{\Gamma \vdash e_1 : T_{11} \to T_2 \qquad \Gamma \vdash e_2 : T_{12} \qquad T_{12} <: T_{11}}{\Gamma \vdash e_1\, e_2 : T_2}$$

Intuitively, we can describe (T-APP') as a function of (T-APP). Such a function includes, at least, giving new variable names when a variable is mentioned more than once, and must relate the new variables with subtyping according to the variance of types (covariant vs contravariant). Our question is: *Can we express, easily, language transformations in a safe calculus?*

Language transformations do not apply just to one language but to several languages. They are beneficial because they can alleviate the burden to language

A. Chatzigeorgiou et al. (Eds.): SOFSEM 2020, LNCS 12011, pp. 547–555, 2020.
https://doi.org/10.1007/978-3-030-38919-2_44

designers, who can use them to automatically generate new language definitions rather than manually defining them, an error prone endeavor.

In this paper, we make the following contributions.

- We present $\mathcal{L}$–Tr (pronounced "Elter"), a formal calculus for language transformations (Sect. 2). We define the syntax (Sect. 2.1), operational semantics (Sect. 2.2), and type system (Sect. 2.3) of $\mathcal{L}$–Tr.
- We prove that $\mathcal{L}$–Tr is type sound (Sect. 2.3).
- We show the applicability of $\mathcal{L}$–Tr to the specification of two transformations: adding subtyping and switching from small-step to big-step semantics (Sect. 3). Our examples show that $\mathcal{L}$–Tr is expressive and offers a rather declarative style to programmers.
- We have implemented $\mathcal{L}$–Tr [8], and we report that we have applied our transformations to several language definitions.

In this paper, we show selected parts of our formalism, and we omit proofs. The full development can be found in the corresponding technical report [7].

2 A Calculus for Language Transformations

We focus on language definitions in the style of operational semantics. To briefly summarize, languages are specified with a BNF grammar and a set of inference rules. BNF grammars have *grammar productions* such as Types $T ::= B \mid T \rightarrow T$. We call Types a *category name*, T is a *grammar meta-variable*, and B and $T \rightarrow T$, as well as, for example, $(\lambda x.e\ v)$, are *terms*. $(\lambda x.e\ v) \longrightarrow e[v/x]$ and $\Gamma \vdash (e_1\ e_2) : T_2$ are *formulae*. An *inference rule* has a set of formulae above the horizontal line, which are called *premises*, and a formula below, the *conclusion*.

2.1 Syntax of $\mathcal{L}$–Tr

Below we show the $\mathcal{L}$–Tr syntax for language definitions.
$cname \in \textsc{CatName}, X \in \textsc{Meta-Var}, opname \in \textsc{OpName}, predname \in \textsc{PredName}$

$$
\begin{array}{ll}
\textsf{Language} & \mathcal{L} ::= (G, R) \\
\textsf{Grammar} & G ::= \{s_1, \ldots, s_n\} \\
\textsf{Grammar Pr.} & s\ ::= cname\ X ::= lt \\
\textsf{Rule} & r\ ::= \dfrac{lf}{f} \\
\textsf{Formula} & f\ ::= predname\ lt \\
\textsf{Term} & t\ ::= X \mid opname\ lt \mid (X)t \mid t[t/X] \\
\textsf{List of Rules} & R ::= \texttt{nil} \mid \textsf{cons}\ r\ R \\
\textsf{List of Formula} & lf\ ::= \texttt{nil} \mid \textsf{cons}\ f\ lf \\
\textsf{List of Terms} & lt\ ::= \texttt{nil} \mid \textsf{cons}\ t\ lt
\end{array}
$$

OpName contains elements such as $\rightarrow$ and λ. PredName contains elements such as $\vdash$ and $\longrightarrow$. We assume that terms and formulae are defined in abstract syntax tree fashion, i.e., they have a top level constructor applied to a list of terms. $\mathcal{L}$–Tr also provides syntax for unary binding $(X)t$ and capture-avoiding substitution $t[t/X]$. $\mathcal{L}$–Tr is then more suited to model static scoping. Lists can be built as usual with the `nil` and `cons` operator.

Below we show the rest of the syntax of $\mathcal{L}$–Tr.

Expression	$e ::= x \mid \mathit{cname} \mid \mathit{str} \mid \hat{t} \mid \hat{f} \mid \hat{r}$
	$\mid \mathtt{nil} \mid \mathtt{cons}\ e\ e \mid \mathtt{head}\ e \mid \mathtt{tail}\ e \mid e@e$
	$\mid \mathtt{map}(e,e) \mid e(e) \mid \mathtt{mapKeys}\ e$
	$\mid \mathtt{just}\ e \mid \mathtt{nothing} \mid \mathtt{get}\ e$
	$\mid \mathit{cname}\ X ::= e \mid \mathit{cname}\ X ::= \ldots\ e$
	$\mid \mathtt{getRules} \mid \mathtt{setRules}\ e$
	$\mid e[p] : e \mid e(\mathtt{keep})[p] : e \mid \mathtt{uniquefy}(e,e,\mathit{str}) \Rightarrow (x,x) : e$
	$\mid \mathtt{if}\ b\ \mathtt{then}\ e\ \mathtt{else}\ e \mid e\ ;\ e \mid e ;_{\mathsf{r}} e \mid \mathtt{skip}$
	$\mid \mathtt{newVar} \mid e\text{'} \mid \mathtt{fold}\ \mathit{predname}\ e$
	$\mid \mathtt{error}$
Boolean Expr.	$b ::= e == e \mid \mathtt{isEmpty}\ e \mid e\ \mathtt{in}\ e \mid \mathtt{isNothing}\ e \mid b\ \mathtt{and}\ b \mid b\ \mathtt{or}\ b \mid \mathtt{not}\ b$
$\mathcal{L}$–Tr Rule	$\hat{r} ::= \frac{e}{e}$
$\mathcal{L}$–Tr Formula	$\hat{f} ::= \mathit{predname}\ e \mid x\ e$
$\mathcal{L}$–Tr Term	$\hat{t} ::= X \mid \mathit{opname}\ e \mid x\ e \mid (X)e \mid e[e/X]$
Pattern	$p ::= x : T \mid \mathit{predname}\ p \mid \mathit{opname}\ p \mid x\ p \mid \mathtt{nil} \mid \mathtt{cons}\ p\ p$
Value	$v ::= t \mid f \mid r \mid \mathit{cname} \mid \mathit{str}$
	$\mid \mathtt{nil} \mid \mathtt{cons}\ v\ v \mid \mathtt{map}(v,v) \mid \mathtt{just}\ v \mid \mathtt{nothing} \mid \mathtt{skip}$

Design Principles: We strive to offer well-crafted operations that map well with language manipulations. There are three features that exemplify our approach the most: (1) selectors $e[p] : e$, (2) the `uniquefy` operation, and (3) the ability to program parts of rules, premises and grammars. We shall describe these feauters, among others, below.

$\mathcal{L}$–Tr has strings, lists, maps and options with their typical operators. The expression $\mathit{cname}\ X ::= e$ augments the current grammar with a new production. $\mathit{cname}\ X ::= \ldots\ e$ (notice the dots) adds the grammar items in the list e to an existing production. `getRules` and `setRules` e retrieve and set the current list of rules, respectively.

Selectors. $e_1[p] : e_2$ selects one by one the elements of the list e_1 that satisfy the pattern p and executes the body e_2 for each of them. It returns a list that collects the result of each iteration. Selectors are useful for selecting elements of a language with great precision. Suppose that we wanted to invert the direction of all subtyping premises in *prems*, $\mathit{prems}[T_1 <: T_2]$: `just` $T_2 <: T_1$ would do just that. Notice the use of options. As we shall see in Sect. 3, it is common for iterations to return `nothing`, and options are then suited. Selectors handle options automatically and remove `nothing`s, so the selector above returns a list of premises rather than a list of options. $e(\mathtt{keep})[p] : e$ works like a selector except that it also returns the elements that failed the pattern-matching.

`uniquefy` is useful because it is often necessary to assign distinct variables when transforming languages. Therefore, $\mathcal{L}$–Tr provides a specific operation for this. $\mathtt{uniquefy}(e_1, e_2, \mathit{str}) \Rightarrow (x, y) : e_3$ takes in input a list of formulae e_1, a map e_2, and a string *str* (we shall discuss x, y, and e_3 shortly). This operation modifies the formulae in e_1 to use different variable names when a variable is mentioned more than once. However, not every variable is subject to the replacement. Only the variables that appear in some specific positions are targeted. The map e_2 and the string *str* contain the information to identify these positions. e_2 maps operator names and predicate names to a list that contains a label (as a string) for each of their arguments. For example, the map $m = \{\vdash \mapsto [\text{"}in\text{"}, \text{"}in\text{"}, \text{"}out\text{"}]\}$ says that Γ and e are inputs in $\Gamma \vdash e : T$, and that T is the output. $\mathcal{L}$–Tr inspects the formulae in e_1 and their terms. Arguments that correspond to the label specified by *str* receive a new variable. To make an example, if *lf* is the list of premises of (T-APP) and m is defined as above, the operation $\texttt{uniquefy}(\mathit{lf}, m, \text{"}out\text{"}) \Rightarrow (x, y) : e_3$ creates the premises of (T-APP') shown in the introduction. The computation continues with the expression e_3 in which x is bound to the new premises and y is bound to a map that summarizes the changes made by `uniquefy`, which associates variables X to the new variables that `uniquefy` has used to replace X. In our example we have the map $\{T_1 \mapsto [T_{11}, T_{12}]\}$ passed to e_3 as y.

Programming rules, premises, and terms is possible in $\mathcal{L}$–Tr thanks to $\mathcal{L}$–Tr terms ($\hat{t}$), $\mathcal{L}$–Tr formulae ($\hat{f}$), and $\mathcal{L}$–Tr rules ($\hat{r}$), which can contain arbitrary expressions, such as if-then-else statements, at any position. This provides a declarative way to create languages, as we shall see in Sect. 3.

The operation `fold` *predname* e creates a list of formulae that interleaves *predname* to any two subsequent elements of the list e. To make an example, the operation $\texttt{fold} = [T_1, T_2, T_3, T_4]$ generates the list of formulae $[T_1 = T_2, T_2 = T_3, T_3 = T_4]$. $\texttt{vars}(e)$ returns the list of the meta-variables in e. `newVar` returns a meta-variable that has not been previously used. The tick operator e' gives a prime to the meta-variables of e_1 (X becomes X'). Some variables have a special treatment in $\mathcal{L}$–Tr. We can refer to the value that a selector iterates over with the variable *self*. If we are in a context that manipulates a rule, we can also refer to the premises and conclusion with variables *premises* and *conclusion*.

The next two sections present the rest of our formalism. We recall that we only show selected parts. The full formalism can be found in [7].

2.2 Operational Semantics of $\mathcal{L}$–Tr

In this section we show a small-step operational semantics for $\mathcal{L}$–Tr. A configuration is denoted with $V; \mathcal{L}; e$, where e is an expression, $\mathcal{L}$ is the language subject of the transformation, and V contains the variables generated by `newVar`. The main reduction relation is $V; \mathcal{L}; e \longrightarrow V'; \mathcal{L}'; e'$, defined as follows. Evaluation contexts E can be found in [7].

$$\frac{V; \mathcal{L}; e \longrightarrow_{@} V'; \mathcal{L}'; e' \quad \vdash \mathcal{L}'}{V; \mathcal{L}; E[e] \longrightarrow V'; \mathcal{L}'; E[e']} \qquad \frac{V; \mathcal{L}; e \longrightarrow_{@} V'; \mathcal{L}'; e' \quad \not\vdash \mathcal{L}'}{V; \mathcal{L}; E[e] \longrightarrow V; \mathcal{L}; \texttt{error}}$$

This relation relies on a step $V; \mathcal{L}; e \longrightarrow_{@} V'; \mathcal{L}'; e'$. $\vdash \mathcal{L}'$ checks that ill-formed elements such as $\vdash T\ T$ have not been inserted. Below, we show the most relevant rules for $V; \mathcal{L}; e \longrightarrow_{@} V'; \mathcal{L}'; e'$.

$$\frac{match(v_1, p) = \theta \qquad \theta' = \begin{cases} \theta^{(r)}_{\mathsf{rule}} & \text{if } v_1 = r \\ \{self \mapsto v_1\} & \text{otherwise} \end{cases}}{V; \mathcal{L}; (\mathtt{cons}\ v_1\ v_2)[p] : e \longrightarrow_{@} V; \mathcal{L}; (cons^*\ e\theta\theta'\ (v_2[p] : e))}$$

(R-SELECTOR-CONS-OK)

$$V; \mathcal{L}; r;_{\mathsf{r}} e \longrightarrow_{@} V; \mathcal{L}; e\theta^{(r)}_{\mathsf{rule}} \qquad \text{(R-RULE-COMP)}$$

$$\frac{X' \notin V \cup vars(\mathcal{L}) \cup range(tick)}{V; (G, R); \mathtt{newVar} \ \longrightarrow_{@} V \cup \{X'\}; \mathcal{L}; X'} \qquad \text{(R-NEWVAR)}$$

$$\frac{(lf', v_2) = uniquefy_{\mathsf{lf}}(lf, v_1, str, \mathtt{map}([], []))}{V; \mathcal{L}; \mathtt{uniquefy}(lf, v_1, str) \Rightarrow (x, y) : e \longrightarrow_{@} V; \mathcal{L}; e[lf'/x, v_2/y]}$$

(R-UNIQUEFY-OK)

(R-SELECTOR-CONS-OK) makes use of the meta-operation $match(v_1, p) = \theta$. If this operation succeeds it returns the substitutions θ with the associations computed during pattern-matching. The body is evaluated with these substitutions and with *self* instantiated. If the element selected is a rule, then the body is instantiated with $\theta^{(r)}_{\mathsf{rule}}$ to refer to that rule as the current rule. If the premises of r are v_1, and the conclusion is v_2 then $\theta^{(r)}_{\mathsf{rule}} \equiv [r/self, v_1/premises, v_2/conclusion]$. The selector returns an option type but the special $cons^*$ discards `nothing`s and unwraps `just`s. (R-RULE-COMP) applies when the first expression has been evaluated to a rule r, and starts the evaluation of the second expression in which r is set as the current rule with $\theta^{(r)}_{\mathsf{rule}}$. (R-NEWVAR) returns a new meta-variable and augments V with it. Meta-variables are chosen among those that are not in the language, have not previously been generated by `newVar`, and also do not clash with ticked variables. (R-UNIQUEFY-OK) defines the semantics for `uniquefy`. It relies on the meta-operation $uniquefy_{\mathsf{r}}(lf, v, str, \mathtt{map}([], []))$, which takes the list of formulae lf, the map v, the string str, and an empty map to start computing the result map. $uniquefy_{\mathsf{r}}$ is a recursive traversal of formuale and terms that seeks variables at the specified positions. This function returns a pair (lf', v_2) where lf' is the modified list of formulae and v_2 maps meta-variables to the new meta-variables that have replaced it.

2.3 Type System of $\mathcal{L}$–Tr

Types and the type environment of $\mathcal{L}$–Tr are defined as follows.

$$\begin{array}{lll} \mathsf{Type} & T ::= & \mathtt{Language} \mid \mathtt{Rule} \mid \mathtt{Formula} \mid \mathtt{Term} \\ & & \mathtt{List}\ T \mid \mathtt{Map}\ T\ T \mid \mathtt{Option}\ T \mid \mathtt{String} \mid \mathtt{OpName} \mid \mathtt{PredName} \\ \mathsf{Type\ Env} & \Gamma ::= & \emptyset \mid \Gamma, x : T \end{array}$$

The typing judgement $\vdash V; \mathcal{L}; e$ type checks configurations.

$$\frac{V \cap vars(\mathcal{L}) = \emptyset \qquad \vdash \mathcal{L} \qquad \emptyset \vdash e : \mathtt{Language}}{\vdash V; \mathcal{L}; e}$$

This judgment checks that the variables of V and those in $\mathcal{L}$ are disjoint. We also check that $\mathcal{L}$ is well-typed and that e is of type $\mathbf{Language}$. The typing judgement $\Gamma \vdash e : T$ means that e has type T under the assignments in Γ.

(T-SELECTOR)

$$\frac{\Gamma \vdash e_1 : \texttt{List}\ T \quad \Gamma \vdash p : T \Rightarrow \Gamma' \qquad \Gamma'' = \begin{cases} \Gamma_{\mathsf{rule}} & \text{if } T = \texttt{Rule} \\ \mathit{self} : T & \text{otherwise} \end{cases} \quad \Gamma, \Gamma', \Gamma'' \vdash e_2 : \texttt{Option}\ T'}{\Gamma \vdash e_1[p] : e_2 : \texttt{List}\ T'}$$

(T-RULE-COMP)

$$\frac{\Gamma \vdash e_1 : \texttt{Rule} \qquad \Gamma, \Gamma_{\mathsf{rule}} \vdash e_2 : \texttt{Rule}}{\Gamma \vdash e_1 ;_{\mathsf{r}} e_2 : \texttt{Rule}}$$

(T-UNIQUEFY)

$$\frac{\Gamma \vdash e_1 : \texttt{List Formula} \quad \Gamma \vdash e_2 : \texttt{Map}\ T'\ (\texttt{List String}) \quad T' \in \{\texttt{OpName}, \texttt{PredName}\} \quad \Gamma, x : \texttt{List Formula}, y : \texttt{Map Term}\ (\texttt{List Term}) \vdash e_3 : T}{\Gamma \vdash \texttt{uniquefy}(e_1, e_2, \mathit{str}) \Rightarrow (x, y) : e_3 : T}$$

(T-SELECTOR) type checks a selector operation. We use $\Gamma \vdash p : T \Rightarrow \Gamma'$ to type check the pattern p and return the type environment for the variables of the pattern. Its definition is standard and is omitted. When we type check the body e_2 we then include Γ'. If the elements of the list are rules then we also include Γ_{rule}, where $\Gamma_{\mathsf{rule}} \equiv \mathit{self} : \texttt{Rule}, \mathit{premises} : \texttt{List Formula}, \mathit{conclusion} : \texttt{Formula}$. This gives a type to the variables that refer to the current rule. Otherwise, we assign *self*. (T-RULE-COMP) type checks a rule composition. In doing so, we type check e_2 with Γ_{rule}. (T-UNIQUEFY) type checks the `uniquefy` operation. The keys of the map are of type `OpName` or `PredName`, and its values are strings. We type check e_3 giving x the type of list of formulae, and y the type of a map from meta-variables to list of meta-variables.

We have proved that $\mathcal{L}$–Tr is type sound.

Theorem 1 (Type Soundness). *For all $V, \mathcal{L}, e$, if $\vdash V; \mathcal{L}; e$ then $V; \mathcal{L}; e \longrightarrow^* V'; \mathcal{L}'; e'$ s.t. (i) $e' = \texttt{skip}$, (ii) $e' = \texttt{error}$, or (iii) $V'; \mathcal{L}'; e' \longrightarrow V''; \mathcal{L}''; e''$.*

3 Examples

We show the applicability of $\mathcal{L}$–Tr with two examples of language transformations: adding subytyping [9] and switching to big-step semantics [6]. In the code we use let-binding and pattern-matching, which can be easily derived ([7]). The code below defines the transformation for adding subtyping. We assume that two maps are already defined, $\mathit{mode} = \{\vdash \mapsto [\text{“}\mathit{inp}\text{”}, \text{“}\mathit{inp}\text{”}, \text{“}\mathit{out}\text{”}]\}$ and $\mathit{variance} = \{\rightarrow\ \mapsto [\text{“}\mathit{contra}\text{”}, \text{“}\mathit{cova}\text{”}]\}$.

$\texttt{getRules}(\texttt{keep})[(\vdash [\Gamma, e, T])]$:

$\texttt{uniquefy}(\mathit{premises}, \mathit{mode}, \text{“}\mathit{out}\text{”}) \Rightarrow (\mathit{uniq}, \mathit{newpremises})$:

$$\frac{\mathit{newpremises}\ @\ \texttt{concat}(\texttt{mapKeys}(\mathit{uniq})[T_f] : \texttt{fold} <: \mathit{uniq}(T_f))}{\mathit{conclusion}}$$

$;_{\mathsf{r}}$

$\texttt{concat}(\mathit{premises}(\texttt{keep})[T_1 <: T_2]$:

$\quad \mathit{premises}[(\vdash [\Gamma, e_v, (c_v\ Ts_v)])]$:

```
8    let vmap = map(Ts_v, variance(c_v)) in
9    if vmap(T_1) = "contra" then T_2 <: T_1
10   else if vmap(T_1) = "inv" and vmap(T_2) = "inv" then T_1 = T_2 else T_1 <: T_2)
     ---------------------------------------------------------------------------------
                                   conclusion
```

Line 1 selects all typing rules, and each of them will be the subject of the transformations in lines 2–10. Line 2 calls **uniquefy** on the premises of the selected rule. We instruct **uniquefy** to give new variables of the typing relation $\vdash$, if they are used more than once in output position. As previously described, **uniquefy** returns the list of new premises, which we bind to *newpremises*, and the map that assigns variables to the list of the new variables generated to replace them, which we bind to *uniq*. The body of **uniquefy** goes from line 3 to 10. Lines 3 and 4 add subtyping premises to the selected rule. The conclusion is left unchanged, with variable *conclusion*. The premises of this rule include the premises just generated by **uniquefy**. Furthermore, we add premises computed as follows. With $\texttt{mapKeys}(uniq)[T_f]$, we iterate over all the variables replaced by **uniquefy**. We take the variables that replaced them and use fold to relate them all with subtyping. In other words, for each $\{T \mapsto [T_1, \ldots, T_n]\}$ in *uniq*, we have the formulae $T_1 <: T_2, \ldots, T_{n-1} <: T_n$. This transformation has created unique outputs and subtyping, but subtyping may be incorrect because if some variable is contravariant its corresponding subtyping premise should be swapped. Lines 6–10, then, adjust the subtyping premises based on the variance of types. Line 6 selects all subtyping premises of the form $T_1 <: T_2$. For each, Line 7 selects typing premises with output of the form $(c_v\ Ts_v)$. We do so to understand the variance of variables. If the first argument of c_v is contravariant, for example, then the first element of Ts_v warrants a swap in a subtyping premise because it is used in contravariant position. We do this by creating a map that associates the variance to each argument of c_v (line 8). The information about variance for c_v is in *variance*. If T_1 or T_2 (from the pattern at line 6) appear in Ts_v then they find themselves with a variance assigned in *vmap*. Lines 9–10 generate a new premise based on the variance of variables. For example, if T_1 is contravariant then we generate $T_2 <: T_1$.

Below, we show the code to turn language definitions into big-step semantics.

```
1 Value[v] : v ⟶ v @
2 getRules(keep)[(op es) ⟶ et] :
3 if isEmpty(Expression[(op _)] : self) then nothing else
4 let v_res = newVar in
5 let emap = createMap((es[e] : newVar), es) in
6   (mapKeys(emap)[e] : if isVar(emap(e)) and not(emap(e) in vars(et))
7                          then nothing else e ⟶ emap(e))
8   @ (if not(et in es) then [(et ⟶ v_res)] else nil) @ premises
  ------------------------------------------------------------------------
9   (op (mapKeys(emap))) ⟶ if not(et in es) then v_res else et
```

Line 1 generates reduction rules such as $\lambda x.e \longrightarrow \lambda x.e$, for each value, as it is standard in big-step semantics. These rules are appended to those generated in lines 2–9. Line 2 selects all the reduction rules. Line 3 leaves out those rules that are not about a top-level expression operator. This skips contextual rules that take a step $E[e] \longrightarrow E[e']$, which do not appear in big-step semantics. To do so, line 3 makes use of *Expression*[(*op* _)] : *self*. As *op* is bound to the operator we are focusing on (from line 2), this selector returns a list with one element if *op* appears in *Expression*, or an empty list otherwise. This is the check we perform at line 3. Line 4 generates a new variable that will store the final value of the step. Line 5 assigns a new variable to each of the arguments in (es). We do so creating a map *emap*. These new variables are the formal arguments of the new rule being generated (lines 8–9). Lines 6–7 make each of these variables evaluate to its corresponding argument in es (line 7). For example, for the β-reduction an argument of es would be $\lambda x.e$ and we therefore generate the premise $e_1 \longrightarrow \lambda x.e$, where e_1 is the new variable that we assigned to this argument at line 5. Line 6 skips generating the reduction premise if it is a variable that does not appear in e_t. For example, in the translation of rule (IF-T) $(\mathit{if}\ \mathit{true}\ e_2\ e_3) \longrightarrow e_2$ we do not evaluate e_3 at all. Line 8 handles the result of the overall small-step reduction. This result is evaluated to a value (v_{res}), unless the target et already appears in the arguments es. The conclusion of the rule syncs with this, and we place v_{res} or e_t in the target of the step accordingly. Line 8 also appends the premises from the original rule, as they contain conditions to be checked.

Our algorithm translates the β-reduction and (IF-T) as follows.

$$(\lambda x.e\ v) \longrightarrow e[v/x] \quad \Rightarrow \quad \frac{e_1' \longrightarrow \lambda x.e \quad e_2' \longrightarrow v \quad e[v/x] \longrightarrow v_{res}}{(e_1'\ e_2') \longrightarrow v_{res}}$$

$$(\mathit{if}\ \mathit{true}\ e_1\ e_2) \longrightarrow e_1 \quad \Rightarrow \quad \frac{e_1' \longrightarrow \mathit{true} \quad e_2' \longrightarrow e_2}{(\mathit{if}\ e_1'\ e_2'\ e_3') \longrightarrow e_2}$$

We have implemented $\mathcal{L}$–Tr and we have applied our algorithms, and extensions of them, to the examples in this paper as well as λ-calculi with lists, pairs, sums, options, let-binding, function composition $(g \circ f)(x)$, and System F. We also considered these calculi in both call-by-value and call-by-name version, as well as lazy evaluation for data types such as pairs and lists. The languages produced by our tool are compiled to λ-prolog, which type checks them successfully and, in fact, can execute them. We have tested the generated languages against simple programs. We have checked that the functionality of subtyping has been added, and that programs evaluate to the expected values in one big-step.

4 Related Work

An excellent classification of language transformations has been provided in [3]. Language workbenches (Rascal, Spoofax, etcetera) implement these types of transformations and similar ones. These transformations are coarse grained in nature because they do not access the components of languages with precision. $\mathcal{L}$–Tr, instead, includes operations to scan rules, and select/manipulate formulae and terms with precision.

Proof assistants are optimized for handling inductive (rule-based) definitions, and can automatically generate powerful inductive reasoning mechanisms from these definitions. $\mathcal{L}$–Tr does not provide these features, and does not assist language designers with their proofs. On the other hand, proof assistants do not have reflective features for programmatically retrieving their own inductive definitions, selected by a pattern, and for manipulating them to form a different specification, which is instead characteristic of $\mathcal{L}$–Tr. A limitation of $\mathcal{L}$–Tr compared to proof assistants (and programming languages) is that $\mathcal{L}$–Tr does not offer recursion but only a simple form of iteration.

Works such as [2] and [1] offer translations from small-step to big-step semantics. Our algorithm differs slightly from that of Ciobâcă [1]. However, we do not provide correctness theorems for our algorithms.

5 Conclusion

We have presented $\mathcal{L}$–Tr, a calculus for expressing language transformations. The calculus is expressive enough to model interesting transformations such as adding subtyping, and switching from small-step to big-step semantics. We have proved the type soundness of $\mathcal{L}$–Tr, and we have implemented the calculus in a tool. As $\mathcal{L}$–Tr manipulates inference systems it can, in principle, be applied to logical systems, and we plan to explore this research venue. Overall, we believe that our calculus offers a rather declarative style for manipulating languages.

References

1. Ciobâcă, Ş.: From small-step semantics to big-step semantics, automatically. In: Johnsen, E.B., Petre, L. (eds.) IFM 2013. LNCS, vol. 7940, pp. 347–361. Springer, Heidelberg (2013). https://doi.org/10.1007/978-3-642-38613-8_24
2. Danvy, O.: Defunctionalized interpreters for programming languages. In: Proceedings of the 13th ACM SIGPLAN International Conference on Functional Programming, ICFP 2008, pp. 131–142. ACM, New York (2008)
3. Erdweg, S., Giarrusso, P.G., Rendel, T.: Language composition untangled. In: LDTA 2012, pp. 7:1–7:8. ACM, New York (2012)
4. Felleisen, M., Findler, R.B., Flatt, M.: Semantics Engineering with PLT Redex. MIT Press, Cambridge (2009)
5. Fowler, M.: Language workbenches: the killer-app for domain specific languages? (2005). http://www.martinfowler.com/articles/languageWorkbench.html
6. Kahn, G.: Natural semantics. In: Brandenburg, F.J., Vidal-Naquet, G., Wirsing, M. (eds.) STACS 1987. LNCS, vol. 247, pp. 22–39. Springer, Heidelberg (1987). https://doi.org/10.1007/BFb0039592
7. Mourad, B., Cimini, M.: A calculus for language transformations (2019). Technical report. arXiv:1910.11924 [cs.PL]
8. Mourad, B., Cimini, M.: L-Tr (2019). http://www.cimini.info/LTR/index.html
9. Pierce, B.C.: Types and Programming Languages. MIT Press, Cambridge (2002)
10. Plotkin, G.D.: A structural approach to operational semantics. DAIMI report FN-19, Computer Science Department of Aarhus University (1981)
11. Rosu, G., Şerbănuţă, T.F.: An overview of the K semantic framework. J. Log. Algebraic Program. **79**(6), 397–434 (2010)

Computing Directed Steiner Path Covers for Directed Co-graphs (Extended Abstract)

Frank Gurski[1(✉)], Stefan Hoffmann[1], Dominique Komander[1], Carolin Rehs[1], Jochen Rethmann[2], and Egon Wanke[1]

[1] Institute of Computer Science, Heinrich Heine University, 40225 Düsseldorf, Germany
frank.gurski@hhu.de

[2] Faculty of Electrical Engineering and Computer Science, Niederrhein University of Applied Sciences, 47805 Krefeld, Germany

Abstract. We consider the Directed Steiner Path Cover problem on directed co-graphs. Given a directed graph $G = (V(G), E(G))$ and a set $T \subseteq V(G)$ of so-called *terminal vertices*, the problem is to find a minimum number of directed vertex-disjoint paths, which contain all terminal vertices and a minimum number of non-terminal vertices (Steiner vertices). The primary minimization criteria is the number of paths. We show how to compute a minimum Steiner path cover for directed co-graphs in linear time. For $T = V(G)$, the algorithm computes a directed Hamiltonian path if such a path exists.

Keywords: Directed co-graphs · Directed Steiner path cover problem

1 Introduction

The Steiner path problem is a restriction of the Steiner tree problem such that the required terminal vertices lie on a path of minimum cost. The related Euclidean bottleneck Steiner path problem was considered in [1] and a linear time solution for the Steiner path problem on trees was given in [11].

While a Steiner tree always exists within connected graphs, it is not always possible to find a Steiner path, which motivates us to consider Steiner path cover problems. The Steiner connectivity problem was considered in [4].

In this article we consider the Directed Steiner Path Cover problem defined as follows. Let G be a directed graph on vertex set $V(G)$ and edge set $E(G)$ and let $T \subseteq V(G)$ be a set of *terminal vertices*. A *directed Steiner path cover* for G is a set of vertex-disjoint simple directed paths in G that contain all terminal vertices of T and possibly also some of the non-terminal (Steiner)

This work was funded in part by the Deutsche Forschungsgemeinschaft (DFG, German Research Foundation) – 388221852.

A. Chatzigeorgiou et al. (Eds.): SOFSEM 2020, LNCS 12011, pp. 556–565, 2020.
https://doi.org/10.1007/978-3-030-38919-2_45

vertices of $V(G) - T$. The *size* of a directed Steiner path cover is the number of its paths, the *cost* is defined as the minimum number of Steiner vertices in a directed Steiner path cover of minimum size.

Name: Directed Steiner Path Cover
Instance: A directed graph G and a set of terminal vertices $T \subseteq V(G)$.
Task: Find a directed Steiner path cover of minimum cost for G.

The Directed Steiner Path Cover problem generalizes the directed Hamiltonian path problem, implying that it is NP-hard. This motivates us to restrict the problem to special inputs. We consider a very natural class of inputs, which is defined as follows.

Directed co-graphs (short for directed complement reducible graphs) can be generated from the single vertex graph by applying disjoint union, order composition and series composition [3]. They also can be characterized by excluding eight forbidden induced subdigraphs, see [6, Fig. 2]. Directed co-graphs are exactly the digraphs of directed NLC-width 1 and a proper subset of the digraphs of directed clique-width at most 2 [9]. Directed co-graphs are also interesting from an algorithmic point of view since several hard graph problems can be solved in polynomial time by dynamic programming along the tree structure of the input graph, see [2,7,8]. Moreover, directed co-graphs are very useful for the reconstruction of the evolutionary history of genes or species using genomic sequence data [12].

In this paper we show how the value of a Steiner path cover of minimum size and cost for the disjoint union, order and series composition of two digraphs can be computed in linear time from the corresponding values of the involved digraphs. Therefore, we define a useful normal form for directed Steiner path covers in digraphs which are defined by the order composition or series composition of two digraphs. Further we sketch an algorithm which constructs a directed Steiner path cover of minimum size and cost for a directed co-graph in linear time.

2 Preliminaries

Definition 1 ([3]). *The class of* directed co-graphs *is recursively defined as follows.*

(i) Every digraph on a single vertex $(\{v\}, \emptyset)$*, denoted by* $\bullet_v$*, is a* directed co-graph.
(ii) If A*,* B *are vertex-disjoint directed co-graphs, then*
  (a) the disjoint union $A \oplus B$*, which is defined as the digraph with vertex set* $V(A) \cup V(B)$ *and edge set* $E(A) \cup E(B)$*,*
  (b) the order composition $A \oslash B$*, defined by their disjoint union plus all possible edges from* $V(A)$ *to* $V(B)$*, and*
  (c) the series composition $A \otimes B$*, defined by their disjoint union plus all possible edges between* $V(A)$ *and* $V(B)$*, are* directed co-graphs.

The recursive generation of a co-graph can be described by a tree called *directed co-tree*. The leaves of the directed co-tree represent the vertices of the digraph and the inner vertices of the directed co-tree correspond to the operations applied on the subgraphs of G defined by the subtrees. For every directed co-graph one can construct a directed co-tree in linear time, see [6].

Next we define a normal form for directed Steiner path covers in digraphs. Therefore we introduce some notations. Let G be a directed co-graph, let $T \subseteq V(G)$ be a set of terminal vertices, and let C be a directed Steiner path cover for G with respect to T. Then, $\mathfrak{s}(C)$ denotes the number of Steiner vertices in the paths of C. We define $p(G,T)$ as the minimum number of paths within a Steiner path cover for G with respect to T. Further let $s(G,T)$ be the minimum number of Steiner vertices in a directed Steiner path cover of size $p(G,T)$ with respect to T. We do not specify set T if it is clear from the context which set is meant.

Lemma 1. *Let C be a directed Steiner path cover for some directed co-graph $G = A \otimes B$ or $G = A \oslash B$ with respect to a set T of terminal vertices. Then, there is a directed Steiner path cover C' with respect to T which does not contain paths p and p' satisfying one of the properties (1)–(4), such that $|C| \geq |C'|$ and $\mathfrak{s}(C) \geq \mathfrak{s}(C')$ holds.*

1. *$p = (x,\ldots)$ or $p = (\ldots,x)$ where $x \notin T$. Comment: No path starts or ends with a Steiner vertex.*
2. *$p = (\ldots,u,x,v,\ldots)$ where $u \in V(A)$, $v \in V(B)$, and $x \notin T$. Comment: On a path, the neighbors u,v of a Steiner vertex x are both contained in the same digraph.*
3. *$p = (\ldots,x)$, $p' = (u,\ldots)$, where $x \in V(A)$, $u \in V(B)$, $p \neq p'$. Comment: There is no path p that ends in A, if there is a path $p' \neq p$ that starts in B.*
4. *$p = (\ldots,x,u,v,y,\ldots)$ where $u,v \notin T$. Comment: The paths contain no edge between two Steiner vertices.*

If $G = A \otimes B$ then cover C' also does not contain paths satisfying properties (5)–(8).

5. *$p = (x,\ldots)$, $p' = (u,\ldots)$, where $x \in V(A)$, $u \in V(B)$, $p \neq p'$. Comment: All paths start in the same digraph.*
6. *$p = (\ldots,x,y,\ldots)$, $p' = (\ldots,u,v,\ldots)$ where $x,y \in V(A)$, $u,v \in V(B)$. Comment: The cover C' contains edges of only one of the digraphs.*
7. *$p = (x,\ldots)$, $p' = (\ldots,u,y,v,\ldots)$, where $x,y \in V(A)$, $u,v \in V(B)$, and $y \notin T$. Comment: If a path starts in A then there is no Steiner vertex in A with two neighbors on the path in B.*
8. *$p = (x,\ldots)$, $p' = (\ldots,u,v,\ldots)$, where $x \in V(A)$ and $u,v \in V(B)$. Comment: If a path starts in A, then no edge of B is contained in the cover.*

The proof of Lemma 1 is omitted due to space restrictions. Since the hypothesis of Lemma 1 is symmetric in A and B, the statement of Lemma 1 is also valid for co-graphs $G = A \otimes B$ if A and B are switched.

Definition 2. *A directed Steiner path cover C for some directed co-graph $G = A \otimes B$ or $G = A \oslash B$ is said to be in* normal form *if it satisfies all properties (1)–(8) given in Lemma 1.*

In the following we assume that a directed Steiner path cover for some directed co-graph $G = A \oslash B$ or $G = A \otimes B$ is always in normal form.

3 Algorithms for the Directed Steiner Path Cover Problem

3.1 Computing the Optimal Number of Paths

Lemma 2. *Let A and B be two vertex-disjoint digraphs and let $T_A \subseteq V(A)$ and $T_B \subseteq V(B)$ be two sets of terminal vertices. Then, the following equations hold:*

1. $p(\bullet_v, \emptyset) = 0$ *and* $p(\bullet_v, \{v\}) = 1$
2. $p(A \oplus B, T_A \cup T_B) = p(A, T_A) + p(B, T_B)$
3. $p(A \otimes B, \emptyset) = 0$
4. $p(A \otimes B, T_A \cup T_B) = \max\{1, p(B, T_B) - |V(A)|\}$ *if* $1 \leq |T_B|$ *and* $|T_A| \leq |T_B|$
5. $p(A \otimes B, T_A \cup T_B) = \max\{1, p(A, T_A) - |V(B)|\}$ *if* $1 \leq |T_A|$ *and* $|T_A| > |T_B|$
6. $p(A \oslash B, T_A \cup T_B) = p(A, T_A)$ *if* $p(A) \geq p(B)$
7. $p(A \oslash B, T_A \cup T_B) = p(B, T_B)$ *if* $p(A) < p(B)$

Proof. 1.–3. Obvious.

4. We show that $p(A \otimes B) \geq \max\{1, p(B) - |V(A)|\}$ applies by an indirect proof. Assume a directed Steiner path cover C for $A \otimes B$ has less than $\max\{1, p(B) - |V(A)|\}$ paths. The removal of all vertices of A from all paths in C gives a directed Steiner path cover of size $|C| + |V(A)| < p(B)$ for B. ↯
   To see that $p(A \otimes B) \leq \max\{1, p(B) - |V(A)|\}$ applies, consider that we can use any vertex of A to combine two paths of the cover of B to one path, since the series composition of A and B creates all directed edges between A and B. If there are more terminal vertices in A than there are paths in the cover of B, i.e. $p(B) < |T_A|$, then we have to split paths of B and reconnect them by terminal vertices of A. This can always be done since $|T_A| \leq |T_B|$.
5. Similar to 4.
6. To see that $p(A \oslash B) \leq p(A)$ applies, consider that we can connect each path of A by each path of B, see Lemma 1(3). Since no edge between B and A is created, no path of B can be extended by a path of A.
   We show that $p(A \oslash B) \geq p(A)$ applies by an indirect proof. Assume a directed Steiner path cover C for $A \oslash B$ contains less than $p(A)$ paths. The removal of all vertices of B from all paths in C gives a Steiner path cover of size $|C| < p(A)$. ↯
7. Similar to 6. □

3.2 Computing the Optimal Number of Steiner Vertices

Remark 1. For two vertex-disjoint directed co-graphs A, B and two sets of terminal vertices $T_A \subseteq V(A)$, $T_B \subseteq V(B)$ it holds that $s(A \oplus B, T_A \cup T_B) = s(A, T_A) + s(B, T_B)$, since the disjoint union does not create any new edges.

Remark 2. Let $G = A \oslash B$ be a directed co-graph, and let C be a directed Steiner path cover of G such that $p = (q_1, u_1, x, q_2, v_1)$ is a path in A, $p_1 = (u_2, q_3)$ and $p_2 = (v_2, q_4)$ are paths in B, all paths are vertex-disjoint paths in C, where $x \notin T$, $u_1, u_2, v_1, v_2 \in T$, and $q_1, \ldots, q_4$ are subpaths. Then, we can split p at vertex x into two paths, combine them with p_1 and p_2 to get (q_1, u_1, u_2, q_3) and (q_2, v_1, v_2, q_4) as new paths and we get a Steiner path cover without increasing the number of paths and one Steiner vertex less than C. If A and B are switched we get (u_2, q_3, q_1, u_1) and (v_2, q_4, q_2, v_1) as new paths and the statement also holds.

Next, we give the central lemma of our work, which is proven by induction on the structure of the directed co-graph.

Lemma 3. *For every directed co-graph G and every directed Steiner path cover C for G with respect to a set T of terminal vertices it holds that $p(G) + s(G) \leq |C| + \mathfrak{s}(C)$.*

Proof. The statement is obviously valid for all directed co-graphs which consist of only one vertex. Let us assume that the statement is valid for directed co-graphs of n vertices. Let A and B are vertex-disjoint directed co-graphs of at most n vertices each.

Let $G = A \oplus B$ be a directed co-graph that consists of more than n vertices. By Lemma 2, and Remark 1, it holds that $p(A \oplus B) + s(A \oplus B) = p(A) + p(B) + s(A) + s(B)$. By the induction hypothesis, it holds that $p(A) + s(A) \leq |C_{|A}| + \mathfrak{s}(C_{|A})$ and $p(B) + s(B) \leq |C_{|B}| + \mathfrak{s}(C_{|B})$, where $C_{|A}$ denotes the cover C restricted to digraph A, i.e. the cover that results from C when all vertices of B are removed. Then, the statement of the lemma follows.

Let $G = A \otimes B$ be a directed co-graph that consists of more than n vertices. Without loss of generality, let $|T_A| \leq |T_B|$.

1. Let $X(A)$ denote the vertices of A used in cover C, and let D denote the cover for B that we obtain by removing the vertices of $X(A)$ from cover C. By induction hypothesis, it holds that $p(B) + s(B) \leq |D| + \mathfrak{s}(D)$.
2. Let $nt(X(A))$ denote the number of non-terminal vertices of $X(A)$. Since covers are in normal form it holds that $\mathfrak{s}(C) = \mathfrak{s}(D) + nt(X(A))$ and $|C| = |D| - |T_A| - nt(X(A))$. Thus, we get $|C| + \mathfrak{s}(C) = |D| + \mathfrak{s}(D) - |T_A|$.

We put these two results together and obtain:

$$p(B) + s(B) - |T_A| \leq |D| + \mathfrak{s}(D) - |T_A| = |C| + \mathfrak{s}(C)$$

To show the statement of the lemma, we first consider the case $p(B) - 1 \leq |V(A)|$. Then, it holds that $p(A \otimes B) = 1$. If $|T_A| \geq p(B) - 1$, then $d := |T_A| - (p(B) - 1)$

many Steiner vertices from B, if available, can be replaced by terminal vertices from A. Otherwise if $|T_A| < p(B) - 1$, then $-d = (p(B) - 1) - |T_A|$ many Steiner vertices from A are used to combine the paths. Thus, it holds that $s(A \otimes B) \leq \max\{0, s(B) - d\}$ since the number of Steiner vertices in an optimal cover is at most the number of Steiner path vertices in a certain cover. Thus, since $p(A \otimes B) = 1$ we get for $s(B) \geq d$:

$$\begin{aligned} p(A \otimes B) + s(A \otimes B) &\leq 1 + s(B) - d = 1 + s(B) - (|T_A| - (p(B) - 1)) \\ &= \cancel{1} + s(B) - |T_A| + p(B) - \cancel{1} \;\leq\; |C| + \mathfrak{s}(C) \end{aligned}$$

If $s(B) < d$ then all Steiner vertices of B can be replaced by terminal vertices of A and since $|T_A| \leq |T_B|$ holds, some of the paths of B can be reconnected by the remaining terminal vertices of A. Thus, $p(A \otimes B) + s(A \otimes B) = 1 \leq |C| + \mathfrak{s}(C)$ holds.

Consider now the case where $p(B) - 1 > |V(A)|$ holds, i.e. not all paths in an optimal cover for B can be combined by vertices of A. By Lemma 2, it holds that $p(A \otimes B) = \max\{1, p(B) - |V(A)|\}$. Thus, for $p(A \otimes B) > 1$ we get:

$$\begin{aligned} p(A \otimes B) + s(A \otimes B) &\leq p(B) - |V(A)| + s(B) + nt(A) \\ &= p(B) + s(B) - |T_A| \;\leq\; |C| + \mathfrak{s}(C) \end{aligned}$$

The non-terminal vertices of A must be used to combine paths of the cover, thus the non-terminal vertices of A become Steiner vertices.

Let $G = A \oslash B$ be a directed co-graph that consists of more than n vertices. By the induction hypothesis, it holds that $p(A) + s(A) \leq |C_{|A}| + \mathfrak{s}(C_{|A})$ and $p(B) + s(B) \leq |C_{|B}| + \mathfrak{s}(C_{|B})$.

First, we consider the case $p(A) > p(B)$. By Lemma 2 it holds $p(A \oslash B) = p(A)$. We can connect every path of A with every path of B. By Remark 2 it holds that if there are more paths in A than in B, for each additional path in A we can remove one Steiner vertex from B. And since an optimal cover has at most as many Steiner vertices as a concrete cover, it holds that $s(A \oslash B) \leq \mathfrak{s}(C_{|A}) + \mathfrak{s}(C_{|B}) - \min\{\mathfrak{s}(C_{|B}), |C_{|A}| - |C_{|B}|\}$. If we sum up both equations, we get

$$p(A \oslash B) + s(A \oslash B) \leq p(A) + \mathfrak{s}(C_{|A}) + \mathfrak{s}(C_{|B}) - \min\{\mathfrak{s}(C_{|B}), |C_{|A}| - |C_{|B}|\}$$

If $\mathfrak{s}(C_{|B}) \geq |C_{|A}| - |C_{|B}|$ holds, and since $\mathfrak{s}(C) = \mathfrak{s}(C_{|A}) + \mathfrak{s}(C_{|B})$ holds, we get

$$p(A \oslash B) + s(A \oslash B) \leq p(A) + \mathfrak{s}(C) - |C_{|A}| + |C_{|B}|.$$

The statement would be shown if $p(A) - |C_{|A}| + |C_{|B}| \leq |C|$ would apply. It holds that $p(A) \leq |C_{|A}|$, since an optimal cover has at most as many paths as a concrete cover, and it holds that $|C_{|B}| \leq |C|$, since $|C| = \max\{|C_{|A}|, |C_{|B}|\}$ and the covers are in normal form. We sum up these equations and we get $p(A) + |C_{|B}| \leq |C_{|A}| + |C|$, which is equivalent to $p(A) - |C_{|A}| + |C_{|B}| \leq |C|$, thus $p(A \oslash B) + s(A \oslash B) \leq |C| + \mathfrak{s}(C)$ has been shown.

If $\mathfrak{s}(C_{|B}) < |C_{|A}| - |C_{|B}|$, then it holds that $p(A \oslash B) + s(A \oslash B) \leq p(A) + \mathfrak{s}(C_{|A})$, and we have to show that $p(A) + \mathfrak{s}(C_{|A}) \leq |C| + \mathfrak{s}(C)$ applies. It holds that

$p(A) \leq |C_{|A}|$, since an optimal cover has at most as many paths as a concrete cover, and it holds that $|C_{|A}| \leq |C|$, since $|C| = \max\{|C_{|A}|, |C_{|B}|\}$ and the covers are in normal form. Furthermore, it holds that $\mathfrak{s}(C_{|A}) \leq \mathfrak{s}(C)$, since a part is only as big as the whole.

The other case $p(A) \leq p(B)$ can be shown in a similar way. □

Remark 3. Let G be a directed co-graph and let C be a directed Steiner path cover for G with respect to some set of terminal vertices T. Then $\mathfrak{s}(C) \geq s(G)$ holds only if $|C| = p(G)$. If $|C| > p(G)$ then $\mathfrak{s}(C)$ might be smaller than $s(G)$.

This fact will be used in the proof of the next lemma.

Lemma 4. *Let A and B be two vertex-disjoint digraphs, and let $T_A \subseteq V(A)$, $T_B \subseteq V(A)$ be sets of terminal vertices. Then, the following equations hold:*

1. $s(\bullet_v, \emptyset) = 0$ *and* $s(\bullet_v, \{v\}) = 0$
2. $s(A \otimes B, T_A \cup T_B) = \max\{0, s(B, T_B) + p(B, T_B) - p(A \otimes B) - |T_A|\}$ *if* $|T_A| \leq |T_B|$
3. $s(A \otimes B, T_A \cup T_B) = \max\{0, s(A, T_A) + p(A, T_A) - p(A \otimes B) - |T_B|\}$ *if* $|T_A| > |T_B|$
4. $s(A \oslash B, T_A \cup T_B) = s(A, T_A) + s(B, T_B)$ *if* $p(A, T_A) = p(B, T_B)$
5. $s(A \oslash B, T_A \cup T_B) = s(A) + s(B) - \min\{s(A), p(B) - p(A)\}$ *if* $p(A) < p(B)$
6. $s(A \oslash B, T_A \cup T_B) = s(A) + s(B) - \min\{s(B), p(A) - p(B)\}$ *if* $p(A) > p(B)$

Proof. 1. Obvious.

2. First, we show $s(A \otimes B) \leq \max\{0, s(B) + p(B) - p(A \otimes B) - |T_A|\}$. By Lemma 3, we know that $s(A \otimes B) + p(A \otimes B) \leq \mathfrak{s}(C) + |C|$ holds for any cover C for co-graph $A \otimes B$ and any set of terminal vertices T. Consider cover C for $A \otimes B$ obtained by an optimal cover D for B in the following way.

Construction 1. *We use the terminal vertices of A to either combine paths of D or to remove a Steiner vertex of D by replacing $v \notin T$ by some terminal vertex of A in a path like $(\ldots, u, v, w, \ldots) \in D$, where $u, w \in T$.*

If $|T_A| \geq s(B) + p(B)$ then all paths of D can be combined and all Steiner vertices of D can be replaced by terminal vertices of A and since $|T_A| \leq |T_B|$ holds, some of the paths can be split and reconnected by the remaining terminal vertices of A. Thus, $\mathfrak{s}(C) + |C| = 1$ and $s(A \otimes B) = 0$.

Otherwise, if $|T_A| < s(B) + p(B)$, then by Construction 1 we get $\mathfrak{s}(C) + |C| = s(B) + p(B) - |T_A|$, and by Lemma 3, we get the statement.

$$\begin{aligned} & s(A \otimes B) + p(A \otimes B) \leq s(B) + p(B) - |T_A| \;=\; \mathfrak{s}(C) + |C| \\ \iff \quad & s(A \otimes B) \leq s(B) + p(B) - p(A \otimes B) - |T_A| \end{aligned}$$

Next, we prove $s(A \otimes B) \geq \max\{0, s(B) + p(B) - p(A \otimes B) - |T_A|\}$. Let $X(A)$ be the vertices of $V(A)$ that are contained in the paths of an optimal cover C for $A \otimes B$. Let D be the cover for B obtained by removing the vertices of $X(A)$ from C. Since the covers are in normal form, the following holds:

$$\begin{aligned} & |X(A)| = nt(X(A)) + |T_A| = |D| - p(A \otimes B) \\ \iff \quad & nt(X(A)) = |D| - p(A \otimes B) - |T_A| \end{aligned}$$

Thus, we get:

$$\begin{array}{rrcl} & s(A \otimes B) - nt(X(A)) & = & \mathfrak{s}(D) = s(A \otimes B) - |D| + p(A \otimes B) + |T_A| \\ \Longleftrightarrow & & & s(A \otimes B) = \mathfrak{s}(D) + |D| - p(A \otimes B) - |T_A| \\ \Rightarrow & & & s(A \otimes B) \geq s(B) + p(B) - p(A \otimes B) - |T_A| \end{array}$$

The implication follows since by Lemma 3 it holds that $\mathfrak{s}(D)+|D| \geq s(B)+p(B)$.

3. Similar to 2.
4. To see that $s(A \oslash B) \leq s(A) + s(B)$ applies, consider optimal covers C and D for A and B, respectively. We construct a cover E for $A \oslash B$ in the following way.

Construction 2. *Connect each path of C by a path of D, see Lemma 1(3).*

Since $|E| = p(A \oslash B)$ holds, we get $s(A \oslash B) \leq \mathfrak{s}(E) = \mathfrak{s}(C)+\mathfrak{s}(D) = s(A)+s(B)$, because an optimal cover has at most as many Steiner vertices as a concrete cover.

To see that $s(A \oslash B) \geq s(A) + s(B)$ applies consider an optimal cover C for $A \oslash B$. Then, it holds that $s(A \oslash B) = \mathfrak{s}(C_{|A}) + \mathfrak{s}(C_{|B}) \geq s(A) + s(B)$, since $|C_{|A}| = p(A) = p(A \oslash B) = p(B) = |C_{|B}|$.

5. We have to distinguish two cases. First, let $s(A) > p(B) - p(A)$.
   To see that $s(A \oslash B) \leq s(A) + s(B) - (p(B) - p(A))$ applies, consider optimal covers C and D for A and B. We construct a cover E for $A \oslash B$ as follows.

Construction 3. *First, we split $p(B)-p(A)$ many paths of C at Steiner vertices as described in Remark 2. Afterwards, we connect each of the resulting paths by a path of D.*

Thus, it holds that $|E| = p(A \oslash B) = p(B)$ and therefore $s(A \oslash B) \leq \mathfrak{s}(C) + \mathfrak{s}(D) - (p(B) - p(A)) = s(A) + s(B) - (p(B) - p(A))$.

Please note, a Steiner path cover C for $A \oslash B$ with $\mathfrak{s}(C_{|A}) > 0$ is not optimal if $|C_{|A}| < |C| = p(A \oslash B)$ holds. By Remark 2 a path of $C_{|A}$ could be splitted at a Steiner vertex and the number of Steiner vertices could be reduced.

To see that $s(A \oslash B) \geq s(A) + s(B) - (p(B) - p(A))$ applies, consider an optimal cover C for $A \oslash B$. Then, it holds that $s(A \oslash B) = \mathfrak{s}(C) = \mathfrak{s}(C_{|A}) + \mathfrak{s}(C_{|B})$, and by the previous note it holds that $|C| = p(A \oslash B) = p(B) = |C_{|A}|$. By Lemma 3 we get $\mathfrak{s}(C_{|A}) + |C_{|A}| \geq s(A) + p(A)$. If we sum up these equations, we get $s(A \oslash B) + p(A \oslash B) = \mathfrak{s}(C_{|A}) + |C_{|A}| + \mathfrak{s}(C_{|B})$. Finally we get:

$$\begin{aligned} s(A \oslash B) &= \mathfrak{s}(C_{|A}) + |C_{|A}| - p(A \oslash B) + \mathfrak{s}(C_{|B}) \\ &\geq s(A) + p(A) - p(B) + \mathfrak{s}(C_{|B}) \ \geq \ s(A) + p(A) - p(B) + s(B) \end{aligned}$$

Consider now the case that $s(A) \leq p(B) - p(A)$. To see that $s(A \oslash B) \leq s(B)$ applies, consider optimal covers C and D for A and B, respectively. We construct a cover E for $A \oslash B$ as follows.

Construction 4. *First, we split as many paths of C at Steiner vertices as possible in a way described in Remark 2. Afterwards, all Steiner vertices of C have been removed and we connect each of the resulting paths by a path of D.*

Thus, it holds that $|E| = p(A \oslash B) = p(B)$ and therefore $s(A \oslash B) \leq \mathfrak{s}(E) = s(B)$.

To see that $s(A \oslash B) \geq s(B)$ applies, consider an optimal cover C for $A \oslash B$. By the above note it holds that $\mathfrak{s}(C_{|A}) = 0$, since C would not be optimal otherwise. Thus, we get $s(A \oslash B) = \mathfrak{s}(C_{|B}) \geq s(B)$, since $|C_{|B}| = p(B)$ holds and by Remark 3.

6. Similar to 5. □

By Lemmas 2 and 4, and since a directed co-tree can be computed in linear time from the input directed co-graph [6], we have shown the following result.

Theorem 1. *Given some directed co-graph G, the values $p(G)$ and $s(G)$ can be computed in linear time with respect to the size of a directed co-expression for G.*

3.3 Computing an Optimal Directed Steiner Path Cover

Next, we sketch an algorithm to compute a solution for the Directed Steiner Path Cover problem for some given directed co-graph G represented by its binary directed co-tree $T(G)$. By performing the rules given in Construction 1–4 bottom-up along $T(G)$, we compute the directed Steiner path cover for G. In order to obtain a linear running time $\mathcal{O}(|V(G)| + |E(G)|)$, we store the paths using double-linked, linear lists. While the paths that contain Steiner vertices are stored in one set and the paths that contain no Steiner vertices are stored in another set. The lists each have a pointer to the first and last element, which are terminal vertices by Lemma 1(1), and they have a pointer to the first and last Steiner vertices. The Steiner vertices are additionally chained as a doubly linked list. Additionally, we store the number of terminal and Steiner vertices for each list. This allows us to perform each construction step in constant time.

Theorem 2. *Given some directed co-graph G, a solution for the* Directed Steiner Path Cover *problem can be computed in linear time with respect to the size of a directed co-expression for G.*

4 Conclusions

Our results for directed co-graphs can be transfered to undirected co-graphs, which are precisely those graphs that can be generated from the single vertex graph by disjoint union and join operations, see [5]. Given some undirected co-graph G, we can solve the Steiner path cover problem in linear time by replacing every edge $\{u, v\}$ of G by two anti-parallel directed edges (u, v) and (v, u) and applying our solution for directed co-graphs.

Since a directed Hamiltonian path exists in digraph G if and only if we have $T = V(G)$ and $p(G) = 1$, our results imply the first linear time algorithm for the directed Hamiltonian path problem on directed co-graphs. This generalizes the known results for undirected co-graphs of Lin et al. [10].

References

1. Abu-Affash, A.K., Carmi, P., Katz, M.J., Segal, M.: The euclidean bottleneck steiner path problem and other applications of (α,β)-pair decomposition. Discret. Comput. Geom. **51**(1), 1–23 (2014)
2. Bang-Jensen, J., Maddaloni, A.: Arc-disjoint paths in decomposable digraphs. J. Graph Theory **77**, 89–110 (2014)
3. Bechet, D., de Groote, P., Retoré, C.: A complete axiomatisation for the inclusion of series-parallel partial orders. In: Comon, H. (ed.) RTA 1997. LNCS, vol. 1232, pp. 230–240. Springer, Heidelberg (1997). https://doi.org/10.1007/3-540-62950-5_74
4. Borndörfer, R., Karbstein, M., Pfetsch, M.: The steiner connectivity problem. Math. Program. **142**(1), 133–167 (2013)
5. Corneil, D., Lerchs, H., Stewart-Burlingham, L.: Complement reducible graphs. Discrete Appl. Math. **3**, 163–174 (1981)
6. Crespelle, C., Paul, C.: Fully dynamic recognition algorithm and certificate for directed cographs. Discrete Appl. Math. **154**(12), 1722–1741 (2006)
7. Gurski, F., Komander, D., Rehs, C.: Computing digraph width measures on directed co-graphs. In: Gąsieniec, L.A., Jansson, J., Levcopoulos, C. (eds.) FCT 2019. LNCS, vol. 11651, pp. 292–305. Springer, Cham (2019). https://doi.org/10.1007/978-3-030-25027-0_20
8. Gurski, F., Rehs, C.: Directed path-width and directed tree-width of directed co-graphs. In: Wang, L., Zhu, D. (eds.) COCOON 2018. LNCS, vol. 10976, pp. 255–267. Springer, Cham (2018). https://doi.org/10.1007/978-3-319-94776-1_22
9. Gurski, F., Wanke, E., Yilmaz, E.: Directed NLC-width. Theor. Comput. Sci. **616**, 1–17 (2016)
10. Lin, R., Olariu, S., Pruesse, G.: An optimal path cover algorithm for cographs. Comput. Math. Appl. **30**, 75–83 (1995)
11. Moharana, S.S., Joshi, A., Vijay, S.: Steiner path for trees. Int. J. Comput. Appl. **76**(5), 11–14 (2013)
12. Nøjgaard, N., El-Mabrouk, N., Merkle, D., Wieseke, N., Hellmuth, M.: Partial homology relations - satisfiability in terms of di-cographs. In: Wang, L., Zhu, D. (eds.) COCOON 2018. LNCS, vol. 10976, pp. 403–415. Springer, Cham (2018). https://doi.org/10.1007/978-3-319-94776-1_34

Counting Infinitely by Oritatami Co-transcriptional Folding

Kohei Maruyama[1(✉)] and Shinnosuke Seki[1,2]

[1] The University of Electro-Communications, 1-5-1 Chofugaoka, Chofu, Tokyo 1828585, Japan
{k.maruyama,s.seki}@uec.ac.jp

[2] École Normale Superiéure de Lyon, 46 allée d'Italie, 69007 Lyon, France

Abstract. A fixed bit-width counter was proposed as a proof-of-concept demonstration of an oritatami model of cotranscriptional folding [Geary et al., Proc. MFCS 2016, LIPIcs 58, 43:1–43:14], and it was embedded into another oritatami system that self-assembles a finite portion of Heighway dragon fractal. In order to expand its applications, we endow this counter with capability to widen bit-width at every encounter with overflow.

1 Introduction

Counting is one of the most essential tasks for computing; as well known, the ability to count suffices to enable Turing universality [9]. Nature has been counting billions of days using molecular "circadian clockwork" which is "as complicated and as beautiful as the wonderful chronometers developed in the 18th century" [8]. Nowadays, developments in molecular self-assembly technology enable us to design molecules to count. Evans has demonstrated a DNA tile self-assembly system that counts accurately *in-vitro* in binary from a programmed initial count until it overflows [3]. In its foundational theory of molecular self-assembly, such binary counters have been proved versatile, being used to assemble shapes of particular size [1,10], towards self-assembly of fractals [7], as an infinite scaffold to simulate all Turing machines in parallel in order to prove undecidability of nondeterminism in the abstract tile-assembly model [2], to name a few.

A fixed bit-width (finite) binary counter has been implemented as a proof-of-concept demonstration of the oritatami model of cotranscriptional folding [5]. As shown in Fig. 1, an RNA transcript folds upon itself while being transcribed (synthesized) from its corresponding DNA template strand. Geary et al. programmed a specific RNA rectangular tile structure into a DNA template in such a way that the corresponding RNA transcript *folds cotranscriptionally* into the programmed tile structure with high probability *in vitro* at

This work is supported in part by KAKENHI Grant-in-Aid for Challenging Research (Exploratory) No. 18K19779 and JST Program to Disseminate Tenure Tracking System No. 6F36, both granted to S. S.

A. Chatzigeorgiou et al. (Eds.): SOFSEM 2020, LNCS 12011, pp. 566–575, 2020.
https://doi.org/10.1007/978-3-030-38919-2_46

Fig. 1. RNA origami. RNA polymerase enzyme (orange) synthesizes the temporal copy (blue) of a gene (gray spiral) out of ribonucleotides of four types A, C, G, and U. (Color figure online)

room temperatures (*RNA origami*) [6]. An oritatami system folds a transcript of abstract molecules called *beads* of finitely many types over the 2-dimensional triangular lattice cotranscriptionally according to a rule set that specifies which types of molecules are allowed to bind once they are placed at the unit distance away. The transcript of the binary counter in [5] is of period 60 as ⓪−①−②−⋯−㊽−㊾−⓪−①⋯ and its period is semantically divided into two half-adder (HA) modules A = ⓪−①−⋯−⑪ and C = ㉚−㉛−⋯−㊶ and two structural modules B and D, which are sandwiched by half-adder modules along the transcript. While being folded cotranscriptionally in zigzags, HA modules increment the current count i by 1, which is initialized on a linear *seed* structure, alike the Evans' counter, whereas structural modules B and D align HA modules properly and also make a turn at an end of the count i; B guides the transcript from a zig to a zag ($\hookrightarrow$) while D does from a zag to a zig ($\hookleftarrow$). This counter was embedded as a component of an oritatami system to self-assemble an arbitrary finite portion of Heighway dragon fractal [7]. Its applications are limited, however, by lack of mechanism to widen bit-width; its behavior is undefined when its count overflows. In this paper, we endow this counter, or more precisely, its structural module B, with the capability to widen the count by 1 bit at every encounter with overflow.

2 Preliminaries

Let Σ be a finite alphabet, whose elements should be regarded as types of abstract molecule, or *beads*. A bead of type $a \in \Sigma$ is called an a-bead. By Σ^* and Σ^ω, we denote the set of finite sequences of beads and that of one-way infinite sequences of beads, respectively. The empty sequence is denoted by λ. Let $w = b_1 b_2 \cdots b_n \in \Sigma^*$ be a sequence of length n for some integer n and bead types $b_1, \ldots, b_n \in \Sigma$. The *length* of w is denoted by $|w|$, that is, $|w| = n$. For two indices i, j with $1 \leq i \leq j \leq n$, we let $w[i..j]$ refer to the subsequence $b_i b_{i+1} \cdots b_{j-1} b_j$; if $i = j$, then $w[i..i]$ is simplified as $w[i]$. For $k \geq 1$, $w[1..k]$ is called a *prefix* of w.

Oritatami systems fold their transcript, which is a sequence of beads, over the triangular grid graph $\mathbb{T} = (V, E)$ cotranscriptionally. For a point $p \in V$, let $\varhexagon_p^d$ denote the set of points which lie in the regular hexagon of radius d centered at the point p. Note that $\bigcirc_p^d$ consists of $3d(d+1)+1$ points. A directed path $P = p_1 p_2 \cdots p_n$ in $\mathbb{T}$ is a sequence of *pairwise-distinct* points $p_1, p_2, \ldots, p_n \in V$ such that $\{p_i, p_{i+1}\} \in E$ for all $1 \le i < n$. Its i-th point is referred to as $P[i]$. Now we are ready to abstract RNA single-stranded structures in the name of conformation. A *conformation* C (over Σ) is a triple (P, w, H) of a directed path P in $\mathbb{T}$, $w \in \Sigma^*$ of the same length as P, and a set of h-interactions $H \subseteq \big\{\{i,j\} \bigm| 1 \le i, i+2 \le j, \{P[i], P[j]\} \in E\big\}$. This is to be interpreted as the sequence w being folded along the path P in such a manner that its i-th bead $w[i]$ is placed at the i-th point $P[i]$ and the i-th and j-th beads are bound (by a hydrogen-bond-based interaction) if and only if $\{i,j\} \in H$. The condition $i+2 \le j$ represents the topological restriction that two consecutive beads along the path cannot be bound. The *length* of C is defined to be the length of its transcript w (that is, equal to the length of the path P). A *rule set* $R \subseteq \Sigma \times \Sigma$ is a symmetric relation over Σ, that is, for all bead types $a, b \in \Sigma$, $(a,b) \in R$ implies $(b,a) \in R$. A bond $\{i,j\} \in H$ is *valid with respect to* R, or simply R-valid, if $(w[i], w[j]) \in R$. This conformation C is *R-valid* if all of its bonds are R-valid. For an integer $\alpha \ge 1$, C is *of arity* α if it contains a bead that forms α bonds but none of its beads forms more. By $\mathcal{C}_{\le\alpha}(\Sigma)$, we denote the set of all conformations over Σ whose arity is at most α; its argument Σ is omitted whenever Σ is clear from the context.

The oritatami system grows conformations by an operation called elongation. Given a rule set R and an R-valid conformation $C_1 = (P, w, H)$, we say that another conformation C_2 is an elongation of C_1 by a bead $b \in \Sigma$, written as $C_1 \xrightarrow{R}_b C_2$, if $C_2 = (Pp, wb, H \cup H')$ for some point $p \in V$ not along the path P and set $H' \subseteq \big\{\{i, |w|+1\} \bigm| 1 \le i < |w|, \{P[i], p\} \in E, (w[i], b) \in R\big\}$ of bonds formed by the b-bead; this set H' can be empty. Note that C_2 is also R-valid. This operation is recursively extended to the elongation by a finite sequence of beads as: for any conformation C, $C \xrightarrow{R}{}^*_\lambda C$; and for a finite sequence of beads $w \in \Sigma^*$ and a bead $b \in \Sigma$, a conformation C_1 is elongated to a conformation C_2 by wb, written as $C_1 \xrightarrow{R}{}^*_{wb} C_2$, if there is a conformation C' that satisfies $C_1 \xrightarrow{R}{}^*_w C'$ and $C' \xrightarrow{R}_b C_2$.

An *oritatami system* (OS) Ξ is a tuple $(\Sigma, R, \delta, \alpha, \sigma, w)$, where Σ and R are defined as above, while the other elements are a positive integer δ called *delay*, a positive integer α called *arity*, an initial R-valid conformation $\sigma \in \mathcal{C}_{\le\alpha}(\Sigma)$ called the *seed*, and a (possibly infinite) *transcript* $w \in \Sigma^* \cup \Sigma^\omega$, which is to be folded upon the seed by stabilizing beads of w one at a time so as to minimize energy collaboratively with the succeeding $\delta-1$ nascent beads. The energy of a conformation $C = (P, w, H)$, denoted by $\Delta G(C)$, is defined to be $-|H|$; the more bonds a conformation has, the more stable it gets. The set $\mathcal{F}(\Xi)$ of conformations *foldable* by the system Ξ is recursively defined as: the seed σ is in $\mathcal{F}(\Xi)$; and provided that an elongation C_i of σ by the prefix $w[1..i]$ be foldable (i.e., $C_0 = \sigma$), its further elongation C_{i+1} by the next bead $w[i+1]$ is foldable if

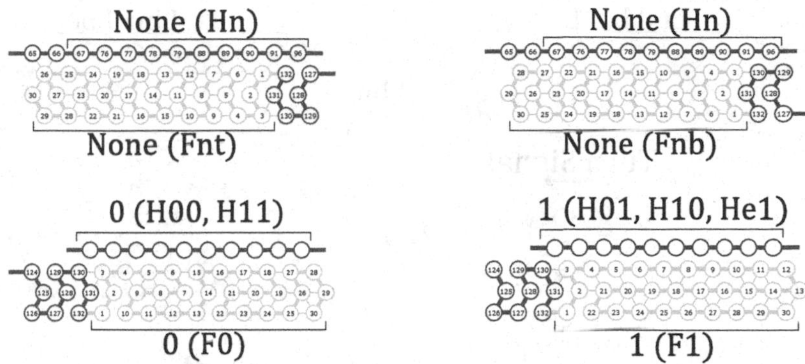

Fig. 2. All the four bricks of module F: the two bricks at the top, Fnt and Fnb, are for zigs while the others, F0 and F1, are for zags.

$$C_{i+1} \in \mathop{\arg\min}_{\substack{C \in \mathcal{C}_{\leq\alpha}\ s.t.\\ C_i \xrightarrow{R}_{w[i+1]} C}} \min\Big\{\Delta G(C') \;\Big|\; C \xrightarrow{R}{}^{*}_{w[i+2\ldots i+k]} C', k \leq \delta, C' \in \mathcal{C}_{\leq\alpha}\Big\}. \quad (1)$$

Then we say that the bead $w[i+1]$ and the bonds it forms are *stabilized* according to C_{i+1}. The easiest way to understand this stabilization process should be the video available at https://www.dailymotion.com/video/x3cdj35, in which the Turing universal oritatami system by Geary et al. [4], whose delay is 3, is running. This video is worth watching because a directed motif called a *glider*, which it features, is utilized for our infinite counter. Note that an arity-α oritatami system cannot fold any conformation of arity larger than α. A conformation foldable by Ξ is *terminal* if none of its elongations is foldable by Ξ.

3 Folding an Infinite Binary Counter

Between two consecutive overflows, the proposed system behaves in the same way as the finite binary counter proposed by Geary et al. [5]. Its transcript folds in a zigzag manner macroscopically (downward in figures throughout this paper). A zig, folding from right to left, increments the current value of the counter by 1. The succeeding zag, folding from left to right, formats the incremented value for the sake of next zig and copies it downward. Unlike the existing counter, when a zig encounters an overflow, it does not abort but rather extends the current value by 1 bit.

The transcript of our counter is periodic. Its period 1-2-3- $\cdots$ -132 is semantically divided into the following four subsequences, called *modules*: 1–30 (Format module or F; colored in green in figures), 31–66 (Left-Turn module or L; blue), 67–96 (Half-Adder module or H; red), and 97–132 (Right-Turn module or R; yellow). The transcript can be hence represented as $(FLHR)^*$ at the module level. Modules are to play their roles in expected environments by folding into

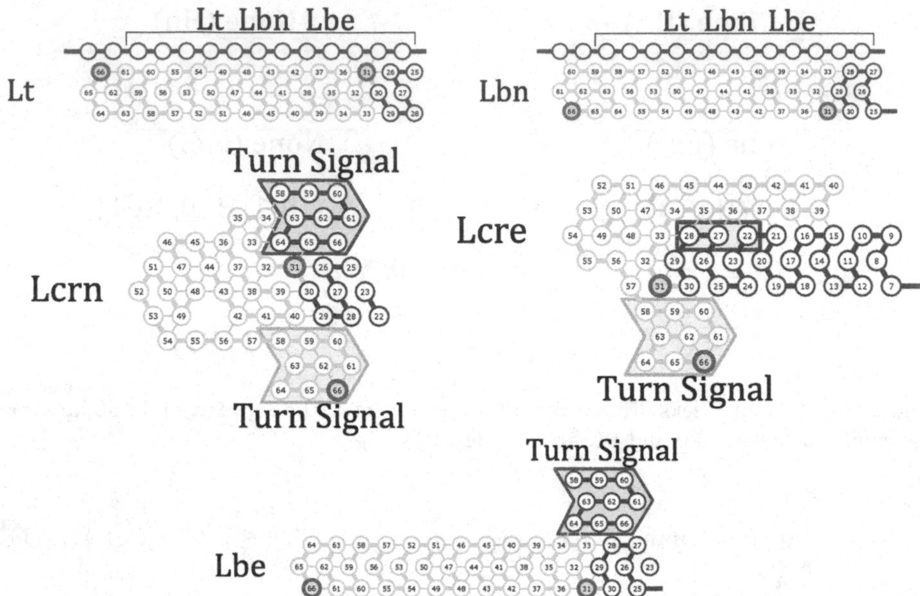

Fig. 3. All the five bricks of module L: `Lt`, `Lbn`, `Lcrn`, `Lcre`, and `Lbe` from top left to bottom right. In zigs, L folds into either `Lt` or `Lbn` depending on where it starts, until the transcript reaches the left end, where L folds either into `Lcrn` if the current value has not been overflowed, or into `Lbe` at an overflow. In the case of overflow, the next L folds into `Lcre`. In zags, L always folds into `Lbn`.

respective conformations which should be pairwise-distinct enough to be distinguishable by other modules transcribed later. Such expected conformations are called a *brick*. For example, module F encounters the four environments shown in Fig. 2 where it takes the four bricks `Fnt`, `Fnb`, `F0`, and `F1`, respectively. Here, by saying (an instance of) a module folds into (or takes) a brick in an environment, what we actually mean is that the rule set is designed so as for the transcript of the module to interact with itself as well as with the environment into that brick according to the oritatami dynamics (1). The whole system is designed to guarantee that each module is transcribed only in one of the environments it expects. This fact is illustrated in the *brick automaton*, which describes pairs of an environment and a brick as a vertex and transitions between them. Since this automaton is closed, it suffices to test whether for all pairs of an environment and a brick, the brick is folded in the environment. This test has been done *in-silico* using our simulator developed for this project. This brick automaton and all the certificates can be found at https://komaruyama.github.io/oritatami-infinit-counter/.

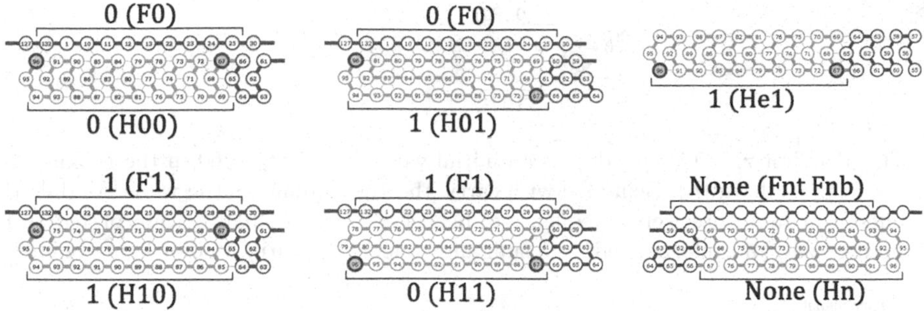

Fig. 4. All the six bricks of module H: `H00`, `H01`, `He1`, `H10`, `H11`, and `Hn` from top left to bottom right. In zags, H always folds into `Hn` while in zigs, it folds into one of the other five bricks.

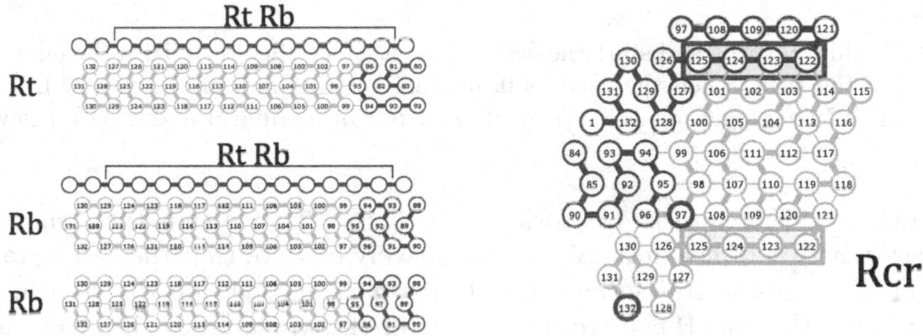

Fig. 5. All the three bricks of module R: `Rt`, `Rb`, and `Rcr`. In zigs, R folds into `Rt` or `Rb`, depending on how high it starts. In zags, R always folds into `Rb` until the transcript reaches the right end, where R folds into `Rcr`.

Seed and Encoding. The initial counter value is encoded as $b_{k-1}b_{k-2}\cdots b_1b_0$ in binary on the seed in the following format

$$\texttt{64-65-66-}\left(\prod_{i=k-1}^{0}\left(w_{Hn}w_{Rb}w_{Fb_i}w_{Lbn}\right)\right)w_{Hn} \tag{2}$$

where w_{Hn}, w_{Rb}, w_{F0}, w_{F1}, and w_{Lbn} are sequences of bead types exposed downward by modules H, R, F, L when they fold into bricks `Hn`, `Rb`, $\texttt{F}b_i$, `Lbn`, respectively, which can be found in Figs. 4, 5, 2, and 3. For example, w_{Hn} = `67-76-77-78-79-88-89-90-91-96`. The seed is examplified for $k = 1$ and $b_0 = 0$ in Fig. 6, where it is colord in purple.

Brick Level Overview. Starting from the seed, this system cyclically transits four phases: zig ($\leftarrow$), left carriage-return ($\hookrightarrow$), zag ($\rightarrow$), and right carriage-return ($\hookleftarrow$). The prefix $(FLHR)^kF$ of the transcript folds into the first zig (recall that k is the bit-width of the initial value). In zigs in general, all the instances of modules F and H fold into bricks of width 10 and height 3, while all the instances

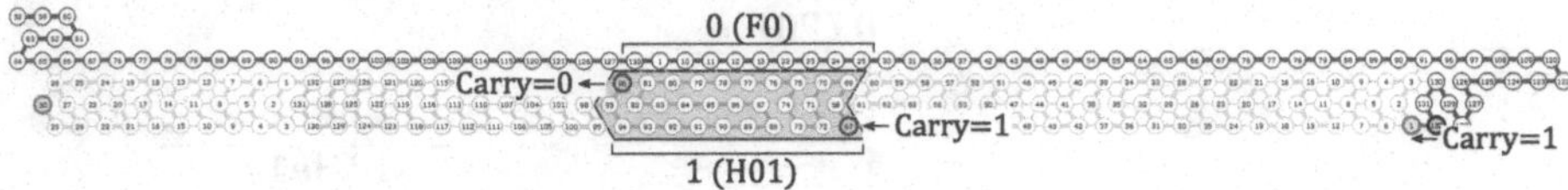

Fig. 6. The first zig. 0 is encoded as an initial value below the seed in the format (2) with $k = 1$ (1-bit width). Being fed with carry, the zig increments the value. Module H outputs 1, or more precisely a sequence of bead types which shall be interpreted as 1 in the next zag and reformatted, as a sum and cancels the carry. (Color figure online)

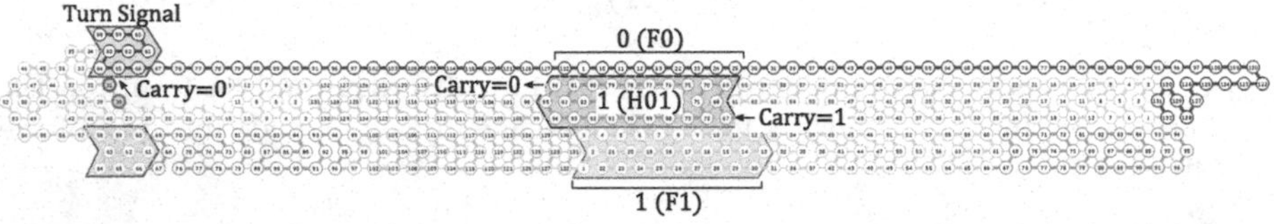

Fig. 7. Module L turns and start the first zag pass. Since there is a Turn Signal at the left end of the seed, when the carry is 0, module L turns and at the end of L forms Turn Signal. In zag pass, module F reads the output of module H and copies it down.

of L and R fold into bricks of width 12 and height 3. Zigs thus turn out to be a linear structure of height 3. We can inductively observe that the i-th instance of H in the prefix is transcribed right below b_{i-1} encoded on the seed in the format (2) so that the H can "read" b_{i-1}. After the whole prefix thus has folded into the first zig, the next L is transcribed right below Turn Signal, which lets the L fold into a special brick for left carriage-return if the zig ended at the top (this occurs when $b_{k-1}b_{k-2}\cdots b_0 < 1^k$) (see Fig. 7). We should note that this special brick `Lcre` is provided with another Turn Signal for the sake of next left carriage-return. Having been thus carriage-returned, the succeeding subsequence $(HRFL)^k H$ of the transcript folds into the first zag. Even in zags instances of F and H fold into bricks of width 10 and height 3, while those of L and R fold into bricks of width 12 and height 3. As a result, zags turn out to be a linear structure of height 3. More importantly, instances of H and F are aligned thus vertically and alternately into columns (see Figs. 6, 7 and 8), i-th of which from the right propagates the $(i-1)$-th bit of the counter value downward. After the whole subsequence has folded into the first zag, an instance of R is transcribed and folded into a special brick `Rcr` for right carriage-return due to the turn signal `125-124-123-122`, which occurs also at the bottom of `Rcr` for the sake of next right carriage-return. This amounts to one cycle of the phase transition.

Increment of the Counter. In a zig, module H plays its primary role as a half-adder and carry transfers through instances of others (F, L, and R) from an instance of H to another for more significant bit. Carry transfers as a height for modules to start. In zigs, modules F, L, and R take the respective two bricks (`Fnt` and `Fnb` for F, `Lt` and `Lbn` for L, and `Rt` and `Rb` for R; see Figs. 2, 3, and 5), both of which start and end at the same height: one at the top while the other

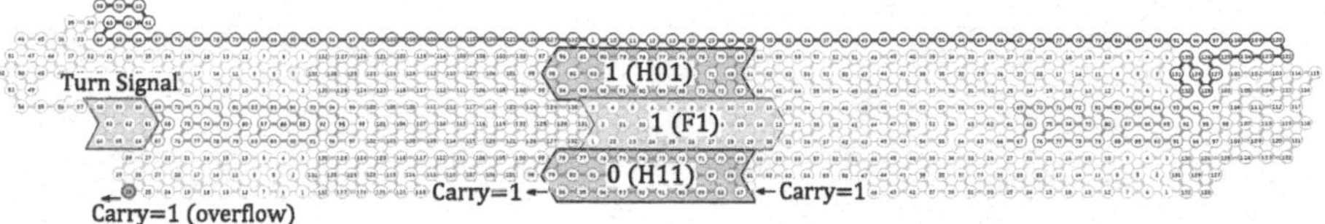

Fig. 8. Reach the left end with carry. Even if transcript of module L sticks to Turn Signal, it can not bind because the distance is long.

at the bottom. A zig is carried by being forced to start at the bottom by the last Rcr or the seed. Until the count overflows, module H encounters only four environments, which encode input 0 as w_{F0} or 1 as w_{F1} and carry or no-carry as of whether the module starts at the bottom or top, where it takes H00, H01, H10, and H11, respectively, as shown in Fig. 4 (Hxc is folded when the input is x and the carry is given if $c = 1$ or not otherwise).

Let us see how the subsequence $(FLHR)^kF$ folds into a zig in order to count up; for $k = 1$ and the current value 0, see Fig. 6. The zig starts at the bottom, that is, being carried, and the carry transfers through the first instances of F and L in the way just explained toward the first instance of H. This H is thus fed with carry and folds into H01 if the bit encoded above is 0, as illustrated in Fig. 6, or H11 if the bit is rather 1. H01 ends at the top, corresponding to canceling the carry out. This absence of carry transfers through the succeeding modules leftward. As a result, the zig ends, or more precisely an instance of F ends folding at the bottom if the current value is overflowed (Fig. 8), or at the top otherwise (Fig. 6). An instance of L is to be transcribed next. It folds either into Lcrn for (normal) carriage-return unless the current value is overflowed, or into Lbe at an overflow.

Bit-Width Expansion at an Overflow. The counter of Geary et al. cannot handle a zig that ends at the bottom, that is, its behavior is undefined at an overflow. In contrast, module L in this infinite counter is designed so as to fold into Lbe in this situation in order to continue counting up (Fig. 9). Observe that the dent on Turn Signal made of 58, 63, and 64 is too far for module L or more precisely its beads 33 and 34 to interact with strongly enough to fold into Lcrn. Lbe is a self-sustaining conformation (glider) so that it can fold even if nothing is around, which occurs at that very moment. For the same reason, the following instances of H, R, and F fold into self-sustaining conformations He1, Rb, and Fnb, respectively (Figs. 9 and 10). Note that He1 is essentially the same as H00 but exposes the opposite side downward, which will be interpreted as the leading bit 1 after expansion in the next zag. When the next instance of L is transcribed, there is nothing around. Nevertheless, it does not fold into Lbe but folds into Lcre for carriage-return; how? It is guided by interaction between the beads 35, 36 along the transcript of L and the Turn Signal 28-27-22 above Fnb (Figs. 3 and 10). This signal is usually hidden geometrically by the previous zag, and hence, does no harm.

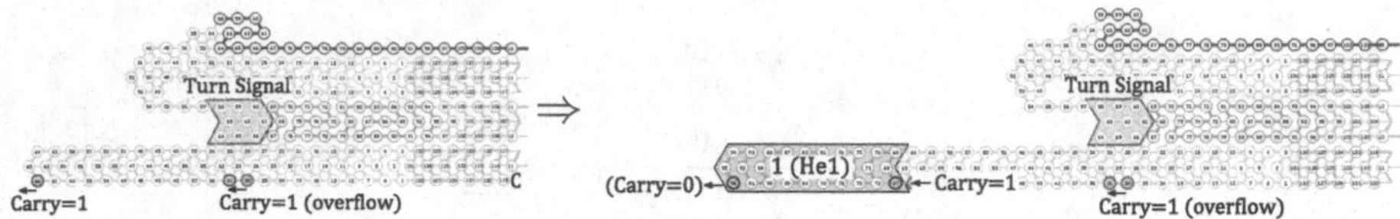

Fig. 9. (Left) Starting from the bottom, the Turn Signal above is too far for this L to fold into `Lcrn`. It rather folds into `Lbe` and initiates bit expansion. (Right) Without anything around, the succeeding H folds into a glider (brick `He1`).

Fig. 10. The succeeding R and F also fold into respective glider-like bricks. (Left) This brick of F (`Fnb`) exposes Turn Signal 28-27-22 (boxed), which is usually "hidden" under the previous zag. (Right) The exposed 28-27-22 (boxed) triggers the folding of next L into a special brick (`Lcre`) for left carriage-return.

Formatting. The value has been successfully incremented but it is not in the format (2) yet. In the upcoming zag, instances of F play their primary role to format 1 bit output by module H (recall that instances of H and F are aligned vertically and alternately). Both of the bricks of L for carriage-return, i.e., `Lcrn` and `Lcre`, end at the bottom so that zags start at the bottom. All modules start and end at the bottom in zags; note that nothing has to be transferred between modules. That is, instances of H, L, and R fold into `Hn`, `Lbn`, and `Rb`, respectively. Below the brick Hxc, an instance of F folds into Fy, where $y = (x + c) \mod 2$.

References

1. Adleman, L., Chang, Q., Goel, A., Huang, M.D.: Running time and program size for self-assembled squares. In: Proceedings of the STOC 2001, pp. 740–748. ACM (2001)
2. Bryans, N., Chiniforooshan, E., Doty, D., Kari, L., Seki, S.: The power of nondeterminism in self-assembly. Theory Comput. **9**, 1–29 (2013)
3. Evans, C.G.: Crystals that count! Physical principles and experimental investigations of DNA tile self-assembly. Ph.D. thesis, Caltech (2014)
4. Geary, C., Étienne Meunier, P., Schabanel, N., Seki, S.: Proving the turing universality of oritatami co-transcriptional folding. In: Proceedings of the ISAAC 2018, pp. 23:1–23:13 (2018)
5. Geary, C., Étienne Meunier, P., Schabanel, N., Seki, S.: Oritatami: a computational model for molecular co-transcriptional folding. Int. J. Mol. Sci. **20**(9), 2259 (2019)
6. Geary, C., Rothemund, P.W.K., Andersen, E.S.: A single-stranded architecture for cotranscriptional folding of RNA nanostructures. Science **345**(6198), 799–804 (2014)
7. Masuda, Y., Seki, S., Ubukata, Y.: Towards the algorithmic molecular self-assembly of fractals by cotranscriptional folding. In: Câmpeanu, C. (ed.) CIAA 2018. LNCS, vol. 10977, pp. 261–273. Springer, Cham (2018). https://doi.org/10.1007/978-3-319-94812-6_22

8. McClung, C.R.: Plant circadian rhythms. Plant Cell **18**, 792–803 (2006)
9. Minsky, M. (ed.): Computation: Finite and Infinite Machines. Prentice-Hall Inc., Upper Saddle River (1967)
10. Rothemund, P.W.K., Winfree, E.: The program-size complexity of self-assembled squares (extended abstract). In: Proceedings of the STOC 2000, pp. 459–468. ACM (2000)

On Synchronizing Tree Automata and Their Work–Optimal Parallel Run, Usable for Parallel Tree Pattern Matching

Štěpán Plachý(✉) and Jan Janoušek(✉)

Faculty of Information Technology, Czech Technical University in Prague, Prague, Czech Republic
{plachste,Jan.Janousek}@fit.cvut.cz

Abstract. We present a way of synchronizing finite tree automata: We define a synchronizing term and a k-local deterministic finite bottom–up tree automaton. Furthermore, we present a work–optimal parallel algorithm for parallel run of the deterministic k-local tree automaton in $\mathcal{O}(\log n)$ time with $\lceil \frac{n}{\log n} \rceil$ processors, for $k \leq \log n$, or in $\mathcal{O}(k)$ time with $\lceil \frac{n}{k} \rceil$ processors, for $k \geq \log n$, where n is the number of nodes of an input tree, on EREW PRAM. Finally, we prove that the deterministic finite bottom–up tree automaton that is used as a standard tree pattern matcher is k-local with respect to the height of a tree pattern.

1 Introduction

Finite tree automaton (FTA) is a standard model of computation for the class of regular tree languages [4,5]. Synchronizing finite string automaton [1,13,14] is a well-studied principle in the theory of formal regular string languages. We are not aware of any existing result dealing with synchronizing finite tree automata.

In this paper we present a new way of synchronizing finite tree automata: We define a synchronizing term and a k-local deterministic finite bottom–up tree automaton (k-local DFTA). In the theory of string languages reading a synchronizing word from any configuration sets a deterministic finite string automaton (DFA) in a well-defined state [1,13]. A stronger property of k-locality, $k \geq 0$, of the DFA occurs when all words of length at least k are synchronizing [13,14]. The new notions are introduced with analogous properties: The run of a DFTA reading a synchronizing term results in a well-defined state in the root of the term and a DFTA is k-local when all terms of height at least k are synchronizing.

Furthermore, we present a new work–optimal parallel algorithm for parallel run of k-local DFTA in $\mathcal{O}(\log n)$ time with $\lceil \frac{n}{\log n} \rceil$ processors, for $k \leq \log n$, or in $\mathcal{O}(k)$ time with $\lceil \frac{n}{k} \rceil$ processors, for $k \geq \log n$, where n is the number of nodes of an input tree, on EREW PRAM. We note that the run of a DFTA can be performed sequentially in $\mathcal{O}(n)$ time. Our parallel algorithm uses several basic parallel techniques and for the sake of work-optimality it further uses

The authors acknowledge the support of the OP VVV MEYS funded project CZ.02.1.01/0.0/0.0/16_019/0000765 "Research Center for Informatics".

A. Chatzigeorgiou et al. (Eds.): SOFSEM 2020, LNCS 12011, pp. 576–586, 2020.
https://doi.org/10.1007/978-3-030-38919-2_47

some nontrivial new operations on the data representation of the input tree. We are not aware of any existing algorithm for work-optimal parallel run of the DFTA. Parallelizing the run of DFTA was studied from the point of view of complexity theory: Lohrey [9] showed that the membership problem for a fixed tree automaton is in the DLOGTIME-uniform NC^1 complexity class, i.e., it can be solved in $\mathcal{O}(\log n)$ time with $\mathcal{O}(n^c)$ processors for some constant $c \geq 1$. Effectively parallelising the run of k-local string DFA was described in [7].

Tree pattern matching (TPM) [6], where a tree pattern to be matched corresponds to a connected subgraph of an input tree, is one of the most important basic algorithmic tree-related problems, with such applications as those searching in data tree representations, for example in compiler optimizations. Many string pattern matching problems were successfully solved by the use of DFA as a useful model of computation. For TPM it is possible to construct analogously DFTA. See [3] for the way of constructing a standard tree pattern matcher as a minimal DFTA, where the DFTA is extended by a simple output function indicating matches of a fixed tree pattern by some specific states of the automaton.

Finally, we prove that the minimal DFTA that is used as a standard tree pattern matcher [3] is k-local with respect to the height of a tree pattern. We are not aware of any existing known result describing a work–optimal parallel algorithm for some TPM problem. Various parallel TPM algorithms were introduced, most notably in Ramesh et al. [11] and in Tarora et al. [12], where a parallel TPM running in $\mathcal{O}(\log n)$ time with $\lceil \frac{nm}{\log n} \rceil$ processors, where m is the number of nodes of the tree pattern, on CREW PRAM is described. Our parallel TPM is work-optimal for any given fixed tree pattern on EREW PRAM.

2 Basic Notions

A *ranked alphabet* $\mathcal{F}$ is a finite nonempty set of symbols each of which has a nonnegative arity (or rank). The arity of a symbol $f \in \mathcal{F}$ is denoted by $arity(f)$ and the set of symbols of arity p is denoted by $\mathcal{F}_p$. The set $T(\mathcal{F}, X)$ of *terms* over a ranked alphabet $\mathcal{F}$ and a set of variables X, $X \cap \mathcal{F} = \emptyset$, is the smallest set defined as $\mathcal{F}_0 \subseteq T(\mathcal{F}, X)$, $X \subseteq T(\mathcal{F}, X)$ and if $f \in \mathcal{F}_p$, $p \geq 1$ and $t_1, \ldots, t_p \in T(\mathcal{F}, X)$ then $f(t_1, \ldots, t_p) \in T(\mathcal{F}, X)$. A term $t \in T(\mathcal{F}, X)$ is called *ground term* if $X = \emptyset$ and the set of all ground terms is denoted by $T(\mathcal{F})$. A *tree language* is a set of ground terms. A term is called *linear* if each variable occurs at most once in the term. All terms in this paper are assumed to be linear. A *ground substitution* σ of a set of variables X over a ranked alphabet $\mathcal{F}$ is a mapping $X \to T(\mathcal{F})$. Ground substitution can be extended to $T(\mathcal{F}, X)$ in such a way that $\forall f \in \mathcal{F}_0 : \sigma(f) = f$ and $\forall p \geq 1, \forall f \in \mathcal{F}_p, \forall t_1, \ldots, t_p \in T(\mathcal{F}, X) : \sigma(f(t_1, \ldots, t_p)) = f(\sigma(t_1), \ldots, \sigma(t_p))$. A *height* of a term $t \in T(\mathcal{F}, X)$ denoted by $height(t)$ is a function inductively defined as $\forall f \in \mathcal{F}_0 : height(f) = 0$, $\forall x \in X : height(x) = 0$ and $\forall p \geq 1$, $\forall f \in \mathcal{F}_p$, $\forall t_1, \ldots, t_p \in T(\mathcal{F}, X) : height(f(t_1, \ldots, t_p)) = 1 + \max_{i=1}^{p} height(t_i)$. A *subterm* $t|_p$ in a term $t \in T(\mathcal{F}, X)$ on a position $p \in \mathbb{N}^*$ is a term defined as, if $|p| = 0$ then $t|_p = t$, otherwise if $t = f(t_1, \ldots, t_r)$ then $t|_{ip'} = t_i|_{p'}$, where $i \leq r, p' \in \mathbb{N}^*$. The set of all subterms of t is denoted by

Subterms(t) and the set of all subterm positions of t is denoted by *SubtPos*(t). A *depth* of the subterm position denoted by *depth*(p) is the length of p. A ground term $t \in T(\mathcal{F})$ *matches a tree pattern* $p \in T(\mathcal{F}, X)$ if $\sigma(p) = t$ for some ground substitution σ.

A *deterministic (bottom–up) finite tree automaton (DFTA)* over a ranked alphabet $\mathcal{F}$ is a 4-tuple $A = (Q, \mathcal{F}, Q_f, \Delta)$, where Q is a finite set of states, $Q_f \subseteq Q$ is a set of final states, and Δ is a transition function of type $f(q_1, \ldots, q_n) \rightarrow q$, where $f \in \mathcal{F}_n, n \geq 0$ and $q, q_1, \ldots, q_n \in Q$. An *extended transition function* $\hat{\Delta}$ of a DFTA $A = (Q, \mathcal{F}, Q_f, \Delta)$ is a mapping $T(\mathcal{F}) \rightarrow Q$ such that $\forall f \in \mathcal{F}_0 : \hat{\Delta}(f) = \Delta(f), \forall p \geq 1, \forall f \in \mathcal{F}_p, \forall t_1, \ldots, t_p \in T(\mathcal{F}) : \hat{\Delta}(f(t_1, \ldots, t_p)) = \Delta(f(\hat{\Delta}(t_1), \ldots, \hat{\Delta}(t_p)))$. A DFTA $A = (Q, \mathcal{F}, Q_f, \Delta)$ is said to *accept* a ground term $t \in T(\mathcal{F})$ if $\hat{\Delta}(t) \in Q_f$. The DFTA A *matches* a tree pattern $p \in T(\mathcal{F}, X)$ if A accepts the ground term t iff t matches p. Such DFTA is called a *tree pattern matcher.* A *language of a set of states* $Q' \subseteq Q$ is a set of ground terms defined as $L(Q') = \{t \in T(\mathcal{F}) : \hat{\Delta}(t) \in Q'\}$. We assume the DFTA to have complete transition function, therefore $L(Q) = T(\mathcal{F})$. A *language of the DFTA* is $L(Q_f)$. The DFTA is *minimal* if $|Q|$ is minimal among automata accepting the same language.

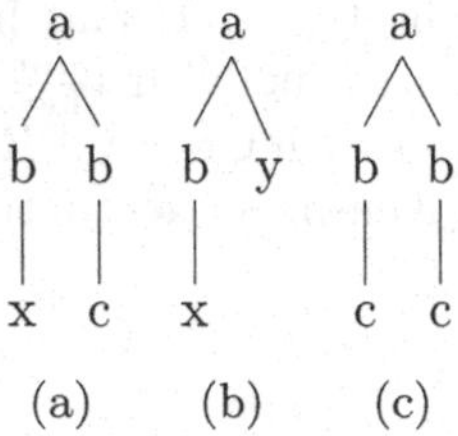

Fig. 1. Example of terms over ranked alphabet $\mathcal{F} = \{a_2, b_1, c_0\}$. Terms *(a)* and *(b)* are terms over $\mathcal{F}$ and set of variables $X = \{x, y\}$, while the term *(c)* is a ground term over $\mathcal{F}$. The term *(c)* can be created by applying a ground substitution σ to either *(a)* or *(b)*, such that $\sigma(x) = c$ and $\sigma(y) = b(c)$, and therefore *(c)* matches *(a)* and *(b)*. The position of constant c in *(a)* is 21.

$A = (Q, \mathcal{F}, Q_f, \Delta)$, where $Q = \{0, 1, 2, 3, 4\}$, $\mathcal{F} = \{a_2, b_1, c_0\}$, $Q_f = \{4\}$ and Δ :

$c \rightarrow 1$	$a(2,3) \rightarrow 4$	$b(q) \rightarrow 2, q \neq 1$
$b(1) \rightarrow 3$	$a(3,3) \rightarrow 4$	$a(q_1, q_2) \rightarrow 0, q_1 \notin \{2,3\} \vee q_2 \neq 3$

Fig. 2. Example of a minimal DFTA matching the tree pattern from Fig. 1a.

Figures 1, 2 and 4 show examples of terms, DFTA matching the pattern in Fig. 1a and its run on a ground term respectively. Note that terms are equivalent to finite ordered rooted labeled ranked trees (further on only trees), where for a term $t \in T(\mathcal{F}, X)$ the set of nodes is $SubtPos(t)$, the root is a node with position being the empty sequence, the label of each node is a ranked symbol or a variable found at its position in the term and each $p'i \in SubtPos(t)$ is the i-th child of its parent p'. We therefore use notions of terms and trees interchangeably. For our algorithms we assume the input tree in an adjacency list form by arrays *label* and *children*, where $label[i]$ contains label of i-th node and $children[i][j]$ contains index of j-th child of i-th node.

As parallel computational model we use the standard EREW PRAM (exclusive read exclusive write parallel random access machine with a shared memory). A parallel algorithm is *work-optimal* if $T(n,p) \cdot p = \mathcal{O}(SU(n))$ for an input of size n and p processors, where $T(n,p)$ is the parallel time complexity of the algorithm and $SU(n)$ is the upper bound time complexity of the best known sequential algorithm for the problem. Several fundamental parallel algorithms are used, including *prefix (suffix) sum*, *segmented prefix (suffix) sum* (or *segmented scan*), *list ranking*, *parenthesis matching*, all running in $\mathcal{O}(\log n)$ time with $n/\log n$ processors, and *Euler tour technique*, which can be constructed in $\mathcal{O}(1)$ time with n processors. For details see [4,8,10].

3 Synchronizing Term and k-Local DFTA

Assuming a DFA without unreachable states and with a complete transition function, we can say that a string is synchronizing if the run of the DFA on the string with any added prefix substring ends up in the same state. An analogous operation for terms and tree patterns is adding a subterm, which can be achieved only by substituting a variable. We define a synchronizing term as follows.

Definition 1. *A term* $t \in T(\mathcal{F}, X)$ *is called* synchronizing *for a DFTA* $A = (Q, \mathcal{F}, Q_f, \Delta)$ *if there exists a state* $q \in Q$ *such that* $\forall \sigma : \hat{\Delta}(\sigma(t)) = q$.

Terms in Fig. 1a and c are synchronizing for the automaton in Fig. 2, while the term in Fig. 1b is not. Note that all ground terms are synchronizing.

A DFA is called k-local if all strings of length at least k are synchronizing. That way we need only a suffix of the input of length k to compute a state. We define a similar property for tree automata such that we need a subtree of only k levels to compute a state.

Definition 2. *A* minimal variable depth $MVD(t)$ *of a term* $t \in T(\mathcal{F}, X)$ *is a function defined as:* $\forall f \in \mathcal{F}_0 : MVD(f) = +\infty$; $\forall x \in X : MVD(x) = 0$; $\forall p \geq 1, \forall f \in \mathcal{F}_p, \forall t_1, \ldots, t_p \in T(\mathcal{F}, X) : MVD(f(t_1, \ldots, t_p)) = 1 + \min_{i=1}^{p} MVD(t_i)$.

Definition 3. *DFTA* $A = (Q, \mathcal{F}, Q_f, \Delta)$ *is* k-local *if all terms* $t \in T(\mathcal{F}, X)$ *such that* $MVD(t) \geq k$ *are synchronizing.*

$A = (Q, \mathcal{F}, Q_f, \Delta)$, where $Q = \{0, 1\}$, $\mathcal{F} = \{b_1, c_0\}$, $Q_f = \{0\}$ and Δ :

$$c \to 0 \qquad b(0) \to 1 \qquad b(1) \to 0$$

Fig. 3. Example of a DFTA accepting ground terms with even number of symbols b. This DFTA is not k-local for any $k \geq 0$.

Figure 3 shows a DFTA that is not k-local for any $k \geq 0$ since its states represent the parity of the number of occurences of symbol b in a subtree, which always depends on the entire subtree of any height.

For any tree pattern a DFTA matching that pattern can be constructed, as shown in [3]. This DFTA is minimal and each state indicates a match of exactly a set of some subterms of the pattern and therefore the state depends only on a subtree no higher than the pattern and is k-local with respect to that height. The automaton in Fig. 2 is a minimal DFTA matching the tree pattern from Fig. 1a and is therefore 3-local.

Theorem 1. *For each term $p \in T(\mathcal{F}, X)$ a minimal DFTA $A = (Q, \mathcal{F}, Q_f, \Delta)$ matching p is k-local, where $k = height(p)$.*

4 Parallel Run of k-local DFTA on EREW-PRAM

The motivating idea for our parallel algorithm, described by Algorithm 1, is dividing the input tree into segments of k levels, or possibly fewer levels in the case of the lowest segment, assigning arbitrary states to each node and performing two passes of the bottom–up computation of the k-local DFTA on each segment in parallel. The first pass would be a synchronization phase that obtains the correct initial states because of k locality. The second pass would perform the run based on the correct states. However, such division is not trivial when the input tree is in adjacency list form. Therefore linear order, defined by Definition 4, is computed first. Table 1 shows an example computation of the algorithm.

Algorithm 1. Parallel run of k-local DFTA

Require: k-local DFTA A, tree of size n given as arrays *label* and *children*
Ensure : *state* array with states of each node

```
Using Euler tour technique on children compute depth array with depth of each node
DMKA ← depthModkSort(children, depth)
step ← computeStep(DMKA, depth)
∀i < n : state[i] ← 0                                   init arbitrary state
computeState(A, DMKA, step, children, state)                    first pass
computeState(A, DMKA, step, children, state)                   second pass
```

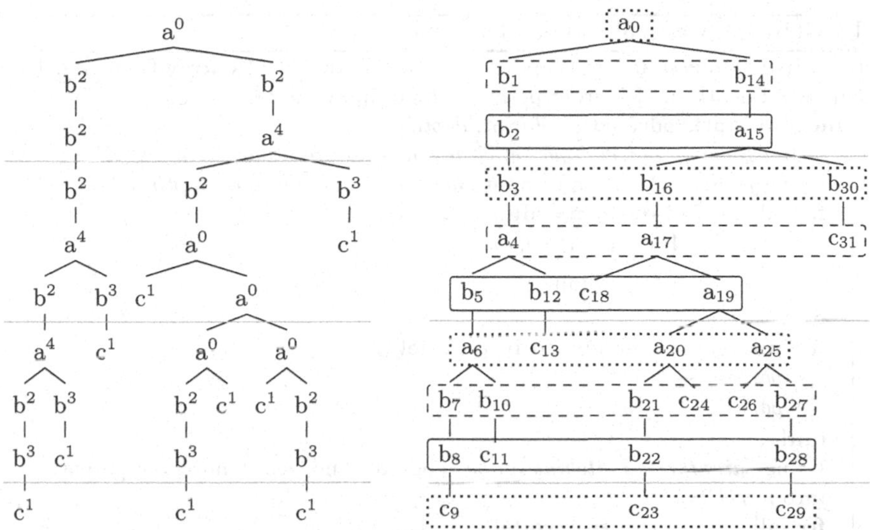

Fig. 4. Figure on the left side shows example of a ground term where numbers in each node represent states after a run of the DFTA in Fig. 2. Separated segments can be computed independently in the parallel algorithm. Figure on the right side shows groups of the depth-mod-k order of the tree for $k = 3$. Nodes are written with their preorder number for further identification.

Definition 4. Depth-mod-k order *of term t is a total preorder of $SubtPos(t)$ where $\forall p_1, p_2 \in SubtPos(t) : p_1 \leq p_2 \Leftrightarrow depth(p_1) \bmod k \leq depth(p_2) \bmod k$.*

An i-th group *of the order is an equivalence class of associated equivalence relation to the preorder such that $p \bmod k = i$ for $p \in SubtPos(t)$ and $i < k$.*

A linear depth-mod-k order *of a term t is a linear order of $SubtPos(t)$ that is a maximal antisymmetric subrelation of the depth-mod-k order of t.*

To compute the linear order we modify a parallel algorithm for computing breadth-first traversal order of tree nodes in [2], where we change the order of levels by modifying some pointers, as described in Algorithm 2. With the Euler tour technique the algorithm constructs a list of parentheses and using parenthesis matching [10] changes pointers to create the resulting order (Fig. 5).

With ordered nodes we can process all nodes in a group in parallel, starting with the lowest group located at the end. As described in Algorithm 4, all processors will traverse a group from right to left, as shown in Fig. 6, until the boundary with the preceding group is reached where some processors might stall if the size of the group is not divisible by the number of processors. To determine when a processor should do computation on a node and when to stall,

Algorithm 2. Computation of depth-mod-k order

```
Require: children array of input tree with n nodes, depth array from Alg. 1
Ensure : depth-mod-k order permutation of input nodes
Function DepthModkSort(children, depth)
    Using Euler tour technique on children compute an array of arcs EA
    Using parallel reduction compute height as maximum of depth array
    for all e ∈ EA do in parallel
        par[e] ← { ')'  if e downgoing
                 { '('  if e upgoing
    end
    for all i ∈ 1, . . . , height do in parallel
        par[−i] ← '('
        par[|EA| − 1 + i] ← ')'
    end
    Using parenthesis matching on par compute indices of matching parentheses
    into nextPar
    for all e ∈ EA do in parallel
        if e downgoing then
            nextPar[e] ← opposite arc of e
        end
    end
    for all i ∈ 1, . . . , height do in parallel
        nextPar[|EA| − 1 + i] ← { nextPar[−i − k]          if i < height − k
                                { nextPar[−i mod k] − 2    if i mod k ≠ k − 1
    end
    for all e ∈ EA do in parallel
        if e upgoing then
            nextPar[e] ← nextPar[nextPar[e]]
        end
    end
    Using list ranking on nextPar create array of lower nodes of corresponding
    arcs, add root of the tree to the beginning and return result
```

while maintaining read exclusivity, we first preprocess the input with Algorithm 3 to calculate in which parallel step a node should be computed. For the traversal at least k sequential steps are needed. Since each processor can stall at most once for each group, the total number of stalls is no more than n with n/k processors.

All operations used can be computed work optimally with at most either $\lceil \frac{n}{\log p} \rceil$ or $\lceil \frac{n}{k} \rceil$ processors. Using the Brent's scheduling principle [8] the whole algorithm can then run work efficiently with the number of processors being minimum of those values. Parallel time changes accordingly.

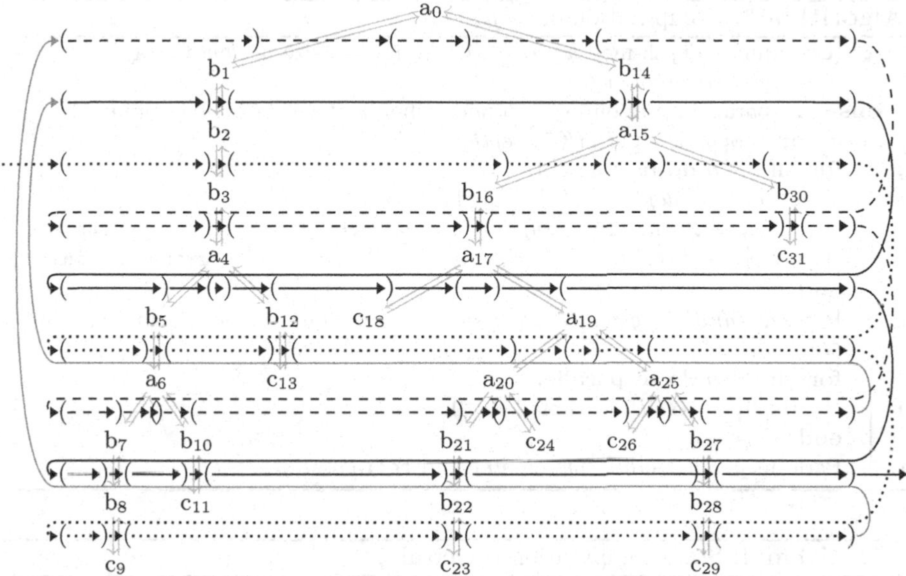

Fig. 5. Linked list of parentheses in Algorithm 2 for tree in Fig. 4 representing nodes, excluding root, in a linear depth-mod-k order (see Fig. 6).

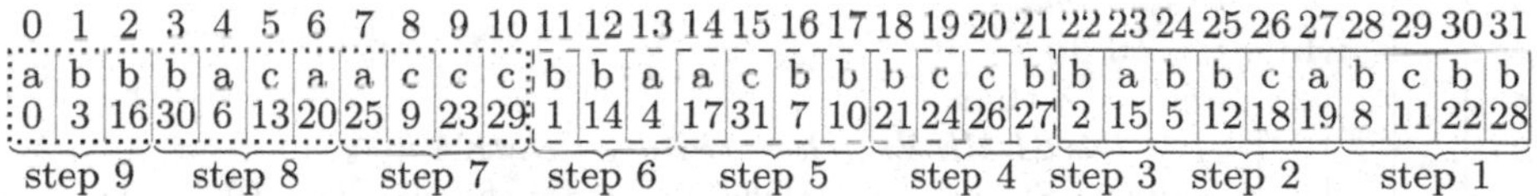

Fig. 6. Nodes in linear depth-mod-k order of the tree in Fig. 4 with parallel steps, computed by Algorithm 3, in which a node should be processed when running Algorithm 4 with 4 processors.

Theorem 2. *Algorithm 1 correctly computes the run of a k-local DFTA on an input tree of size n in time $\mathcal{O}(\max(k, \log n))$ with $\lceil \frac{n}{\max(k,\log n)} \rceil$ processors on EREW PRAM.*

For the future work properties of synchronizing DFTA, such as those analogous to Černý conjecture [13] for DFA, can be studied as well as constructing efficient parallel algorithms for various problems solvable by k-local DFTA.

Algorithm 3. Computation of *step* array

Require: linear depth-mod-k order array of tree of size n, *depth* array, processor count p
Ensure : parallel iteration for each node when a state should be computed

Function `computeStep`(*DMKA, depth*)
  for all $i < n$ **do in parallel**
    $group[i] \leftarrow depth[DMKA[i]] \bmod k$
    $GE[i] \leftarrow i = n - 1 \vee group[i] \neq group[i+1]$ `group end`
    $GEI[i] \leftarrow i \cdot GE[i]$ `group end index`
  end
  Perform *parallel segmented suffix sum* on GEI with segment bound flags in GE
  for all $i < n$ **do in parallel**
    $step[i] \leftarrow (GEI[i] - i) \bmod p = 0$
  end
  Perform *parallel suffix sum* on *step* and **return** *step*

Algorithm 4. State computation traversal

Require: k-local DFTA, depth-mod-k order array of size n, *step* array, *children* array, *state* array, processor id *pid*
Ensure : recalculated *state* array

Procedure `computeState`($A = (Q, \mathcal{F}, Q_f, \Delta)$, *DMKA, step, children, state*)
  do in parallel
    $j \leftarrow 1, i \leftarrow n - pid$
    while $i >= 0$ **do**
      if $step[i] = j$ **then**
        $f \leftarrow label[DMKA[i]], ch \leftarrow children[DMKA[i]]$
        $childrenStates \leftarrow state[ch[0]] \dots state[ch[arity(f) - 1]]$
        $state[DMKA[i]] \leftarrow \Delta(f, childrenStates)$
        $i \leftarrow i - p$
      end
      $j \leftarrow j + 1$
    end
  done

Table 1. Computation of Algorithm 1 with the 3-local DFTA from Fig. 2, the input tree from Fig. 4 and 4 processors. Arrays in parenteses are shown permuted by the linear depth-mod-k order (array *DMKA*). Array *GE* indicates group end. Arrays *GEI* (group end index) and *step* are shown before and after their respective suffix sum operation. Array *state* is shown on initialization and after the first and the second pass.

(DMKA)	0	1	2	3	4	5	6	7	8	9	10	11	12	13	14	15	16	17	18	19	20	21	22	23	24	25	26	27	28	29	30	31
DMKA	0	3	16	30	6	13	20	25	9	23	29	1	14	4	17	31	7	10	21	24	26	27	2	15	5	12	18	19	8	11	22	28
(label)	a	b	b	b	a	c	a	a	c	c	c	b	b	a	a	c	b	b	b	c	c	b	b	a	b	b	c	a	b	c	b	b
(children)	1	4	17	31	7		21	26				2	15	5	18		8	11	22			28	3	16	6	13		20	9		23	29
	14				10		24	27						12	19									30				25				
(depth)	0	3	3	3	6	6	6	6	9	9	9	1	1	4	4	4	7	7	7	7	7	7	2	2	5	5	5	5	8	8	8	8
(group)	0	0	0	0	0	0	0	0	0	0	0	1	1	1	1	1	1	1	1	1	1	1	2	2	2	2	2	2	2	2	2	2
GE	0	0	0	0	0	0	0	0	0	0	1	0	0	0	0	0	0	0	0	0	0	1	0	0	0	0	0	0	0	0	0	1
GEI	0	0	0	0	0	0	0	0	0	0	10	0	0	0	0	0	0	0	0	0	0	21	0	0	0	0	0	0	0	0	0	31
	10	10	10	10	10	10	10	10	10	10	10	21	21	21	21	21	21	21	21	21	21	21	31	31	31	31	31	31	31	31	31	31
step	0	0	1	0	0	0	1	0	0	0	1	0	0	1	0	0	0	1	0	0	0	1	0	1	0	0	0	1	0	0	0	1
	9	9	9	8	8	8	8	7	7	7	7	6	6	6	5	5	5	5	4	4	4	4	3	3	2	2	2	2	1	1	1	1
(state)	0	0	0	0	0	0	0	0	0	0	0	0	0	0	0	0	0	0	0	0	0	0	0	0	0	0	0	0	0	0	0	0
	0	2	2	3	4	1	0	0	1	1	1	2	2	0	0	1	2	3	2	1	1	2	2	0	2	2	1	0	2	1	2	2
	0	2	2	3	4	1	0	0	1	1	1	2	2	4	0	1	2	3	2	1	1	2	2	4	2	3	1	0	3	1	3	3

References

1. Béal, M.-P., Perrin, D.: Symbolic dynamics and finite automata. In: Rozenberg, G., Salomaa, A. (eds.) Handbook of Formal Languages, pp. 463–506. Springer, Heidelberg (1997). https://doi.org/10.1007/978-3-662-07675-0_10
2. Chen, C.C.-Y., Das, S.: Breadth-first traversal of trees and integer sorting in parallel. Inf. Process. Lett. **41**, 39–49 (1992)
3. Cleophas, L.G.W.A.: Tree algorithms: two taxonomies and a toolkit. Ph.D. thesis, Department of Mathematics and Computer Science (2008)
4. Comon, H., et al.: Tree automata techniques and applications (2007). http://www.grappa.univ-lille3.fr/tata. Accessed 12 Oct 2007
5. Gécseg, F., Steinby, M.: Tree languages. In: Handbook of Formal Languages, vol. 3, pp. 1–68. Springer, New York (1997). https://doi.org/10.1007/978-3-642-59136-5
6. Hoffmann, C.M., O'Donnell, M.J.: Pattern matching in trees. J. ACM **29**(1), 68–95 (1982)
7. Holub, J., Štekr, S.: On parallel implementations of deterministic finite automata. In: Maneth, S. (ed.) CIAA 2009. LNCS, vol. 5642, pp. 54–64. Springer, Heidelberg (2009). https://doi.org/10.1007/978-3-642-02979-0_9
8. JaJa, J.F.: An Introduction to Parallel Algorithms. Addison Wesley Longman Publishing Co., Inc., Redwood City (1992)
9. Lohrey, M.: On the parallel complexity of tree automata. In: Middeldorp, A. (ed.) RTA 2001. LNCS, vol. 2051, pp. 201–215. Springer, Heidelberg (2001). https://doi.org/10.1007/3-540-45127-7_16
10. Prasad, S.K., Das, S.K., Chen, C.C.-Y.: Efficient EREW PRAM algorithms for parentheses-matching. IEEE Trans. Parallel Distrib. Syst. **5**(9), 995–1008 (1994)
11. Ramesh, R., Ramakrishnan, I.V.: Parallel tree pattern matching. J. Symb. Comput. **9**(4), 485–501 (1990)

12. Tarora, K., Hirata, T., Inagaki, Y.: A parallel algorithm for tree pattern matching. Syst. Comput. Japan **24**(5), 30–39 (1993)
13. Černý, J.: Poznámka k homogénnym experimentom s konečnými automatmi. Matematicko-fyzikálny časopis **14**(3), 208–216 (1964)
14. Rosenauerová, B., Černý, J., Pirická, A.: On directable automata. Kybernetika **7**(4), 289–298 (1971)

On the Hardness of Energy Minimisation for Crystal Structure Prediction

Duncan Adamson[1,2(✉)], Argyrios Deligkas[2], Vladimir V. Gusev[2], and Igor Potapov[1]

[1] Department of Computer Science, Univiersity of Liverpool, Liverpool, England
duncan.adamson@liverpool.ac.uk
[2] Leverhulme Research Centre for Functional Materials Design, Liverpool, England

Abstract. Crystal Structure Prediction (CSP) is one of the central and most challenging problems in materials science and computational chemistry. In CSP, the goal is to find a configuration of ions in 3D space that yields the lowest potential energy. Finding an efficient procedure to solve this complex optimisation question is a well known open problem in computational chemistry. Due to the exponentially large search space, the problem has been referred in several materials-science papers as "NP-Hard" without any formal proof. This paper fills a gap in the literature providing the first set of formally proven NP-Hardness results for a variant of CSP with various realistic constraints. In particular, this work focuses on the problem of *removal*: the goal is to find a substructure with minimal energy, by removing a subset of the ions from a given initial structure. The main contributions are NP-Hardness results for the CSP *removal* problem, new embeddings of combinatorial graph problems into geometrical settings, and a more systematic exploration of the energy function to reveal the complexity of CSP. These results contribute to the wider context of the analysis of computational problems for weighted graphs embedded into the 3-dimensional Euclidean space, where our NP-Hardness results holds for complete graphs with edges which are weighted proportional to the distance between the vertices.

1 Introduction

One of the central and most challenging problems in materials science and computational chemistry is the problem of predicting the structure of a crystal given the set of ions composing it [14]. The goal is to find a structure of ions that achieves the lowest energy. This problem, *Crystal Structure Prediction* (CSP), has remained open due to the complexity of solving it optimally [14] and the combinatorial explosion following a brute-force approach. Current approaches to this problem are based on heuristic techniques [9,12], however they cannot guarantee optimality while remaining computationally demanding.

In generic formulations of CSP there are many degrees of freedom due to the numerous parameters: the number of ions, their positions, and the unique interactions between each type of ion. The search space remains exponential in

A. Chatzigeorgiou et al. (Eds.): SOFSEM 2020, LNCS 12011, pp. 587–596, 2020.
https://doi.org/10.1007/978-3-030-38919-2_48

size even for greatly simplified versions of CSP. Due to this, CSP has, incorrectly, been referred to in several computational-chemistry papers as "NP-Hard and very challenging" [11]. However the argument that the search must be done in a set of exponential size implies NP-Hardness does not hold.

The two results which are often mentioned in context of the NP-Hardness of CSP are [3] and [13]. In [3], within the context of the Ising model, the authors show NP-Hardness in the model of placing ± 1 charges on a graph with degree at most 6 taking into account only the local interactions between connected vertices. In [13], provides a reduction to TSP, showing the problem belongs to NP however not Hardness.

In this work, several variants of CSP are considered, providing alternative reasons for the hardness of closely related problems, focusing on the problem of *removal*. Inspiration comes from hard combinatorial problems in graph theory and proposes several new embeddings of NP-Hard graph problems into numerical versions of CSP which can be seen as an optimisation problem for weighted geometric graphs with a non-linear objective function. The input is a configuration of the ions, with the goal to remove a subset of the ions such that the interaction energy among the remaining atoms is minimised. The problem of removing vertices of a graph whose deletion results in a subgraph satisfying some specific property have been intensively studied in the combinatorial graph theory. [8] shows that for a large class of properties this problem is NP-Complete, extended in [16] and [15] to further properties showing NP-Completeness for bipartite graphs and for non-trivial hereditary properties.

The *removal* problem can be seen as a variant of combinatorial CSP problem, where the positions of the ions correspond to points in a discrete grid. The idea is to find an optimal structure by placing many copies of the ions used to build a new structure in unrealistic positions in the discrete space. Due to the nature of the energy function, when the goal is to minimise the potential energy, the excess ions must be removed. In this variant of the *removal* problem for which NP-Hardness is shown, the initial configuration (from where the ions are removed) is part of the input and has only vacant positions or positions with a single ions in the discrete three-dimensional-Euclidean space.

Our Contributions. This work provides the first NP-Hardness results for CSP [7] with realistic constraints, providing new embeddings of combinatorial graph problems in geometrical settings, as well as exploring the energy function in a more systematic way that could reveal the computational complexity of CSP. Moreover, these results can be seen as part of a more general problem of removing vertices from a weighted graph embedded into 3D Euclidean space. Three versions of this problem are considered:

- **k-Charge Removal:** Remove exactly k charges minimising the total energy;
- **Minimal At-Least-k-Charge Removal:** A generalisation of k-charge removal where the removed set is a *minimal* set of at least k charges minimising the total energy;

- **At-Least-k-Charge Removal:** A generalisation of min-at-least-k-charge removal where the removed set is of least charges but not necessarily minimal, minimising the total energy.

One challenge of the Euclidean graphs considered here is that these graphs are complete, with edges weighted proportional to the distance between the vertices. Many classical NP-Hard problems are much harder to embed into this setting. Even for some existing hardness results, in both the geometric and more restricted Euclidean setting, to bring these problems into a bounded number of dimensions often requires non-trivial technical proofs as dimension often is part of the input [2,10]. Often these constructions utilise the results on geometric graphs embedded into the plane [5,6], with many problems in this field open.

This work will be organised as follows: Sect. 2 provides relevant notation and definitions, Sect. 3 presents NP-Hardness for the general case of the problems under both energy function in $\mathcal{F}$ and the *Coulomb* (electrostatic) potential. Section 4 restricts the problem to only 2 species under the *Buckingham-Coulomb* (interatomic) potential, and is shown the remain NP-Hard in Theorem 5. The full version of this paper, containing the omitted proofs is available at arXiv [1].

Theorem	Summary	Setting
Theorem 1	NP-Completeness by reduction from the clique problem	All problems, under any energy function in $\mathcal{F}$, charges of $\pm c$ for a given c and an unbounded number of ion species
Theorem 2	NP-Completeness by extension of Theorem 1	All problems, under any energy function in $\mathcal{F}$, any bounded set of charges and an unbounded number of ion species
Theorem 3	Reduction to max-weight-k-clique	k-charge removal or minimal-at-least-k-charge removal under any computable energy function, charges of $\pm c$ for a given c, and a unbounded number of ion species
Theorem 4	NP-Completeness by reduction from the knapsack problem	Minimal-at-least-k-charge removal and at-least-k-charge removal, under the Coulomb potential energy function, unbounded number of charges and unbounded number of ion species
Theorem 5	NP-Completeness by reduction from independent set on penny graphs	All problems, under the Buckingham-Coulomb potential energy function, charges of ± 1, and two species of ion

2 Notation and Definitions

Unit Cell. A *crystal* is a solid material whose ions, are arranged in a highly ordered arrangement, forming a *crystal structure* that extends in all directions. A crystal structure is described by its *unit cell*; a region of $\mathbb{R}^3$ bounded by a parallelepiped representing a period containing ions in a specific arrangement. The unit cells are stacked in $\mathbb{R}^3$ tiling the whole space forming a crystal. The unit cell is a parallelepiped alongside the *arrangement* of ions with their specie. Each unit cell contains a set of n *ions* within the parallelepiped. Each ion, i, has a *specie*, e.g. Ti or Sr, and a non-zero charge q_i. The specie for an ion i will be denoted $S(i)$. All unit cells are neutrally charged, i.e., $\sum_{1 \le i \le n} q_i = 0$. An arrangement defines a position for every ion in the unit cell.

Energy. The energy of a crystal is computed by summing the pairwise interactions between all pairs of ions. A positive value for the pairwise interaction means the two ions are repelling, while a negative value means they are attracting. Each pair of species has a unique set of parameters (called *force fields*) which are applied to the common energy function U alongside the Euclidean distance between the ions. In general, energy is defined via series as a crystal is infinite.

In this paper interaction will be restricted to a single unit cell. The primary reason is that the energy between ions in different unit cells quickly converges, making the energy within a single unit cell a good approximation of the total.

Each arrangement has n ions and a corresponding *potential energy* PE, calculated with respect to the given energy function U. The goal is to minimise the potential energy. Pairwise interaction between two ions i and j with respect to the energy function U is $U(i,j)$, denoted U_{ij} when it is clear from the context. The value of U_{ij} is defined by the force field of the ions and the Euclidean distance between them, which is included as one of the parameters. The total potential energy for an arrangement of n ions is given by $PE = \sum_{1 \le i,j \le n, i \ne j} U_{ij}$.

This paper will consider a general class of energy functions, called the *controllable* potential functions, denoted $\mathcal{F}$. All functions in $\mathcal{F}$ are computable in polynomial time for any input. Intuitively, for every $f \in \mathcal{F}$ there exists a set of force field parameters that counteract the distance parameter r. Formally, a function $f : \mathbb{R}^n \mapsto \mathbb{R}$ belongs to $\mathcal{F}$ if and only if for any given $a \in \mathbb{R}$ and any fixed $r \in \mathbb{R}^+$ there exists a set $\{x_1 \dots x_{n-1}\} \in \mathbb{R}^{n-1}$ such that $f(x_1, \dots, x_{n-1}, r\} = a$.

The most popular function for CSP, which will be focused on in this paper, is the *Buckingham-Coulomb* potential [4], which is the sum of the Buckingham and Coulomb potentials. The Coulomb potential for a pair of ions i, j is $U_{ij}^C = \frac{q_i q_j}{r_{ij}}$, where r_{ij} is the Euclidean distance between the ions. The Buckingham potential for a pair of ions i, j, U_{ij}^B, is defined by four parameters. These are the distance and the three force field parameters, $A_{S(i),S(j)}$, $B_{S(i),S(j)}$, $C_{S(i),S(j)}$, which are dependent on the specie of the ions. It should be noted that all three parameters are positive values. The energy is calculated as $U_{ij}^B = \frac{A_{S(i),S(j)}}{e^{B_{S(i),S(j)} r_{ij}}} - \frac{C_{S(i),S(j)}}{r_{ij}^6}$. Therefore the Buckingham-Coulomb potential is given by:

$$U_{ij}^{BC} = U_{ij}^{B} + U_{ij}^{C} = \frac{A_{S(i),S(j)}}{e^{B_{S(i),S(j)}r_{ij}}} - \frac{C_{S(i),S(j)}}{r_{ij}^{6}} + \frac{q_i q_j}{r_{ij}}.$$

Proposition 1. *There exists a set of parameters for the Buckingham-Coulomb function such that it is in $\mathcal{F}$.*

Crystals as Geometric Graphs. Using the above definitions, it can be shown how crystals may be viewed as geometric graphs. Recall that each ion corresponds to a charged point in $\mathbb{R}^3$. Each ion is represented with a weighted vertex, also placed into $\mathbb{R}^3$ at the same position as the ion, giving a total of n vertices. The vertex corresponding to the ion i, denoted v_i, is assigned a weight of q_i. $wt(v_i)$ will denote the weight of a given vertex v_i, i.e. $wt(v_i) = q_i$. For notation, V^+ will denote the set of vertices with a positive weight in V, and V^- for the set of vertices with a negative weight in V. Between each pair of vertices there is an edge, weighted by the pairwise interaction of the corresponding ions U_{ij}. Note that U_{ij} will be determined by the length of the edge, which will be a straight line in the space. The energy of a crystal graph $G = \{V, E\}$ can be computed as $PE = \sum_{\{v_i,v_j\}\in E} U_{ij}$. Geometric graphs created from a unit cell will be referred to as *crystal graphs.*

The *Charge Removal* Problem. The Charge removal problem takes as input a crystal graph G corresponding to a "dense" initial arrangement of ions, with the goal of removing some subset of vertices X. In the most general case this may be any subset, provided the final graph is *charge neutral*, meaning it satisfies $\sum_{v_i\in R} wt(v_i) = 0$. It will be assumed that the initial graph is charge neutral, and therefore that X is also neutral. This work will consider three variants of this problem where there are further conditions on the set, summarised in Table 1. Note that the second of these, *At-Least-k-Charges*, becomes the general case when $k = 0$. This work considers three restrictions on the removed set, which are defined in Table 1. The base version of the problem is stated as:

Instance: A crystal graph G, with edges weighted by a given common energy function U.
Goal: The set of charges R satisfying P from G such that $G' = G \setminus R$ created by the removal of R from G which minimises $\sum_{\{v_i,v_j\}\in E'} U_{ij}$.

From this problem, a decision version may be obtained by asking if there exists a removal that leaves G' with no-more total energy than some goal g, i.e. $\sum_{v_i,v_j\in V',i\neq j} U_{ij} \leq g$. In the case there is some restriction on the output, there may also be additional input - in all the cases considered here this will be a natural k. In the remainder of this work, the problems under the restrictions in Table 1 and will be denoted as follows:

- The k-Charge-Removal Problem (K-CHARGE REMOVAL).
- The At-Least-k-Charge-Removal Problem (AT-LEAST-K-CHARGE REMOVAL).
- The Minimal At-Least-k-Charge Removal Problem (MINIMAL-AT-LEAST-K-CHARGE REMOVAL).

Table 1. Summary of restrictions for the charge removal problem.

Restriction	Summary
k-Charges	A *neutral* set of charges R where $\left\lvert \sum_{v_i \in R+} wt(v_i) \right\rvert = k$
At-Least-k-charges	A neutral set of charges R where $R \subseteq V$ and $\sum_{v_i \in R+} wt(v_i) \geq k$
Minimal-At-Least-k-charges	A *minimal* set of at-least-k-charges R - where minimal means that there does not exist any neutral subset $R' \subset R$ where R' is also a set of at least k-charges

Proposition 2. *A solution to* K-CHARGE REMOVAL *or* AT-LEAST-K-CHARGE REMOVAL *can be verified in polynomial time.*

Proposition 3. *A set of k-charges may be verified as minimal in polynomial time if and only if the set of allowed values for charges is polynomially bounded.*

Proposition 2 follows from noting that for a given graph with precomputed weights for the edges, the requisite edges and vertices may be summed to verify that it is either a set of k or of at-least-k charges, and that the energy is bellow the required bound in the decision case. Proposition 3 is shown by reduction from the subset sum problem to the problem of verifying if the set is *minimal*, as defined in Table 1.

3 NP-Hardness for an Unbounded Number of Ion Species

This section will focus on results for an unbounded number of ion species. Theorems 1, 2 and 4 will show NP-Hardness for various settings via a series of reductions under the general class of potential function in the case of Theorems 2 and 2, and under the Coulomb energy in Theorem 4. Theorem 3 will show a novel way of encoding the problem into the well studied max-weight clique problem. While these results will apply to all restrictions, it should be noted that in the case the charges are not bounded, although MINIMAL-AT-LEAST-K-CHARGE REMOVAL will remain NP-Hard it will not be in NP.

Theorem 1. K-CHARGE REMOVAL, MINIMAL-AT-LEAST-K-CHARGE REMOVAL *and* AT-LEAST-K-CHARGE REMOVAL *are NP-Complete for energy functions in* $\mathcal{F}$ *for charges of* $\pm c$*, for any natural number* c*.*

Theorem 2. K-CHARGE REMOVAL *remains NP-Hard for set of allowed charges with unique magnitude and an energy function within* $\mathcal{F}$*.*

Theorem 3. K-CHARGE REMOVAL *can be reduced to* MAX-WEIGHT K-CLIQUE *in polynomial time, under the restriction that charges are limited* $\pm c$ *and the energy function is computable within polynomial time.*

Theorems 1 and 2 come by a reduction from the Max-Clique problem. Theorem 1 provides a construction for the decision version of the charge removal problem from an instance of Max-Clique using constant charges such that under any of the restrictions on the removed vertices a solution the the charge removal instance will imply a solution to the Max-Clique problem. This is extended in Theorem 2, where it is shown that this construction may be extended with a set of *dummy* vertices, the removal of which may be done at no cost while maintaining the total set a charge neutral. Theorem 3 provides a novel encoding of the charge removal problem into the well known maximum weight clique problem.

Theorem 4. AT-LEAST-K-CHARGE REMOVAL *remains NP-Hard when the energy function is limited to the Coulomb potential.*

Theorem 4 compliments Proposition 3 by showing that, even in the case the removal does not have to be verified as minimal, the complexity of finding a solution may still be NP-Hard for the Coulomb potential function.

4 NP-Hardness for a Bounded Number of Species

In Sect. 3 NP-Hardness was shown for the case that there was an unbounded number of species, and NP-completeness in the case that there is a bounded number of charges. This will now be strengthened by considering instances with only two unique species. Only the Buckingham-Coulomb potential function with charges of ± 1 will be considered in this section. All three problems will again be considered, noting that for charges of ± 1 K-CHARGE REMOVAL is equivalent to MINIMAL-AT-LEAST-K-CHARGE REMOVAL. NP-Hardness will be shown by a reduction from INDEPENDENT-SET on penny graphs adapting it to the Euclidean settings of crystal graph of ions within a unit cell. The Independent Set problem, denoted INDEPENDENT-SET, takes as input a graph, G, and a natural number k. The goal is to find an *independent set*, i.e. a set of vertices such that no two are adjacent, of size k in G, or report that one does not exist. Penny graphs are the class of graphs where each vertex may be drawn as a unit circle such that no two circles overlap, and an edge between two vertices exist if and only if the corresponding circles are tangent, i.e. they intersect at only a single point. Finding an independent set on this class of graphs is known to be NP-Hard [5].

Sketch of the construction of the K-CHARGE REMOVAL **instance:** Starting with an instance of INDEPENDENT-SET on a maximum degree 3 planar graph, containing the graph and a natural number k a penny graph, G, is created using Theorem 1.2 from Cerioli et al., using a radius of $\frac{n}{2}$ for the pennies. Graphs created in this manner will be denoted *long orthogonal penny graphs.* The K-CHARGE REMOVAL instance is created by placing a positive ion above the centre of each penny, and a negative ion bellow.

Ion Species: The positive and negative species are assigned charges of magnitude 1. From these species there are parameters for the interaction between two ions of the positive specie, two ions of the negative specie, and between one ion of the positive specie and one of the negative specie. For brevity, 1 and 2 will denote the positive and negative specie respectively. Under this construction, the interaction between the two ions of the positive specie is the same as between two ions of the negative specie. Therefore the parameters that may be set are $A_{11}, B_{11}, C_{11}, A_{12}, B_{12}$, and C_{12}.

Notation: Let $k' = n - k$, being the number of charges that are required to be removed to be left with an independent set of size k.charge Note that as the charge of each ion has a magnitude of one, a removal of k' can only be achieved by removing k' positive and k' negative ions. The goal energy for the construction is set as $g = (k-1)(\frac{A_{12}}{e^{B_{12}}} - C_{12} - 1)$. To simplify the equations regarding the interaction between planes, let $\widehat{r}$ denote $\sqrt{r^2+1}$. An independent set is *left* if the ions left after a removal of k' charges have labels corresponding to an independent set in G. To ensure that an independent set is left of size k if and only if one exists, the following three inequalities must be satisfied:

$$\frac{A_{11}}{e^{B_{11}n}} - \frac{C_{11}}{n^6} + \frac{1}{n} + \frac{A_{12}}{e^{B_{12}\widehat{n}}} - \frac{C_{12}}{\widehat{n}^6} - \frac{1}{\widehat{n}} \geq \left|\frac{A_{12}}{e^{B_{12}}} - C_{12} - 1\right| \tag{1}$$

$$n^2\left|\frac{A_{11}}{e^{B_{11}r}} - \frac{C_{11}}{r^6} + \frac{1}{r} + \frac{A_{12}}{e^{B_{12}\widehat{r}}} - \frac{C_{12}}{\widehat{r}^6} - \frac{1}{\widehat{r}}\right| \leq \left|\frac{A_{12}}{e^{B_{12}}} - C_{12} - 1\right|, \quad r \geq \sqrt{2}n \tag{2}$$

$$\frac{A_{11}}{e^{B_{11}r}} - \frac{C_{11}}{r^6} + \frac{1}{r} + \frac{A_{12}}{e^{B_{12}\widehat{r}}} - \frac{C_{12}}{\widehat{r}^6} - \frac{1}{\widehat{r}} > 0, \quad r \geq \sqrt{2}n \tag{3}$$

Theorem 5 formally states the correctness of this reduction, via Lemmas 1–4.

Lemma 1. *Inequalities (1) and (2) are sufficient to ensure that an independent set is left if one exists.*

Lemma 2. *There exists, for any structure created from a long orthogonal penny graph, some parameters such that Inequalities (1, 2) and (3) are satisfied.*

Lemma 3. *Given k pairs, the energy will be less than* $(k-1)(\frac{A_{12}}{e^{B_{12}}} - C_{12} - 1)$ *only if the pairs correspond to an independent set of size k, for* $\frac{A_{12}}{e^{B_{12}}} - C_{12} - 1 < 0$.

Lemma 4. *It is always preferable to remove pairs from the construction from a long orthogonal penny graph under Inequalities (1–3).*

Lemmas 1 and 2 show that the inequalities ensure that leaving an independent set is preferable, and are satisfiable for the Buckingham-Coulomb potential. Lemma 3 provides bounds, which may be calculated exactly using the construction provided by Lemma 2. Lemma 4 proves that when removing either member of a pair vertices, it is always preferable to select the the other member for removal.

Theorem 5. K-CHARGE REMOVAL, MINIMAL-AT-LEAST-K-CHARGE REMOVAL *and* AT-LEAST-K-CHARGE REMOVAL *are NP-Complete when limited to only two species of ion and restricted to the Buckingham-Coulomb potential function.*

Proof. Lemma 1 shows that, under Inequalities (1) and (2), the optimal solution will be to leave an independent set. Lemma 2 provides a construction such that the inequalities are satisfiable. Lemma 3 shows the upper bound is reachable if and only if an independent set has been left. It follows from Lemma 4 that it is preferable to remove a set of pairs over any other set of charges. Therefore there will be a satisfiable instance of K-CHARGE REMOVAL or any generalisation if and only if the instance of INDEPENDENT SET on a max degree 3 planar graph is satisfiable. Conversely if the INDEPENDENT SET instance is satisfiable, the corresponding K-CHARGE REMOVAL instance can be satisfied by leaving the vertices corresponding to the independent set in the penny graph construction. Hence under these restriction all three problems will be NP-Complete. □

Conclusions and Future Work: Motivated by analyses of computational complexity for CSP Problem we defined a class of functions for which the k-charge removal problem is NP-Complete in general. We have also shown that the problem remains NP-Complete under both the restriction that we have only two species of ions and the Buckingham-Coulomb energy function and the restriction we only use the Coulomb potential on an unbounded number of ion species. One obvious question would be if approximation results can be gained for this problem. From a chemistry stand point, while we have made progress towards physical constructions there is still a lot that could be done. As such investigation into the restrictions of having more realistic physical values remains an important unexplored direction. Another question would be if we can investigate the convergence of these interactions, particularly the Coulomb potential, over a periodic structure to more fully understand the energy function.

References

1. Adamson, D., Deligkas, A., Gusev, V., Potapov, I.: On the Hardness of Energy Minimisation for Crystal Structure Prediction. arXiv.org/abs/1910.12026
2. Ageev, A.A., Kelmanov, A.V., Pyatkin, A.V.: Np-hardness of the euclidean max-cut problem. Doklady Math. **89**, 343–345 (2014)
3. Barahona, F.: On the computational complexity of Ising spin glass models. J. Phys. A: Math. Gen. **15**(10), 3241–3253 (1982)
4. Buckingham, R.A.: The classical equation of state of gaseous helium, neon and argon. Proc. Royal Soc. London Ser. A. Math. Phys. Sci. **168**(933), 264–283 (1938)

5. Cerioli, M.R., Faria, L., Ferreira, T.O., Protti, F.: A note on maximum independent sets and minimum clique partitions in unit disk graphs and penny graphs: complexity and approximation. RAIRO **45**(3), 331–346 (2011)
6. Clark, B.N., Colbourn, C.J., Johnson, D.S.: Unit disk graphs. Disc. Math. **86**(1), 165–177 (1990)
7. Zanella, M., et al.: Accelerated discovery of two crystal structure types in a complex inorganic phase field. Nature **546**(7657), 280 (2017)
8. Krishnamoorthy, M., Deo, N.: Node-deletion np-complete problems. SIAM J. Comput. **8**(4), 619–625 (1979)
9. Lyakhov, A.O., Oganov, A.R., Mario, V.: How to predict very large and complex crystal structures. Comput. Phys. Commun. **181**(9), 1623–1632 (2010)
10. Mahajan, M., Nimbhorkar, P., Varadarajan, K.: The planar k-means problem is np-hard. In: WALCOM: Algorithms and Computation, pp. 274–285 (2009)
11. Oganov, A.R.: Crystal structure prediction: reflections on present status and challenges. Faraday Discuss. **211**, 643–660 (2018)
12. Oganov, A.R., Glass, C.W.: Crystal structure prediction using ab initio evolutionary techniques: principles and applications. J. Chem. phys. **124**(24), 244704 (2006)
13. Wille, L.T., Vennik, J.: Computational complexity of the ground-state determination of atomic clusters. J. Phys. A: Math. Gen. **18**(8), L419–L422 (1985)
14. Woodley, S.M., Catlow, R.: Crystal structure prediction from first principles. Nat. Mater. **7**(12), 937 (2008)
15. Yannakakis, M.: Node-and edge-deletion np-complete problems. In: STOC, vol. 1978, pp. 253–264 (1978)
16. Yannakakis, M.: Node-deletion problems on bipartite graphs. SIAM J. Comput. **10**(2), 310–327 (1981)

Practical Implementation of a Quantum Backtracking Algorithm

Simon Martiel and Maxime Remaud(✉)

Atos, Quantum R&D, 78340 Les Clayes-sous-Bois, France
{simon.martiel,maxime.remaud}@atos.net

Abstract. In previous work, Montanaro presented a method to obtain quantum speedups for backtracking algorithms, a general meta-algorithm to solve constraint satisfaction problems (CSPs). In this work, we derive a space efficient implementation of this method. Assume that we want to solve a CSP with m constraints on n variables and that the domain in which these variables take their value is of cardinality d. Then, we show that the implementation of Montanaro's backtracking algorithm can be done by using $\mathcal{O}(n \log d)$ data qubits. We detail an implementation of the predicate associated to the CSP with an additional register of $\mathcal{O}(\log m)$ qubits. We explicit our implementation for graph coloring and SAT problems, and present simulation results. Finally, we discuss the impact of the usage of static and dynamic variable ordering heuristics in the quantum setting.

Keywords: Backtracking algorithm · Quantum walk · CSP · Graph coloring · SAT

1 Introduction

Quantum computing. Quantum computing is one of the most promising emerging computation technology. Theory promises algorithmic speedups ranging from quadratic, for unstructured problems, up to exponential for some particular key problems. Besides some problems such as integer factoring and its obvious applications in cryptology, very few applicable large scale algorithms have been derived or studied. In 2015, Montanaro [16] presented a general method to obtain speedups of backtracking-based algorithms, relying on Belovs' previous work [5]. The algorithm uses Quantum Walks, a well developed quantum algorithmic tool intensively studied in the scope of search algorithms [1,2,7,13,14,19,20].

Constraint satisfaction problems. CSPs form a very general class of problems, that encompass a large set of practical problems. The most famous examples are the Boolean satisfiability problem (SAT) [12] and the graph coloring problem [15]. CSPs have been widely studied and a well known meta-algorithm taking advantage of their structure to solve them is backtracking. For example, the DPLL [8,9] backtracking-based algorithm has been introduced in 1962 and is currently the procedure at the basis of some of the most efficient SAT solvers

A. Chatzigeorgiou et al. (Eds.): SOFSEM 2020, LNCS 12011, pp. 597–606, 2020.
https://doi.org/10.1007/978-3-030-38919-2_49

[10,11,21]. Since backtracking algorithms explore a tree whose vertices are partial solutions to the associated CSP [21], one can think of using a quantum walk to explore such a tree faster.

Quantum backtracking. In 2017, Ambainis and Kokainis [3] have dealt with Montanaro's algorithm in depth. Using these works, it has been shown how to get a quantum speedup of the two most efficient forms of enumeration (a lattice algorithm) [4] and of branch-and-bound algorithms [17]. Thus, Montanaro's algorithm is of high interest in computing science and cryptography. In 2018, its complexity has been reviewed when applied to the graph coloring problem and SAT, assuming access to a very large amount of qubits, since the circuit depth was aggressively optimized [6]. The aim of our work is to investigate a memory efficient implementation of this backtracking algorithm, in order to be able to validate the algorithm on small instances via classical emulation. We manage to reduce the number of data qubits to $\mathcal{O}(n \log d)$ and ancillae to $\mathcal{O}(\log m)$.

The paper is organized as follows. Section 2 introduces definitions. We discuss the choice of heuristic in Sect. 3. Then, Sect. 4 is about the use of the predicate, how to implement it and how to check some generic constraints. We conclude by presenting some of the results we got for graph coloring, thanks to a simulator, in Sect. 5.

2 Preliminaries

Thereafter, we will denote by $\mathcal{P}$ a constraint satisfaction problem defined by a triple $\langle X, D, C \rangle$, where $X = \{x_1, \dots, x_n\}$ is a set of n variables, $D = [\![1, d]\!]$ is a set of d values and C is a set of m constraints. If a variable has no value, we will denote it by "$*$".

Let x be an assignment of values to the n variables of a CSP $\mathcal{P}$. It will be said to be a solution to $\mathcal{P}$ if it verifies all the constraints in C; complete if $\forall x_i \in X,\ x_i \neq *$, partial otherwise; valid if it is partial and can be extended to a solution; invalid if it is partial and not a solution.

A standard approach for solving a CSP $\mathcal{P}$ is the technique of backtracking. For this, we assume that we have access to a predicate P which can receive an assignment (complete or partial) of values x as argument and returns "true" if x is a solution, "indeterminate" if it is valid, "false" otherwise. We also assume access to a heuristic h which specifies which variable should be instantiated next.

Montanaro's backtracking algorithm [16] is based on a quantum walk on trees. The idea is summarized as follows. Consider a rooted tree $\mathcal{T}$ with T vertices, labeled $r, 1, \dots, T-1$, the vertex r being the root of $\mathcal{T}$. Hereafter, A (resp., B) will denote the set of vertices at an even (resp., odd) distance from r and $x \longrightarrow y$ will mean that y is a child of x in the tree. The quantum walk operates on the space spanned by $\{|x\rangle;\ x \in \{r\} \cup [\![1, T-1]\!]\}$, and starts in the state $|r\rangle$. It is based on a set of diffusion operators D_x, where D_x is the identity if x is a solution, otherwise, diffuses on the subspace spanned by $\{|x\rangle\} \cup \{|y\rangle : x \longrightarrow y\}$. A step of the walk consists in applying the operator $R_B R_A$, where:

$$R_A = \bigoplus_{x \in A} D_x \text{ and } R_B = |r\rangle\langle r| + \bigoplus_{x \in B} D_x$$

Thanks to these operators, an algorithm for detecting a solution in a tree can be established (Algorithm 1). It is the phase estimation of the operator $R_B R_A$ which allows the quantum walker to go through the paths leading to a solution in $\mathcal{T}$. By applying the detection algorithm to wisely chosen vertices of $\mathcal{T}$, it is possible to construct an hybrid algorithm for finding a solution (described in Subsection 2.1 of [16]). Montanaro has shown that it finds a solution in time complexity $\mathcal{O}(\sqrt{T}n^{3/2}\log n \log 1/\delta)$ and $\mathcal{O}(\sqrt{T}n \log^3 n \log 1/\delta)$ if there exists a unique solution. For more details, we refer to [16].

Algorithm 1. Detecting a solution (Algorithm 2 of [16])

Require: Operators R_A, R_B, a failure probability δ, upper bounds on the depth n and the number of vertices T. Let $\beta, \gamma > 0$ be universal constants to be determined.

Repeat $K = \lceil \gamma \log 1/\delta \rceil$ times:
  Apply phase estimation to the operator $R_B R_A$ with precision $\beta/\sqrt{Tn}$.
  If the eigenvalue is 1 **then** accept
If number of acceptances $\geq 3K/8$ **then return** "Solution exists"
  else return "No solution"

3 Variable Ordering Heuristics

In order to be able to test the algorithm with the means we have, we initially took a close look at the choice of the heuristic. Indeed, the largest qubit overhead in the implementation comes from the heuristic implementation, since one has to store variable indexes inside the quantum memory ($\mathcal{O}(n \log n)$ qubits are needed). Moreover, even though the depth overhead of the heuristic is asymptotically negligible compared to the rest of the algorithm, it seems that for instances of reasonable size, this overhead is not negligible.

Since we are interested in optimizing the number of qubits, we chose to deal with static variable ordering (SVO) heuristics, which can be classically precomputed at the start of the meta-algorithm. The benefits are twofold: only $\mathcal{O}(n \log d)$ data qubits are required to store an assignment and we do not have to produce a reversible implementation of a dynamic variable ordering (DVO) heuristic.

In [6,18], the implementation of Montanaro's algorithm has been optimized in depth and its complexity has been studied by considering the use of a DVO heuristic. However, the benefit of implementing a DVO heuristic is unclear in the range of parameters for which it is claimed in [6] that a graph could be colored in one day (up to approximately 150 vertices). Let T (resp., T') be the number of vertices in the backtracking tree associated with an SVO (resp., DVO) heuristic and c_T (resp., $c_{T'}$) the number of calls to $R_B R_A$ made by Montanaro's algorithm. If we denote the depth of the overhead due to the implementation of a DVO heuristic by d_h and the depth of the operators R_A and R_B without heuristic by d_R, we have that using a dynamic heuristic is more efficient than using a static one if $c_T d_R \geq c_{T'}(d_R + d_h)$, i.e. if $\frac{c_T}{c_{T'}} \geq 1 + \frac{d_h}{d_R}$. This is asymptotically true but thanks to the script computing the algorithm's complexity to solve graph coloring given in [18], one can compute an estimate of $\frac{d_h}{d_R}$. In the considered range of parameters, we can see that using a DVO heuristic would be better than using an SVO one if $c_T \geq \frac{4}{3}c_{T'}$. Even if no DVO heuristic has been proved to verify such a bound, DSATUR could be a good candidate. Nevertheless, the time-saving trick allowed by a quantum DVO heuristic may not be obvious and it might be worth using an SVO heuristic, at least for "small" instances.

The modified algorithm for R_A and R_B is presented as Algorithm 2.

Algorithm 2. Implementation of the operator R_A

Require: A basis state $|\ell\rangle|v_1\rangle \dots |v_n\rangle \in \mathbb{C}^{n+1} \otimes (\mathbb{C}^{d+1})^{\otimes n}$ corresponding to a partial assignment $x_1 = v_1, \cdots, x_\ell = v_\ell$. Ancilla registers : $\mathcal{H}_{\text{anc}}$, $\mathcal{H}_{\text{children}}$, storing a tuple (a, S), where $a \in \{*\} \cup D$, $S \subseteq D$, initialized to $a = *$, $S = \emptyset$.

1: If ℓ is odd, swap a with v_ℓ.
2: Compute $P(x)$.
3: If $P(x)$ is true, go to step 8.
4: If $a \neq *$, subtract 1 from ℓ.
5: For each $w \in D$, if $P(v_1, \cdots, v_\ell, w)$ is not false, set $S = S \cup \{w\}$.
6: If $\ell = 0$, $i = n$, else, $i = 1$. Perform $I - 2\,|\phi_{i,S}\rangle\langle\phi_{i,S}|$ on $\mathcal{H}_{\text{anc}}$.
7: Revert steps 5 and 4.
8: Revert steps 2 and 1.

R_B is similar, except that: step 1 is preceded by the check "If $\ell = 0$, return"; "odd" is replaced with "even" in step 1; and the check "If $\ell = 0$ is removed from step 6.

---

4 Generic Implementation

Thereafter, we will use the following notations: $\nu = \lceil \log(n+1) \rceil$, $\delta = \lceil \log(d+1) \rceil$ and $\mu = \lceil \log(m+1) \rceil$.

The most straightforward way of implementing the predicate $\mathcal{P}$ is to compute a logical AND of the result of the evaluation of each constraint over the current variable assignment. This would lead to a circuit using $m+1$ work qubits. We present another solution, a quantum counter, using only μ qubits.

In the case of the partial predicate, an index ℓ stored in a quantum register indicates that the first $\ell + 1$ variables have been assigned a value. Therefore, if is sufficient to check the constraints that depend on at least one variable in $\{x_{i_1}, \cdots, x_{i_\ell}\}$. To find out if a constraint has to be checked, we use a system of comparison of values, one of which is quantum and the other classical. The overall process requires $\mathcal{O}(m)$ additions on ν qubits.

Similarly, if we apply the detection algorithm to a vertex located at the ℓ-th level of the backtracking tree, it is unnecessary to check the constraints that depend solely on the variables in $\{x_{i_1}, \cdots, x_{i_{\ell-1}}\}$.

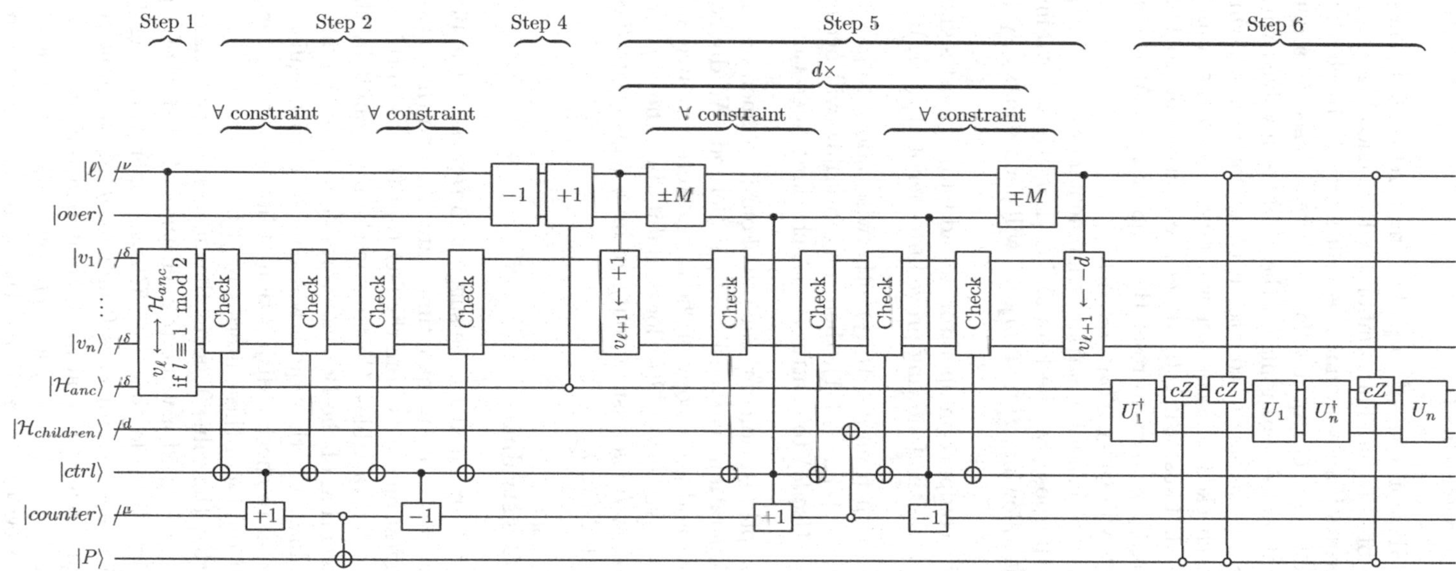

Fig. 1. Circuit corresponding to Algorithm 2 for R_A. Steps 7 and 8 are not represented since they consist in reversing steps 5, 4, 2 and 1. Step 3 is realized by controlling cZ (controlled Z) operations in Step 6 with $|P\rangle$. In the same way, it is sufficient to control the three cZ in Step 6 to control the whole operator, due to the reversibility of the other steps. Circuit for R_B is quite similar. "If $\ell \equiv 1 \mod 2$" is replaced by "If $\ell \equiv 0 \mod 2$" in Step 1. Step 6 is simplified (the two last cZ operations, U_n and $U_n^\dagger$ are removed). An additional ancillary qubit set to 0 is necessary. It is flipped before Step 1 if $\ell = 0$ and used to control the remaining cZ operation.

4.1 How to Implement a Predicate

Let $\mathcal{P} = \langle X, D, C \rangle$ be a CSP. For all $i \in [\![1, n]\!]$, the binary representation of the value assigned to x_i is denoted by v_i. The symbol $*$ will be encoded by 0.

The step 2 in Fig. 1 verifies if an assignment is solution to $\mathcal{P}$, by checking for all $i \in [\![1, m]\!]$ the constraint $C_i \in C$. An ancillary register is used as a counter. In order to check C_i, a subroutine depending on the set Y_i of involved variables in C_i will be used and will increase the counter if C_i is not verified. Once all the constraints have been checked, the counter will be equal to the number of constraints that are violated. Thus, if it is 0, we set the result qubit to 1. Then, we reverse the C_i checking operations to reset the counter to 0. At the end of the circuit, the result qubit will carry $P(v_1, \cdots, v_n)$.

We also want to be able to verify if a partial assignment $x_1 = v_1, \cdots, x_{\ell+1} = v_{\ell+1}$ is valid for $\mathcal{P}$. For this purpose, we check for all $i \in [\![1, m]\!]$ the constraint C_i if $\max_{x_i \in Y_i}\{i\} - 1 \leq \ell$. In the following, $\forall i \in [\![1, m]\!]$, M_i will denote $\max_{x_i \in Y_i}\{i\} - 1$ and $M_0 = 0$.

In order to compute the comparison operator, we add a bit (most significant one) to $|\ell\rangle$ (call it $|a\rangle$) and use the following procedure (denoted by M):

1. If $M_i - M_{i-1} > 0$, subtract it from $|\ell - M_{i-1}\rangle$, otherwise, add it to $|\ell - M_{i-1}\rangle$;
2. If $M_i > \ell$ (i.e., if $\max_{x_i \in Y_i}\{i\} > \ell + 1$), then some overflow will occur, and thus, $|a\rangle$ will be flipped. Use $|a\rangle$ to control the C_i checking operation.

Within the scope of an optimization of the depth of our implementation, note that constraint checking operations can easily be parallelized. We divide C in $k \in \mathbb{N}$ sets of $\frac{m}{k}$ constraints and use a copy of the v_i registers and a counter for checking each one. This involves using $\mathcal{O}(kn \log d)$ qubits but the depth of the predicate would be divided by k (we just fan-out/fan-in v_i registers to k copies, which can be done in depth $\mathcal{O}(\log k)$).

4.2 How to Check a Constraint

For the specific case of Boolean variables, we suggest to represent the bit 0 by the quantum state $|10\rangle$, the bit 1 by the quantum state $|11\rangle$ and the unassigned symbol $*$ by $|00\rangle$. Our suggestion stems from the fact that thanks to this choice, it is simpler to manipulate the variables. Controlling the right qubit by the left one allows us to do the negation of the Boolean variable without having a side effect on the unassigned values. The right qubit will allow us to distinguish 1 from $*$ and 0, which will be useful to check a disjunction of literals (by incrementing the counter if all the right qubits of these literals are 0).

Since we have chosen that the value 0 means that $v_i = *$, we can easily check if the variable x_i has been assigned a value (e.g., if the i-th vertex of a graph G has been colored). In order to do this, the "Check" operation (Fig. 1) is a δ-Toffoli gate checking that each bit of v_i is 0 and targeting a qubit set to 0 (*ctrl*). Thus, this qubit can be used to control a counter increment by 1.

If we want to check if the variables x_j and x_k have different values (e.g., if the coloring of the edge between the j-th and the k-th vertices of a graph is well colored), we can use the same idea as above. In fact, we want to increase the couter if $v_j = v_k$, i.e. if $v_j \oplus v_k = 0$. For that, we apply a bit-wise XOR to the values of the two variables (the result is stocked in place of v_k) and then we apply the δ-Toffoli gate, controlled by the bits of $v_j \oplus v_k$.

4.3 General Structure

Thanks to the precedent subsections, step 2 and most of step 5 of Algorithm 2 can be realized. The rest of the implementation is most straightforward, circuit of the operator R_A is given in appendix (Fig. 1). Note that, contrary to what is stated in [6], we can simply add one control qubit to three (resp., one) controlled-Z gates in step 6 to control the whole operator R_A (resp., R_B).

5 Simulation Results

Now, we present some results of our simulations for the graph coloring problem. Please note that despite these graphs being small and our implementation being space optimized, up to 30 qubits are necessary.

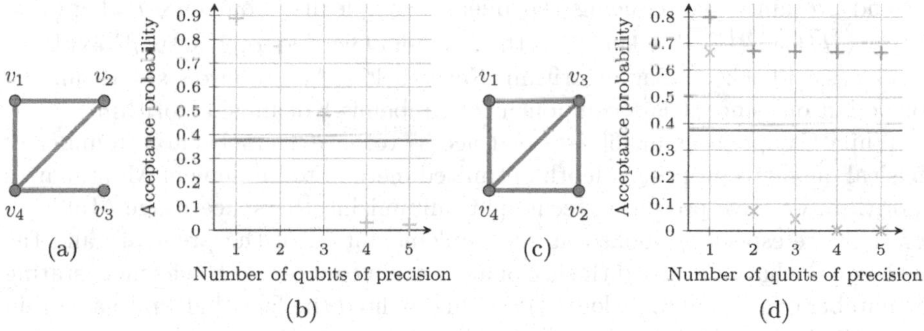

Fig. 2. (a): a first graph. (b): some results of our simulations for graph in Fig. (a). (c): a second graph. (d): some results of our simulations for graph in Fig. (c). (Color figure online)

For the first example, we used our implementation of Montanaro's algorithm for graph coloring on the graph in Fig. 2.(a) with 2 colors. Obviously, no 2-coloring exists for this graph, but it is quite interesting to wonder which precision the Algorithm 1 requires in the phase estimation step to output "No solution", on such a small graph. Figure 2.(b) presents the probabilities of acceptance depending on the precision used. We can see that for a precision smaller than 4, the Algorithm 1 will fail with high probability, since the threshold probability fixed in [16] is 0.375 (blue line).

For the second example, we applied the algorithm on the graph in Fig. 2.(c) with 3 colors and colored the vertices in increasing order of indices. We can

Table 1. Comparison of our $R_B R_A$ circuit with CKM one.

Circuit	Input qubits	Ancillae	Toffoli	Depth
DVO-CKM [6]	$\mathcal{O}(n \log n)$	$\mathcal{O}(n^2 d \log d)$	$\mathcal{O}(md \log d)$	$\mathcal{O}(n \log n)$
SVO-CKM	$\mathcal{O}(n \log d)$	$\mathcal{O}(n^2 \log d)$	$\mathcal{O}(m \log d)$	$\mathcal{O}(n \log n)$
Our	$\mathcal{O}(n \log d)$	$\mathcal{O}(\log m)$	$\mathcal{O}(md \log n)$	$\mathcal{O}(md \log n \log \log n)$

see that v_2 must be the same color as v_1 if we want to get a solution. We distinguished the cases where $v_1 = v_2$ and $v_1 \neq v_2$ ($v_1 \neq *$, $v_2 \neq *$, $v_3 = v_4 = *$) and looked at the results where the algorithm has been applied on these partial colorings. Figure 2.(d) presents the probabilities of acceptance depending on the precision used: plus signs when $v_1 = v_2$, crosses otherwise. Algorithm 1 quickly discriminates the two cases, since 2 qubits of precision are already sufficient.

6 Conclusion

In this paper, we presented our implementation of Montanaro's algorithm, but an improved quantum algorithm for backtracking has been introduced by Ambainis and Kokainis [3], reducing the queries complexity from $\mathcal{O}(\sqrt{Tn} \log \frac{1}{\delta})$ to $\mathcal{O}(n^{3/2}\sqrt{T'} \log^2 \frac{n \log T}{\delta})$, where T' is the number of vertices of $\mathcal{T}$ actually explored by a classical backtracking algorithm. Nevertheless, Montanaro's algorithm can not be left out since it is a component of Ambainis-Kokainis' algorithm.

While Campbell et al. [6] assumed access to an extremely large number of physical qubits to propose a depth optimized method to implement Montanaro's algorithm, we have presented techniques minimizing the space usage (Table 1). For that, we especially looked at the implementation of the predicate and the heuristic. We have proposed the use of a quantum counter for the former (stating the number of ancillae to $\mathcal{O}(\log m)$) and highlighted the fact that up to a certain point, the latter might not be quantumly implemented (improving the number of input qubits from $\mathcal{O}(n \log n)$ to $\mathcal{O}(n \log d)$). However, these propositions are not asymptotically competitive, although our implementation of the predicate could be parallelized to be efficient and could lead to a trade-off between the space usage and the time usage. As far as the heuristic is concerned, it would be interesting to establish a precise resource estimation and define to what extent an SVO heuristic would present benefits compared to a DVO one.

Acknowledgments. This work was supported by Atos. The implementation was developed in python using Atos' pyAQASM library. All simulations were performed on the Atos Quantum Learning Machine. We acknowledge support from the French ANR project ANR-18-CE47-0010 (QUDATA), the QuantERA ERA-NET Cofund in Quantum Technologies implemented within the European Union's Horizon 2020 Program (QuantAlgo project), and the French ANR project ANR-18-QUAN-0017 (QuantAlgo Project).

References

1. Ambainis, A.: Quantum walks and their algorithmic applications. Int. J. Quantum Inf. **01**(04), 507–518 (2003). https://doi.org/10.1142/S0219749903000383
2. Ambainis, A.: Quantum walk algorithm for element distinctness. SIAM J. Comput. **37**(1), 210–239 (2007). https://doi.org/10.1137/S0097539705447311
3. Ambainis, A., Kokainis, M.: Quantum algorithm for tree size estimation, with applications to backtracking and 2-player games. In: Proceedings of the 49th STOC. ACM (2017). https://doi.org/10.1145/3055399.3055444
4. Aono, Y., Nguyen, P.Q., Shen, Y.: Quantum lattice enumeration and tweaking discrete pruning. In: Peyrin, T., Galbraith, S. (eds.) ASIACRYPT 2018. LNCS, vol. 11272, pp. 405–434. Springer, Cham (2018). https://doi.org/10.1007/978-3-030-03326-2_14
5. Belovs, A., Childs, A.M., Jeffery, S., Kothari, R., Magniez, F.: Time-efficient quantum walks for 3-distinctness. In: Fomin, F.V., Freivalds, R., Kwiatkowska, M., Peleg, D. (eds.) ICALP 2013. LNCS, vol. 7965, pp. 105–122. Springer, Heidelberg (2013). https://doi.org/10.1007/978-3-642-39206-1_10
6. Campbell, E., Khurana, A., Montanaro, A.: Applying quantum algorithms to constraint satisfaction problems. Quantum **3**, 167 (2018). https://doi.org/10.22331/q-2019-07-18-167
7. Childs, A., Cleve, R., Deotto, E., Farhi, E., Gutmann, S., Spielman, D.: Exponential algorithmic speedup by a quantum walk. In: Proceedings of the 35th STOC. ACM (2003). https://doi.org/10.1145/780542.780552
8. Davis, M., Logemann, G., Loveland, D.: A machine program for theorem-proving. Commun. ACM **5**(7), 394–397 (1962). https://doi.org/10.1145/368273.368557
9. Davis, M., Putnam, H.: A computing procedure for quantification theory. JACM **7**(3), 201–215 (1960). https://doi.org/10.1145/321033.321034
10. Eén, N., Sörensson, N.: An extensible SAT-solver. In: Giunchiglia, E., Tacchella, A. (eds.) SAT 2003. LNCS, vol. 2919, pp. 502–518. Springer, Heidelberg (2004). https://doi.org/10.1007/978-3-540-24605-3_37
11. Gomes, C.P., Kautz, H., Sabharwal, A., Selman, B.: Satisfiability solvers. In: Handbook of Knowledge Representation. Elsevier (2008). https://doi.org/10.1016/S1574-6526(07)03002-7
12. Gu, J., Purdom, P.W., Franco, J., Wah, B.W.: Algorithms for the satisfiability (SAT) problem: a survey. In: Handbook of Combinatorial Optimization. Springer (1999) https://doi.org/10.1007/978-1-4757-3023-4_7
13. Kempe, J.: Quantum random walks: an introductory overview. Contemp. Phys. **44**(4), 307–327 (2003). https://doi.org/10.1080/00107151031000110776
14. Magniez, F., Nayak, A., Roland, J., Santha, M.: Search via quantum walk. In: Proceedings of the 39th STOC. Theory of Computing (2007). https://doi.org/10.1145/1250790.1250874
15. Malaguti, E., Toth, P.: A survey on vertex coloring problems. Int. Trans. Oper. Res. **17**(1), 1–34 (2010). https://doi.org/10.1111/j.1475-3995.2009.00696.x
16. Montanaro, A.: Quantum walk speedup of backtracking algorithms. Theory Comput. **14**(15), 1–24 (2018). https://doi.org/10.4086/toc.2018.v014a015
17. Montanaro, A.: Quantum speedup of branch-and-bound algorithms. arXiv:1906.10375 (2019)
18. Montanaro, A.: Data from Quantum algorithms for CSPs. c9pb. Accessed Jul 2019. https://doi.org/10.5523/bris.19va21gun3c7629f291kmd6w37

19. Santha, M.: Quantum walk based search algorithms. In: Agrawal, M., Du, D., Duan, Z., Li, A. (eds.) TAMC 2008. LNCS, vol. 4978, pp. 31–46. Springer, Heidelberg (2008). https://doi.org/10.1007/978-3-540-79228-4_3
20. Szegedy, M.: Quantum speed-up of Markov chain based algorithms. In: FOCS 2004. IEEE (2004). https://doi.org/10.1109/FOCS.2004.53
21. van Beek, P.: Backtracking search algorithms. In: Handbook of Constraint Programming. Elsevier (2006). https://doi.org/10.1016/S1574-6526(06)80008-8

Simplified Emanation Graphs: A Sparse Plane Spanner with Steiner Points

Bardia Hamedmohseni[1], Zahed Rahmati[1(✉)], and Debajyoti Mondal[2]

[1] Department of Mathematics and Computer Science, Amirkabir University of Technology, Tehran, Iran
{hamedmohseni,zrahmati}@aut.ac.ir

[2] Department of Computer Science, University of Saskatchewan, Saskatoon, Canada
dmondal@cs.usask.ca

Abstract. Emanation graphs of grade k, introduced by Hamedmohseni, Rahmati, and Mondal, are plane spanners made by shooting 2^{k+1} rays from each given point, where the shorter rays stop the longer ones upon collision. The collision points are the Steiner points of the spanner.

We introduce a method of simplification for emanation graphs of grade $k = 2$, which makes it a competent spanner for many possible use cases such as network visualization and geometric routing. In particular, the simplification reduces the number of Steiner points by half and also significantly decreases the total number of edges, without increasing the spanning ratio. Exact methods of simplification is provided along with comparisons of simplified emanation graphs against Shewchuk's constrained Delaunay triangulations on both synthetic and real-life datasets. Our experimental results reveal that the simplified emanation graphs outperform constrained Delaunay triangulations in common quality measures.

Keywords: Network visualization · Mesh generation · Plane spanners

1 Introduction

Let $G = (V, E)$ be a geometric graph in the Euclidean plane. For a pair of vertices u, v, we denote by $d_G(u, v)$ and $d_E(u, v)$, the minimum graph distance and the Euclidean distance between u and v, respectively. The spanning ratio of G is the maximum value of $\frac{d_G(u,v)}{d_E(u,v)}$ over all pairs of vertices $\{u, v\} \in V$. A graph is called a t-*spanner* if its spanning ratio is less than or equal to t.

Many applications use t-*spanners*, and in general, planar geometric graphs, in different applied areas of computational geometry and data visualization. Nachmanson *et al.* [10] introduced a system called GraphMaps for interactive visualization of large graphs based on constrained Delaunay triangulations. Mondal and Nachmanson [9] introduced and used a specific mesh called the competition mesh to improve GraphMaps. Given a set of points P, a *competition mesh*

The full version of this paper can be found in arXiv [8].

Work of D. Mondal is supported in part by NSERC.

A. Chatzigeorgiou et al. (Eds.): SOFSEM 2020, LNCS 12011, pp. 607–616, 2020.
https://doi.org/10.1007/978-3-030-38919-2_50

is constructed by shooting from each point, four axis-aligned rays at the same speed, where the shorter rays stop the longer ones upon collision (the rays that are not stopped are clipped by the axis-aligned bounding box of P). This can also be seen as a variation of a motorcycle graph [3]. The points corresponding to the collisions are called Steiner points.

The ray shooting idea that the competition mesh used, encouraged the introduction of a new, general *t-spanner* called *the emanation graph* by Hamedmohseni, Rahmati, and Mondal [6]. An emanation graph of grade k, is obtained by shooting 2^{k+1} rays around each given point. Given a set P of n points in the plane, an emanation graph M_k is constructed by shooting 2^{k+1} rays from each point $p \in P$ with equal $\frac{\pi}{2^k}$ angles between them. Each ray stops as soon as it hits another ray of a larger length or upon reaching the bounding box $R(P)$. When two parallel rays collide they both stop and when two rays with equal length collide at a point, one of them is randomly stopped. The competition mesh is thus the emanation graph of grade 1. Figure 1(left) depicts an emanation graph of grade 2 with six points in the plane.

An emanation graph of grade 1 is a $\sqrt{10}$-spanner with at most $4n$ Steiner points [6]. Emanation graphs of larger grades allow many redundant edges and Steiner points, i.e., elements that can be removed without increasing the spanning ratio. Redundant edges make a spanner visually cluttered and unsuitable for the visualization purposes unless we further refine the layout. In this paper we propose a simplification for the emanation graphs of grade 2 (Fig. 1 (right)). Our simplified version of the emanation graph has the potential to be used in tools such as GraphMaps for interactive visualization of large graphs, and serve as an alternative spanner with better properties. We now briefly review the literature related to the emanation graphs and other geometric spanners.

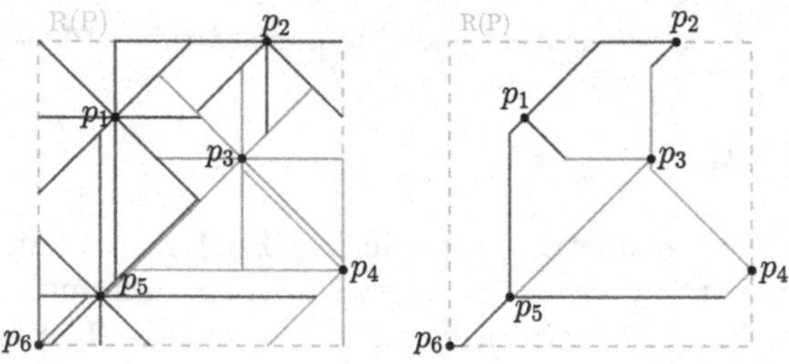

Fig. 1. (Left) A non-simplified and complete emanation graph of grade $k = 2$ for an example point set, and (Right) its simplified version.

Related Work: The literature on geometric spanners is rich and there are many approaches to construct geometric spanners and meshes. We refer the reader to [2] and [11] for the surveys on geometric spanners and mesh generation, respectively. Emanation graph was motivated by a well-studied question in this context: Given a set P of n points in the plane, can we compute a planar spanner

$G = (V, E)$ of P with small size, degree, spanning ratio and few Steiner points? There are fewer approaches known for constructing planar spanners with Steiner points compared to the vast literature on planar geometric spanners that do not use them.

Comparing emanation graphs with traditional spanners such as Delaunay triangulation and its variants reveals interesting differences. While Delaunay meshes generally have better spanning ratios, there is no guarantee on the minimum angle between edges incident to the same node, *i.e.* angular resolution of the resulting graph. Shewchuk [12] has thoroughly examined the *angular constraints* on Delaunay triangulations and introduced a Delaunay mesh generation algorithm which adds Steiner points to the original vertex set to increase the graph's angular resolution; however, this algorithm does not guarantee to exit for angular constraints over 34°, meaning that it may run forever. For an emanation graph, the angular resolution is determined by it's grade k, and all emanation graphs of grade $k = 2$ have 45° angular resolution.

Contributions: We provide a simplification method for emanation graphs which works by building a *Simplified Emanation Graph (SEG)* from scratch, instead of removing extra edges from the original version. Then we compare SEG with *constrained Delaunay triangulations* and demonstrate it's advantages under various quality metrics. Good geometric properties of emanation graphs mostly belong to grades $k \leq 2$, *e.g.* much fewer vertex degrees and sufficiently good spanning ratio. Yet the current form of these graphs output a cluttered and visually complex layout. We provide a simplification method for graphs of grade $k = 2$. Thus whenever we refer to Simplified Emanation Graph (SEG) in this paper, we refer to grade $k = 2$. This simplification process greatly reduces the total number of edges while the good properties of the original graph such as the spanning ratio and angular resolution are preserved.

We compared emanation graphs with Delaunay triangulation on both real-life geospatial data and synthetic point sets. The synthetic point sets were created from small world graphs by FMMM algorithm [4], which is a well-known force directed algorithm to create network visualization. The experimental results reveal the reduction of total number of edges, total edge length and the average vertex degree in less than half, while the number of Steiner points and the spanning ratios are comparable.

2 Simplification Method

In an emanation graph, it is common to find two paths of shortest length between a pair of vertices, *e.g.* p_1 and p_3 in Fig. 1 (left). Our simplification attempts to remove such redundancy.

We iterate on the vertices and find only one *nearest neighbor* for every 2^{k+1} directions. Although this appears to be similar to the construction of Θ-graphs, but there are also significant differences in the technique for finding appropriate sweep lines. After selecting this nearest neighbor, we check whether they can connect or that their connection is somehow interfered by a ray of another vertex. For the ease of explanation, the rightward ray of a vertex is labeled r_1 and its other rays are numbered counter-clockwise (Fig. 2). During the computation of the neighbors of p, we will refer to two important vertex types p_s (the 'top' neighbor to connect to p) and p_c (the candidate vertices to check while searching the correct neighbor). We use emanated rays $\{r_1, r_2, r_3, ...\}$ and their angular bisectors labeled b_1, a_1, a_2, b_2, respectively, as guidelines to sweep appropriate regions (cones) to search for p_s and p_c. We use the notation $C_{a_1 a_2}$ to refer to the cone shaped region between the two guidelines a_1 and a_2, and denote by l_g a sweep line orthogonal to the guideline g, starting from p.

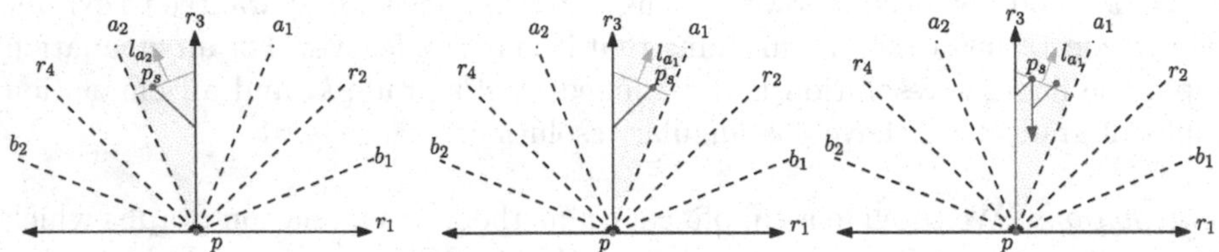

Fig. 2. (left)–(middle) Illustration for the selection of p_s. Both sweep lines start at the same time from p and stop as soon as one finds a vertex p_s. (right) An example, where a successful connection between p and p_s has been made, but a horizontal sweep cannot find p_s.

While describing the computation, instead of iterating on directions, we rotate the plane by $(\frac{\pi}{2^k})$-degrees at each step, and then find a proper *top neighbor* for each vertex. The top neighbor of each vertex p is labeled p_s: The first vertex found sweeping up p's top cones $C_{a_1 r_3}$ and $C_{r_3 a_2}$. Two sweep lines l_{a_1} and l_{a_2}, orthogonal to a_1 and a_2, respectively, are used simultaneously to sweep $C_{a_1 r_3}$ and $C_{r_3 a_2}$ as drawn in Fig. 2 (left)–(middle).

Note that using a single horizontal sweep line may not hit the correct neighbor p_s to be connected to p, *e.g.* the first point q hit by the horizontal sweep line maybe a vertex near p_s in the same cone and one of the downward rays of p_s may block the connection between q and p (contradicting that q is the correct neighbor). Figure 2 (right) illustrates an example for such cases.

To find whether p_s should be connected to p, we need to check whether there is a vertex p_c with $|p|_y \leq |p_c|_y < |p_s|_y$ whose ray reaches the potential connection between p_s and p faster than that of the rays of p_s and p. If so, then p_s and p should not be connected. The notation $|p|_y$ refers to the y coordinate of vertex p. We now show how to check each candidate p_c vertex in the four cones $C_{b_1 r_2}, C_{r_2 a_1}, C_{a_2 r_4}$, and $C_{r_4 b_2}$. Cones $C_{r_1 b_1}$ and $C_{b_2 r_5}$ and their vertices are skipped as no vertex in these areas can reach p's connection to p_s in time.

We use sweep lines with angles specific to each cone to find the *first* p_c *vertex* in that cone, *i.e.* the vertex winning the competition of reaching p's connection to p_s among all the points in the underlying cone. Such a selection of the first candidate vertex p_c ensures that its ray is not interfered by another point inside

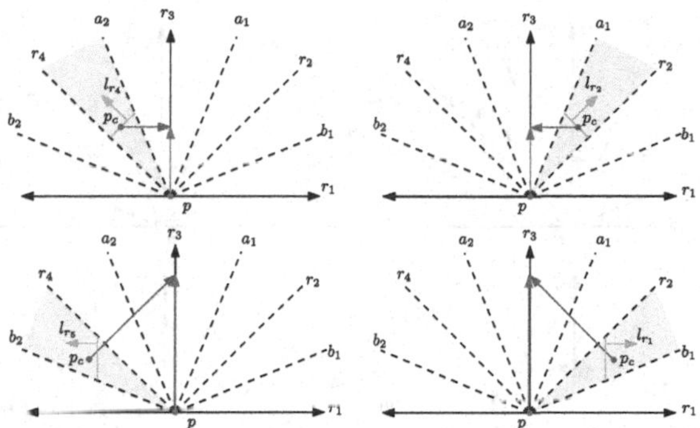

Fig. 3. Sweep lines used to select p_c in each cone around p, drawn in yellow color. They start from p and stop upon finding a vertex. (Color figure online)

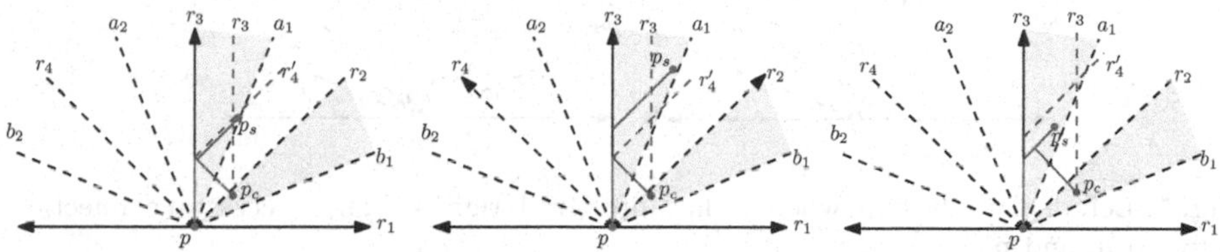

Fig. 4. Left and middle depict two different cases where p_c has interfered, right shows a successful connection between p_s and p.

this cone. Figure 3 illustrates the sweep lines for each cone. Depending on their geometric properties, every vertex in a cone has one ray which is the most competent, for example in a vertex $p_c \in C_{b_1 r_2}$, it's r_4, the north-western ray, may interfere with p, thus to find the most competent vertex inside $C_{b_1 r_2}$ we use a vertical sweep line l_{r_1} starting from p. In other words, if p_c is the correct neighbor to be connected to p, then to reach the ray of p, any subsequent point in the cone will need to have a longer ray than that of p_c. The same method applies to the other cases.

After finding our candidate p_c vertices, we must check for special conditions in each and every one of them individually in order to know whether they can block the connection between p_s and p. These conditions are thoroughly explained later. For every vertex p and each rotation, we find p_s and a list of possibly interfering vertices p_c using the selection methods provided above. During these iterations we skip pairs that are already connected, therefore, if p is already connected to p_s, we do not check if p_s can connect to p. This almost halves the total number of edges and Steiner points by avoiding redundant paths between two connected vertices in the original emanation graph. Theorem 2 provides a bound on total Steiner points in a SEG.

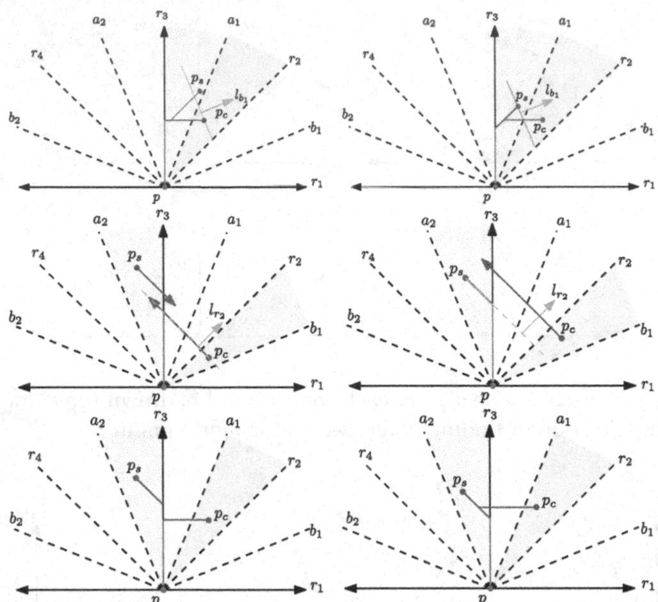

Fig. 5. Left depicts the case where p_c has interfered, right shows a successful connection between p_s and p.

For each $p_c \in C_{b_1a_1}$, there are four different cases in which p_c does not interfere the connection between p_s and p. Figures 4 and 5 illustrate the example for each case, where the rightmost section in each figure depicts the case when p can successfully connect to p. The conditions are as follows:

1. $p_s \in C_{a_1r_3}$ and $p_c \in C_{b_1r_2}$ and $|p_s|_x < |p_c|_x$ and p_s is below r'_4 of p_c: the continued refraction of r_4 of p_c after hitting r_3 of p; see Fig. 4.
2. $p_s \in C_{a_1r_3}$ and $p_c \in C_{r_2a_1}$ and p_s is swept before p_c by the sweep line l_{b_1} orthogonal to b_1 of p; see Fig. 5 (top).
3. $p_s \in C_{r_3a_2}$ and $p_c \in C_{b_1r_2}$ and p_s is swept before p_c by a sweep line orthogonal to r_2 of p; see Fig. 5 (middle).
4. $p_s \in C_{r_3a_2}$ and $p_c \in C_{r_2a_1}$ and $|p_sp|_x < |p_sp_c|_y$; see Fig. 5 (bottom).

Explaining cases where $p_c \in C_{a_2b_2}$ and $|p_c|_x < |p|_x$ is straightforward, as every condition needs to be vertically mirrored, relative to p. We thus describe the conditions regarding these mirrored cases without any additional figure.

1. $p_s \in C_{r_3a_2}$ and $p_c \in C_{b_2r_4}$ and $|p_s|_x > |p_c|_x$ and p_s is below r'_2 of p_c: continued refraction of r_2 of p_c after hitting r_3 of p.
2. $p_s \in C_{r_3a_2}$ and $p_c \in C_{r_4a_2}$ and p_s is swept before p_c by a sweep line orthogonal to b_2 of p.
3. $p_s \in C_{a_1r_3}$ and $p_c \in C_{b_2r_4}$ and p_s is swept before p_c by a sweep line orthogonal to r_4 of p.
4. $p_s \in C_{a_1r_3}$ and $p_c \in C_{r_4a_2}$ and $|pp_s|_x < |p_cp_s|_y$.

Properties of SEG: The following two lemmas discuss a few properties that SEG provides as a spanner. These properties along with ones that result into a visually less cluttered image, highlight the purpose of SEG opposed to it's normal version and in comparison to other commonly used spanners. Yet the lemma proofs are omitted from this paper and provided in our full paper [8].

Lemma 1. *A SEG on a set of n points can be constructed in time $O(n \cdot \text{polylog}(n))$.*

Lemma 2. *A SEG of grade $k = 2$ is a max-degree-8 geometric spanner with at most $4n$ Steiner points.*

Table 1. Results of our comparisons on 3 random and two real data sets. CIT marks the results related to World's most populated cities data set while AIR refers to the data set of US Airlines, lines that are unmarked are related to our random experimentation, based on averages of 1000 instances. *SEG* stands for Simplified Emanation Graph of grade $k = 2$ and DEL C $= \alpha$ is a α° constrained Delaunay Triangulation.

Configuration	Point Count	Data Set	Steiner Points	Max Degree	Average Degree	Edge Count	Max Edge Len	Edge Length	Total Edge Length	Min Angle	Spanning Ratio
SEG	***100***		***197***	***6.20***	***2.55***	***379***	***80.33***	***19.36***	***7319***	***45***	***1.88***
DEL C = 0	100		0	9.55	5.66	283	304.15	50.50	14301	0.57	1.37
DEL C = 22.5	100		87.73	9.05	5.40	506	89.40	29.34	14803	22.68	1.44
DEL C = 33	100		315.60	8.18	5.56	1156	56.95	18.75	21392	33.07	1.59
SEG	***500***		***1085***	***6.85***	***2.63***	***2087***	***69.35***	***12.58***	***20261***	***45***	***2.07***
DEL C = 0	500		0	10.31	5.91	1478	317.07	28.97	42820	0.27	1.39
DEL C = 22.5	500		253	9.48	5.75	2165	76.87	21.06	45576	22.55	1.60
DEL C = 33	500		1017	8.69	5.79	4398	47.44	14.35	62963	33.02	1.84
SEG	***1000***		***2177***	***7.01***	***2.66***	***4231***	***58.10***	***9.52***	***40289***	***45***	***2.16***
DEL C = 0	1000		0	10.72	5.95	2974	284.05	21.16	62933	0.20	1.40
DEL C = 22.5	1000		472	9.75	5.83	4296	64.75	15.91	68346	22.53	1.96
DEL C = 33	1000		1933	8.90	5.86	8601	39.35	10.95	94099	33.01	2.16
SEG	***235***	***AIR***	***485***	***7***	***2.69***	***970***	***61.17***	***9.95***	***9651***	***45***	***1.96***
DEL C = 0	235	AIR	0	11	5.89	692	291.34	26.64	18432	0.06	1.39
DEL C = 22.5	235	AIR	221	9	5.57	1270	66.82	15.5	19682	22.53	1.48
DEL C = 33	235	AIR	729	8	5.66	2727	56.96	10.15	27691	33	1.73
SEG	***1000***	***CIT***	***1161***	***7.00***	***2.94***	***3177***	***580***	***20.39***	***64796***	***45***	***2.28***
DEL C = 0	1000	CIT	0	12	5.95	2975	2024	50.93	151541	0.09	1.41
DEL C = 22.5	1000	CIT	1358	10	6.28	7414	373	27.64	192302	22.55	1.49
DEL C = 33	1000	CIT	4676	9	6.03	17139	166	18.34	308560	33.00	1.6

3 Experimental Comparison

In this section we compare SEG with graphs generated with Delaunay triangulation: constrained [12] and normal.

We generated three sample data sets [7], each containing 1000 random Newman_ Watts_Strogatz small world graphs using NetworkX [5]. All the graphs

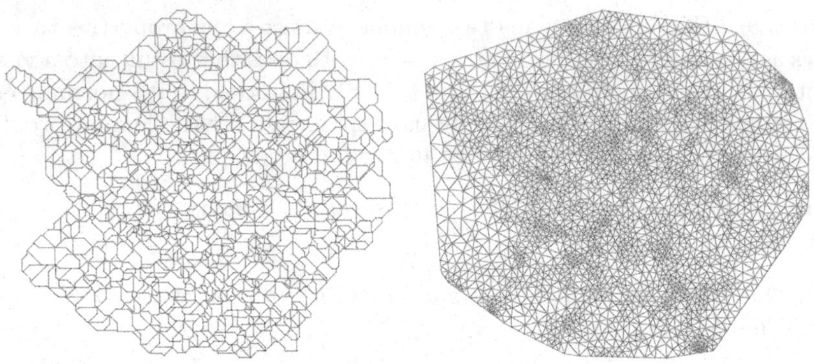

Fig. 6. (left) A SEG of grade 2 on a sample of size 1000. (right) The corresponding 33° constrained Delaunay triangulation.

in a data set contains the same number of nodes. Thus the three data sets contain graphs of size 100, 500, and 1000. We generated the layout for all these graphs using the fast multi-pole multilevel (FMMM) layout [4]. Aside from experimenting on randomly generated data, we also tried SEG on two commonly used data sets: *Locations of 1000 Most Populated Cities* and *US Airports* [1].

Figure 6 depicts SEG and the corresponding constrained Delaunay triangulations for a sample of size 1000.

Although one would like to have angular constraints higher than 33° and close to what emanation graph gives, but the algorithm for constrained Delaunay triangulation doesn't guarantee an exit for larger angular resolutions. We used Triangle [13] to compute the Delaunay triangulations. The metrics we chose to compare our samples are *Steiner Point Count, Vertex Degree, Edge Count, Edge Length, Angle and Spanning Ratio.* Results are depicted in Table 1, separated by different configurations and the number of vertices. Every row of the table shows the mean performance over all 1000 instances of the graphs. In comparison with 33° constrained Delaunay triangulation, SEG provides:

- Much better angular resolution (45° compared to 33°)
- Less than half the number of edges
- Less than half the total edge length
- Less than half the average vertex degree
- Slightly worse spanning ratio (within a factor of 1.18 when $n = 100$ and $n = 500$; and the comparable when $n = 1000$)
- Comparable number of Steiner points (less than half the number of Steiner points for $n = 100$; but slightly worse for $n = 1000$)

4 Discussion

In this paper we present an algorithm to simplify emanation graphs of grade $k = 2$, and experimentally evaluate its aesthetic qualities compared to the Delaunay triangulation and constrained Delaunay triangulation. Our experimental result shows the potential of the simplified emanation graph to be considered as a good alternative to these traditional spanners.

A theoretical open question is to prove a tight upper bound on spanning ratio of the simplified version. Furthermore, one can implement simplified emanation graphs in visualization systems such as GraphMaps [10] to compare the visual results with that of generated by the Delaunay and constrained Delaunay triangulations.

Another interesting avenue for future research is to look for local drawing methods for emanation graph, which output a roughly exact drawing based on user's view-port and zoom level, without computing all other nodes outside user's view-port. Also, extending simplified emanation graphs to a triangulated mesh by triangulating the faces maybe considered as a possible extension of this paper.

References

1. Gephi sample data sets: Us airlines (2019). https://github.com/gephi/gephi/wiki/Datasets. Accessed 6 June 2019
2. Bose, P., Smid, M.H.M.: On plane geometric spanners: a survey and open problems. Comput. Geom. **46**(7), 818–830 (2013)
3. Eppstein, D., Goodrich, M.T., Kim, E., Tamstorf, R.: Motorcycle graphs: canonical quad mesh partitioning. Comput. Graph. Forum **27**(5), 1477–1486 (2008)
4. Hachul, S., Jünger, M.: Large-graph layout with the fast multipole multilevel method. University of Cologne, Computer Science Department, Technical report. Cologne (2005)
5. Hagberg, A.A., Schult, D.A., Swart, P.J.: Exploring network structure, dynamics, and function using networkx. In: Proceedings of the 7th Python in Science Conference (SciPy) (2008)
6. Hamedmohseni, B., Rahmati, Z., Mondal, D.: Emanation graph: a new t-spanner. In: Proceedings of the 30th Canadian Conference on Computational Geometry (CCCG), pp. 311–317 (2018)
7. Hamedmohseni, B., Rahmati, Z., Mondal, D.: Simplified emanation graph - implementations and tests (2019). https://github.com/sneyes/SEG/tree/master. Accessed 6 June 2019
8. Hamedmohseni, B., Rahmati, Z., Mondal, D.: Simplified emanation graphs: A sparse plane spanner with steiner points. https://arxiv.org/abs/1910.10376 (2019)
9. Mondal, D., Nachmanson, L.: A new approach to GraphMaps, a system browsing large graphs as interactive maps. In: Telea, A., Kerren, A., Braz, J. (eds.) Proceedings of the 13th International Joint Conference on Computer Vision, Imaging and Computer Graphics Theory and Applications (VISIGRAPP), pp. 108–119. SciTePress (2018)
10. Nachmanson, L., Prutkin, R., Lee, B., Riche, N.H., Holroyd, A.E., Chen, X.: GraphMaps: browsing large graphs as interactive maps. In: Di Giacomo, E., Lubiw, A. (eds.) GD 2015. LNCS, vol. 9411, pp. 3–15. Springer, Cham (2015). https://doi.org/10.1007/978-3-319-27261-0_1
11. Owen, S.J.: A survey of unstructured mesh generation technology. In: Proceedings of the 7th International Meshing Roundtable (IMR), pp. 239–267 (1998)

12. Shewchuk, J.R.: Triangle: engineering a 2D quality mesh generator and delaunay triangulator. In: Lin, M.C., Manocha, D. (eds.) WACG 1996. LNCS, vol. 1148, pp. 203–222. Springer, Heidelberg (1996). https://doi.org/10.1007/BFb0014497
13. Triangle: A two-dimensional quality mesh generator and delaunay triangulator (2013). https://www.cs.cmu.edu/~quake/triangle.html. Accessed 6 June 2019

Simultaneous FPQ-Ordering and Hybrid Planarity Testing

Giuseppe Liotta[1], Ignaz Rutter[2], and Alessandra Tappini[1](✉)

[1] Dipartimento di Ingegneria, Università degli Studi di Perugia, Perugia, Italy
giuseppe.liotta@unipg.it, alessandra.tappini@studenti.unipg.it
[2] Department of Computer Science and Mathematics, University of Passau, Passau, Germany
rutter@fim.uni-passau.de

Abstract. We study the interplay between embedding constrained planarity and hybrid planarity testing. We consider a constrained planarity testing problem, called 1-Fixed Constrained Planarity, and prove that this problem can be solved in quadratic time for biconnected graphs. Our solution is based on a new definition of fixedness that makes it possible to simplify and extend known techniques about Simultaneous PQ-Ordering. We apply these results to different variants of hybrid planarity testing, including a relaxation of NodeTrix Planarity with fixed sides, that allows rows and columns to be independently permuted.

1 Introduction

A *flat clustered graph* (G, S) consists of a graph G and a set S of vertex disjoint subgraphs of G called *clusters*. An edge connecting two vertices in different clusters is an *inter-cluster edge* while an edge with both end-vertices in a same cluster is an *intra-cluster edge*. A *hybrid representation* of (G, S) is a drawing of the graph that adopts different visualization paradigms to represent the clusters and to represent the inter-cluster edges. For example, Fig. 1(a) depicts a flat clustered graph and Fig. 1(b) shows a NodeTrix representation of this graph.

A NodeTrix *representation* is a hybrid representation of a flat clustered graph where the clusters are depicted as adjacency matrices and the inter-cluster edges are drawn according to the node-link paradigm. NodeTrix representations have been introduced to visually explore non-planar networks by Henry et al. [7] in one of the most cited papers of the InfoVis conference. They have been intensively studied in the last few years, see e.g. [4,6].

PolyLink representations are a generalization of NodeTrix representations. In a PolyLink representation every vertex of each cluster has two copies that lie on opposite sides of a convex polygon (in a NodeTrix representation the

Work partially supported by: MIUR, grant 20174LF3T8 AHeAD: efficient Algorithms for HArnessing networked Data; Dip. Ingegneria Univ. Perugia, grants RICBASE2017WD-RICBA18WD: "Algoritmi e sistemi di analisi visuale di reti complesse e di grandi dimensioni"; German Science Found. (DFG), grant Ru 1903/3-1.

A. Chatzigeorgiou et al. (Eds.): SOFSEM 2020, LNCS 12011, pp. 617–626, 2020.
https://doi.org/10.1007/978-3-030-38919-2_51

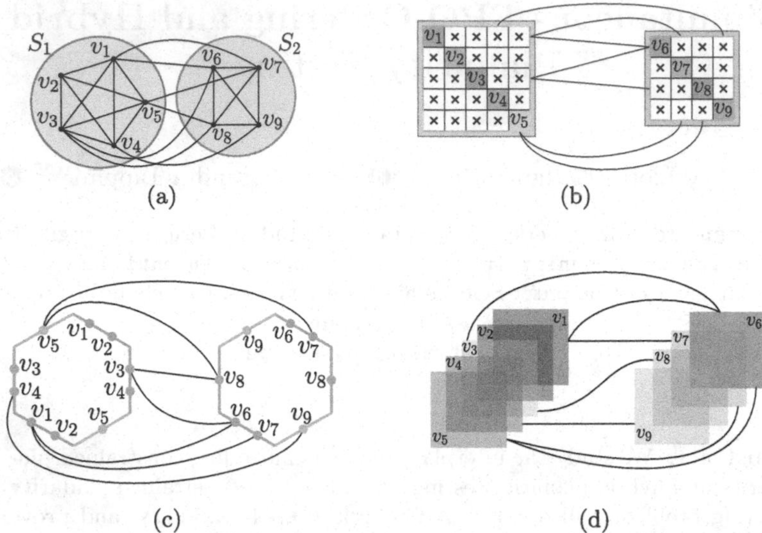

Fig. 1. (a) A flat clustered graph (G, S). Clusters S_1 and S_2 are highlighted. (b) A NodeTrix representation of (G, S). (c) A PolyLink representation of (G, S). (d) An intersection-link representation of (G, S).

polygon is a square); see Fig. 1(c). *Intersection-link representations* are another example of hybrid representations: Each vertex of (G, S) is a simple polygon and two polygons overlap if and only if there is an intra-cluster edge connecting them [1]. Figure 1(d) is an intersection-link representation with unit squares.

Given a flat clustered graph (G, S) and a hybrid representation paradigm, it makes sense to ask whether (G, S) is *hybrid planar*, that is, whether (G, S) admits a drawing in the given paradigm such that no two inter-cluster edges cross. In general terms, hybrid planarity testing is a more challenging problem than "traditional" planarity testing. Hybrid representations allow multiple copies for each vertex, which facilitates the task of avoiding crossings but makes the problem of testing the graph for planarity combinatorially more complex. Hybrid planarity testing can be studied in both the "fixed sides" and the "free sides" scenarios. Let $e = (u, v)$ be an inter-cluster edge where u is a vertex of cluster $C_u \in S$ and v is a vertex of $C_v \in S$. The fixed sides scenario specifies the sides of the geometric objects representing C_u and C_v to which e is incident; the free sides scenario allows the algorithm to choose the sides of incidence of e. For example, NodeTrix Planarity with fixed sides specifies whether e is incident to the top, bottom, left, or right copy of u (v) in the matrix representing C_u (C_v).

This paper studies different variants of hybrid planarity testing in the fixed sides scenario. It adopts a unified approach that models these problems as instances of a suitably defined constrained planarity testing problem on a graph G. The constrained planarity problem specifies for each vertex v which cyclic orders for the edges of G incident to v are allowed. Choosing an order for a vertex of G influences the allowed orders for other vertices of G; such dependencies between different allowed orders are expressed by a directed acyclic graph (DAG) whose nodes are FPQ-trees (a variant of PQ-trees). Our contribution is as follows:

- We introduce and study 1-Fixed Constrained Planarity and show that this problem can be solved in quadratic time for biconnected graphs, by modeling it as an instance of Simultaneous FPQ-Ordering. 1-Fixed Constrained Planarity generalizes the partially PQ-constrained planarity testing problem studied by Bläsius and Rutter [2]. Our solution exploits a new definition of fixedness that simplifies and extends results of [2].
- We show that a relaxation of NodeTrix Planarity with fixed sides, that allows to independently permute the rows and the columns of the matrices, can be modeled as an instance of 1-Fixed Constrained Planarity, and hence it can be solved in quadratic time if the multi-graph obtained by collapsing the clusters to single vertices is biconnected. We recall that NodeTrix Planarity with fixed sides is NP-complete in general, but it is linear-time solvable if the rows and the columns of each matrix cannot be permuted [4]. Thus it makes sense to further explore the conditions under which the problem is polynomially tractable.
- We introduce PolyLink representations and we show that biconnected instances of PolyLink Planarity with fixed sides can be solved in quadratic time. As a byproduct, we obtain that a special instance of intersection-link planarity, *clique planarity with fixed sides*, can be solved in quadratic time. Note that clique planarity is NP-complete in general [1]. We remark that PolyLink Planarity is equivalent to NodeTrix Planarity with free sides if the polygons have maximum size four and each side is associated with the same set of vertices.

For reasons of space, the results about PolyLink Planarity and some missing proofs and details can be found in [9].

2 Preliminaries

PQ-Trees: A *PQ-tree* is a data structure that represents a family of permutations on a set of elements [3]. In a PQ-tree, each element is represented by a leaf node, and each non-leaf node is either a *P-node* or a *Q-node*. The children of a P-node can be arbitrarily permuted, while the order of the children of a Q-node is fixed up to a reversal. Three main operations are defined on PQ-trees [2,3]. Let T be a PQ-tree and let L be the set of its leaves. Given $S \subseteq L$, the *projection of T to S*, denoted as $T|_S$, is a PQ-tree T' that represents the orders of S allowed by T, such that T' contains only the leaves of T that belong to S. T' is obtained form T by removing all the leaves not in S and simplifying the result, where simplifying means, that former inner nodes now having degree 1 are removed iteratively and that degree-2 nodes together with both incident edges are iteratively replaced by single edges. The *reduction of T with S*, denoted as $T + S$, is a PQ-tree T' that represents only the orders represented by T where the leaves of S are consecutive. A Q-node in $T+S$ can determine the orientation of several Q-nodes of T, while if we consider a P-node μ' in $T+S$, there is exactly one P-node μ in T that depends on μ'. We say that μ' *stems from* μ. Given two PQ-trees T_1 and T_2, the *intersection of* T_1 *and* T_2, denoted as $T_1 \cap T_2$, is a PQ-tree T' representing the orders of L represented by both T_1 and T_2. If T_1 and T_2 have the same leaves, their intersection is obtained by applying to T_2 a sequence of reductions with subsets of leaves whose orders are given by T_1 [2].

Simultaneous PQ-Ordering: An instance of SIMULTANEOUS PQ-ORDERING [2] is a DAG of PQ-trees that establishes relations between each parent node and its children nodes. Informally, the DAG imposes that the order of the leaves of a parent node must be "in accordance with" the order of the leaves of its children. More formally, let $N = \{T_1, \dots, T_k\}$ be a set of PQ-trees whose leaves are $L(T_1), \dots, L(T_k)$, respectively. Let $\mathcal{I} = (N, Z)$ be a DAG with vertex set N and such that every arc in Z is a triple $(T_i, T_j; \varphi)$ where T_i is the tail vertex, T_j is the head vertex, and $\varphi : L(T_j) \to L(T_i)$ is an injective mapping from the leaves of T_j to the leaves of T_i $(1 \leq i, j \leq k)$. Given two cyclic orders O_i and O_j defined by T_i and T_j, respectively, we say that O_i *extends* $\varphi(O_j)$ if $\varphi(O_j)$ is a suborder of O_i. The SIMULTANEOUS PQ-ORDERING problem asks whether there exist cyclic orders $O_1, \dots O_k$ of $L(T_1), \dots, L(T_k)$, respectively, such that for each arc $(T_i, T_j; \varphi) \in Z$, O_i extends $\varphi(O_j)$. Let $(T_i, T_j; \varphi)$ be an arc in Z. An internal node μ_i of T_i is *fixed by* an internal node μ_j of T_j (and μ_j *fixes* μ_i in T_i) if there exist leaves $x, y, z \in L(T_j)$ and $\varphi(x), \varphi(y), \varphi(z) \in L(T_i)$ such that (i) removing μ_j from T_j makes x, y, and z pairwise disconnected in T_j, and (ii) removing μ_i from T_i makes $\varphi(x)$, $\varphi(y)$, and $\varphi(z)$ pairwise disconnected in T_i.

An instance $\mathcal{I} = (N, Z)$ of SIMULTANEOUS PQ-ORDERING is *normalized* if, for each arc $(T_i, T_j; \varphi) \in Z$ and for each internal node $\mu_j \in T_j$, tree T_i contains exactly one node μ_i that is fixed by μ_j. Every instance of SIMULTANEOUS PQ-ORDERING can be normalized by means of an operation called the *normalization* [2], which is defined as follows. Consider each arc $(T_i, T_j; \varphi) \in Z$ and replace T_j with $T_i|_{\varphi(L(T_j))} \cap T_j$ in $\mathcal{I}$, that is, replace tree T_j with its intersection with the projection of its parent T_i to the set of leaves of T_i obtained by applying mapping φ to the leaves $L(T_j)$ of T_j. Consider a normalized instance $\mathcal{I} = (N, Z)$. Let μ be a P-node of a PQ-tree T with parents $T_1, \dots, T_p$ and let $\mu_i \in T_i$ be the unique node in T_i, with $1 \leq i \leq p$, fixed by μ. The *fixedness* of μ is defined as $\text{fixed}(\mu) = \omega + \sum_{i=1}^{p}(\text{fixed}(\mu_i) - 1)$, where ω is the number of children of T containing a node that fixes μ. A P-node μ is *k-fixed* if $\text{fixed}(\mu) \leq k$. Also, instance $\mathcal{I}$ is *k-fixed* if all the P-nodes of any PQ-tree $T \in N$ are k-fixed.

FPQ-Trees: An *FPQ-tree* is a PQ-tree where, for some of the Q-nodes, the reversal of the permutation described by their children is not allowed. To distinguish these Q-nodes from the regular Q-nodes, we call them *F-nodes* [8]. The study of Bläsius and Rutter on SIMULTANEOUS PQ-ORDERING also considers the case in which the permutations described by some of the Q-nodes are totally

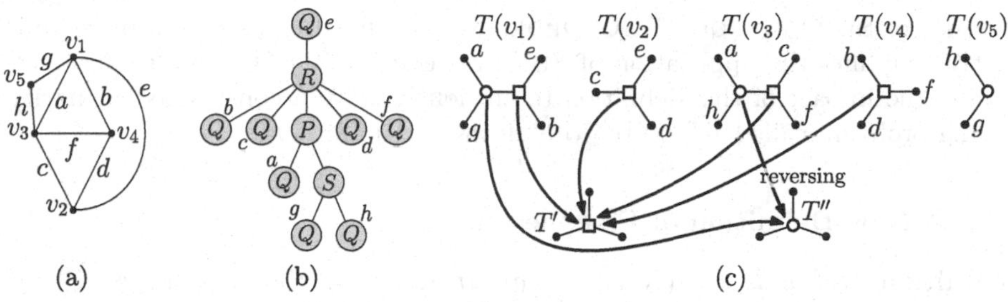

Fig. 2. (a) A biconnected planar graph G. (b) An SPQR-decomposition tree of G. (c) The embedding DAG $\mathcal{D}$ of G. P-nodes are depicted as circles, while Q-nodes are boxes.

fixed, hence the results given in [2] for SIMULTANEOUS PQ-ORDERING also hold when the nodes of the input DAG are FPQ-trees. In the rest of the paper we talk about SIMULTANEOUS FPQ-ORDERING to emphasize the presence of F-nodes, since they play an important role in our applications of hybrid planarity testing.

Embedding DAG: Let G be a biconnected planar graph and let $\mathcal{T}$ be an SPQR-tree of G [5]. We can associate with each vertex v of G an FPQ-tree $T(v)$ called the *embedding tree of* v, whose leaves correspond to the edges incident to v [2]; $T(v)$ encodes all the cyclic orders of the edges incident to v that are described by $\mathcal{T}$. The cyclic orders around a vertex in a planar embedding of a graph depend on the cyclic orders of the edges around other vertices. Such dependencies can be conveniently modeled as a DAG of FPQ-trees, called the *embedding DAG of G* and denoted as $\mathcal{D}$ (see also [9]). Figure 2(b) shows an SPQR-tree of the graph in Fig. 2(a), and (c) is the corresponding embedding DAG. The injective function φ for each arc of $\mathcal{D}$ associates the leaves of a source FPQ-tree to the leaves of a sink FPQ-tree. For example, there is a mapping between the leaves a, b, and e of $T(v_1)$ and the leaves of the sink FPQ-tree T'; while a suitable mapping between $T(v_1)$ and the sink FPQ-tree T'' maps g, a and b to the leaves of T''. Let v be a vertex of G. The *embedding DAG of* v, denoted as $\mathcal{D}(v)$, is the subgraph of $\mathcal{D}$ induced by $T(v)$ and by the FPQ-trees that are connected to $T(v)$. Note that $\mathcal{D}$ and $\mathcal{D}(v)$ are instances of SIMULTANEOUS FPQ-ORDERING.

3 Fixedness and 1-Fixed Constrained Planarity

Bläsius and Rutter in [2] show that normalized instances of SIMULTANEOUS FPQ-ORDERING can be solved in quadratic time if they are 2-fixed. In their applications, instances are normalized (or have a very simple structure) so that it is easy to verify whether an instance is 2-fixed. The difficulty of applying their result to other contexts is that if the instances are not normalized, it is quite technical to understand the structure of the normalized instance and to check whether it is 2-fixed. We present a new definition of fixedness that does no longer require the normalization as a preliminary step to check whether an instance of Simultaneous FPQ-Ordering is 2-fixed. This definition significantly simplifies the application of Simultaneous FPQ-Ordering. Also, we discuss the impact of this definition to efficiently solve a constrained planarity testing problem, called 1-Fixed Constrained Planarity.

3.1 A New Definition of Fixedness

Definition 1. *Let $\mathcal{I} = (N, Z)$ be an instance of* Simultaneous FPQ-Ordering *and let μ be a P-node of an FPQ-tree that belongs to a node v of $\mathcal{I}$. The* fixedness *of μ is denoted as* $\text{fixed}(\mu)$. *Let ω be the number of children of v fixing μ. If v is a source, we define* $\text{fixed}(\mu) = \omega$. *If v is not a source, let p be the number of parent nodes $T_1, \dots, T_p$ of v in $\mathcal{I}$. For $i = 1, \dots, p$, let F_i be the set of P-nodes of T_i that is fixed by μ. If $|F_i| = 0$ for some $i = 1, \dots, k$, then* $\text{fixed}(\mu) = 0$, *otherwise* $\text{fixed}(\mu) = \omega + \sum_{i=1}^{p} \max_{\nu \in F_i}(\text{fixed}(\nu) - 1)$. *The P-node μ is* k-fixed *if* $\text{fixed}(\mu) \leq k$. *Instance $\mathcal{I}$ is* k-fixed *if all P-nodes of FPQ-trees $T \in N$ are k-fixed.*

We remark that Definition 1 coincides with the notion of fixedness given in [2] if we restrict ourselves to normalized instances. Namely, in a normalized instance, $|F_i| = 1$ for $i = 1, \dots, p$, and the maximum vanishes.

Lemma 1. *Let $\mathcal{I}$ be an instance of* Simultaneous FPQ-Ordering *and let $\mathcal{I}'$ be the normalization of $\mathcal{I}$. Then* $\text{fixed}(\mathcal{I}') \leq \text{fixed}(\mathcal{I})$.

By Lemma 1, it suffices to check the 2-fixedness of a non-normalized instance of Simultaneous FPQ-Ordering to conclude that it can be solved in quadratic time by exploiting [2, Theorems 3.11, 3.16]. We now further simplify the applicability of the result.

Let $\mathcal{I} = (N, A)$ be an instance of Simultaneous FPQ-Ordering. We denote by $\text{source}(\mathcal{I})$ the set of sources of $\mathcal{I}$. A solution of an instance $\mathcal{I} = (N, A)$ of Simultaneous FPQ-Ordering determines a tuple of cyclic orders $(O_v)_{v \in N}$. In many cases, we are only interested in the cyclic orders at the sources, and we therefore define $\text{sol}(\mathcal{I}) = \{(O_v)_{v \in \text{source}(\mathcal{I})} \mid \mathcal{I} \text{ has a solution } (O'_v)_{v \in N} \text{ with } O_v = O'_v \text{ for } v \in \text{source}(\mathcal{I})\}$. We say that an instance $\mathcal{I}$ has *P-degree* k if every node whose FPQ-tree contains a P-node has at most k parents. Let $\mathcal{I}$ and $\mathcal{I}'$ be two instances of Simultaneous FPQ-Ordering such that there exists a bijective mapping M between the sources of $\mathcal{I}$ and the sources of $\mathcal{I}'$ with $L(M(T)) = L(T)$ for each source T of $\mathcal{I}$. We call $\mathcal{I}$ and $\mathcal{I}'$ *joinable*. The *join DAG* of $\mathcal{I}$ and $\mathcal{I}'$ is the instance $\mathcal{I} \bowtie \mathcal{I}'$ obtained by replacing, for each source node T of $\mathcal{I}$ (and each corresponding source node $M(T)$ of $\mathcal{I}'$), the nodes T (and $M(T)$) by $T \cap M(T)$ and identifying the respective nodes of $\mathcal{I}$ and $\mathcal{I}'$. By construction, it is $\text{sol}(\mathcal{I} \bowtie \mathcal{I}') = \text{sol}(\mathcal{I}) \cap \text{sol}(\mathcal{I}')$.

Lemma 2. *Let $\mathcal{I}$ and $\mathcal{I}'$ be joinable instances of* Simultaneous FPQ-Ordering *with P-degree at most* 2 *and such that their associated DAGs each have height* 1. *If both $\mathcal{I}$ and $\mathcal{I}'$ are* 1*-fixed, then $\mathcal{J} = \mathcal{I} \bowtie \mathcal{I}'$ is* 2*-fixed.*

3.2 1-Fixed Constrained Planarity

Let $G = (V, E)$ be a biconnected planar graph, let $v \in V$ be a vertex, and let $E(v)$ be the edges of G incident to v. A *1-fixed constraint* $\mathcal{C}(v)$ for v is a 1-fixed instance of SIMULTANEOUS FPQ-ORDERING such that it has P-degree at most 2 and it has a single source whose FPQ-tree has the edges in $E(v)$ as its leaves. The following property is implied by [2, Section 4.1].

Property 1. For each vertex v of G, $\mathcal{D}(v)$ is a 1-fixed constraint.

Let $\mathcal{E}$ be an embedding of G and let $\mathcal{E}(v)$ be the cyclic order that $\mathcal{E}$ induces on the edges around v. We say that embedding $\mathcal{E}$ *satisfies* constraint $\mathcal{C}(v)$ if there exists a solution for $\mathcal{C}(v)$ such that the order of the source is $\mathcal{E}(v)$.

Given a graph G and a 1-fixed constraint for each vertex of G, the 1-FIXED CONSTRAINED PLANARITY testing problem asks whether G is 1*-fixed constrained planar*, i.e., it admits a planar embedding that satisfies all the constraints.

Theorem 1. *Let $G = (V, E)$ be a biconnected planar graph with n vertices, and for each $v \in V$ let $\mathcal{C}(v)$ be a 1-fixed constraint.* 1-FIXED CONSTRAINED PLANARITY *can be tested in $O(n^2)$ time.*

Proof. Let $\mathcal{D}$ be the embedding DAG of G, where sol($\mathcal{D}$) corresponds bijectively to the rotation systems of the planar embeddings of G [2]. The embedding DAG $\mathcal{D}(v)$ of a vertex $v \in V$ is such that sol($\mathcal{D}(v)$) corresponds bijectively to the cyclic orders that the planar embeddings of G induce around v. Let $\mathcal{C}$ denote the instance of SIMULTANEOUS FPQ-ORDERING that is the disjoint union $\bigcup_{v \in V} \mathcal{C}(v)$, and observe further that sol($\mathcal{C}$) are precisely the rotations at vertices that satisfy all the constraints $\mathcal{C}(v)$. Observe further that $\mathcal{D}$ and $\mathcal{C}$ are joinable, and sol($\mathcal{D} \bowtie \mathcal{C}$) are exactly the rotation systems of planar embeddings of G that satisfy all the constraints $\mathcal{C}(v)$, $v \in V$. By Property 1, both $\mathcal{D}$ and $\mathcal{C}$ are 1-fixed, have height 1 and P-degree at most 2. Therefore, by Lemma 2 $\mathcal{J} = \mathcal{D} \bowtie \mathcal{C}$ is 2-fixed and by Lemma 1 also the normalization of $\mathcal{J}$ is 2-fixed. It follows that the normalization of $\mathcal{J}$ can be solved in $O(n^2)$ time [2, Theorems 3.11, 3.16]. The overall result follows from the fact that $\mathcal{D}$ and $\mathcal{C}$ have size linear in n and their normalization can be computed in linear time [2, Lemma 3.12].

4 Hybrid Planarity Testing Problems

We recall that in a NODETRIX representation each cluster is represented as an adjacency matrix, while the inter-cluster edges are simple curves connecting the corresponding matrices and not crossing any other matrix [4,6,7]. A NODETRIX *graph* is a flat clustered graph with a NODETRIX representation. For example, Fig. 1(b) is a NODETRIX representation of the graph in Fig. 1(a); note that for every vertex there are four segments, one for each side of the matrix, to which inter-cluster edges can be connected. A NODETRIX representation is *with fixed sides* if the sides of the matrices to which inter-cluster edges must be incident are given as a part of the input. NODETRIX PLANARITY with fixed sides is

NP-hard [4], and it is fixed parameter tractable with respect to the maximum size of clusters and to the treewidth of the graph obtained by collapsing each cluster into a single vertex, as shown in [6,8]. NodeTrix Planarity with fixed sides is known to be solvable in linear time when rows and columns are not allowed to be permuted [4]. This naturally raises the question about whether a polynomial-time solution exists also for less constrained versions of NodeTrix Planarity.

We study the scenario in which the permutations of rows and columns can be chosen independently. Namely, we introduce a relaxed version of NodeTrix Planarity with fixed sides, called Row-Column Independent NodeTrix Planarity (RCI-NT Planarity for short). RCI-NT Planarity asks whether a flat clustered graph admits a planar NodeTrix representation in the fixed sides scenario, but it allows to permute the rows and the columns independently of one another. A graph for which the RCI-NT Planarity test is positive is said to be RCI-NT *planar.*

The Equipped Frame Graph: We model RCI-NT Planarity as an instance of 1-Fixed Constrained Planarity defined on a (multi-)graph associated with (G, S), that we call the *equipped frame graph* of G, denoted as G_F, and that is obtained from G by collapsing each cluster into a single vertex. More precisely, G_F has $n_F = |S|$ vertices, each one corresponding to one of the matrices defined by S. There is an edge between two vertices u and v of G_F if and only if there is an edge in G between matrices M_u and M_v corresponding to u and to v, respectively. A NodeTrix graph is biconnected if its equipped frame graph is biconnected and, from now on, we consider biconnected NodeTrix graphs.

Each vertex v of G_F is associated with a *constraint DAG* $\mathcal{H}(v)$ whose nodes are FPQ-trees. More precisely, the source vertex of $\mathcal{H}(v)$ is an FPQ-tree T_M consisting of an F-node with four incident P-nodes; each of such P-nodes describes

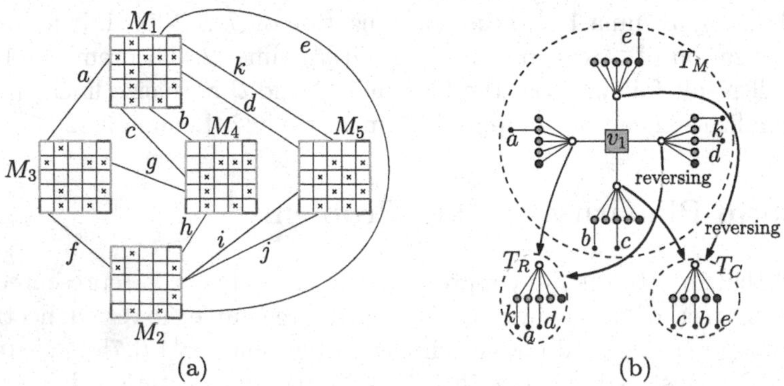

Fig. 3. (a) An RCI-NT planar graph (G, S) that is not NodeTrix planar with fixed sides. (b) The constraint DAG $\mathcal{H}(v_1)$ associated with vertex v_1 of the equipped frame graph of G, corresponding to matrix M_1. P-nodes are circles, F-nodes are shaded boxes.

possible permutations for the rows or for the columns of the matrix M_v. Two P-nodes encode the permutations of the rows (on the left and right hand-side of M_v), and the other two P-nodes encode the permutations of the columns (on the top and bottom hand-side of M_v). The source of $\mathcal{H}(v)$ has two adjacent vertices; one of these adjacent vertices is associated with an FPQ-tree T_R, and the other one is associated with an FPQ-tree T_C. T_R specifies permutations for the rows of M_v, and T_C specifies permutations for the columns of M_v, that must be respected by the P-nodes of the FPQ-tree in the root of $\mathcal{H}(v)$. We say that T_R and T_C define the *coherence* between the permutations of the rows and the permutations of the columns, respectively. Figure 3(a) shows a NodeTrix graph (G, S) and Fig. 3(b) shows the constraint DAG $\mathcal{H}(v_1)$ associated with vertex v_1 of the equipped frame graph of G. Note that G is RCI-NT planar but it is not NodeTrix planar with fixed sides: If we require the rows and the columns of M_1 to have the same permutation, it is easy to check that either a crossing between b and c or one between d and k occurs. Two arcs of Fig. 3(b) are labeled *reversing* because, for any given permutation of the rows (columns), the rows (columns) are encountered in opposite orders when walking around M_1. Note that $\mathcal{H}(v)$ is an instance of Simultaneous FPQ-Ordering.

Property 2. For each vertex v of G_F, $\mathcal{H}(v)$ is a 1-fixed constraint.

Let $\mathcal{D}$ be the embedding DAG of G_F. Each vertex v of G_F is associated with its constraint DAG $\mathcal{H}(v)$ and its embedding DAG $\mathcal{D}(v)$.

Lemma 3. *A biconnected* NodeTrix *graph with fixed sides is* RCI-NT *planar if and only if its equipped frame graph is* 1*-fixed constrained planar.*

Testing RCI-NT Planarity: Based on Lemma 3, we shall test whether (G, S) is RCI-NT planar by testing whether G_F is 1-fixed constrained planar. Observe that $\mathcal{H}(v)$ and $\mathcal{D}(v)$ have the same leaves, since they describe possible cyclic orders for the same set of inter-cluster edges, namely those incident to the matrix M_v associated with v in G_F, hence $\mathcal{H}(v)$ and $\mathcal{D}(v)$ are joinable instances of Simultaneous FPQ-Ordering. G_F is 1-fixed constrained planar if and only if it admits a planar embedding such that, for each vertex v the cyclic order of the edges incident to v satisfies both the constraints given by $\mathcal{H}(v)$ and the ones given by $\mathcal{D}(v)$. These constraints are described by the join DAG $\mathcal{J}(v) = \mathcal{H}(v) \bowtie \mathcal{D}(v)$. Properties 1, 2, and Lemma 2 imply that, for each vertex v of G, $\mathcal{J}(v)$ is 2-fixed.

We now exploit Theorem 1, and hence we can test in $O(n_F^2)$ time whether G_F is 1-fixed constrained planar, where n_F is the number of vertices of G_F. By Lemma 3, and since constructing G_F may require $O(n^2)$ time, the following holds.

Theorem 2. *Let (G, S) be a biconnected* NodeTrix *graph.* RCI-NT Planarity *can be tested in $O(n^2)$-time, where n is the number of vertices of G.*

References

1. Angelini, P., Da Lozzo, G., Di Battista, G., Frati, F., Patrignani, M., Rutter, I.: Intersection-link representations of graphs. JGAA **21**(4), 731–755 (2017)
2. Bläsius, T., Rutter, I.: Simultaneous PQ-ordering with applications to constrained embedding problems. ACM Trans. Algorithms **12**(2), 16:1–16:46 (2016)
3. Booth, K.S., Lueker, G.S.: Testing for the consecutive ones property, interval graphs, and graph planarity using PQ-tree algorithms. J. C. Syst. Sci. **13**(3), 335–379 (1976)
4. Da Lozzo, G., Di Battista, G., Frati, F., Patrignani, M.: Computing NodeTrix representations of clustered graphs. J. Graph Algorithms Appl. **22**(2), 139–176 (2018)
5. Di Battista, G., Tamassia, R.: On-line planarity testing. SIAM J. Comput. **25**(5), 956–997 (1996)
6. Di Giacomo, E., Liotta, G., Patrignani, M., Rutter, I., Tappini, A.: NodeTrix planarity testing with small clusters. Algorithmica **81**(9), 3464–3493 (2019)
7. Henry, N., Fekete, J., McGuffin, M.J.: NodeTrix: a hybrid visualization of social networks. IEEE Trans. Vis. Comput. Graph. **13**(6), 1302–1309 (2007)
8. Liotta, G., Rutter, I., Tappini, A.: Graph planarity testing with hierarchical embedding constraints. CoRR abs/1904.12596 (2019)
9. Liotta, G., Rutter, I., Tappini, A.: Simultaneous FPQ-Ordering and hybrid planarity testing. CoRR abs/1910.10113 (2019)

Two-Player Competitive Diffusion Game: Graph Classes and the Existence of a Nash Equilibrium

Naoka Fukuzono[1(✉)], Tesshu Hanaka[2(✉)], Hironori Kiya[1(✉)], Hirotaka Ono[1(✉)], and Ryogo Yamaguchi[3(✉)]

[1] Graduate School of Informatics, Nagoya University, Furo-cho, Chikusa-ku, Nagoya, Japan
fukuzono.naoka@h.mbox.nagoya-u.ac.jp, kiya.hironori@f.mbox.nagoya-u.ac.jp, ono@i.nagoya-u.ac.jp

[2] Department of Information and System Engineering, Chuo University, 1-13-27Kasuga, Bunkyo-ku, Tokyo, Japan
hanaka.91t@g.chuo-u.ac.jp

[3] Development Bank of Japan, Tokyo, Japan

Abstract. The competitive diffusion game is a game-theoretic model of information spreading on a graph proposed by Alon et al. (2010). In the model, a player chooses an initial vertex of the graph, from which information by the player spreads through the edges connected with the initial vertex. If a vertex that is not yet influenced by any information receives information by a player, it is influenced by the information and it diffuses it to adjacent vertices. A vertex that simultaneously receives two or more types of information does not diffuse any type of information from then on. The objective of a player is to maximize the number of vertices influenced by the player's information. In this paper, we investigate the existence of a pure Nash equilibrium of the two-player competitive diffusion game on chordal and its related graphs. We show that a pure Nash equilibrium always exists on block graphs, split graphs and interval graphs, all of which are well-known subclasses of chordal graphs. On the other hand, we show that there is an instance with no pure Nash equilibrium on (strongly) chordal graphs; the boundary of the existence of a pure Nash equilibrium is found.

Keywords: Nash equilibrium · Competitive diffusion game · Algorithmic game theory · Chordal graph

1 Introduction

The competitive diffusion game is a game-theoretic model of information spreading on a graph proposed by Alon et al. [1]. It is introduced in order to study several competitive diffusion phenomena on social network services (SNS),

This work was partially supported by JSPS KAKENHI Grant Numbers JP17K19960, 17H01698, 19K21537.

A. Chatzigeorgiou et al. (Eds.): SOFSEM 2020, LNCS 12011, pp. 627–635, 2020.
https://doi.org/10.1007/978-3-030-38919-2_52

such as Facebook and Twitter. For example, viral marketing is a typical commercial activity utilizing information diffusion phenomena on a social network. A game-theoretical setting happens when several companies want to sell interoperable products via viral marketing.

In the model, each player has its own information and wants to spread it to vertices in a graph. To this end, a player chooses an initial vertex of the graph, from which information by the player spreads through the edges connected with the initial vertex. If a vertex that is not yet influenced by any information receives information by a player, it is influenced by the information and it diffuses the information to adjacent vertices. A vertex that simultaneously receives two or more types of information by multiple players does not diffuse any type of information from then on. The objective of a player is to maximize the number of vertices affected by the player's information. These settings are interpreted in real world situations as follows: A graph is a social network and each vertex represents a person (potential customer) and an edge represents that two persons corresponding end vertices are friends in SNS. Players are commercial companies that want to sell interoperable products via viral marketing. Each company asks a person on the SNS to advertise its own product by paying some amount of money. The person receiving money recommends the product of the company to his/her friends. A person receives a recommendation of a product from a friend, he/she decides to buy the product and newly recommends the product of the company to his/her friends. Sometimes a person simultaneously receives two types of recommendations. Then he/she gets confused, and he/she does not buy any of the products and recommend anything. This is a simplest model and we can consider more generalized models.

In analyses of game-theoretic models, one of typical approaches is to focus on Nash equilibria. This is because finding a Nash equilibrium is related to predict behaviours of rational players. Although it is known that there exists a mixed-strategy Nash equilibrium for every finite game, a pure Nash equilibrium does not always exist. In fact, in the competitive diffusion game, there is a graph under which no pure Nash equilibrium exists even for the two-player case [1], whereas a graph in some restricted graph classes such as cycles always has a pure Nash equilibrium for any number of players [3]. If a game has a pure Nash equilibrium, it implies that it is relatively easy to analyze. From such a viewpoint, several studies try to find classes of graphs under which a pure Nash equilibrium always exists. For more details, see the following subsection.

1.1 Related Work

There are many studies that focus on the existence of a pure Nash equilibrium of the two-player competitive diffusion game. For example, Alon et al. give a graph with diameter 3 has no pure Nash equilibrium [1]. Takehara et al. give a stronger example, a graph with diameter 2 has no pure Nash equilibrium [14]. On the other hand, Small and Mason show that a pure Nash equilibrium always exists on trees [12]. Roshanbin shows that a pure Nash equilibrium always exists on cycles and grid graphs [11], and Sukenari et al. show that a pure Nash equilibrium

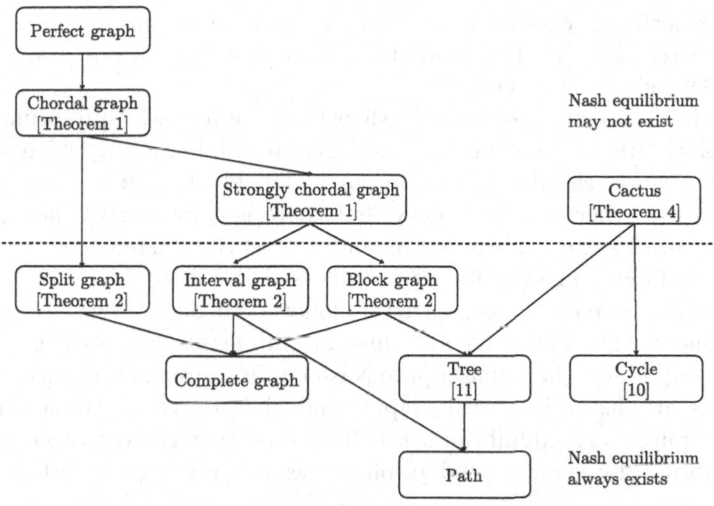

Fig. 1. Graph classes and the existence of a pure Nash equilibrium. Connections between two graph classes imply that the above one is a super class of the below one.

always exists on torus grid graphs [13]. These results are about the two-player competitive diffusion game. For three or more players, the situation is different. For example, in most of the cases, a path always has a pure Nash equilibrium. The exception is the case where the number of players is 3 and the number of vertices is at least 6. On the other hand, a cycle always has a pure Nash equilibrium for the case where the number of players and the number of vertices are arbitrary [3].

If the number k of players is bounded by a constant, it can be done in polynomial time to check whether a given graph has a pure Nash equilibrium or not, because the number of combinations of strategies is $O(n^k)$, where n is the number of vertices. On the other hand, it is not trivial to check the existence of a pure Nash equilibrium for general k. Etesami and Basar show that the decision problem of the existence a pure Nash equilibrium for general k is NP-complete [5]. Furthermore, Ito et al. show that the decision problem of the existence a pure Nash equilibrium is W[1]-hard when it is parameterized by k [8].

1.2 Our Results

In this paper, we investigate the existence of a pure Nash equilibrium of the two-player competitive diffusion game on chordal and its related graphs. A graph is called *chordal* if every induced cycle in the graph should have exactly three vertices. The class of chordal graphs is a well-known graph class in many research fields, and they are also called rigid circuit graphs or triangulated graphs. Particularly in the algorithm theory, it is considered very important, because many NP-hard optimization problems in general graphs can be solved in polynomial time if the input graph is chordal. This is a motivation that we focus on chordal graphs. Furthermore, chordal graphs and its related graph classes are intensively and extensively studied, and there are many well-known graph classes. For example, trees are also chordal.

We obtain the following results: We show that a pure Nash equilibrium always exists on block graphs, split graphs and interval graphs, all of which are well-known subclasses of chordal graphs. In particular, block graphs is a super class of trees. On the other hand, we show that there is a (strongly) chordal graph that has no pure Nash equilibrium; the boundary of the existence of a pure Nash equilibrium is found. The results are summarized in Fig. 1.

The rest of the paper is organized as follows. In Sect. 2, we define several notations and terminology, and introduce graph classes. Section 3 is the main part of this paper. We show that a pure Nash equilibrium always exists on block graphs, split graphs and interval graphs, and give a (strongly) chordal graph that has no pure Nash equilibrium. Finally, we present the existence of a pure Nash equilibrium for some related graph classes and settings in Sect. 4.

2 Preliminaries

In this paper, we use the standard graph notations. Let $G = (V, E)$ be an undirected conneted graph where $|V| = n$ and $|E| = m$. For $V' \subseteq V$, we denote by $G[V']$ a subgraph induced by V'. We denote by $N(v)$ the set of neighbors of v. A vertex set C is called a *clique* if $G[C]$ is a complete graph. Moreover, a clique C is *maximal* if there is no clique C' such that $C \subsetneq C'$.

2.1 Competitive Diffusion Game

Let p_1 and p_2 be players 1 and 2, respectively. Also, let $G = (V, E)$ be an undirected connected graph. Then the two-player competitive diffusion game on G proceeds as follows (see also Fig. 2).

Time 1. Each player chooses one vertex $v \in V$. The vertex v is called an *initial vertex*. If v is chosen by only one player, we say v is *dominated* by p. Otherwise, that is, if two players choose v, the vertex v is a *neutral* vertex. In the subsequent time, no player can dominate the neutral vertex. A vertex is called *undominated* if it is neither a dominated vertex nor a neutral vertex.

Time t $(t \geq 2)$. A vertex $v \in V$ is dominated by a player p at time t if (i) v is neither neutral nor dominated by any player by time $(t - 1)$, and (ii) v has a neighbor dominated by p, but does not have a neighbor dominated by the other player p'. If v satisfies (i) and has both neighbors of p and p', then v becomes a neutral vertex at time t. When no player can dominate a vertex any more, the game ends.

When player p chooses a vertex $s \in V$ at time 1, we call a vertex s the *strategy* of p. For two players p_1 and p_2, a *strategy profile* $\mathbf{s} = (s_1, s_2)$ is a pair of strategies of p_1 and p_2. If p_1 changes the strategy s_1 to s_1', we denote it by

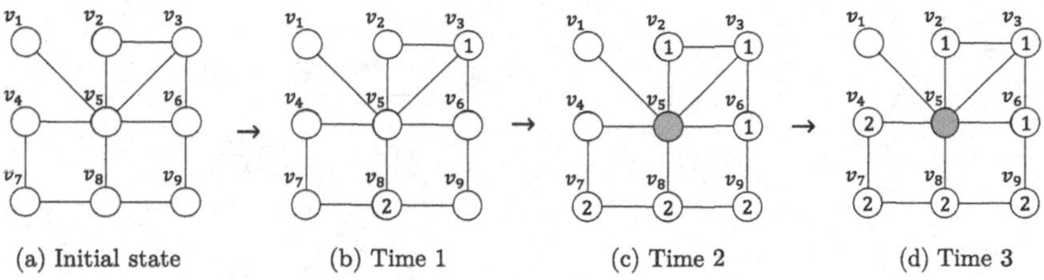

Fig. 2. An example how two-player competitive diffusion game goes. Vertices with 1 and 2 stand for vertices dominated by p_1 and p_2, respectively. White vertices are undominated vertices and a grey vertex is a neutral vertex. (a) A graph is an initial state. (b) At time 1, p_1 chooses v_3 and p_2 chooses v_8. (c) At time 2, v_2 and v_6 are dominated by p_1 and v_7 and v_9 are dominated by p_2. Vertex v_5 becomes a neutral vertex. (d) At time 3, v_4 is dominated by p_2. Since no player can dominate a vertex any more, the game ends. In the end of the game, v_1 is an undominated vertex. The utility of p_1 is $U_1(v_3, v_8) = 3$ and the utility of p_2 is $U_2(v_3, v_8) = 4$.

$(s_1, s_2) \to (s'_1, s_2)$. Similarly, if p_2 changes the strategy s_2 to s'_2, we denote it by $(s_1, s_2) \to (s_1, s'_2)$. For a strategy profile (s_1, s_2), we say s_i *dominates* vertices dominated by p_i in the end of a game. For a strategy profile s, the utility $U_i(\mathrm{s})$ of p_i is the sum of vertices dominated by p_i at the end of a game. In Fig. 2, the utility of p_1 is $U_1(v_3, v_8) = 3$ and the utility of p_2 is $U_2(v_3, v_8) = 4$.

Then we define a pure Nash equilibrium in the two-player competitive diffusion game.

Definition 1. *A strategy profile* $\mathbf{s} = (s_1, s_2)$ *is called a* pure Nash equilibrium *if there is no vertex* $v \in V$ *such that* $U_1(v, s_2) > U_1(s_1, s_2)$ *or* $U_2(s_1, v) > U_1(s_1, s_2)$*, that is, if no player can increase the utility by changing the strategy.*

We call the two-player competitive diffusion game *2-CDG* for short. Also, we simply use term "Nash equilibrium" instead of pure Nash equilibrium" from here on. If both p_1 and p_2 choose $v \in V$ as initial vertices, the utilities of p_1 and p_2 are 0 because v is a neutral vertex and other vertices are undominated in the end of a game. Then, a player has an incentive to change the strategy from v to another vertex because if two players choose different vertices, the utilities of them are at least 1. This implies that the strategy profile (v, v) for any $v \in V$ cannot be a Nash equilibrium. Thus, we suppose that two players choose different vertices as initial vertices.

2.2 Graph Classes

In this subsection, we define several graph classes. A graph $G = (V, E)$ is a *chordal graph* if every cycle of length at least 4 has a chord, or equivalently every induced cycle has exactly 3 vertices [4]. A graph $G = (V, E)$ is a *strongly chordal graph* if it is chordal graph and includes no n-sun (for $n \geq 3$) as an induced subgraph [6]. A cycle is a graph with closed circuits. A graph $G = (V, E)$ is a *block graph* if all the two connected components are cliques [7]. A graph $G = (V, E)$ is

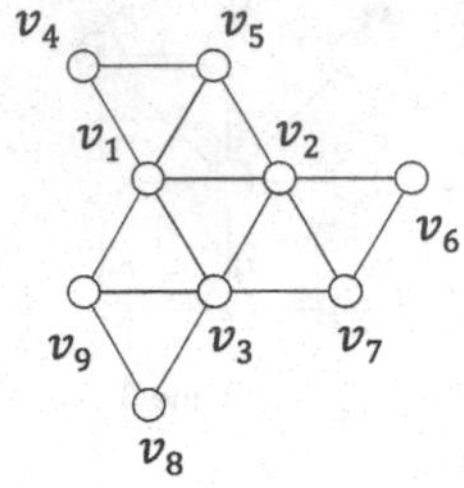

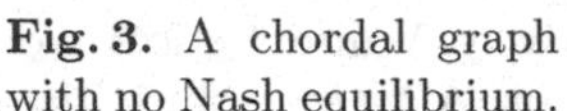

Fig. 3. A chordal graph with no Nash equilibrium.

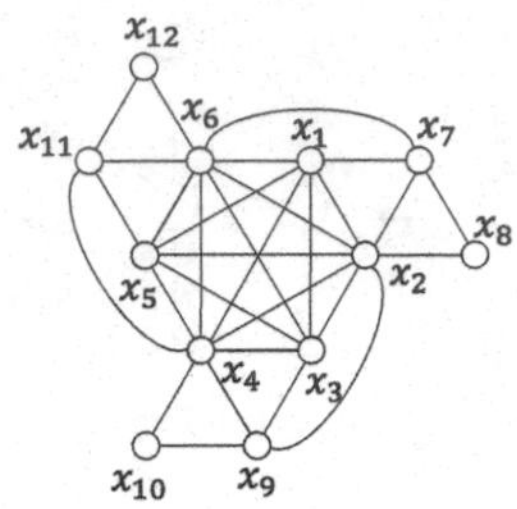

Fig. 4. A strongly chordal graph with no Nash equilibrium.

a *split graph* if V can be partitioned in an independent set I and a clique C [10]. A graph $G = (V, E)$ is a *interval graph* if it has an intersection model consisting of intervals on a real line corresponding to a vertex such that there is an edge in G if and only if two lines are intersect [9]. For more information about graph classes, see [2].

3 The Existence of a Nash Equilibrium

In this section, we investigate the existence of Nash equilibrium in 2-CDG. The results are summarized as the following two theorems (also see Fig. 1).

Theorem 1. *There is a chordal graph with* 9 *vertices and diameter* 3 *that has no Nash equilibrium in 2-CDG. There is a strongly chordal graph with* 12 *vertices and diameter* 3 *that has no Nash equilibrium in 2-CDG.*

Theorem 2. *If a graph G belongs to classes of split graphs, block graphs or interval graphs, 2-CDG on G always has a Nash equilibrium.*

We can show Theorem 1 by the concrete examples in Figs. 3 and 4, though we omit the details explaining that they really have no Nash equilibrium. For Theorem 2, we need more careful arguments, though we omit the detailed proof again. Instead, we briefly give several key ideas or overviews of the proof. Sections 3.1, 3.2 and 3.3 explain how we can show the existence of a Nash equilibrium in split graphs, block graphs and interval graphs, respectively.

3.1 Split Graph

In the proof for split graphs, we introduce the following two lemmas. The former gives a simple utility expression and the latter restricts pairs of strategies as possible Nash equilibria. Combining these, we can show that contradiction arises if no Nash equilibrium exists in a split graph. Both of the proofs are omitted.

Lemma 1. *Let $G = (C \cup I, E)$ be a split graph where C forms a clique and I is an independent set. If the strategy profile of p_1 and p_2 is (u, v) where $u, v \in C$, the utilities of p_1 and p_2 are $U_1(u, v) = |N(u)| - |N(u) \cap N(v)| + 1$ and $U_2(u, v) = |N(v)| - |N(v) \cap N(u)| + 1$, respectively.*

Lemma 2. *On any split graph $G = (C \cup I, E)$, if both p_1 and p_2 choose vertices in C, there is no strategy to choose a vertex in I that increases own utilities.*

3.2 Block Graph

The proof is based on the characterization of two vertices in a block graph G that give a Nash equilibrium. To describe them, we introduce several new terminology and notations in the following. We suppose that G is a block graph that is not complete, since the 2-CDG on a complete graph trivially has a Nash equilibrium.

A vertex v is called a *cut vertex* if $G[V \setminus \{v\}]$ has at least two components. On a block graph, a non cut vertex is contained in exactly one maximal clique C and all the neighbors is in C. For a maximal clique C, we suppose that p_1 and p_2 select $x \in C$ and $y \in C$, respectively. If x is not a cut vertex, the utility of p_1 is $|\{x\}| = 1$ since x is adjacent to only vertices in C on block graph G and every vertex in $C \setminus \{x, y\}$ becomes a neutral vertex. Suppose that x is a cut vertex in G. Then, let $D_x(C)$ be the set of vertices in connected components not containing C in $G[V \setminus \{x\}]$. Since $y \in C$, the set of vertices dominated by x is $D_x(C) \cup \{x\}$. Note that every vertex in $C \setminus \{x, y\}$ is neutral. Moreover, every vertex in C is a neutral vertex and every vertex in $V \setminus (D_x(C) \cup D_y(C) \cup C)$ is an undominated vertex. Thus, the utility of p_1 is $|D_x(C)| + 1$ if p_1 chooses $x \in C$ and p_2 chooses a vertex in $C \setminus \{x\}$. For a maximal clique C on block graph G, we denote by $w(C, u)$ the number of vertices dominated by u when either p_1 or p_2 chooses $u \in C$ and the other chooses a vertex in $C \setminus \{u\}$. Note that if u is a cut vertex, then $w(C, u) = |D_u(C)| + 1$, and otherwise $w(C, u) = |\{u\}| = 1$. Also, we suppose that a maximal clique $C = \{u_1^C, \dots, u_k^C\}$ satisfies that $w(C, u_1^C) \geq \cdots \geq w(C, u_k^C)$.

By using these, we have the following lemma, which proves the corresponding part of Theorem 2. We omit the proof due to space limitation.

Lemma 3. *Let $\mathcal{C}$ be the set of maximal cliques in G. Also, let C^* be a maximal clique such that $w(C^*, u_2^{C^*}) = \max_{C \in \mathcal{C}} w(C, u_2^C)$. Then the strategy profile $(u_1^{C^*}, u_2^{C^*})$ is a Nash equilibrium.*

3.3 Interval Graph

We assume that an interval graph $G = (V, E)$ is given by corresponding intervals $\mathcal{I} = \{I_1, \dots, I_n\}$, where each interval $I_i = [a_i, b_i]$ $(i = 1, \dots, n)$ of two integer $a_i \leq b_i$ corresponds to vertex i. The endpoint a_i of I_i is called the *initial endpoint* and the other endpoint b_i is called the *terminal endpoint*. We assume that $\{I_1, \dots, I_n\}$ are sorted in nonincreasing order of the initial endpoints a_i's. On an interval graph, there is an edge if and only if two intervals are intersect. An interval graph is called a *proper interval graph* if no interval properly contains any other interval, and known to have a unit interval representation, in which each interval has unit length [2].

Here, we prove the existence of a Nash equilibrium of 2-CDG on any proper interval graph as Lemma 4, instead of interval graph. It is because the proof for proper interval graphs is essentially same but simpler. For interval graphs, we just give a corresponding lemma (Lemma 5) without proof.

Lemma 4. *Suppose that $G = (V, E)$ is a proper interval graph and $U_1(x, y) \geq U_2(x, y)$ for any strategy profile (x, y). Let $\{x^*, y^*\}$ be an edge satisfying $U_2(x^*, y^*) = \max_{\{x,y\} \in E} U_2(x, y)$. Then (x^*, y^*) is a Nash equilibrium in 2-CDG on G.*

Proof. Prove by contradiction. Suppose that strategy profile (x^*, y^*) is not a Nash equilibrium; a player has an incentive to change the strategy. Without loss of generality, we suppose $a_{x^*} \leq a_{y^*}$, that is, the initial point of a_x is on the left side of a_y by the unit interval representation. We first observe that p_2 does not change the strategy since $U_2(x^*, y^*) = \max_{\{x,y\} \in E} U_2(x, y)$ holds. Thus p_1 is the player that changes the strategy x^*. Let D_{x^*} be the set of vertices such that every vertex v satisfies $a_v \leq a_{x^*}$. Moreover, let D_{y^*}be the set of vertices such that every vertex v satisfies $a_{y^*} \leq a_v$. Since $\{x^*, y^*\} \in E$, any vertex v satisfying $a_{x^*} \leq a_v \leq a_{y^*}$ is in $N(x^*) \cap N(y^*)$. Then we observe that p_1 dominates vertices in $D_{x^*} \setminus (N(x^*) \cap N(y^*))$. Also, p_2 dominates vertices in $D_{y^*} \setminus (N(x^*) \cap N(y^*))$. Since $U_1(x^*, y^*) \geq U_2(x^*, y^*)$, p_1 does not change the strategy to any vertex v satisfying $a_{y^*} \leq a_v$. Moreover, any vertex v satisfying $a_v \leq a_{x^*}$ can dominate only $D_{x^*} \setminus ((N(x^*) \cap N(y^*)) \cup \{x^*\})$ and the utility of p_1 does not increase when p_1 changes the strategy to v. Finally, we consider the case that p_1 changes to x' satisfying $a_{x^*} < a_{x'} < a_{y^*}$. When p_1 changes the strategy from x^* to x', at most it can dominate vertices in $D_{x'} \setminus ((N(x') \cap N(y^*) \cup \{x^*\})$. As the result, we have $U_1(x^*, y^*) \geq U_1(x', y^*)$. Thus, p_1 does not change the strategy from x^* to any vertex in V, which implies (x^*, y^*) is a Nash equilibrium, a contradiction. □

This lemma is extended to interval graphs by focusing on a subset of vertices. Let $G = (V, E)$ be an interval graph and we assume that no two identical intervals exists in V, without loss of generality. For $x \in V$, we call x an *essential* vertex if no $y \in V$ properly contains x. The following lemma is the extension of Lemma 4. Note that in a proper interval graph V itself is the set of essential vertices.

Lemma 5. *Suppose that $G = (V, E)$ is an interval graph, V' is the set of essential vertices in G and $U_1(x, y) \geq U_2(x, y)$ for any strategy profile (x, y). Let $\{x^*, y^*\}$ be an edge satisfying $U_2(x^*, y^*) = \max\{U_2(x, y) \mid x, y \in V', \{x, y\} \in E\}$. Then (x^*, y^*) is a Nash equilibrium in 2-CDG on G.*

4 Some Other Results

In this section, we briefly present two related results, though we omit all of the proofs and detailed explanation due to the space limitation. One is about vertex-weighted cycles (Theorem 3), and the other is about cacti (Theorem 4). In the vertex-weighted model, the utility is defined by not the number of influenced vertices but the total weights of influenced vertices. A cactus is a connected graph in which any two simple cycles have at most one vertex in common. Intuitively, given a block graph, we can obtain a cactus by removing the internal edges in all the cliques with size at least 4. Theorem 3 and the results of unweighted case [3] contrast, and Theorem 4 and the block graph part of Theorem 2 contrast.

Theorem 3. *In 2-CDG, any vertex weighted cycle of length at most* 5 *always has a Nash equilibrium, whereas for any length at least* 6 *there are a cycle that has no Nash equilibrium.*

Theorem 4. *In 2-CDG, any vertex-weighted cactus whose induced cycles consist of at most* 5 *always has a Nash equilibrium. On the other hand, there is an unweighted cactus containing a cycle with length at least 6 that has no Nash equilibrium.*

References

1. Alon, N., Feldman, M., Procaccia, A.D., Tennenholtz, M.: A note on competitive diffusion through social networks. Inf. Process. Lett. **110**(6), 221–225 (2010)
2. Brandstadt, A., Spinrad, J.P., et al.: Graph Classes: A Survey, vol. 3. SIAM, Philadelphia (1999)
3. Bulteau, L., Froese, V., Talmon, N.: Multi-player diffusion games on graph classes. Internet Math. **12**(6), 363–380 (2016)
4. Dirac, G.A.: On rigid circuit graphs. Abh. Math. Semin. Univ. Hambg **25**(1), 71–76 (1961)
5. Etesami, S.R., Basar, T.: Complexity of equilibrium in competitive diffusion games on social networks. Automatica **68**, 100–110 (2016)
6. Farber, M.: Characterizations of strongly chordal graphs. Discrete Math. **43**(2–3), 173–189 (1983)
7. Harary, F.: A characterization of block-graphs. Can. Math. Bull. **6**(1), 1–6 (1963)
8. Ito, T., et al.: Competitive diffusion on weighted graphs. In: Dehne, F., Sack, J.-R., Stege, U. (eds.) WADS 2015. LNCS, vol. 9214, pp. 422–433. Springer, Cham (2015). https://doi.org/10.1007/978-3-319-21840-3_35
9. Lekkeikerker, C., Boland, J.: Representation of a finite graph by a set of intervals on the real line. Fundamenta Mathematicae **51**(1), 45–64 (1962)
10. Roberts, F.S., Spencer, J.H.: A characterization of clique graphs. J. Comb. Theory Ser. B **10**(2), 102–108 (1971)
11. Roshanbin, E.: The competitive diffusion game in classes of graphs. In: Gu, Q., Hell, P., Yang, B. (eds.) AAIM 2014. LNCS, vol. 8546, pp. 275–287. Springer, Cham (2014). https://doi.org/10.1007/978-3-319-07956-1_25
12. Small, L., Mason, O.: Nash equilibria for competitive information diffusion on trees. Inf. Process. Lett. **113**(7), 217–219 (2013)
13. Sukenari, Y., Hoki, K., Takahashi, S., Muramatsu, M.: Pure Nash equilibria of competitive diffusion process on toroidal grid graphs. Discrete Appl. Math. **215**, 31–40 (2016)
14. Takehara, R., Hachimori, M., Shigeno, M.: A comment on pure-strategy nash equilibria in competitive diffusion games. Inf. Process. Lett. **112**(3), 59–60 (2012)

Foundations of Data Science and Engineering – Short Papers

Automatic Text Generation in Slovak Language

Dominik Vasko, Samuel Pecar(✉), and Marian Simko

Faculty of Informatics and Information Technologies,
Slovak University of Technology in Bratislava, Ilkovicova 2,
842 16 Bratislava, Slovakia
{xvasko,samuel.pecar,marian.simko}@stuba.sk

Abstract. Automatic text generation can significantly help to ease human effort in many every-day tasks. Recent advancements in neural networks supported further research in this area and also brought significant improvement in quality of text generation. Unfortunately, most of the research deals with English language and possibilities of text generation of Slavic languages was not fully explored yet. Our work is concerned with automatic text generation and language modeling for Slovak language. Since Slovak language has more complicated grammatical structure and morphology, the task of text generation is also more challenging. We experimented with the neural approaches in natural language generation and performed several experiments with text generation in both Slovak and English language for two different domains. Additionally, we performed an experiment with human annotators to assess the quality of generated texts. Our experiments showed promising results and we can consider using neural networks for text generation as sufficient also for text generation in Slovak language.

Keywords: Natural language processing · Language modeling · Text generation

1 Introduction

Natural text generation belongs to one of the most popular tasks in natural language processing (NLP). It can be also considered as one of the essential parts for many higher level tasks in NLP, such as summarization or dialogue systems. During the history of AI, there have been many approaches for text generation. We are concerned with the approach based on the probability distribution in natural languages, which also employs neural network architectures.

In this paper, we present a study of language modeling along with a qualitative evaluation of models on two different languages – Slovak and English. To the best of our knowledge, our study is the first attempt to experiment with text generation in Slovak language. The main contribution of our paper is exploration of techniques for natural language generation in Slovak language

A. Chatzigeorgiou et al. (Eds.): SOFSEM 2020, LNCS 12011, pp. 639–647, 2020.
https://doi.org/10.1007/978-3-030-38919-2_53

and experiments with human annotators to assess the quality of generated texts in the same domains (Wikipedia articles and European Parliament speeches) for two different languages (Slovak and English).

We train neural networks, which learn the probability distribution of characters or words from text corpora. Based on already seen words or characters, these networks can predict the probability of a following character or word. The process of generating samples is performed in two steps. First, the network learns the probability distribution on examples (input/output pairs); this is the training part. Second, an initial word or character is given to the network and the output is interpreted; this is the inference part. Based on the second step, we can build up long text samples, which can be used for manual evaluation.

Language modeling can be utilized as part of many NLP tasks, where models would be hard to train even on superb hardware. Other reasons for using language models is the ease of obtaining corpora. Separate inputs and outputs are not needed since the output is only shifted by one token relative to the input.

It is also important to compare similar architectures on different languages. Since the most research for language modeling is done on English datasets, performance for other languages might vary. Furthermore, to ensure that the generated texts are meaningful and can be understood, a qualitative evaluation by humans is needed, contrary to evaluation by other automated methods, which might yield negative results due to the complexity of languages.

2 Related Work

Language models (LM) are often used as a part of more complex tasks in natural language processing, such as text summarization [12], machine translation [2] or automatic speech recognition [8].

There are many applications for LMs, especially in sequence-to-sequence tasks, which often employ an encoder-decoder architecture. In encoder-decoder architectures, the input sequences have to be represented as one value, which is used to generate a new sequence. Therefore, there is a lot of motivation to train language models that perform outstandingly on a wide range of languages.

Statistical LM based on recurrent neural models proved to outperform approaches, such as the ones based on grammars or dense neural networks [1], since they are made to model longer continuous sequences with dependencies.

Different improvements over classical recurrent neural network (RNN) architectures can be used to obtain better language models. One of those improved architectures is the long short-term memory (LSTM) network. It deals with the shortcomings of classical RNNs, especially vanishing gradients [6], which allow LSTMs to outperform RNNs. The authors have tried and succeeded in the task of language modeling due to the fact that these models are better suited for very long dependencies [15]. This all comes with the cost of more parameters to train, which increases the whole training time and memory usage. An improved modification of the RNN, which addresses the same issues are the gated recurrent unit (GRU) networks. They are similar to LSTM but have much less parameters to learn and achieve comparable results [3].

Improvement over classical RNN for LM was achieved in many tasks such as simplification, summarization or translation with the use of a mechanism called attention. This mechanism guides the selection of better predictions based on the previous context. It has been shown that it increases not just the performance but also the convergence of the network, while using less parameters. We employed a customized attention mechanism based on a previous work of Salton [13].

A work by Cotterell et al. on the topic of comparison of language modeling for multiple languages suggests the differences due to different levels of morphology complexity [4]. Slovak language being morphologically richer means that it is supposed to be more difficult to model. Although the published experiment was evaluated only automatically and only with the use of one dataset (European Parliament speeches), it clearly shows that some languages with richer morphology are more difficult to model, as a result of their morphology (it was also reported that after lemmatization the differences between languages mostly disappeared).

3 Model

In this section, we describe general architecture of the proposed models along with their properties and training process. In total, we used two similar architectures with the difference of using/not using an attention mechanism (see Fig. 1).

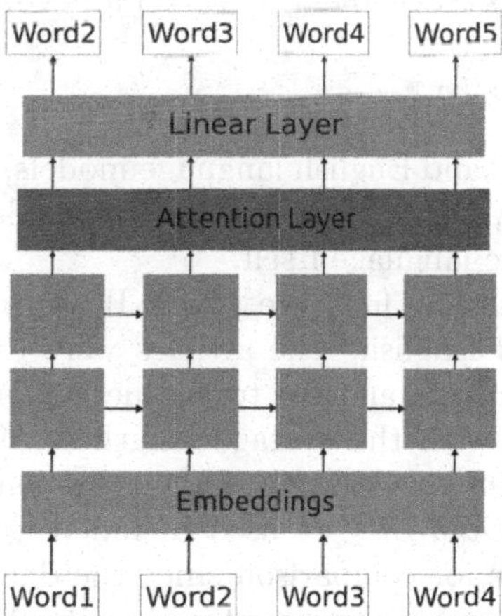

Fig. 1. Our LSTM neural network architecture. Attention layer is red. (Color figure online)

During the preprocessing stage, the whole dataset is divided into sequences of the same length, which are organized into batches. Since our model utilizes embeddings, every character or word is represented with a corresponding number (index), which is used to look up word representations in the embedding layer.

Then, the pre-processed batches are fed into the network, starting with the embedding layer. Embeddings help with the representation of inputs, instead of one-hot encoding which uses discrete values for every single input, continuous values can be used with much lower dimensionality. As a result, less data is needed to learn relationships. Weights of the embedding layer are learned and updated during the training phase. The same weights are used in the output, where the conversion from vector to index is performed.

After obtaining the vector representations, the whole input sequence is fed to the first LSTM, which consequently outputs another value. The output from the first LSTM layer is again fed to the second LSTM. Two stacked LSTM layers were used. After the last LSTM layer, the output value is forwarded to a dense linear layer and decoded into a vector, which contains probability values for every character or word. The size of this vector is the same as the size of dictionary. This probability is obtained after application of the softmax function.

For our second architecture, we employed an attention layer, which calculates a context vector for every time step. This context vector is concatenated to the output of the current time step [13].

4 Evaluation

We performed an evaluation of our models in several steps. First, we trained multiple language models on data described below for each language with proposed architecture and generated articles using these models. Random generated articles were selected and were given to participants (speaking both Slovak and English), who manually assessed them based on a few selected measures.

4.1 Dataset

Since we train both Slovak and English language models, semantically equivalent text corpora were needed, meaning the content of the datasets was the same or very similar except for the language itself.

We used two datasets. The first one (*Data-Wiki*) included the Wikipedia[1] for Slovak and Simplified English. The articles were extracted to a single file and consequently preprocessed and fed to the networks The Slovak Wikipedia contains 225,080 articles with the average length of 892 characters while the simple English had 188,198 articles with the average length of 655 characters.

The second dataset (*Data-EP*) we used included Euro-parliament speeches [5]. It is more appropriate for comparison since the data in both languages are the same. The whole dataset was manually translated to every EU language. After dividing the speeches we ended up with about 130,000 training examples for the word level model and 330,000 examples for the character level model.

4.2 Training

First, two character level models were trained on the *Data-Wiki* dataset. We used the same architecture for both models, two layers of LSTM layers with hidden and embedding sizes of 200 units with a dense layer at the end.

[1] https://dumps.wikimedia.org/.

Next, four additional models were trained, character and word level, with an additional attention layer [13]. Those models were trained on the *Data-EP* dataset. The architecture was bigger then the first one with two layers of LSTM cells with hidden size of 512 features, embedding sizes of 32 features for characters and 512 features for words. Adam was used as the optimization algorithm.

Both datasets were divided into chunks of length 200 characters or 90 words in the case of Euro-parliament speeches, and 320 characters in the case of Wikipedia articles. The models used those chunks as individual training examples and after each example the internal state of LSTM networks was reset. The generated texts were around the same length as the input sequences.

The models were trained based on the following two rules. First, we started with a learning rate of 0.001. If the model became over-fitted, we used the model with the best validation perplexity. Second, if the validation perplexity plateaued for 4 epochs, the learning rate was decreased tenfold. Character level models were trained for around 40 epochs, while word level model for around 10 epochs.

4.3 Article Generation

After the training of the models, we generated some samples. The output of our model is a vector with probabilities for each character or word, which sum up to 1. We used softmax with a different base values ranging from $\beta = 0.75 - 0.9$ to select the best candidates. Furthermore, sampling from multinomial distribution was used to increase the diversity of our outputs. This candidate was then fed to the network again. We repeated this step until we had a long enough sequence. Tables 1 and 2 show some generated samples from our models.

Table 1. Sample from an article generated by the Slovak and English model

Saint-Maritre je francúzska obec, ktorá sa nachádza v departemente Orne, v regióne Dolná Normandia. Obec má rozlohu. Najvyšší bod je položený a najnižší bod Počet obyvateľov obce je ().
Sixth Airport College, Pennsylvania State won includes several movies in Switzerland, USA from 1983 from the Operation Guide Rails, a regional album by the German rock band in America in April 2010.

4.4 Results

In this section, we sum results of our experiments for text generation. The evaluation consisted of two parts. In the first part, we assessed text generation automatically by computing perplexities of trained models. In the second part, the generated text samples were evaluated manually by human participants.

In Table 3, we show the results of automatic evaluation. These results are in line with previous research [4] done on the topic of differences in language

Table 2. Sample from an article generated by the Slovak and English models trained on speeches from Euro-parliament

(PL) Vážený pán predsedajúci, chcel by som začať jedným slovom Jean Mondelsonovi, pánovi Kallasovi, a pánovi Markovi Medina, ktoré dostal jednoznačnú odpoveď. Správa pána Casparyho o situácii v Gaze neumožňuje zverejniť iniciatívy, ktoré sa týkajú dotácií na boj proti terorizmu a neustále úspešnému preskúmaniu zodpovednosti v regiónoch Európy. Komisia sa stále venuje schengenskej otázke, ktorá sa venuje otázke energetiky a splnenie nariadenia, ktoré majú tiež v tejto oblasti prebiehať pri príležitosti procesu dosiahnutia väčšej spolupráce s krajinami Mercosuru.
(CS) Mr President, I would like to commend the Commissioner will have a rule of law. It is up to that of a more democratic tracht and lead it. One of the European Union could have the case that we do not want to see what happened for the people and the rapporteur. We have also required to ensure that we trafficking in the European Union, in the same resources will have to improve the crisis and the financial situation and on the highest possible procedure.

modeling for different languages. Particularly that morphologically richer languages are harder to learn with the word-level models. This also explains the gap between the Slovak and English models.

In manual evaluation we evaluated the generated samples from two datasets. We assessed the following metrics: *existence of words*, *syntax*, *morphology* and *semantics* of the article. The participants had to subjectively select from value on a scale from 1 to 10, smaller values meaning that the metric was bad, e.g most of the words did not exist. Each participant was shown 6 articles (3 in English and 3 in Slovak). The generated samples of the Wikipedia models were presented to 14 participants, the generated samples of the Euro-parliament speeches to 21 participants. The results of manual evaluation can be seen in Tables 4 and 5, where the average rating of the metrics are shows.

Table 3. Perplexities of trained models on the validation and testing set. Perplexity is calculated as $e^{\text{cross-entropy loss}}$. The models were trained on the *Data-EP* and *Data-Wiki* datasets. The *SK* and *EN* denote the language and the *W* or *CH* represents input type (words or characters)

Models	Validation perplexity	Test perplexity
$Data\text{-}EP_{EN-CH}$	2.1344	2.1364
$Data\text{-}EP_{SK-CH}$	2.2321	2.2313
$Data\text{-}EP_{EN-W}$	26.0888	26.2095
$Data\text{-}EP_{SK-W}$	30.8373	30.3086
$Data\text{-}Wiki_{EN-CH}$	3.1365	–
$Data\text{-}Wiki_{SK-CH}$	4.6751	–

Table 4. Evaluation results on the *Data-Wiki* dataset for character level input.

	Data-Wiki$_{EN-CH}$	Data-Wiki$_{SK-CH}$
Word existence	7.88	6.98
Morphology	6.67	5.43
Syntax	5.95	4.23
Semantics	7.75	8.33

Table 5. Results of human evaluation on the *Data-EP* dataset for both character level and word level input.

	Data-EP$_{EN-CH}$	Data-EP$_{EN-W}$	Data-EP$_{SK-CH}$	Data-EP$_{SK-W}$
Word existence	8.68	7.89	8.58	8.84
Morphology	6.37	8.63	7.95	6.95
Syntax	6.74	8.63	7.47	6.37
Semantics	5.21	7.58	6.47	6.21

In most cases, English language models outperformed Slovak ones in perplexity. We can also observe the results of manual evaluation are different then automatic one. The participants viewed the Slovak character level language models better then English ones. While for the word level models the English ones performed better. In the case of the models trained on EP speeches, every single English model outperformed the Slovak ones. This can be due to several reasons. Character level language models can model better intra-word probabilities and also lack of the ability to generalize enough to successfully and model syntax level dependencies. The meaning in English relies more on the syntax, while in Slovak more on the morphology of the words.

We can also see that going from the least difficult to more difficult metric (from word existence to semantics) both in English and Slovak models had no problem creating words from characters and most of the words did exist. On the other hand, the articles themselves were making less sense. This could be related to the fact that the models were trained on short examples.

When evaluating models trained on the *Data-EP* dataset, we aimed to assess overall authenticity of generated samples (see Table 6). We included additional binary measure to assess if the generated sample as whole is seen as generated or original (non-generated, genuine). The generated samples (articles) were

Table 6. Evaluation results of manual assessment in English and Slovak

	English		Slovak	
	Condition			
	Positive	Negative	Positive	Negative
Prediction positive	TP = 103	FP = 42	TP = 145	FP = 26
Prediction negative	FN = 65	TN = 105	FN = 23	TN = 121
F1 score	0.6581	–	0.8555	–

evaluated by 21 participants. Each participant had to evaluate 30 articles, half of them were generated and the other half were original. True positives (TP) are generated articles which were classified as generated, true negatives (TN) original articles classified as original, false positives (FP) original articles classified as generated and false negatives (FN) generated articles classified as originals.

5 Conclusions and Future Work

Evaluation done by humans showed that the resulting language models are quite different, when using two different languages and equivalent models. The difference reflects mainly in the quality of resulting generated texts. While the perplexities of English models were much lower, manual evaluation showed that Slovak language models trained on character level outperformed English ones and English language models outperformed Slovak ones on word level.

Potential further improvements of our model could involve using convolutional neural networks (CNNs) instead of embeddings, as CNNs were reported to yield better results [7]. The use of softmax at the dense layer of our model can also create a bottleneck [16]. Methods such as the use of more sophisticated embeddings or pre-trained embeddings has shown to increase performance of neural networks for different tasks. Such approach could also be used in our model [10]. Furthermore, training a language model on a larger corpora and transferring the weights into our model using transfer learning [11] could be used to compensate the lack of the data, especially in Slovak language.

It would also be possible to further extend the application of very similar neural models on different languages for tasks such as text summarization [9] or text simplification, where a similar level of model architecture can be applied.

The results show the evaluators had more problems identifying generated texts in English language. This suggests higher vulnerability of English language for potential misuse for harmful purposes [14]. Interestingly, richer morphology of Slovak language introduces more safety checks for people that can prevent from being misused.

Acknowledgments. This work was partially supported by the Slovak Research and Development Agency under the contract No. APVV-17-0267 and No. APVV SK-IL-RD-18-0004 and the Scientific Grant Agency of the Slovak Republic, grant No. VG 1/0667/18 and grant No. VG 1/0725/19 and the education and research development project "STU as a digital leader", project no. 002STU-2-1/2018 by the Ministry of Education, Science, Research and Sport of the Slovak Republic and by the student grant provided by Softec Pro Society.

References

1. Bengio, Y., Ducharme, R., Vincent, P., Jauvin, C.: A neural probabilistic language model. J. Mach. Learn. Res. **3**(Feb), 1137–1155 (2003)
2. Brants, T., Popat, A.C., Xu, P., Och, F.J., Dean, J.: Large language models in machine translation. In: Proceedings of the 2007 Joint Conference on Empirical Methods in Natural Language Processing and Computational Natural Language Learning (EMNLP-CoNLL) (2007). http://aclweb.org/anthology/D07-1090

3. Chung, J., Gulcehre, C., Cho, K., Bengio, Y.: Empirical evaluation of gated recurrent neural networks on sequence modeling. In: NIPS 2014 Workshop on Deep Learning, December 2014
4. Cotterell, R., Mielke, S.J., Eisner, J., Roark, B.: Are all languages equally hard to language-model? In: Proceedings of the 2018 Conference of the NAACL: Human Language Technologies, Volume 2 (Short Papers), pp. 536–541. ACL, New Orleans, June 2018. https://doi.org/10.18653/v1/N18-2085
5. Galuščáková, P., Garabík, R., Bojar, O.: English-Slovak parallel corpus (2012). http://hdl.handle.net/11858/00-097C-0000-0006-AAE0-A
6. Hochreiter, S., Schmidhuber, J.: Long short-term memory. Neural Comput. **9**(8), 1735–1780 (1997). https://doi.org/10.1162/neco.1997.9.8.1735
7. Józefowicz, R., Vinyals, O., Schuster, M., Shazeer, N., Wu, Y.: Exploring the limits of language modeling (2016). http://arxiv.org/abs/1602.02410
8. Mikolov, T., Karafiát, M., Burget, L., Černockỳ, J., Khudanpur, S.: Recurrent neural network based language model. In: Eleventh Annual Conference of the International Speech Communication Association (2010)
9. Pecar, S.: Towards opinion summarization of customer reviews. In: Proceedings of ACL 2018, Student Research Workshop, pp. 1–8. ACL, Melbourne, July 2018. https://doi.org/10.18653/v1/P18-3001
10. Peters, M., et al.: Deep contextualized word representations. In: Proceedings of the 2018 Conference of the NAACL: Human Language Technologies, Volume 1 (Long Papers), pp. 2227–2237. ACL, New Orleans, June 2018. https://doi.org/10.18653/v1/N18-1202
11. Pikuliak, M., Simko, M., Bielikova, M.: Towards combining multitask and multilingual learning. In: Catania, B., Královič, R., Nawrocki, J., Pighizzini, G. (eds.) SOFSEM 2019. LNCS, vol. 11376, pp. 435–446. Springer, Cham (2019). https://doi.org/10.1007/978-3-030-10801-4_34
12. Rush, A.M., Chopra, S., Weston, J.: A neural attention model for abstractive sentence summarization. In: Proceedings of the 2015 Conference on Empirical Methods in Natural Language Processing, pp. 379–389. ACL (2015). https://doi.org/10.18653/v1/D15-1044, http://aclweb.org/anthology/D15-1044
13. Salton, G., Ross, R., Kelleher, J.: Attentive language models. In: Proceedings of the Eighth International Joint Conference on Natural Language Processing (Volume 1: Long Papers), pp. 441–450. Asian Federation of Natural Language Processing, Taipei, November 2017. https://www.aclweb.org/anthology/I17-1045
14. Simko, J., Hanakova, M., Racsko, P., Tomlein, M., Moro, R., Bielikova, M.: Fake news reading on social media: an eye-tracking study. In: Proceedings of the 30th ACM Conference on Hypertext and Social Media, pp. 221–230. ACM (2019)
15. Sundermeyer, M., Schlüter, R., Ney, H.: LSTM neural networks for language modeling. In: 13th Annual Conference of the International Speech Communication Association (2012)
16. Yang, Z., Dai, Z., Salakhutdinov, R., Cohen, W.W.: Breaking the softmax bottleneck: a high-rank RNN language model. CoRR abs/1711.03953 (2017). http://arxiv.org/abs/1711.03953

Connecting Galaxies: Bridging the Gap Between Databases and Applications

Henrietta Dombrovskaya[1], Jeff Czaplewski[1], and Boris Novikov[2(✉)]

[1] Braviant Holdings, Chicago, IL 60602, USA
{hettie.dombrovskaya,jeff.czaplewski}@braviantholdings.com

[2] National Research University Higher School of Economics, Saint Petersburg, Russia
borisnov@acm.org

Abstract. An incompatibility of object-oriented application code and relational database engine often causes performance problems, known as Impedance Mismatch, which negatively affect business-critical application functions. The incompatibility can also over-complicate application design and increase the costs of development.

We address these issues, applying a concept of the API contracts to the interaction between the application and the database. We introduce a new technique providing for the transfer of complex objects rather than low-level records. We describe the implementation of the proposed solution in industrial settings and show how suggested techniques streamline the application development, at the same time providing significant performance gains.

Keywords: Database connectivity · Complex objects · Impedance mismatch · Performance

1 Introduction

It is hardly possible to find an application area that does not use databases. However, in the overwhelming number of cases, the system design is driven by application developers, whose perceptions regarding databases are oversimplified.

Recently there have been many discussions in the database research community regarding the relevance of the data management scientific paradigms to the industry. Several authors [14,18] have observed that industry does not seem to use most of the advantages provided by relational databases. The recent move toward NoSQL databases [17] is powered by a desire to run fast and free of any rigorous RDBMS requirements.

There is some rationale behind these practices. The object-oriented approach to the application design reduces the role of a database to a persistence layer where NoSQL storage managers, including key-value stores and document stores like Mongo DB [16], may indeed be more appropriate.

A. Chatzigeorgiou et al. (Eds.): SOFSEM 2020, LNCS 12011, pp. 648–656, 2020.
https://doi.org/10.1007/978-3-030-38919-2_54

However, we also can observe movement in the opposite direction. Some of the most popular NoSQL systems are gradually implementing more advanced database features, such as joins, declarative (SQL-like) query languages, and even transactions. The latter observation shows that advanced DBMS features are still needed [20].

Modern high-end, high-performance object-relational databases can present complex objects by their own means; however, the most widely used data transfer interfaces do not allow the transferring of such objects. The connectivity standards (such as JDBC) provide only a simplified database view, hiding advanced DBMS features from a client application.

On the application side Object Relational Mappers (ORMs), disassemble complex objects into atomic components, generating queries, which are either too generic or too small, producing inefficient communications and leaving no room for query optimization. As a result, application response time can degrade and directly affect business performance [1,2,10,12,13].

This observation suggests that the most significant impact of object-relational impedance mismatch is in the data exchange rather than model conversions.

In this paper, we introduce a technique for bi-directional transfer of complex objects between applications and DBMS. These techniques rely on the well-known concept of API contract and avoid explicit conversions to a low-level representation. Both sides convert their internal representations into complex hierarchical objects represented in JSON format suitable for transfer.

This approach requires some compromises. Application developers have to accept the fact that relationships may exist in the persistence layer and thus delegate certain parts of logic to the database. On the other side, database developers have to provide and accept non-first-normal-form (nested relations) as a unit of data transfer. As a reward, application developers get the convenience of APIs, while the database side benefits from the coarse granularity of requests. Our solution was developed and fully implemented at Braviant Holdings, with the database of choice being PostgreSQL and the applications running on Java. The resulting solution appeared to be very generic and can be implemented in a variety of platforms.

The rest of the paper is organized as follows. Section 2 describes the proposed model, Sect. 3 outlines the implementation. Section 4 shows the performance results. Sections 5 provides an overview of the related research.

2 Models and Conversions

The primary reason for the inefficiency of interactions between a database and an application is the inability to preserve complex object structure, which leads to the need to disassemble objects into primitive types. A typical interaction between different layers of the system is shown in Fig. 1a.

The format for data transfer differs inevitably from both internal formats. We name application objects A-objects, database objects D-objects, and transfer objects T-objects. Both A-objects and D-objects may include references to other respective objects, thus producing more complex objects. However, to avoid the need for hyperlinks, T-objects must be hierarchical. Any data model that supports hierarchical structures (e.g., XML and JSON) can serve as a T-model.

Two conversions are required for each interaction between an application and a database. To transfer data from a DB to an application, D-objects are converted into T-objects, then T-objects are converted to A-objects.

The main distinguishing feature of the approach presented in this paper is that the model mapping is symmetrical, and conversion to the transfer model is performed on both sides. This feature enables us to define data transfer in terms of complex hierarchical objects that may contain nested collections, as shown in Fig. 1b.

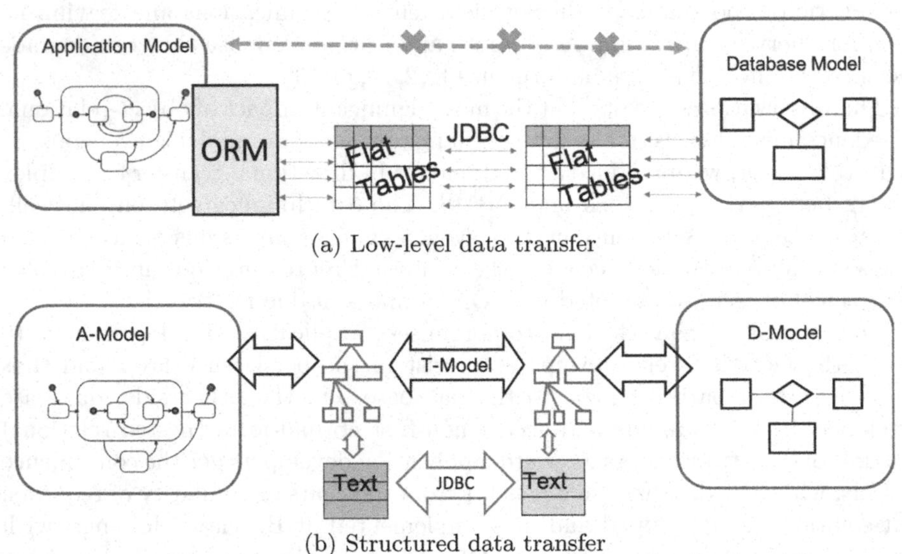

(a) Low-level data transfer

(b) Structured data transfer

Fig. 1. Data transfer architectures

Resulting JSON objects are considered as plain text strings, which can be transferred via low-level JDBC connections. The immediate advantages are:

1. The data structure is preserved during the data exchange.
2. All information representing an object can be transferred within a single transmission even if data are stored in several tables.
3. Several objects may be transferred in a single bulk collection.

The combination of items 2 and 3 results in a reduction of the number of needed interactions, contributing to overall performance improvement. Of course, this cannot prevent poor application design exploiting too fine-grained interactions, but provides for effective and natural coarse-grained interactions in terms of complex objects.

To preserve the flexibility of an application and to facilitate adjustments and improvements, and at the same time to build a scalable system, it is essential to separate the database logic and the application logic. Although this division seems to be very subjective, we were able to specify rules in the Logic Split methodology, originally introduced in [6]. At Braviant Holdings, we designed our system so that each database function returns all the data elements needed for a specific endpoint exposed by the application level microservice. Since relationships between the data are known to the database engine, query execution has been optimized for maximum efficiency. For example, our search functions allow arbitrary sets of search criteria that are passed to search functions as JSON objects. Depending on the specific set of search attributes, the select statement may utilize different execution plans. Moreover, on the database side, we can generate the SQL statements completely differently, joining different tables, in order to achieve the best performance for each specific set of attributes. In each case, the result set has an identical structure.

3 Implementation

3.1 Application Considerations

Our current solution calls for a contract to be established between the application layer and the database much the way you would see a contract over a RESTful web service. Through this contract, we've simplified the persisting of objects by serializing the objects into JSON payloads that the database can consume. This results in one DB call to persist an object regardless of its structure or complexity. Likewise, when retrieving objects, we deserialize the result coming back from the database to our model in a single database call. We are also able to pass additional parameters as a part of the contract to tell the database that we may want additional pieces of the model similar to an ODATA web service request. We have found that there are several advantages to this approach.

One of the advantages is the simplified implementation of the data access layer on the application side. Previously we had tried using Hibernate as our ORM but found that when persisting and retrieving objects, there were certain scenarios where Hibernate did not work. In many cases, Hibernate would make multiple queries to retrieve a complex object as well as to persist one. We also found ourselves writing many native queries and thus embedding database specifics into the application layer. To correctly retrieve the data that we needed, the results of our query was an untyped generic array, basically `Object[]`. Converting this array to our model resulted in a very brittle implementation. In contrast, our current solution uses a contract to determine the inputs and outputs of every call to the database. This allows application developers to code to the contract and easily mock out any dependencies when testing as the calls to and from the database will abide by the contract.

An additional advantage is ease of development. When developing our data access layer, we found several well-established libraries that would meet our needs on the application side. A pattern was quickly established for serializing and deserializing the objects as well as the correct JDBC APIs to use when interacting with the DB in a variety of situations. As each new DB interaction arose, we were able to reuse the same pattern for implementation. This allowed us to spend more time designing the JSON payload to ensure it was meeting the current and future needs of the business. Reusing the same pattern of interactions also reduced our implementation time, minimized the possibilities for defects, and allowed minimal code changes to impact our entire DB access implementation.

3.2 Database Considerations

Although conceptually any relational database may be used to store complex objects, a restriction to first normal form inevitably leads to a need to use multiple relations for its representation. To be able to conform to the contract, we need to assemble the object into the single units suitable for transfer to or from the application.

Fortunately, all high-end DBMSs allow nested collections as values of relation attributes. Any action on such a collection is supported with strong type checking. However, these kinds of complex objects are supported in PostgreSQL only internally. To externalize complex objects, we rely on a notation designed for data transfer, i.e., JSON.

The following steps build a JSON representation of T-object:

- Construct an internal complex (nested collection) object. This process is fully supported by type checking.
- Convert the complex internal object (or set of objects) into JSON format with a single invocation of the `to_json` function (and then convert the output into text string for JDBC transfer).

Update requests require the mapping of T-objects to D-objects. Regardless of whether an operation is insert, update, or delete, each of the data manipulation functions receives a T-object (JSON object) as an input parameter. Inside each function, T-object specific parsing is performed, and each element of the T-object is converted to the corresponding element of the matching D-object. While we know what type of operation we are going to perform on the top level, different operations may be needed on the lower level.

To be able to specify what exactly should be done, the contract between a database and an application includes the following:

- If a primary key value is omitted, an operation is an insert, and a function assigns a new primary key
- If a primary key is present, an operation is an update, which updates the values for all the keys passed
- No key means no changes; we do not automatically set any key values to NULL.
- To delete a database object, we utilize a special key ‘command’, which allows only one value - ‘delete’.

We wrap our queries into function calls for two main reasons. First, we want to simplify interactions between an application and a database, which would allow us to abstract D-objects from A-objects. The second reason is to ensure strong type dependencies. Our functions return complex nested objects of predefined structure, each structure being a PostgreSQL user-defined type. Any change made to the user-defined type results in cascade drop of all dependent functions, prompting to recreate them reflecting the change.

The ideal solution to abstract D-objects from A-objects would be to define D-objects as object types with accompanying methods. However, PostgreSQL does not support either object types with a state shared between methods, nor Oracle PL/SQL-style packages with the static state. Since we do not need to pass internal state between methods, PL/pgSQL functions work well, and naming rules are enforced to group related functions together.

4 Field Measurements

The framework described in this paper was developed to support new services of the Braviant Holdings OLTP system. Although the foundation for this work, Logic Split methodology [6] and [7] was laid earlier, the two differ significantly.

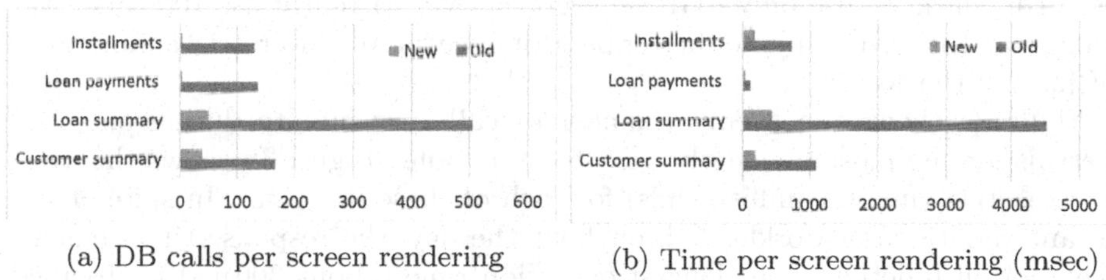

(a) DB calls per screen rendering (b) Time per screen rendering (msec)

Fig. 2. The *Logic Split* performance gain on frequent application functions

The Logic Split methodology was implemented to fix the flaws of the original application design, which used the Active Record ORM. When we succeeded in replacing multiple (in some cases hundreds) database calls with one function execution, the performance increase was dramatic. Figure 2 represents the results that were achieved during 2014–2016.

Note, that the PostgreSQL cluster was running on 80 thread 2.4 GHz CPUs with 512 GB RAM almost entirely used by DB cache. Also, due to the complexity of the legacy system, we were unable to improve any updates.

The most important advantage of our current solution is the ease of use. In 2014, the major complaint was that the application development using the Logic Split was time-consuming and unnatural. In contrast, the new methodology is very easily adopted and does not impose any risks on projects timelines.

At Braviant Holdings, our goal was to design a system with a specific focus on performance and scalability. We host our databases on Amazon RDS, which

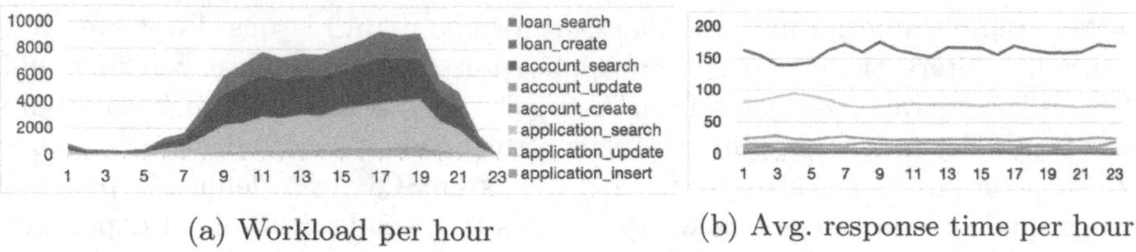

(a) Workload per hour (b) Avg. response time per hour

Fig. 3. Average execution time and avg. operations/min per hour

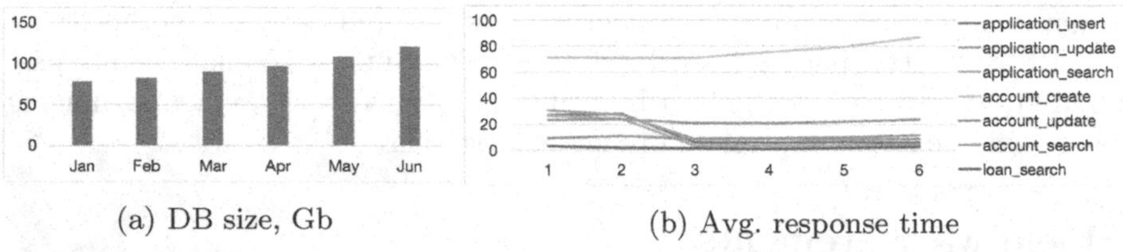

(a) DB size, Gb (b) Avg. response time

Fig. 4. Scalability: Monthly DB growth and response times

means that the available resources are limited to one of the typical configurations. Specifically, our OLTP database runs on xLarge instance with 4 vCPU and 16 GB RAM (of which we use only 6 GB on average). We currently use 100 times less memory and achieve similar or better performance for all data retrieval functions, not just for the top 10.

Figure 3a shows the average number of calls per minute during one-hour intervals for the most frequently executed functions. Figure 3b shows the average execution time (in milliseconds) for each of these functions. In spite of significant variations of workload throughout the day, the response time remains stable for all functions. The longest execution time (about 300 ms) is observed for `account_update` function, which updates multiple rows in several different tables. The most frequently executed function is `account_search`. The number of executions of this function varies from 100 to over 3000 per hour, with average execution times varying from 220 ms to 320 ms.

Another measure of scalability is the dependency of average execution times on the overall size of the database. Figure 4 shows that the execution times of some most commonly used functions remain stable while the database is growing. There is a room for scale-up and no need in scale-out in the foreseeable future.

5 Related Work

Many ORMs recognize the existence of ORIM (Object-Relational Impedance Mismatch) and have made some improvements to accommodate bulk data processing. However, the absence of bulk processing is not considered a part of ORIM. Hibernate [11] is a widely used ORM that provides two SQL-like languages: HSQL and JPQL. Both of them allow constructing declarative queries, but the usage of these features is optional.

LINQ (Language Integrated Query) [15] is a set of features that define a query as a class of objects, but an option of performing bulk operations does not mean developers are inclined to "thinking sets" [4]. SQLAlchemy Toolkit [19] is an extremely powerful tool for building database applications. It includes a variety of options that allow exact control over database interactions. However, since the toolkit was created by database professionals, it calls for a similar level of DB expertise from the application developers. Another important direction of research is related to finding algorithms that would help identify and fix the inefficient application code [3,5]. In [5], the authors present a tool to trace the occurrences of non-performing code, by coupling the application trace with the database trace. This tool helps to identify the opportunities for bulk processing, indexes optimization, reduction of fetched data volumes, etc. The AppSleuth [3] is another example of a tool, that focuses on application code improvement. It identifies *delinquent design patterns*, such as fetching one record at a time rather than a pre-selected set of records. Several research groups focus their efforts on a holistic approach to application optimization. One of the most advanced projects of that kind is DBridge [8,9], which explores different methods of holistic application optimization, based on the analysis of the application source code. The most important advantage of DBridge tool is that the code optimization process is automated and that the system guarantees the consistency of the results.

6 Conclusion

We described an approach based on a contract between the application and the database considered as a service providing for a transfer of complex hierarchical objects represented in JSON format. Although the processing of JSON requires some additional effort on the database side, it significantly simplifies application development and makes mapping between models straightforward.

Further, this approach allows developers to define strong type dependencies, and also allows the application to manipulate object sets, rather than individual objects. Coarse-grained interactions, in turn, allows the use of DBMS optimization capabilities to build a scalable and robust system.

The framework described in this paper has been in use at Braviant Holdings for over two years in a real production environment. Measurements show efficiency, effectiveness, and scalability of our techniques.

References

1. How one second could cost amazon $1.6 billion in sales, March 2015. http://www.fastcompany.com/1825005/how-one-second-could-cost-amazon-16-billion-sales/
2. Baxter, R.: Improving site speed - talk about the business benefit, October 2017. https://builtvisible.com/improving-site-speed-talk-about-the-business-benefit/
3. Cao, W., Shasha, D.: AppSleuth: a tool for database tuning at the application level. In: Proceedings of the 16th International Conference on Extending Database Technology, EDBT 2013, pp. 589–600. ACM, New York (2013)

4. Celko, J.: Joe Celko's Thinking in Sets: Auxiliary, Temporal, and Virtual Tables in SQL. Data Management Systems, 1st edn. The Morgan Kaufmann, San Francisco (2008)
5. Chaudhuri, S., Narasayya, V., Syamala, M.: Bridging the application and DBMS divide using static analysis and dynamic profiling. In: Proceedings of the 2009 ACM SIGMOD International Conference on Management of Data, SIGMOD 2009, pp. 1039–1042. ACM, New York (2009)
6. Dombrovskaya, H., Lee, R.: Talking to the database in a semantically rich way. In: Amer-Yahia, S., Christophides, V., Kementsietsidis, A., Garofalakis, M.N., Idreos, S., Leroy, V. (eds.) Proceedings of the 17th International Conference on Extending Database Technology, EDBT 2014, Athens, Greece, 24–28 March 2014, pp. 676–687. OpenProc.org (2014)
7. Dombrovskaya, H., Rangarajan, S., Marks, J.: FastFunction: replacing a herd of lemmings with a cheetah a ruby framework for interaction with PostgreSQL databases. In: 32nd IEEE International Conference on Data Engineering, ICDE 2016, Helsinki, Finland, 16–20 May 2016, pp. 1275–1286. IEEE Computer Society (2016)
8. Emani, K.V., Deshpande, T., Ramachandra, K., Sudarshan, S.: DBridge: translating imperative code to SQL. In: Proceedings of the 2017 ACM International Conference on Management of Data, SIGMOD 2017, pp. 1663–1666. ACM, New York (2017)
9. Emani, K.V., Ramachandra, K., Bhattacharya, S., Sudarshan, S.: Extracting equivalent SQL from imperative code in database applications. In: Proceedings of the 2016 International Conference on Management of Data, SIGMOD 2016, pp. 1781–1796. ACM, New York (2016)
10. Google's push to speed up your web site, October 2017. https://searchengineland.com/googles-push-to-speed-up-your-web-site-42177/
11. Hibernate web site (2017). http://www.hibernate.org/about/
12. How loading time affects your bottom line, March 2015. https://blog.kissmetrics.com/loading-time/
13. Speed is a killer? Why decreasing page load time can drastically increase conversion, March 2015. https://blog.kissmetrics.com/speed-is-a-killer/
14. Kumar, A.: Ml/Ai systems and applications: is the SIGMOD/VLDB community losing relevance? August 2018. https://wp.sigmod.org/?p=2454
15. Language-Integrated Query (LINQ) (2017). https://docs.microsoft.com/en-us/dotnet/csharp/programming-guide/concepts/linq/
16. MongoDB (2018). https://www.mongodb.com/
17. NoSQL (2018). http://nosql-database.org/
18. Pavlo, A.: What are we doing with our lives?: nobody cares about our concurrency control research. In: Proceedings of the 2017 ACM International Conference on Management of Data, SIGMOD 2017, p. 3. ACM, New York (2017)
19. SQLAlchemy (2015). http://www.sqlalchemy.org/
20. Stonebraker, M.: The "NOSQL" discussion has nothing to do with SQL, November 2009. https://cacm.acm.org/blogs/blog-cacm/50678-the-nosql-discussion-has-nothing-to-do-with-sql/fulltext

GRaCe: A Relaxed Approach for Graph Query Caching

Francesco De Fino(✉), Barbara Catania, and Giovanna Guerrini

University of Genova, Genoa, Italy
francesco.defino@dibris.unige.it

Abstract. SPARQL query optimization is an important issue for RDF data stores that can benefit from the usage of caching frameworks. Most caching approaches rely on a precise match semantics, that limits the number of cache hits and, as a consequence, the potential benefit. Others propose relaxed matches for the entire query, which is precisely executed over the cached result set. In this paper, to overcome these limitations we propose GRaCe, a Graph Relaxed Caching approach for RDF data stores. GRaCe supports relaxed cache matches and a relaxed query semantics, thus increasing the number of cache hits. Experimental results show that a relaxed cache can significantly reduce query execution time in all the scenarios where a relaxed query result is tolerated.

1 Introduction

Motivation. An increasing number of data sources are represented in RDF and queried through SPARQL. RDF stores, SPARQL endpoints, and, in general, Semantic Web query engines heavily depend on the ability of efficiently executing SPARQL queries. SPARQL query optimization often relies on caching frameworks, by which the results of previously executed queries are stored in a cache to be reused for further query processing. Most caching approaches adopt an exact semantics for matches, thus limiting the number of hits and, as a consequence, the number of cache-based optimizations. Other approaches (e.g., [3,14]) admit relaxed matches, based on query containment, for the whole query, which is then precisely executed over the cached result set. In this way, however, when the result set is huge or the matched query is quite far from the input one, the performance improvement could be limited. Both approaches return a precise result to the user. There are situations, however, in which a precise answer is not needed. This happens, e.g., in very interactive environments where the user can tolerate a loss in accuracy for a gain in performance. In such cases, relaxed cache matches and relaxed query semantics could be an interesting alternative approach towards increasing the number of cache hits thus reducing the query execution time, at the price of reducing answer accuracy, in a controlled way.

Contribution. In this paper, we propose GRaCe, a Graph Relaxed Caching approach for RDF data stores. In GRaCe, SPARQL queries are executed in a relaxed way. Thus, a superset of the query result can be returned to the user for efficiency purposes. An execution plan for a SPARQL query is composed

A. Chatzigeorgiou et al. (Eds.): SOFSEM 2020, LNCS 12011, pp. 657–666, 2020.
https://doi.org/10.1007/978-3-030-38919-2_55

of (optimal) atomic plans for subqueries, each corresponding to either a traditional query execution or a (precise or relaxed) cache match. The best plan is determined by relying on a dynamic programming planner (DPP), that, besides execution cost as in [11], also considers the degree of relaxation due to relaxed cache matches. The proposed framework is parametric with respect to the specific condition used for relaxation and query containment. In this paper, however, we focus on relaxed matches based on class and predicate generalization constraints, derived from RDF schema information. For designing GRaCe we thus: (i) propose a sufficient condition for query containment for a relevant subset of SPARQL queries, extending that in [11] to cope with relaxation based on Subproperty and Subclass RDFS entailment rules; (ii) design a relaxed cache selection algorithm, obtained as a variation of the A* algorithm over a tree representing in a compact way all the cached queries; (iii) propose a DPP taking relaxation into account in computing query (sub)plan execution costs.

Related Work. Existing caching approaches for SPARQL queries rely on a precise semantics, thus, differently from GRaCe, they return the exact query result. They can be classified depending on whether they support exact or relaxed (i.e., containment based) matches and the approach used for efficiently checking equality or containment - well known NP-complete problems - for specific types of SPARQL queries. Among exact-match caching approaches, we mention the one in [10], relying on a hashing approach for identifying matches, and the one in [11], based on the usage of canonical labels for checking graph isomorphism. Among relaxed matches approaches, the work in [3] relies on tight simulation for checking containment. Cached queries are stored using their spanning polytree and search is done by using an indexing system based on polytree signatures. The problem of SPARQL query containment under the RDFS entailment regime is studied in [1], where the problem is reduced to the expressive logic of μ-calculus. In [14], subgraph/supergraph relationships are detected by relying on subgraph isomorphism. In [8], sufficient conditions for checking graph similarity are provided for different types of SPARQL queries but no specific caching approach is proposed. More recently, f-graph queries are introduced to solve the containment problem in polynomial time [9]. Among the approaches described so far, the one in [11] uses cache matches in the context of more complex but precise query execution plans, identified through the usage of a dynamic programming planner.

Organization. The remainder of this paper is organized as follows. In Sect. 2, we formalize the problem we want to address. The cache selection algorithm and the cache-based query processing planner are described in Sects. 3 and 4, respectively. Some preliminary experimental results are reported in Sect. 5. Finally, Sect. 6 presents some concluding remarks and outlines future work. Due to space constraints, the paper presents the basic ideas underlying the proposed system. Additional details can be found in [2].

2 Problem Statement and GRaCe Architecture

The focus of our work is the design of a cache-based SPARQL execution engine, relying on relaxation for performance issues. To this aim, we consider SPARQL queries executed over RDF datasets [12]. For the sake of simplicity, we only deal with Basic Graph Pattern (BGP) SPARQL queries, i.e., set of triple patterns, each corresponding to a *subject*, *predicate*, *object* triple, where *subject* and *predicate* can be either a URI or a variable and *object* can be either a URI, a variable, or a literal.

A *graph-based relaxed cache* $\mathcal{C}$ is a set of pairs $(Q_1, r_1), ..., (Q_n, r_n)$, where each Q_i is a query and r_i is a set of triples returned as result for Q_i in previous executions. The *cache selection algorithm* we consider relies on a relaxed approach: given a query Q and a cache $\mathcal{C}$, it selects a query Q_i in $\mathcal{C}$ such that $Q \sqsubseteq Q_i$, i.e., *Q is contained in Q_i* (thus, for each RDF dataset d, $Q(d) \subseteq Q_i(d)$). We say that *Q_i is a relaxed cache hit for/generalizes Q*.

Since more than one cache item might contain the query, the issue arises of selecting the "best match" through a *relaxation cost function* $relax_{\mathcal{C}}()$ that quantifies the distance between Q and Q_i. When $Q \sqsubseteq Q_i$, such distance is computed taking into account RDF schema information and in particular the entailment rules related to *subPropertyOf* and *subClassOf* properties (see Fig. 1 [4]). It can be easily proved that, given a query Q containing the triple pattern in bold in one of the rules in Fig. 1 and assuming the RDF schema contains the other triple in the premise of the same rule, a query Q_i obtained from Q by replacing the triple pattern in bold with that appearing as the consequence of the same rule, generalizes Q, i.e., $Q \sqsubseteq Q_i$.

$$\textit{Subproperty}(sp) \quad (1)\,\frac{(\mathbf{a;sp;b})(b;sp;c)}{(a;sp;c)} \qquad (2)\,\frac{(a;sp;b)(\mathbf{x;a;y})}{(x;b;y)}$$

$$\textit{Subclass}(sc) \quad (3)\,\frac{(\mathbf{a;sc;b})(b;sc;c)}{(a;sc;c)} \qquad (4)\,\frac{(a;sc;b)(\mathbf{x;type;a})}{(x;type;b)}$$

Fig. 1. RDFS entailment rules

The GRaCe architecture is then obtained by extending classical cache-based query processing architectures, like that proposed in [11], by taking into account relaxation in several places. More precisely: (i) the *cache* might associate queries with relaxed results, generated from previous executions based on at least one relaxed cache hit; (ii) the *cache selection algorithm* detects the best cache hits based on a relaxation function; (iii) the *planner* chooses the best execution plan for the query at hand, taking into account relaxation costs besides processing time; (iv) the *cache update* module is extended to cope with cache redundancy, due to the presence in cache of queries generalizing other queries in cache (not addressed in this paper).

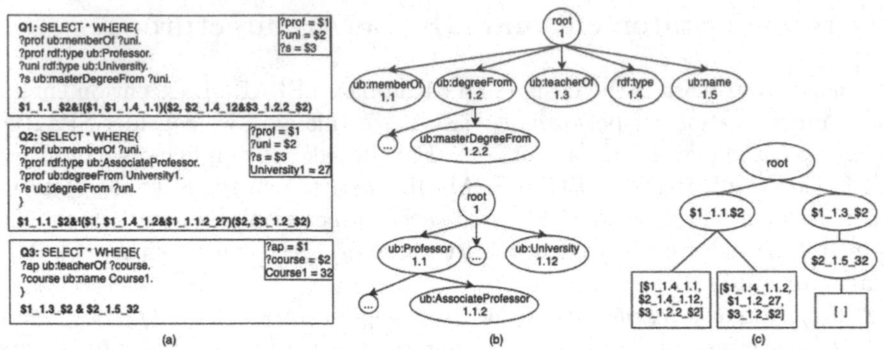

Fig. 2. (a) SPARQL queries and corresponding canonical labels; (b) Predicate and class taxonomies; (c) Graph cache tree

3 Cache Selection Algorithm

3.1 Graph Query Matching and Cache Data Structure

Cached queries are represented as a tree by exploiting a *canonical labeling* extending the labeling in [11]. All isomorphic forms of a SPARQL query are assigned the same label, generated through an extension of the Bliss algorithm [7]. Each canonical label represents a graph query as a string (see Fig. 1(a)): each variable, resource, and predicate is represented by an id and triple patterns are listed according to the ordering generated by the Bliss algorithm, applied over a specific vertex-coloured representation of the input graph. In [11], a variation of this basic algorithm is also provided with an ad-hoc treatment for star subqueries (i.e., the set of triple patterns sharing the same variable and in join with at most one other triple patterns). The canonical label first lists the canonical label of the non-star subquery (*skeleton canonical label*) and then lists (after symbol "!") the canonical representation of each star subquery in a predefined order.

When $subPropertyOf$ and $subClassOf$ relationships form two taxonomy trees, the approach in [11] can be extended to deal with entailment rules presented in Fig. 1. A label according to the Dewey numbering scheme is assigned to each node in the taxonomy trees, and such labels are used inside canonical labeling as predicate and class ids (see Fig. 2(a) and (b)).

It can be shown that, when $Q \sqsubseteq Q_i$, the canonical label of Q_i differs from that of Q only for Dewey identifiers of corresponding but generalized predicates and classes. Subgraph isomorphism can then be checked by relying on the approach in [11] and replacing predicate/class equality with generalization tested on Dewey identifiers.

A set of extended canonical labels can be represented as a *graph cache tree* as follows. Each node but the root (which is a dummy node) is associated with the canonical representation of a triple pattern appearing in the skeleton canonical label of a cached query. There is an edge from node n_1 to node n_2 if the triple pattern associated with n_2 follows the one associated with n_1 in the canonical

$$relax_C(Q, Q_i) = \begin{cases} \sum_{j=1}^{h} relax_d(t_j, g_j) + (k-h) \cdot (h_{pred}^{max} + 2h_{res}^{max} + 1) & k > h \\ \sum_{j=1}^{h} relax_d(t_j, g_j) & h = k \\ k(h_{pred}^{max} + 2h_{res}^{max} + 1) & k < h \end{cases}$$

Fig. 3. $relax_C()$ function, applied over $Q \equiv t_1, ..., t_k$, $Q_i \equiv g_1, ..., g_h$

representation of a cached query. Each skeleton canonical label of a cached query thus corresponds to a path in the cached tree. Since many queries in the cache may share the same skeleton canonical representation, each node corresponding to the final triple pattern of a skeleton label has a child for each cached query sharing that skeleton label, corresponding to the canonical representation of one of its star subqueries (see Fig. 2(c)).

3.2 Selection Algorithm

The best cache hit for a query Q is identified through a relaxation function $relax_C()$ that, given two queries $Q \equiv t_1, ..., t_k$ and $Q_i \equiv g_1, ..., g_h$, returns a value quantifying the relaxation distance between Q and Q_i (see Fig. 3). When $Q \sqsubseteq Q_i$, $k \geq h$ holds and $relax_C()$ is defined in terms of function $relax_d()$ that, for each pair of corresponding triple patterns t_j and g_j in Q and Q_i, such that t_j generalizes g_j, returns the sum of the distances between corresponding predicates/subjects/objects inside taxonomies. In all the other cases, an upper bound, based on the maximum height in the input taxonomies, is returned.

The best relaxed cache hit can then be detected by searching the cache tree through a customized version of the A* search [6], computing costs based on function $relax_C()$. Differently from the state-of-the-art A* algorithm, our search: (i) takes as input the canonical representation of an input query Q; (ii) compares the i-th triple pattern appearing in the label with labels of nodes at level i in the tree, checking generalization; (iii) considers a path as a candidate answer if it corresponds to a query that generalizes the query at hand. The cost of a path, i.e., of a cached query Q_i, corresponds to the result of $relax_C(Q, Q_i)$.

4 Extending the Planner

In order to choose the best query execution plan for a given SPARQL query, we rely on a dynamic programming planner (DPP) approach that selects the optimal query plan by decomposing the problem taking into account all the connected subgraphs of the input query. Differently from state of the art approaches, e.g., [11], we exploit both query execution and relaxation costs in the identification of the optimal query plan.

For each connected subgraph, three different *atomic plans* can be considered: (i) *precise plan, without cache match* (T): the execution corresponds to a traditional approach over the input dataset; (ii) *precise plan, with cache match* (P): the execution coincides with a precise cache match; (iii) *relaxed plan, with cache match* (R): the execution coincides with a relaxed cache match.

The cost C of each atomic plan is computed as the sum of two sub-costs, C_{exe} and C_{relax}.

C_{exe} is the cost of executing the query over the input dataset. This cost is set to 1 for P and R plans since no query has to be executed in this case. For T plans, a standard estimation of the execution cost based on database statistics can be used. However, since our main aim is to investigate cache usage performance, we simply set this cost to the number of triple patterns in the input query Q. The rationale is that the higher the number of triple patterns, the higher the number of potential joins and therefore the overall execution cost.

C_{relax} represents the loss in precision due to a relaxed plan. This cost is 0 for T and P plans. For R plans, this cost can be precisely defined as $r(Q_i, Q, D) = |Q_i(D) - Q(D)|$, i.e., as the difference between the result cardinality of the relaxed query Q_i and that of the given query Q. While the cardinality of the relaxed cached query is known (since it has already been executed), the cardinality of the input query is not and can only be estimated (see, e.g., [13]). By estimating the difference based on function $relax_d()$ we can simply obtain a value for $r(Q_i, Q, D)$. We refer the interested reader to [2] for additional details.

The DPP relies on an efficient approach for enumerating all possible subgraphs, following the approach in [11]. For each identified subgraph Q_j of the input query, the cost of each atomic plan P_j is computed as $\alpha C^j_{exe} + \beta C^j_{relax}$, where α and β allow the system to weight performance w.r.t. relaxation costs.

5 Experimental Evaluation

5.1 Experimental Setup

GRaCe has been developed in Java, version HotSpot(TM) 64-Bit Server VM under Java SE 10.0.1, relying on Apache Jena 3.12.0. The experiments were performed on a machine with CPU Intel Core i3-2350M 2.30 GHz, 8 GB of RAM size, running Ubuntu 18.04 LTS.

Our evaluation relies on the Lehigh University Benchmark (LUBM) [5]. LUBM features an OWL ontology for the university domain, enables scaling of datasets to an arbitrary size, and includes a class and a predicate taxonomies that we exploit for checking query containment. The considered LUBM dataset describes 4 universities, extended with all triples inferred according to the rules presented in Fig. 1, computed with the Jena reasoning module. The total number of triples is 985879. The LUBM dataset has been stored under Apache Jena TDB[1] and T plans have been executed by Jena querying facilities. While LUBM comes with a given workload, such queries have not been designed for relaxation

[1] https://jena.apache.org/documentation/tdb/.

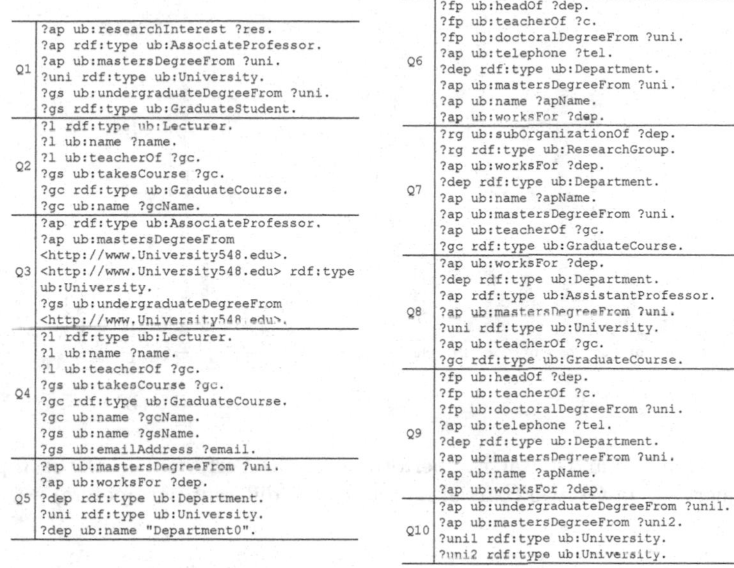

Fig. 4. SPARQL queries over LUBM Benchmark

purposes. We therefore designed our own workload of 10 queries $Q_1, ..., Q_{10}$ (see Fig. 4), differing for the total number of relaxation steps that can be applied to them. We then created a cache containing 100 entries, corresponding to a cache tree of about 4000 nodes. The result of each cached query is computed by a T plan and stored together with the query in the cache. The cache has been generated so that for queries $Q_1, ..., Q_5$ a precise cache hit for the whole query can be found while for queries $Q_6, ..., Q_{10}$ only a relaxed cache hit can be detected.

The aim of the performed experiments is to: (i) analyze the benefits obtained by the usage of a relaxed cache during query processing; (ii) analyze the behaviour of the DPP for the selection of the optimal query plan. In the experiments, we consider three GRaCE versions: one relying on the DPP described in Sect. 4 (denoted by GRaCe) and two versions in which the DPP only considers *total cache hits*, i.e., cache hits for the whole query, either precise (denoted by GRaCe_p^-) or relaxed (denoted by GRaCe_r^-). As usual, each query is executed 10 times in GRaCe, GRaCE_p^-, or GRaCE_r^-, depending on the experiments, and the average execution time is computed.

5.2 Experimental Results

Relaxed Cache Benefits. For analyzing relaxed cache benefits, we executed queries $Q_1, ..., Q_5$ on GRaCe_p^- and queries $Q_6, ..., Q_{10}$ on GRaCe_r^-. For each query Q_i, we compared the performance of the execution of a T plan with respect

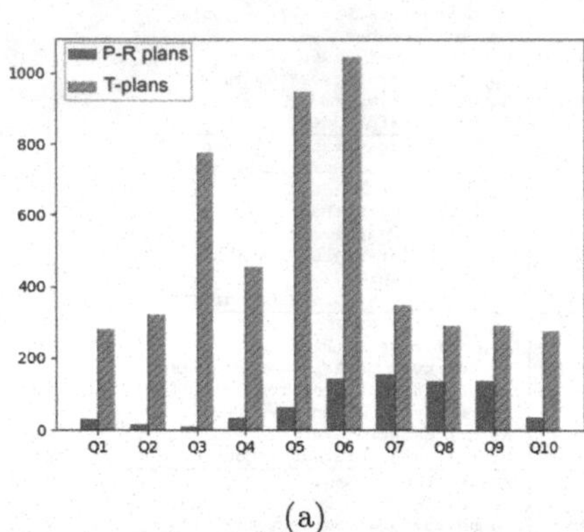

(a)

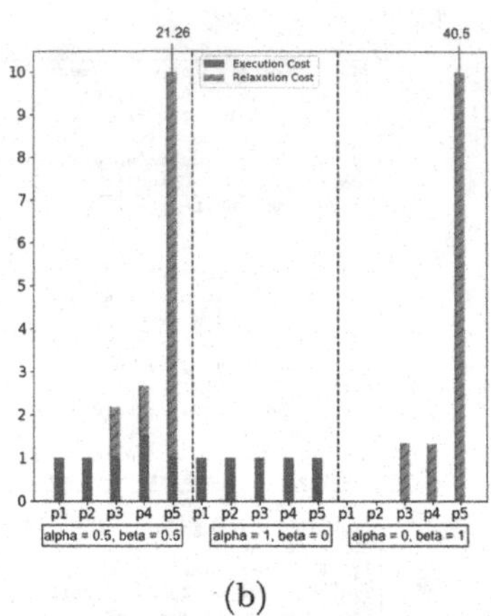

(b)

Fig. 5. (a) GRaCE_p^- and GRaCE_r^- performance; (b) Performance of the top-5 best plans for query Q_6 in GRaCe (partial matches with different parameters)

to the performance of a P or R plan (see Fig. 5(a)). As expected, P and R plans have better performance than T plans since no SPARQL query has to be executed. The performance of a P or R plan depends on the number of visited cache tree nodes and the number of generalization checks to be performed. Thus, when the cache hit is precise, the cost for selecting the best cache hit is lower since more pruning is applied during the cache tree visit.

DPP Behaviour. In the second experiment, we consider GRaCe for analyzing the behaviour of the DPP. To this aim, we consider query Q_6 (see Table 4). Tables 1, 2, and 3 report the 5 plans with the lowest costs, generated by the DPP for different values of α and β. Numbers in subgraph descriptions correspond to triple pattern positions inside the query. For each plan, we report the considered query subgraphs (which results are then joined) and the type of the corresponding selected plan. Figure 5(b) shows the total estimated cost for each plan, pointing out C_{exe} and C_{relax} costs. We can see that, when $\alpha = 0.5, \beta = 0.5$, plans $p1$ and $p2$, that correspond to precise match hits, have the lowest cost since in this case relaxation is not applied and cached results are just retrieved and joined. When $\alpha = 1, \beta = 0$, the relaxation cost is 0 ($\beta = 0$); since in this case all plans correspond to a relaxed match hit (see Table 2), any of them can be selected by the DPP. Finally, when, $\alpha = 0, \beta = 1$, only the amount of relaxation is relevant for estimating the cost, thus, either plan $p1$ or $p2$ is selected since, in both cases, no relaxation is applied (see Table 3).

Table 1. Top-5 best plans for query Q_6, $\alpha = 0.5, \beta = 0.5$

Plan	Subgraphs	Subplans
p1	(1)(2)(3)(5)(8) ⋈ (1)(4)(5)(6)(7)(8)	$P \bowtie P$
p2	(1)(2)(3)(5)(8) ⋈ (1)(4)(5)(6)(7)	$P \bowtie P$
p3	(4)(7) ⋈ (1)(2)(3)(5)(6)(8)	$T \bowtie R$
p4	(2)(3) ⋈ (4)(7) ⋈ (5)(6)(8)	$R \bowtie T \bowtie R$
p5	(1)(2)(3)(5)(8) ⋈ (4)(6)	$P \bowtie R$

Table 2. Top-5 best plans for query Q_6, $\alpha = 1, \beta = 0$

Plan	Subgraphs	Subplans
p1	(1)(3)(4)(5)(6)(7)(8)	R
p2	(1)(2)(3)(4)(5)(7)(8)	R
p3	(1)(2)(5)(6)(8)	R
p4	(3)(4)(6)(7)(8)	R
p5	(1)(3)(6)	R

Table 3. Top-5 best plans for query Q_6, $\alpha = 0, \beta = 1$

Plan	Subgraphs	Subplans
p1	(1)(2)(3)(4)(5)(6)(7)(8)	T
p2	(1)(2)(3)(5) ⋈ (1)(4)(5)(7)(8)	$T \bowtie P$
p3	(4)(7) ⋈ (1)(5)(6)(8)	$T \bowtie R$
p4	(2)(3) ⋈ (4)(7) ⋈ (5)(6)(8)	$R \bowtie T \bowtie R$
p5	(1)(2)(3)(5)(8) ⋈ (4)(6)	$P \bowtie R$

6 Concluding Remarks

In this paper, we have presented a Graph Relaxed Caching approach for RDF data stores. To the best of our knowledge, our framework is the first relaxed caching framework for speeding up query processing taking into account relaxation during plan cost estimation. Experimental results show that the usage of a relaxed cache can significantly increase performance and it is suitable when relaxed query results are acceptable. Future work focuses on two main issues: (i) the definition of efficient cache replacement algorithms, taking into account potential cache redundancy, i.e., the presence in cache of queries generalizing each other; (ii) the extension of GRaCe to deal with SPARQL endpoints, and in general Semantic Web query engines.

References

1. Chekol, M.W., Euzenat, J., Genevès, P., Layaïda, N.: SPARQL query containment under RDFS entailment regime. In: Gramlich, B., Miller, D., Sattler, U. (eds.) IJCAR 2012. LNCS (LNAI), vol. 7364, pp. 134–148. Springer, Heidelberg (2012). https://doi.org/10.1007/978-3-642-31365-3_13
2. De Fino, F.: Relaxation meets caching: towards smart caching approaches for graph query processing. Ph.D. thesis. University of Genova, Italy (2020, in preparation)
3. Fard, A., et al.: Effective caching techniques for accelerating pattern matching queries. In: Big Data 2014, pp. 491–499 (2014)
4. Frosini, R., et al.: Flexible query processing for SPARQL. Semant. Web **8**(4), 533–563 (2017)

5. Guo, Y., Pan, Z., Heflin, J.: LUBM: a benchmark for OWL knowledge base systems. J. Web Semant. **3**(2–3), 158–182 (2005)
6. Hart, P.E., Nilsson, N.J., Raphael, B.: A formal basis for the heuristic determination of minimum cost paths. IEEE Trans. Syst. Sci. Cybern. **4**(2), 100–107 (1968)
7. Junttila, T., Kaski, P.: Engineering an efficient canonical labeling tool for large and sparse graphs. In: International Workshop on Algorithm Engineering and Experiments (ALENEX), pp. 135–149 (2007)
8. Lorey, J., Naumann, F.: Caching and prefetching strategies for SPARQL queries. In: Cimiano, P., Fernández, M., Lopez, V., Schlobach, S., Völker, J. (eds.) ESWC 2013. LNCS, vol. 7955, pp. 46–65. Springer, Heidelberg (2013). https://doi.org/10.1007/978-3-642-41242-4_5
9. Mailis, T., et al.: An efficient index for RDF query containment. In: SIGMOD Conference 2019, pp. 1499–1516 (2019)
10. Martin, M., Unbehauen, J., Auer, S.: Improving the performance of semantic web applications with SPARQL query caching. In: Aroyo, L., Antoniou, G., Hyvönen, E., ten Teije, A., Stuckenschmidt, H., Cabral, L., Tudorache, T. (eds.) ESWC 2010. LNCS, vol. 6089, pp. 304–318. Springer, Heidelberg (2010). https://doi.org/10.1007/978-3-642-13489-0_21
11. Papailiou, N., et al.: Graph-aware, workload-adaptive SPARQL query caching. In: SIGMOD Conference 2015, pp. 1777–1792 (2015)
12. Prud'hommeaux, E., Seaborne, A.: SPARQL query language for RDF, W3C recommendation (2008). https://www.w3.org/TR/rdf-sparql-query/
13. Stocker, M., et al.: SPARQL basic graph pattern optimization using selectivity estimation. In: WWW 2008, pp. 595–604 (2008)
14. Wang, J., et al.: GC: a graph caching system for subgraph/supergraph queries. PVLDB **11**(12), 2022–2025 (2018)

Modelling of the Fake Posting Recognition in On-Line Media Using Machine Learning

Kristína Machová[(✉)], Marián Mach, and Gabriela Demková

Department of Cybernetics and Artificial Intelligence,
Technical University, Letná 9, 042 00 Košice, Slovakia
kristina.machova@tuke.sk, gabriela.demkova@student.tuke.sk

Abstract. Discuss content in the online web space has a significant impact on social life in recent years, especially in the political world. The impact of social networks has its advantages and disadvantages. An important disadvantage is a rising of the antisocial content in online communities. The antisocial content represents a serious and actual problem that is reinforced by a simplifying the process of creating and disseminating of antisocial posts. A typical example is a spreading of fake reviews. Detection of fake reviews is becoming one of the most important areas of research in last years. It is easier to track the impact of fake reviews than to detect them. The aim of this paper is to create suitable models for the fake reviews recognition using machine learning algorithms particularly decision tree, random forests, support vector machine and naïve Bayes classifier. Using a confusion matrix, several indicators of binary classification efficiency were quantified in the process of these models testing.

Keywords: Social media mining · Model for fake reviews identification · Machine learning methods · Antisocial posting

1 Introduction

In today's technology, nearly three and a half billion people have access to the Internet. At its beginning, the web was used to spread knowledge and education, first among academics and later among the general public. When the social networks began to emerge later, their goal was similar. Over time and with their rapid development, they have become not only a communication channel but also a means of sharing photos, videos, articles, opinions, even though a mobile phone. Many people do things in their lives just to share it on social networks. Unfortunately, this communication tool also has a dark side. It has become the home of fake reviews, gossip, or nonsense, which unfortunately users continue to share without validation. Every day, falsehood and deceit are spread through social networks for a variety of reasons as a financial gain or a gain the favor of the greatest number of people. And consumer users are just helping.

The concept "fake reviews" is neologism, which is very often used to refer to a fictive message. The fictive information is distributed mainly by social media, but it can be also distributed through the conventional media. Fake reviews is written and

A. Chatzigeorgiou et al. (Eds.): SOFSEM 2020, LNCS 12011, pp. 667–675, 2020.
https://doi.org/10.1007/978-3-030-38919-2_56

published in order to mislead and sometimes to harm the reputation of a company, entity or person, and to profit from it either financially or politically. They usually have sensational headlines to increase readability of posts.

The effort to manipulate people's minds is very old. In ancient times, tribal leaders, princes, kings, and pharaohs wanted to manipulate power. In these times, it was enough to influence those who had some power. With the arrival of the city states, it was necessary to manipulate wider groups of people as senators or ambassadors. With the oncoming of democratic regimes, it was necessary to persuade the masses of the people about their truth by means of books or daily newspapers. The problem at that time was that the reader had to buy a book or newspaper with lies. When the radio and television came, the manipulation was easier, because they allowed the information to be spread among masses without necessity to pay for the content so much. The mass manipulation began to fall slowly with the oncoming of the Internet. Suddenly it was also very easy and quick to find out what was true.

After some time, the mass manipulation took on a new form, for example the form of fake reviews. Social networks provided an ideal environment for the fake reviews. If misleading information comes from multiple sources in a similar period, it is not difficult to believe that it is serious information. Most social networks users give them only a quick look. Time and space to confront the source of information is significantly low. Large number of users do not verify the truth of information because the information is already enjoyed by thousands of users. Everybody who has a social network account can create a professionally looking posts, which are spread quickly and for free [1].

2 Fake Reviews Detection

2.1 Fake Reviews in Online Space

The fake reviews under mines serious media companies and make it difficult for journalists to report significant reviews stories. BuzzFeed, an American Internet media and reviews agency, found that the top twenty fake reports of the 2016 US presidential election received more engagement on Facebook than the first twenty election stories from nineteen major social media. Also, well known publishers have been anonymously attacked by sites that published fake reviews because it was difficult to detect sources of them. During and after the presidential election, Donald Trump began using the term "fake reviews" to describe negative information about his presidency.

To be able to detect fake reviews, at first the concept has to be characterized. There are more types of fake reviews:

- Satire or parody - no intention to cause harm, but there is a possible craziness
- False link - subtitles and headlines do not support content
- Misconception - use of misleading information to confront a problem or an individual
- False context - if a true content is shared with a false context information
- Fraudulent content - if original sources are supplied with false resources
- Manipulated content - when original information or images intend for fraud are manipulated, for example in a "modified" photo
- Invented content - the new content is 100% false, designed to deceive and damage

According the source, the fake reviews on traditional media has psychology and social foundations. On the other hand, the fake reviews on social media can be characterized by malicious accounts and its echo chamber [2].

2.2 State of the Art

There are different approaches to fake reviews detection. The first approach is based on a *content analysis* using knowledge (an external source can be used for check-up the truth of reviews) and the style analysis (a spreading of fake and misleading information requires the special writing style). The second approach is based on a *social context analysis* of a stance, attitude and a propagation of reviews. Detecting false messages on social networks is a relatively new area of research. The survey [2] addresses related research areas, open issues, and future research directions from a data mining perspective. The fake reviews detection can be oriented on data, feature, model and application.

Social media such as Facebook and Twitter are undoubtedly main channels for spreading misleading information. Facebook has attempted to implement detection tools. The first is, that users have the option to mark reviews they consider to be a fake reviews. To identify the source, badges are created that mark the lie and allow users to learn more about the story. When enough users label a story to be fake, the frequency of the shared article decreases. To prevent the spread of fake reviews, the company reduces the number of tagged posts and thus reduces their spreading. Repeated offenders, who spread often misleading messages, have removed advertising rights, thereby reducing their distribution as well as earning.

Another possibility is to implement an artificial intelligence to detect fake messages. The artificial intelligence can learn quickly and efficiently to determine words and phrases most relevant to fake reviews - using the trial-and-error method. The artificial intelligence can be used to identify inappropriate posts and to recognize extremism, violence, hatred, threats and other forms of misleading information in online discussions.

The work [3] represents an approach to fake reviews detection using spam indicators. It integrates content and usage information to detect fake products reviews. The model is based also on a reviewer's behavioural trails interlinked by specific spam indicators as an extreme evaluation, a big number of post in a short time period and a similarity of posts of the reviewer.

Another study is based on the detection of online fake reviews using a text analysis approach based on n-gram models and machine learning techniques. It compares six different classification techniques, namely, K-nearing neighbors (KNN), logistic regression, linear support vector machine (SVM), decision trees and stochastic gradient descent. To reduce the size of the lexical profile of texts, two methods were used TF and TF-IDF weightings. The authors collected 12,600 false and 12,600 true reviews on the 2016 political situation. The study showed linear models are better than nonlinear ones. The highest accuracy was achieved using the SVM algorithm and the lowest accuracy was achieved using the KNN algorithm [4].

The work [5] is focused on an uncovering fake reviews by classifying it using naive Bayes classifier and random forests. The reviews were obtained from Amazon and included the seller's website, product name, rating, reviewer ID, review topic, review content, date added, review impact (how many people consider it useful), and whether the purchase was verified. The experiments showed that the random forests model achieved better results than the naive Bayes classifier.

Another work [6] answers interesting questions, for example if the performance of the classification methods for fake reviews filtering is affected when they are used in real-world scenarios that require online learning. Their datasets were from various domains as a trip advising, a recommendation of hotels, or an evaluation of restaurants. Some of data were ordered chronologically and some were not. They used naïve Bayes multinomial and Bernoulli, K-nearest neighbors, decision trees, random forests and support vector machines. The best results were achieved by support vector machines (F1 measure = 0.899).

3 Used Machine Learning Methods

To solve the problem of fake reviews detection, we have selected following machine learning algorithms: naive Bayes, decision trees, random forests and support vector machines. We have chosen them for number of reasons, for example - they are the most reliable, understandable, often used with success, etc. [7, 8].

Naive Bayes is a probabilistic classifier based on Bayes' theorem and the independence assumption between features. It is very natural selection when the final decision about class fake/non-fake review depends on conditional probability of words in the given review on the class. The naive Bayes is often applied as a baseline for text classification. This method is successful on extremely short text. On the other hand, its performance can be outperformed by support vector machines [9] in the case of big lexical profile of data texts.

Another selection is a model which uses a tree of decisions to predict a fake/non-fake label for a new post sample. The tree is learned on the training set using a standard top-down approach, which starts with a full dataset in one root (parent) node A question divides a node to sub-nodes - each representing answers to the question [10]. The advantages of decision trees in comparison with two above methods are their intuitive interpretation and the non-linear solution. The decision trees were successfully used for part-of-speech tagging [11] and text documents categorization and parsing [12].

Random Forests [13] represents a composed learning. It builds a set of de-correlated decision trees. It averages results of the set of decision trees to final decision about class. The classification result is determined by voting. To ensure the condition that individual tree models must be independent, the random forests technique uses a random selection of attributes for each tree generation. Advantages of Random forests are following.

- Since these trees are independent, it is suitable and easier to generate them in parallel.
- The random selection of a training set for training of each decision tree enable to validate (to test) it on data, which was not used for training. It facilitates validation.
- This approach is fast and accurate - so it has been used very often in recent years.

The Support Vector Machine (SVM) model separates the sample space into two or more classes with the widest margin possible, which enables to find the best separating hyperplane. SVM is originally a linear classifier. However, the classifier can perform a non-linear classification [14] using non-linear function or kernel function. Kernel method maps features into a higher dimensional separable space. The objective is to maximize the width of the margin, which is known as the primal problem of support vector machines [15].

4 Models Building

The main aim of this work was to find the most accurate machine learning algorithm for learning the model that could detect the fake reviews. We used CRISP-DM methodology [16] for the data mining process. We had a dataset that contained the title, the text and the label of the posts in the form of marks False and True. The attributes Title and Text of the posts have been pre-processed. Input to the modelling was in two forms - a document term matrix with and without TF-IDF weighting. We have verified the created models on the test set using several indicators of binary classification.

4.1 Data Source

We have chosen a dataset that was freely available at https://www.kaggle.com/. Dataset, *Fake News Detection*, contained 4009 records and 4 attributes. The message tag was specified using the *Label* attribute, which takes two values, 0 - indicates fake reviews, and 1 - indicates true post. The attribute represents the target attribute. The proportion of false and true posts was 51.31% to 46.69%. We suppose, that it reflects a reality of a discussions, when authors want to respond to fake reviews to balanced it. Dataset also contained attributes: *URLs*, which indicates the location of the post on the Internet, the *Headline* and *Body* of a post. The dataset contains reviews about the new USA President Donald Trump. The dataset was divided into two datasets. First data set contains only bodies of all reviews and the second dataset contains only headlines of reviews. These two datasets were used for testing. We wanted to figure out which machine learning methods are better for extremely short texts and which methods can be used for all reviews.

4.2 Data Preprocessing

The data pre-processing is one of the most time consuming phases of the process of data mining. The quality of pre-processed input data affects the quality of output. The data pre-processing contained following steps:

- removing unsuitable symbols and unnecessary gaps
- conversion to lowercase and delete punctuation
- remove "stop words"
- stemming
- create document term matrix (DTM) using "Bag of words" representation
- create DTM with weighting scheme TF-IDF [16].

5 Models Testing

Four models were trained using following four machine learning methods: naive Bayes (NB), decision tree (DT), random forests (RF) and support vector machine (SVM). The input data, process of models learning, and testing are illustrated in Fig. 1.

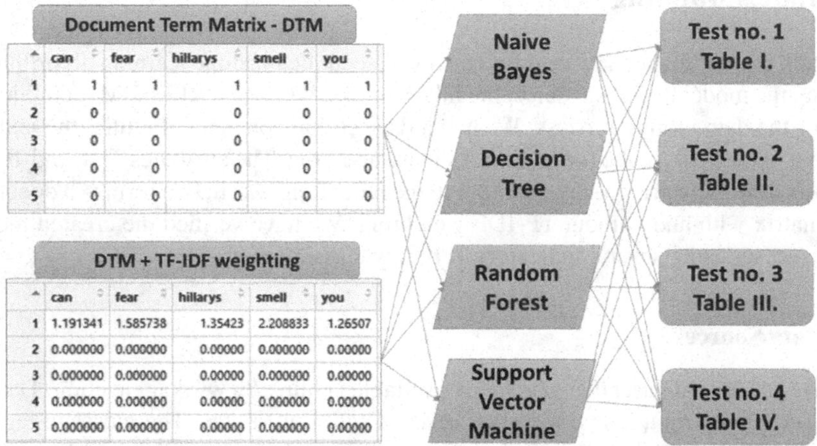

Fig. 1. Illustration of input data and generated model testing

The input of the training was consisted from tests of posts (attribute Body) from dataset described above. The data were pre-processed to the form of document term matrix (DTM) and processed by four above mentioned machine learning methods under two different conditions: with and without TF-IDF waiting. Using a confusion matrix, following indicators of binary classification were quantified: Accuracy, Interval of accuracy (Table 1), Precision, Recall and F1 measure as a harmonic average of precision and recall (see Table 2).

Table 1. Accuracy and Interval of accuracy of models learned from the body of posts

Body of posts	DTM		DTM + TF-IDF	
	Accuracy	Interval of accuracy	Accuracy	Interval of accuracy
NB	0.844	(0.822, 0.864)	0.904	(0.886, 0.920)
DT	0.881	(0.862, 0.899)	0.904	(0.886, 0.920)
RF	**0.978**	**(0.969, 0.988)**	**0.983**	**(0.973, 0.989)**
SVM	0.782	(0.758, 0.805)	0.944	(0.930, 0.957)

Experiments presented in both Tables 1 and 2 showed that in the case when the input of the learning are whole texts of reviews or posts the best model is model learned by random forests algorithm. The random forests method is best in all results of the

Table 2. Recall, Precision and F1 measure of models learned from the body of posts

Body of posts	DTM			DTM + TF-IDF		
	Recall	Precision	F1	Recall	Precision	F1
NB	0.910	0.818	0.862	0.880	0.936	0.907
DT	0.938	0.854	0.894	0.920	0.902	0.911
RF	**0.964**	**0.995**	**0.979**	**0.972**	**0.995**	**0.983**
SVM	0.570	0.938	0.709	0.916	0.978	0.946

monitored parameters of effectivity: Recall, Precision, F1 measure and Accuracy. Also Intervals of accuracy are narrowest and smallest for random forests model. A smaller range of the Interval of accuracy means a better model.

Our best result achieved by SVM in F1 measure is 0.946 what is better than in the similar work [6].

Other four models were trained using following four machine learning methods: naive Bayes (NB), decision tree (DT), random forests (RF) and support vector machine (SVM). The input of the training was created with headlines of posts (attribute Headline) from dataset described above. The data were pre-processed to the form of document term matrix (DTM) without and with TF-IDF waiting. Using a confusion matrix, following indicators of binary classification were quantified: Accuracy, Interval of accuracy (Table 3), Precision, Recall and F1 measure (Table 4).

Table 3. Accuracy and Interval of accuracy of models learned from the headlines of posts

Headline of posts	DTM		DTM + TF-IDF	
	Accuracy	Interval of accuracy	Accuracy	Interval of accuracy
NB	**0.802**	**(0.778, 0.824)**	**0.812**	**(0.788, 0.833)**
DT	0.551	(0.523, 0.580)	0.550	(0.521, 0.578)
RF	0.749	(0.724, 0.773)	0.760	(0.735, 0.784)
SVM	0.762	(0.737, 0.786)	0.775	(0.750, 0.798)

Experiments presented in Table 3 shows that in the case when the input of the learning were extremely short texts as headlines or titles of posts the best model was model learned by Naïve Bayes learning method when Accuracy and Interval od accuracy was taken into account. The results in Table 4 are not so clear. When Recall and Precision was taken into account naïve Bayes model and decision tree model were the best alternately according the way of pre-processing: DTM or DTM with TF-IDF waiting. But when F1 measure was taken into account the naïve Bayes model was the best. We can close the evaluation by claim, that naïve Bayes model is best for Headlines of post, because F1 measure is

Table 4. Recall, Precision and F1 measure of models learned from the headlines of posts

Headline of posts	DTM			DTM + TF-IDF		
	Recall	Precision	F1	Recall	Precision	F1
NB	0.792	**0.829**	**0.810**	**0.820**	0.826	**0.823**
DT	***1.000***	0.543	0.704	0.200	***0.898***	0.327
RF	0.830	0.734	0.779	0.710	0.827	0.764
SVM	0.718	0.814	0.763	0.766	0.802	0.783

harmonic means of Precision and Recall and so the F1 measure takes into account both types of mistakes – numbers of falls positive and falls negative classifications.

6 Conclusions

The approach to the fake reviews detection in online discussions was introduced. The approach was based on models for fake reviews classification generated by machine learning algorithms: naive Bayes, decision trees, random forests and support vector machines. Generated models have been tested. Experiments showed that input data representation is important, as in most cases models that worked with the document term matrix with a TF-IDF weighting (DTM + TF-IDF) achieved better results. The naive Bayes model appeared to be the best for a smaller data input sample for example in the form of headlines or titles of posts. On the other hand, the random forests model appeared to be the best for larger data input samples as whole texts of posts. For future, the presented 4 approaches are planned to be explore by evaluating their robustness on a progressively unbalanced dataset and on other datasets.

This work has produced results that could be further developed, as the problem of fake reviews steadily increases. These issues should be discussed, their dangers highlighted, and they can be resolved by finding and detecting them. For future, we would like to use a deep learning method to get better results mainly in the accuracy measure. The problem of fake reviews detection could be also analyzed from the point of sentiment or opinion polarity [17].

Acknowledgements. The work presented in this paper was supported by the Slovak Research and Development Agency under the contract APVV-017-0267 and APVV-16-0213.

References

1. Vítek, F.: Fake news – where did it begin and where do we go?, May 2019. http://mocnedata.sk/2018-fake-news/
2. Shu, K., Sliva, A., Wang, S., Tang, J., Liu, H.: Fake news detection on social media: a data mining perspective. Newsletter **19**(1), 22–36 (2017)

3. Dematis, I., Karapistoli, E., Vakali, A.: Fake review detection via exploitation of spam indicators and reviewer behavior characteristics. In: Tjoa, A.M., Bellatreche, L., Biffl, S., van Leeuwen, J., Wiedermann, J. (eds.) SOFSEM 2018. LNCS, vol. 10706, pp. 581–595. Springer, Cham (2018). https://doi.org/10.1007/978-3-319-73117-9_41
4. Ahmed, H., Traore, I., Saad, S.: Detection of online fake news using N-gram analysis and machine learning techniques. In: Traore, I., Woungang, I., Awad, A. (eds.) ISDDC 2017. LNCS, vol. 10618, pp. 127–138. Springer, Cham (2017). https://doi.org/10.1007/978-3-319-69155-8_9
5. Chowdhary, N.S., Pandit, A.A.: Fake review detection using classification. Int. J. Comput. Appl. **180**(50), 16–21 (2018)
6. Cardoso, E.F., Silva, R.M., Almeida, T.A.: Towards automatic filtering of fake reviews. Neurocomputing **309**, 1–41 (2018)
7. Russell, S.J., Norvig, P.: Artificial Intelligence. A Modern Approach, 3rd edn, pp. 1–932. Prentice Hall, Pearson Education, Upper Saddle River (2010). ISBN-13: 978-0-13-604259-4
8. Han, J., Kamber, M.: Data Mining: Concepts and Techniques, 3rd edn, pp. 1–703. Morgan Kaufmann, Elsevier, Burlington (2012)
9. Pang, B., Lee, L., Vaithyanathan, S.: Thumbs up? Sentiment classification using machine learning techniques. In: Proceedings of the Conference on Empirical Methods in Natural Language Processing (EMNLP 2002), Philadelphia, pp. 79–86 (2002)
10. Kingsford, C., Salzberg, S.L.: What are decision trees? Nat. Biotechnol. **26**(1), 1011–1013 (2008)
11. Orphanos, G., Kalles, D., Papagelis, T., Christodoulakis, D.: Decision trees and NLP: a case study in POS tagging. Academia, 1–7 (1999)
12. Magerman, D.M.: Statistical decision-tree models for parsing. In: Proceeding ACL 1995 Proceedings of the 33rd Annual Meeting on Association for Computational Linguistics, pp. 276–283 (1995)
13. Breiman, L.: Random forests. Mach. Learn. **45**(1), 5–32 (2001)
14. James, G., Witten, D., Hastie, T., Tibshirani, R.: An Introduction to Statistical Learning: With Applications in R. STS, vol. 103, pp. 1–426. Springer, New York (2013). https://doi.org/10.1007/978-1-4614-7138-7
15. Ben-Hur, A., Horn, D., Siegelmann, H.T., Vapnik, V.: Support vector clustering. J. Mach. Learn. Res. **2**(2), 125–137 (2001)
16. Paralič, J., et al.: Mining Knowledge from Texts. Equilibria, Košice (2010)
17. Mikula, M., Machová, K.: Combined approach for sentiment analysis in Slovak using a dictionary annotated by particle swarm optimization. Acta Electrotechnica et Informatica **18**(2), 27–34 (2018)

Two-Step Memory Networks for Deep Semantic Parsing of Geometry Word Problems

Ishadi Jayasinghe(✉) and Surangika Ranathunga

Department of Computer Science and Engineering, University of Moratuwa, Katubedda, Moratuwa 10400, Sri Lanka
{ishadij.12,surangika}@cse.mrt.ac.lk

Abstract. Semantic parsing of geometry word problems (GWPs) is the first step towards automated geometry problem solvers. Existing systems for this task heavily depend on language-specific NLP tools, and use hard-coded parsing rules. Moreover, these systems produce a static set of facts and record low precision scores. In this paper, we present the two-step memory network, a novel neural network architecture for deep semantic parsing of GWPs. Our model is language independent and optimized for low-resource domains. Without using any language-specific NLP tools, our system performs as good as existing systems. We also introduce on-demand fact extraction, where a solver can query the model about entities during the solving stage that alleviates the problem of imperfect recalls.

Keywords: Semantic parsing · Memory networks · Low-resource domains

1 Introduction

A Geometry Word Problem (GWP) usually consists of a set of sentences that describes some geometric entities and a question regarding those entities. An example of a description of a GWP is given in Table 1 (*Sentences* section). Ample amount of previous research available in this area of automated solving of GWPs attests to the importance of this domain [2,3]. However, most of the existing systems for this task require the GWPs to be manually parsed to machine-understandable formats. In this research, we tackle this problem of automating the parsing of GWPs, which paves the way towards end-to-end automatic solving of GWPs.

We use deep semantic parsing here, as it focuses on building more formal representations that support automated reasoning [1]. This task of deep parsing of geometry questions is inherently difficult due to several reasons. One critical reason is the lack of data [3], so using usual deep learning models for this task is not viable. Hence, existing systems such as GEOS and GEOSV2 have resorted to using hard-coded parsing rules [2,3]. These rule-based approaches bring forth

A. Chatzigeorgiou et al. (Eds.): SOFSEM 2020, LNCS 12011, pp. 676–685, 2020.
https://doi.org/10.1007/978-3-030-38919-2_57

a couple of limitations. Firstly, these systems are limited only for one language, and even within that language, the scalability and the maintenance are difficult due to hard-coded rules. Secondly, these systems record low precision scores [3].

Table 1. Task definition

Sentences	
Line AB is parallel to line CD	
AB is a chord of the circle with center O	
Initial fact extraction	
Unary rules	Line(AB), Line(CD), Circle(O)
Binary rules	Parallel(AB, CD), IsChordOf(AB, O)
Relation completion	
Parallel(Line(AB), Line(CD))	
IsChordOf(Line(AB), Circle(O)))	
On-demand fact extraction	
AB→ {Line(AB), Chord(AB)}	
CD→ Line(CD)	
O→ Circle(O)	

A solver that uses deductive reasoning, rather than a numeric approach, is more interpretable [2]. Hence, the importance of entities for such a system is dynamic. For example, during a proof where the goal is to prove the entities AB and CD are equal, and if the solver deduces a relationship such as $AB = BC$, proving that $BC = CD$ would achieve the target. Thus, the solver will have to explore about BC. However, due to imperfect recalls, the facts relevant to BC can be missing in the extracted fact set. This can inhibit solving the problem. Considering the requirements above, we model this task as a question answering (Q/A) task. Through this way of modeling, in addition to the capability to extract facts similar to the other existing systems, we also facilitate the solver to extract facts about interested entities (BC in the example above) during the deduction process. This effectively increases the recall of the system, thus lessening the severity of the impact caused by imperfect recalls during the initial fact extraction phase. We term this feature as **on-demand fact extraction**.

This Q/A task is built on Memory networks [6]. We enhance this memory network model to suit for low resource datasets facilitating (*i*) the introduction of deep learning to this task (which in turn brings forth advantages such as inter-language usability, and low-cost maintainability), (*ii*) on-demand fact extraction resulting in an overall high recall, and (*iii*) the capability to accommodate multiple sentences. We show this model performs comparable to existing systems, and is better than memory networks in all the tasks. We also show that the **position encoding** scheme we introduce improves results on all the tasks[1].

2 Related Work

GEOS focuses on solving geometry questions in an end-to-end manner [3]. Using a hardcoded set of parsing rules, GEOS generates a set of facts that are scored for their accuracy using a discriminative model. Although GEOS produces a large number of inaccurate rules at the end of text parsing, one reason this is not affecting GEOS is due to this limited scope. Hence, if this parser is used for problems where proofs are needed, not only might the wrong facts would lead to a wrong proof, proving might even become impossible if some of the necessary facts are missing in the extracted set.

RNNs such as LSTMs are frequently used in a wide array of tasks [4]. However, it is known that their memory is limited and they have trouble in modeling long-range dependencies [5,6]. Even though increasing the layer size is a solution (as in the machine translation model developed by Xu et al. [7]), it is not possible on low resource domains.

Memory networks [6] were proposed as a means to query from multiple sentences. Consider we have N number of sentences as memories, with the memory for sentence i denoted as m_i. The memory network first selects the most matching memory m_{o1} with the query x. Then it goes for a second iteration to select the most matching memory m_{o2}, but now, not only with the query x but also with the memory m_{o1}. This process is repeated for the configured number of hops (2 in this case). Finally, the most matching word ($\hat{a}$) from the vocabulary W is selected based on its similarity score with the query x and the selected memories (m_{o1} and m_{o2} in this case).

End-to-end Memory Networks [5] were introduced to make memory networks end-to-end trainable by replacing the hard attention mechanism in memory networks by a soft attention mechanism. Sukhbaatar et al. [5] also introduced modifications to reduce the trainable parameter count.

3 Two-Step Memory Networks

3.1 Task Definition

Table 1 illustrates the intended use of our model. Our model should initially produce unary rules (rules with one child), binary rules (rules with two children), and, finally, completed relations (merging unary and binary rules) for a given set of sentences. Also, it should be capable of extracting facts related to a given entity as shown under *On-demand Fact Extraction* in the table.

We model this task as a question-answering (Q/A) task. First, a list of keywords from the training data is automatically extracted along with their valences.

[1] Our work can be accessed from this repository: https://github.com/IshJ/Two-step-memory-networks.

For example, if there are two rules such as {"parallel AB CD", "line AB"}, the words *parallel*, and *line* are extracted along with 2 and 1 as their valences respectively. These words are used as the query words. Two models are trained for unary and binary rules.

3.2 Limitations of Existing Memory Networks

End-to-end memory networks store a given sentence as the sum of the embedding vectors of its words resulting in a lossy sentence representation. Even though this has not caused a significant loss of information with the experiments carried out by Sukhbaatar et al. [5], we found this representation to be too lossy for our task. As discussed above, increasing the network capacity to retain more data is not viable here. On the other hand, if we can store only the important memories, we will not need to compress information as much as above. This is our motivation behind coming up with a model that can selectively store only the essential data within a limited capacity.

3.3 Model Formulation

Here, we first need to select the sentences that are relevant to the query at hand. Taking x as the query and m_i as the ith sentence, the probability is calculated for the relevance of sentence i to the query x:

$$p_i = Sigmoid(x^T m_i) \tag{1}$$

Figure 1 shows an example for the resulting probability distribution. Lighter colors indicate higher probabilities. We do not concern about handling inter-sentence dependencies in this research. So we are not going for a multi-hops approach for selecting the relevant sentences. As a hyper-parameter of the model, we maintain a threshold for the lowest probability a sentence should get to be considered as a relevant sentence. This is hard attention. Even though this makes the network incapable of being trained end-to-end using standard backpropagation methods [7], we adopt it to keep the computations simple.

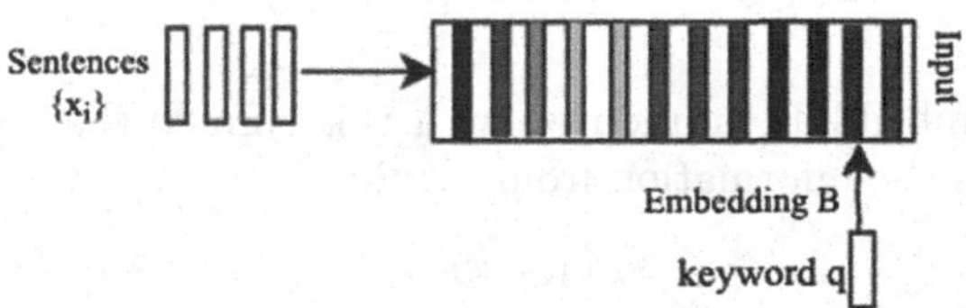

Fig. 1. Architecture of the sentence selection model.

After selecting the relevant sentences, each of the selected sentences is fed to the **word-level memory network** (Fig. 2). Here, the model attends each selected sentence in the word level.

First, the words of the query are embedded with the embedding matrices A and B respectively. Then the dot product is calculated between each word embedding and the query. These scores are converted to a probability distribution using the $Softmax$ operation:

$$p_i = Softmax(x^T m_i) \tag{2}$$

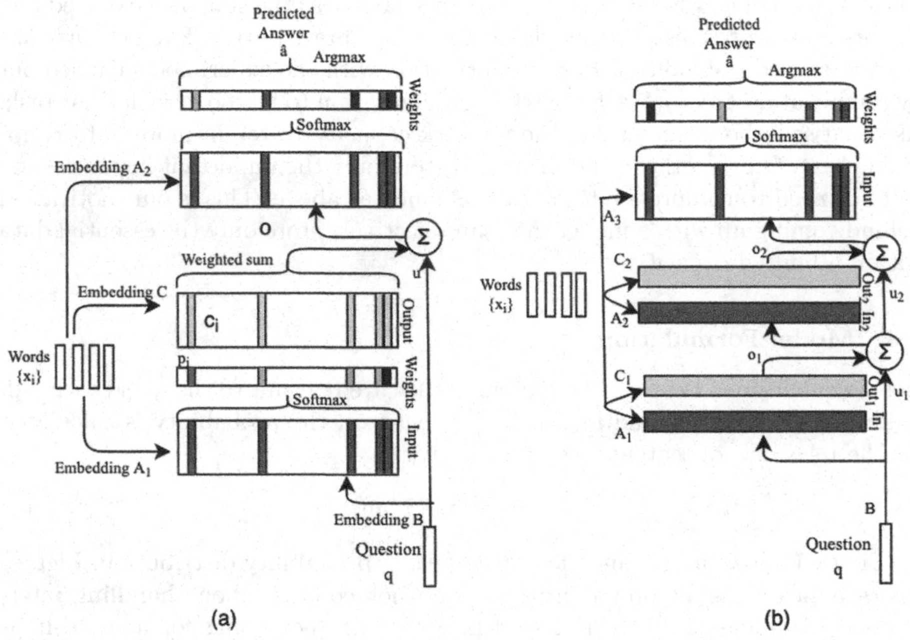

Fig. 2. Architecture of the word-level memory network

After that, the words are again embedded using the embedding matrix C, and then the weighted sum of these embedding vectors is calculated:

$$o = \sum_i p_i c_i \tag{3}$$

Now the query embedding is updated with this sum so that the query for the next hop contains the information from this hop:

$$x_{k+1} = x_k + o_k \tag{4}$$

In the single hop scenario (Fig. 2a), the words are embedded using a second input embedding matrix (A_2). From the dot products calculated with the updated query, a probability distribution is produced. This process is similar to the process described above (Eq. 2). Now, as the answer, the word with the highest probability is selected:

$$\hat{a} = Argmax(Softmax(x_{k+1}^T m_i)) \tag{5}$$

In the multi-hop scenario (Fig. 2b), equations from 2–4 are executed in a loop for the given number of hops. After each hop, the query is updated with the weighted sum of the word embeddings, thus allowing the forward-passing of information. Finally, similar to the single hop scenario, the answer is selected based on the probabilities calculated in the final hop.

Focus controlling (FC) with position encoding: If we consider the sentence "In the given figure, line AB is parallel to line CD", it contains two rules for the same keyword *line*. If we query the sentence just with *line*, we will not be able to get both rules. To overcome this limitation, we introduce a mechanism to "tell" the network where to focus. We do this by a customized **position encoding** (PE) scheme. This scheme also serves the purpose of giving the model a sense of the order of the words. We build our scheme based on the PE scheme introduced by Sukhbaatar et al. [5]. We first extend the embedding matrices with a size of $2 * max\_len$, where max_len refers to the maximum sentence length in the training dataset. After that, instead of the memory m_i being simply the embedding of w_i, we modify the memory m_i to be $embedding(w_i) + embedding(max\_len + d_i)$ where d_i refers to the distance between w_i and the matched keyword in the sentence. Words before the query word get negative values for d_i, whereas the words after the query word get positive values for d_i. According to this scheme, the memory in the location of the query always gets added the same vector; $embedding(max\_len)$. So does the words around the query word. Hence, we can expect the model to learn to focus more on memories near the query word.

Unary Rule Extraction: This task only needs single word answers. So, we use the model in Fig. 2b as it is.

Binary Rule Extraction: Here, we focus on rules such as "parallel AB CD". Unlike extracting unary rules, not only do we have to extract two words but also the second word depends on the first word retrieved. Due to this dependency, we cannot model this problem as a multi-class classification problem. Also, as explained earlier, introducing an RNN is not viable. Therefore we come up with a layer-wise retrieving mechanism; assuming we have k number of hops and $layer_k$ as the final layer, we retrieve the first literal (AB in this case) from $layer_{k-1}$ and the second literal (CD) from $layer_k$. We define the loss function as the summation of individual losses (categorical cross-entropies) from the two layers;

$$Loss = \sum_{i}^{l} (t_{i_k} log(s_{i_k}) \ + t_{i_{k-1}} log(s_{i_{k-1}}) \ ), i = 1, 2...l \tag{6}$$

Here, l refers to the memory size (or the length of the sentence). t_{i_k} refers to the ground truth value for the ith location for k^{th} layer. Usually (if label smoothing or any such technique is not used), t_{i_k} is 1 for the location of the second literal (CD) and 0 for the other locations. Similarly, $t_{i_{k-1}}$ is 1 for the location of the first literal (AB) and 0 for the other locations. s_{i_k} refers to the probability computed by the model for the ith location for the kth layer.

On-demand Fact Extraction: We model this task similar to unary rule extraction. Here, we query with the interested entity. For example, if we consider the sentence "AB is a tangent to circle O", and if we query the sentence with AB, the system will produce *tangent*. Through FC, as described above, we retrieve multiple rules for a single entity.

4 Experiments

GEOSV2 have used a publicly unavailable dataset that is larger than the dataset used in GEOS. Thus, we compare the scores of our model with the scores of GEOS and GEOSV2 when they are trained on the dataset used in GEOS (Table 2). We use the training and the practice datasets in the table as our training and evaluation datasets respectively.

Table 2. Statistics of the dataset. Introduced by Seo et al. [3].

	Total	Training	Practice	Official
Questions	186	67	64	55
Sentences	326	121	110	105
Binary relations	337	110	108	119
Unary relations	437	141	150	146

4.1 Unary Rule Extraction

The keywords with valence 1 are used for this task. *Unary Rule Extraction* in Table 3 indicates the results of this task. We handle multiple occurrences of the same keyword through FC. Figure 3 shows how the probability distribution changes based on the focused location during multiple rule extraction for the same keyword. We can see a significant improvement in the F1-score when it comes to two-step memory networks from end-to-end memory networks.

Table 3. Precision, Recall, and F1 scores for the tasks of unary rule extraction, binary rule extraction, and on-demand fact extraction

Task	Model	P	R	F1
Unary rule extraction	End-to-end MN	0.51	0.25	0.33
	Two-step MN without FC	0.52	0.42	0.46
	Two-step MN with dynamic FC	0.55	0.58	0.56
	Two-step MN with fixed FC	**0.68**	**0.72**	**0.70**
Binary rule extraction	End-to-end MN	**0.83**	0.19	0.30
	Two-step MN without FC	0.36	0.60	0.45
	Two-step MN with fixed FC	0.49	**0.62**	**0.55**
On-demand fact extraction	End-to-end MN	0.49	0.20	0.29
	Two-step MN without FC	0.49	0.56	0.52
	Two-step MN with dynamic FC	**0.73**	**0.80**	**0.76**

Dynamic Versus Fixed FC: Dynamic FC refers to changing the position encoding based on the location of the keyword. Consider the sentence ".. line AB is parallel to line CD". We will use $\{line, 5\}$ and $\{line, 10\}$ for querying under this setting. Fixed FC refers to keeping the position of the keyword fixed ignoring its multiple occurrences. Interestingly, when it comes to results, for unary rule extraction, we can see that dynamic FC has lower scores compared to fixed FC (3rd and 4th rows of *Unary Rule Extraction* in Table 3). The disadvantage caused by being unable to produce multiple rules for the same keyword can be seen to be overridden by the advantage of being able to ignore the query word position during training. As expected, having FC in either setting above is better than having no FC (2nd and 3rd rows of *Unary Rule Extraction* in Table 3).

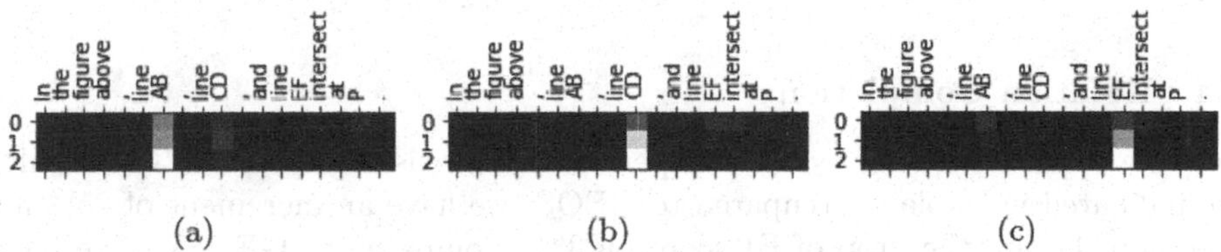

Fig. 3. Extracting the rules (a) "line AB", (b) "line CD", and (c) "line EF" with the keyword *line*.

4.2 Binary Rule Extraction

The (*Binary Rule Extraction* in Table 3) indicates the results for this task. We can see a significant improvement in the F1-score with the two-step memory network (two-step MN) and further improvements with FC. Figure 4 shows an example for this task.

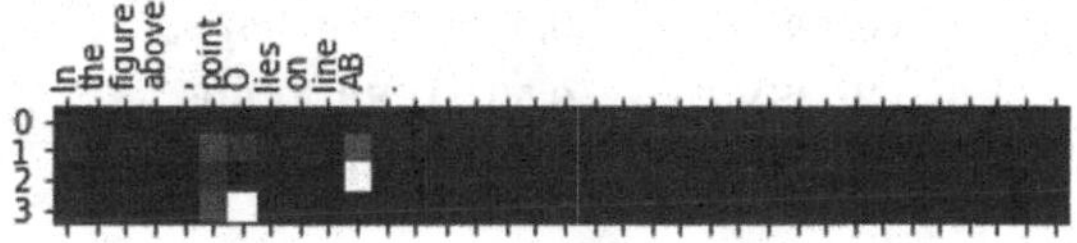

Fig. 4. Binary rule extraction for the keyword *lies*.

4.3 On-Demand Fact Extraction

Here, as the training set, we reversed the unary rules so that the entity would be the querying word and the property would be the answer. During the testing phase, using a regex expression, we first scan the sentence to retrieve all the candidates, and then we query the sentence with each of those candidates (Fig. 5).

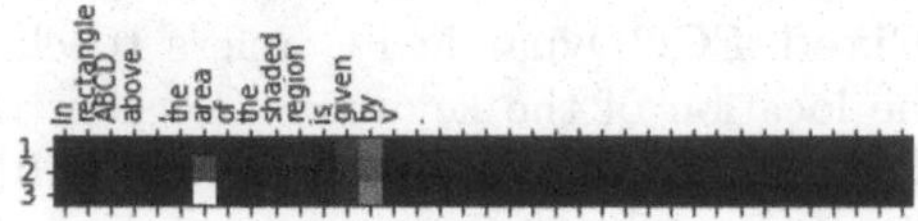

Fig. 5. Fact extraction for the entity v. We can see how the answer *area* is refined over hops.

It is important to note that we use this regex rule only for evaluating the system. We do not need this rule for training the system and inferencing afterwards. *On-demand Fact Extraction* in Table 3 shows our results under this task. Similar to the above tasks, we can see substantial improvements with the improvements we introduce.

4.4 Relation Completion

Here we see whether our system outperforms the existing systems. The results are indicated in Table 4. Compared to GEOS, we have an increment of 25% for precision despite the drop of F1-score by 3%. Compared to GEOSV2, we have a precision increment of 20% while we record a drop of 3% with F1-score. Both GEOS and GEOSV2 have low precision scores despite the high recalls, which occurs when a system produces a lot of false positives. As our system records a higher precision only with a slight drop of the F1-score, we claim that the rules generated by our system are more reliable. With on-demand fact extraction (that records a high F1-score of 76%, a recall of 80%, and a precision of 73%), we effectively increase the recall of our system.

Table 4. Precision, Recall, and F1 score for relation completion

	P	R	F1
GEOS	0.57	0.82	0.67
GEOSV2	0.59	**0.83**	**0.69**
Two-step MN	**0.71**	0.60	0.65

5 Conclusion and Future Work

In this paper, we present the two-step memory network, a novel neural network architecture, for deep semantic parsing of GWPs. Our system is competitive with GEOS and GEOSV2. Also, our system can generate on-demand rules, which alleviates the problem of having a low recall. Unlike existing systems that are heavily dependent on NLP tools and hardcoded parsing rules, our system does not use any language dependent tools and optimized for low-resource languages. Also, since we do not use any domain specific rules, we conjecture that our system can be used for other similar parsing tasks too.

Even though we have provisioned for multiple sentences, we did not experiment this. Research in this line will make way to modeling inter-sentence dependencies such as co-reference resolution.

Acknowledgments. This research was funded by a Senate Research Committee (SRC) Grant of University of Moratuwa, Sri Lanka and LK Domain Registry, Sri Lanka.

References

1. Miikkulainen, R.: Subsymbolic case-role analysis of sentences with embedded clauses. Cogn. Sci. **20**(1), 47–73 (1996)
2. Sachan, M., Xing, E.: Learning to solve geometry problems from natural language demonstrations in textbooks. In: Proceedings of the 6th Joint Conference on Lexical and Computational Semantics (* SEM 2017), pp. 251–261 (2017)
3. Seo, M., et al.: Solving geometry problems: combining text and diagram interpretation. In: Proceedings of EMNLP 2015, pp. 1466–1476 (2015)
4. Socher, R., et al.: Recursive deep models for semantic compositionality over a sentiment treebank. In: Proceedings of EMNLP 2013, pp. 1631–1642 (2013)
5. Sukhbaatar, S., et al.: End-to-end memory networks. In: NIPS, pp. 2440–2448 (2015)
6. Weston, J., et al.: Memory networks. arXiv preprint arXiv:1410.3916 (2014)
7. Xu, K., et al.: Show, attend and tell: neural image caption generation with visual attention. In: International Conference on Machine Learning, pp. 2048–2057 (2015)

Foundations of Software Engineering – Short Papers

A Case Study on a Hybrid Approach to Assessing the Maturity of Requirements Engineering Practices in Agile Projects (REMMA)

Mirosław Ochodek, Sylwia Kopczyńska(✉), and Jerzy Nawrocki

Poznan University of Technology, Poznań, Poland
{miroslaw.ochodek,sylwia.kopczynska}@cs.put.poznan.pl

Abstract. *Context:* Requirements Engineering (RE) is one of the key processes in software development. With the advent of agile software development methods, new challenges have emerged for traditional, prescriptive maturity models aiming to support the improvement of RE process. One of the main problems is that frequently the guidelines prescribed by agile approaches have to be adapted to a project's context to provide benefits. Therefore, it might be naive to believe that it is possible to propose a prescriptive method of RE process improvement that will suit all agile projects without any alteration. *Objective:* The aim of the paper is to evaluate a hybrid approach to assessing the maturity of agile RE (REMMA), which combines elements of prescriptive and problem-oriented improvement methods. *Method:* The usefulness, ease of use, and cost-effectiveness of REMMA were investigated through a case study performed in one of the biggest software houses in Central Europe. *Results:* The results of the case study suggest that the method seems easy to use, affordable, and is perceived as a useful tool to support the process of improving RE practices in agile projects. Its feature of taking into account the dependencies between practices and the necessity to adapt them to a certain project context was regarded as well suited for the agile context. *Conclusions:* REMMA, which includes two main components: a maturity model for agile RE (a set of state-of-the-art agile RE practices) and an assessment method that makes it possible to evaluate how well the agile RE practices are implemented, seems to be a useful tool supporting improvement of RE in agile projects.

Keywords: Requirements Engineering · Process assessment · Process maturity · Process improvement · Agile

1 Introduction

Requirements Engineering (RE) is one of the key processes in software development. It has been observed that when the RE process is orchestrated properly it can favorably influence the whole software development process [3,6]. Conversely, problems related to requirements were often identified as main causes of the failures of IT projects, e.g., [4,8,9,18].

A. Chatzigeorgiou et al. (Eds.): SOFSEM 2020, LNCS 12011, pp. 689–698, 2020.
https://doi.org/10.1007/978-3-030-38919-2_58

With the advent of agile software development methods, such as eXtreme Programming (XP) [1], or Scrum [14], new challenges have emerged for RE. The approach to planning projects has changed from traditional—predictive to adaptive [7]. The need for adaptivity is deeply-rooted in the Agile Manifesto [2] that advise teams to regularly reflect on how to become more effective and make use of the lessons they learned. Although this principle makes the team responsible for improving the software development process, many agile software development methods introduce additional roles responsible for driving the improvement process, e.g., Coach in XP [1], or Scrum Master [14] (we will refer to them as agile coaches). The agile coaches are supposed to have decent knowledge of agile methods to help teams find the best solutions to their problems, as well as be able to convince management to allocate necessary resources to support the improvement process.

Among the different tools that agile coaches can use to support the improvement process, they can employ one of the existing agile maturity models (e.g., AMM [11], SAMI [15]). These tools might be useful for discovering problems in their projects, as they *prescribe* sets of guidelines related to the proper usage of practices in agile projects.

One of such tools is the approach to assessing the maturity of Requirements Engineering practices in agile projects (REMMA). It is a unique approach. It does not only support the improvement of RE by indicating some practices that need to be tweaked but also it allows incorporating information about how the specific context of a project affects the usage of agile RE practices. Thus, REMMA is a hybrid method in the sense that it combines elements characteristic to both prescriptive and inductive approaches to process improvement. REMMA has been proposed by Ochodek and Kopczyńska in [10]. Although it has been developed using Design Science Research, so its components resulted from several empirical studies, it lacks empirical validation. Therefore, in the paper we present a study with which we would like to fill this gap.

We carried out an exploratory case study in a software house context to understand to what extent the method satisfies its requirements: is perceived as useful (i.e., provides results that correctly reflect the current state of implementation of RE practices in a project), easy to use (can be used by project team), and cost-effective (project team can afford using the method).

2 Case Study

We conducted an exploratory *case study* to *characterize* and *understand* an application of REMMA to assess the alignment of practices within the context of an organization simultaneously running several agile software development projects using the guidelines by Runeson et al. [13].

To formulate the research questions we referred to the theoretical framework of the Technology Acceptance Model (TAM) [17], which was used while creating REMMA to define quality attributes that the method should exhibit. The following determinants of *Perceived usefulness* (PU) and *perceived ease of use*

(PEOU) were selected—job relevance, output quality, and result demonstrability, and self-efficacy (an individual's belief that he or she can perform the maturity assessment in his or her project), and perception of external control, which is the degree to which an individual believes that organizational and technical resources exist to support the use of the artifact (e.g., management support). Finally, the authors wanted the method to be cost-effective concerning the effort required to perform the assessment. Thus, the following research questions were formulated:

- **RQ1** What is the cost of performing the assessment?
- **RQ2** How easy to use is the method?
- **RQ3** How useful are the results of the assessment?

Case and Subject Selection. In our study, we decided to look for an organization that could provide us with some '*typical cases*' [19]. We decided to conduct the study in a software house (we refer to it as Company) that at the moment of conducting the study employed nearly 100 Python programmers, which made it one of the largest software development houses in Central Europe. The company ran approximately seven concurrent projects. It had three locations in Poland. One of their main goals is continuously improving and working towards becoming a truly agile organization.

We were allowed to carry out the case study in three of the company's project teams. We were also able to talk to a person that had worked at the company for 3 months as a Scrum Master (Coach) with the goal to improve their processes (we will refer to this person as Agile Coach). This gave us the chance to investigate the RE practices from multiple perspectives (Agile Coach, project team members) and triangulate the observations made during the case study. The Agile Coach had cross-sectional knowledge of all the projects and was no longer employed at the time. This was important for us and helped reduce potential biases. Prior to the case study, we had asked the Agile Coach to select the projects (cases) to analyze. He chose three projects, A, B, and C that he thought corresponded to outstanding, good and poor projects with respect to how agile practices were implemented at the time he was working for the company.

All projects developed web-based applications using Scrum for small- or medium-sized organizations that were located in different countries. We describe the projects in more detail in Table 1, to the extent that we are allowed.

Data Collection and Analysis Procedure. We decided to use semi-structured interviews to collect data during the study. The interview guide for the study was prepared by one of the researchers (the primary interviewer), based on research questions. As a following step, the guide was reviewed by the secondary interviewer. We additionally decided to extend the model by adding more context factors than those presented in the detailed paper on REMMA [10]. Based on our experience we defined the following factors: adherence to Scrum, use of XP, team location, team communication type, type of budget, and staff retention. Finally, we decided to test the prepared instrumentation during a pilot study. The primary interviewer conducted an interview session with the secondary interviewer as an interviewee. The pilot interview pertained to the three software projects that the secondary interviewer had participated in during the previous year.

After incorporating all the suggestions and remarks from the pilot study, we proceeded to the actual interview sessions. We began by interviewing the Agile Coach in two sessions. Afterward, we met with the representatives of each project team. We organized separate sessions for each project team. Each session was conducted in Polish, and was constituted by five phases:

1. *General Questions*—the goal was to better understand the company and project context and to inquire about the interviewees' experience;
2. *Assessment*—the primary interviewer facilitated the process of interviewees collectively filling in the assessment form about their project using a prototype version of a software tool supporting the REMMA assessment;
3. *Presentation of the method*—both interviewers explained the mechanism of practice alignment assessment with REMMA by giving a small presentation;
4. *Presentation of results*—both interviewers discussed the assessment results with the interviewees. Then the interviewers presented results of the assessment performed by the Agile Coach and asked interviewees to discuss the commonalities and differences between these two assessments;
5. *Questions about REMMA and Closing up*—the interviewers asked about the interviewees' impressions of using REMMA and closed up the interview.

Before the sessions, we obtained support for the study from the CEO. During the sessions, we observed that the participants were committed, sometimes even enthusiastic. Both interviewers were present during all sessions. The primary interviewer focused on asking questions while the secondary interviewer tried to reflect and ask follow-up questions. All recordings from the sessions were transcribed by the primary interviewer (7.5 h of the audio recordings) and reviewed by the secondary interviewer. Both interviewers analyzed the collected data according to the guidelines of Charmaz [5]. First, initial coding was carried out incident-by-incident, individually by each researcher. Then in vivo codes underwent constant comparison, and finally axial coding was performed by both reviewers working together. At the end, we discussed the results. The part of our coding scheme that we perceive as relevant to answering the research questions is presented in Fig. 1. To make the codes more understandable to the reader, we substituted their names with exemplary quotations they tagged. We also changed the names of categories to questions. Finally, we used font weight to reflect the number of projects in which the underlying code existed.

3 Results

Cost of Performing Assessment (RQ1). The Agile Coach needed 40 min to assess all three projects while projects' teams assessed their projects in 20 min (C), 28 min (B), and 38 min (A). All interviewees stated that from their perspective using REMMA did not require much effort. Moreover, they regarded it as an acceptable investment for their current projects. They also claimed that they could afford to perform this assessment regularly, e.g., every sprint (Fig. 1).

Table 1. Description of the projects and interviewees presented according to the guidelines of Petersen and Wohlin [12].

	A	B	C
Product	The web-application has been developed by Company for 0.8 years. It is a new version of an existing customer-tailored system used by more than 20 people. The total team size is more than 20FTE**.	The product has been developed by the company for 1.5 years. It is a web application used by users worldwide, available online. Team B (10FTE) have developed the product from its beginning and has been the only team working on new features for the product since then.	Team C (6FTE) developed a product-line-type web application, customized for each customer. The team took over the development from a previous team that had worked on it for several years.
	Company cooperates based on nearshore and offshore outsourcing contract relationships with all clients. Products are developed mainly in Python.		
Process	Scrum	Scrum, plus some elements of TDD	introducing Scrum
	Company is responsible only for development and alpha testing. Sprints take 2 weeks(A,B), 2-3 weeks(C).		
Practices, Tools, and Techniques	There are no specific tools and techniques used, only tools supporting development and project management.		
People (from the company)	7FTE including devs, testers, Scrum Master, Proxy Product Owner***	15FTE including devs, testers, Scrum Master, Proxy Product Owner	7FTE including devs, testers, Proxy Product Owner
Organization Size* & Location	medium-sized other country, same timezone	small other country, different timezone (> 5h)	medium-sized other country, different timezone (> 5h)
Market	One customer (the customer will use the product and sell it to multiple other customers)	One customer (a publicly available application)	One customer (the customer will use the product and sell it to multiple other customers)
Customer (domain)	Marketing	Social Media	Medicine
Interviewees (#)	2	2	1
Roles	(1) Proxy Product Owner, Test Engineer (2) Scrum Master	(1) Proxy Product Owner, Test Engineer (2) Scrum Master, Tech Lead	Proxy Product Owner, Developer
Experience in IT projects and in agile projects [years]	(1) 2, 1 (2) 6, 1	(1) 1.5, 1.5 (2) 1.5, 2	3, 2

*) *According to EU recommendation 2003/361/EC* [16] **) *FTE = Full-Time Employee;* ***) *Proxy Product Owner = a person who works on-site during working hours of the software development team on behalf of the actual product owner. He or she is a single point of contact with the product owner (PO), and can autonomously make certain limited decisions regarding requirements. As a result, to some extent, this compensates for the customer's low availability to the team.*

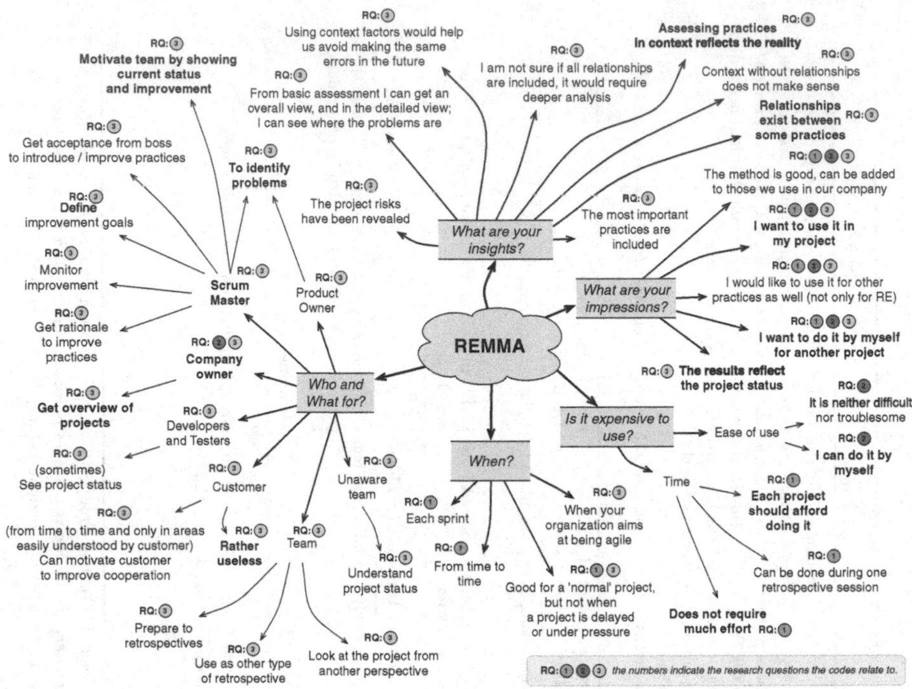

Fig. 1. Interviewees opinions on REMMA. The size and emphasis reflects the number of projects in which the statement was made.

REMMA was also indicated as well suited for retrospectives. Conversely, one person asserted that he would not use REMMA when there are multiple problems in the project as *"investing in anything other than source code development activities would be treated as a waste of resources"*. Taking into consideration the quantitative data related to the effort required to perform assessments, and the positive feedback of team members, we might come to the conclusion that the method seems affordable for an agile project, which is in a relatively favorable situation. Nevertheless, using REMMA might not be the best option if a project is in a dramatic situation.

Perceived Ease of Use of the Method (RQ2). (1) Self-efficacy. At first, we focus on identifying possible problems with understanding REMMA and any spontaneous questions asked during the interview. There was a question regarding the interpretation of the practice assessment scale and the rationale behind it, and about the source of the practices. There appeared ten questions concerning several terms used in the names of practices, e.g., "*what is an elevator test?*". These required the interviewers to read aloud the description of the ambiguous practice from the REMMA assessment form. The remaining practices and context factors seemed self-explanatory. As the Fig. 1 shows, the majority of interviewees claimed that they would be able to conduct the assessment by themselves. The Agile Coach concluded the interview by stating that *"from my perspective, it is a super tool for the Scrum Master, who can perform the assessment by himself without any great effort"*. **(2) Perception of external control.** Taking into account the previously discussed cost-efficiency of REMMA, we can state that participants considered using the method in their projects. In addition, they suggested that the management might be interested in using the method to "*get an overview of the projects.*"

Usefulness of REMMA (RQ3). Output quality. *Project A*: The results of both the team members' and Agile Coach's assessments showed that all critical and the majority of important practices were performed in this project at the de facto standard level in basic assessment. There were only two important practices identified as never used. Both assessments indicated that there are certain important practices to be improved, and additional practices to be introduced, in the *Knowledge sharing* area. Project A was believed to be the best Agile project in the company concerning its process quality. The team members seemed very proud of their success in applying Scrum "*everybody says that we do it by the book and even better*". *Project B*: The Agile Coach assessed the project as middling in applying the Agile principles. Around 80% of critical and ca. 75% of important practices were at least normally used. Two critical, one important, and two additional practices were performed at discretionary-use level. Moreover, two important and three additional practices were never used. There were also five negative influencers (practices insufficiently used) and one practice unfit for the context. The Agile Coach triggered some positive changes during his presence in the company. When we asked to comment on the differences between their assessment and the one of the Agile Coach, the team members easily identified the improvement steps that led them to the up-to-date status. *Project C*: It had the highest number of practices that required improvement. This state was correctly reflected in the results of the assessment—the project obtained the lowest score compared to the others. Only around 20% and 36% of practices were assessed to be de facto standard or normally used by team members and the Agile Coach respectively. In both assessments, there were 8–9 practices that negatively impacted the usage of other practices and 8 practices unfit for the context. Both assessments also indicated that *Customer Involvement, Knowledge sharing* areas had the lowest result. The assessment of the *Planning* area by the Agile Coach showed that the practices were at least normally used. While discussing the Agile Coach's assessment with the team, they confirmed that the Agile Coach had tried to introduce them, but they were rejected soon after.

We can state that all the interviews approved the results of assessment provided by the method. They also confirmed the appropriateness of each major increase or decrease in influence and contextual assessments. Moreover, we analyzed the codes developed during the analysis of transcripts; we assigned them into categories of the project's problems, strong points, and context factors. We wanted to find out if they were mirrored in the assessments. We observed that 100% of problems and strong points, described by interviewees, were reflected in the assessments results. In addition, we obtained examples of practice implementation for 75% of cases.

Result Demonstrability. Some of the remarks from interviewees concerned the usability of our prototype tool used to support the assessments. Most of them were related to graphical details, such as colors, or layout. We also received comments regarding different forms of presenting the results. Interviewees suggested that they should be provided in a form supporting a top-down approach to analysis. They stated that, first of all, that the method should be able to provide an overall result of assessment—accurately expressed by a single number. For instance, in REMMA the overall assessment is provided in the form of TL and PIP measures, or a chart summarizing the results of the basic assessment. The interviewees supported this idea with the argument that some people may not be interested in the details of assessment. Second, team members might want to receive a quick report on the current status of their project, without needing to analyze the results. On the other hand, agile coaches and team members who are working on improvement may like to receive more detailed reports. Such reports should help them identify potential problems in their project.

Job Relevance. All of the interviewees stated that they were interested in using REMMA in their projects. Moreover, two interviewees were eager to use it in another project right after the interview. During the sessions, we tried to identify which project roles could potentially benefit from applying REMMA in certain situations (see Fig. 1). The participants indicated that Scrum Master would be the role that would use the tool most extensively for supporting the whole improvement process (identification of problems, defining goals, monitoring the process). However, they also believed that a company owner (management) would consider the tool valuable, as it provides comprehensive information about project status. Overall, REMMA was recognized by project stakeholders as an appropriate tool to make the development team more aware of the current status of a project or to support retrospectives. According to the interviewees, the teams that are unaware of how agile practices shall be implemented may find REMMA particularly valuable.

4 Conclusions

To validate the approach to assessing the maturity of Requirements Engineering practices in agile projects (REMMA) proposed method we conducted an exploratory case study in one of the biggest software houses in Central Europe. The results of the study made us expect that the proposed approach might be useful for agile development teams to identify strengths and weaknesses of RE and provoke improvement. Besides, according to the study participants the method seems cost-effective and simple enough to be regularly used in agile software projects.

Finally, we are aware of the limitations of the case study as a validation method, but we believe that the promising results might provoke further implementation and evaluation of the method by practitioners in various organizations. Moreover, we hope that the paper will facilitate the discussion about the role of project context in the assessment of the agile-projects maturity.

Acknowledgements. We thank the employees of Company for the participation in the study. We especially thank Maciej Dziergwa, Oliwia Gogolewska, Jakub Jurkiewicz, Sebastian Kalinowski, Michał Kwiatkowski, Klaudia Prasek, and Dariusz Śmigiel.

References

1. Beck, K., Andres, C.: Extreme Programming Explained: Embrace Change. Addison-Wesley Professional, Boston (2000)
2. Beck, K., et al.: The Agile Manifesto. http://agilemanifesto.org. Accessed 28 Aug 2015
3. Brodman, J.G., Johnson, D.L.: Return on Investment (ROI) from software process improvement as measured by US industry. Softw. Process: Improv. Pract. **1**(1), 35–47 (1995)
4. Charette, R.N.: Why software fails. IEEE Spectr. **42**(9), 36 (2005)
5. Charmaz, K.: Constructing Grounded Theory. SAGE Publications, Thousand Oaks (2006)
6. Damian, D., Zowghi, D., Vaidyanathasamy, L., Pal, Y.: An industrial case study of immediate benefits of requirements engineering process improvement at the Australian center for Unisys software. Empir. Softw. Eng. **9**, 45–75 (2004)
7. Elshandidy, H., Mazen, S.: Agile and traditional requirements engineering: a survey. Int. J. Sci. Eng. Res. **4**(9), 473–482 (2013)
8. Kappelman, L.A., McKeeman, R., Zhang, L.: Early warning signs of it project failure: the dominant dozen. Inf. Syst. Manag. **23**(4), 31–36 (2006)
9. May, L.: Major causes of software project failures. CrossTalk-J. Defense Softw. Eng. **11**(7), 9–12 (1998)
10. Ochodek, M., Kopczyńska, S., Nawrocki, J.: A hybrid approach to assessing the maturity of Requirements Engineering practices in agile projects (REMMA). http://remma.cs.put.poznan.pl/about
11. Patel, C., Ramachandran, M.: Agile maturity model (AMM): a software process improvement framework for agile software development practices. Int. J. Softw. Eng. IJSE **2**(1), 3–28 (2009)
12. Petersen, K., Wohlin, C.: Context in industrial software engineering research. In: Proceedings of ESEM, pp. 401–404. IEEE (2009)
13. Runeson, P., Host, M., Rainer, A., Regnell, B.: Case Study Research in Software Engineering: Guidelines and Examples. Wiley, Hokoben (2012)
14. Schwaber, K., Sutherland, J.: The Scrum Guide^{TM}. The Definitive Guide to Scrum: The Rules of the Game. Scrum.org (2013)
15. Sidky, A.: A structured approach to adopting agile practices: the agile adoption framework. Ph.D. thesis, Virginia Polytechnic Institute and State University (2007)
16. The Commission of the European Communities: Commission Recommendation of 6 May 2003 concerning the definition of micro, small and medium-sized enterprises (2003/361/EC)

17. Venkatesh, V., Bala, H.: Technology acceptance model 3 and a research agenda on interventions. Decis. Sci. **39**(2), 273–315 (2008)
18. Verner, J., Cox, K., Bleistein, S., Cerpa, N.: Requirements engineering and software project success: an industrial survey in Australia and the US. Australas. J. Inf. Syst. **13**(1), 1–14 (2005)
19. Yin, R.: Case Study Research: Design and Methods. SAGE Publications, Thousand Oaks (2003)

Does Live Regression Testing Help?

Marek Bruchatý and Karol Rástočný(✉)

Institute of Informatics, Information Systems and Software Engineering, Faculty of Informatics and Information Technologies, Slovak University of Technology in Bratislava, Ilkovičova 2, Bratislava, Slovakia
{xbruchaty,karol.rastocny}@stuba.sk

Abstract. Regression testing is an expensive, yet crucial part of the software development process. As regression test suites grow in size, the time required for their execution increases proportionally, and their execution is often either delegated to a specialized testing environment out of developers reach, or they are omitted completely. This could have a variety of negative effects on the developers' productivity, including interruptions and slowdown of developers' workflow. We propose a method of live regression unit testing to address these issues via incorporating Regression Test Selection and Test Case Prioritization techniques and an automatized change detection mechanism to run the regression testing in the background automatically. By combining the test results with source code changes and code coverage information, we are able to precisely identify source code changes responsible for test failures. By the paired two-sample t-test we proved, that our method is able to increase the speed of fault detection and to fix changes responsible for incorrect behaviour almost 2 times (p-value = 0.001, $\alpha = 0.05$).

Keywords: Regression testing · Regression test selection · Test case prioritization

1 Introduction and Related Work

Frequent re-execution of test suites with large numbers of test cases can be very resource-intensive, mainly due to the testing time requirements. Regression test selection (RTS) and Test case prioritization (TCP) are two of the most common practices of regression test optimization (RTO) addressing these issues. Although extensive research exists regarding both optimization practises, their application to near real-time environment was not sufficiently examined. A method of automatic and optimized regression testing designed for individual development environments can have a profound positive effect on the effectiveness of individual developers regarding source code development.

Previous research conducted on both of RTS and TCP practices resulted in a variety of new techniques addressing the issues of applying selection and optimization to a wide variety of programming languages and different types of software projects [7]. As described in the work of Biswas et al. [3], RTS

A. Chatzigeorgiou et al. (Eds.): SOFSEM 2020, LNCS 12011, pp. 699–707, 2020.
https://doi.org/10.1007/978-3-030-38919-2_59

techniques can be categorized by the type of programs they are intended for, mainly divided on the basis of the programming paradigm these programs are created in. The Object Oriented (OO) approach introduces several important concepts fundamental for the OO programming paradigm, e.g. polymorphism. These fundamental concepts directly affect the design of methods used for RTS. One of often used techniques are Firewall-based techniques that are based on the concept of a firewall, originally introduced by Leung and White [5]. Firewall-based techniques use an abstract firewall placed around certain parts of the program, that contain code modifications. Tests, that exercise at least some code from the parts of the program enclosed by the firewall are then added to the test suite for rerun. Firewall techniques are applicable on different testing levels, e.g. unit tests or integration tests.

As described by Rothermel et al. [9], TCP is a problem of finding the best possible prioritization/reordering of a test suite, such that a selected awarding function always yields more preferable award value to tests placed sooner in the ordered test suite. Many goals of prioritization can be defined [9], e.g. reveal faults earlier in the test suite, accelerate the code coverage of the system. As described by Rothermel et al. [9] there is a strict distinction between TCP intended for the initial testing of software and regression testing of software. In case of regressional TCP, information gathered from previous runs are used to prioritize the subsequent reruns. Therefore at least one successful test suite run need to be performed in order for this technique to work. Rothermel et al. [9] also defines two varieties of test case prioritization: general TCP and version-specific TCP. General TCP is aimed at defining a test case order that will be used over a sequence of subsequent versions of the program [4,9]. General TCP can be therefore performed for any release of the program. Version-specific TCP is aimed on finding a test case order for a specific version of the program [4,9].

Great insight to the effectiveness and possibilities of combining RTS and TCP techniques can be acquired from the work of Beszédes et al. [2]. They were able to reduce the average test case selection size to below 10% of the original size, while still preserving about 50% of test suite inclusiveness [2].

Despite the fact, that hybrid optimization techniques are promoted as a viable method for further optimization in papers published on the subject of RTS and TCP [1], none of these works provide any usable information about their effectiveness in a real production environment and on wider variety of software projects. To the best of our knowledge, no available research suggests that a hybrid approach can be successfully used in a near real-time environment. Exemplary implementation of the near real-time testing is Live Unit Testing[1] in Microsoft Visual Studio. Although no empirical evidence confirms nor disproves the effectiveness of hybrid RTS in a near real-time environment. For this reason we propose a hybrid RTO technique and evaluate effectiveness of developers in their change implementation tasks.

[1] https://docs.microsoft.com/en-us/visualstudio/test/live-unit-testing.

2 Proposed Method

We propose the RTO technique, that targets the problem of regression testing inefficiency during development. Our aim is to optimize the process of regression testing performed by individual developers during the development of a source code and adjacent test cases. This type of regression testing is a frequently occurring process, that is routinely performed by developers as a form of prevention from introducing software bugs to the modified system [6]. It provides insurance for the developers, that changes introduced to the modified system are not causing any problems and that they coincide with existing functional requirements. Successfully satisfied functional requirements outlined by a test suite can be used by the author of the changes as a form of validation for the modified source code. Based on the outcomes of a regression testing process, changes compliant with all necessary requirements can be then marked as completed. Completed source code changes are changes that are verified for not having any harmful effects on the original system.

The proposed method incorporating the techniques of RTS and TCP to form one unified continuous testing platform, that can be integrated into an integrated development environment (IDE). The proposed platform also oversees the execution of test cases and collection of testing data. Our platform is intended to work in the background, without the need of any oversight from the developer. Minimization of unnecessary interactions between the user and the testing platform can be achieved by test automation. The proposed method uses changes of the source code as triggers, designed to initiate specific processes of the testing platform. Based on the severity of detected changes, different steps including multiple optimization techniques can be applied. Once the final test suite is ready to run, our proposed testing tool executes all prepared test cases and gather the execution data. This entire process is initiated by the developer indirectly, only as a side effect of applying changes to the source code. The use of triggers assumes, that the regression testing will take place every time the developer introduces a modification to the source code. Our method proposes a form of real-time test execution where the source code is tested while it is being developed. By creating this method, we want to accomplish a level of near real-time testing, to be able to use the gathered test execution information to provide a quick feedback for the developer. By building our method on top of an IDE, we are able to utilize the provided features and use them to notify the developer about detected faults immediately after they are detected. By informing the developer about detected fault that were introduced to the system with recent changes, we suppose that we would be able to help the developer to identify failing tests and the defective changes more effectively.

The proposed method utilizes following workflow:

1. A developer performs a change of the original source code;
2. The testing platform detects the newly applied change to the source code;
3. The testing platform collects data about the changed source code and effected test cases;

4. The testing platform executes an optimized regression testing and collects testing results;
5. The testing platform alerts the developer about its findings if necessary (if test failure/s occur);
6. The developer uses the alert/s to quickly localize the defective change/s of the source code responsible for possible test failure/s.

2.1 Regression Test Selection Method

Our RTS method uses a change-based test selection approach that uses code coverage information to determine the areas of the system affected by the modifications of the source code. Due to the fact, that our method is prone to encounter possible performance issues caused by the amount of computation required for its reliable work we decided, that the high level of inclusiveness that can be achieved with advanced test selection techniques is inferior to the performance and high efficiency that rises from the use of more minimalistic selection techniques like the change-based approach.

Our method uses the code coverage information obtained from the available test suites to identify units of the source code affected by recent changes. The coverage information collected at the procedural level, i.e. on level of methods, is stored in a coverage database. The coverage database stores the relations between test cases and code units as well as records about the outcomes of previously performed test runs, i.e. recent test failures. By using the coverage information to identify all effected code units, we are able to determine which tests are necessary to add to the test suite prepared for the upcoming test run. If combined with previously failed test cases, we are capable of assembling a test suite with a high level of inclusiveness while maintaining a smaller test suite size. Our selection method is based on a method applied by Beszédes et al. [2], who have achieved positive results on which are our estimates for inclusiveness based on. This test selection method offers a good foundation for test optimization on which test case prioritization can be applied.

2.2 Test Case Prioritization Method

Our method of TCP acts as a second layer of RTO, following the RTS process. The output of the RTS process – a reduced regression test suite is supplied as an input to the TCP process. The TCP process then executes a set of sub-processes that collectively produces a version of the test suite designed to fail fast if a newly introduced source code bug is detected.

During the first two steps of this method focused on information acquisition, code coverage information supplied by a chosen testing framework is used to identify the coverage of individual test cases. At this stage, a set of important test cases is already known from the previous step of RTS. The coverage information is thus used with additional information collected from previous test suite runs to evaluate each test case in terms of its significance with regards to source code changes. Previously gathered information includes the defect frequency, i.e. the number of previously unsuccessful test runs for a particular test case, and the time of the last test run for a particular test case. An awarding value from the interval [0; 1] is then computed and assigned to all of the considered test cases (higher value = higher probability of test failure). Test cases within the test suite are then reordered in a descending order based on the awarding value they have had received. This ensures, that the test suite will be executed in an order ranging from a test case most susceptible to failure to the least one.

3 Evaluation

The aim of our experimental evaluation is to find out if the proposed method increases the developers' efficiency of finding source code changes responsible for test failures. Higher efficiency means, that the developer spends less time on finding the failing test cases, as well as the actual changes of the source code responsible for the failures.

3.1 Experiment Design

This experiment relies on the implementation of the proposed method as a working prototype[2], on which we are able to conduct necessary experiments. For this purpose, we have decided to use the IntelliJ IDEA platform to create a IDE plugin[3] able to provide all necessary features based upon the proposed method. This choice also enables the subjects (IDE users) to interact with the prototype directly within the IDE environment.

Goal. We suppose that by incorporating the prototype to the IDE, the developer would be able to efficiently detect changes responsible for test failures and fix them in shorter period of time compared to a situation where the prototype is not available.

Procedure. We propose an experiment in which we test each test subject two times using two different software projects. Two sets of three tasks are created, one set for each of the two projects. These tasks contain instructions for the participants that instructs them to modify the existing source code of a specific software project in a certain way. One task is purposely designed so that it contradicts with precisely one test case from the existing test suite. Source code changes required by this task will therefore produce a failed test case when running the available regression test suite. A participant is required to fulfill all three of the tasks defined for each project and confirm their compliance with the existing regression test suite. The existing test suite is marked as the source of truth and therefore every task that contradicts with it should be ignored. If such conflict occurs, the participant is required to find the failed test case and

[2] https://github.com/marekbruchaty/livetest.
[3] https://www.youtube.com/watch?v=PFdGbQaFUPk.

then locate the source code changes responsible for the test case failure. Once the participant is aware of the source code changes responsible for the test case failure, he/she is required to revert all changes responsible for the failure and retest the source code again to confirm its compliance with the test suite. If the changes carried out do not result in a test case failure, the participant can continue to the next task. A task ignored by the participant due to it being in conflict with the existing test suite is also considered as successfully completed.

The experiment consists of two independent measurements conducted on each of the participants, one for each of the two software projects. The key difference between these measurement is, that the participant carries out one measurement without the use of the proposed IDE plugin and need to run the regression test suite manually, whereas during the other measurement the testing plugin is enabled, and the participant can interact with it. Once the participant finishes with the first test, he/she is tested for the second time using the second configuration. To further reduce the transfer of knowledge between the two measurements, we reverse the order in which the two measurements are conducted for the second half of the subjects.

Different projects are used to prevent the participants to gain an advantage by learning the source code between the two measurements. Both projects have source code with similar complexity and comparable test suites.

Screen recording is used to record the behaviour of all participants during the testing process and to extract individual time intervals critical for the experimental evaluation. These individual time intervals are recorded for the evaluation of the experiment:

- Time used to determine the state of the test suite, measured from the time of the last executed modification to the time when the state of the sets suite is known;
- Time used to find the failing test case, measured from the time when the test suite state is determined to the time when the failing test case is discovered;
- Time used to fix the failing test suite, measured from the discovery of the failing test case to the time when the test suite does not contain any other failed test cases.

Participants. The participants of this experiment were chosen from the ranks of developers, with various levels of expertise with python programming language. Since it is hard to obtain a large test group of this type, we decided to perform the experiment on a smaller test group of 10 developers. While it is necessary to include different expertise levels in the experiment since this method is intended for use in any common development scenario, our subject group is mainly composed of junior or medium-experienced developers. All developers chosen for this experiment have a good knowledge of the IDE used during the experiment and are familiar at least with the basic knowledge required to navigate the environment, to use the editor, and to run test cases. Statistical evaluation method appropriate for the composition and the size of the sample group was chosen accordingly.

Data Analysis. We have designed the experiment to produce an output containing two sets of sample values, one set for each tested source code. Samples from the first set are paired with the values from the second set, based on the repeated measurement carried out by the same test subject.

The paired two-sample t-test, also known as the dependent samples t-test was chosen as a statistical test for our hypotheses. It is used to objectively evaluate the data obtained from the experiment and determine, if the differences between the two sets of measurements are statistically significant. The paired two-sample t-test operates on two sets of input data, as the name suggests. It assumes, that a relationship exists between two particular sample values, measured separately in each set. Our evaluation method repeats the same experiment with a particular test subject repeatedly in a slightly different configuration. Our data is therefore paired based on the tested subject and is ideal for this type of statistical test. The use of the paired t-test gives us an advantage over the use of unpaired t-test because the use of paired observations reduces the intersubject variability caused by a variety of differences between the tested subjects. The paired design therefore tends to increase the signal-to-noise ration that directly determines the statistical significance. Therefore, paired two-sample t-test can be theoretically marked as the more powerful compared to the unpaired t-test, capable to identify a significant difference between measurements using less resources, if one exists [8]. This design allows us to test our hypotheses on a relatively low subject group with 10 participants.

Based on our subject analysis, we have formulated 5 alternative hypotheses:

- H_1: The time needed to find the state of the test suite after applying source code changes is shorter when using the proposed method.
- H_2: The time needed to find the test case responsible for test suite failure is shorter using the proposed method.
- H_3: The time needed to fix the source code and retest it after discovering the failed test case is shorter when using the proposed method.
- H_4: The time needed to fix the source code and retest it after finding the state of the test suite is shorter when using the proposed method.
- H_5: The time needed to fix the source code and retest it after applying source code change is shorter when using the proposed method.

3.2 Evaluation Results

We measured time intervals for each one of our hypotheses. Each of these time intervals contains two sets of paired observations, one pair for each of the test subjects – observations with and without the use of the Livetest plugin. We have analysed all sets of measured time intervals by the Shapiro-Wilk test and we found all sets approximately normally distributed within their observation groups.

We have computed the two-sample paired t-tests for each of the paired groups of observations associated with each of the proposed alternative hypotheses. All t values computed using the paired t-test fell into the expected critical region predicted by each of the alternative hypotheses. All p values are therefore smaller than the significance level defined by the chosen α (0.05) value, see Table 1. We are therefore able to reject the null hypotheses for all of the alternative hypotheses and assume, that the use of our method improves the developers' effectiveness for each of the specified intervals.

Table 1. Two sample t-test results

Hypothesis	Difference	t (observed)	$\|t\|$ (critical)	df	p value	α
H_1	11.550	8.697	2.262	9	<0.0001	0.05
H_2	2.800	3.579	2.262	9	0.006	0.05
H_3	9.700	3.750	2.262	9	0.005	0.05
H_4	15.600	3.612	2.262	9	0.006	0.05
H_5	28.400	5.059	2.262	9	0.001	0.05

4 Findings and Discussion

The most substantial differences were observed regarding the hypotheses H_1 and H_3, as well as the hypothesis H_5. The mean difference between the two sets of observations measured for the H_1 hypothesis is the largest among the performed tests. Introduction of the Livetest plugin to the developers' workflow decreased the variance in the time needed by the developer to identify the state of the test suite after modifying the source code. We attribute this result to the fact, that the developer is not required to manually run the appropriate test suite and the testing process is initiated automatically after a source code change is detected. The test suite rerun is also initiated faster by the Livetest plugin as if initiated manually by the developer. The test subjects were often reviewing the source code changes before they approached the rerun of the test suite. The Livetest plugin however executes the appropriate test cases right after the developer finishes modifying the source code, regardless of the fact that he/she is still reviewing the changes or not. This greatly increases the test suite state discoverability. The conducted experiments have used only a small test suite with small number of test cases that can be rerun in a fraction of a second. The effects and the differences in means of the two sets of observations can increase if the test suite grows in size.

The observations related to the H_3 hypothesis also show a difference between the means of the observations. We attribute these results to the easily discoverable state of the test suite achieved by using the gutter icons and the detailed popup messages containing the effected test cases and their state as well as the highlighting of the source code changes responsible for the test suite failure.

While the measurements associated with the H_2 hypothesis show slightly less dominant difference between the means of the observations, the observations related to the hypotheses H_1 and H_3 outbalance the observations for the hypothesis H_2 and therefore results in a substantial difference between the means of the observations associated with the hypothesis H_5. The overall developers' effectiveness in regard to the implementation of source code changes and reparation of possible test suite failures is therefore significantly higher.

The overall reduction of the variance of the observed time intervals when using the Livetest plugin indicates, that the plugin successfully helped to guide the developer through the testing process and eliminated redundant actions the developer. This time can be further decreased by refining the test suite optimization process as well as the design of the software prototype.

Additional experiments need to be performed to correctly identify all the impacts and possible areas of improvements for the proposed method. The evaluation on a larger test group with greater diversity of experience with unit and regression testing is one of the most important improvements. Secondly, experiments with a large software project containing several thousand unit test should also provide valuable information regarding the true effectiveness of the proposed method.

Acknowledgements. This work was partially supported by the Slovak Research and Development Agency under the contract No. APVV-15-0508, and by the Scientific Grant Agency of the Slovak Republic, grant No. VG 1/0759/19.

References

1. Ansari, A., Khan, A., Khan, A., Mukadam, K.: Optimized regression test using test case prioritization. Procedia Comput. Sci. **79**, 152–160 (2016). Proceedings of Int, p. 2016. Conf. on Communication, Computing and Virtualization (ICCCV)
2. Beszédes, A., Gergely, T., Schrettner, L., Jász, J., Langó, L., Gyimóthy, T.: Code coverage-based regression test selection and prioritization in Webkit. In: 2012 28th IEEE International Conference on SW Maintenance (ICSM), pp. 46–55 (2012)
3. Biswas, S., Mall, R., Satpathy, M., Sukumaran, S.: Regression test selection techniques: a survey. Informatica **35**, 289–321 (2011)
4. Elbaum, S., Malishevsky, A.G., Rothermel, G.: Test case prioritization: a family of empirical studies. IEEE Trans. SW Eng. **28**(2), 159–182 (2002)
5. Elbaum, S., Rothermel, G., Penix, J.: Techniques for improving regression testing in continuous integration development environments. In: Proceedings of the 22nd ACM SIGSOFT International Symposium on Foundations of SW Engineering, FSE 2014, pp. 235–245. ACM, New York (2014)
6. Kandil, P., Moussa, S., Badr, N.: Cluster-based test cases prioritization and selection technique for agile regression testing. J. Softw. Evol. Process **29**(6), 19 (2017)
7. Kazmi, R., Jawawi, D.N.A., Mohamad, R., Ghani, I.: Effective regression test case selection: a systematic literature review. ACM Comput. Surv. **50**(2), 29:1–29:32 (2017)
8. NCSS, LLC: Paired t-test (2019). https://ncss-wpengine.netdna-ssl.com/wp-content/themes/ncss/pdf/Procedures/NCSS/Paired_T-Test.pdf. Accessed 27 Aug 2019
9. Rothermel, G., Untch, R.H., Chu, C., Harrold, M.J.: Prioritizing test cases for regression testing. IEEE Trans. SW Eng. **27**(10), 929–948 (2001)

Foundations of Algorithmic Computational Biology – Short Paper

Dense Subgraphs in Biological Networks

Mohammad Mehdi Hosseinzadeh(✉)

Università degli Studi di Bergamo, Bergamo, Italy
m.hosseinzadeh@unibg.it

Abstract. A fundamental problem in analysing biological networks is the identification of dense subgraphs, since they are considered to be related to relevant parts of networks, like communities. Many contributions have been focused mainly in computing a single dense subgraph, but in many applications we are interested in finding a set of dense, possibly overlapping, subgraphs. In this paper we consider the Top-k-Overlapping Densest Subgraphs problem, that aims at finding a set of k dense subgraphs, for some integer $k \geq 1$, that maximize an objective function that consists of the density of the subgraphs and the distance among them. We design a new heuristic for the Top-k-Overlapping Densest Subgraphs and we present an experimental analysis that compares our heuristic with an approximation algorithm developed for Top-k-Overlapping Densest Subgraphs (called DOS) on biological networks. The experimental result shows that our heuristic provides solutions that are denser than those computed by DOS, while the solutions computed by DOS have a greater distance. As for time-complexity, the DOS algorithm is much faster than our method.

Keywords: Biological networks · Graph algorithms · Heuristics · Dense subgraph

1 Introduction

Analyzing biological interactions is an important task to study complex biological data, which are usually represented as a network. Network mining is a fundamental method that helps us in understanding biological networks and their underlying properties.

One of the most studied problems in this area is the identification of dense subgraphs [8]. Identification of dense subgraphs in a network enables for example to obtain a better understanding of the processes organization in a biological system. For instance, in protein-protein interaction networks, finding dense subgraphs gives a comprehensive view of protein interaction patterns, thus providing information to understand biological processes at various resolutions [15]. Molecular processes require usually the interaction of many proteins. Proteins densely interact together, forming large molecular machines. Identification of

This paper was supported by STARS Supporting Talented Research.

A. Chatzigeorgiou et al. (Eds.): SOFSEM 2020, LNCS 12011, pp. 711–719, 2020.
https://doi.org/10.1007/978-3-030-38919-2_60

such densely interconnected groups in protein-protein interaction is crucial to understand and explore the organization of biological networks [15].

Most of the contributions focused on the identification of a subgraph of interest, like a clique [12] or a relaxed clique (for example an s-club, a t-clique, a k-core, and an s-plex) [1,13,16]. Several of these definitions lead to optimization problems that are NP-hard (for example finding a clique of at least a given cardinality [12]). The dense subgraph problem, which asks for a subgraph of maximum average-degree density, is instead polynomial-time solvable [11] and it can be approximated within factor $\frac{1}{2}$ in linear-time [2,4].

In several applications, finding a single subgraph is not enough to understand the property of a network. The Top-k-Overlapping Densest Subgraphs problem, introduced in [9], given an input graph, asks for a set of $k \geq 1$ subgraphs that may overlap (since communities may share vertices, for example hubs [9,14]). Top-k-Overlapping Densest Subgraphs aims at maximizing an objective function that includes the overall density of the k subgraphs and a distance between them. The distance allows to differentiate the returned subgraphs and it is controlled by a parameter $\lambda > 0$. For values of λ close to zero, the density plays a dominant role in the objective function, hence we can have subgraphs with a substantial overlap. For larger values of λ, the returned subgraphs may be disjoint.

Top-k-Overlapping Densest Subgraphs has been shown to be approximable within factor $\frac{1}{10}$ [9]. This approximation factor has been improved to $\frac{1}{2}$ when $k < |V|$ and to $\frac{2}{3}$ when k is a constant [6]. Furthermore, Top-k-Overlapping Densest Subgraphs has been shown to be NP-hard, when $k = 3$ and $\lambda = 3|V|^3$ [6].

1.1 Related Works

An approach similar to Top-k-Overlapping Densest Subgraphs was proposed in [3]. The problem defined in [3], given a graph $G = (V, E)$, asks for a set of k subgraphs of maximum density, such that the maximum pairwise Jaccard coefficient of the subgraphs in the solution is bounded by a value.

A dynamic variant of this problem, whose goal is finding a set of k disjoint subgraphs in a dynamic graph, has been recently considered in [17].

Other approaches related to Top-k-Overlapping Densest Subgraphs include covering or partitioning an input graph in dense subgraphs, like Minimum Clique Partition [10] or Minimum s-Club Covering [7]. However, notice that these approaches require that all the vertices of the graph belong to some dense subgraph of the solution, which is not the case for Top-k-Overlapping Densest Subgraphs, where a solution consists of collection of k subgraphs, with no constraint on the covering of the graph.

1.2 Contribution

In this paper, we design in Sect. 3 a new heuristic for the Top-k-Overlapping Densest Subgraphs, inspired by the approximation algorithm in [6]. The heuristic iteratively adds a subgraph by applying three phases and by picking the densest subgraph computed by these phases. Then, in Sect. 4 we compare our heuristic with the approximation algorithm of [9], called DOS (Dense Overlapping Subgraphs). We compare the two algorithms on biological networks ranging from 453 to 3518 vertices, and our heuristic always produces denser subgraphs while DOS graphs have greater distance. On the other hand, the algorithm [9] is much faster than our method.

2 Definitions

In this section, we give the definitions that we use in the rest of the paper and we present the formal definition of Top-k-Overlapping Densest Subgraphs.

We consider only undirected graphs. Consider a graph $G = (V, E)$, and a subset $W \subseteq V$, then $G[W]$ denotes *subgraph* of G induced by W, that is $G[W] = (W, E_W)$, where E_W is defined as:

$$E_W = \{\{u, v\} : \{u, v\} \in E \wedge u, v \in W\}. \tag{1}$$

Let $G[W_1]$ and $G[W_2]$, with $W_1, W_2 \subseteq V$, be two subgraphs of $G = (V, E)$, then $G[W_1]$ and $G[W_2]$ are called *overlapping* when $W_1 \cap W_2 \neq \emptyset$. $G[W_1]$ and $G[W_2]$ are *distinct* when $W_1 \neq W_2$.

Next, we introduce formally the definition of density, which is fundamental to define Top-k-Overlapping Densest Subgraphs.

Definition 1. *Let $G = (V, E)$ be a graph and let $G[W] = (W, E_W)$, with $W \subseteq V$, be a subgraph of G. Then the density of $G[W]$, denoted by $dens(G[W])$, is defined as follows:*

$$dens(G[W]) = \frac{|E_W|}{|W|}. \tag{2}$$

Given a graph $G = (V, E)$, a subgraph $G[U]$, with $U \subseteq V$, is a *densest subgraph* of G if it has maximum density among the subgraphs of G. Consider the example of Fig. 1. The subgraph induced by $\{v_4, v_5, v_6, v_7, v_8, v_9\}$ is the densest subgraph and has density $\frac{11}{6}$.

Let $G = (V, E)$ be a graph and let $\mathcal{W} = \{G[W_1], \ldots, G[W_k]\}$ be a collection of subgraphs of G. The density of $\mathcal{W}$, denoted by $dens(\mathcal{W})$, is defined as follows:

$$dens(\mathcal{W}) = \sum_{i=1}^{k} dens(G[W_i]). \tag{3}$$

Top-k-Overlapping Densest Subgraphs asks for a collection $\mathcal{W}$ of k subgraphs of the input graph G, with $k \geq 1$, such that $\mathcal{W}$ maximizes an objective function that contains the density of $\mathcal{W}$ and the distance between the subgraphs in $\mathcal{W}$. We present the definition of distance between two subgraphs of G introduced in [9] and used in Top-k-Overlapping Densest Subgraphs.

Definition 2. *Let $G = (V, E)$ be a graph and let $G[W]$, $G[Z]$, with $W, Z \subseteq V$, be two subgraphs of G. The distance between $G[W]$ and $G[Z]$, denoted by $d : 2^{G[V]} \times 2^{G[V]} \to \mathbb{R}_+$ is defined as follows:*

$$d(G[W], G[Z]) = \begin{cases} 2 - \frac{|W \cap Z|^2}{|W||Z|} & \text{if } W \neq Z, \\ 0 & \text{otherwise.} \end{cases} \tag{4}$$

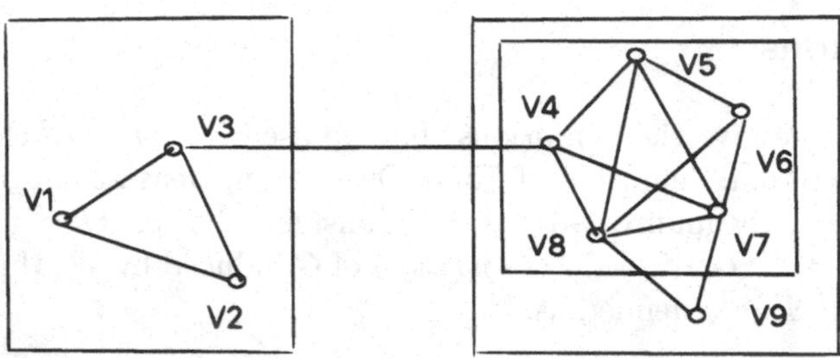

Fig. 1. A graph and a solution $\mathcal{W}$ of Top-k-Overlapping Densest Subgraphs, for $k = 3$, consisting of the three subgraphs included in boxes. The density of the subgraph induced by $\{v_4, v_5, v_6, v_7, v_8, v_9\}$ is $\frac{11}{6}$. The distance between the subgraphs induced by $\{v_4, v_5, v_6, v_7, v_8, v_9\}$ and $\{v_1, v_2, v_3\}$ is 2, as they do not overlap. The distance between the subgraphs induced by $\{v_4, v_5, v_6, v_7, v_8, v_9\}$ and $\{v_4, v_5, v_6, v_7, v_8\}$ is $2 - \frac{25}{30} = \frac{7}{6}$, since they overlap for five vertices.

Notice that $1 < d(G[W], G[Z]) \leq 2$, for any two distinct subgraphs of G. Next, we present the formal definition of the Top-k-Overlapping Densest Subgraphs problem.

Problem 1. Top-k-Overlapping Densest Subgraphs
Input: a graph $G = (V, E)$, a parameter $\lambda > 0$.
Output: a collection $\mathcal{W} = \{G[W_1], \ldots, G[W_k]\}$ of k subgraphs of G, with $k \geq 1$, such that the following objective function is maximized:

$$dens(\mathcal{W}) + \lambda \sum_{i=1}^{k-1} \sum_{j=i+1}^{k} d(G[W_i], G[W_j]) \tag{5}$$

2.1 Greedy and Constrained Greedy Algorithm for Densest-Subgraph

Although a densest subgraph of a given graph can be computed in polynomial time with the Goldberg's Algorithm [11], as in [9] we apply the faster approximation algorithm proposed in [2,4], denoted by Greedy, that has a linear-time complexity and achieves an approximation factor of $\frac{1}{2}$.

Given an input graph $G = (V, E)$, Greedy iteratively removes from G a vertex having lowest degree and stops when all the vertices of the graph have been removed. Greedy returns a densest subgraph among those computed starting from G by iteratively removing vertices.

We consider also a variant of the Greedy algorithm, called Constrained-Greedy. The Constrained-Greedy, given an input graph $G = (V, E)$ and a subset $C \subseteq V$, called the *constrained set*, computes a subgraph of G that contains C. Similar to Greedy, Constrained-Greedy iteratively removes a vertex of $V \setminus C$ having lowest degree (hence vertices in C can not be removed), until it obtains a graph on the set C of vertices. Then Greedy-Constrained returns a densest subgraph among those computed starting from G by iteratively removing vertices.

3 A Heuristic for Top-k-Overlapping Densest Subgraphs

In this section, we present our heuristic for Top-k-Overlapping Densest Subgraphs, called Iterative Dense Subgraphs (IDS). IDS is based on the Greedy and the Constraint-Greedy algorithms (see Sect. 2.1) to find k distinct subgraphs. The heuristic makes k iterations to compute $\mathcal{W} = \{G[W_1], ..., G[W_k]\}$, and at iteration i, with $1 \leq i \leq k$, it adds a subgraph denoted by $G[W_i]$ to $\mathcal{W}$.

The first iteration of IDS applies Greedy on the input graph G to compute subgraph $G[W_1]$. In iteration i, with $2 \leq i \leq k$, given the subgraph $G[W_{i-1}]$ added to $\mathcal{W}$ in iteration $i-1$, IDS applies Steps 1–3 (described later) to compute subgraph $G[W_i]$ and add it to $\mathcal{W} = \{G[W_1], ..., G[W_{i-1}]\}$. Each Step s, with $s = 1, 2, 3$, of iteration i, $2 \leq i \leq k$, computes a subgraph $G[W_{i,s}]$.

IDS adds to $\mathcal{W}$ the densest among the subgraphs $G[W_{i,1}]$, $G[W_{i,2}]$, $G[W_{i,3}]$, and the subgraphs $G[W_{j,1}]$, $G[W_{j,2}]$, $G[W_{j,3}]$, with $2 \leq j \leq i-1$, computed from iterations 2 to $i-1$ and not already added to $\mathcal{W}$. Hence notice that the subgraphs computed by each iteration, and not already added to $\mathcal{W}$, are saved and considered in next iterations of IDS. Steps 1–3 of iteration i, with $2 \leq i \leq k$, are based on sets $Q_{i,s} \subseteq V$, with $s = 1, 2, 3$, of at most q_i vertices ($|Q_{i,s}|$ will be defined as the maximum value not greater than q_i), where q_i is defined as follows:

$$q_i = \lambda \frac{\sqrt{|W_{i-1}|}}{2}. \tag{6}$$

Sets $Q_{i,s}$ are used by IDS to differentiate the subgraphs computed in Step 1–3 from subgraph $G[W_{i-1}]$. In particular, $Q_{i,s}$, with $2 \leq i \leq k$ and $s = 1, 2, 3$, is used by IDS in two following ways:

(I) A set $Q_{i,1}$ of at most q_i vertices that will not be contained in the subgraph $G[W_{i,1}]$ computed by Step 1 of the ith iteration. In this case we select $Q_{i,1}$ as a set of at most q_i vertices with lower degree.

(II) A set $Q_{i,s}$, with $s = 2, 3$, of at most q_i vertices that will be contained in the subgraphs $G[W_{i,2}]$ and $G[W_{i,3}]$ respectively, computed by Step 2 and Step 3 of the ith iteration. In this case we select $Q_{i,s}$ as a set of at most q_i vertices with higher degree.

Recall that, $G[W_{i-1}]$ is the subgraph added at iteration $i-1$ to the solution $\mathcal{W} = \{G[W_1], ..., G[W_{i-1}]\}$. The input of the i-th iteration of IDS consists of G, $\mathcal{W} = \{G[W_1], ..., G[W_{i-1}]\}$ and a set $Q_{i,s}$, with $s = 1, 2, 3$ (the definition of $Q_{i,s}$ depends on the specific step we consider).

Next, we give some details about Steps 1–3 of each iteration i, $2 \leq i \leq k$, of heuristic IDS.

Step 1: Given a set $Q_{i,1} \subseteq W_{i-1}$ of at most q_i vertices of lower degree, Step 1 computes a subgraph $G[W_{i,1}]$, distinct from the subgraphs in $\mathcal{W}$ such that $W_{i,1} \cap Q_{i,1} = \emptyset$. Step 1 computes subgraph $G[W_{i,1}]$ by applying the Greedy algorithm on input $G[W_{i-1} \setminus Q_{i,1}]$.

Step 2: Given a set $Q_{i,2} \subseteq (V \setminus W_{i-1})$ of at most q_i vertices of higher degree, Step 2 computes a subgraph $G[W_{i,2}]$, distinct from the subgraphs in $\mathcal{W}$ such that $(W_{i-1} \cup Q_{i,2}) \subseteq W_{i,2}$. Step 2 computes subgraph $G[W_{i,2}]$ by applying the Constrained-Greedy algorithm (described in Sect. 2.1) on G, with constrained set $W_{i-1} \cup Q_{i,2}$, that is $(W_{i-1} \cup Q_{i,2}) \subseteq W_{i,2}$.

Step 3: Given a set $Q_{i,3} \subseteq V$ of at most q_i vertices of higher degree not covered by any subgraph of $\mathcal{W}$, Step 3 computes a subgraph $G[W_{i,3}]$, distinct from the subgraphs in $\mathcal{W}$ such that $Q_{i,3} \subseteq W_{i,3}$. Step 3 computes subgraph $G[W_{i,3}]$ by applying Constrained-Greedy algorithm on G, with constrained set $Q_{i,3}$, that is $Q_{i,3} \subseteq W_{i,3}$.

Notice that some of the Steps 1–3, at an iteration i, with $2 \leq i \leq k$, may not be applied. For example, if each vertex of V is covered by the subgraphs in $\mathcal{W}$, then Step 3 cannot be applied. Moreover, notice that if Steps 1–3 of iteration i of IDS are not able to compute a subgraph of G distinct from those in $\mathcal{W}$, we apply a post-processing phase in order to have $|\mathcal{W}| = k$. This post-processing phase, starting from subgraph $G[W_{i-1}]$, greedily adds to $\mathcal{W}$ a set of densest subgraphs obtained by removing a single vertex from $G[W_{i-1}]$ or by adding a single vertex to $G[W_{i-1}]$.

Next, we discuss the time complexity of IDS.

Lemma 1. *Given an input graph $G = (V, E)$, the time complexity of* IDS *is $O(k^2|V| + k|E|)$.*

Proof. First, recall that the time complexity of Greedy is $O(|V| + |E|)$ [4]. It follows that also the time complexity of Constrained-Greedy is $O(|V| + |E|)$.

IDS makes k iterations to compute $\mathcal{W}$. In the first iteration, IDS applies Greedy, hence this iteration has time complexity $O(|V| + |E|)$. Consider now iteration i, with $2 \leq i \leq k$, in which IDS applies Steps 1–3. In Step s, with $s = 1, 2, 3$, the set $Q_{i,s}$ is computed in time $O(|V|)$, by applying the Counting Sort algorithm [5] to sort the vertices by their degree, then either Greedy or Constrained-Greedy is applied, which requires time $O(|V| + |E|)$. Finally, the set of covered vertices is updated in time $O(|V|)$ (this set is required for Step 3) and the comparison between a subgraph $G[W_{i,s}]$ and the subgraphs in $\mathcal{W}$, in order to verify if $G[W_{i,s}]$ is distinct from the subgraphs in $\mathcal{W}$, requires time $O(k|V|)$. It follows that each iteration i, with $2 \leq i \leq k$, requires time $O(k|V| + |E|)$. The overall time complexity of IDS is then $O(k^2|V| + k|E|)$. □

4 Experimental Results

In this section, we show the experimental results of the comparison of our proposed heuristic IDS with DOS [9]. DOS is an iterative approximation algorithm for Top-k-Overlapping Densest Subgraphs that achieve the approximation ratio of $\frac{1}{10}$. Given an input $G = (V, E)$, DOS has time complexity $O(k|E| + |V|(t + k))$, where $t = \min\{2^k, |V|\}$. We consider the value of λ equal to range from 0.25 to 2 (similarly to [9]).

The DOS was implemented in Python, while we implemented IDS in MATLAB R2018b. IDS and DOS were run on a computer with processor 2.9 GHz Intel Core i5 and 8 GB of RAM, MacOS version 10.14.3.

Table 1. Performance of IDS and DOS on real-world networks with $k = 10$. For each network, we report the size of the network (number of vertices $|V|$ and edges $|E|$), the density (Den.), distance (Dis.) and the objective function value (Obj. val.).

				IDS			DOS		
Set	$\|V\|$	$\|E\|$	λ	Den.	Dis.	Obj. val.	Den.	Dis.	Obj. val.
c.elegans	453	2K	0.25	74.96	51.63	**87.87**	51.92	86.55	73.56
			1	74.81	51.83	**126.64**	36.65	89.39	126.04
			2	74.16	56.41	186.98	32.57	89.85	**212.28**
Diseasome	516	1K	0.25	48.10	56.13	**62.14**	37.78	88.84	59.99
			1	45.91	56.99	102.90	33.29	89.51	**122.81**
			2	44.31	62.30	168.92	28.28	89.97	**208.21**
Yeast-protein-inter	2114	4K	0.25	29.45	54.71	**43.13**	19.86	88.86	42.08
			1	26.67	63.08	89.75	17.16	89.67	**106.83**
			2	26.05	73.44	172.94	15.27	89.93	**195.13**
Worm	3518	13K	0.25	77.39	49.10	**89.66**	49.15	87.11	70.93
			1	76.61	53.12	**129.73**	32.10	89.69	121.79
			2	75.48	55.68	186.85	24.08	89.93	**203.95**

Following [9], we set $k = 10$ for the Top-k-Overlapping Densest Subgraphs problem. We present the results of both algorithms on real-world networks in Table 1. We consider four biological networks (see [18]), with the size ranging from 453 to 3518 vertices. The c.elegans network represents substrates vertices type and metabolic reactions edges type. Diseasome network represents disease map whose vertices are diseases and edges are various molecular relationships between the disease-associated cellular components. Yeast-protein-inter and worm networks are networks whose vertices are protein and edges are interaction between proteins.

5 Conclusion

In this paper, we propose a new heuristic (IDS) for Top-k-Overlapping Densest Subgraphs, a problem recently introduced for finding a collection of dense subgraphs in a network. We compare IDS with DOS [9], an approximation algorithm designed for Top-k-Overlapping Densest Subgraphs, on four biological networks. The experimental results show that always IDS performs better than DOS in terms of density, while DOS is faster. A future direction of research is to expand the comparison to synthetic and other real datasets.

References

1. Alba, R.D.: A graph-theoretic definition of a sociometric clique. J. Math. Sociol. **3**, 113–126 (1973)
2. Asahiro, Y., Iwama, K., Tamaki, H., Tokuyama, T.: Greedily finding a dense subgraph. In: Karlsson, R., Lingas, A. (eds.) SWAT 1996. LNCS, vol. 1097, pp. 136–148. Springer, Heidelberg (1996). https://doi.org/10.1007/3-540-61422-2_127
3. Balalau, O.D., Bonchi, F., Chan, T.H., Gullo, F., Sozio, M.: Finding subgraphs with maximum total density and limited overlap. In: Cheng, X., Li, H., Gabrilovich, E., Tang, J. (eds.) Proceedings of the Eighth ACM International Conference on Web Search and Data Mining, WSDM 2015, pp. 379–388. ACM (2015). https://doi.org/10.1145/2684822.2685298
4. Charikar, M.: Greedy approximation algorithms for finding dense components in a graph. In: Jansen, K., Khuller, S. (eds.) APPROX 2000. LNCS, vol. 1913, pp. 84–95. Springer, Heidelberg (2000). https://doi.org/10.1007/3-540-44436-X_10
5. Cormen, T.H., Leiserson, C.E., Rivest, R.L., Stein, C.: Introduction to Algorithms, 3rd edn. MIT Press, Cambridge (2009)
6. Dondi, R., Hosseinzadeh, M.M., Mauri, G., Zoppis, I.: Top-k overlapping densest subgraphs: approximation and complexity. In: Proceeding in 20th Italian Conference on Theoretical Computer Science (2019, to appear)
7. Dondi, R., Mauri, G., Sikora, F., Zoppis, I.: Covering a graph with clubs. J. Graph Algorithms Appl. **23**(2), 271–292 (2019). https://doi.org/10.7155/jgaa.00491
8. Fratkin, E., Naughton, B.T., Brutlag, D.L., Batzoglou, S.: MotifCut: regulatory motifs finding with maximum density subgraphs. Bioinformatics **22**(14), 156–157 (2006). https://doi.org/10.1093/bioinformatics/btl243
9. Galbrun, E., Gionis, A., Tatti, N.: Top-k overlapping densest subgraphs. DataMin. Knowl. Discov. **30**(5), 1134–1165 (2016). https://doi.org/10.1007/s10618-016-0464-z
10. Garey, M.R., Johnson, D.S.: Computers and Intractability: A Guide to the Theory of NP-Completeness. WH Freeman & Co., Stuttgart (1979)
11. Goldberg, A.V.: Finding a Maximum Density Subgraph. University of California Berkeley, CA (1984)
12. Karp, R.M.: Reducibility among combinatorial problems. In: Miller, R.E., Thatcher, J.W., Bohlinger, J.D. (eds.) Complexity of Computer Computations. IRSS, pp. 85–103. Plenum Press, New York (1972). https://doi.org/10.1007/978-1-4684-2001-2_9
13. Komusiewicz, C.: Multivariate algorithmics for finding cohesive subnetworks. Algorithms **9**(1), 21 (2016)
14. Leskovec, J., Lang, K.J., Dasgupta, A., Mahoney, M.W.: Community structure in large networks: natural cluster sizes and the absence of large well-defined clusters. Internet Math. **6**(1), 29–123 (2009). https://doi.org/10.1080/15427951.2009.10129177

15. Ma, X., Zhou, G., Shang, J., Wang, J., Peng, J., Han, J.: Detection of complexes in biological networks through diversified dense subgraph mining. J. Comput. Biol. **24**(9), 923–941 (2017)
16. Mokken, R.: Cliques, clubs and clans. Qual. Quant. Int. J. Methodol. **13**(2), 161–173 (1979)
17. Nasir, M.A.U., Gionis, A., Morales, G.D.F., Girdzijauskas, S.: Fully dynamic algorithm for top-k densest subgraphs. In: Lim, E., et al. (eds.) Proceedings of the 2017 ACM on Conference on Information and Knowledge Management, CIKM 2017, pp. 1817–1826. ACM (2017). https://doi.org/10.1145/3132847.3132966
18. Rossi, R.A., Ahmed, N.K.: The network data repository with interactive graph analytics and visualization. In: Proceedings of the Twenty-Ninth AAAI Conference on Artificial Intelligence (2015). http://networkrepository.com

Author Index

Printed in the United States
By Bookmasters